Constructibility

Since their inception, the Perspectives in Logic and Lecture Notes in Logic series have published seminal works by leading logicians. Many of the original books in the series have been unavailable for years, but they are now in print once again.

In this volume, the 6th publication in the Perspectives in Logic series, Keith Devlin gives a comprehensive account of the theory of constructible sets at an advanced level. The book provides complete coverage of the theory itself, rather than the many and diverse applications of constructibility theory, although applications are used to motivate and illustrate the theory.

The book is divided into two parts: Part A (Elementary Theory) deals with the classical definition of the L_α-hierarchy of constructible sets and may be used as the basis of a graduate course on constructibility theory; and Part B (Advanced Theory) deals with the J_α-hierarchy and the Jensen "fine-structure theory".

KEITH J. DEVLIN works in the Department of Mathematics at the University of Lancaster.

PERSPECTIVES IN LOGIC

The *Perspectives in Logic* series publishes substantial, high-quality books whose central theme lies in any area or aspect of logic. Books that present new material not now available in book form are particularly welcome. The series ranges from introductory texts suitable for beginning graduate courses to specialized monographs at the frontiers of research. Each book offers an illuminating perspective for its intended audience.

The series has its origins in the old *Perspectives in Mathematical Logic* series edited by the Ω-Group for "Mathematische Logik" of the Heidelberger Akademie der Wissenschaften, whose beginnings date back to the 1960s. The Association for Symbolic Logic has assumed editorial responsibility for the series and changed its name to reflect its interest in books that span the full range of disciplines in which logic plays an important role.

For more information, see www.aslonline.org/books_perspectives.html

PERSPECTIVES IN LOGIC

Constructibility

KEITH J. DEVLIN

Lancaster University

ASSOCIATION FOR SYMBOLIC LOGIC

CAMBRIDGE
UNIVERSITY PRESS

CAMBRIDGE
UNIVERSITY PRESS

University Printing House, Cambridge CB2 8BS, United Kingdom

One Liberty Plaza, 20th Floor, New York, NY 10006, USA

477 Williamstown Road, Port Melbourne, VIC 3207, Australia

4843/24, 2nd Floor, Ansari Road, Daryaganj, Delhi – 110002, India

79 Anson Road, #06–04/06, Singapore 079906

Cambridge University Press is part of the University of Cambridge.

It furthers the University's mission by disseminating knowledge in the pursuit of education, learning, and research at the highest international levels of excellence.

www.cambridge.org
Information on this title: www.cambridge.org/9781107168350
10.1017/9781316717219

Association for Symbolic Logic
Richard A. Shore, Publisher
Department of Mathematics, Cornell University, Ithaca, NY 14853
http://www.aslonline.org

A catalogue record for this publication is available from the British Library.

ISBN 978-1-107-16835-0 Hardback

Preface to the Series

Perspectives in Mathematical Logic

(Edited by the Ω-group for "Mathematische Logik" of the Heidelberger Akademie der Wissenschaften)

On Perspectives. *Mathematical logic arose from a concern with the nature and the limits of rational or mathematical thought, and from a desire to systematise the modes of its expression. The pioneering investigations were diverse and largely autonomous. As time passed, and more particularly since the mid-fifties, interconnections between different lines of research and links with other branches of mathematics proliferated. The subject is now both rich and varied. It is the aim of the series to provide, as it were, maps or guides to this complex terrain. We shall not aim at encyclopaedic coverage; nor do we wish to prescribe, like Euclid, a definitive version of the elements of the subject. We are not committed to any particular philosophical programme. Nevertheless we have tried by critical discussion to ensure that each book represents a coherent line of thought; and that, by developing certain* themes, *it will be of greater interest than a mere assemblage of results and techniques.*

The books in the series differ in level: some are introductory, some highly specialised. They also differ in scope: some offer a wide view of an area, others present a single line of thought. Each book is, at its own level, reasonably self-contained. Although no book depends on another as prerequisite, we have encouraged authors to fit their book in with other planned volumes, sometimes deliberately seeking coverage of the same material from different points of view. We have tried to attain a reasonable degree of uniformity of notation and arrangement. However, the books in the series are written by individual authors, not by the group. Plans for books are discussed and argued about at length. Later, encouragement is given and revisions suggested. But it is the authors who do the work; if, as we hope, the series proves of value, the credit will be theirs.

History of the Ω-Group. *During 1968 the idea of an integrated series of monographs on mathematical logic was first mooted. Various discussions led to a meeting at Oberwolfach in the spring of 1969. Here the founding members of the group (R. O. Gandy, A. Levy, G. H. Müller, G. E. Sacks, D. S. Scott) discussed the project in earnest and decided to go ahead with it. Professor F. K. Schmidt and Professor Hans Hermes gave us encouragement and support. Later Hans Hermes joined the group. To begin with all was fluid. How ambitious should we be? Should we write the books ourselves? How long would it take? Plans for authorless books were promoted, savaged and scrapped. Gradually there emerged a form and a method. At the end of an infinite discussion we found our name, and that of the series. We established our centre in Heidelberg. We agreed to meet twice a year together with authors, consultants and assistants, generally in Oberwolfach. We soon found the value of collaboration: on the one hand the permanence of the founding group gave coherence to the over-all plans; on the other hand the stimulus of new contributors kept the project alive and flexible. Above all, we found how intensive discussion could modify the authors' ideas and our own. Often the battle ended with a detailed plan for a better book which the author was keen to write and which would indeed contribute a perspective.*

Oberwolfach, September 1975

Acknowledgements. *In starting our enterprise we essentially were relying on the personal confidence and understanding of Professor Martin Barner of the Mathematisches Forschungsinstitut Oberwolfach, Dr. Klaus Peters of Springer-Verlag and Dipl.-Ing. Penschuck of the Stiftung Volkswagenwerk. Through the Stiftung Volkswagenwerk we received a generous grant (1970 – 1973) as an initial help which made our existence as a working group possible.*

Since 1974 the Heidelberger Akademie der Wissenschaften (Mathematisch-Naturwissenschaftliche Klasse) has incorporated our enterprise into its general scientific program. The initiative for this step was taken by the late Professor F. K. Schmidt, and the former President of the Academy, Professor W. Doerr.

Through all the years, the Academy has supported our research project, especially our meetings and the continuous work on the Logic Bibliography, in an outstandingly generous way. We could always rely on their readiness to provide help wherever it was needed.

Assistance in many various respects was provided by Drs. U. Felgner and K. Gloede (till 1975) and Drs. D. Schmidt and H. Zeitler (till 1979). Last but not least, our indefatigable secretary Elfriede Ihrig was and is essential in running our enterprise.

We thank all those concerned.

Heidelberg, September 1982

R. O. Gandy *H. Hermes*
A. Levy *G. H. Müller*
G. E. Sacks *D. S. Scott*

Author's Preface

This book is intended to give a fairly comprehensive account of the theory of constructible sets at an advanced level. The intended reader is a graduate mathematician with some knowledge of mathematical logic. In particular, we assume familiarity with the notions of formal languages, axiomatic theories in formal languages, logical deductions in such theories, and the interpretation of languages in structures. Practically any introductory text on mathematical logic will supply the necessary material. We also assume some familiarity with Zermelo-Fraenkel set theory up to the development or ordinal and cardinal numbers. Any number of texts would suffice here, for instance *Devlin* (1979) or *Levy* (1979).

The book is not intended to provide a complete coverage of the many and diverse applications of the methods of constructibility theory, rather the theory itself. Such applications as are given are there to motivate and to exemplify the theory.

The book is divided into two parts. Part A ("Elementary Theory") deals with the classical definition of the L_α-hierarchy of constructible sets. With some pruning, this part could be used as the basis of a graduate course on constructibility theory. Part B ("Advanced Theory") deals with the J_α-hierarchy and the Jensen "fine-structure theory".

Chapter I is basic to the entire book. The first seven or eight sections of this chapter should be familiar to the reader, and they are included primarily for completeness, and to fix the notation for the rest of the book. Sections 9 through 11 may well be new to the reader, and are fundamental to the entire development. Thus a typical lecture course based on the book would essentially commence with section 9 of Chapter I. After Chapter II, where the basic development of constructibility theory is given, the remaining chapters of Part A are largely independent, though it would be most unnatural to cover Chapter IV without first looking at Chapter III. Likewise, in Part B, after the initial chapter (Chapter VI) there is a large degree of independence between the chapters. (Indeed, given suitable introduction by an instructor, Chapter IX could be read directly after Chapter IV.)

Constructibility theory is plagued with a large number of extremely detailed and potentially tedious arguments, involving such matters as investigating the exact logical complexity of various notions of set theory. In order to try to strike a balance between the need to have a readable book of reasonable length, and the requirements of a beginning student of the field, as our development proceeds we give progressively less detailed arguments, relying instead upon the developing ability of the reader to fill in any necessary details. Thus the experienced reader may well find that it is necessary to skip over some of the earlier proofs, whilst the novice will increasingly need to spend time supplying various details. This is particularly true of Chapter II and the latter parts of Chapter I upon which Chapter II depends.

As this is intended as an advanced reference text, we have not provided an extensive selection of exercises. Those that are given consist largely of extensions or enlargements of the main development. Together with filling in various details in our account, these should suffice for a full understanding of the main material, which is their only purpose. The exercises occur at the end of each chapter (except for Chapter I), with an indication of the stage in the text which must be reached in order to attempt them.

Chapters are numbered by Roman numerals and results by normal numerals. A reference to "II.5" means section 5 of Chapter II, whilst "V.3.7" would refer to result 7 in section 3 of Chapter V. The mention of the chapter number would be supressed within that chapter. The end of a proof is indicated by the symbol □. If this occurs directly after the statement of a result, it should be understood that either the proof of the result is obvious (possibly in view of earlier remarks) or else (according to context) that the proof is a long one that will stretch over several pages and involve various lemmas. During the course of some of the longer proofs, many different symbols are introduced. In order to help the reader to keep track of them, at the points where new symbols are defined the symbol concerned appears in the outer margin of the book.

Finally, I would like to express my gratitude to all of those who have helped me in the preparation of this book. There are the members of the Ω-Group, who gave me the benefit of their views during the early stages of planning. Gert Müller kept a watchful eye on matters managerial, and Azriel Levy took on the task of editor, reading through various versions of the manuscript and making countless suggestions for improvements. Others who read through all or parts of the final manuscript are (in order of the number of errors picked up) Stevo Todorčević, Klaus Gloede, Jakub Jasinski, Włodek Bzyl, Martin Lewis, and Dieter Donder. Not to forget Ronald Jensen. Although he played no part in the writing of this book, it is clear (or will be if you get far enough into the book) that without his work there would have been practically nothing to write about!

Financial support during the preparation of the manuscript was provided by the Heidelberger Akademie der Wissenschaften.

Keith J. Devlin

Contents

Part A. Elementary Theory

Part A

Elementary Theory

Chapter I
Preliminaries

The fundamental set theory of this book is Zermelo-Fraenkel set theory. In this chapter we give a brief account of this theory, insofar as we need it. Sections 1 through 5 cover the early development of the theory up to ordinal and cardinal numbers. The remaining six sections deal with some special topics of direct relevance to the subject matter of this book, and the coverage is therefore a little more complete than in the previous sections.

1. The Language of Set Theory

The language of set theory, LST, is the first-order language with predictates $=$ (equality) and $\in$ (set membership), logical symbols $\wedge$ (and), $\neg$ (not), and $\exists$ (there exists), variables $v_0, v_1, \ldots,$ and (for convenience) brackets (,).

The *primitive* (or *atomic*) *formulas* of LST are strings of the forms

$$(v_m = v_n), \quad (v_m \in v_n).$$

The *formulas* of LST are generated from the primitive formulas by means of the following schemas: if Φ, Ψ are formulas, so too are the strings

$$(\Phi \wedge \Psi), \quad (\neg \Phi), \quad (\exists v_n \Phi).$$

(We generally use capital Greek letters to denote formulas of LST.)

The notions of *free* and *bound* variables are defined as usual. A *sentence* is a formula with no free variables.

We write $x \notin y$ for $\neg (x \in y)$ and $x \neq y$ for $\neg (x = y)$. (We generally use, x, y, z, etc. to denote arbitrary variables of LST.)

The defined logical symbols $\vee$, $\rightarrow$, $\leftrightarrow$, $\forall$ are introduced in the usual way, and are frequently treated as if they were basic symbols of LST (i.e. having the same status as $\wedge$, $\neg$, $\exists$). Likewise for the bounded quantifiers $(\exists v_m \in v_n)$ and $(\forall v_m \in v_n)$ (where $m \neq n$), introduced by the schemas:

$$(\exists v_m \in v_n)\, \Phi \quad \text{replaces} \quad \exists v_m((v_m \in v_n) \wedge \Phi);$$

$$(\forall v_m \in v_n)\, \Phi \quad \text{replaces} \quad \forall v_m((v_m \in v_n) \rightarrow \Phi).$$

The symbols $\subseteq$ and $\exists!$ are defined thus:

$$y \subseteq z \quad \text{abbreviates } (\forall x \in y)(x \in z);$$

$$\exists! x\Phi \quad \text{abbreviates } \exists y \forall x (y = x \leftrightarrow \Phi).$$

(Thus $\exists! x\Phi$ means "there is a unique x such that Φ".) We also write

$$y \subset z \quad \text{to mean } y \subseteq z \wedge y \neq z.$$

The above abbreviations are never regarded as a fundamental part of the language LST, however, unlike the bounded quantifiers, etc.

One final remark. In writing formulas, we strive for legibility at the expense of strict adherence to the syntax of LST. This particularly applies to our use of parentheses, which are omitted wherever possible. Also, when nesting of clauses is required, we sometimes use both (square) brackets as well as parentheses, for clarity. Out notation for the interpretation of variables in formulas is also chosen with clarity in mind. If we write, say, $\Phi(v_i, v_j)$, we mean that the free variables of Φ are amongst the variables v_1, v_j. If we subsequntly write $\Phi(x, y)$, where x and y are specific sets, we mean that Φ is a valid assertion when x interprets v_i and y interprets v_j. (Of course, we have also decided to use x, y, z, etc. to denote arbitrary variables of LST. But in any given case, the context should indicate the intended meaning.[1]

2. The Zermelo-Fraenkel Axioms

The theory ZF is the LST theory whose axioms are the usual axioms for first-order logic (for the langugage LST), together with the following axioms (i)–(vii):

(i) Extensionality: $\forall x \forall y [\forall z (z \in x \leftrightarrow z \in y) \rightarrow (x = y)]$
(ii) Union: $\forall x \exists y \forall z [z \in y \leftrightarrow (\exists u \in x)(z \in u)]$
(iii) Infinity: $\exists x [\exists y (y \in x) \wedge (\forall y \in x)(\exists z \in x)(y \in z)]$
(iv) Power Set: $\forall x \exists y \forall z [z \in y \leftrightarrow z \subseteq x]$
(v) Foundation: $\forall x [\exists y (y \in x) \rightarrow \exists y (y \in x \wedge (\forall z \in y)(z \notin x))]$
(vi) Comprehension (schema): $\forall \vec{a} \forall x \exists y \forall z [z \in y \leftrightarrow z \in x \wedge \Phi(z, \vec{a})]$,
where Φ is any LST formula whose free variables are amongst z, $\vec{a}$, and where the variables $\vec{a}$, x, y, z are all distinct.

(We use $\vec{x}$, $\vec{a}$, etc. to denote finite strings of variables, $\forall \vec{a}$ to abbreviate $\forall a_1, \ldots, \forall a_n$ and $\Phi(z, \vec{a})$ to abbreviate $\Phi(z, a_1, \ldots, a_n)$. In more complicated situations,

1 Strictly speaking there is no clash of notation here. As far as formal set theory is concerned there are simply *variables* (to denote "sets"). But as usual, to avoid incomprehensible use of quantifiers and formulas to define specific sets, we argue in a loose, *semantic* fashion whenever possible, and then it can be useful to distinguish between "formal variables" and "sets which interpret those variables".

we often use expressions such as $\vec{x}_0, \ldots, \vec{x}_n$. Here, $\vec{x}_0$ will denote some sequence $x_{00}, \ldots, x_{0k}$, $\vec{x}_1$ will be another sequence $x_{10}, \ldots, x_{1l}$, possibly of a different length, according to context, and so on.)

(vii) Collection (schema):

$$\forall \vec{a} [\forall x \exists y \Phi(y, x, \vec{a}) \rightarrow \forall u \exists v (\forall x \in u)(\exists y \in v) \Phi(y, x, \vec{a})],$$

where Φ is any LST-formula whose free variables are amongst $y, x, \vec{a}$, and where the variables $\vec{a}, x, y, u, v$ are all distinct.

In (iii), the exact formulation of the Axiom of Infinity is not important, and different texts often give different formulations. The main point is to guarantee the existence of at least one infinite set. Axiom (vi) (the Comprehension Axiom schema) is sometimes referred to as the Subset Selection schema. The German word *Aussonderungsaxiom* is also quite common for this axiom scheme. In Axiom (vii) (Collection), notice that we have placed the variable y before the variable x. This is purely a stylistic convention, of course, and reflects the fact that in our representation of a function as a set of ordered pairs, we shall take the first member of each ordered pair as the value of the function and the second element as the argument. Axiom schemas (vi) and (vii) are often replaced by a single schema: the *Axiom of Replacement*.

Notice that by virtue of the two axiom schemas, the above list of axioms for ZF is infinite. We shall soon be able to prove that no finite collection of LST sentences suffices to axiomatise ZF.

By the Axiom of Infinity, there exists at least one set. The Axiom of Comprehension then yields the existence of the empty set $\emptyset$. Many texts include as an axiom of ZF the *Null Set Axiom*, which is the assertion that there exists a set having no elements, viz.:

$$\exists x \forall y (y \notin x).$$

Zermelo-Fraenkel set theory includes one further axiom:

(viii) Axiom of Choice (AC):

$$\forall x [(\forall y \in x)(y \neq \emptyset) \wedge (\forall y, y' \in x)(y \neq y' \rightarrow \forall w (w \in y \leftrightarrow w \notin y') \\ \rightarrow (\exists z)(\forall y \in x)(\exists ! v \in y)(v \in z)].$$

We denote Zermelo-Fraenkel set theory (which includes AC) by ZFC. This nomenclature is now fairly standard, despite the rather unfortunate fact that it means that the letters ZF do not stand for "Zermelo-Fraenkel" set theory, but just a part of that theory. To try to avoid any confusion, throughout the book we shall stick to the abbreviated notations ZF and ZFC. Hence, we shall have the "equation"

$$\mathrm{ZFC} = \mathrm{ZF} + \mathrm{AC}.$$

ZFC is our basic set theory. On occasions it will be important to note that AC is not being used in an argument, and in such cases we shall write, for example,

$$\mathrm{ZF} \vdash \Phi$$

or else

$$\Phi \rightarrow_{\mathrm{ZF}} \Psi$$

to mean, respectively, that Φ is provable in ZF or that Ψ is provable from Φ together with the axioms of ZF.

3. *Elementary Theory of* ZFC

3.1 (Sets and Classes). The basic objects of discussion of ZFC (i. e. the objects over which the variables range) are called *sets*. The *universe* is the collection of all sets, and is denoted by V. If $\Phi(v_0, v_1, \dots, v_n)$ is an LST formula and $x_1, \dots, x_n$ are sets, the collection of all sets x for which $\Phi(x, x_1, \dots, x_n)$ is a *class*, denoted by

$$\{x \mid \Phi(x, x_1, \dots, x_n)\}.$$

Every set, y, is a class (consider the formula $\Phi(x, y) \equiv (x \in y)$), but not every class is a set (consider the formula $\Phi(x) \equiv (x \notin x)$, which would lead at once to the Russell paradox if the class it defined were a set). We often write

$$\{x \in y \mid \Phi(x, x_1, \dots, x_n)\}$$

in place of

$$\{x \mid x \in y \wedge \Phi(x, x_1, \dots, x_n)\}.$$

(By the Axiom of Comprehension, this class is always a set.) We generally use capital Roman letters X, Y, Z etc. to denote classes, with lower case Roman letters being reserved for sets (as well as for variables of LST, which denote sets, of course). A class which is not a set is called a *proper class*. Proper classes do not fall under the scope of the axioms of ZFC, but their usage is convenient. We assume the reader is familiar both with the use of proper classes in set theory and the means by which such usage may be avoided if required. A particular example occurs in VI.1, where we discuss the rudimentary functions. It is convenient, though avoidable, to develop the relevant theory in terms of "functions" defined on the whole of V, even though, as proper classes these cannot be *functions* in the sense of set theory at all.

Our set-theoretic notation is standard. The set consisting of precisely the elements $x_1, \dots, x_n$ is denoted by

$$\{x_1, \dots, x_n\};$$

$\{x\}$ is the *singleton* of x, and $\{x, y\}$ is the *unordered pair* of x, y. Many texts include as an axiom of ZF the *Pairing Axiom*, which asserts that for every pair of elements x, y, the set $\{x, y\}$ exists, i.e.

$$\forall x \forall y \exists z \forall u (u \in z \leftrightarrow u = x \vee u = y).$$

However, as this "axiom" is easily proved from the axioms we listed earlier, we did not take it as a basic axiom.

The *ordered pair* of x and y is defined by

$$(x, y) = \{\{x\}, \{x, y\}\},$$

and has the property that

$$(x, y) = (x', y') \quad \text{iff } x = x' \text{ and } y = y'.$$

The *union* of x (i.e. the set of all members of all members of x) is denoted by $\bigcup x$, and is guaranteed to exist by the Union Axiom. We write $x \cup y$ instead of $\bigcup \{x, y\}$. The *intersection* of x, $\bigcap x$, is defined by

$$y \in \bigcap x \quad \text{iff } (\forall z \in x)(y \in z),$$

and is a set whenever $x \neq \emptyset$. (By our definition, $\bigcap \emptyset = V$, but this is not a case that will ever concern us.) We write $x \cap y$ for $\bigcap \{x, y\}$. The *difference* of x and y is defined by

$$x - y = \{z \in x \mid z \notin y\}.$$

The *power set* of x (i.e. the set of all subsets of x) is denoted by $\mathscr{P}(x)$, and is guaranteed to exist by the Power Set Axiom.

3.2 (Ordinals). A class M is said to be *transitive* if

$$x \in y \in M \rightarrow x \in M.$$

If $\mathrm{Trans}(v_0)$ denotes the LST formula

$$(\forall v_1 \in v_0)(\forall v_2 \in v_1)(v_2 \in v_0),$$

then a set x will be transitive iff $\mathrm{Trans}(x)$.

An *ordinal number* (or simply, an *ordinal*) is a transitive set which is linearly ordered by $\in$. We use $\alpha, \beta, \gamma, \ldots$ to denote ordinals. We denote by $\mathrm{On}(v_0)$ the LST-formula

$$\mathrm{Trans}(v_0) \wedge (\forall v_1 \in v_0)(\forall v_2 \in v_0)(v_1 = v_2 \vee v_1 \in v_2 \vee v_2 \in v_1).$$

It is not hard to show that a set x will be an ordinal iff $\mathrm{On}(x)$.

If α, β are ordinals, either $\alpha = \beta$ or $\alpha \in \beta$ or $\beta \in \alpha$. So the class

$$\mathrm{On} = \{x \mid \mathrm{On}(x)\}$$

is totally ordered by $\in$. We often write $\alpha < \beta$ instead of $\alpha \in \beta$, and $\alpha \leqslant \beta$ instead of $(\alpha < \beta \vee \alpha = \beta)$. It is easily seen that $\alpha < \beta$ is equivalent to $\alpha \subset \beta$. Moreover, for any ordinal α,

$$\alpha = \{\beta \mid \beta < \alpha\}.$$

By the Axiom of Foundation, the relation $<$ is in fact a well-ordering of On (i.e. every non-empty subset of On has a $<$-least element).

If A is a set of ordinals, then $\cup A$ is also an ordinal. In fact, $\cup A$ is the least ordinal δ such that $(\forall \alpha \in A)(\alpha \leqslant \delta)$. This least δ is also called the *supremum* of A, denoted by $\sup(A)$. Thus $\sup(A)$ and $\cup A$ coincide.

The first ordinal (under the canonical well-ordering $\in$) is the null set, $\emptyset$, but when considered as an ordinal it is usually denoted by 0. The next ordinal is the set $\{0\}$, denoted by 1. Then comes the ordinal $\{0, 1\}$, denoted by 2, followed by $3 = \{0, 1, 2\}$, and so on. If α is an ordinal, so too is $\alpha \cup \{\alpha\}$, and there is no ordinal γ strictly between α and $\alpha \cup \{\alpha\}$. We call $\alpha \cup \{\alpha\}$ the *successor* of α, denoted by $\alpha + 1$. Any ordinal of the form $\alpha + 1$ is called a *successor ordinal*. An ordinal α is a successor ordinal iff $\mathrm{succ}(\alpha)$, where $\mathrm{succ}(v_0)$ is the LST-formula

$$\mathrm{On}(v_0) \wedge (\exists v_1 \in v_0)(\forall v_2 \in v_0)(v_2 \in v_1 \vee v_2 = v_1).$$

A non-zero ordinal which is not a successor ordinal is called a *limit ordinal*. If $\lim(v_0)$ is the LST-formula

$$\mathrm{On}(v_0) \wedge (\exists v_1 \in v_0)(v_1 = v_1) \wedge (\forall v_1 \in v_0)(\exists v_2 \in v_0)(v_1 \in v_2),$$

then an ordinal α will be a limit ordinal iff $\lim(\alpha)$. Using the Axiom of Infinity, together with other ZF axioms, it can be shown that a limit ordinal exists. The least limit ordinal is denoted by ω. The elements of the set ω are precisely the finite ordinal numbers, and are called the *natural numbers*. We usually denote natural numbers by m, n, i, j, k, etc. Notice that ω is definable by the formula

$$\lim(v_0) \wedge (\forall v_1 \in v_0)(\mathrm{succ}(v_1) \vee (\forall v_2 \in v_1)(v_2 \neq v_2)).$$

We usually write $\exists \alpha \Phi(\alpha)$ in place of

$$\exists v_0 [\mathrm{On}(v_0) \wedge \Phi(v_0)],$$

and $\forall \alpha \Phi(\alpha)$ in place of

$$\forall v_0 [\mathrm{On}(v_0) \rightarrow \Phi(v_0)].$$

If $(X, <)$ is a well-ordered set, there is a unique ordinal number α such that $(X, <)$ is isomorphic to α (with the usual ordering). This α is called the *order-type* of $(X, <)$, denoted by $\mathrm{otp}(X, <)$.

3.3 (Relations and Functions). Let $n > 0$ be a natural number. The *n-tuple* $(x_1, \ldots, x_n)$ of sets $x_1, \ldots, x_n$ is defined thus:

$$\text{if } n = 1, \qquad (x_1) = x_1;$$
$$\text{if } n > 1, \quad (x_1, \ldots, x_n) = (x_1, (x_2, \ldots, x_n))$$
$$(= (x_1, (x_2, (x_3, \ldots, x_n))) = \text{etc.})$$

If $X_1, X_2, \ldots, X_n$ are classes, their *Cartesian product* is the class:

$$X_1 \times X_2 \times \ldots \times X_n = \{(x_1, \ldots, x_n) \mid x_1 \in X_1 \wedge \ldots \wedge x_n \in X_n\}.$$

We write X^2 in place of $X \times X$, X^3 for $X \times X \times X$, etc.

Let X be any class. An *n-ary relation* on X is a class $R \subseteq X^n$. We often write $R(\vec{x})$ in place of $(\vec{x}) \in R$.

Suppose R is an $(n+1)$-ary relation on a class X, where $n > 0$. The *domain* of R is the class

$$\operatorname{dom}(R) = \{(\vec{x}) \mid \exists y R(y, \vec{x})\}.$$

The *range* of R is the class

$$\operatorname{ran}(R) = \{y \mid \exists \vec{x} R(y, \vec{x})\}.$$

If $Z \subseteq X$, we set

$$R \restriction Z = \{(y, \vec{x}) \in R \mid \vec{x} \in Z\}.$$

(Notice that accoding to our conventions concerning finite strings of variables, $\vec{x} \in Z$ means $x_1 \in Z, \ldots, x_n \in Z$. If we want to mean that $(x_1, \ldots, x_n) \in Z$ we would write $(\vec{x}) \in Z$.)

We define

$$R''Z = \operatorname{ran}(R \restriction Z).$$

Let X be a class, $n > 0$. An *n-ary function* over X is an $(n+1)$-ary relation R on X such that

$$(\forall (\vec{x}) \in \operatorname{dom}(R))(\exists! y) R(y, \vec{x}).$$

We often write $R(\vec{x}) = y$ instead of $R(y, \vec{x})$ in such cases. Thus $R(\vec{x})$ is the unique y such that $R(y, \vec{x})$. We say that R is *total* on X iff $\operatorname{dom}(R) = X$.

Let f be an n-ary function over V. We write

$$f: X \to Y$$

to denote that (f is a function and) $\operatorname{dom}(f) = X$ and $\operatorname{ran}(f) \subseteq Y$. We say that f is *one-one* (or *injective*), and write

$$f: X \xrightarrow{(1-1)} Y,$$

iff for all $x_1, x_2 \in X$,

$$x_1 \neq x_2 \rightarrow f(x_1) \neq f(x_2).$$

We say f is *onto* Y (or is a *surjection* to Y), and write

$$f\colon X \xrightarrow{\text{onto}} Y,$$

iff $\operatorname{ran}(f) = Y$. We say f is a *bijection* iff it is both one-one and onto, and write

$$f\colon X \leftrightarrow Y.$$

If f is bijective there is a unique function $f^{-1}\colon Y \rightarrow X$ (called the *inverse* of f) such that

$$(\forall x \in X)(f^{-1}(f(x)) = x),$$

$$(\forall y \in Y)(f(f^{-1}(y)) = y).$$

Regardless of whether or not f is bijective, we set, for any $Z \subseteq Y$,

$$f^{-1}{''}Z = \{x \in X \mid f(x) \in Z\}.$$

The set $f^{-1}{''}Z$ is called the *preimage of Z under f*.

Notice that by our definition of the n-tuples $(x_1, \ldots, x_n)$, every function is a set of ordered pairs, regardless of whether or not the function is unary.

If X and Y are structures of the same type, we write

$$f\colon X \cong Y$$

if f is a bijection from X to Y which preserves the structure (i.e. if f is an *isomorphism*).

We denote the composition of functions f, g by $f \circ g$, as usual. Thus if $f\colon Y \rightarrow Z$ and $g\colon X \rightarrow Y$, we define $f \circ g\colon X \rightarrow Z$ by

$$(\forall x \in X)(f \circ g(x) = f(g(x))).$$

For any sets x, y we define

$${}^{x}y = \{f \mid f\colon x \rightarrow y\}.$$

The *identity function* is the unary function

$$\mathrm{id} = \{(y, x) \mid y = x\}.$$

Of course, being a proper class, id is not strictly speaking a function at all, but for any set X, $\mathrm{id} \restriction X$ will be a function, so this definition is convenient.

A function whose domain is an ordinal is called a *sequence*; if α is that ordinal domain, we say that the sequence is an α-sequence. If f is an α-sequence, and if $f(\xi) = x_\xi$ for all $\xi < \alpha$, we often write $f = (x_\xi \mid \xi < \alpha)$.

If $f: I \to V$, and if we denote $f(i)$ by x_i for each $i \in I$, we often write $\{x_i \mid i \in I\}$ in place of $f'' I$, $\bigcup_{i \in I} x_i$ in place of $\cup (f'' I)$, and $\bigcap_{i \in I} x_i$ in place of $\cap (f'' I)$. Similarly, given a sequence $f = (x_\nu \mid \nu < \tau)$, we sometimes write $\bigcup_{\nu<\tau} x_\nu$ for $\cup (f'' \tau)$ and $\bigcap_{\nu<\tau} x_\nu$ for $\cap (f'' \tau)$. And if $h = (\alpha_\nu \mid \nu < \tau)$ is a sequence of ordinals, we would write $\sup_{\nu<\tau} \alpha_\nu$ for $\sup (h'' \tau)$.

The inverse functions to the ordered pair function are defined thus:

if $u = (x, y)$, then $(u)_0 = x$ and $(u)_1 = y$;

if u is not an ordered pair, then $(u)_0 = (u)_1 = \emptyset$.

Similarly we define inverse functions $(u)_0^n, \ldots, (u)_{n-1}^n$ to the n-tuple function.

3.4 (Induction and Recursion). By using the Axiom of Foundation (together with other axioms of ZF), every instance of the following *schema of proof* by $\in$*-induction* can be proved in ZF:

$$\forall x [(\forall y \in x)\, \Phi(y) \to \Phi(x)] \to \forall x\, \Phi(x).$$

More generally, if X is any class, a relation $R \subseteq X^2$ is said to be *well-founded* iff:

(i) $(\forall x \in \mathrm{dom}(R))\,[\{y \mid R(y, x)\}$ is a set$]$;
(ii) $\forall a\, [a \neq \emptyset \wedge a \subseteq X \to (\exists x \in a)(\forall y \in a) \neg R(y, x)]$.

If R is such a relation, then every instance of the following *schema of proof by induction on* R is a theorem of ZF:

$$(\forall x \in X)[(\forall y \in X)(R(y, x) \to \Phi(y)) \to \Phi(x)] \to (\forall x \in X)\, \Phi(x).$$

It follows from the above that every instance of the following *schema of definition by recursion* is provable in ZF. Let G be a total $(n+2)$-ary function over V, and let H be a total unary function over V such that the relation $\{(z, y) \mid z \in H(y)\}$ is well-founded. Then there is a unique, total $(n+1)$-ary function F over V such that

$$F(y, \vec{a}) = G(y, \vec{a}, F \upharpoonright (H(y) \times \{(\vec{a})\}))$$

(Actually, some care is required in formulating this result precisely. Given formulas which determine G and H, possibly with reference to certain set parameters, one can explicitly write down a third formula such that in ZF it is provable that the class determined by this formula has all of the properties required of F above. We assume the reader is quite familiar with all of this, though in fact we shall not really need to know the exact formulation. As is usually the case in set theory, all that we require is the knowledge that ZF allows definitions "of a recursive nature".)

A particular case of the above recursion principle is when $H = \mathrm{id}$, when it is called the *principle of* $\in$*-recursion*. There are also recursion principles applicable to functions defined not on all of V but on some class, the most common example being when the class concerned is the class of all ordinals; $\in$-recursion restricted to the ordinals is known as *ordinal recursion*, and will be used frequently in this book.

The total, unary function TC (*transitive closure*) is defined by the $\in$-recursion

$$\mathrm{TC}(x) = x \cup \bigcup \{\mathrm{TC}(y) \mid y \in x\}.$$

(Intuitively, $\mathrm{TC}(x) = x \cup (\bigcup x) \cup (\bigcup\bigcup x) \cup (\bigcup\bigcup\bigcup x) \cup \ldots$. It is not hard to show that $\mathrm{TC}(x)$ is the $\subseteq$-smallest transitive set y such that $x \subseteq y$. We call the set $\mathrm{TC}(x)$ the *transitive closure* of x.

The relation $\{(x, y) \mid x \in \mathrm{TC}(y)\}$ is well-founded. Hence we can carry out definitions by recursion on this relation. This form of recursive definition will also be quite common in this book.

3.5 (The Cumulative Hierarchy). The *cumulative hierarchy of sets* is defined by the ordinal recursion

$$V_0 = \emptyset;$$

$$V_\alpha = \bigcup \{\mathscr{P}(V_\beta) \mid \beta < \alpha\}.$$

It is not hard to see that

$$V = \bigcup \{V_\alpha \mid \alpha \in \mathrm{On}\}.$$

(The proof makes central use of the Axiom of Foundation, and indeed the above "equation" may be taken as an alternative formulation of this axiom.)

The *rank* of a set x is the least ordinal α such that $x \in V_{\alpha+1}$. We may define the rank function directly by means of the $\in$-recursion

$$\mathrm{rank}(\emptyset) = 0$$

$$\mathrm{rank}(x) = \bigcup \{\mathrm{rank}(y) + 1 \mid y \in x\}.$$

Notice that $x \in y$ implies $\mathrm{rank}(x) < \mathrm{rank}(y)$.

4. Ordinal Numbers

The notion of an ordinal number plays a central role in set theory, and we have referred to ordinals several times already. In this section we consider, very briefly, the arithmetic of ordinal numbers.

Let α, β be ordinals. We define the *ordinal sum* $\alpha + \beta$ by recursion on β, thus:

$$\alpha + 0 = \alpha;$$
$$\alpha + (\beta + 1) = (\alpha + \beta) + 1;$$
$$\alpha + \delta = \bigcup\{(\alpha + \beta) \mid \beta < \delta\}, \qquad \text{if } \lim(\delta).$$

The ordinal sum is not commutative; for example, $1 + \omega = \omega$ but $\omega + 1 > \omega$.

The *ordinal product* $\alpha \cdot \beta$ is defined by the following recursion on β:

$$\alpha \cdot 0 = 0;$$
$$\alpha \cdot (\beta + 1) = (\alpha \cdot \beta) + \alpha;$$
$$\alpha \cdot \delta = \bigcup\{\alpha \cdot \beta \mid \beta < \delta\}, \qquad \text{if } \lim(\delta).$$

The ordinal product is not commutative; for example $2 \cdot \omega = \omega$ but $\omega \cdot 2 = \omega + \omega > \omega$.

(Both ordinal sum and ordinal product can be defined in an alternative fashion, but we shall not go into that here.)

Notice that $(\omega \cdot \alpha \mid \alpha \in \mathrm{On})$ enumerates 0 and all the limit ordinals.

For $\alpha > 0$, the *ordinal power* α^β is defined by the following recursion on β:

$$\alpha^0 = 1;$$
$$\alpha^{(\beta+1)} = (\alpha^\beta) \cdot \alpha;$$
$$\alpha^\delta = \bigcup\{\alpha^\beta \mid \beta < \delta\}, \qquad \text{if } \lim(\delta).$$

We shall not be concerned with any of the properties of ordinal exponentiation.

5. Cardinal Numbers

A *cardinal number* (or simply, a *cardinal*) is an ordinal α such that there is no $\beta < \alpha$ for which there is a function $f: \beta \xrightarrow{\text{onto}} \alpha$.

Clearly, $0, 1, 2, 3, \ldots, n, \ldots, \omega$ are all cardinals. All other cardinals are said to be *uncountable*. We generally use κ, λ, μ, to denote cardinals.

Using AC it can be shown that for every set x there is a unique cardinal κ for which there is a bijection $f: \kappa \leftrightarrow x$. We call κ the *cardinality* of x, denoted by $|x|$.

Clearly, if κ is a cardinal, then, recalling that κ is the set $\{\alpha \mid \alpha < \kappa\}$, we have $|\kappa| = \kappa$.

If κ is a cardinal, the least cardinal greater than κ is called the (*cardinal*) *successor* of κ, and is denoted by κ^+. For convenience, we extend this notation so that for any *ordinal* α, α^+ denotes the least *cardinal* greater than α. (So in particular, we have $\alpha^+ = |\alpha|^+$.) Any infinite cardinal of the form κ^+ is called a *successor cardinal*. An infinite cardinal which is not a successor cardinal is called a *limit cardinal*.

Clearly, if lim (β) and $(\kappa_\alpha \mid \alpha < \beta)$ is a strictly increasing sequence of cardinals, then $\sup_{\alpha<\beta} \kappa_\alpha$ is a limit cardinal.

The canonical, monotone enumeration of the infinite cardinals, $(\omega_\alpha \mid \alpha \in \mathrm{On})$, is defined by the following recursion:

$$\omega_0 = \omega;$$

$$\omega_{\alpha+1} = \omega_\alpha^+;$$

$$\omega_\delta = \sup_{\alpha<\delta} \omega_\alpha, \quad \text{if } \lim(\delta).$$

Of course, each cardinal ω_α is also an ordinal. In order to distinguish between the two cases when ω_α is being used as a cardinal and when it is being used as an ordinal, many texts use the symbol $\aleph_\alpha$ ("aleph-α") to denote ω_α regarded as a cardinal, and reserve the notation ω_α for pure ordinal use. However, in this book we shall have very little occasion to use ω_α as an ordinal (in the strict sense), so we shall rely upon the single notation ω_α in all cases.

Notice that ω_α is a limit cardinal iff $\alpha = 0$ or lim (α).

We write

$$\forall \kappa\, \Phi(\kappa)$$

in place of

$$\forall \alpha\, [\alpha \text{ is a cardinal} \rightarrow \Phi(\alpha)].$$

and

$$\exists \kappa\, \Phi(\kappa)$$

in place of

$$\exists \alpha\, [\alpha \text{ is a cardinal} \wedge \Phi(\alpha)].$$

Let $(\kappa_\alpha \mid \alpha < \beta)$ be a sequence of cardinals. The *cardinal sum* $\sum_{\alpha<\beta} \kappa_\alpha$ is defined thus:

$$\sum_{\alpha<\beta} \kappa_\alpha = |\{(\xi, \alpha) \mid \xi < \kappa_\alpha \wedge \alpha < \beta\}|.$$

Clearly, $\sum_{\alpha<\beta} \kappa_\alpha$ is the cardinality of the union of any disjoint collection $\{A_\alpha \mid \alpha < \beta\}$ of sets such that $|A_\alpha| = \kappa_\alpha$.

Clearly,

$$\sum_{\alpha<\beta} \kappa_\alpha = \sup_{\alpha<\beta} \kappa_\alpha.$$

We write $\kappa_0 + \kappa_1$ instead of $\sum_{\alpha<2} \kappa_\alpha$. Provided that at least one of κ_0, κ_1 is infinite, we have

$$\kappa_0 + \kappa_1 = \max(\kappa_0, \kappa_1).$$

We usually rely upon context to distinguish between ordinal and cardinal addition, rather than introduce additional notation. For instance, $\alpha + \beta$ would usually mean ordinal addition, whereas $\kappa + \lambda$ would mean cardinal addition.

The *cardinal product*, $\prod_{\alpha<\beta} \kappa_\alpha$, is defined thus:

$$\prod_{\alpha<\beta} \kappa_\alpha = |\{f \mid f\colon \beta \to \bigcup_{\alpha<\beta} \kappa_\alpha \wedge (\forall \alpha < \beta)(f(\alpha) \in \kappa_\alpha)\}|.$$

Clearly, $\prod_{\alpha<\beta} \kappa_\alpha$ is the cardinality of the *cartesian product*

$$\bigtimes_{\alpha<\beta} A_\alpha = \{f \mid f\colon \beta \to \bigcup_{\alpha<\beta} A_\alpha \mathbin{\&} (\forall \alpha < \beta)(f(\alpha) \in A_\alpha)\}$$

of any family of sets A_α, $\alpha < \beta$, such that $|A_\alpha| = \kappa_\alpha$.

We write $\kappa_0 \cdot \kappa_1$ instead of $\prod_{\alpha<2} \kappa_\alpha$. If at least one of κ_0, κ_1 is infinite and neither is 0, then

$$\kappa_0 \cdot \kappa_1 = \max(\kappa_0, \kappa_1).$$

The *cardinal power*, κ^λ, is defined by

$$\kappa^\lambda = \prod_{\alpha<\lambda} \kappa.$$

(Again, context and notation are used to distinguish between cardinal and ordinal exponentiation.) Recalling that for any sets x, y,

$${}^x y = \{f \mid f\colon x \to y\},$$

we see that if $|x| = \lambda$ and $|y| = \kappa$ then

$$\kappa^\lambda = |{}^x y|.$$

In particular,

$$\kappa^\lambda = |{}^\lambda \kappa| = |\{f \mid f\colon \lambda \to \kappa\}|.$$

By considering characteristic functions of subsets of x, we see easily that

$$|\mathscr{P}(x)| = 2^{|x|},$$

for any set x. Consequently, for any cardinal κ,

$$2^\kappa = |\mathscr{P}(\kappa)|.$$

By means of the well-known Cantor diagonal argument, it follows that

$$2^\kappa > \kappa .$$

Hence

$$2^\kappa \geqslant \kappa^+ .$$

For finite κ, we only have equality in the above in the cases $\kappa = 0$ and $\kappa = 1$. But for infinite κ, the axioms of ZFC set theory do not provide enough information to decide whether or not $2^\kappa = \kappa^+$. (The precise situation is rather complicated, and certainly outside the scope of this book.) The statement

$$2^\omega = \omega_1 ,$$

which is neither provable nor refutable in ZFC, is known as the *continuum hypothesis* (CH). The statement

$$\forall \alpha (2^{\omega_\alpha} = \omega_{\alpha+1}) ,$$

which is likewise neither provable nor refutable in ZFC, is known as the *generalised continuum hypothesis* (GCH). (The word "continuum" is used here because 2^ω is the cardinality of the real number continuum.)

5.1 Lemma.

(i) $\sum$ *and* $\prod$ *are commutative and associative operations on cardinal numbers.*
(ii) $\prod$ *distributes over* $\sum$; *i.e.*

$$\prod_{\alpha<\gamma} \sum_{\beta<\delta} \kappa_{\alpha\beta} = \sum_{f \in {}^\gamma\delta} \prod_{\alpha<\gamma} \kappa_{\alpha f(\alpha)} .$$

(iii) $\kappa . \sum_{\alpha<\beta} \kappa_\alpha = \sum_{\alpha<\beta} \kappa . \kappa_\alpha .$

(iv) $\sum_{\alpha<\beta} \kappa = |\beta| . \kappa .$

Proof. An easy exercise. □

5.2 Lemma. *If* $2 \leqslant \kappa \leqslant \lambda$ *and* $\lambda \geqslant \omega$, *then* $\kappa^\lambda = 2^\lambda$.

Proof. Clearly, $2^\lambda \leqslant \kappa^\lambda$. Conversely,

$$\kappa^\lambda \leqslant \lambda^\lambda \leqslant (2^\lambda)^\lambda = 2^{\lambda \cdot \lambda} = 2^\lambda . \quad \square$$

5.3 Lemma ("The König Inequality"). *If* $\kappa_\alpha < \lambda_\alpha$ *for all* $\alpha < \beta$, *then*

$$\sum_{\alpha<\beta} \kappa_\alpha < \prod_{\alpha<\beta} \lambda_\alpha .$$

Proof. Define

$$f\colon \bigcup_{\alpha<\beta} (\kappa_\alpha \times \{\alpha\}) \to \bigtimes_{\alpha<\beta} \lambda_\alpha$$

by setting

$$[f(\xi, \alpha)](\nu) = \begin{cases} \xi + 1, & \text{if } \nu = \alpha \\ 0, & \text{if } \nu \neq \alpha \end{cases}.$$

Clearly, f is one-one. Hence

$$\sum_{\alpha<\beta} \kappa_\alpha \leqslant \prod_{\alpha<\beta} \lambda_\alpha .$$

We show that equality is impossible. Let

$$h\colon \bigcup_{\alpha<\beta} (\kappa_\alpha \times \{\alpha\}) \to \bigtimes_{\alpha<\beta} \lambda_\alpha .$$

We show that h cannot be onto. For $\gamma < \beta$, define

$$h_\gamma\colon \bigcup_{\alpha<\beta} (\kappa_\alpha \times \{\alpha\}) \to \lambda_\gamma$$

by

$$h_\gamma(\xi, \alpha) = [h(\xi, \alpha)](\gamma).$$

Since $\kappa_\gamma < \lambda_\gamma$, $h_\gamma \upharpoonright (\kappa_\gamma \times \{\gamma\})$ cannot map onto λ_γ, so we can pick

$$a_\gamma \in \lambda_\gamma - h_\gamma''(\kappa_\gamma \times \{\gamma\}).$$

Define $g \in \bigtimes_{\alpha<\beta} \lambda_\alpha$ by

$$g(\gamma) = a_\gamma \qquad (\gamma < \beta).$$

Clearly, $g \notin \operatorname{ran}(h)$, so h is not onto, and we are done. □

5.4 Lemma. *Let κ_α be cardinals for $\alpha < \beta$, and set*

$$\kappa = \sum_{\alpha<\beta} \kappa_\alpha .$$

For any cardinal λ,

$$\lambda^\kappa = \prod_{\alpha<\beta} \lambda^{\kappa_\alpha}.$$

Proof. Let

$$X = \bigcup_{\alpha<\beta} (\kappa_\alpha \times \{\alpha\}).$$

Thus $|X| = \kappa$. Let $f \in {}^X\lambda$. For $\alpha < \beta$, define $f_\alpha \in {}^{(\kappa_\alpha)}\lambda$ by

$$f_\alpha(\nu) = f(\nu, \alpha).$$

Then

$$(f_\alpha \mid \alpha < \beta) \in \bigtimes_{\alpha<\beta} ({}^{\kappa_\alpha}\lambda).$$

Since the mapping

$$f \mapsto (f_\alpha \mid \alpha < \beta)$$

from ${}^X\lambda$ to $\bigtimes_{\alpha<\beta} ({}^{\kappa_\alpha}\lambda)$ is clearly one-one and onto, the desired equality is proved. □

A subset A of a limit[2] ordinal α is said to be *unbounded* in α iff for no $\gamma < \alpha$ do we have $A \subseteq \gamma$. Equivalently, $A \subseteq \alpha$ is unbounded in α iff

$$(\forall \nu \in \alpha)(\exists \tau \in A)(\tau \geqslant \nu).$$

(i.e. iff $\sup(A) = \alpha$.)

Let γ, α be limit ordinals. A function $f\colon \gamma \to \alpha$ is said to be *cofinal* iff f is order-preserving and $\operatorname{ran}(f)$ is an unabounded subset of α.

Let α be a limit ordinal. The *cofinality* of α is the least ordinal γ such that there is a cofinal function $f\colon \gamma \to \alpha$. We denote the cofinality of α by $\operatorname{cf}(\alpha)$. It is easily seen that $\operatorname{cf}(\alpha)$ is always a cardinal.

A limit ordinal α is said to be *regular* iff $\operatorname{cf}(\alpha) = \alpha$; otherwise it is *singular*. Every regular ordinal is a cardinal. Also, $\operatorname{cf}(\alpha)$ is always a regular cardinal. The cardinal ω is regular; the cardinal ω_ω is singular of cofinality ω.

5.5 Lemma. *Let κ be an infinite cardinal. Then $\operatorname{cf}(\kappa)$ is the least α such that there are cardinals $\kappa_\xi < \kappa$, $\xi < \alpha$ such that*

$$\kappa = \sum_{\xi<\alpha} \kappa_\xi.$$

Proof. Let $\lambda = \operatorname{cf}(\kappa)$, and let α be least such that $\kappa = \sum_{\xi<\alpha} \kappa_\xi$ for some $\kappa_\xi < \kappa$. We must show that $\lambda = \alpha$.

Let $(\gamma_\xi \mid \xi < \lambda)$ be cofinal in κ. For each $\xi < \lambda$, $|\gamma_\xi| < \kappa$. But $\kappa = \bigcup_{\xi<\lambda} \gamma_\xi$. It follows easily that $\kappa = \sum_{\xi<\lambda} |\gamma_\xi|$. Hence $\alpha \leqslant \lambda$.

Suppose that $\alpha < \lambda$. Pick $\kappa_\xi < \kappa$ for $\xi < \alpha$ so that $\kappa = \sum_{\xi<\alpha} \kappa_\xi$. Since $\alpha < \lambda$, $(\kappa_\xi \mid \xi < \alpha)$ is not cofinal in κ, so for some $\gamma < \kappa$, we have $\kappa_\xi \leqslant \gamma$ for all $\xi < \alpha$. Hence $\sum_{\xi<\alpha} \kappa_\xi \leqslant \sum_{\xi<\alpha} |\gamma| = |\alpha| \cdot |\gamma| < \kappa$, which is a contradiction. Thus $\alpha = \lambda$. □

2 On a formal level, this and the following definitions can be applied to any ordinal, but the notions are trivial in the case of successor ordinals.

5.6 Lemma. *Let κ be an infinite cardinal. Then κ is regular iff*

$$(\forall \lambda < \kappa)({}^{\lambda}\kappa = \bigcup_{\alpha<\kappa} {}^{\lambda}\alpha).$$

Proof. ($\rightarrow$) Let $\lambda < \kappa$. If $f \in \bigcup_{\alpha<\kappa} {}^{\lambda}\alpha$, then $f \in {}^{\lambda}\kappa$. Conversely, if $f \in {}^{\lambda}\kappa$, then since $\mathrm{cf}(\kappa) = \kappa > \lambda$, $\mathrm{ran}(f) \subseteq \alpha$ for some $\alpha < \kappa$, so $f \in {}^{\lambda}\alpha$. Thus ${}^{\lambda}\kappa = \bigcup_{\alpha<\kappa} {}^{\lambda}\alpha$.

($\leftarrow$) Let $\lambda < \kappa$, $f\colon \lambda \rightarrow \kappa$. For some $\alpha < \kappa, f \in {}^{\lambda}\alpha$, so f cannot be cofinal. Hence κ is regular. □

5.7 Lemma. *Let κ be an infinite cardinal. Then κ^{+} is regular.*

Proof. Suppose $\mathrm{cf}(\kappa^{+}) \leqslant \kappa$. By 5.5 there are cardinals $\kappa_\alpha < \kappa^{+}$ such that

$$\kappa^{+} \leqslant \sum_{\alpha<\kappa} \kappa_\alpha .$$

Then

$$\kappa^{+} \leqslant \sum_{\alpha<\kappa} \kappa_\alpha \leqslant \sum_{\alpha<\kappa} \kappa = \kappa \cdot \kappa = \kappa,$$

a contradiction. □

5.8 Lemma. *Let κ be an infinite cardinal. Then $\kappa^{\mathrm{cf}(\kappa)} > \kappa$.*

Proof. Let $\kappa = \sum_{\alpha<\mathrm{cf}(\kappa)} \kappa_\alpha$, where $\kappa_\alpha < \kappa$. By 5.3 we have

$$\kappa = \sum_{\alpha<\mathrm{cf}(\kappa)} \kappa_\alpha < \prod_{\alpha<\mathrm{cf}(\kappa)} \kappa = \kappa^{\mathrm{cf}(\kappa)}. \quad \square$$

The following result shows that under certain circumstances $\mathrm{cf}(\kappa)$ may be the least cardinal λ such that $\kappa^{\lambda} > \kappa$.

5.9 Lemma. *Let κ be an infinite cardinal. If $\lambda < \mathrm{cf}(\kappa)$ and $(\forall \mu < \kappa)(2^{\mu} \leqslant \kappa)$, then $\kappa^{\lambda} = \kappa$.*

Proof. By an argument as in 5.6 we see that if $\lambda < \mathrm{cf}(\kappa)$,

$$\kappa^{\lambda} = |{}^{\lambda}\kappa| = |\bigcup_{\alpha<\kappa} {}^{\lambda}\alpha| \leqslant \sum_{\alpha<\kappa} |\alpha|^{\lambda}.$$

But for $\alpha < \kappa$ we have

$$|\alpha|^{\lambda} \leqslant (2^{|\alpha|})^{\lambda} = 2^{|\alpha| \cdot \lambda} \leqslant \kappa .$$

Hence

$$\kappa \leqslant \kappa^{\lambda} \leqslant \kappa . \quad \square$$

We define the *weak power* of κ by λ thus:

$$\kappa^{<\lambda} = \sum_{\mu<\lambda} \kappa^{\mu} .$$

5.10 Lemma. *Let κ be an infinite cardinal.*

(i) *If κ is regular and $(\forall \mu < \kappa)(2^\mu \leqslant \kappa)$, then $\kappa^{<\kappa} = \kappa$.*
(ii) *Assume* GCH. *Then κ is regular iff $\kappa^{<\kappa} = \kappa$.*

Proof. (i) By 5.9, $\kappa^\lambda = \kappa$ for all $\lambda < \text{cf}(\kappa) = \kappa$. Thus

$$\kappa^{<\kappa} = \sum_{\lambda<\kappa} \kappa^\lambda = \sum_{\lambda<\kappa} \kappa = \kappa.$$

(ii) If κ is regular then (since GCH implies $2^\mu = \mu^+ \leqslant \kappa$ for all $\mu < \kappa$) by (i) we have $\kappa^{<\kappa} = \kappa$. If κ is singular, then by 5.8, $\kappa^{<\kappa} \geqslant \kappa^{\text{cf}(\kappa)} > \kappa$. □

5.11 Lemma. *Let κ be an infinite cardinal.*

(i) $\kappa^\kappa \geqslant \kappa^{<\kappa} \geqslant 2^{<\kappa} \geqslant \kappa$.
(ii) *If $\kappa = \lambda^+$, then $\kappa^{<\kappa} = 2^{<\kappa} = 2^\lambda$.*

Proof. (i) The only non-trivial inequality is $2^{<\kappa} \geqslant \kappa$. And this follows quite easily from the fact that for every $\lambda < \kappa$, $2^{<\kappa} \geqslant 2^\lambda > \lambda$.

(ii) We have $2^{<\kappa} = 2^\lambda$ and $\kappa^{<\kappa} = \kappa^\lambda$. Hence

$$2^\lambda = 2^{<\kappa} \leqslant \kappa^{<\kappa} = \kappa^\lambda \leqslant (2^\lambda)^\lambda = 2^\lambda. \quad \square$$

5.12 Lemma. GCH $\leftrightarrow (\forall \kappa \geqslant \omega)(2^{<\kappa} = \kappa)$.

Proof. ($\rightarrow$) By GCH,

$$2^{<\kappa} = \sum_{\lambda<\kappa} 2^\lambda = \sum_{\lambda<\kappa} \lambda^+ = \kappa,$$

for any infinite κ.

($\leftarrow$) For any infinite κ,

$$2^\kappa = 2^{<\kappa^+} = \kappa^+. \quad \square$$

6. *Closed Unbounded Sets*

Let α be a limit ordinal. Recall that a set $A \subseteq \alpha$ is *unbounded* in α iff

$$(\forall \nu \in \alpha)(\exists \tau \in A)(\tau \geqq \nu).$$

A set $A \subseteq \alpha$ is *closed* in α iff $\bigcup (A \cap \gamma) \in A$ for all $\gamma < \alpha$. Equivalently, if we define a *limit point* of A to be any limit ordinal γ such that $A \cap \gamma$ is unbounded in γ, then A will be closed in α iff it contains all its limit points below α. Still another formulation is that A is closed in α iff, whenver $\lim(\tau)$ and $(\alpha_\nu \mid \nu < \tau)$ is a strictly increasing sequence of elements of A which is not cofinal in α, then $\bigcup_{\nu<\tau} \alpha_\nu \in A$.

We use the abbreviation "club" to mean "closed and unbounded". Club sets play an important role in our development.

6.1 Lemma. *Let κ be an infinite cardinal, $\mathrm{cf}(\kappa) > \omega$. If A, B are club subsets of κ, then $A \cap B$ is club in κ.*

Proof. That $A \cap B$ is closed in κ is obvious. To establish unboundedness, let $\alpha < \kappa$ be given. Pick $\beta_0 \in A$, $\beta_0 > \alpha$. By recursion now, let β_{2n+1} be the least member of B than β_{2n} and let β_{2n+2} be the least member of A greater than β_{2n+1}. Let $\beta = \bigcup_{n<\omega} \beta_n$. Since $\beta = \bigcup_{n<\omega} \beta_{2n}$ and A is closed we have $\beta \in A$, and similary $\beta = \bigcup_{n<\omega} \beta_{2n+1}$ implies $\beta \in B$. Thus $\beta \in A \cap B$, $\beta > \alpha$, and we are done. □

By generalising the above proof we obtain:

6.2 Lemma. *Let κ be an uncountable regular cardinal. If $\lambda < \kappa$ and A_ν, $\nu < \lambda$, are club subsets of κ, then $\bigcap_{\nu<\lambda} A_\nu$ is a club subset of κ.* □

Let α be an infinite ordinal. A non-decreasing function $f\colon \alpha \to \mathrm{On}$ is said to be *continuous* if for every limit ordinal $\delta < \alpha$,

$$f(\delta) = \bigcup_{\beta<\delta} f(\beta).$$

A *normal* function on α is a (strictly) increasing, continuous function $f\colon \alpha \to \mathrm{On}$.

6.3 Lemma. *Let α be a limit ordinal. If f is an increasing function from α into* On, *then $f(\gamma) \geqslant \gamma$ for all $\gamma \in \alpha$.*

Proof. By induction on γ. If the result holds below γ, then for all $\beta < \gamma$ we have $f(\gamma) > f(\beta) \geqslant \beta$, and hence $f(\gamma) \geqslant \gamma$. □

6.4 Lemma. *Let κ be an uncountable regular cardinal.*

(i) *If $A \subseteq \kappa$ is club, then the enumeration of A in increasing oder (as ordinals) is a normal function from κ to κ.*

(ii) *If $f\colon \kappa \to \kappa$ is a normal function, then* $\mathrm{ran}(f)$ *is a club subset of κ.*

Proof. Trivial. □

6.5 Lemma. *Let κ be an uncountable regular cardinal, f a normal function from κ to κ. Then the set $\{\alpha \in \kappa \mid f(\alpha) = \alpha\}$ is club in κ.*

Proof. Let

$$A = \{\alpha \in \kappa \mid f(\alpha) = \alpha\}.$$

If $\gamma < \kappa$ is a limit point of A, then

$$f(\gamma) = \bigcup_{\alpha<\gamma} f(\alpha) = \bigcup_{\alpha \in A \cap \gamma} f(\alpha) = \bigcup_{\alpha \in A \cap \gamma} \alpha = \gamma$$

(since $\bigcup (A \cap \gamma) = \gamma$), so $\gamma \in A$. Hence A is closed in κ.

To show that A is unbounded in κ, let $\alpha_0 \in \kappa$ be given. By recursion, define $\alpha_{n+1} = f(\alpha_n)$. Set $\alpha = \bigcup_{n<\omega} \alpha_n$. Then by 6.3, we have $\alpha \geqslant \alpha_0$ and, by continuity,

$$f(\alpha) = \bigcup_{n<\omega} f(\alpha_n) = \bigcup_{n<\omega} \alpha_{n+1} = \alpha,$$

so $\alpha \in A$. Hence A is unbounded in κ. □

6.6 Lemma. *Let κ be an uncountable regular cardinal, and let $h\colon \kappa \to \kappa$. Set*

$$A = \{\gamma \in \kappa \mid (\forall \nu < \gamma)(h(\nu) < \gamma)\}.$$

Then A is club in κ.

Proof. It is immediate that A is closed in κ. To prove unboundedness, given $\alpha_0 \in \kappa$, define, by recursion, α_{n+1} to be the least γ such that $h''\alpha_n \subseteq \gamma$, and set $\alpha = \bigcup_{n<\omega} \alpha_n$. Clearly, $\alpha \geqslant \alpha_0$ and $\alpha \in A$. □

7. The Collapsing Lemma

In this section we prove a simple lemma which is of extreme importance in constructibility theory: the Mostowski-Shepherdon Collapsing Lemma. We start with a definition.

A set X is said to be *extensional* if:

$$(\forall u, v \in X)(u \neq v \to (\exists x \in X)(x \in u \leftrightarrow x \notin v))$$

(In other words, a set X is extensional iff the structure $\langle X, \in \rangle$ is a model of the Axiom of Extensionality.)

7.1 Theorem (The Collapsing Lemma). *Let X be an extensional et. Then there is a unique transitive set M and a unique bijection $\pi\colon X \leftrightarrow M$ such that*

$$\pi\colon \langle X, \in \rangle \cong \langle M, \in \rangle.$$

Moreover, if $Y \subseteq X$ is transitive, then $\pi \restriction Y = \mathrm{id} \restriction Y$. (The transitive set M is called the *transitive collapse* or *transitivisation* of X.)

Proof. We first of all see what a function π as in the theorem must look like. This will amount to a proof of the uniqueness of π and M. So suppose that

$$\pi\colon \langle X, \in \rangle \cong \langle M, \in \rangle,$$

where M is transitive. Let $x, y \in X$, $y \in x$. By isomorphism, $\pi(y) \in \pi(x)$. Hence

$$\{\pi(y) \mid y \in X \wedge y \in x\} \subseteq \pi(x).$$

Now let $x \in X$, $z \in \pi(x)$. Since M is transitive, $z \in M$. So for some $y \in X$, $z = \pi(y)$. Then $\pi(y) \in \pi(x)$, so as π is an isomorphism, $y \in x$. Thus

$$\pi(x) \subseteq \{\pi(y) \mid y \in X \wedge y \in x\}.$$

Hence

$$\pi(x) = \{\pi(y) \mid y \in X \wedge y \in x\}.$$

This shows us what form π must have, as well as providing a proof of the uniqueness of such a π (and hence also of $M = \operatorname{ran}(\pi)$). (More precisely, since π, M necessarily have the above structure, given any two candidates π_1, M_1 and π_2, M_2, a trivial $\in$-induction shows that $\pi_1(x) = \pi_2(x)$ for all $x \in X$, so $\pi_1 = \pi_2$ and $M_1 = M_2$.)

We prove existence. Define π on X by the $\in$-recursion:

$$\pi(x) = \{\pi(y) \mid y \in X, y \in x\}.$$

Set $M = \operatorname{ran}(\pi)$. We show that π, M are as required.

We prove first that π is one-one. Suppose not, and pick $x_1 \in X$ of least rank such that for some $x_2 \in X$, $x_2 \neq x_1$ and $\pi(x_2) = \pi(x_1)$. Since X is extensional there is a $y \in X$ such that either $y \in x_1$, $y \notin x_2$ or else $y \notin x_1$, $y \in x_2$.

Suppose first that $y \in x_1$, $y \notin x_2$. By definition of $\pi(x_1)$ we have $\pi(y) \in \pi(x_1)$. Hence as $\pi(x_1) = \pi(x_2)$, we have $\pi(y) \in \pi(x_2)$. So by definition of $\pi(x_2)$ there must be a $z \in X$, $z \in x_2$, such that $\pi(y) = \pi(z)$. Since $y \notin x_2$, $y \neq z$. But $\operatorname{rank}(y) < \operatorname{rank}(x_1)$ so the existence of such a z contradicts the choice of x_1.

Now suppose that $y \notin x_1$, $y \in x_2$. By definition of $\pi(x_2)$, we have $\pi(y) \in \pi(x_2)$. Hence as $\pi(x_1) = \pi(x_2)$, we have $\pi(y) \in \pi(x_1)$. So by definition of $\pi(x_1)$ there must be a $z \in X$, $z \in x_1$, such that $\pi(y) = \pi(z)$. Since $y \notin x_1$, $y \neq z$. But $\operatorname{rank}(z) < \operatorname{rank}(x_1)$, so the existence of such a y (for z) contradicts the choice of x_1.

Having established that π is one-one, we show next that π is an $\in$-isomorphism. Let $x, y \in X$. If $x \in y$, then by definition of $\pi(y)$ we have $\pi(x) \in \pi(y)$. Conversely, suppose $\pi(x) \in \pi(y)$. Then by definition of $\pi(y)$, $\pi(x) = \pi(z)$ for some $z \in X$, $z \in y$. But π is one-one. Hence $x = z$, giving $x \in y$. Thus π is an $\in$-isomorphism.

Finally, suppose $Y \subseteq X$ is transitive. Then for $x \in Y$ we have

$$y \in x \rightarrow y \in Y,$$

so for $x \in Y$ we can write the definition of π as

$$\pi(x) = \{\pi(y) \mid y \in x\}.$$

Suppose now that $\pi \upharpoonright Y \neq \mathrm{id} \upharpoonright Y$. Pick $x \in Y$ of minimal rank such that $\pi(x) \neq x$. Then for $y \in x$, we must have $\pi(y) = y$, so

$$\pi(x) = \{\pi(y) \mid y \in x\} = \{y \mid y \in x\} = x,$$

a contradiction. The proof is complete. □

By an analogous argument we may also prove the following result:

7.2 Theorem (The Representation Lemma). *Let X be a set (or, more generally, a class), E a binary relation on X such that:*

(i) $(\forall u, v \in X)(u \neq v \rightarrow (\exists x \in X)(xEu \leftrightarrow \neg xEv))$;
(ii) *E is well-founded.*

Then there is a unique transitive set (in the general case a class) M and a unique map π such that

$$\pi: \langle X, E\rangle \cong \langle M, \in\rangle.$$

(M is called the *transitivisation* of $\langle X, E\rangle$.) □

8. Metamathematics of Set Theory

In this section we establish various metamathematical results about the theory ZF. We commence with the well-known *reflection principle* of Montague and Lévy. Loosely speaking, this says that any valid sentence Φ of LST "reflects down" to some initial section V_α of V (i.e. is valid in V_α).

We first of all prove a "generalised reflection principle". We need some preliminaries.

Let M be any class. For each formula Φ of LST we define a new formula Φ^M, called the *relativisation* of Φ to M. Φ^M is also a formula of LST. This may not be immediately clear from our definition, since it will appear that we are using a unary predicate letter M. But since M is a class, it must be defined by some formula of LST, and by replacing all mention of "M" in our definition by this formula we obtain a formula of LST.

The idea is that Φ^M should make the same assertion as Φ, but referring only to the sets in M. In particular, all quantifiers in Φ^M should range only over M. The formal definition of Φ^M proceeds by unravelling the logical construction of Φ and using the following rules.

If Φ is primitive, then $\Phi^M = \Phi$.

If Φ is of the form $\Phi_1 \wedge \Phi_2$, then $\Phi^M = \Phi_1^M \wedge \Phi_2^M$.

If Φ is of the form $\neg \Phi_0$, then $\Phi^M = \neg (\Phi_0^M)$.

If Φ is of the form $\exists v_n \Phi_0$, then $\Phi^M = (\exists v_n \in M)(\Phi_0^M)$. More precisely, if Θ is the LST formula such that

$$M = \{x \mid \Theta(x)\},$$

then Φ^M is the formula

$$\exists v_n [\Theta(v_n) \wedge \Phi_0^M].$$

(This is the really significant clause of course, being the only one which involves M.)

8.1 Theorem (Generalised Reflection Principle). *Let $(W_\alpha \mid \alpha \in \mathrm{On})$ be a hierarchy of transitive sets, definable by a formula, Ψ, of* LST *in the sense that*

$$W_\alpha = \{x \mid \Psi(x, \alpha)\},$$

and suppose that:

(i) $\alpha < \beta \to W_\alpha \subseteq W_\beta$;

(ii) $\lim(\delta) \to W_\delta = \bigcup_{\alpha<\delta} W_\alpha$.

Let

$$W = \bigcup_{\alpha \in \mathrm{On}} W_\alpha.$$

Let $\Phi(\vec{v})$ be an LST*-formula with free variables amongst $\vec{v}$. Then the following sentence is a theorem of* ZF:

$$(\forall \alpha)(\exists \beta > \alpha)[\lim(\beta) \wedge (\forall \vec{v} \in W_\beta)[\Phi^W(\vec{v}) \leftrightarrow \Phi^{W_\beta}(\vec{v})]].$$

Proof. Let $\Phi(\vec{v})$ be a given formula of LST. Let[3] $\Phi_0(\vec{x}_0), \ldots, \Phi_n(\vec{x}_n)$ be a sequence of LST-formulas such that $\Phi_n = \Phi$ and for each $i = 0, \ldots, n$, either Φ_i is a primitive formula or else is obtained from previous formulas in the sequence by a direct application of negation, conjunction, or existential quantification. (The existence of such a sequence follows from the definition of a formula of LST.) We define ordinal-valued functions $f_i(\vec{x}_i)$, $i = 0, \ldots, n$, as follows. If Φ_i is primitive or of the form $\neg \Phi_j$ for some $j < i$, or of the form $\Phi_j \wedge \Phi_k$ for some $j, k < i$, let $f_i(\vec{x}_i) = 0$. If $\Phi_i(\vec{x}_i)$ is of the form $\exists y \Phi_j(y, \vec{x}_i)$ for some $j < i$, let $f_i(\vec{x}_i)$ be the least ordinal γ such that

$$(\exists y \in W)\, \Phi_j^W(y, \vec{x}_i) \to (\exists y \in W_\gamma)\, \Phi_j^W(y, \vec{x}_i).$$

Given α now, let $\beta > \alpha$ be a limit ordinal such that for each $i = 0, \ldots, n$,

$$(\forall \vec{x}_i \in W_\beta)(f_i(\vec{x}_i) < \beta).$$

Using the Axiom of Collection, it is easy to show that such a β exists. We prove by induction on $i = 0, \ldots, n$ that for $\vec{x}_i \in W_\beta$,

$$\Phi_i^W(\vec{x}_i) \leftrightarrow \Phi_i^{W_\beta}(\vec{x}_i).$$

If Φ_i is primitive (in particular, if $i = 0$) this is immediate. And if Φ_i is $\neg \Phi_j$ or $\Phi_j \wedge \Phi_k$ (where $j, k < i$), the induction step is trivial. So suppose that $\Phi_i(\vec{x}_i)$ is $\exists y \Phi_j(y, \vec{x}_i)$, where $j < i$. Let $\vec{x}_i \in W_\beta$.

Assume first that $\Phi_i^W(\vec{x}_i)$. Thus

$$(\exists y \in W)\, \Phi_j^W(y, \vec{x}_i).$$

---

3 For the convention regarding expressions such as $\vec{x}_0, \ldots, \vec{x}_n$, see page 4.

Since $f_i(\vec{x}_i) < \beta$, it follows that

$$(\exists y \in W_\beta)\, \Phi_j^W(y, \vec{x}_i).$$

But for $y, x_i \in W_\beta$, the induction hypothesis gives

$$\Phi_j^W(y, \vec{x}_i) \leftrightarrow \Phi_j^{W_\beta}(y, \vec{x}_i).$$

Hence $\Phi_i^{W_\beta}(\vec{x}_i)$.

Now assume that $\Phi_i^{W_\beta}(\vec{x}_i)$. Thus

$$(\exists y \in W_\beta)\, \Phi_j^{W_\beta}(y, \vec{x}_i).$$

Using the induction hypothesis, it follows that

$$(\exists y \in W_\beta)\, \Phi_j^W(y, \vec{x}_i).$$

But $W_\beta \subseteq W$. Hence

$$(\exists y \in W)\, \Phi_j^W(y, \vec{x}_i).$$

In other words, $\Phi_i^W(\vec{x}_i)$.

The proof is complete. □

8.2 Corollary (The Reflection Principle). *Let $\Phi(\vec{v})$ be any formula of* LST *with free variables amongst $\vec{v}$. Then the following sentence is provable in* ZF:

$$(\forall \alpha)(\exists \beta > \alpha)[\lim(\beta) \wedge (\forall \vec{v} \in V_\beta)[\Phi(\vec{v}) \leftrightarrow \Phi^{V_\beta}(\vec{v})]]. \quad \square$$

Using 8.2, it is quite easy to show that ZF is not finitely axiomatisable. For suppose there were a finite set $\{\Phi_1, \ldots, \Phi_n\}$ of LST-sentences which yielded all of the ZF axioms. Let

$$\Phi = \Phi_1 \wedge \ldots \wedge \Phi_n.$$

Thus Φ is a single axiom for the theory ZF. By the Reflection Principle there is an ordinal α such that Φ^{V_α}. Let α be the least such. Then V_α is a model of ZF. Hence we can apply the Reflection Principle *within* V_α to find an ordinal $\beta \in V_\alpha$ such that, *within* V_α, Φ^{V_β}. But this implies that Φ^{V_β} is valid in the real world, and since $\beta < \alpha$, this contradicts the choice of α, and we are done. Notice that we have made various assumptions in this argument. Firstly, that the set "V_β" as constructed within the "universe" V_α is the same as the real V_β, constructed in V. Secondly, we assumed that if Φ^{V_β} is true inside V_α, then Φ^{V_β} really is true. These are easily verified examples of the important general concept of *absoluteness*, discussed below.

Let M be a transitive class and let $\Phi(\vec{x})$ be an LST-formula. We say $\Phi(\vec{x})$ is *downward absolute (D-absolute) for M* iff

$$(\forall \vec{x} \in M)(\Phi(\vec{x}) \rightarrow \Phi^M(\vec{x})).$$

We say $\Phi(\vec{x})$ is *upward absolute* (*U-absolute*) *for M* iff

$$(\forall \vec{x} \in M)(\Phi^M(\vec{x}) \rightarrow \Phi(\vec{x})).$$

Finally, we say $\Phi(\vec{x})$ is *absolute for M* iff it is both D-absolute and U-absolute for M.

In cases where it is clear which class M is concerned, we often drop the explicit mention of M, and say, for example, simply that "$\Phi(\vec{x})$ is absolute".

It is clear that primitive LST-formulas will be absolute for any class M. For other formulas we usually need to assume that the class M is transitive; in which case absoluteness is related to the logical complexity of the formula. In this connection, a classification of the logical complexity of formulas due to Lévy is useful.

In order to describe the *Lévy hierarchy* in the simplest fashion, it is convenient to regard both universal quantifiers and the two types of bounded quantifier as integral parts of LST, rather than as mere abbreviations.

Let Φ be an LST-formula. We say that Φ is Σ_0 (or Π_0) iff it contains no unbounded quantifiers. Thus the only quantifiers in Φ will be of the form $(\forall x \in y)$ or $(\exists x \in y)$. Now let $n \geqslant 1$. Recursively, we say that Φ is Σ_n iff it is of the form $\exists \vec{x}\, \Psi(\vec{x})$ where $\Psi(\vec{x})$ is Π_{n-1}, and that Φ is Π_n iff it is of the form $\forall \vec{x}\, \Psi(\vec{x})$ where $\Psi(\vec{x})$ is Σ_{n-1}.

Thus, to say that a formula is Σ_n is to say that the formula consists of a Σ_0 formula preceded by n blocks of like quantifiers, commencing with a block of existential quantifiers and alternating between blocks of existential quantifiers and blocks of universal quantifiers.

A formula Φ is said to be Σ_n^{ZF} iff there is a Σ_n formula Ψ such that

$$ZF \vdash \Phi \leftrightarrow \Psi.$$

Similarly Π_n^{ZF}. For $n \geqslant 1$, a formula Φ is said to be Δ_n^{ZF} iff it is both Σ_n^{ZF} and Π_n^{ZF}.

If T is some subtheory of ZF, we define Σ_n^T, Π_n^T, Δ_n^T in a similar fashion, with T in place of ZF.

The following simple result is fundamental to much of our later work.

8.3 Lemma. *Let T be some subtheory of* ZF *(possibly* ZF *itself). Let M be a transitive class such that Ψ^M for every axiom Ψ of T. Let Φ be any formula of* LST.

(i) *If Φ is Σ_0^T, then Φ is absolute (for M).*
(ii) *If Φ is Σ_1^T, then Φ is U-absolute.*
(iii) *If Φ is Π_1^T, then Φ is D-absolute.*
(iv) *If Φ is Δ_1^T, then Φ is absolute.*

Proof. (i) Let Φ have its free variables amongst $\vec{v}$, and let $\Psi(\vec{v})$ be a Σ_0 formula of LST such that

$$T \vdash \Phi \leftrightarrow \Psi.$$

Since T is a subtheory of ZF,

$$\forall \vec{v}\,[\Phi \leftrightarrow \Psi]$$

is a valid assertion. And since Θ^M for every axiom Θ of T,

$$(\forall \vec{v}\,[\Phi \leftrightarrow \Psi])^M$$

is a valid assertion, which is to say that

$$(\forall \vec{v} \in M)\,[\Phi^M \leftrightarrow \Psi^M]$$

is valid. Hence it suffices to prove the result for Ψ. In other words, there is no loss of generality in assuming that Φ is itself a Σ_0 formula.

If Φ is primitive, the result is immediate. We proceed now by induction on the construction of Φ. If Φ is of the form $\Phi_1 \wedge \Phi_2$ or of the form $\neg \Phi_0$, the result for Φ follows trivially from the result for Φ_0, Φ_1, Φ_2. Suppose that $\Phi(y, \vec{v})$ has the form $(\exists x \in y)\,\Psi(x, y, \vec{v})$, where the result is valid for Ψ. Let $y, \vec{v} \in M$. If $\Phi(y, \vec{v})^M$, then $[(\exists x \in y)\,\Psi(x, y, \vec{v})]^M$ so for some $x \in M$ such that $x \in y$ we have $\Psi(x, y, \vec{v})^M$. So by induction hypothesis, $\Psi(x, y, \vec{v})$. Hence $(\exists x \in y)\,\Psi(x, y, \vec{v})$, which means that $\Phi(y, \vec{v})$. Conversely, suppose $\Phi(y, \vec{v})$. Thus $(\exists x \in y)\,\Psi(x, y, \vec{v})$, so for some $x \in y$, $\Psi(x, y, \vec{v})$. But M is transitive, so as $y \in M$ we have $x \in M$ also. (Note the importance of transitivity here.) Hence by the induction hypothesis we can conclude that $\Psi(x, y, \vec{v})^M$, and hence that $[(\exists x \in y)\,\Psi(x, y, \vec{v})]^M$, i.e. $\Phi(y, \vec{v})^M$. The case where Φ has the form $(\forall x \in y)\,\Psi(x, y, \vec{v})$ is handled similarly. This proves (i).

(ii) As in part (i) there is no loss in generality in assuming that Φ is a Σ_1 formula. Let $\Phi = \Phi(\vec{v}) = \exists \vec{u}\,\Psi(\vec{u}, \vec{v})$, where Ψ is Σ_0. Assume $\Phi(\vec{v})^M$, where $\vec{v} \in M$. Thus for some $\vec{u} \in M$, $\Psi(\vec{u}, \vec{v})^M$. By part (i), it follows that $\Psi(\vec{u}, \vec{v})$. Thus $\exists \vec{u}\,\Psi(\vec{u}, \vec{v})$, i.e. $\Phi(\vec{v})$.

(iii) As before we may assume that Φ is Π_1. Let $\Phi = \Phi(\vec{v}) = \forall \vec{u}\,\Psi(\vec{u}, \vec{v})$, where Ψ is Σ_0. Assume $\Phi(\vec{v})$, where $\vec{v} \in M$. Then for all $\vec{u}$, $\Psi(\vec{u}, \vec{v})$. In particular, for all $\vec{u} \in M$, $\Psi(\vec{u}, \vec{v})$. But by part (i), if $\vec{u} \in M$, then $\Psi(\vec{u}, \vec{v})$ implies $\Psi(\vec{u}, \vec{v})^M$. Hence for all $\vec{u} \in M$, $\Psi(\vec{u}, \vec{v})^M$. In other words, $\Phi(\vec{v})^M$.

(iv) By parts (ii) and (iii). □

Let us see where many of the simpler formulas of LST lie in the Lévy hierarchy. Notice that when we speak of, for example

the formula "x is a finite sequence",

we mean the "obvious" rendering of this statement as a formula in LST. In most cases the rendering is indeed *obvious*. If there is any significant doubt, we shall indicate the manner in which the statement is expressed in LST. In the case of our first lemma there is no such problem.

8.4. Lemma. *The following formulas are* Σ_0: $x = y$, $x \in y$, $x \subseteq y$, $y = \{x_1, \ldots, x_n\}$, $y = (x_1, \ldots, x_n)$, $y = (x)_i^n$ *(for* $i = 0, \ldots, n-1$*)*, $y = x \cup z$, $y = x \cap z$, $y = \bigcup x$, $y = \bigcap x$, $y = x - z$, "*x is an n-tuple*", "*x is a relation on y*", "*x is a function*", $y = \mathrm{dom}(x)$, $y = \mathrm{ran}(x)$, $y = x(z)$, $y = x''z$, $y = x \upharpoonright z$, $y = x \times z$, $y = x^{-1}$, $y = x \cup \{x\}$, $\mathrm{On}(x)$, $\lim(x)$, $\mathrm{succ}(x)$, "*x is a natural number*", "*x is a sequence*", $x : y \to z$, $x : y \leftrightarrow z$. □

In the case of the next lemma, we shall indicate the fashion in which the statement is to be expressed in LST, since, unlike the previous lemma, there are several possibilities.

8.5 Lemma. *The formula "x is finite" is* Σ_1.

Proof. x is finite $\leftrightarrow \exists n \exists f [n$ is a natural number $\wedge f: n \leftrightarrow x]$. □

The following lemma gives various closure properties for the levels in the Lévy hierarchy. The proofs are all trivial.

8.6 Lemma. *Let T be any* LST *theory.* (*By convention, T therefore includes all the axioms for predicate logic for* LST.) *Let* Φ, Ψ *be formulas of* LST.

(i) *If* Φ, Ψ *are* Σ_0^T, *so too are* $\Phi \wedge \Psi$, $\Phi \vee \Psi$, $\neg \Phi$.
(ii) *If* Φ *is* Σ_n^T, $\neg \Phi$ *is* Π_n^T; *if* Φ *is* Π_n^T, $\neg \Phi$ *is* Σ_n^T.
(iii) Φ *is* Δ_n^T *iff both* Φ *and* $\neg \Phi$ *are* Σ_n^T.
(iv) *If* Φ, Ψ *are* Σ_n^T, *so are* $\Phi \wedge \Psi$, $\Phi \vee \Psi$, $\exists x \Phi$, $(\exists x \in z) \Phi$.
(v) *If* Φ, Ψ *are* Π_n^T, *so are* $\Phi \wedge \Psi$, $\Phi \vee \Psi$, $\forall x \Phi$, $(\forall x \in z) \Phi$.
(vi) *If* Φ, Ψ *are* Δ_n^T, *so are* $\Phi \wedge \Psi$, $\Phi \vee \Psi$, $\neg \Phi$.
(vii) $m < n \rightarrow \Sigma_m^T, \Pi_m^T \subseteq \Delta_n^T$. □

8.7 Lemma. *The formula*

$$\mathrm{WF}(x, y) \leftrightarrow \text{"}x \text{ is a well-founded relation on } y\text{"}$$

is Δ_1^{ZF}.

Proof. It is easily seen that the formula

"x is a binary relation on y"

is Σ_0. Consequently, we need only concentrate upon the clause of WF which relates to well-foundedness. Let Φ denote this clause. Now, if E is a binary relation on a set X, the obvious rendering of $\Phi(E, X)$ is:

$$\forall A [A \subseteq X \wedge A \neq \emptyset \rightarrow (\exists a \in A)(\forall x \in A) \neg (x E a)].$$

(This is the *definition* of well-foundedness.) This shows at once that in its canonical rendering in LST, the formula $\mathrm{WF}(x, y)$ is Π_1. But in ZF it is easy to prove, for E, X as above, the equivalence

$$\Phi(E, X) \leftrightarrow \exists f [f: X \rightarrow \mathrm{On} \wedge (\forall x, y \in X)(x E y \rightarrow f(x) < f(y))].$$

(This involves a fairly routine application of the Recursion Principle.) This shows that Φ, and hence WF, are Σ_1^{ZF}, so we are done. □

Given an LST formula $\Phi(y, \vec{z})$, we denote by $\Phi((x)_0, \vec{z})$ the LST-formula

$$(\exists u \in x)(\exists a \in u)(\exists b \in u)[x = (a, b) \wedge \Phi(a, \vec{z})].$$

Similarly for $\Phi((x)_1, \vec{z})$.

Again, we denote by $\Phi(x(y), \vec{z})$ the LST-formula

$$(x \text{ is a function}) \wedge (\exists w \in x)(\exists u \in w)(\exists v \in u)[w = (v, y) \wedge \Phi(v, \vec{z})].$$

The following lemma is an immediate consequence of these definitions.

8.8 Lemma. *If $\Phi(x, \vec{z})$ is a Σ_0 formula of* LST, *then so too are $\Phi((x)_0, \vec{z})$, $\Phi((x)_1, \vec{z})$, and $\Phi(x(y), \vec{z})$.* □

8.9 Lemma (Contraction of Quantifiers). *Let T be any* LST *theory whose axioms include the axioms of null set and pairing* (see sections 2 and 3). *Then*:

(i) *Let $n \geqslant 1$ and let $\Phi(\vec{z})$ be a Σ_n formula. Then there is a Σ_0 formula $\Psi(\vec{x}, \vec{z})$ such that*

$$T \vdash \Phi(\vec{z}) \leftrightarrow \exists x_1 \forall x_2 \exists x_3 \ldots - x_n \Psi(x_1, \ldots, x_n, \vec{z}).$$

(ii) *Let $n \geqslant 1$ and let $\Phi(\vec{z})$ be a Π_n formula. Then there is a Σ_0 formula $\Phi(\vec{x}, \vec{z})$ such that*

$$T \vdash \Phi(\vec{z}) \leftrightarrow \forall x_1 \exists x_2 \forall x_3 \ldots - x_n \Psi(x_1, \ldots, x_n, \vec{z}).$$

Proof. We prove (i). The proof of (ii) is similar. Consider first the case $n = 1$. A general Σ_1 formula has the form

$$\Phi(\vec{z}) \colon \exists y_1 \exists y_2 \ldots \exists y_m \Theta(y_1, \ldots, y_m, \vec{z}),$$

where Θ is Σ_0. If $m = 1$ now there is nothing further to prove. Suppose that $m = 2$. (All other cases $m > 2$ are handled similarly.) Let $\Psi(x, \vec{z})$ be the formula:

$$(x \text{ is an ordered pair}) \wedge \Theta((x)_0, (x)_1, \vec{z}).$$

By 8.8, Ψ is Σ_0. And clearly, by our assumptions on T,

$$T \vdash \Phi(\vec{z}) \leftrightarrow \exists x \Psi(x, \vec{z}).$$

That deals with the case $n = 1$. We consider next the case $n = 2$, and leave it to the reader to see that the same idea works for all cases $n \geqslant 2$. Suppose $\Phi(\vec{z})$ is the formula

$$\exists u_1 \exists u_2 \ldots \exists u_p \forall v_1 \forall v_2 \ldots \forall v_q \Theta(\vec{u}, \vec{v}, \vec{z}),$$

where Θ is Σ_0. Let $\Psi(x, y, \vec{z})$ be the formula

$$(x \text{ is a } p\text{-tuple}) \wedge [(y \text{ is a } q\text{-tuple}) \rightarrow \Theta((x)_0^p, \ldots, (x)_{p-1}^p, (y)_0^q, \ldots, (y)_{q-1}^q, \vec{z})].$$

By 8.8, Ψ is Σ_0. Moreover,

$$T \vdash \Phi(\vec{z}) \leftrightarrow \exists x \forall y \Psi(x, y, \vec{z}).$$

The proof is complete. □

For the case where the theory, T, concerned is ZF, the following lemma extends the closure rules given in 8.6.

8.10 Lemma.

(i) *If* Φ *is a* Σ_n *formula of* LST, *then* $(\forall x \in y)\,\Phi$ *is* Σ_n^{ZF}.
(ii) *If* Φ *is a* Π_n *formula of* LST, *then* $(\exists x \in y)\,\Phi$ *is* Π_n^{ZF}.

Proof. We prove (i) and (ii) simultaneously by induction on n. For $n = 0$ there is nothing to prove. Suppose now that (i) and (ii) hold for n. We prove (i) and (ii) for $n + 1$.

(i) Let Φ be Σ_{n+1}. By 8.9 there is a Π_n formula Ψ such that

$$\mathrm{ZF} \vdash \Phi \leftrightarrow \exists z\,\Psi.$$

Hence,

$$\mathrm{ZF} \vdash (\forall x \in y)\,\Phi \leftrightarrow (\forall x \in y)\,\exists z\,\Psi.$$

But, by using the Axiom of Collection,

$$\mathrm{ZF} \vdash (\forall x \in y)\,\exists z\,\Psi \leftrightarrow \exists u\,(\forall x \in y)\,(\exists z \in u)\,\Psi.$$

Thus

$$\mathrm{ZF} \vdash (\forall x \in y)\,\Phi \leftrightarrow \exists u\,(\forall x \in y)\,(\exists z \in u)\,\Psi.$$

By induction hypothesis, $(\exists z \in u)\,\Psi$ is Π_n^{ZF}. Hence, using 8.7, $(\forall x \in y)(\exists z \in u)\,\Psi$ is Π_n^{ZF}. Thus $\exists u\,(\forall x \in y)(\exists z \in u)\,\Psi$ is $\Sigma_{n+1}^{\mathrm{ZF}}$, which means that $(\forall x \in y)\,\Phi$ is $\Sigma_{n+1}^{\mathrm{ZF}}$, as required.

(ii) Now suppose that Φ is Π_{n+1}. Then $\neg\,\Phi$ is $\Sigma_{n+1}^{\mathrm{ZF}}$. Hence by the above, $(\forall x \in y)\neg\,\Phi$ is $\Sigma_{n+1}^{\mathrm{ZF}}$. It follows that $\neg\,(\exists x \in y)\,\Phi$ is $\Sigma_{n+1}^{\mathrm{ZF}}$, and hence that $(\exists x \in y)\,\Phi$ is Π_{n+1}^{ZF}. □

9. The Language $\mathscr{L}_V$

We develop, within set theory, a formal "language", $\mathscr{L}_V$, which consists of an analogue of LST (which analogue we shall denote by $\mathscr{L}$), together with an individual constant "symbol" for each set (in V). The purpose of the subscript V in "$\mathscr{L}_V$" is to indicate that these constants are present. Later on we shall consider sublanguages $\mathscr{L}_X$ of $\mathscr{L}_V$ for any set X, where we only allow constants which denote elements of X. In particular, $\mathscr{L}_\emptyset$ is the same as $\mathscr{L}$, the formal analgoue of LST.

It should be emphasised that the entire development of $\mathscr{L}_V$ takes place within set theory. In particular, all the "symbols" and "formulas" of $\mathscr{L}_V$ will be sets. We shall require that the various syntactic and semantic notions of $\mathscr{L}_V$ have certain absoluteness properties, and in order to see that this is the case we shall need to examine the logical complexity of the (real) LST formulas which define the various notions of $\mathscr{L}_V$. So, as we proceed with our development of $\mathscr{L}_V$ within set theory, we shall make regular metamathematical digressions to examine the logical structure of the various notions. To try to minimise any confusion, we shall use lower case Greek letters $\varphi, \psi, \theta, \ldots$ to denote "formulas" of $\mathscr{L}_V$ (with upper case Greek letters $\Phi, \Psi, \Theta, \ldots$ for formulas of LST as before). However, since $\mathscr{L}$ will have the same structure as LST, it would be an unnecessary complication to use separate symbols for the variables and connectives of these two languages, so we shall leave this distinction to the reader, who will always be aided by the context.

The basic symbols of $\mathscr{L}_V$ will be certain sets, and the formulas of $\mathscr{L}_V$ will be certain finite sequences of these sets. Accordingly, we must begin by establishing some notations concerning finite sequence. (Incidentally, the exact fashion in which $\mathscr{L}_V$ is defined is not important, and we have just chosen a reasonably convenient method.)

The sequence with domain $\{0\}$ and value x is denoted by $\langle x \rangle$. The finite sequence with domain $\{0, \ldots, n-1\}$ and values $x_0, \ldots, x_{n-1}$ is denoted by $\langle x_0, \ldots, x_{n-1} \rangle$. (Notice that $\langle x_0, \ldots, x_{n-1} \rangle$ is not the same as the n-tuple $(x_0, \ldots, x_{n-1})$.)

If s, t are sequences, $s^\frown t$ denotes the *concatenation* of s and t, i.e. if $s = \langle x_0, \ldots, x_{n-1} \rangle$ and $t = \langle y_0, \ldots, y_{m-1} \rangle$, then

$$s^\frown t = \langle x_0, \ldots, x_{n-1}, y_0, \ldots, y_{m-1} \rangle .$$

If s is a finite sequence, $\|s\|$ denotes the greatest element of $\mathrm{dom}(s)$, i.e. $\|s\| = \mathrm{dom}(s) - 1$.

The *variables* of the language $\mathscr{L}_V$ are the sets $(2, n)$, for $n \in \omega$, and we shall denote $(2, n)$ by the symbol v_n.

Let $\mathrm{Vbl}(x)$ be the following LST formula:

$$[x \text{ is an ordered pair}] \wedge [(x)_0 = 2] \wedge [(x)_1 \text{ is a natural number}].$$

Clearly,

$$\mathrm{Vbl}(x) \leftrightarrow x \text{ is a variable of } \mathscr{L}_V.$$

In the above equivalence, note the use of the symbol x. Vbl is an LST formula, and x is a variable of LST. Being a variable of LST, x denotes a set. $\mathrm{Vbl}(x)$ says that the set denoted by x has the form that we have decided to refer to as a "variable" of $\mathscr{L}_V$. With a little experience, any initial difficulties the reader may encounter due to points such as this should be easily overcome.

For each set x, $\mathscr{L}_V$ has an *individual constant symbol*, namely the set $(3, x)$, which we shall denote by $\mathring{x}$.

Let $\mathrm{Const}(x)$ be the LST formula:

$$[x \text{ is an ordered pair}] \wedge [(x)_0 = 3].$$

Clearly,

$$\mathrm{Const}(x) \leftrightarrow x \text{ is a constant of } \mathscr{L}_V.$$

The *primitive formulas* of $\mathscr{L}_V$ are the sequences of the forms

$$\langle 0, 4, x, y, 1\rangle \quad \text{and} \quad \langle 0, 5, x, y, 1\rangle,$$

where x and y are variables or constants of $\mathscr{L}_V$. The sequence $\langle 0, 4, x, y, 1\rangle$ will be denoted by $(x \in y)$, and the sequence $\langle 0, 5, x, y, 1\rangle$ by $(x = y)$. (Thus we are using the number 0 to correspond to the open bracket symbol of LST and the number 1 to correspond to the close bracket symbol. The number 4 indicates a membership formula, and the number 5 indicates an equality formula.)

Let PFml(x) be the LST formula:

$$[x \text{ is a function}] \wedge [\mathrm{dom}(x) = 5] \wedge [x(0) = 0] \wedge [x(1) = 4 \vee x(1) = 5] \\ \wedge [\mathrm{Vbl}(x(2)) \vee \mathrm{Const}(x(2))] \wedge [\mathrm{Vbl}(x(3)) \vee \mathrm{Const}(x(3))] \\ \wedge [x(4) = 1].$$

Clearly,

$$\mathrm{PFml}(x) \leftrightarrow x \text{ is a primitive formula of } \mathscr{L}_V.$$

9.1 Lemma. *The* LST *formulas* Vbl(x), Const(x), PFml(x) are all Σ_0 (*when written out fully in* LST).

Proof. Immediate. □

The *formulas* of $\mathscr{L}_V$ are built up from the primitive formulas by means of the following schemas:

$$(\varphi \wedge \psi) = \langle 0, 6\rangle ^\frown \varphi ^\frown \psi ^\frown \langle 1\rangle$$
$$(\neg \varphi) = \langle 0, 7\rangle ^\frown \varphi ^\frown \langle 1\rangle$$
$$(\exists u \varphi) = \langle 0, 8, u\rangle ^\frown \varphi ^\frown \langle 1\rangle,$$

where φ, ψ are formulas of $\mathscr{L}_V$ and u is a variable of $\mathscr{L}_V$. (Note again the use of 0 and 1 as brackets, with the numbers 6, 7, 8 indicating the operations of conjunction, negation, and existential quantification, respectively.)

We shall presently write down a Σ_1 formula of LST which says "x is a formula of $\mathscr{L}_V$". But before we can do this we require several preliminary notions.

The following LST formula, Finseq(x), says that "x is a finite sequence":

$$[x \text{ is a sequence}] \wedge (\forall u \in \mathrm{dom}(x))[u \text{ is a natural number}] \\ \wedge (\exists u \in \mathrm{dom}(x))(\forall u \in \mathrm{dom}(x))[u \in v \vee u = v].$$

9.2 Lemma. *The* LST *formula* Finseq(x) *is* (*when written out fully in* LST) Σ_0.

Proof. All we need to observe is that expressions such as

$$(\forall u \in \mathrm{dom}(x))[\ldots u \ldots]$$

can be replaced by

$$(\forall z \in x)[\ldots (z)_1 \ldots],$$

which is Σ_0 by 8.8. □

We now write down LST formulas which describe the construction of the "formulas" of $\mathscr{L}_V$.

Let $F_\in(\theta, x, y)$ be the LST formula

$$\begin{aligned}&\mathrm{Finseq}(\theta) \wedge [\mathrm{dom}(\theta) = 5] \wedge [\theta(0) = 0] \wedge [\theta(1) = 4] \wedge [\theta(2) = x]\\ &\qquad \wedge [\theta(3) = y] \wedge [\theta(4) = 1].\end{aligned}$$

Clearly, if $x, y \in \mathrm{Vbl} \cup \mathrm{Const}$, then

$$F_\in(\theta, x, y) \leftrightarrow \theta \text{ is the } \mathscr{L}_V \text{ formula } (x \in y).$$

Let $F_=(\theta, x, y)$ be the LST formula

$$\begin{aligned}&\mathrm{Finseq}(\theta) \wedge [\mathrm{dom}(\theta) = 5] \wedge [\theta(0) = 0] \wedge [\theta(1) = 5] \wedge [\theta(2) = x]\\ &\qquad \wedge [\theta(3) = y] \wedge [\theta(4) = 1].\end{aligned}$$

Thus if $x, y \in \mathrm{Vbl} \cup \mathrm{Const}$, then

$$F_=(\theta, x, y) \leftrightarrow \theta \text{ is the } \mathscr{L}_V \text{ formula } (x = y).$$

Let $F_\wedge(\theta, \varphi, \psi)$ be the LST formula

$$\begin{aligned}&\mathrm{Finseq}(\theta) \wedge \mathrm{Finseq}(\varphi) \wedge \mathrm{Finseq}(\psi)\\ &\quad \wedge [\mathrm{dom}(\theta) = \mathrm{dom}(\varphi) + \mathrm{dom}(\psi) + 3] \wedge [\theta(0) = 0] \wedge [\theta(1) = 6]\\ &\quad \wedge [\theta(\|\theta\|) = 1] \wedge (\forall i \in \mathrm{dom}(\varphi))[\theta(i + 2) = \varphi(i)]\\ &\quad \wedge (\forall i \in \mathrm{dom}(\psi))[\theta(\mathrm{dom}(\varphi) + i + 2) = \psi(i)].\end{aligned}$$

Thus if $\varphi, \psi \in \mathrm{Fml}$, then

$$F_\wedge(\theta, \varphi, \psi) \leftrightarrow \theta \text{ is the } \mathscr{L}_V \text{ formula } (\varphi \wedge \psi).$$

Let $F_\neg(\theta, \varphi)$ be the LST formula

$$\begin{aligned}&\mathrm{Finseq}(\theta) \wedge \mathrm{Finseq}(\varphi) \wedge [\mathrm{dom}(\theta) = \mathrm{dom}(\varphi) + 3] \wedge [\theta(0) = 0]\\ &\quad \wedge [\theta(1) = 7] \wedge [\theta(\|\theta\|) = 1] \wedge (\forall i \in \mathrm{dom}(\varphi))[\theta(i + 2) = \varphi(i)].\end{aligned}$$

Thus if $\varphi \in \mathrm{Fml}$, then

$$F_\neg(\theta, \varphi) \leftrightarrow \theta \text{ is the } \mathscr{L}_V \text{ formula } (\neg \varphi).$$

Finally, let $F_\exists(\theta, u, \varphi)$ be the LST formula

$$\begin{aligned}&\mathrm{Finseq}(\theta) \wedge \mathrm{Finseq}(\varphi) \wedge [\mathrm{dom}(\theta) = \mathrm{dom}(\varphi) + 4] \wedge [\theta(0) = 0]\\ &\qquad \wedge [\theta(1) = 8] \wedge [\theta(2) = u] \wedge [\theta(\|\theta\|) = 1]\\ &\qquad \wedge (\forall i \in \mathrm{dom}(\varphi))[\theta(i + 3) = \varphi(i)].\end{aligned}$$

Thus if $\varphi \in \mathrm{Fml}$ and $u \in \mathrm{Vbl}$, we have

$$F_\exists(\theta, u, \varphi) \leftrightarrow \theta \text{ is the } \mathscr{L}_V \text{ formula } (\exists u\, \varphi).$$

9.3 Lemma. *The* LST *formulas* $F_\in, F_=, F_\wedge, F_\neg, F_\exists$ *are all* Σ_0 *(when written out fully in* LST*).*

Proof. In view of the remark made in the proof of 9.2, this is clear from the nature of the formulas concerned. □

Now, if φ is a formula of $\mathscr{L}_V$, there must be a finite sequence $\psi_0, \ldots, \psi_n$ of $\mathscr{L}_V$ formulas such that $\psi_n = \varphi$ and for each i, ψ_i is either a primitive formula or else is obtained from one or two formulas in the list $\psi_0, \ldots, \psi_{i-1}$ by an application of one of the schemas for generating formulas. The sequence $\psi_0, \ldots, \psi_n$ thus describes the way that φ is built up as a formula. We write down an LST formula, Build(φ, ψ) which says that ψ is just such a sequence $\psi_0, \ldots, \psi_n$. Build(φ, ψ) is as follows:

$$\begin{aligned} &\mathrm{Finseq}(\psi) \wedge [\psi_{\|\psi\|} = \varphi] \wedge (\forall i \in \mathrm{dom}(\psi))[\mathrm{PFml}(\psi_i) \\ &\qquad \vee (\exists j, k \in i) F_\wedge(\psi_i, \psi_j, \psi_k) \vee (\exists j \in i) F_\neg(\psi_i, \psi_j) \\ &\qquad \vee (\exists j \in i)(\exists u \in \mathrm{ran}(\varphi))(\mathrm{Vbl}(u) \wedge F_\exists(\psi_i, u, \psi_j))]. \end{aligned}$$

9.4 Lemma. *The* LST *formula* Build(φ, ψ) *is* Σ_0 *(when written out fully in* LST*).*

Proof. The main point to check is that expressions such as

$$(\forall i \in \mathrm{dom}(\psi))(\exists j, k \in i) F_\wedge(\psi_i, \psi_j, \psi_k)$$

are Σ_0. Well, this one can be written as

$$\begin{aligned} &(\forall i \in \mathrm{dom}(\psi))(\exists j, k \in i)(\exists a, b, c \in \mathrm{ran}(\psi))[a = \psi_i \wedge b = \psi_j \wedge c \\ &\wedge c = \psi_k \wedge F_\wedge(a, b, c)], \end{aligned}$$

which is immediately recognisable as Σ_0 now. The other cases are handled similarly. □

Clearly,

$$\varphi \text{ is a formula of } \mathscr{L}_V \leftrightarrow (\exists f)\,\mathrm{Build}(\varphi, f),$$

which presents us with a Σ_1 formula of LST to define the formulas of $\mathscr{L}_V$. Now, our main purpose in analysing the logical complexity of the syntactic notions of $\mathscr{L}_V$ is to enable us to prove various absoluteness results. In the case of Σ_0 notions, such as in Lemmas 9.1 through 9.4, there is no problem, since then 8.3 (i) guarantees absoluteness for all transitive classes. But for notions which are not Σ_0, such as the notion of being a formula of $\mathscr{L}$, it is not enough to know that the concept is Σ_1, for that will only guarantee U-absoluteness (see 8.3 (ii)). For full absoluteness we require (see 8.3 (iv)) an equivalent Π_1 definition as well. Moreover (see 8.3 again), in order that any absoluteness results have the widest possible application, it is important that the equivalence of the Σ_1 and Π_1 definitions be proved in the simplest theory possible, thereby giving absoluteness for all transitive models of that theory. We now develop such a theory: we call it *Basic Set Theory* (BS).

BS is the LST theory having the following axioms:

(1) *Extensionality*: $\forall x \forall y [\forall z (z \in x \leftrightarrow z \in y) \rightarrow (x = y)]$;
(2) *Induction schema*: $\forall \vec{a} [\forall x ((\forall y \in x) \Phi(y, \vec{a}) \rightarrow \Phi(x, \vec{a})) \rightarrow \forall x \Phi(x, \vec{a})]$, where Φ is any formula of LST with free variables amongst $x, \vec{a}$;
(3) *Pairing*: $\forall x \forall y \exists z \forall w [(w \in z) \leftrightarrow (w = x \vee w = y)]$;
(4) *Union*: $\forall x \exists y \forall z [(z \in y) \leftrightarrow (\exists u \in x)(z \in u)]$;
(5) *Infinity*: $\exists x [\mathrm{On}(x) \wedge (x \neq 0) \wedge (\forall y \in x)(\exists z \in x)(y \in z)]$;
(6) *Cartesian Product*: $\forall x \forall y \exists z \forall u [(u \in z) \leftrightarrow (\exists a \in x)(\exists b \in y)(u = (a, b))]$;
(7) Σ_0-*Comprehension (schema)*: $\forall \vec{a} \forall x \exists y \forall z [(z \in y) \leftrightarrow (z \in x \wedge \Phi(\vec{a}, z))]$, where $\Phi(\vec{a}, z)$ is a Σ_0 formula of LST.

Clearly, BS is a subtheory of ZF (i.e. all the axioms of BS are theorems of ZF). Indeed, axioms (1), (4) and (5) are axioms of ZF, though in the present formulation of the Axiom of Infinity we axiomatically guarantee the existence of an infinite ordinal, rather than any infinite set as we did with ZF. Axioms (3) and (6), which guarantee the existence of the unordered pair $\{x, y\}$ and the Cartesian product $x \times y$ of any two sets x and y, respectively, are easily proved theorems of ZF. Axiom (7) is just the restriction of the usual Comprehension Axiom Schema to the Σ_0 formulas of LST. In the absence of full Comprehension, we replace the Axiom of Foundation of ZF by the induction schema (2).

Notice that BS allows for the construction of all finite sets, i.e. for any n,

$$\mathrm{BS} \vdash \forall x_1 \ldots \forall x_n \exists y \forall z [(z \in y) \leftrightarrow (z = x_1 \vee \ldots \vee z = x_n)].$$

We now write down a formula of LST, Seq(u, a, n), which says that u is the set of all m-sequences of members of a for all $m < n$. Now, the "obvious" formula which says this is:

$$(\forall x \in u)(\exists m \in n)(x \text{ is an } m\text{-sequence of members of } a)$$
$$\wedge (\forall x)(\forall m \in n)(x \text{ is an } m\text{-sequence of members of } a \rightarrow x \in u).$$

But this formula is Π_1, whereas we shall require a Σ_1 definition. (Though we shall show that our Σ_1 definition is in fact BS-provably equivalent to a Π_1 definition.) We obtain our desired Σ_1 formula by regarding the members of the set u being built up in stages, constructing first the 1-sequences, then the 2-sequences, and so on. (The function f in the following formula enumerates these sets of finite sequences.)

Let Seq(u, a, n) be the following formula of LST:

$$(\exists f)[\mathrm{Finseq}(f) \wedge (n \text{ is a natural number}) \wedge (\mathrm{dom}(f) = n)$$
$$\wedge (u = \bigcup \mathrm{ran}(f)) \wedge (\forall i \in \mathrm{dom}(f))(\forall x \in f(i))(\mathrm{Finseq}(x) \wedge (\mathrm{dom}(x) = i)$$
$$\wedge (\forall j \in i)(x(j) \in a)) \wedge (\forall i \in \mathrm{dom}(f))(\forall j \in i)(\forall x \in f(j))(\forall p \in a)$$
$$\cdot (i = j + 1 \rightarrow x \cup \{(p, i)\} \in f(i))].$$

It is easily seen that this formula is Σ_1 (when written out fully in LST).

9.5 Lemma. *The* LST *formula* Seq (u, a, n) *is* Δ_1^{BS}.

Proof. The only unbounded quantifier in the above formula is $(\exists f)$. This quantifier can, without any loss of generality, be restricted to range over the set of n-sequences of finite sequences from a. (That is to say, if such an f exists, it will have to lie in this bounding set.) Consequently, it is clear from the definition of BS that:

$$\mathrm{BS} \vdash (\forall a)(\forall n \in \omega)(\exists u)\, \mathrm{Seq}(u, a, n).$$

But we obviously have:

$$\mathrm{BS} \vdash \forall a\, \forall n\, \forall u\, \forall v\, [\mathrm{Seq}(u, a, n) \wedge \mathrm{Seq}(v, a, n) \rightarrow u = v].$$

Hence,

$$\mathrm{BS} \vdash \mathrm{Seq}(u, a, n) \leftrightarrow [(n \text{ is a natural number}) \wedge \forall z\, [\mathrm{Seq}(z, a, n) \rightarrow z = u]].$$

This proves the lemma, since the expression on the right of this equivalence is Π_1. □

We are now able to write down an LST formula Fml(x) such that:

$$\mathrm{Fml}(x) \leftrightarrow x \text{ is a formula of } \mathscr{L}_V.$$

As we mentioned earlier, the obvious way to do this is by the formula

$$(\exists f)\, \mathrm{Build}(x, f).$$

So let us take this as our formula Fml(x). By 9.4, Fml(x) is Σ_1.

9.6 Lemma. *The* LST *formula* Fml(x) *is* Δ_1^{BS}.

Proof. Consider the quantifier $\exists f$ in the expression

$$(\exists f)\, \mathrm{Build}(x, f).$$

We may clearly bind this quantifier by the set

$$A(x) = \bigcup_{n \in \mathrm{dom}(x)} {}^{n+1}\Big[\bigcup_{n \in \mathrm{dom}(x)} {}^{m+1}\mathrm{ran}(x)\Big].$$

(Because, as is easily seen,

$$(\exists f)\, \mathrm{Build}(x, f) \rightarrow (\exists f \in A(x))\, \mathrm{Build}(x, f).)$$

Moreover, it is easily checked that

$$\mathrm{BS} \vdash \forall x\, \exists y\, [y = A(x)].$$

Hence

$$\mathrm{BS} \vdash \mathrm{Fml}(x) \leftrightarrow \mathrm{Finseq}(x) \wedge \forall u \forall v [\mathrm{Seq}(u, \mathrm{ran}(x), \mathrm{dom}(x) + 2) \wedge \mathrm{Seq}(v, u, \mathrm{dom}(x) + 2) \rightarrow (\exists f \in v) \, \mathrm{Build}(x, f)].$$

This provides us with a Π_1^{BS} equivalent to $\mathrm{Fml}(x)$, so we are done. □

The above trick of finding a convenient bound for a quantifier will be used frequently during our development of $\mathscr{L}_V$.

Given any class X, $\mathscr{L}_X$ is the "sublanguage" of $\mathscr{L}_V$ obtained by omitting from $\mathscr{L}_V$ all constants $\mathring{z}$ for $z \notin X$. We write $\mathscr{L}$ instead of $\mathscr{L}_\emptyset$. Thus $\mathscr{L}$ is a formal analogue of LST within set theory. We shall be particularly concerned with the languages $\mathscr{L}_u$ where u is a set.

Let $\mathrm{Const}(x, u)$ be the LST formula

$$\mathrm{Const}(x) \wedge (x)_1 \in u.$$

The LST formulas $\mathrm{PFml}(x, u)$ and $\mathrm{Fml}(x, u)$ are defined in exactly the same way as the formulas $\mathrm{PFml}(x)$ and $\mathrm{Fml}(x)$ except that $\mathrm{Const}(x)$ is replaced everywhere by $\mathrm{Const}(x, u)$. Clearly,

$$\mathrm{Fml}(x, u) \leftrightarrow x \text{ is a formula of } \mathscr{L}_u.$$

By means of arguments as before, we have:

9.7 Lemma.

(i) *The* LST *formulas* $\mathrm{Const}(x, u)$ *and* $\mathrm{PFml}(x, u)$ *are* Σ_0.
(ii) *The* LST *formula* $\mathrm{Fml}(x, u)$ *is* Δ_1^{BS}. □

Our next task is to write down an LST formula $\mathrm{Fr}(\varphi, x)$ such that

$$\mathrm{Fr}(\varphi, x) \leftrightarrow \mathrm{Fml}(\varphi) \wedge [x \text{ is the set of variables occuring free in } \varphi].$$

Now, given a formula φ, how would one go about checking whether a set x is the set of free variables of φ? One way would be to concentrate on a sequence ψ for which $\mathrm{Build}(\varphi, \psi)$, and proceed along the members of ψ, keeping track of the free variables at each stage. This approach leads to the following formula, which we take as our $\mathrm{Fr}(\varphi, x)$:

$$\begin{aligned}
&\exists \psi \exists f [\mathrm{Build}(\varphi, \psi) \wedge \mathrm{Finseq}(f) \wedge (\mathrm{dom}(f) = \mathrm{dom}(\psi)) \wedge (x = f(\|f\|)) \\
&\wedge (\forall i \in \mathrm{dom}(f)) [(\exists j, k \in i) [F_\wedge(\psi_i, \psi_j, \psi_k) \wedge (f(i) = f(j) \cup f(k))] \\
&\qquad \vee (\exists j \in i) [F_\neg(\psi_i, \psi_j) \wedge (f(i) = f(j))] \\
&\qquad \vee (\exists j \in i)(\exists u \in \mathrm{ran}(\varphi)) [\mathrm{Vbl}(u) \wedge F_\exists(\varphi_i, u, \psi_j) \wedge (f(i) = f(j) - \{u\})] \\
&\qquad \vee [\mathrm{PFml}(\psi_i) \wedge [[\mathrm{Vbl}((\psi_i)_2) \wedge \mathrm{Vbl}((\psi_i)_3) \wedge f(i) = \{(\psi_i)_2, (\psi_i)_3\}] \\
&\qquad\qquad \vee [\mathrm{Vbl}((\psi_i)_2) \wedge \mathrm{Const}((\psi_i)_3) \wedge f(i) = \{(\psi_i)_2\}] \\
&\qquad\qquad \vee [\mathrm{Const}((\psi_i)_2) \wedge \mathrm{Vbl}((\psi_i)_3) \wedge f(i) = \{(\psi_i)_3\}] \\
&\qquad\qquad \vee [\mathrm{Const}((\psi_i)_2) \wedge \mathrm{Const}((\psi_i)_3) \wedge f(i) = \emptyset]]]].
\end{aligned}$$

Clearly, the LST formula $\mathrm{Fr}(\varphi, x)$ is Σ_1.

9.8 Lemma. *The* LST *formula* $\mathrm{Fr}(\varphi, x)$ *is* Δ_1^{BS}.

Proof. Clearly,

$$\mathrm{BS} \vdash \mathrm{Fr}(\varphi, x) \leftrightarrow [\mathrm{Fml}(\varphi) \wedge \forall z [\mathrm{Fr}(\varphi, z) \rightarrow z = x]].$$

This gives us a Π_1^{BS} characterisation of $\mathrm{Fr}(\varphi, x)$. □

We now formulate an LST formula $\mathrm{Sub}(\varphi', \varphi, v, t)$, which will say that φ' and φ are formulas of $\mathcal{L}_V$, v is a variable, t is a constant, and φ' is the result of subsituting t for every free occurrence of v in φ. To arrive at this formula, we adopt a procedure similar to the one used above for $\mathrm{Fr}(\varphi, x)$. Pick a sequence ψ such that $\mathrm{Build}(\varphi, \psi)$. Proceed through ψ, substituting t for every free occurrence of v at each stage. If the quantifier v is ever encountered, delete any substitutions made previously within the scope of this quantifier. In order to write this out in an intelligible fashion, we consider first the restriction of $\mathrm{Sub}(\varphi', \varphi, v, t)$ to primitive formulas φ. Let $S(\varphi', \varphi, v, t)$ denote this restricted formula. That is, let $S(\varphi', \varphi, v, t)$ be the following LST formula:

$$\begin{aligned}
&\mathrm{PFml}(\varphi') \wedge \mathrm{PFml}(\varphi) \wedge \mathrm{Vbl}(v) \wedge \mathrm{Const}(t) \\
&\wedge [[F_=(\varphi, \varphi_2, \varphi_3) \wedge [[\varphi_2 \neq v \wedge \varphi_3 \neq v \wedge (\varphi' = \varphi)] \\
&\qquad\qquad \vee [\varphi_2 = v \wedge \varphi_3 \neq v \wedge F_=(\varphi', t, \varphi_3)] \\
&\qquad\qquad \vee [\varphi_2 \neq v \wedge \varphi_3 = v \wedge F_=(\varphi', \varphi_2, t)] \\
&\qquad\qquad \vee [\varphi_2 = v \wedge \varphi_3 = v \wedge F_=(\varphi', t, t)]]] \\
&\quad \vee [F_\in(\varphi, \varphi_2, \varphi_3) \wedge [\ldots\ \ldots\ \ldots]]],
\end{aligned}$$

where the expression denoted … … … in the above is just as in the $F_=$ part, but with $F_\in$ in place of $F_=$.

Notice that $S(\varphi', \varphi, v, t)$ is Σ_0. Let $\mathrm{Sub}(\varphi', \varphi, v, t)$ be the following LST formula:

$$\begin{aligned}
&\mathrm{Fml}(\varphi') \wedge \mathrm{Fml}(\varphi) \wedge \mathrm{Vbl}(v) \wedge \mathrm{Const}(t) \wedge \exists\psi\, \exists\theta\, [\mathrm{Build}(\varphi, \psi) \\
&\wedge \mathrm{Finseq}(\theta) \wedge (\mathrm{dom}(\theta) = \mathrm{dom}(\psi)) \wedge (\theta_{\|\theta\|} = \varphi') \\
&\wedge (\forall i \in \mathrm{dom}(\psi)) [(\exists j, k \in i)(F_\wedge(\psi_i, \psi_j, \psi_k) \wedge F_\wedge(\theta_i, \theta_j, \theta_k)) \\
&\qquad\qquad \vee (\exists j \in i)(F_\neg(\psi_i, \psi_j) \wedge F_\neg(\theta_i, \theta_j)) \\
&\qquad\qquad \vee (\exists j \in i)(\exists u \in \mathrm{ran}(\varphi))(\mathrm{Vbl}(u) \wedge (u \neq v) \\
&\qquad\qquad\qquad \wedge F_\in(\psi_i, u, \psi_j) \wedge F_\exists(\theta_i, u, \theta_j)) \\
&\qquad\qquad \vee (\exists j \in i)(F_\exists(\psi_i, v, \psi_j) \wedge (\theta_i = \psi_i)) \\
&\qquad\qquad \vee S(\theta_i, \psi_i, v, t)]].
\end{aligned}$$

This formula is clearly Σ_1. Moreover:

9.9 Lemma. *The* LST *formula* $\mathrm{Sub}(\varphi', \varphi, v, t)$ *is* Δ_1^{BS}.

Proof. Clearly,

$$\mathrm{BS} \vdash \mathrm{Sub}(\varphi', \varphi, v, t) \leftrightarrow \mathrm{Fml}(\varphi) \wedge \mathrm{Vbl}(v) \wedge \mathrm{Const}(t) \wedge \forall \psi\, [\mathrm{Sub}(\psi, \varphi, v, t) \rightarrow \psi = \varphi'],$$

which gives us the lemma at once, since the expression on the right of the above equivalence is clearly Π_1^{BS}. □

We are now able to define the notion of satisfaction ("truth") for the languages $\mathscr{L}_u$. We shall write down an LST formula $\mathrm{Sat}(u, \varphi)$ such that

> $\mathrm{Sat}(u, \varphi) \leftrightarrow \varphi$ is a sentence of $\mathscr{L}_u$ which is true in the structure $\langle u, \in \rangle$ under the canonical interpretation (i.e. with x interpreting $\mathring{x}$ for each x in u).

The standard way to define satisfaction is as follows. Let f be a function with domain ω such that $f(0)$ is the set of all primitive formulas of $\mathscr{L}_u$ and, in general, $f(i+1)$ is the set of all formulas of $\mathscr{L}_u$ which are obtained from formulas in $f(i)$ by a single application of one of the three formula building schemas. Let g be a function with domain ω such that $g(i)$ is the set of all formulas in $f(i)$ which have no free variables and which are true in $\langle u, \in \rangle$. Both f and g can be defined by simple recursions. The function g then provides us with all the sentences of $\mathscr{L}_u$ which are true in $\langle u, \in \rangle$. (The function f is required in order to handle negation in passing from $g(i)$ to $g(i+1)$.) Our formula $\mathrm{Sat}(u, \varphi)$ will be obtained by considering the above process taken sufficiently far to check whether φ is in $g(i)$ or not, when i is chosen so that $\varphi \in f(i)$.

Let $E(\varphi, u)$ be the following LST formula:

$$(\exists x, y \in u)[(x \in y) \wedge F_{\in}(\varphi, \mathring{x}, \mathring{y})] \vee (\exists x \in u) F_{=}(\varphi, \mathring{x}, \mathring{x}).$$

Clearly, $E(\varphi, u)$ says that φ is a primitive sentence of $\mathscr{L}_u$ which is true in the structure $\langle u, \in \rangle$. Provided that we are careful when we write it out in LST, the formula $E(\varphi, u)$ is Σ_0. For example, in rendering the clause $(\exists x \in u) F_{=}(\varphi, \mathring{x}, \mathring{x})$ in LST we must proceed thus:

$$(\exists x \in u)(\exists y \in \mathrm{ran}(\varphi))(y = \mathring{x} \wedge F_{=}(\varphi, y, y)).$$

It should now be clear that $E(\varphi, u)$ is Σ_0.

The following LST formula, $S(u, \varphi)$, expresses in LST the notion that φ is a sentence of $\mathscr{L}_u$ which is true in $\langle u, \in \rangle$. (However, as $S(u, \varphi)$ will not be Σ_1, this is not our sought after formula $\mathrm{Sat}(u, \varphi)$, but rather a precursor to it.)

$$\begin{aligned}&(u \neq \emptyset) \wedge \mathrm{Fml}(\varphi, u) \wedge \exists f \exists g\, [\mathrm{Finseq}(f) \wedge \mathrm{Finseq}(g) \wedge (\mathrm{dom}(f) = \mathrm{dom}(g))\\ &\wedge (\varphi \in g(\|g\|)) \wedge \forall \psi (\psi \in f(0) \leftrightarrow \mathrm{PFml}(\psi, u)) \wedge \forall \psi (\psi \in g(0) \leftrightarrow E(\psi, u))\\ &\wedge (\forall j \in \mathrm{dom}(f))(\forall i \in j)(\forall \psi)[\psi \in f(i+1) \leftrightarrow (\psi \in f(i)) \vee (\exists \theta, \theta' \in f(i)) F_{\wedge}(\psi, \theta, \theta')\\ &\quad \vee (\exists \theta \in f(i)) F_{\neg}(\psi, \theta) \vee (\exists \theta \in f(i))(\exists v \in \mathrm{ran}(\psi))(\mathrm{Vbl}(v) \wedge F_{\in}(\psi, v, \theta))] \wedge\end{aligned}$$

$$(\forall j \in \mathrm{dom}(g))(\forall i \in j)(\forall \psi)[\psi \in g(i+1) \leftrightarrow (\psi \in g(i)) \vee (\exists \theta, \theta' \in g(i)) F_\wedge(\psi, \theta, \theta') \vee (\exists \theta \in f(i))(\theta \notin g(i) \wedge F_\neg(\psi, \theta)) \vee (\exists \theta \in f(i))(\exists v \in \mathrm{ran}(\psi))(\exists x \in u)(\exists \theta' \in g(i))[\mathrm{Vbl}(v) \wedge F_\exists(\psi, v, \theta) \wedge \mathrm{Sub}(\theta', \theta, v, \mathring{x})]]].$$

It is easily seen that the above formula does define the satisfaction relation. But it is not a Σ_1 formula. The problem is the quantifier $(\forall \psi)$, which appears four times, and the unbounded quantifiers involved in the Δ_1^{BS} formula $\mathrm{Sub}(\theta', \theta, v, \mathring{x})$, which occurs inside the scope of a number of other quantifiers. However, it is easily seen that the truth of $S(u, \varphi)$ is not affected by binding all unbounded quantifiers involved (including the $\exists f$ and the $\exists g$) by the set

$$w(u, \varphi) = \Big(\bigcup_{m \in \mathrm{dom}(\varphi)} {}^{m+1}[9 \cup \{v_i \mid i \in \omega\} \cup \{\mathring{x} \mid x \in u\}]\Big) \cup \Big(\bigcup_{n \in \mathrm{dom}(\varphi)} {}^{n+1}\Big[\bigcup_{m \in \mathrm{dom}(\varphi)} {}^{m+1}[9 \cup \{v_i \mid i \in \omega\} \cup \{\mathring{x} \mid x \in u\}]\Big]\Big).$$

(The first set in the above union includes all $\mathscr{L}_u$ formulas of lengths at most that of φ, and the second set includes all finite sequences of such formulas whose domain is at most $\mathrm{dom}(\varphi)$.) Let $S'(u, \varphi, w)$ be the formula obtained from $S(u, \varphi)$ by binding all quantifiers not already bound by w. Then for $\mathrm{Sat}(u, \varphi)$ we take the following LST formula:

$$\exists w \exists x \exists y \exists a \exists b \exists t\, [(a = \{\mathring{x} \mid x \in u\}) \wedge (\text{“}t = \omega\text{”}) \wedge (b = \{v_i \mid i \in t\}) \wedge \mathrm{Seq}(x, 9 \cup a \cup b, \mathrm{dom}(\varphi) + 1) \wedge \mathrm{Seq}(y, x, \mathrm{dom}(\varphi) + 1) \wedge (w = x \cup y) \wedge S'(u, \varphi, w)],$$

where the formula “$t = \omega$” is written out thus:

$$\mathrm{On}(t) \wedge \lim(t) \wedge (\forall i \in t)[(\exists j \in i)(i = j + 1) \vee (\forall j \in i)(j \neq j)].$$

By our previous remarks, $\mathrm{Sat}(u, \varphi)$ is equivalent to $S(u, \varphi)$, so indeed

$$\mathrm{Sat}(u, \varphi) \leftrightarrow \varphi \text{ is a sentence of } \mathscr{L}_u \text{ which is true in } \langle u, \in \rangle.$$

Moreover, $\mathrm{Sat}(u, \varphi)$ is clearly Σ_1, and in fact:

9.10 Lemma. *The* LST *formula* $\mathrm{Sat}(u, \varphi)$ *is* Δ_1^{BS}.

Proof. Clearly

$$\mathrm{BS} \vdash \neg \mathrm{Sat}(u, \varphi) \leftrightarrow \neg [\mathrm{Fml}(\varphi, u) \wedge \mathrm{Fr}(\varphi, \emptyset)] \vee \exists \theta [F_\neg(\theta, \varphi) \wedge \mathrm{Sat}(u, \theta)].$$

Hence $\neg \mathrm{Sat}(u, \varphi)$ is Σ_1^{BS}, whence $\mathrm{Sat}(u, \varphi)$ is Π_1^{BS}. □

We often write $\models_u \varphi$ instead of $\mathrm{Sat}(u, \varphi)$.

As we have remarked earlier, the collection of sets which constitute the “formulas” of $\mathscr{L}$ provides us with an analogue of the formulas of the genuine language

LST. Given any formula Φ of LST we can construct a set φ which, according to the "syntax" of $\mathscr{L}$ developed above, has the same logical structure as Φ. In this context, the following result indicates how the formal notion of satisfaction just defined corresponds to the genuine notion of truth.

9.11 Lemma. *Let $\Phi(v_0, \ldots, v_n)$ be any formula of* LST, *and let $\varphi(v_0, \ldots, v_n)$ be its counterpart in $\mathscr{L}$ (in the sense described above). Then*

$$\mathrm{ZF} \vdash \forall u (\forall x_0 \in u) \ldots (\forall x_n \in u)[\Phi^u(x_0, \ldots, x_n) \leftrightarrow \mathrm{Sat}(u, \varphi(\mathring{x}_0, \ldots, \mathring{x}_n))].$$

Proof. By induction on the construction of Φ and φ. (The easy details are left as an exercise for the reader.) □

Notice that the above result is a theorem schema for ZF, which takes us from a given LST formula Φ and the genuine notion of truth to a "formula" φ of $\mathscr{L}$ and the mathematically defined notion of satisfaction.

By analogy with LST, we define a "Lévy hierarchy" for the formulas of $\mathscr{L}_V$. For reasons of technical convenience we only allow for single quantifiers rather than blocks of like quantifiers as we did for LST.

A formula φ of $\mathscr{L}_V$ is said to be Σ_0 (or Π_0) if, whenever a quantifier $\exists v_n$ occurs in φ it does so in the context

$$\exists v_n (v_n \in x \wedge \ldots)$$

for some $x \in \mathrm{Vbl} \cup \mathrm{Const}$. The following LST formula, $\mathrm{Fml}^{\Sigma_0}(\varphi)$, clearly defines this notion:

$$\begin{aligned}\mathrm{Fml}(\varphi) \wedge (\forall i \in \mathrm{dom}(\varphi))[(\varphi_i = 0 \wedge \varphi_{i+1} = 8 \wedge \mathrm{Vbl}(\varphi_{i+2}))\\ \rightarrow (\varphi_{i+3} = 0 \wedge \varphi_{i+4} = 6 \wedge \varphi_{i+5} = 0 \wedge \varphi_{i+6} = 4\\ \wedge \varphi_{i+7} = \varphi_{i+2} \wedge (\mathrm{Const}(\varphi_{i+8}) \vee \mathrm{Vbl}(\varphi_{i+8})) \wedge \varphi_{i+9} = 1)].\end{aligned}$$

Notice that except for the part $\mathrm{Fml}(\varphi)$, this formula is Σ_0. Likewise for the LST formula $\mathrm{Fml}^{\Sigma_0}(\varphi, u)$, which says that φ is a Σ_0 formula of $\mathscr{L}_u$. The following lemma is immediate:

9.12 Lemma. *The* LST *formulas* $\mathrm{Fml}^{\Sigma_0}(\varphi)$ *and* $\mathrm{Fml}^{\Sigma_0}(\varphi, u)$ *are* Δ_1^{BS}. □

A formula φ of $\mathscr{L}_V$ is said to be Σ_1 if it is of the form $\exists v_n \psi$, where ψ is Σ_0, and is said to be Π_1 if it is of the form $\neg \exists v_n \psi$ where ψ is Σ_0. In general, an $\mathscr{L}_V$ formula is said to be Σ_{n+1} if it is of the form $\exists v_m \psi$ where ψ is Π_n, and is said to be Π_{n+1} if it is of the form $\neg \psi$ where ψ is Σ_{n+1}.

9.13 Lemma. *Fix $n \geqslant 1$. Then there are Δ_1^{BS} formulas* $\mathrm{Fml}^{\Sigma_n}(\varphi)$, $\mathrm{Fml}^{\Pi_n}(\varphi)$, $\mathrm{Fml}^{\Sigma_n}(\varphi, u)$, $\mathrm{Fml}^{\Pi_n}(\varphi, u)$ *of* LST *such that:*

$\mathrm{Fml}^{\Sigma_n}(\varphi) \leftrightarrow \varphi$ *is a* Σ_n *formula of* $\mathscr{L}_V$;
$\mathrm{Fml}^{\Pi_n}(\varphi) \leftrightarrow \varphi$ *is a* Π_n *formula of* $\mathscr{L}_V$;
$\mathrm{Fml}^{\Sigma_n}(\varphi, u) \leftrightarrow \varphi$ *is a* Σ_n *formula of* $\mathscr{L}_u$;
$\mathrm{Fml}^{\Pi_n}(\varphi, u) \leftrightarrow \varphi$ *is a* Π_n *formula of* $\mathscr{L}_u$.

Proof. These are all more or less the same as $\mathrm{Fml}^{\Sigma_0}(\varphi)$, considered earlier. For example, $\mathrm{Fml}^{\Sigma_1}(\varphi)$ is:

$$\begin{aligned}&\mathrm{Fml}(\varphi) \wedge [\varphi_0 = 0 \wedge \varphi_1 = 8 \wedge \mathrm{Vbl}(\varphi_2)\\ &\quad \wedge (\forall i \in \mathrm{dom}(\varphi))[(i > 2 \wedge \varphi_i = 0 \wedge \varphi_{i+1} = 8 \wedge \mathrm{Vbl}(\varphi_{i+2}))\\ &\quad \rightarrow (\varphi_{i+3} = 0 \wedge \varphi_{i+4} = 6 \wedge \varphi_{i+5} = 0 \wedge \varphi_{i+6} = 4 \wedge \varphi_{i+7} = \varphi_{i+2}\\ &\quad \wedge (\mathrm{Const}(\varphi_{i+8}) \vee \mathrm{Vbl}(\varphi_{i+8})) \wedge \varphi_{i+9} = 1)]].\end{aligned}$$

(As n increases, the length and complexity of $\mathrm{Fml}^{\Sigma_n}(\varphi)$, etc. also increases, of course, but the overall pattern is much the same.) □

Occasionally we shall wish to consider extensions of the languages $\mathscr{L}_u$ in which there are a finite number of additional predicates. Specifically, let k be some natural number and let $A_1 \subseteq u^{n(1)}, \ldots, A_k \subseteq u^{n(k)}$. The language $\mathscr{L}_u(\mathring{A}_1, \ldots, \mathring{A}_k)$ has the same structure as $\mathscr{L}_u$ except that there are the k extra predicate letters $\mathring{A}_1, \ldots, \mathring{A}_k$, where $\mathring{A}_i$ is $n(i)$-ary for each i. More precisely, for each $i = 1, \ldots, k$, amongst the primitive formulas of $\mathscr{L}_u(\mathring{A}_1, \ldots, \mathring{A}_k)$ we allow the sequences

$$\langle 0, 8 + i, x_1, \ldots, x_{n(i)}, 1\rangle,$$

where $x_1, \ldots, x_{n(i)} \in \mathrm{Vbl} \cup \mathrm{Const}_u$. We usually write $\mathring{A}_i(x_1, \ldots, x_{n(i)})$ in place of the sequence $\langle 0, 8 + i, x_1, \ldots, x_{n(i)}, 1\rangle$. With this modification to the primitive formulas, the development of the rest of the language $\mathscr{L}_u(\mathring{A}_1, \ldots, \mathring{A}_k)$ proceeds exactly as for $\mathscr{L}_u$. Consequently, all of the results obtained in this section for the languages $\mathscr{L}_u$ hold in this more general situation. (Note that the interpretation of $\mathscr{L}_u(\mathring{A}_1, \ldots, \mathring{A}_k)$ in the structure $\langle u, \in, A_1, \ldots, A_k\rangle$ is the obvious, canonical one.)

We shall require the following formal analogue of the metamathematical notion of absoluteness (Section 8).

Let $\varphi(\vec{x})$ be any $\mathscr{L}(\mathring{A}_1, \ldots, \mathring{A}_k)$ formula. Let $\mathbf{M}, \mathbf{N}$ be structures appropriate for this language, $\mathbf{M}$ a substructure of $\mathbf{N}$. We say that φ is *U-absolute* for $\mathbf{M}, \mathbf{N}$ iff

$$(\forall \vec{x} \in M)(\models_{\mathbf{M}} \varphi(\vec{\mathring{x}}) \text{ implies } \models_{\mathbf{N}} \varphi(\vec{\mathring{x}})).$$

We say that φ is *D-absolute* for $\mathbf{M}, \mathbf{N}$ iff

$$(\forall \vec{x} \in M)(\models_{\mathbf{N}} \varphi(\vec{\mathring{x}}) \text{ implies } \models_{\mathbf{M}} \varphi(\vec{\mathring{x}})).$$

We say that φ is *absolute* for $\mathbf{M}, \mathbf{N}$ iff it is both U-absolute and D-absolute for $\mathbf{M}, \mathbf{N}$. Analogous to 8.3 we have:

9.4 Lemma. *Let* $\mathbf{M}, \mathbf{N}$ *be* $\mathscr{L}(\mathring{A}_1, \ldots, \mathring{A}_k)$ *structures,* $\mathbf{M}$ *a substructure of* $\mathbf{N}$. *Suppose further that both M and N are transitive sets. Let $\varphi(\vec{x})$ be a formula of* $\mathscr{L}(\mathring{A}_1, \ldots, \mathring{A}_k)$.

(i) *If φ is Σ_0, then φ is absolute for* $\mathbf{M}, \mathbf{N}$.
(ii) *If φ is Σ_1, then φ is U-absolute for* $\mathbf{M}, \mathbf{N}$.
(iii) *If φ is Π_1, then φ is D-absolute for* $\mathbf{M}, \mathbf{N}$.

Proof. Similar to the proof of 8.3. (The details are left as an exercise for the reader.) □

The following lemma concerns the relationship between the two languages LST and $\mathscr{L}$, and is related to Lemma 9.11.

9.15 Lemma. *Let $\Phi(\vec{x})$ be a Σ_0 formula of* LST, *and let $\varphi(\vec{x})$ be its counterpart in $\mathscr{L}$. Then*

$$\mathrm{ZF} \vdash \text{“}\textit{For any transitive set } M, (\forall \vec{x} \in M)[\Phi(\vec{x}) \leftrightarrow \models_M \varphi(\mathring{\vec{x}})]\text{”}.$$

Proof. By an easy induction on the length of Φ. (The details are left as an exercise to the reader.) □

We shall make considerable use of 9.15 and generalisations thereof in Chapter II.

10. Definability

Consider a structure of the form

$$\mathbf{M} = \langle M, \in, A_1, \ldots, A_k \rangle,$$

where M is a non-empty set and $A_i \subseteq M^{n(i)}$ for $i = 1, \ldots, k$. (In such cases we often omit specific reference to $\in$, as is always the case with $=$, of course.) By the **M**-*language* we mean the language $\mathscr{L}_M(\mathring{A}_1, \ldots, \mathring{A}_k)$ introduced at the end of the previous section. As we indicated there, all of the various definitions and results of section 9 hold for **M**-languages. For instance, there is a Δ_1^{BS} formula Sat$(\mathbf{M}, \varphi)$ of LST such that Sat$(\mathbf{M}, \varphi)$ iff φ is a sentence of the **M**-langugage which is true in **M** (under the standard interpretation). Note that we usually write $\models_{\mathbf{M}} \varphi$ instead of Sat$(\mathbf{M}, \varphi)$.

Let $N \subseteq M$. A set $R \subseteq M^m$ is said to be $\Sigma_n^{\mathbf{M}}(N)$ iff there is a Σ_n formula $\varphi(v_0, \ldots, v_{m-1})$ of the **M**-language, whose constants are all members of the set $\{\mathring{a} \mid a \in N\}$, such that

$$(\forall x_0, \ldots, x_{m-1} \in M)[(x_0, \ldots, x_{m-1}) \in R \leftrightarrow \models_{\mathbf{M}} \varphi(\mathring{x}_0, \ldots, \mathring{x}_{m-1})].$$

Similary for $\Pi_n^{\mathbf{M}}(N)$. A set $R \subseteq M^m$ is $\Delta_n^{\mathbf{M}}(N)$ iff it is both $\Sigma_n^{\mathbf{M}}(N)$ and $\Pi_n^{\mathbf{M}}(N)$.

We write $\Sigma_n^{\mathbf{M}}$ instead of $\Sigma_n^{\mathbf{M}}(\emptyset)$ and $\Sigma_n(\mathbf{M})$ instead of $\Sigma_n^{\mathbf{M}}(M)$. Similarly for Π and Δ.

A set $R \subseteq M^m$ is said to be **M**-*definable* iff it is $\Sigma_n(\mathbf{M})$ for some n.

Notice that the above notions are all formally defined within set theory, and are not metamathematical notions. For example, there is an LST formula $\Phi(R, M)$ such that

$$\Phi(R, M) \leftrightarrow M \text{ is a non-empty set } \wedge R \subseteq M \wedge R \text{ is } M\text{-definable}.$$

(As an exercise, the reader may like to investigate the logical complexity of such a formula.)

It φ is a formula of the **M**-language, the interpretations (in **M**) of the constants $\mathring{x}$ which occur in φ are called the *parameters* of φ.

Let A be some class of m-tuples, **M** a given structre. We say that the class A is $\Sigma_n^{\mathbf{M}}(N)$ iff $A \cap M^m$ is $\Sigma_n^{\mathbf{M}}(N)$, etc.

A related notion is the following. Let $\mathscr{F}$ be a class of structures of the form $\mathbf{M} = \langle M, \in, A_1, \ldots, A_k \rangle$, where k is fixed and each A_i is $n(i)$-ary, for fixed $n(i)$, $i = 1, \ldots, k$. Let A be a class of m-tuples. We say that A is *uniformly* $\Sigma_n^{\mathbf{M}}$ for $\mathbf{M} \in \mathscr{F}$ iff there is a *single* Σ_n formula $\varphi(v_0, \ldots, v_{m-1})$ of $\mathscr{L}(\mathring{A}_1, \ldots, \mathring{A}_k)$ such that for each $\mathbf{M} \in \mathscr{F}$,

$$A \cap M^m = \{(x_0, \ldots, x_{m-1}) \mid \models_{\mathbf{M}} \varphi(\mathring{x}_0, \ldots, \mathring{x}_{m-1})\}.$$

Similary for *uniformly* $\Pi_n^{\mathbf{M}}$ and *uniformly* $\Delta_n^{\mathbf{M}}$. We shall presently give some examples of these important (to us) concepts. In order to do so, however, we require some preliminary ideas.

A set M is said to be *amenable* iff it is transitive and satisfies the following conditions:

(i) $(\forall x, y \in M)(\{x, y\} \in M)$;
(ii) $(\forall x \in M)(\bigcup x \in M)$;
(iii) $\omega \in M$;
(iv) $(\forall x, y \in M)(x \times y \in M)$;
(v) if $R \subseteq M$ is $\Sigma_0(M)$, then $(\forall x \in M)(R \cap x \in M)$.

(Intuitively speaking, an amenable set is thus a transitive "model" of the theory BS of section 9. The idea behind this definition is that it will enable us to prove, within set theory, semantic analogues of the logical complexity results of section 9.)

Notice that if M is amenable, then $x \in M$ whenever $x \subseteq M$ is finite.

10.1 Lemma. *The predicate "x is finite" is uniformly* Σ_1^M *for amenable* M.

Proof. Let $\Phi(x, n, f)$ be the Σ_0 LST formula

$$(n \text{ is a natural number}) \wedge (f: n \leftrightarrow x).$$

Clearly, for any set x,

$$x \text{ is finite} \leftrightarrow \exists n \, \exists f \, \Phi(x, n, f).$$

Let φ be the analogue to Φ *in* $\mathscr{L}$. We prove that for any amenable set M,

$$(\forall x \in M)[\exists n \, \exists f \, \Phi(x, n, f) \leftrightarrow \models_M \exists n \, \exists f \, \varphi(\mathring{x}, n, f)],$$

which proves the lemma, of course.

Let M be amenable, $x \in M$. Suppose first that

$$\models_M \exists n \, \exists f \, \varphi(\mathring{x}, n, f).$$

Then by 9.11,

$$[\exists n \, \exists f \, \Phi(x, n, f)]^M.$$

But Σ_1 formulas of LST are U-absolute for transitive classes. Hence,

$$\exists n \exists f \Phi(x, n, f),$$

as required. Now suppose that this last formula is true. Pick n, f such that

$$\Phi(x, n, f).$$

Since M is transitive and $\omega \in M$ we have $\omega \subseteq M$, so certainly $n \in M$. Hence $x \times n \in M$. But $f \subseteq x \times n$ and f is finite, so $f \in M$. Then by the D-absoluteness of all Σ_0 formulas of LST, we have

$$[\Phi(x, n, f)]^M.$$

Thus

$$[\exists n \exists f \Phi(x, n, f)]^M,$$

and by 9.11 we conclude that

$$\models_M \exists n \exists f \varphi(\mathring{x}, n, f),$$

and we are done. □

If $R(x_0, \ldots, x_m)$ and $S(x_0, \ldots, x_m)$ are relations on M, then, extending our convention that $R(x_0, \ldots, x_m)$ means $(x_0, \ldots, x_m) \in R$, etc., we write:

$$\begin{array}{ll}
(R \wedge S)(x_0, \ldots, x_m) & \text{iff } (x_0, \ldots, x_m) \in R \cap S, \\
(R \vee S)(x_0, \ldots, x_m) & \text{iff } (x_0, \ldots, x_m) \in R \cup S, \\
(\neg R)(x_0, \ldots, x_m) & \text{iff } (x_0, \ldots, x_m) \in M^{m+1} - R, \\
(\exists x_0 R)(x_1, \ldots, x_m) & \text{iff } (\exists x_0 \in M)((x_0, x_1, \ldots, x_m) \in R), \\
((\exists x_0 \in z) R)(z, x_1, \ldots, x_m) & \text{iff } (\exists x_0 \in z)((x_0, x_1, \ldots, x_m) \in R), \\
\text{etc.} &
\end{array}$$

By means of, in particular, quantifier contraction along the lines of 8.9, we can easily prove:

10.2 Lemma. *Let M be an amenable set, and let $\mathbf{M} = \langle M, A_1, \ldots, A_k \rangle$. Let R, S be m-ary relations on M.*

(i) *If R, S are $\Sigma_0^{\mathbf{M}}(N)$, so too are $R \wedge S$, $R \vee S$, $\neg R$.*
(ii) *If R is $\Sigma_n^{\mathbf{M}}(N)$, $\neg R$ is $\Pi_n^{\mathbf{M}}(N)$.*
(iii) *If R is $\Pi_n^{\mathbf{M}}(N)$, $\neg R$ is $\Sigma_n^{\mathbf{M}}(N)$.*
(iv) *R is $\Delta_n^{\mathbf{M}}(N)$ iff both R and $\neg R$ are $\Sigma_n^{\mathbf{M}}(N)$.*
(v) *If R, S are $\Sigma_{n+1}^{\mathbf{M}}(N)$, so are $R \wedge S$, $R \vee S$, $\exists x R$, $(\exists x \in z) R$.*
(vi) *If R, S are $\Pi_{n+1}^{\mathbf{M}}(N)$, so are $R \wedge S$, $R \vee S$, $\forall x R$, $(\forall x \in z) R$.*
(vii) *If R, S are $\Delta_{n+1}^{\mathbf{M}}(N)$, so are $R \wedge S$, $R \vee S$, $\neg R$.* □

The following simple lemma employs the same trick used in the proofs of both 9.8 and 9.9.

10.3 Lemma. *Let M be an amenable set, and let $\mathbf{M} = \langle M, A_1, \ldots, A_k \rangle$. Let $f \subseteq M \times M$ be a function. (We say f is a function over M in this case.) If f is $\Sigma_n^{\mathbf{M}}(N)$ and* dom(f) *is $\Pi_n^{\mathbf{M}}(N)$, then both f and* dom(f) *are $\Delta_n^{\mathbf{M}}(N)$.*

Proof. Since

$$x \in \mathrm{dom}(f) \leftrightarrow \exists y [y = f(x)],$$

we see at once that dom(f) is $\Sigma_n^{\mathbf{M}}(N)$. To see that f is $\Pi_n^{\mathbf{M}}(N)$, note the equivalence

$$y = f(x) \leftrightarrow [x \in \mathrm{dom}(f)] \wedge \forall z [z = f(x) \rightarrow y = z]. \quad \square$$

10.4 Corollary. *Let $\mathbf{M}$ be as above. If $f: M \rightarrow M$ is $\Sigma_n^{\mathbf{M}}(N)$, then f is in fact $\Delta_n^{\mathbf{M}}(N)$.* $\square$

10.5 Lemma. *Let M be amenable, and let $\mathbf{M} = \langle M, A_1, \ldots, A_k \rangle$. Let $n \geqslant 1$ and let f be a $\Sigma_n^{\mathbf{M}}(N)$ m-ary function over M (i.e. $f \subseteq M^{m+1}$). Let g be a $\Sigma_n^{\mathbf{M}}(N)$ unary function over M and let R be a $\Sigma_n^{\mathbf{M}}(N)$ unary relation on M. Then h, S are $\Sigma_n^{\mathbf{M}}(N)$, where:*

(i) *h is the m-ary function defined by*

$$h(\vec{x}) = g \circ f(\vec{x});$$

(ii) *S is the m-ary relation defined by*

$$S(\vec{x}) \leftrightarrow R(f(\vec{x})).$$

Proof. By 10.2 and the observations

$$y = h(\vec{x}) \leftrightarrow \exists z [y = g(z) \wedge z = f(\vec{x})],$$
$$S(\vec{x}) \leftrightarrow \exists z [R(z) \wedge z = f(\vec{x})]. \quad \square$$

10.6 Lemma. *Let M be amenable, and let $\mathbf{M} = \langle M, A_1, \ldots, A_k \rangle$. If $R(x)$ is a $\Sigma_n^{\mathbf{M}}(N)$ unary relation on M, so too is $Q(x)$, where*

$$Q(x) \leftrightarrow [x \text{ is an ordered pair } \wedge R((x)_0)].$$

Similarly for $(x)_1$, etc. (We usually write $R((x)_0)$ in place of $Q(x)$ as defined above, etc.)

Proof. $Q(x) \leftrightarrow x$ is an ordered pair $\wedge\ (\exists u \in x)(\exists y \in u)(y = (x)_0 \wedge R(y))$. $\square$

10.7 Lemma (Contraction of Parameters). *Let M be as above. Let $n \geqslant 1$, and let R be a $\Sigma_n(\mathbf{M})$ relation on M. Then there is a single element $p \in M$ such that R is $\Sigma_n^{\mathbf{M}}(\{p\})$.*

Proof.. Let R be $\Sigma_n^{\mathbf{M}}(\{p_1, \ldots, p_m\})$. Set

$$p = (p_1, \ldots, p_m).$$

Using the method of 10.6 it is easily seen that R is $\Sigma_n^{\mathbf{M}}(\{p\})$. $\square$

We note also the following consequence of 9.11:

10.8 Lemma. *Fix $n \geqslant 1$. Let M be amenable, and let $\mathbf{M} = \langle M, A_1, \ldots, A_k \rangle$. If $R(\vec{z})$ is a $\Sigma_n^{\mathbf{M}}(N)$ relation on M, there is a $\Sigma_0^{\mathbf{M}}(N)$ relation $S(\vec{x}, \vec{z})$ on M such that*

$$R(\vec{z}) \leftrightarrow (\exists x_1 \in M)(\forall x_2 \in M)(\exists x_3 \in M) \ldots (- x_n \in M) S(\vec{x}, \vec{z}). \quad \square$$

For later use we make the following definitions. Let $\mathbf{M} = \langle M, A_1, \ldots, A_k \rangle$, $\mathbf{N} = \langle N, B_1, \ldots, B_k \rangle$. We say that $\mathbf{N}$ is an *elementary substructure* of $\mathbf{M}$, and write $\mathbf{N} \prec \mathbf{M}$, iff $N \subseteq M$, B_i is the restriction of A_i to N for $i = 1, \ldots, k$, and for all sentences φ of $\mathscr{L}_N(\mathring{A}_1, \ldots, \mathring{A}_k)$.

$$\models_{\mathbf{N}} \varphi \quad \text{iff } \models_{\mathbf{M}} \varphi .$$

(Notice that the sentence φ may contain constants denoting elements of N.) For $n \geqslant 0$, we say that $\mathbf{N}$ is a Σ_n *elementary substructure* of $\mathbf{M}$, and write $\mathbf{N} \prec_n \mathbf{M}$, iff the above holds when φ is restricted to be a Σ_n sentence. We shall write $X \prec \mathbf{M}$ to mean that X is the domain of a (necessarily unique for X) elementary substructure of $\mathbf{M}$, and analogously $X \prec_n \mathbf{M}$. We write $\pi : \mathbf{N} \prec \mathbf{M}$ (respectively $\pi : \mathbf{N} \prec_n \mathbf{M}$) iff π is an isomorphism from $\mathbf{N}$ to an elementary (respectively Σ_n elementary) substructure of $\mathbf{M}$.

11. Kripke-Platek Set Theory. Admissible Sets

We have already worked with one subtheory of ZF, namely the Basic Set Theory, BS. In this section we consider another, much stronger subtheory: *Kripke-Platek Set Theory*, KP. This is a particularly important subtheory of ZF for various reasons. One reason, of relevance to us, is that KP is the weakest subtheory of ZF which suffices for the construction of the constructible hierarchy of sets, introduced in Chapter II.

The theory KP is the LST theory whose axioms are the axioms of BS, together with the Σ_0 *Collection Schema*:

$$\forall \vec{a} [\forall x \exists y \Phi(y, x, \vec{a}) \rightarrow \forall u \exists v (\forall x \in u)(\exists y \in v) \Phi(y, x, \vec{a})],$$

where Φ is a Σ_0 formula of LST.

By an *admissible set* we mean an amenable set M (see section 10) such that for any $\Sigma_0(M)$ relation $R \subseteq M \times M$, if

$$(\forall x \in M)(\exists y \in M) R(y, x)$$

then for any $u \in M$ there is a $v \in M$ such that

$$(\forall x \in u)(\exists y \in v) R(y, x).$$

Clearly, the notion of an admissible set is related to the theory KP in the same way that the notion of an amenable set is related to the theory BS (i.e. admissible sets are transitive "models" of the theory KP.)

For κ an uncountable cardinal, we define

$$H_\kappa = \{x \mid |\mathrm{TC}(x)| < \kappa\}.$$

Using the following lemma, we shall be able to show that H_κ is an admissible set for any uncountable cardinal κ.

11.1 Lemma. *Let $\varphi(\vec{x})$ be a Σ_1 formula of $\mathscr{L}$. Let κ, λ be uncountable cardinals, $\lambda < \kappa$. If $\vec{x} \in H_\lambda$ are such that $\models_{H_\kappa} \varphi(\vec{x})$, then $\models_{H_\lambda} \varphi(\vec{x})$.*

Proof. Let

$$W = \mathrm{TC}(\{\vec{x}\}).$$

Clearly, $W \in H_\lambda$. Pick $M \prec H_\kappa$ with $W \subseteq M$ and $|M| = |W| < \lambda$. (That this can always be done follows from the Löwenheim-Skolem-Tarski Theorem. We assume the reader is familiar with this theorem.) Let

$$\pi: M \cong N$$

be the collapsing isomorphism (see 7.1), where N is transitive. Then $|N| = |M| < \lambda$, so $N \in H_\lambda$ and $N \subseteq H_\lambda$. Now, $\pi^{-1}: N \prec H_\kappa$ and (see 7.1) $\pi \upharpoonright W = \mathrm{id} \upharpoonright W$, so $\models_N \varphi(\vec{x})$. But φ is Σ_1, so by 9.14, φ is U-absolute for N, H_λ. Thus $\models_{H_\lambda} \varphi(\vec{x})$, as required. □

11.2 Lemma. *If κ is an uncountable cardinal, then H_κ is admissible.*

Proof. It is easily seen that H_κ is amenable for any uncountable cardinal κ. (Exercise: Check this.) Moreover, it is also easy to see that in the case where κ is regular, H_κ is in fact admissible. We are therefore left with proving admissibility in the case where κ is singular.

So assume that κ is singular, and let $R \subseteq H_\kappa \times H_\kappa$ be $\Sigma_0(H_\kappa)$. We must show that if

$$(\forall x \in H_\kappa)(\exists y \in H_\kappa) R(y, x)$$

and if $u \in H_\kappa$, then there is a $v \in H_\kappa$ such that

$$(\forall x \in u)(\exists y \in v) R(y, x).$$

Let $\varphi(y, x, \vec{a})$ be a Σ_0 formula of $\mathscr{L}$ and $\vec{a} \in H_\kappa$ be such that

$$R(y, x) \leftrightarrow \models_{H_\kappa} \varphi(\mathring{y}, \mathring{x}, \mathring{\vec{a}}).$$

Let $u \in H_\kappa$ be given. We seek a $v \in H_\kappa$ such that

$$\models_{H_\kappa} (\forall x \in \mathring{u})(\exists y \in \mathring{v})\, \varphi(y, x, \mathring{\vec{a}}).$$

Let

$$W = \mathrm{TC}(\{u, \vec{a}\}).$$

Then $W \in H_\kappa$, so as κ is singular there is a regular cardinal $\lambda < \kappa$ such that $W \in H_\lambda$. Now, by the assumptions on R,

$$\models_{H_\kappa} \forall x \exists y \varphi(y, x, \mathring{\vec{a}}).$$

So for all $x \in H_\kappa$,

$$\models_{H_\kappa} \exists y \varphi(y, \mathring{x}, \mathring{\vec{a}}).$$

So by 11.1, for all $x \in H_\lambda$,

$$\models_{H_\lambda} \exists y \varphi(y, \mathring{x}, \mathring{\vec{a}}).$$

Thus

$$\models_{H_\lambda} \forall x \exists y \varphi(y, x, \mathring{\vec{a}}).$$

But λ is regular, so as we observed above, H_λ is admissible. Thus as $u \in H_\lambda$, there is a $v \in H_\lambda$ such that

$$\models_{H_\lambda} (\forall x \in \mathring{u})(\exists y \in \mathring{v}) \varphi(y, x, \mathring{\vec{a}}).$$

But the sentence involved here is Σ_0, and hence (by 9.14) absolute for H_λ, H_κ. Thus

$$\models_{H_\kappa} (\forall x \in \mathring{u})(\exists y \in \mathring{v}) \varphi(y, x, \mathring{\vec{a}}),$$

and we are done. □

We shall obtain a few elementary results about the theory KP. Our first two show that KP entails stronger versions of the Collection and Comprehension Axioms than were allowed for in the axioms.

11.3 Lemma (Σ_1-Collection Principle). *Let $\Phi(y, x, \vec{a})$ be a Σ_1 formula of* LST. *Then*

$$\mathrm{KP} \vdash \forall \vec{a} [\forall x \exists y \Phi(y, x, \vec{a}) \rightarrow \forall u \exists v (\forall x \in u)(\exists y \in v) \Phi(y, x, \vec{a})].$$

Proof. Let $\Psi(z, y, x, \vec{a})$ be a Σ_0 formula of LST such that

$$\mathrm{KP} \vdash \Phi(y, x, \vec{a}) \leftrightarrow \exists z \Psi(z, y, x, \vec{a}).$$

(By 8.9, such a formula can always be found.) Argue in KP from now on.

Let $\vec{a}$ be given, and assume

$$\forall x \exists y \Phi(y, x, \vec{a}).$$

Then

$$\forall x \exists y \exists z \Psi(z, y, x, \vec{a}).$$

Hence

$$\forall x \exists w \Psi((w)_0, (w)_1, x, \vec{a}).$$

Given u now we must find a v such that

$$(\forall x \in u)(\exists y \in v) \Phi(y, x, \vec{a}).$$

But by 8.8, the formula $\Psi((w)_0, (w)_1, x, \vec{a})$ is Σ_0, so by Σ_0-Collection there is a t such that

$$(\forall x \in u)(\exists w \in t) \Psi((w)_0, (w)_1, x, \vec{a}).$$

Let $v = \bigcup\bigcup t$. Then

$$(\forall x \in u)(\exists y \in v)(\exists z) \Psi(z, y, x, \vec{a}).$$

Hence

$$(\forall x \in u)(\exists y \in v) \Phi(y, x, \vec{a}),$$

as required. □

11.1 Lemma (Δ_1-Comprehension Principle). *Let $\Phi(z, \vec{a})$ be a Δ_1^{KP} formula of* LST. *Then*

$$\mathrm{KP} \vdash \forall \vec{a} \forall x \exists y \forall z [z \in y \leftrightarrow z \in x \wedge \Phi(z, \vec{a})].$$

Proof. By 8.9 we can find Σ_0 formulas Θ, Ψ of LST such that

$$\mathrm{KP} \vdash \Phi(z, \vec{a}) \leftrightarrow \forall v \Theta(v, z, \vec{a}),$$
$$\mathrm{KP} \vdash \Phi(z, \vec{a}) \leftrightarrow \exists v \Psi(v, z, \vec{a}).$$

We argue in KP from now on.

Let $\vec{a}$, x be given. We seek a y such that

$$\forall z [z \in y \leftrightarrow z \in \wedge \Phi(z, \vec{a})].$$

Now,

$$\forall z [\Phi(z, a) \vee \neg \Phi(z, a)].$$

Hence

$$\forall z \exists v [\Psi(v, z, \vec{a}) \vee \neg \Theta(v, z, \vec{a})].$$

By Σ_0-Collection there is thus a set u such that

(*) $$(\forall z \in x)(\exists v \in u)[\Psi(v, z, \vec{a}) \vee \neg \Theta(v, z, \vec{a})].$$

By Σ_0-Comprehension, let

$$y = \{z \in x \mid (\exists v \in u) \Psi(v, z, \vec{a})\}.$$

We finish by showing that

$$y = \{z \in x \mid \Phi(z, \vec{a})\}.$$

Certainly, for any $z \in x$,

$$(\exists v \in u)\, \Psi(v, z, \vec{a}) \rightarrow \exists v\, \Psi(v, z, \vec{a}) \\ \leftrightarrow \Phi(z, \vec{a}),$$

so what we must prove is that for any $z \in x$,

$$\Phi(z, \vec{a}) \rightarrow (\exists v \in u)\, \Psi(v, z, \vec{a}).$$

By (*), we know that there is a $v \in u$ such that

$$\Psi(v, z, \vec{a}) \vee \neg\, \Theta(v, z, \vec{a}).$$

But

$$\Phi(z, \vec{a}) \leftrightarrow \forall v\, \Theta(v, z, \vec{a}).$$

Hence for the $v \in u$ chosen above, we must have

$$\Psi(v, z, \vec{a}).$$

We are done. □

The following lemma is a useful alternative to the Σ_1-Collection Principle (11.3).

11.5 Lemma (Localised Σ_1-Collection Schema). *If Φ is a Σ_1 formula of* LST, *then*

$$\mathrm{KP} \vdash \forall \vec{a}\, [(\forall x \in u)\, \exists y\, \Phi(y, x, \vec{a}) \rightarrow \exists v\, (\forall x \in u)(\exists y \in v)\, \Phi(y, x, \vec{a})].$$

Proof. Argue in KP. Assume

$$(\forall x \in u)\, \exists y\, \Phi(y, x, \vec{a}).$$

Then

$$\forall x\, \exists y\, (x \notin u \vee \Phi(y, x, \vec{a})).$$

So by Σ_1-Collection there is a v such that

$$(\forall x \in u)(\exists y \in v)(x \notin u \vee \Phi(y, x, \vec{a})).$$

But this is logically equivalent to

$$(\forall x \in u)(\exists y \in v)\, \Phi(y, x, \vec{a}),$$

and we are done. □

The next lemma extends 8.6 (iv), (v) for the theory KP, and is a special case of 8.10.

11.6 Lemma. (i) *If* $\Phi(y, \vec{x})$ *is a* Σ_1 *formula of* LST, *then* $(\forall z \in y)\, \Phi(z, \vec{x})$ *is* Σ_1^{KP}.
(ii) *If* $\Phi(y, \vec{x})$ *is a* Π_1 *formula of* LST, *then* $(\exists z \in y)\, \Phi(z, \vec{x})$ *is* Π_1^{KP}.

Proof. We prove (i); (ii) then follows by taking negations. Let $\Psi(w, y, \vec{x})$ be a Σ_0 formula such that

$$\mathrm{KP} \vdash \Phi(y, \vec{x}) \leftrightarrow \exists w\, \Psi(w, y, \vec{x}).$$

By 8.9, such a Ψ can be found, of course. We argue in KP from now on.
We have:

$$\begin{aligned}(\forall z \in y)\, \Phi(z, \bar{x}) &\leftrightarrow (\forall z \in y)(\exists w)\, \Psi(w, z, \vec{x})\\ &\rightarrow (\exists v)(\forall z \in y)(\exists w \in v)\, \Psi(w, z, \vec{x}) \qquad \text{(by 11.5)}\\ &\rightarrow (\forall z \in y)(\exists w)\, \Psi(w, z, \vec{x}) \qquad \text{(by logic)}\\ &\leftrightarrow (\forall z \in y)\, \Phi(z, \vec{x}).\end{aligned}$$

This provides us with the Σ_1 equivalent

$$(\exists v)(\forall z \in y)(\exists w \in v)\, \Psi(w, z, \vec{x})$$

to $(\forall z \in y)\, \Phi(z, \vec{x})$. □

Now, both in the case of BS and KP, as well as considering these as LST theories, we introduced analoguous, set-theoretic notions defined within set theory proper, namely the notions of amenable and admissible sets, respectively. This is to enable us to obtain, within set theory itself, "localised" analogues of some of our later results concerning the logical complexity of the constructible hierarchy, and related notions. By and large, the importance of this will become clear as we progress through Chapter II, but in the meantime, by way of an illustration, we formulate our next result not as a theorem schema for KP, as we did with the previous four lemmas, but rather as a (ZF) theorem about admissible sets. Hopefully, the reader should have no difficulty in reformulating both the statement and the proof of this lemma along the lines of the previous KP-results.

11.7 Lemma. *Let* M *be an admissible set, and let* F *be a* $\Sigma_1(M)$ *function over* M. *If* $u \in M$ *and* $u \subseteq \operatorname{dom}(F)$, *then* $F \restriction u$, $F''u \in M$.

Proof. By 10.3, $F \restriction u$ is $\Delta_1(M)$. So by Δ_1-Comprehension (11.4),

$$w \in M \rightarrow w \cap (F \restriction u) \in M.$$

Now,

$$(\forall x \in u)\, \exists y\, [y = F(x)],$$

so by 11.5 (or rather the consequence/analogue of 11.5 for admissible sets) there is a $v \in M$ such that

$$(\forall x \in u)(\exists y \in v)\, [y = F(x)].$$

Thus $F \upharpoonright u \subseteq v \times u$. But $w = v \times u \in M$, by the Cartesian Product Axiom. Hence

$$F \upharpoonright u = w \cap (F \upharpoonright u) \in M.$$

By Σ_0-Comprehension now,

$$F'' u = v \cap \{y \mid (\exists x \in u)[(y, x) \in F \upharpoonright u]\} \in M. \quad \square$$

So far we have stated results either as theorem schemas for KP or as theorems within ZF about admissible sets. It is convenient to state the next result as a theorem schema *in terms of classes* (as we often do for ZF). Thus, a Σ_1^{KP} *class* is a class of the form

$$\{x \mid \Phi(x)\}$$

where Φ is a Σ_1^{KP} formula of LST, etc. And a Σ_1^{KP} *function over* V is a class of the form

$$\{(y, \vec{x}) \mid \Phi(y, \vec{x})\},$$

such that Φ is a Σ_1 formula of LST and

$$\mathrm{KP} \vdash \forall \vec{x}\,[\exists y\, \Phi(y, \vec{x}) \rightarrow \exists!\, y\, \Phi(y, \vec{x})].$$

11.8 Lemma (The Recusion Theorem). *Let G be a total, $(n+2)$-ary, Σ_1^{KP} function over V. Then there is a total, $(n+1)$-ary, Σ_1^{KP} function, F, over V such that:*

$$\mathrm{KP} \vdash F(y, \vec{x}) = G(y, \vec{x}, (F(z, \vec{x}) \mid z \in y)).$$

Proof. Let $\Phi(\sigma, \vec{x})$ be the LST formula

$$[\text{“}\sigma \text{ is a function”}] \wedge [\text{“dom}(\sigma) \text{ is transitive”}]$$
$$\wedge\ [(\forall y \in \mathrm{dom}(\sigma))(\sigma(y) = G(y, \vec{x}, \sigma \upharpoonright y))].$$

Since G is total, by 10.3, G is in fact a Δ_1^{KP} class. Hence Φ is Δ_1^{KP}. Thus $\Psi(z, y, \vec{x})$ is a Σ_1^{KP} formula, where

$$\Psi(z, y, \vec{x}) = (\exists \sigma)[\Phi(\sigma, \vec{x}) \wedge \sigma(y) = z].$$

Claim 1. $\mathrm{KP} \vdash (\forall \vec{x}, y)(\exists z)\, \Psi(z, y, \vec{x})$.

Proof of claim: Argue in KP. Suppose otherwise. Pick $\vec{x}, y$ so that

$$\neg (\exists z)\, \Psi(z, y, \vec{x}).$$

By the Axiom of Foundation, we can ensure that y is chosen here so that

$$(\forall y' \in y)(\exists z)\, \Psi(z, y, \vec{x}).$$

By 11.5, we can find a set v such that

$$(\forall y' \in y)(\exists \sigma \in v)(y' \in \mathrm{dom}(\sigma) \wedge \Phi(\sigma, \vec{x})).$$

By Δ_1-Comprehension, set

$$w = v \cap \{\sigma \mid \Phi(\sigma, \vec{x})\}.$$

Let $\varrho = \bigcup w$. Then ϱ is a function. To see this, it clearly suffices to show that if $z \in \mathrm{dom}(\sigma_1) \cap \mathrm{dom}(\sigma_2)$, where $\Phi(\sigma_1, \vec{x})$ and $\Phi(\sigma_2, \vec{x})$, for $\sigma_1, \sigma_2 \in v$, then $\sigma_1(z) = \sigma_2(z)$. But this follows from the nature of Φ by $\in$-induction: if $\sigma_1(z') = \sigma_2(z')$ for all $z' \in z$, then $\sigma_1 \upharpoonright z = \sigma_2 \upharpoonright z$, and therefore

$$\sigma_1(z) = G(z, \vec{x}, \sigma_1 \upharpoonright z) = G(z, \vec{x}, \sigma_2 \upharpoonright z) = \sigma_2(z).$$

And clearly, $\mathrm{dom}(\varrho)$ is transitive. It is now clear that $\Phi(\varrho, \vec{x})$. Let

$$\tau = \varrho \cup \{(G(y, \vec{x}, \varrho \upharpoonright y), y)\}.$$

Clearly, $\Phi(\tau, \vec{x})$. But

$$\tau(y) = G(y, \vec{x}, \varrho \upharpoonright y).$$

Hence $\Psi(\tau(y), y, \vec{x})$, contrary to the choice of $\vec{x}$, y. The claim is proved.

Let F be the class

$$\{(z, y, \vec{x}) \mid \Psi(z, y, \vec{x})\}.$$

Claim 2. $\mathrm{KP} \vdash F$ is a function.

Proof of claim: Just as the proof that ϱ was a function in claim 1.

Clearly, F is a required for the lemma. □

11.9 Corollary. *The function* TC (*transitive closure*) *is* Σ_1^{KP} (*and hence* Δ_1^{KP}). □

Using 11.9, together with an argument much as in 11.8, we get:

11.10 Lemma (TC-Recursion Theorem). *Let G be a total, $(n+2)$-ary, Σ_1^{KP} function over V. Then there is a total, $(n+1)$-ary, Σ_1^{KP} function, F, over V such that*

$$\mathrm{KP} \vdash F(y, \vec{x}) = G(y, \vec{x}, (F(z, \vec{x}) \mid z \in \mathrm{TC}(y))). \quad \square$$

Chapter II
The Constructible Universe

In Zermelo-Fraenkel set theory, the notion of what consitutes a set is not really defined, but rather is taken as a basic concept. The Zermelo-Fraenkel axioms describe the *properties* of sets and the set-theoretic universe. For instance, if X is an infinite set, the Power Set Axiom tells us that there is a set, $\mathscr{P}(X)$, which consists of *all* subsets of X. But the other axioms do not tell us very much about the members of $\mathscr{P}(X)$, or give any indication as to how big a set this is. The Axiom of Comprehension says that $\mathscr{P}(X)$ will contain all sets which are *describable* in a certain, well-defined sense, and AC will provide various choice sets and well-orderings. But the word "all" in the phrase "all subsets of X" is not really explained. Of course, as mathematicians we are (are we not?) quite happy with the notion of $\mathscr{P}(X)$, and so long as there are no problems, Zermelo-Fraenkel set theory can be taken as a perfectly reasonable theory. But as we know, ZFC set theory does have a major drawback: there are several easily formulated questions which cannot be answered on the basis of the ZFC axioms alone. A classic example is the status of the *continuum hypothesis*, $2^{\omega} = \omega_1$. It can be argued that this cannot be decided in ZFC because the ZFC axioms do not say just what constitutes a subset of ω; hence we cannot relate the size of $\mathscr{P}(\omega)$ to the infinite cardinal numbers ω_α, $\alpha \in \mathrm{On}$. (The formal *proof* of the undecidability of CH is rather different from the above "plausibility argument".)

One way of overcoming the difficulty of undecidable questions is to extend the theory ZFC, to obtain a richer theory which provides more information about sets. (An alternative solution is simply to accept as a fact of life that some questions have no answer.) One highly successful extension of ZFC is the *constructible set theory* of Gödel. In this theory the notion of a "set" is made precise (at least relative to the ordinals). The idea is as follows.

The fundamental picture of the set-theoretic universe which the Zermelo-Fraenkel axioms supply is embodied in the cumulative hierarchy of sets. We commence with the null set, $\emptyset$, and obtain all other sets by iteratively applying the (undescribed) power set operation, $\mathscr{P}$. Thus:

$$V_0 = \emptyset; \qquad V_{\alpha+1} = \mathscr{P}(V_\alpha); \qquad V_\lambda = \bigcup_{\alpha<\lambda} V_\alpha, \qquad \text{if } \lim(\lambda).$$

Then

$$V = \bigcup_{\alpha \in \mathrm{On}} V_\alpha.$$

In *constructible set theory* we do not take as basic the (so called "unrestricted") power set operator, $\mathscr{P}$. Rather we say that a set can only be said to exist if it is definable over existing sets in much the same way that classes are obtained. Recall from I.10 that a subset y of a set x is said to be *x-definable* iff there is a formula $\varphi(v_0)$ of $\mathscr{L}_x$ such that

$$y = \{a \in x \mid \models_x \varphi(\mathring{a})\}.$$

To obtain the *constructible universe* of sets, we start with the empty set and iterate the operation of taking all the definable subsets at each stage. This provides us with a universe of sets in which the notion of what constitutes a set is very precisely defined (relative to the ordinals).

Now, although we can regard constructible set theory as an *alternative* to Zermelo-Fraenkel set theory, as axiomatic theories the former is an extension of the latter: in fact constructible set theory is just ZFC together with one additional axiom – the *Axiom of Constructibility*. In this volume we are taking ZFC as our basic set theory, and we shall study the notion of constructibility in its own right. Indeed, many mathematicians feel that constructible set theory is *not* a reasonable *fundamental* set theory in the sense that ZFC is, and that constructibility should *only* be studied as an interesting notion within the ZFC framework. In any event, the notion is an interesting and fruitful one, as we hope to demonstrate in the ensuing pages.

In this chapter we define the constructible universe and develop its elementary theory.

1. Definition of the Constructible Universe

Let X be any set. By

$$\mathrm{Def}(X)$$

we mean the set of all subsets of X which are X definable (in the sense of I.10). That is, $\mathrm{Def}(X)$ consists of all sets, a, such that for some formula $\varphi(v_0)$ of $\mathscr{L}_X$,

$$a = \{x \in X \mid \models_X \varphi(\mathring{x})\}.$$

The function Def is a well-defined set-theoretic function, and indeed has the definition:

$$\begin{aligned} v = \mathrm{Def}(u) \leftrightarrow (\forall x \in v)(\exists \varphi)&[\mathrm{Fml}(\varphi, u) \wedge \mathrm{Fr}(\varphi, \{v_0\}) \\ &\wedge (x = \{z \in u \mid \exists \psi (\mathrm{Sub}(\psi, \varphi, v_0, \mathring{z}) \wedge \mathrm{Sat}(u, \psi))\})] \\ &\wedge (\forall \varphi)[(\mathrm{Fml}(\varphi, u) \wedge \mathrm{Fr}(\varphi, \{v_0\})) \\ &\rightarrow (\exists x \in v)(x = \{z \in u \mid \exists \psi (\mathrm{Sub}(\psi, \varphi, v_0, \mathring{z}) \wedge \mathrm{Sat}(u, \psi))\})]. \end{aligned}$$

(We shall presently examine the logical complexity of this definition.)

By recursion on $\alpha \in \mathrm{On}$ we define

$$L_0 = \emptyset; \quad L_{\alpha+1} = \mathrm{Def}(L_\alpha); \quad L_\lambda = \bigcup_{\alpha<\lambda} L_\alpha, \quad \text{if } \lim(\lambda).$$

$(L_\alpha \mid \alpha \in \mathrm{On})$ is the *constructible hierarchy*, and is clearly a well-defined function (in the class sense) of ZF set theory (see later for more details). Hence L is a well-defined class (again, more details later), where we set:

$$L = \bigcup_{\alpha \in \mathrm{On}} L_\alpha .$$

L is the *constructible universe.* A set x is said to be *constructible* iff $x \in L$.

Our first lemma below establishes various simple and basic results about the constructible hierarchy.

1.1 Lemma.

(i) $\alpha \leqslant \beta$ *implies* $L_\alpha \subseteq L_\beta$.
(ii) *Each* L_α *is transitive. (Hence* L *is transitive.)*
(iii) $L_\alpha \subseteq V_\alpha$ *for all* α.
(iv) $\alpha < \beta$ *implies* $\alpha, L_\alpha \in L_\beta$. *(Hence* $\mathrm{On} \subseteq L$.*)*
(v) *For all* α, $L \cap \alpha = L_\alpha \cap \mathrm{On} = \alpha$.
(vi) *For* $\alpha \leqslant \omega$, $L_\alpha = V_\alpha$.
(vii) *For* $\alpha \geqslant \omega$, $|L_\alpha| = |\alpha|$.

Proof. (i) and (ii). We prove by simultaneous induction on α that;

(a) $\quad \gamma < \alpha \rightarrow L_\gamma \subseteq L_\alpha$;

(b) $\quad L_\alpha$ is transitive.

For $\alpha = 0$ this is trivial. For limit α, we have $L_\alpha = \bigcup_{\gamma<\alpha} L_\gamma$, so (a) and (b) are immediate consequences of the induction hypothesis. (In particular, note that any union of transitive sets is transitive.) In order to prove that (a) and (b) for $\alpha + 1$ follow from (a) and (b) for α, let us start with (a) for $\alpha + 1$. It clearly suffices to prove that $L_\alpha \subseteq L_{\alpha+1}$. Let $x \in L_\alpha$. Then by (b) for α, $x \subseteq L_\alpha$, so by Σ_0-absoluteness,

$$x = \{y \in L_\alpha \mid \models_{L_\alpha} \text{“}\mathring{y} \in \mathring{x}\text{”}\} \in \mathrm{Def}(L_\alpha) = L_{\alpha+1}.$$

To prove (b) for $\alpha + 1$ now, let $x \in y \in L_{\alpha+1}$. Since $y \in L_{\alpha+1} = \mathrm{Def}(L_\alpha) \subseteq \mathscr{P}(L_\alpha)$, we have $y \subseteq L_\alpha$. But then $x \in L_\alpha$, so by (a) for $\alpha + 1$ just proved, we have $x \in L_{\alpha+1}$, and we are done.

(iii) By induction on α. For $\alpha = 0$ we have

$$L_0 = V_0 = \emptyset .$$

At limit stages λ, we have

$$L_\lambda = \bigcup_{\alpha<\lambda} L_\alpha \quad \text{and} \quad V_\lambda = \bigcup_{\alpha<\lambda} V_\alpha,$$

so if $L_\alpha \subseteq V_\alpha$ for all $\alpha < \lambda$, then $L_\lambda \subseteq V_\lambda$. Finally, if $L_\alpha \subseteq V_\alpha$, then

$$L_{\alpha+1} = \mathrm{Def}(L_\alpha) \subseteq \mathscr{P}(L_\alpha) \subseteq \mathscr{P}(V_\alpha) = V_{\alpha+1}.$$

(iv) By (i) it suffices to prove that α, $L_\alpha \in L_{\alpha+1}$ for all α. Well, for any α,

$$L_\alpha = \{x \in L_\alpha \mid \models_{L_\alpha} \text{“}\mathring{x} = \mathring{x}\text{”}\} \in \mathrm{Def}(L_\alpha) = L_{\alpha+1}.$$

To prove that $\alpha \in L_{\alpha+1}$ we proceed by induction on α. Assume that $\gamma \in L_{\gamma+1}$ for all $\gamma < \alpha$. Then by (i), $\gamma \in L_\alpha$ for all $\gamma < \alpha$, i.e. $\alpha \subseteq L_\alpha$. Thus by (ii), $\alpha = L_\alpha \cap \mathrm{On}$. But $\mathrm{On}(v_0)$ is a Σ_0 formula and is thus absolute for L_α. Hence

$$\alpha = \{x \in L_\alpha \mid \models_{L_\alpha} \mathrm{On}(\mathring{x})\} \in \mathrm{Def}(L_\alpha) = L_{\alpha+1}.$$

(Actually we are being a bit sloppy here. As we defined it, $\mathrm{On}(x)$ is a formula of LST, and thus not available for use as above. However, if we take instead the corresponding $\mathscr{L}$-formula, as described in I.9.11, then by I.9.15 we see that for any x and any transitive set M which contains x:

$$x \text{ is an ordinal} \leftrightarrow \models_M \mathrm{On}(\mathring{x}).$$

In future we shall not bother too much about fine points of this nature.)

(v) That $L_\alpha \cap \mathrm{On} = \alpha$ was proved during the proof of (iv). In view of (ii), this proves all of the equalities in (v).

(vi) For $\alpha = 0$ we have $L_0 = \emptyset = V_0$. Let $\alpha < \omega$ and assume that $L_\alpha = V_\alpha$. We prove that $L_{\alpha+1} = V_{\alpha+1}$. By (ii) it suffices to prove that $V_{\alpha+1} \subseteq L_{\alpha+1}$. Let $x \in V_{\alpha+1}$. Then $x \subseteq V_\alpha = L_\alpha$, and there are $a_1, \ldots, a_n \in L_\alpha$ such that

$$x = \{a_1, \ldots, a_n\}.$$

(Because V_α is finite for each $\alpha < \omega$.) Hence

$$x = \{z \in L_\alpha \mid \models_{L_\alpha} (\mathring{z} = \mathring{a}_1 \vee \ldots \vee \mathring{z} = \mathring{a}_n)\} \in \mathrm{Def}(L_\alpha) = L_{\alpha+1}.$$

Thus by induction, $L_\alpha = V_\alpha$ for all $\alpha < \omega$. It follows at once that

$$L_\omega = \bigcup_{\alpha<\omega} L_\alpha = \bigcup_{\alpha<\omega} V_\alpha = V_\omega.$$

(vii) By (v) we have $|\alpha| \leqslant |L_\alpha|$ for all α. By induction on $\alpha \geqslant \omega$ we prove that $|L_\alpha| \leqslant |\alpha|$ for all $\alpha \geqslant \omega$. For $\alpha = \omega$ this holds by (vi), since

$$|L_\omega| = |V_\omega| = \omega.$$

Suppose next that $\lim(\lambda)$ and we know that $|L_\alpha| \leqslant |\alpha|$ for all $\alpha < \lambda$. Then

$$|L_\lambda| = |\bigcup_{\alpha<\lambda} L_\alpha| \leqslant \sum_{\alpha<\lambda} |L_\alpha| \leqslant \sum_{\alpha<\lambda} |\alpha| = |\lambda|.$$

Finally, suppose that $|L_\alpha| \leqslant |\alpha|$. We prove that $|L_{\alpha+1}| \leqslant |\alpha|$ $(= |\alpha + 1|)$. Well, since $\mathscr{L}$ is countable, the set of formulas of $\mathscr{L}_{L_\alpha}$ is easily seen to have cardinality $|L_\alpha|$. But this at once implies that

$$|L_{\alpha+1}| = |\mathrm{Def}(L_\alpha)| \leqslant |L_\alpha| \leqslant |\alpha|,$$

and we are done. □

Let M be a transitive proper class, and let T be a theory in LST. We say that M is an *inner model* of T iff Φ^M for every axiom Φ of T. (The name "inner model" arises from the case where T is the theory ZF, in which case M is a sort of "inner universe" of set theory. But it is convenient to formulate the definition to cover all LST theories T.) The following result is fundamental to all work on constructibility theory.

1.2 Theorem. *The class L is an inner model of* ZF. *More precisely, for every axiom Φ of* ZF,

$$\mathrm{ZF} \vdash \Phi^L.$$

Proof. For each axiom Φ of ZF in turn, we argue in ZF to prove Φ^L.

I. *Extensionality.* We must prove

$$[(\forall x, y)[(\forall z)(z \in x \leftrightarrow z \in y) \to (x = y)]]^L.$$

Thus, given $x, y \in L$, we must prove

$$[(\forall z)(z \in x \leftrightarrow z \in y) \to (x = y)]^L.$$

This is the same as

$$(\forall z \in L)(z \in x \leftrightarrow z \in y) \to (x = y).$$

But since L is transitive, $x, y \subseteq L$, so this is the same as

$$(\forall z)(z \in x \leftrightarrow z \in y) \to (x = y).$$

And this is true by virtue of the (real) Axiom of Extensionality itself.

II. *Union.* We must prove

$$[\forall x \, \exists y \, \forall z (z \in y \leftrightarrow (\exists u \in x)(z \in u))]^L.$$

Thus, given an $x \in L$ we must find a $y \in L$ such that

$$[\forall z (z \in y \leftrightarrow (\exists u \in x)(z \in u)]^L,$$

i.e. such that

$$(\forall z \in L)(z \in y \leftrightarrow (\exists u \in x)(z \in u)).$$

By the Axiom of Union itself, let

$$y = \bigcup x .$$

Since $x \in L$ there is an ordinal α such that $x \in L_\alpha$. Since L_α is transitive, $y \subseteq L_\alpha$. Moreover,

$$y = \{z \in L_\alpha \mid \models_{L_\alpha} (\exists v_1 \in \mathring{x})(\mathring{z} \in v_1)\},$$

so

$$y \in \operatorname{Def}(L_\alpha) = L_{\alpha+1} \subseteq L .$$

But since $y = \bigcup x$,

$$(\forall z)(z \in y \leftrightarrow (\exists u \in x)(z \in u)),$$

so in particular

$$(\forall z \in L)(z \in y \leftrightarrow (\exists u \in x)(z \in u)),$$

as required.

III. *Infinity*. We must show that

$$[\exists x [\exists y (y \in x) \wedge (\forall y \in x)(\exists z \in x)(y \in z)]^L .$$

But by 1.1 (iv), $\omega \in L_{\omega+1} \subseteq L$, so this is immediate.

IV. *Power Set*. We must show that

$$[\forall x \, \exists y \, \forall z (z \in y \leftrightarrow z \subseteq x)]^L .$$

So, given $x \in L$ we must find a $y \in L$ such that

$$(\forall z \in L)(z \in y \leftrightarrow z \subseteq x) .$$

By the Axioms of Power Set and Comprehension, let

$$y = \{z \in \mathscr{P}(x) \mid z \in L\} .$$

We prove that $y \in L$, in which case y is clearly as required.

For each $z \in y$, let $f(z)$ be the least α such that $z \in L_\alpha$. By the Axiom of Collection, let α exceed all $f(z)$ for $z \in y$. Thus $y \subseteq L_\alpha$. But then

$$y = \{z \in L_\alpha \mid \models_{L_\alpha} (\mathring{z} \subseteq \mathring{x})\} \in \operatorname{Def}(L_\alpha) = L_{\alpha+1} .$$

Thus $y \in L$.

V. *Foundation.* We must prove that

$$[\forall x[\exists y(y \in x) \rightarrow \exists y(y \in x \wedge (\forall z \in y)(z \notin x))]]^L.$$

Let $x \in L$ be given, $x \neq \emptyset$. We must find a $y \in L$ such that $y \in x$ and

$$[(\forall z \in y)(z \notin x)]^L.$$

By the Axiom of Foundation itself there is a $y \in x$ such that

$$(\forall z \in y)(z \notin x).$$

But L is transitive, so $y \in L$. Clearly, y is as required.

VI. *Comprehension.* Let $\Phi(v_0, \ldots, v_n)$ be a formula of LST. We must prove that

$$[\forall x \forall v_1 \ldots \forall v_n \exists y \forall z[(z \in y) \leftrightarrow (z \in x) \wedge \Phi(z, v_1, \ldots, v_n)]]^L.$$

Let $x, a_1, \ldots, a_n \in L$ be given. We seek a $y \in L$ such that

$$(\forall z \in L)[(z \in y) \leftrightarrow (z \in x) \wedge \Phi^L(z, a_1, \ldots, a_n)].$$

Pick α so that $x, a_1, \ldots, a_n \in L_\alpha$. Applying the Generalised Reflection Principle (I.8.1) to the constructible hierarchy, we can find a $\beta > \alpha$ such that

$$(\forall \vec{z} \in L_\beta)[\Phi^{L_\beta}(\vec{z}) \leftrightarrow \Phi^L(\vec{z})].$$

Let $\varphi(v_0, \ldots, v_n)$ be the $\mathscr{L}$-formula corresponding to Φ, and set

$$y = \{z \in L_\beta \mid \models_{L_\beta} [\varphi(\mathring{z}, \mathring{a}_1, \ldots, \mathring{a}_n) \wedge (\mathring{z} \in \mathring{x})]\}.$$

Then $y \in L_{\beta+1} \subseteq L$. But by I.9.11,

$$y = \{z \in x \mid \Phi^{L_\beta}(z, a_1, \ldots, a_n)\}.$$

So by choice of β,

$$y = \{z \in x \mid \Phi^L(z, a_1, \ldots, a_n)\}.$$

Since $y \in L$ we are done.

VII. *Collection.* We must show that if $\Phi(v_0, \ldots, v_n)$ is any LST formula, then

$$[\forall v_2 \ldots v_n [\forall x \exists y \Phi(y, x, v_2, \ldots, v_n) \\ \rightarrow \forall u \exists v (\forall x \in u)(\exists y \in v) \Phi(y, x, v_2, \ldots, v_n)]]^L.$$

Let $a_2, \ldots, a_n \in L$ be given, and suppose that

$$(\forall x \in L)(\exists y \in L) \Phi^L(y, x, a_2, \ldots, a_n).$$

We must show that, if we are given a $u \in L$ then there is a $v \in L$ such that

$$(\forall x \in u)(\exists y \in v)\, \Phi^L(y, x, a_2, \ldots, a_n).$$

(Since $u, v \subseteq L$ here, by the transitivity of L, we do not need to bind x, y by L.) Well, for each $x \in u$, let $f(x)$ be the least ordinal γ such that

$$(\exists y \in L_\gamma)\, \Phi^L(y, x, a_2, \ldots, a_n).$$

By the Axiom of Collection (in V), let α exceed all $f(x)$ for $x \in u$. Let $v = L_\alpha$. By 1.1 (iv), $v \in L$. Clearly,

$$(\forall x \in u)(\exists y \in v)\, \Phi^L(y, x, a_2, \ldots, a_n),$$

so we are done.

The theorem is proved. □

We shall in fact prove that

$$\mathrm{ZF} \vdash (\mathrm{AC})^L,$$

so L is an inner model of ZFC. This in turn will enable us to prove that AC cannot be disproved in ZF set theory. But first we must establish some further technical results about the constructible hierarchy. This is the business of the next section.

2. *The Constructible Hierarchy. The Axiom of Constructibility*

Recall from I.10 that a transitive set M is *amenable* iff:

(i) $(\forall x, y \in M)(\{x, y\} \in M)$;
(ii) $(\forall x \in M)(\bigcup x \in M)$;
(iii) $\omega \in M$;
(iv) $(\forall x, y \in M)(x \times y \in M)$;
(v) if $R \subseteq M$ is $\Sigma_0(M)$, then $(\forall x \in M)(R \cap x \in M)$.

(Intuitively, M is a "model" of the theory BS of I.9.) Our first lemma enables us to apply the results of I.9 and I.10 to the limit levels of the constructible hierarchy.

2.1 Lemma. *For each limit ordinal $\alpha > \omega$, L_α is amenable.*

Proof. (i) Let $x, y \in L_\alpha$. Since

$$L_\alpha = \bigcup_{\beta < \alpha} L_\beta,$$

there is a $\beta < \alpha$ such that $x, y \in L_\beta$. Then

$$\{x, y\} = \{z \in L_\beta \mid \models_{L_\beta} (\mathring{z} = \mathring{x} \vee \mathring{z} = \mathring{y})\} \in L_{\beta+1} \subseteq L_\alpha .$$

(ii) Let $x \in L_\alpha$. For some $\beta < \alpha$, $x \in L_\beta$. Since L_β is transitive, $\bigcup x \subseteq L_\beta$, and we have

$$\bigcup x = \{z \in L_\beta \mid \models_{L_\beta} (\exists u \in \mathring{x})(\mathring{z} \in u)\} \in L_{\beta+1} \subseteq L_\alpha .$$

(iii) By 1.1 (iv), $\omega \in L_\alpha$.

(iv) Let $x, y \in L_\alpha$. For some $\beta < \alpha$, $x, y \in L_\beta$. Since L_β is transitive, $x, y \subseteq L_\beta$. Let $a \in x$, $b \in y$. Then $a, b \in L_\beta$, so clearly (see the proof of (i)) $\{a\}, \{a, b\} \in L_{\beta+1}$, and hence $(a, b) = \{\{a\}, \{a, b\}\} \in L_{\beta+2}$. Thus $x \times y \subseteq L_{\beta+2}$ and we have

$$x \times y = \{z \in L_{\beta+2} \mid \models_{L_{\beta+2}} (\exists a \in \mathring{x})(\exists b \in \mathring{y})[\mathring{z} = (a, b)]\} \in L_{\beta+3} \subseteq L_\alpha .$$

(In fact $x \times y \in L_{\beta+2}$. Why?)

(v) Let $R \subseteq L_\alpha$ be $\Sigma_0(L_\alpha)$, $u \in L_\alpha$. We show that $R \cap u \in L_\alpha$. Let $\varphi(v_0, \ldots, v_n)$ be a Σ_0 formula of $\mathscr{L}$ and let $a_1, \ldots, a_n \in L_\alpha$ be such that

$$(\forall x \in L_\alpha)[x \in R \leftrightarrow \models_{L_\alpha} \varphi(\mathring{x}, \mathring{a}_1, \ldots, \mathring{a}_n)].$$

Pick $\beta < \alpha$ such that $u, a_1, \ldots, a_n \in L_\beta$. Since L_β is transitive, $u \subseteq L_\beta$, so

$$R \cap u = \{x \mid x \in u \wedge x \in R\} = \{x \in L_\beta \mid x \in u \wedge x \in R\} .$$

Now, being Σ_0, φ is absolute for L_β, L_α (by I.9.14), so for $x \in L_\beta$,

$$\models_{L_\beta} \varphi(\mathring{x}, \mathring{a}_1, \ldots, \mathring{a}_n) \leftrightarrow \models_{L_\alpha} \varphi(\mathring{x}, \mathring{a}_1, \ldots, \mathring{a}_n) .$$

Hence

$$\begin{aligned} R \cap u &= \{x \in L_\beta \mid x \in u \wedge \models_{L_\alpha} \varphi(\mathring{x}, \mathring{a}_1, \ldots, \mathring{a}_n)\} \\ &= \{x \in L_\beta \mid x \in u \wedge \models_{L_\beta} \varphi(\mathring{x}, \mathring{a}_1, \ldots, \mathring{a}_n)\} \\ &= \{x \in L_\beta \mid \models_{L_\beta} [\mathring{x} \in \mathring{u} \wedge \varphi(\mathring{x}, \mathring{a}_1, \ldots, \mathring{a}_n)]\} \in L_{\beta+1} \subseteq L_\alpha . \end{aligned}$$

The proof is complete. □

Towards the end of Chapter I, we mentioned on more than one occasion that it would be necessary to carry through two parallel developments concerning logical complexity, one of a metamathematical nature, involving the language LST, the other within set theory, utilising the language $\mathscr{L}$. We are now at the point where we must begin this process.

In I.9, we investigated the logical complexity of the basic syntactical and semantical notions of the language $\mathscr{L}_V$, showing that each concept could be defined by means of a formula of LST which is Δ_1^{BS}. We shall make *direct* use of these results. However, we shall require analogous results obtained within set theory. More precisely, working within the theory ZF (in fact KP will suffice, as we shall see), we shall need to examine the construction of the constructible hierarchy with regards to its definability properties along the lines of I.10. As a starting point, let us observe that now that we have the language $\mathscr{L}$ available, we can use it to analyse the syntax and semantics of $\mathscr{L}_V$ instead of working in LST. For this, it is convenient to agree to identify each formula of LST with the class it determines. With which convention it should be clear that each of the BS-complexity results of I.9 provides (by means of the replacement of LST by $\mathscr{L}$) a uniform definability result for amenable sets. For example, by repeating the proof of I.9.10 for $\mathscr{L}$ in place of LST, we obtain a proof of the fact that the class Sat ($= \{(u, \varphi) \mid \mathrm{Sat}(u, \varphi)\}$) is uniformly Δ_1^M for amenable sets M. That is, there is a Σ_1 formula $\psi(x, y)$ of $\mathscr{L}$ and a Π_1 formula $\theta(x, y)$ of $\mathscr{L}$ such that for any amenable set M, if $u, \varphi \in M$, then

$$\mathrm{Sat}(u, \varphi) \leftrightarrow \models_M \psi(\mathring{u}, \mathring{\varphi}) \leftrightarrow \models_M \theta(\mathring{u}, \mathring{\varphi}).$$

(The formulas ψ and θ are just the $\mathscr{L}$ analogues of the LST formulas described in I.9.10.)

Let $\mathrm{Seq}(y, x)$ be the LST formula which says that y is the set of all finite sequence from x. More precisely (cf. I.9.5), let $\mathrm{Seq}(y, x)$ be the LST formula:

$$\begin{aligned}(\exists f)[&(f \text{ is a function}) \wedge (\mathrm{dom}(f) = \omega) \wedge (f(0) = \emptyset) \wedge (y = \textstyle\bigcup \mathrm{ran}(f)) \\ &\wedge (\forall n \in \omega)(\forall s \in f(n+1))(\exists t \in f(n))(\exists a \in x)(s = t \cup \{(a, n)\}) \\ &\wedge (\forall n \in \omega)(\forall s \in f(n))(\forall a \in x)(\exists t \in f(n+1))(t = s \cup \{(a, n)\})].\end{aligned}$$

2.2 Lemma.

(i) *The* LST *formula* $\mathrm{Seq}(y, x)$ *is* Δ_1^{KP}.

(ii) *The class* Seq *is uniformly* $\Delta_1^{L_\alpha}$ *for limit* $\alpha > \omega$.

Proof. (i) As it stands, $\mathrm{Seq}(y, x)$ is Σ_1. Or rather, it is Σ_1 provided we eliminate explicit mention of ω by means of the prefix (to the entire formula)

$$\begin{aligned}\exists w\,[&\mathrm{On}(w) \wedge (\forall u \in w)(u \text{ is a natural number}) \wedge (\forall u \in w)(\exists v \in w)(u \in v) \\ &\wedge \ldots\ \ldots\ \ldots],\end{aligned}$$

thereafter replacing each mention of ω by w.

Now, in KP, using the Recursion Theorem (I.11.8), for any set x we can construct a function f as in $\mathrm{Seq}(y, x)$, so

$$\mathrm{KP} \vdash \forall x \exists y \,\mathrm{Seq}(y, x).$$

Clearly, any such y must be unique. Thus,

$$\mathrm{KP} \vdash \mathrm{Seq}(y, x) \leftrightarrow \forall z\,[\mathrm{Seq}(z, x) \rightarrow z = y].$$

This shows that $\mathrm{Seq}(y, x)$ is Δ_1^{KP}.

(ii) The ideas employed in the proof of this part of the lemma will be used several times in what follows, so we shall first of all consider in general how we can get from a Σ_1^{KP} definability result to a $\Sigma_1^{L_\alpha}$ definability result.

Suppose then that $\Phi(f, x)$ is a Σ_0 formula of LST, determining the class $A = \{x \mid \exists f \Phi(f, x)\}$. We wish to prove that the class A is $\Sigma_1^{L_\alpha}$ for some limit ordinal $\alpha > \omega$. Consider the $\mathscr{L}$-analogue of $\exists f \Phi(f, x)$, which will be of the form $\exists f \varphi(f, x)$, where φ is a Σ_0 formula of $\mathscr{L}$. We prove that for any $x \in L_\alpha$,

$$x \in A \quad \text{iff } \models_{L_\alpha} \exists f \varphi(f, \mathring{x}).$$

Now by I.9.15, if $x, f \in L_\alpha$, we have

$$\Phi(f, x) \leftrightarrow \models_{L_\alpha} \varphi(\mathring{f}, \mathring{x}).$$

Consequently, for $x \in L_\alpha$,

$$\models_{L_\alpha} \exists f \varphi(f, \mathring{x}) \quad \text{implies } \exists f \Phi(f, x).$$

This leaves us with the proof that

$$\exists f \Phi(f, x) \quad \text{implies } \models_{L_\alpha} \exists f \varphi(f, \mathring{x}).$$

So, in practice what we must prove is that if there is an f such that $\Phi(f, x)$, then there is such an f in L_α. (In all the cases we shall encounter, any such f will be unique, so what we shall prove is that if $\Phi(f, x)$, where $x \in L_\alpha$, then $f \in L_\alpha$.) Now let us see how this works in the case of the problem in hand.

Let $\varphi(y, x)$ be the $\mathscr{L}$-analogue of the LST-formula $\mathrm{Seq}(y, x)$. Let $\alpha > \omega$, $\lim(\alpha)$. We prove that for any $x, y \in L_\alpha$,

$$\mathrm{Seq}(y, x) \leftrightarrow \models_{L_\alpha} \varphi(\mathring{y}, \mathring{x}).$$

This will show that the class Seq is uniformly $\Sigma_1^{L_\alpha}$ for limit $\alpha > \omega$. We shall also prove that for any $x \in L_\alpha$ there is a (necessarily unique) $y \in L_\alpha$ such that $\mathrm{Seq}(y, x)$, from which fact it follows as in part (i) that Seq is also $\Pi_1^{L_\alpha}$ (uniformly for limit $\alpha > \omega$).

Let $x \in L_\alpha$. Pick $\gamma < \alpha$ so that $\gamma > \omega$ and $x \in L_\gamma$. If $a \in x$, then we have $(a, n) = \{\{a\}, \{a, n\}\} \in L_{\gamma+1}$ for any $n \in \omega$, so if s is any finite sequence from x, then $s \in L_{\gamma+2}$. Thus $\mathrm{Seq}(y, x)$, where

$$y = \{s \in L_{\gamma+2} \mid \models_{L_{\gamma+2}} \text{“}\mathring{s} \text{ is a finite sequence from } \mathring{x}\text{”}\}.$$

But $y \in L_{\gamma+3} \subseteq L_\alpha$. Consider now the function f which figures in the formula $\mathrm{Seq}(y, x)$. If it exists (i.e. if $\mathrm{Seq}(y, x)$), then clearly,

$$f = \{(s, n) \mid s = {}^n x \wedge n \in \omega\}.$$

It is easily seen that for any $n \in \omega$, ${}^n x \in L_{\gamma+3}$, so $({}^n x, n) \in L_{\gamma+5}$, giving $f \in L_{\gamma+6}$.

Thus $f \in L_\alpha$, which implies (see the above discussion):

$$\mathrm{Seq}(y, x) \leftrightarrow \models_{L_\alpha} \varphi(\mathring{y}, \mathring{x}).$$

The proof is complete.[4] □

Let Pow(y, x) be the LST formula which says that y is the set of all finite subsets of x. More precisely, let Pow(y, x) be as follows:

$$\exists z\,[\mathrm{Seq}(z, x) \wedge y = \{\mathrm{ran}(u) \mid u \in z\}].$$

2.3 Lemma.
(i) *The* LST *formula* Pow(y, x) *is* Δ_1^{KP}.
(ii) *The class* Pow *is uniformly* $\Delta_1^{L_\alpha}$ *for limit* $\alpha > \omega$.

Proof. (i) As it stands, Pow(y, x) is Σ_1. Moreover,

$$\mathrm{KP} \vdash \forall x\, \exists!\, y\, \mathrm{Pow}(y, x),$$

so as in 2.2 it follows that Pow(y, x) is in fact Δ_1^{KP}.

(ii) This follows from part (i) by a straightforward application of the technique discussed above. (The details are left as an exercise for the reader.) □

We shall now write down an LST formula $A(v, u)$ such that

$$A(v, u) \leftrightarrow v = \mathrm{Def}(u).$$

Namely:

$$\begin{aligned}
&(\forall x \in v)(\exists \varphi)[\mathrm{Fml}(\varphi, u) \wedge \mathrm{Fr}(\varphi, \{v_0\}) \wedge (x \subseteq u)\\
&\wedge (\forall z \in u)(z \in x \leftrightarrow \exists \psi\,(\mathrm{Sub}(\psi, \varphi, v_0, \mathring{z}) \wedge \mathrm{Sat}(u, \psi)))]\\
&\wedge (\forall \varphi)[(\mathrm{Fml}(\varphi, u) \wedge \mathrm{Fr}(\varphi, \{v_0\}))\\
&\qquad \rightarrow (\exists x \in v)[(x \subseteq u) \wedge (\forall z \in u)(z \in x\\
&\qquad\qquad \leftrightarrow \exists \psi\,(\mathrm{Sub}(\psi, \varphi, v_0, \mathring{z}) \wedge \mathrm{Sat}(u, \psi)))]].
\end{aligned}$$

Our task now is to modify this formula in order to obtain a Σ_1 formula equivalent to it. Broadly speaking, the idea is to find a single bound for all of the unbounded quantifiers in $A(v, u)$, much as we did when we formulated the formula Sat(u, φ) prior to I.9.10. What happened there was that we commenced with a formula $S(u, \varphi)$, which embodied the canonical definition of the notion

"φ is a sentence of $\mathscr{L}_u$ which is true in $\langle u, \in \rangle$",

and then found a bound for all unbounded quantifiers in $S(u, \varphi)$. Now, the binding set used there is not large enough to handle the quantifiers involved in $A(v, u)$

4 This is not quite accurate, since we did not bother with the quantifier $\exists w$ mentioned in the proof of 2.2. This was because we knew already that $\omega \in L_\alpha$. In practice we shall always restrict our attention only to the "significant" quantifier(s).

(though it will clearly suffice for those quantifiers involved in those parts of $A(v, u)$ concerning Sat). So, as there is clearly no point in rebinding quantifiers which are already bound, let us at once amend $A(v, u)$ by replacing each occurrence of the formula $\mathrm{Sat}(u, \varphi)$ in $A(v, u)$ by $S(u, \varphi)$, denoting the resulting formula by $B(v, u)$. Since $S(u, \varphi)$ is equivalent to $\mathrm{Sat}(u, \varphi)$, $B(v, u)$ will be equivalent to $A(v, u)$. We now seek a bound for all the unbounded quantifiers in $B(v, u)$. (This bound will have to be large enough to rebind all of the quantifiers we have just freed in passing from $A(v, u)$ to $B(v, u)$, of course.) Let $C(w, v, u)$ be the formula obtained from $B(v, u)$ by binding all unbounded quantifiers by w. (Thus $C(w, v, u)$ is a Σ_0 formula.) We must now see what sort of set we can take for the bound w.

The unbounded quantifiers involved in $B(v, u)$ fall into three types: those that range over formulas ($\exists \varphi$ and $\forall \varphi$ as in $A(v, u)$), those that range over finite sequences of formulas (such quantifiers occur in Fml, Fr, Sub, and $S(u, \varphi)$), and those ranging over finite sequences of finite sets of variables (these occur in Fr). Hence, all unbounded quantifiers in $B(v, u)$ can (without loss of meaning) be bound by the set

$$\begin{aligned} K(u) = {} & [\text{the set of finite sequences of members of the set } 9 \cup \{v_i \mid i \in \omega\} \cup \{\mathring{x} \mid x \in u\}] \\ & \cup [\text{the set of finite sequences of finite sequences of members of the set } 9 \cup \{v_i \mid i \in \omega\} \cup \{\mathring{x} \mid x \in u\}] \\ & \cup [\text{the set of finite sequences of finite subsets of the set } \{v_i \mid i \in \omega\}]. \end{aligned}$$

Let $K(w, u)$ be the LST formula which says "$w = K(u)$", namely:

$$\begin{aligned} (\exists a, b, c, d, e, f)[[(\forall z \in d)\,\mathrm{Vbl}(z) \wedge (\forall i \in \omega)(v_i \in d)] \\ \wedge [(\forall z \in e)\,\mathrm{Const}(z, u) \wedge (\forall z \in u)(\mathring{z} \in e)] \\ \wedge [\mathrm{Seq}(a, 9 \cup d \cup e)] \\ \wedge [\mathrm{Seq}(b, a)] \\ \wedge [\mathrm{Pow}(f, d) \wedge \mathrm{Seq}(c, f)] \\ \wedge [w = a \cup b \cup c]]. \end{aligned}$$

Provided we remove the explicit mention of ω as in 2.2, we see that the formula $K(w, u)$ is Σ_1. If we now let $D(v, u)$ be the formula

$$\exists w[K(w, u) \wedge C(w, v, u)],$$

then clearly,

$$D(v, u) \leftrightarrow v = \mathrm{Def}(u).$$

Moreover, we have:

2.4 Lemma.

(i) *The* LST *formula* $D(v, u)$ *is* Σ_1 *and* Δ_1^{KP}.

(ii) *The class* D *is uniformly* $\Delta_1^{L_\alpha}$ *for limit* $\alpha > \omega$.

Proof. As in 2.2 and 2.3. (The details are left as an exercise for the reader.) □

Noting that if $\lim(\alpha)$ and $\alpha > \omega$, the set L_α is closed under the function Def (this observation forms part of the proof of 2.4), we often state part (ii) of 2.4 in the following form:

2.5 Corollary. *The function* Def *is uniformly* $\Delta_1^{L_\alpha}$ *for limit* $\alpha > \omega$. □

We now write down an LST formula, $E(f, \alpha)$ such that

$$E(f, \alpha) \leftrightarrow f = (L_\gamma \mid \gamma \leqslant \alpha).$$

Namely:

$$\begin{aligned} &\mathrm{On}(\alpha) \wedge (f \text{ is a function}) \wedge (\mathrm{dom}(f) = \alpha + 1) \wedge (f(0) = \emptyset) \\ &\quad \wedge (\forall \gamma \in \mathrm{dom}(f))[((\lim(\gamma) \wedge \gamma > 0) \rightarrow (f(\gamma) = \bigcup_{\delta < \gamma} f(\delta))) \\ &\quad \wedge (\mathrm{succ}(\gamma) \rightarrow D(f(\gamma), f(\gamma - 1)))]. \end{aligned}$$

Clearly, $E(f, \alpha)$ says what we want it to, and our next task is to modify this formula to obtain an equivalent Σ_1 formula, just as we just did for $D(v, u)$. Let $F(w, f, \alpha)$ be the Σ_0 formula obtained from $E(f, \alpha)$ by replacing the clause $D(f(\gamma), f(\gamma - 1))$ by $C(w, f(\gamma), f(\gamma - 1))$, and rendering the clause

$$f(\gamma) = \bigcup_{\delta < \gamma} f(\delta)$$

in the form

$$(\forall x \in f(\gamma))(\exists \delta \in \gamma)(x \in f(\delta)) \wedge (\forall \delta \in \gamma)(f(\delta) \subseteq f(\gamma)).$$

Comparing the present situation with that which led up to 2.4, we see that all the unbounded quantifiers which figure in $E(f, \alpha)$ (namely as part of the clause $D(f(\gamma), f(\gamma - 1))$) can be bound by the set

$$K(\bigcup \mathrm{ran}(f)).$$

So if we let $G(f, \alpha)$ be the Σ_1 formula

$$\exists w[K(w, \bigcup \mathrm{ran}(f)) \wedge F(w, f, \alpha)],$$

we have

$$G(f, \alpha) \leftrightarrow f = (L_\gamma \mid \gamma \leqslant \alpha).$$

Moreover:

2.6 Lemma.

(i) *The* LST *formula* $G(f, \alpha)$ *is* Δ_1^{KP}.

(ii) *The class* G *is uniformly* $\Delta_1^{L_\alpha}$ *for limit* $\alpha > \omega$.

Proof. (i) The proof boils down to proving that

$$\mathrm{KP} \vdash \forall \alpha \, \exists f G(f, \alpha).$$

But this follows from 2.4 (i) together with the KP-Recursion Theorem (I.11.8), which enables us to construct, within KP, the function $(L_\gamma \mid \gamma \leqslant \alpha)$ for any ordinal α.

(ii) Here we quickly reduce to proving that for any limit ordinal $\alpha > \omega$, if $\delta < \alpha$ then $(L_\gamma \mid \gamma \leqslant \delta) \in L_\alpha$. In fact it is not hard to see that if $\delta > \omega$, then $(L_\gamma \mid \gamma \leqslant \delta) \in L_{\delta+4}$, so we are done. (We leave all the details to the reader.) □

Let $H(x, \alpha)$ be the LST formula which says that "$x = L_\alpha$", namely:

$$\exists f [G(f, \alpha) \wedge (x = f(\alpha))].$$

The following lemma follows easily from 2.6 using the by now familiar arguments:

2.7 Lemma.
(i) *The* LST *formula* $H(x, \alpha)$ *is* Δ_1^{KP}.
(ii) *The class* H *is uniformly* $\Delta_1^{L_\alpha}$ *for limit* $\alpha > \omega$. □

Noting that if $\alpha > \omega$ is a limit ordinal, then L_α is closed under the function $\gamma \mapsto L_\gamma$, we have, by the above:

2.8 Lemma. *The function* $\gamma \mapsto L_\gamma$ *is uniformly* $\Delta_1^{L_\alpha}$ *for limit* $\alpha > \omega$. □

The following absoluteness results may now be proved.

2.9 Lemma. *Let* M *be an inner model of* KP. *For any* $\alpha \in \mathrm{On}$, $L_\alpha \in M$ *and* $(L_\alpha)^M = L_\alpha$. *(This equality means that if* $[H(x, \alpha)]^M$, *then* $x = L_\alpha$.) *Hence* $(L)^M = L$.

Proof. As we observed above, the KP-Recursion Theorem enables us to construct $(L_\gamma \mid \gamma \leqslant \alpha)$ for any ordinal α. Hence if $\alpha \in \mathrm{On}$, then we have $(L_\gamma^M \mid \gamma \leqslant \alpha) \in M$, so in particular $L_\alpha^M \in M$. But by 2.7 (i) and I.8.3 (iv), we have the absoluteness result $L_\alpha^M = L_\alpha$. The lemma is proved. □

2.10 Lemma. *Let* M *be an admissible set, and let* $\lambda = \sup (M \cap \mathrm{On})$. *For any* $\alpha \in \lambda$, $(L_\alpha)^M = L_\alpha$. *Hence* $(L)^M = L_\lambda$.

Proof. Analogous to the proof of 2.9. □

2.11 Lemma. *For any* α, $(L_\alpha)^L = L_\alpha$. *Hence* $(L)^L = L$.

Proof. Directly from 2.9. □

2.12 Lemma. *Let* $\alpha > \omega$ *be a limit ordinal. For all* $\gamma < \alpha$, $(L_\gamma)^{L_\alpha} = L_\gamma$. *Hence* $(L)^{L_\alpha} = L_\alpha$.

Proof. Much as for 2.10, except that we use the closure properties of L_α rather than admissibility. The details are left to the reader. □

2.13 Lemma. *The* LST *formula*

"*x is constructible*"

is Σ_1^{KP}.

Proof.

$$\begin{aligned} x \textit{ is constructible} &\leftrightarrow x \in L \\ &\leftrightarrow \exists \alpha (x \in L_\alpha) \\ &\leftrightarrow \exists \alpha \exists u (u = L_\alpha \wedge x \in u). \end{aligned}$$

The result follows from 2.7 (i) now. □

The *Axiom of Constructibility* is the assertion that all sets are constructible:

$$\forall x (x \in L).$$

This is usually abbreviated as:

$$V = L.$$

For the most part we shall be treating the assertion $V = L$ as a particularly interesting set-theoretical statement, not as a fundamental *axiom* of set theory in the sense of the axioms of ZFC. Thus the use of the word "axiom" in this connection is somewhat different from the more common usage. From our standpoint it would perhaps be more suitable to refer to $V = L$ as the "Hypothesis of Constructibility". However, we shall stick to the accepted usage of the phrase "Axiom of Constructibility".

2.14 Lemma. *The* LST *formula* $V = L$ *is* Π_2^{KP} .

Proof. $V = L \leftrightarrow \forall x (x \in L)$, which is Π_2^{KP} by virtue of 2.13. □

2.15 Theorem. $\mathrm{ZF} \vdash (V = L)^L$. *Hence* L *is an inner model of the theory* $\mathrm{ZF} + (V = L)$.

Proof. By 2.11, $(L)^L = L$. But clearly, $(V)^L = L$. Hence,

$$(V)^L = (L)^L.$$

In other words,

$$(V = L)^L. \quad \square$$

3. The Axiom of Choice in L

In this section we shall show that

$$\mathrm{ZF} \vdash (\mathrm{AC})^L.$$

We do this in a very strong fashion. We exhibit a formula $\Phi(v_0, v_1)$ of LST such that (suitably expressed)

$$\mathrm{ZF} \vdash [\text{"}\Phi \text{ well-orders the universe"}]^L.$$

In order to describe the formula Φ, it is necessary to look once more at the definition of the constructible hierarchy. Recall that in passing from L_α to $L_{\alpha+1}$, we allow any elements of L_α to figure as parameters in definitions of the new sets appearing in $L_{\alpha+1}$. The following lemma shows that we may be rather more restrictive than this, and provides us with a slightly more convenient characterisation of $L_{\alpha+1}$ in terms of L_α.

3.1 Lemma. *Let $x \in L_{\alpha+1}$. Then there is a formula $\varphi(v_0, \ldots, v_n)$ of $\mathscr{L}$ (so in particular φ contains no individual constant symbols) and ordinals $\gamma_1, \ldots, \gamma_n < \alpha$ such that*

$$x = \{z \in L_\alpha \mid \models_{L_\alpha} \varphi(\mathring{z}, \mathring{L}_{\gamma_1}, \ldots, \mathring{L}_{\gamma_n})\}.$$

Proof. By induction on α. For $\alpha = 0$ there is nothing to prove, since $\emptyset$ is the only possible set x. Let $\alpha > 0$ now, and suppose that the lemma is valid below α. If $x \in L_{\alpha+1}$ there is an $\mathscr{L}$-formula $\psi(v_0, \ldots, v_n)$ and elements $p_1, \ldots, p_n$ of L_α such that

$$x = \{z \in L_\alpha \mid \models_{L_\alpha} \psi(\mathring{z}, \mathring{p}_1, \ldots, \mathring{p}_n)\}.$$

Pick $\gamma < \alpha$ so that $p_1, \ldots, p_n \in L_{\gamma+1}$. By induction hypothesis, for each $i = 1, \ldots, n$ there is an $\mathscr{L}$-formula $\psi_i(v_0, \ldots, v_{k(i)})$ and ordinals $\gamma_1^i, \ldots, \gamma_{k(i)}^i < \gamma$ such that

$$p_i = \{z \in L_\gamma \mid \models_{L_\gamma} \psi_i(\mathring{z}, \mathring{L}_{\gamma_1^i}, \ldots, \mathring{L}_{\gamma_{k(i)}^i})\}.$$

For each i, let $\psi_i'(v_0, \ldots, v_{k(i)}, v_{k(i)+1})$ be the $\mathscr{L}$-formula obtained from $\psi_i(v_0, \ldots, v_{k(i)})$ by binding all unbounded quantifiers by $v_{k(i)+1}$. Then clearly,

$$p_i = \{z \in L_\alpha \mid \models_{L_\alpha} [(\mathring{z} \in \mathring{L}_\gamma) \wedge \psi_i'(\mathring{z}, \mathring{L}_{\gamma_1^i}, \ldots, \mathring{L}_{\gamma_{k(i)}^i}, \mathring{L}_\gamma)]\}.$$

Hence,

$$\begin{aligned} x = \{z \in L_\alpha \mid \models_{L_\alpha} & (\exists p_1) \ldots (\exists p_n)[\psi(\mathring{z}, p_1, \ldots, p_n) \\ & \wedge (\forall v)[(v \in p_1) \leftrightarrow (v \in \mathring{L}_\gamma \wedge \psi_1'(v, \mathring{L}_{\gamma_1^1}, \ldots, \mathring{L}_{\gamma_{k(1)}^1}, \mathring{L}_\gamma))] \wedge \ldots\ldots \\ & \wedge (\forall v)[(v \in p_n) \leftrightarrow (v \in \mathring{L}_\gamma \wedge \psi_n'(v, \mathring{L}_{\gamma_1^n}, \ldots, \mathring{L}_{\gamma_{k(n)}^n}, \mathring{L}_\gamma))]]\}. \end{aligned}$$

The lemma is proved. □

We shall now fix a simple, effective well-ordering of the formulas of $\mathscr{L}$. The precise definition is not important. For definiteness, we say that if φ and ψ are formulas of $\mathscr{L}$, so in particular, φ and ψ are both finite sequences of sets, then $\varphi \dashv \psi$ iff either φ is an initial segment of ψ or else $k(\varphi(i)) < k(\psi(i))$, where i is the least integer such that $\varphi(i) \neq \psi(i)$ and where the function k is defined on the set

$$9 \cup \{v_n \mid n \in \omega\}$$

by

$$k(x) = \begin{cases} x, & \text{if } x \in 9 \\ n + 9, & \text{if } x = v_n \ (= (2, n)). \end{cases}$$

We also define $<^*$ to be the lexicographic well-ordering of the finite sequences of ordinals, i.e. if s and t are finite sequences of ordinals, then $s <^* t$ iff

(i) $\mathrm{dom}(s) < \mathrm{dom}(t)$, or else
(ii) $\mathrm{dom}(s) = \mathrm{dom}(t)$ and $s(i) < t(i)$, where i is least such that $s(i) \neq t(i)$.

Using 3.1, we now define a well-ordering of the class L. Let $x, y \in L$. We set $x <_L y$ iff either:

(A) The least α such that $x \in L_{\alpha+1}$ is smaller than the least β such that $y \in L_{\beta+1}$; or else
(B) there is an α such that x and y both lie in $L_{\alpha+1} - L_\alpha$ and either:
(B1) the $\dashv$-least formula $\varphi(v_0, \ldots, v_n)$ of $\mathscr{L}$ for which there are ordinals $\gamma_1, \ldots, \gamma_n < \alpha$ such that

$$x = \{z \in L_\alpha \mid \models_{L_\alpha} \varphi(\mathring{z}, \mathring{L}_{\gamma_1}, \ldots, \mathring{L}_{\gamma_n})\}$$

$\dashv$-precedes the $\dashv$-least formula $\psi(v_0, \ldots, v_m)$ of $\mathscr{L}$ for which there are ordinals $\delta_1, \ldots, \delta_m < \alpha$ such that

$$y = \{z \in L_\alpha \mid \models_{L_\alpha} \psi(\mathring{z}, \mathring{L}_{\delta_1}, \ldots, \mathring{L}_{\delta_m})\}; \qquad \text{or else}$$

(B2) the formulas φ and ψ in (B1) coincide, but the $<^*$-least n-sequence $\langle \gamma_1, \ldots, \gamma_n \rangle$ of ordinals $\gamma_i < \alpha$ which defines x as in (B1) $<^*$-precedes the $<^*$-least n-sequence $\langle \delta_1, \ldots, \delta_n \rangle$ of ordinals $\delta_i < \alpha$ which defines y.

It is easily seen that $<_L$ is indeed a well-ordering of L. Our task now is to investigate the logical complexity of this well-ordering.

The following LST-formula, $N(\alpha, x, \varphi, t)$ says that φ is a formula of $\mathscr{L}$, t is a finite sequence of ordinals less than α, the free variables of φ are $v_0, \ldots, v_n$, where $n = \mathrm{dom}(t)$, and $x = \{z \in L_\alpha \mid \models_{L_\alpha} \varphi(\mathring{z}, \mathring{L}_{t(0)}, \ldots, \mathring{L}_{t(n-1)})\}$:

$$\begin{aligned} \exists u\, \exists f\, \exists n\, \exists \psi\, [&\mathrm{Fml}(\varphi, \emptyset) \wedge \mathrm{Finseq}(t) \wedge (\mathrm{dom}(t) = n) \wedge (\forall i \in n)(t(i) \in \alpha) \\ &\wedge \mathrm{Fr}(\varphi, u) \wedge (f: n+1 \leftrightarrow u) \wedge (\forall i \in n+1)(f(i) = v_i) \\ &\wedge \mathrm{Finseq}(\psi) \wedge (\mathrm{dom}(\psi) = n+1) \wedge (\psi(0) = \varphi) \\ &\wedge (\forall i \in n)\, \mathrm{Sub}(\psi(i+1), \psi(i), v_{i+1}, \mathring{L}_{t(i)}) \wedge (x \subseteq L_\alpha) \\ &\wedge (\forall z \in L_\alpha)(z \in x \leftrightarrow \exists \theta(\mathrm{Sub}(\theta, \psi(n), v_0, \mathring{z}) \wedge \mathrm{Sat}(L_\alpha, \theta)))]. \end{aligned}$$

The following LST-formula, $M(\alpha, x, \varphi)$, says that φ is the $\dashv$-least formula of $\mathscr{L}$ such that $N(\alpha, x, \varphi, t)$ for some t:

$$(\exists t) N(\alpha, x, \varphi, t) \wedge (\forall \varphi')[(\exists t') N(\alpha, x, \varphi', t') \rightarrow (\varphi \quad \varphi' \vee \varphi = \varphi')].$$

The next formula of LST, $P(\alpha, x, \varphi, t)$, says that t is the $<^*$-least suitable sequence of ordinals less than α such that $N(\alpha, x, \varphi, t)$:

$$N(\alpha, x, \varphi, t) \wedge (\forall t')[N(\alpha, x, \varphi, t') \rightarrow (t \leqslant^* t')].$$

We are now able to write down a formula of LST which expresses the relation $x <_L y$ outlined above. We shall not bother to replace the relations $\dashv$ and $<^*$ by their LST-definitions, since it should be obvious how this can be done, and is thus not worth causing further complications. Let $X(x, y)$ be the following LST-formula (to express $x <_L y$):

$$(\exists \alpha)[(x \in L_\alpha) \wedge (y \notin L_\alpha)] \vee (\exists \alpha) Q(x, y, \alpha),$$

where $Q(x, y, \alpha)$ is the LST-formula

$$\begin{aligned}
&[(x \in L_{\alpha+1}) \wedge (y \in L_{\alpha+1}) \wedge (x \notin L_\alpha) \wedge (y \notin L_\alpha)] \\
&\quad \wedge [(\exists \varphi, \psi)[M(\alpha, x, \varphi) \wedge M(\alpha, y, \psi) \wedge (\varphi \dashv \psi)] \\
&\quad \vee (\exists \varphi)[M(\alpha, x, \varphi) \wedge M(\alpha, y, \varphi) \\
&\qquad\qquad \wedge (\exists s, t)[P(\alpha, x, \varphi, s) \wedge P(\alpha, y, \varphi, t) \wedge (s <^* t)]]].
\end{aligned}$$

Now, it is easily seen that any unbounded quantifiers in the formula $Q(x, y, \alpha)$ may be bound by $L_{\max(\omega, \alpha+4)}$. (This includes the quantifiers which are required in order to define $\dashv$ and $<^*$.) So if $R(x, y, \alpha, w)$ is the formula obtained from $Q(x, y, \alpha)$ by binding all quantifiers not already bound by w, we see that the relation $x <_L y$ is expressed by the formula

$$(\exists \alpha)[(x \in L_\alpha) \wedge (y \notin L_\alpha)] \vee (\exists \alpha)(\exists w)[w = L_{\max(\omega, \alpha+4)} \wedge R(x, y, \alpha, w)].$$

We denote this formula by $\mathrm{WO}(x, y)$. It is clearly Σ_1. Moreover,

3.2 Lemma.

(i) *The* LST *formula* $\mathrm{WO}(x, y)$ *is* $\Delta_1^{\mathrm{KP}+(V=L)}$.

(ii) $\mathrm{KP} \vdash$ "$\{(x, y) \mid \mathrm{WO}(x, y)\}$ *is a well-ordering of* L".

Proof. We prove (ii) first. From the way we evolved the formula $\mathrm{WO}(x, y)$ above, it is clear that what we must prove is that (working in KP) if $x, y \in L$, $x \neq y$, then either $\mathrm{WO}(x, y)$ or else $\mathrm{WO}(y, x)$. But if x, y are as stated, then either $x <_L y$ or else $y <_L x$, of course. So we are reduced to proving that if $x <_L y$, then the sets required to exist by virtue of the existential quantifiers involved in the formula $\mathrm{WO}(x, y)$ can all be constructed (from x, y) in KP. This is easily seen, and is left as an exercise for the reader.

We now prove (i). We know that the formula $\mathrm{WO}(x, y)$ is Σ_1. But by (ii) we have

$$\mathrm{KP} \vdash (\forall x, y \in L)[\mathrm{WO}(x, y) \leftrightarrow \neg [(x = y) \vee \mathrm{WO}(y, x)]].$$

Hence $\mathrm{WO}(x, y)$ is also $\Pi_1^{\mathrm{KP}+(V=L)}$. □

Let $\mathrm{wo}(x, y)$ be the analogue of $\mathrm{WO}(x, y)$ in $\mathscr{L}$. Then:

3.3 Lemma.

(i) *If* $x, y \in L_\alpha$, *then*

$$\mathrm{WO}(x, y) \leftrightarrow \models_{L_\gamma} \mathrm{wo}(\mathring{x}, \mathring{y}),$$

where $\gamma = \max(\omega, \alpha + 5)$.

(ii) *If $\alpha > \omega$ is a limit ordinal, then*

$$\{(x, y) \mid \models_{L_\alpha} \mathrm{wo}(\mathring{x}, \mathring{y})\} \quad \textit{is a well-ordering of } L_\alpha.$$

(iii) *The relation $x <_L y$ is uniformly $\Delta_1^{L_\alpha}$ for limit $\alpha > \omega$.*

Proof. (i) This follows from 3.2 by the kind of argument outlined in 2.2. The heart of the proof is to show that the existential quantifiers involved in wo (x, y) can be bound by L_γ, where γ is as stated. The details are left as an exercise for the reader.

(ii) This follows immediately from (i).

(iii) This also follows from (i). □

We often write $x <_L y$ in place of both WO (x, y) and wo (x, y).

The following lemma is clear from the definition of $<_L$, and will often be used without mention.

3.4 Lemma.

(i) *If $x <_L y$ and $y \in L_\alpha$, then $x \in L_\alpha$.*

(ii) *If $x \in L_\alpha$ and $y \notin L_\alpha$, then $x <_L y$.*

(iii) *If $x \in y \in L$, then $x <_L y$.* □

For later use we also prove the following result.

3.5 Lemma. *Let* pr *be the predecessor function defined on L by*

$$\mathrm{pr}(x) = \{z \mid z <_L x\}.$$

(i) $x \in L \to \mathrm{pr}(x) \in L$.

(ii) *if $\alpha > \omega$ is a limit ordinal, then $x \in L_\alpha \to \mathrm{pr}(x) \in L_\alpha$.*

(iii) pr *is uniformly $\Delta_1^{L_\alpha}$ for limit $\alpha > \omega$.*

Proof. (i) follows directly from (ii).

(ii) Let $x \in L_\alpha$. Choose $\beta < \alpha$ so that $x \in L_\beta$. We know that

$$z <_L x \to z \in L_\beta.$$

Moreover by 3.3 (i),

$$(\forall a, b \in L_\beta)[\mathrm{WO}(a, b) \leftrightarrow \models_{L_\gamma} \mathrm{wo}(\mathring{a}, \mathring{b})],$$

where $\gamma = \max(\omega, \beta + 5)$. Hence

$$\mathrm{pr}(x) = \{z \in L_\gamma \mid \models_{L_\gamma} \mathrm{wo}(\mathring{z}, \mathring{x})\} \in L_{\gamma+1} \subseteq L_\alpha.$$

(iii) Let $w(z, x, 1)$ be the $\mathscr{L}$-formula obtained from wo (z, x) by binding any unbounded quantifiers by 1. By 3.3 (i),

$$y = \mathrm{pr}(x) \leftrightarrow \models_{L_\alpha} (\exists \beta)[(\mathring{x} \in L_\beta) \wedge (\mathring{y} \subseteq L_\beta) \\ \wedge (\forall z \in L_\beta)(z \in \mathring{y} \leftrightarrow w(z, \mathring{x}, L_{\max(\omega, \beta+5)}))].$$

So pr is uniformly $\Sigma_1^{L_\alpha}$ for limit $\alpha > \omega$. Hence by I.10.4, pr is uniformly $\Delta_1^{L_\alpha}$ for limit $\alpha > \omega$. □

The following result is also fundamental to much of the work on constructibility.

3.6 Lemma. *There is a* Σ_1 *formula* Enum(α, x) *of* LST, *absolute for* L, *such that*

$$\mathrm{KP} \vdash \text{“}\mathit{If}\ F = \{(x, \alpha) \mid \mathrm{Enum}(\alpha, x)\},\ \mathit{then}\ F\colon \mathrm{On} \leftrightarrow L\text{”}.$$

Proof. Intuitively, Enum(α, x) says that x is the α-the member of L under the well-ordering $<_L$. Thus, Enum(α, x) is the formula:

$$\begin{aligned}&(\exists f)[(f \text{ is a function}) \wedge (\mathrm{dom}(f) = \alpha + 1)\\ &\quad \wedge (\forall \xi, \zeta \in \alpha + 1)(\xi < \zeta \rightarrow f(\xi) <_L f(\zeta))\\ &\quad \wedge (\exists z)[(z = \mathrm{pr}(x)) \wedge (\forall y \in z)(\exists \beta \in \alpha)(y = f(\beta))] \wedge (f(\alpha) = x)].\end{aligned}$$

It is easily seen that this formula is as stated in the lemma. □

As an illustration of the use of 3.6 we show that $V = L$ can be reduced to an “axiom of constructibility for sets of ordinals”.

3.7 Lemma. $\mathrm{KP} \vdash \forall a(a \subseteq \mathrm{On} \rightarrow a \in L) \rightarrow (V = L)$.

Proof. (In KP.) Assume all sets of ordinals are constructible. We prove by $\in$-induction that

$$\forall x(x \in L).$$

Let x be given, and suppose that

$$y \in x \rightarrow y \in L.$$

By Σ_1-Collection, let

$$a = \{\alpha \mid \alpha \in \mathrm{On} \wedge (\exists y \in x)\,\mathrm{Enum}(\alpha, y)\}.$$

By hypothesis, $a \in L$. Hence, using the induction hypothesis,

$$x' = \{y \mid (\exists \alpha \in a)\,\mathrm{Enum}(\alpha, y)^L\} \in L.$$

But by the absoluteness of Enum, $x = x'$, so we have $x \in L$, as required. □

3.8 Theorem. $\mathrm{ZF} \vdash (\mathrm{AC})^L$.

Proof. An immediate consequence of 3.6. (The function F well-orders L.) □

Notice that we made no use of AC in the above proof. This will be important to us in the next section.

4. Constructibility and Relative Consistency Results

The construction of inner models such as L provides us with a useful method for obtaining relative consistency results. The idea is as follows. Suppose Φ is some statement in LST, and that there is a class M such that

$$\mathrm{ZF} \vdash (\mathrm{ZF} + \Phi)^M.$$

Then the consistency of the theory ZF + Φ follows from the consistency of ZF. Indeed, given a proof of an inconsistency in ZF + Φ we could, in a highly effective manner, produce from it a proof of an inconsistency in ZF. To see this, let $\Psi_0, \ldots, \Psi_n$ be a proof (in the formal sense) of an inconsistency in ZF + Φ. Thus, for each $i = 0, \ldots, n$, Ψ_i is a formula of LST which is either an axiom of the theory ZF + Φ or else follows from some of $\Psi_0, \ldots, \Psi_{i-1}$ by an application of a rule of logic, and Ψ_n is a statement such as $(0 = 1)$. Consider the sequence $\Psi_0^M, \ldots, \Psi_n^M$. If Ψ_i is an axiom of ZF + Φ, then Ψ_i^M is a theorem of ZF, by the assumption on M. And if Ψ_i follows from some of $\Psi_0, \ldots, \Psi_{i-1}$ by means of a rule of logic, then Ψ_i^M follows from the corresponding members of $\Psi_0^M, \ldots, \Psi_{i-1}^M$ by means of the same rule. Hence Ψ_n^M is a theorem of ZF. But since Ψ_n is an inconsistency, so too is Ψ_n^M.

As a particular instance of the above considerations, we have

4.1 Theorem. *If* ZF *is a consistent theory, so too is* ZFC.

Proof. By 1.2 and 3.8,

$$\mathrm{ZF} \vdash (\mathrm{ZF} + \mathrm{AC})^L. \quad \square$$

Similarly, using 2.15, we obtain

4.2 Theorem. *If* ZF *is a consistent theory, so too is* ZFC + $(V = L)$. □

A consequence of this last result is that any statement Φ which we can prove in the theory ZFC + $(V = L)$ will have automatically been shown to be consistent with ZFC. Thus proofs of results in the theory ZFC + $(V = L)$ have a significance in terms of ZFC set theory, regardless of the light in which $V = L$ is viewed.

We end this short section by giving a characterisation of L in terms of inner models.

4.3 Theorem (The Minimal Model Property). *L is the smallest inner model of* ZF.

Proof. By 1.1, L is a transitive proper class. By 1.2, L is thus an inner model of ZF. Let M be any other inner model of ZF. By 2.9, $(L)^M = L$. *Thus $L \subseteq M$.* □

In fact the above proof tells us more, namely:

4.4 Theorem (The Minimal Model Property for KP). *L is the smallest inner model of* KP. □

5. *The Condensation Lemma. The* GCH *in* L

Recall from I.10 the definitions of the notions of *elementary substructure*, Σ_n*-elementary substructure*, *elementary embedding*, etc., together with the notation we established concerning these notions.

The following lemma is often useful in this connection.

5.1 Lemma. *Let* $\mathbf{M} = \langle M, \in, A_1, \ldots, A_n \rangle$, *where* M *is an amenable set, and let* $\mathbf{N} = \langle N, \ldots \rangle$ *be a substructure of* $\mathbf{M}$. *Let* $n > 0$. *The following are equivalent*:

(i) $\mathbf{N} \prec_n \mathbf{M}$;

(ii) *if* A *is a non-empty* $\Sigma_n^{\mathbf{M}}(N)$ *subset of* M, *then* $A \cap N \neq \emptyset$.

Proof. Before we start, we recall that in the definition of Σ_n in the case of the language $\mathscr{L}_V$ we do not allow repeated quantifiers.

(i) $\rightarrow$ (ii). Let A be a non-empty $\Sigma_n^{\mathbf{M}}(N)$ subset of M, and let $\varphi(x, y)$ be a Π_{n-1} formula of the $\mathbf{M}$-language, with parameters from N, such that

$$A = \{x \in M \mid \models_{\mathbf{M}} \exists y\, \varphi(\mathring{x}, y)\}.$$

Since $A \neq \emptyset$,

$$\models_{\mathbf{M}} \exists x \exists y \varphi(x, y).$$

So, as M is amenable,

$$\models_{\mathbf{M}} \exists z\, \varphi((z)_0, (z)_1).$$

So, as $\mathbf{N} \prec_n \mathbf{M}$, by (i),

$$\models_{\mathbf{N}} \exists z\, \varphi((z)_0, (z)_1).$$

So for some $x \in N$,

$$\models_{\mathbf{N}} \exists y\, \varphi(\mathring{x}, y).$$

But $\exists y \varphi(\mathring{x}, y)$ is a Σ_n formula of the $\mathbf{M}$-language with parameters from N, so by (i) again,

$$\models_{\mathbf{M}} \exists y\, \varphi(\mathring{x}, y).$$

Hence $x \in A$. But $x \in N$ also. Thus, as required,

$$A \cap N \neq \emptyset.$$

(ii) $\rightarrow$ (i). We prove by induction on the length of formulas that for any sentence φ of the $\mathbf{M}$-language with parameters from N which is at most Σ_n,

$$\models_{\mathbf{N}} \varphi \quad \text{iff} \quad \models_{\mathbf{M}} \varphi.$$

If φ is primitive the result is trivial. If φ is of the form $\neg \psi$ or else of the form $\psi_1 \wedge \psi_2$, the induction step is immediate. There remains the case where φ is of the form $\exists x \psi(x)$. Suppose first that

$$\models_{\mathbf{N}} \varphi .$$

Thus,

$$\models_{\mathbf{N}} \exists x \psi(x) .$$

So for some $x \in N$,

$$\models_{\mathbf{N}} \psi(\mathring{x}) .$$

Now, $\psi(\mathring{x})$ is shorter than φ and is at most Π_{n-1}, so by induction hypothesis,

$$\models_{\mathbf{M}} \psi(\mathring{x}) .$$

Thus

$$\models_{\mathbf{M}} \exists x \psi(x) ,$$

i.e.

$$\models_{\mathbf{M}} \varphi .$$

Conversely, assume now this last fact. Then

$$\models_{\mathbf{M}} \exists x \psi(x) .$$

Let

$$A = \{x \in M \mid \models_{\mathbf{M}} \psi(\mathring{x})\} .$$

Then A is non-empty, and is a $\Sigma_n^{\mathbf{M}}(N)$ subset of M. (In fact A is $\Pi_{n-1}^{\mathbf{M}}(N)$.) So by (ii),

$$A \cap N \neq \emptyset .$$

Let $x \in A \cap N$. Then

$$\models_{\mathbf{M}} \psi(\mathring{x}) .$$

But $\psi(\mathring{x})$ has parameters from N, is shorter than φ, and is at most Π_{n-1}. So by induction hypothesis,

$$\models_{\mathbf{N}} \psi(\mathring{x}) .$$

Thus,

$$\models_{\mathbf{N}} \exists x \psi(x) ,$$

i.e.

$$\models_{\mathbf{N}} \varphi .$$

The proof is complete. □

The following theorem is arguably the most important single result in constructibility theory (as far as applications are concerned).

5.2 Theorem (The Condensation Lemma). *Let α be a limit ordinal. If*

$$X \prec_1 L_\alpha,$$

then there are unique π and β such that $\beta \leqslant \alpha$ and:

(i) $\pi \colon \langle X, \in\rangle \cong \langle L_\beta, \in\rangle$;
(ii) *if $Y \subseteq X$ is transitive, then $\pi \restriction Y = \mathrm{id} \restriction Y$;*
(iii) $\pi(x) \leqslant_L x$ *for all $x \in X$.*

Proof. Notice first that by an easy induction on m we can prove that $L_m \subseteq X$ for all $m < \omega$. Indeed, if $x \in L_{m+1}$, then x is of the form

$$x = \{a_1, \ldots, a_k\}$$

for some $a_1, \ldots, a_k \in L_m$, and

$$\models_{L_\alpha} \exists x [(\mathring{a}_1 \in x) \wedge \ldots \wedge (\mathring{a}_k \in x) \wedge (\forall z \in x)((z = \mathring{a}_1) \vee \ldots \vee (z = \mathring{a}_k))],$$

so if $L_m \subseteq X$ this sentence is true in X, which means that $x \in X$. Since $L_m \subseteq X$ for all $m < \omega$, we have $L_\omega \subseteq X$. Thus in the case $\alpha = \omega$, we have $X = L_\alpha$, and the theorem is trivially valid. So from now on we shall assume that $\alpha > \omega$.

Note first that X is extensional. For suppose that $x, y \in X$, $x \neq y$. Then

$$\models_{L_\alpha} \exists z (z \in \mathring{x} \leftrightarrow z \notin \mathring{y}),$$

so as $X \prec_1 L_\alpha$,

$$\models_X \exists z (z \in \mathring{x} \leftrightarrow z \notin \mathring{y}),$$

which means that for some $z \in X$,

$$z \in x \leftrightarrow z \notin y.$$

Since X is extensional, by the Collapsing Lemma (I.7.1) there is a unique π and a unique transitive set M such that

$$\pi \colon X \cong M.$$

We shall show that $M = L_\beta$ for a (unique) ordinal $\beta \leqslant \alpha$.

By 2.7 there is a Σ_0 formula $\Phi(z, v, \gamma)$ of LST such that

(a) $$\forall \gamma \forall v [v = L_\gamma \leftrightarrow \exists z \Phi(z, v, \gamma)],$$

and moreover, if $\varphi(z, v, \gamma)$ is the $\mathscr{L}$-analogue of $\Phi(z, v, \gamma)$, then (using I.9.15)

(b) $$(\forall \gamma < \alpha)(\forall v)[v = L_\gamma \leftrightarrow v \in L_\alpha \wedge \models_{L_\alpha} \exists z \varphi(z, \mathring{v}, \mathring{\gamma})].$$

Now, $\pi^{-1}: M \prec_1 L_\alpha$, so if $\models_M \mathrm{On}(x)$ then $\pi^{-1}(x) \in \alpha$. Moreover, by (b) we have

(c) $$(\forall \gamma < \alpha)[\models_{L_\alpha} \exists v \exists z \varphi(z, v, \mathring{\gamma})].$$

Hence, applying π^{-1}, we get

(d) $$(\forall \gamma \in M)[\models_M \exists v \exists z \varphi(z, v, \mathring{\gamma})].$$

So,

(e) $$(\forall \gamma \in M)(\exists v \in M)(\exists z \in M)[\models_M \varphi(\mathring{z}, \mathring{v}, \mathring{\gamma})].$$

Thus as M is transitive, I.9.15 gives

(f) $$(\forall \gamma \in M)(\exists v \in M)(\exists z \in M)\Phi(z, v, \gamma).$$

Thus by (a)

(g) $$(\forall \gamma \in M)(L_\gamma \in M).$$

Now, M is transitive, so

$$M \cap \mathrm{On} = \beta$$

for some ordinal β. Thus (g) becomes

(h) $$(\forall \gamma \in \beta)(L_\gamma \in M).$$

So, as M is transitive, we conclude that

(i) $$\bigcup_{\gamma < \beta} L_\gamma \subseteq M.$$

Again, since

$$L_\alpha = \bigcup_{\gamma < \alpha} L_\gamma,$$

we have

(j) $$(\forall x \in L_\alpha)[\models_{L_\alpha} \exists \gamma \exists v \exists z(\varphi(z, v, \gamma) \wedge (\mathring{x} \in v))].$$

Applying π^{-1},

(k) $$(\forall x \in M)[\models_M \exists \gamma \exists v \exists z(\varphi(z, v, \gamma) \wedge (\mathring{x} \in v))].$$

Thus,

(l) $$(\forall x \in M)(\exists \gamma \in M)(\exists v \in M)(\exists z \in M)[\models_M \varphi(\mathring{z}, \mathring{v}, \mathring{\gamma}) \wedge (\mathring{x} \in \mathring{v})].$$

So by I.9.15,

(m) $(\forall x \in M)(\exists \gamma \in M)(\exists v \in M)(\exists z \in M)[\Phi(z, v, \gamma) \wedge (x \in v)]$.

Hence by (a),

(n) $(\forall x \in M)(\exists \gamma \in M)(x \in L_\gamma)$.

Thus by definition of β,

(o) $(\forall x \in M)(\exists \gamma \in \beta)(x \in L_\gamma)$.

In other words,

(p) $M \subseteq \bigcup_{\gamma<\beta} L_\gamma$.

Combining (i) and (p) we conclude that

(q) $M = \bigcup_{\gamma<\beta} L_\gamma$.

But lim(α), so

$$(\forall v \in \alpha)[\models_{L_\alpha} \exists \tau (\mathring{v} < \tau)],$$

which implies that

$$(\forall v \in M)[\models_M \exists \tau (\mathring{v} < \tau)].$$

Thus,

$$(\forall v \in \beta)(\exists \tau \in \beta)(v < \tau).$$

Hence lim(β), and (q) becomes

$$M = L_\beta.$$

That completes the proof of part (i) of the theorem.

Part (ii) follows immediately from I.7.1. We are left with the proof of part (iii).

Suppose that $\pi(x) >_L x$ for some $x \in X$. Let x_0 be the $<_L$-least such x. Since $x_0 \in X$, $\pi(x_0) \in L_\beta$. But $x_0 <_L \pi(x_0)$. So by 3.4(i), $x_0 \in L_\beta$. Hence $x_0 = \pi(x_1)$ for some $x_1 \in X$. Thus

$$\pi(x_1) = x_0 <_L \pi(x_0).$$

But $<_L$ is uniformly $\Sigma_1^{L_\lambda}$ for limit $\lambda > \omega$ and π^{-1}: $L_\beta \prec_1 L_\alpha$ (and moreover α and β are limit ordinals), so the above inequality yields

$$x_1 <_L x_0.$$

But this means that $x_1 <_L \pi(x_1)$, which contadicts the choice of x_0. The proof is complete. □

Using the Condensation Lemma, we shall prove that the GCH is valid in L. We require the following lemma, which though stated for limit levels of the constructible hierarchy is really a result about structures with definable well-orders.

5.3 Lemma. *Let α be a limit ordinal, and let $X \subseteq L_\alpha$. Let M be the set of all elements of L_α which are definable in L_α from elements of X. (i.e. $a \in M$ iff for some formula $\varphi(v_0)$ of $\mathscr{L}_X$, a is the unique element of L_α such that $\models_{L_\alpha} \varphi(\mathring{a})$.)*
Then

$$X \subseteq M \prec L_\alpha,$$

and moreover M is the smallest elementary substructure of L_α which contains all elements of X.

Proof. If $\alpha = \omega$ then as in the proof of 5.2 we see at once that $M = L_\alpha$, and that the only elementary submodel of L_α is L_α itself. So we may assume that $\alpha > \omega$ from now on.

If $x \in X$, then x is definable in L_α by means of the formula

$$(v_0 = \mathring{x})$$

so $X \subseteq M$. To show that $M \prec L_\alpha$ we prove that for any formula $\varphi(v_0)$ of $\mathscr{L}_X$,

$$\models_{L_\alpha} \exists x \varphi(x) \quad \text{implies } (\exists x \in M)[\models_{L_\alpha} \varphi(\mathring{x})].$$

(This is *Tarski's Criterion* for being an elementary submodel.) Let $\psi(v_0)$ be the following $\mathscr{L}_X$-formula:

$$\varphi(v_0) \wedge (\forall v_1)(v_1 <_L v_0 \rightarrow \neg \varphi(v_1)).$$

If $\models_{L_\alpha} \exists x \varphi(x)$ then $\models_{L_\alpha} \exists x \psi(x)$. But there is clearly just one $x \in L_\alpha$ such that $\models_{L_\alpha} \psi(\mathring{x})$. Hence the formula $\psi(v_0)$ defines x from elements of X in L_α. Thus $x \in M$. Since $\models_{L_\alpha} \varphi(\mathring{x})$, we are done.

Suppose now that $X \subseteq N \prec L_\alpha$. We show that $M \subseteq N$. Let $x \in M$. For some formula $\varphi(v_0)$ of $\mathscr{L}_X$, x is the unique element of L_α such that $\models_{L_\alpha} \varphi(\mathring{x})$. Now, $\models_{L_\alpha} \exists v_0 \varphi(v_0)$, so as $X \subseteq N \prec L_\alpha$, we have $\models_N \exists v_0 \varphi(v_0)$. So for some $y \in N$, $\models_N \varphi(\mathring{y})$. But $N \prec L_\alpha$, so $\models_{L_\alpha} \varphi(\mathring{y})$. Hence $y = x$, and we are done. □

5.4 Corollary. *Let α be a limit ordinal. For any $X \subseteq L_\alpha$ there is a unique smallest $M \prec L_\alpha$ such that $X \subseteq M$. For this M,*

$$|M| = \max(|X|, \omega).$$

Proof. Let M be as in 5.3. Since $\mathscr{L}_X$ has $\max(|X|, \omega)$ many formulas, we clearly have $|M| = \max(|X|, \omega)$. □

Now, one striking difference between the constructible hierarchy and the cumulative hierarchy of sets is the rate of growth. By definition, if $x \in V_\alpha$, then at level $V_{\alpha+1}$, all subsets of x appear. But the same is not true for the constructible hierarchy. For instance, $L_{\omega+2}$ will contain some subsets of ω, but not all of them. More will appear at level $L_{\omega+3}$, still more at level $L_{\omega+4}$, etc. However, as our next lemma shows, there is a bound to this "gradual growth" process.

5.5 Lemma. *Assume $V = L$. Let κ be a cardinal. If x is a bounded subset of κ (or more generally if $x \subseteq L_\alpha$ for some $\alpha < \kappa$), then $x \in L_\kappa$.*

Proof. For $\kappa \leqslant \omega$ the result is trivial, since then $L_\kappa = V_\kappa$. So assume $\kappa > \omega$. Pick $\alpha < \kappa$ so that $\alpha \geqslant \omega$ and $x \subseteq L_\alpha$, and let λ be a limit ordinal such that $\lambda \geqslant \kappa$ and $x \in L_\lambda$. By 5.4, let $M \prec L_\lambda$ be such that $L_\alpha \cup \{x\} \subseteq M$ and $|M| = |L_\alpha|$. By the Condensation Lemma, let $\pi: M \cong L_\gamma$. Since $L_\alpha \cup \{x\}$ is a transitive subset of M, $\pi \upharpoonright L_\alpha \cup \{x\} = \mathrm{id} \upharpoonright L_\alpha \cup \{x\}$. In particular, $\pi(x) = x$. Thus $x \in L_\gamma$. Now by 1.1 (vii), $|L_\alpha| = |\alpha|$ and $|L_\gamma| = |\gamma|$. Hence,

$$|\gamma| = |L_\gamma| = |\pi'' M| = |M| = |L_\alpha| = |\alpha| < \kappa.$$

Thus $\gamma < \kappa$. But then $L_\gamma \subseteq L_\kappa$, so $x \in L_\kappa$, and we are done. □

5.6 Theorem. *$V = L$ implies* GCH.

Proof. By 5.5, $\mathscr{P}(\kappa) \subseteq L_{\kappa^+}$ for all infinite cardinals κ. So by 1.1 (vii),

$$(\forall \kappa)(2^\kappa \leqslant |L_{\kappa^+}| = \kappa^+).$$

The result follows at once. □

5.7 Corollary. $\mathrm{ZF} \vdash (\mathrm{GCH})^L$.

Proof. We know that

$$\mathrm{ZF} \vdash [\mathrm{ZFC} + (V = L)]^L.$$

By 5.6,

$$\mathrm{ZFC} + (V = L) \vdash \mathrm{GCH}.$$

The result follows at once. □

5.8 Corollary. *If* ZF *is consistent, so too is* ZFC + GCH.

Proof. By the discussion in section 4. □

We finish this section by proving two special cases of the Condensation Lemma for later use. First a technical lemma.

5.9 Lemma. *Let $\alpha > \omega$ be a limit ordinal, $N \subseteq L_\alpha$. Let $A(x)$ be a non-empty $\Sigma_0^{L_\alpha}(N)$ predicate on L_α. Let x be the $<_L$-least element of L_α such that $A(x)$. Then x is Σ_1-definable from elements of N in L_α.*

Proof. We can define x in L_α by the predicate

$$A(x) \wedge (\exists u)[u = \mathrm{pr}(x) \wedge (\forall z \in u) \neg A(z)].$$

By 3.5, this is Σ_1. □

5.10 Lemma. *Assume $V = L$. If $X \prec_1 L_{\omega_1}$, then $X = L_\alpha$ for some $\alpha \leqslant \omega_1$.*

Proof. By the condensation lemma, there are π, α such that $\alpha \leqslant \omega_1$ and $\pi\colon X \cong L_\alpha$. If $Y \subseteq X$ is transitive, then $\pi \mid Y = \mathrm{id} \restriction Y$. So if we can show that X itself is transitive we shall be done.

Let $x \in X$. Then $x \in L_{\omega_1} = \bigcup_{\gamma<\omega_1} L_\gamma$, so $x \in L_\gamma$ for some $\gamma < \omega_1$. But $|L_\gamma| \leqslant |\gamma| + \omega = \omega$, so as L_γ is transitive, x is countable. There is thus a function $f\colon \omega \xrightarrow{\text{onto}} x$. Let f be in fact the $<_L$-least such function. By 5.5, $f \in L_{\omega_1}$. So by 5.9, f is Σ_1-definable from x in L_{ω_1}. But $x \in X \prec_1 L_{\omega_1}$. Thus $f \in X$. But clearly $\omega \subseteq X$. Thus $f(n) \in X$ for all $n < \omega$. Thus $x = f''\omega \subseteq X$. Hence X is transitive, and we are done. □

5.11 Lemma. *Assume $V = L$. Let $\kappa > \omega_1$ be a cardinal. If $\omega_1 \in X \prec_1 L_\kappa$, then $X \cap L_{\omega_1} = L_\alpha$ for some $\alpha \leqslant \omega_1$.*

Proof. Since $\omega_1 \in X$ and $X \prec_1 L_\kappa$, we have $L_{\omega_1} \in X$ (by 2.8). Clearly,

$$X \cap L_{\omega_1} = \{x \in X \mid \models_X \text{“}\mathring{x} \in \mathring{L}_{\omega_1}\text{”}\}.$$

But $X \prec_1 L_\kappa$. So, using an obvious extension of our established notation, for any Σ_1 sentence φ of $\mathscr{L}_{X \cap L_{\omega_1}}$, we have

$$\models_{L_{\omega_1}} \varphi \quad \text{iff} \quad \models_{L_\kappa} \varphi^{L_{\omega_1}} \quad \text{iff} \quad \models_X \varphi^{L_{\omega_1}} \quad \text{iff} \quad \models_{X \cap L_{\omega_1}} \varphi.$$

Thus $X \cap L_{\omega_1} \prec_1 L_{\omega_1}$, and we are done by 5.10. (Note that in fact the above argument works for any φ, Σ_1 or not, so that we actually have $X \cap L_{\omega_1} \prec L_{\omega_1}$.) □

In connection with 5.11, let us just mention that if $\kappa > \omega_1$ is a cardinal and $X \prec_2 L_\kappa$, then we automatically have $\omega_1, L_{\omega_1} \in X$, since L_{ω_1} is definable in L_κ by the Σ_2-formula (in free variable u)

$$(\exists v)(u = L_v) \wedge (\forall x \in u)(\exists f \in u)(f\colon \omega \xrightarrow{\text{onto}} x) \wedge (\forall f) \neg (f\colon \omega \xrightarrow{\text{onto}} u).$$

6. Σ_n *Skolem Functions*

The notion of a Σ_n skolem function for a structure L_α plays an important role in some of the deeper parts of constructibility theory (the so-called *fine structure theory*). In this section we introduce the basic ideas, and in section 7 we give an application, but a detailed study will not be begun until Chapter VI.

Let $(\varphi_i \mid i < \omega)$ enumerate all Π_{n-1} formulas of $\mathscr{L}$ with free variables v_0, v_1. Fix α a limit ordinal grater than ω. For each $i < \omega$ and each $x \in L_\alpha$, if

$$\models_{L_\alpha} \exists y\, \varphi_i(y, \mathring{x}),$$

let $h_i(x)$ be some element y of L_α such that

$$\models_{L_\alpha} \varphi_i(\mathring{y}, \mathring{x}).$$

This defines an ω-sequence $(h_i \mid i < \omega)$ of partial functions h_i from L_α to L_α. By the usual methods of contracting quantifiers and parameters, if $X \subseteq L_\alpha$ is closed under ordered pairs, and if Y is the closure of X under the functions $h_i, i < \omega$, then $Y \prec_n L_\alpha$. (The argument is given, in effect, in 6.2 below.) Now, as far as generating Σ_n elementary substructures is concerned, the exact definition of the functions h_i is not important. But we shall require something rather more than this, though at this stage the reader will simply have to postpone a proper motivation until later, since the situation we are leading towards is rather complicated. What we require is a particularly "nice", canonical definition of the functions h_i. In particular, we want the functions h_i to be definable over L_α in as logically simple as fashion as possible. Now, the most obvious canonical definition of the functions h_i would be to make use of the canonical well-ordering $<_L$ of L, setting $h_i(x) \simeq$ the $<_L$-least $y \in L_\alpha$ such that $\models_{L_\alpha} \varphi_i(\mathring{y}, \mathring{x})$. (The symbol $\simeq$ is standard when partial functions are concerned: to write $f(x) \simeq g(x)$ means that $f(x)$ is defined iff $g(x)$ is defined, in which case they are equal.) It is easily seen that each function h_i so defined will be Π_n. It turns out that we want functions h_i that are Σ_n (and in fact rather more than that), so this obvious definition is not adequate for our purposes. To be precise, what we require is the following. Let h be the partial function defined on $\omega \times L_\alpha$ by

$$h(i, x) \simeq h_i(x).$$

Then the function h should be $\Sigma_n(L_\alpha)$.

The construction of a function h as outlined above requires some considerable effort, and will be postponed until much later. For the moment we investigate the general properties such a function will have.

Let M be an amenable set, and let

$$\mathbf{M} = \langle M, \in, A_1, \ldots, A_k \rangle.$$

Let $n \geqslant 1$. A Σ_n *skolem function* for $\mathbf{M}$ is a $\Sigma_n(\mathbf{M})$ function h such that $\mathrm{dom}(h) \subseteq \omega \times M$, $\mathrm{ran}(h) \subseteq M$, for which there is a $p \in M$ such that h is $\Sigma_n^{\mathbf{M}}(\{p\})$, and whenever A is a non-empty $\Sigma_n^{\mathbf{M}}(\{p, x\})$ subset of M for some $x \in M$, there is an $i \in \omega$ such that $h(i, x) \in A$. (In this situation we say that p is a *good parameter* for h.)

6.1 Lemma. *Let M be an amenable set, and let*

$$\mathbf{M} = \langle M, \in, A_1, \ldots, A_k \rangle.$$

Let $n \geqslant 1$, and let h be a Σ_n skolem function for $\mathbf{M}$. Then:

(i) *if $x \in M$, then*

$$x \in h''(\omega \times \{x\}) \prec_n \mathbf{M};$$

(ii) *if $q \in M$ and if $X \subseteq M$ is closed under ordered pairs, then*

$$X \cup \{q\} \subseteq h''(\omega \times (X \times \{q\})) \prec_n \mathbf{M}.$$

Proof. Let p be a good parameter for h. We prove (i) first, Set

$$N = h''(\omega \times \{x\}).$$

Since $\{x\}$ is a $\Sigma_0^{\mathbf{M}}(\{x\})$ subset of M,

$$h(i, x) \in \{x\}$$

for some $i < \omega$, so $x \in N$. To show that $N \prec_n \mathbf{M}$ we use 5.1. Let A be a non-empty $\Sigma_n^{\mathbf{M}}(N)$ subset of M. We must show that $A \cap N \neq \emptyset$. Pick $y_1, \ldots, y_m \in N$ so that $A \in \Sigma_n^{\mathbf{M}}(\{y_1, \ldots, y_m\})$. By definition of N there are $j_1, \ldots, j_m < \omega$ such that

$$y_1 = h(j_1, x), \ldots, y_m = h(j_m, x).$$

Let $\varphi(v_0, \ldots, v_m)$ be a Σ_n formula of the $\mathbf{M}$-language, having no individual constants, such that for any $a \in M$,

$$a \in A \quad \text{iff } \models_{\mathbf{M}} \varphi(\mathring{a}, \mathring{y}_1, \ldots, \mathring{y}_m);$$

and let $\psi(v_0, v_1, v_2, v_3)$ be a Σ_n formula of the $\mathbf{M}$-language, also having no individual constants, such that for any $u, v \in M$, $i \in \omega$,

$$v = h(i, u) \quad \text{iff } \models_{\mathbf{M}} \psi(\mathring{v}, \mathring{i}, \mathring{u}, \mathring{p}).$$

Then, for any $a \in M$,

$$a \in A \quad \text{iff } \models_{\mathbf{M}} (\exists y_1, \ldots, y_m)[\psi(y_1, \mathring{j}_1, \mathring{x}, \mathring{p}) \wedge \ldots \wedge \psi(y_m, \mathring{j}_m, \mathring{x}, \mathring{p}) \wedge \varphi(\mathring{a}, y_1, \ldots, y_m)].$$

The predicate A is thus seen to be $\Sigma_n^{\mathbf{M}}(\{x, p\})$. (The parameters $j_1, \ldots, j_m$ can be ignored since, being integers they can be replaced by their set-theoretic definitions, i.e. $0 = \emptyset$, $1 = \{\emptyset\}$, etc.) So by the definition of the Σ_n skolem function concept there is an $i < \omega$ such that

$$h(i, x) \in A.$$

Thus $A \cap N \neq \emptyset$, and we are done.

We turn to the proof of (ii). Let

$$N = h''(\omega \times (X \times \{q\})).$$

As in part (i) we get $X \cup \{q\} \subseteq N$, and we must show that $N \prec_n \mathbf{M}$. Again, we begin by picking an arbitrary $\Sigma_n^{\mathbf{M}}(N)$ subset A of M, and show that if $A \neq \emptyset$ then

$A \cap N \neq \emptyset$. Pick $y_1, \ldots, y_m \in N$ so that $A \in \Sigma_n^{\mathbf{M}}(\{y_1, \ldots, y_m\})$. Pick $j_1, \ldots, j_m < \omega$ and $x_1, \ldots, x_m \in X$ so that

$$y_1 = h(j_1, (x_1, q)), \ldots, y_m = h(j_m, (x_m, q)).$$

Set $x = (x_1, \ldots, x_m)$. Since X is closed under ordered pairs, $x \in X$. As in part (i) it is easily seen that A is $\Sigma_n^{\mathbf{M}}(\{p, (x, q)\})$. It follows that there is an $i < \omega$ such that $h(i, (x, q)) \in A$, giving $A \cap N \neq \emptyset$. □

6.2 Corollary. *Let* $\mathbf{M}$, *n*, *h be as in* 6.1. *If* $X \subseteq M$ *and if* $h''(\omega \times X)$ *is closed under ordered pairs, then*

$$X \subseteq h''(\omega \times X) \prec_n \mathbf{M}.$$

Proof. Let

$$Y = h''(\omega \times X).$$

By 6.1 it suffices to prove that $h''(\omega \times Y) = Y$. Well, since $X \subseteq Y$ we clearly have

$$Y = h''(\omega \times X) \subseteq h''(\omega \times Y).$$

Conversely, suppose $z \in h''(\omega \times Y)$. Pick $i \in \omega$, $y \in Y$ so that $z = h(i, y)$. For some $j \in \omega$, $x \in X$ we have $y = h(j, x)$. Thus $z = h(i, h(j, x))$, which shows that z is Σ_n-definable from p and x in $\mathbf{M}$. Thus $\{z\}$ is a $\Sigma_n^M(\{x, p\})$ subset of M. So for some $k \in \omega$, $h(k, x) \in \{z\}$. Thus $z \in h''(\omega \times X) = Y$, and we are done. □

If $\alpha > \omega$ is a limit ordinal, then L_α has a Σ_n skolem function for every $n \geqslant 1$. For $n > 1$, the proof of this fact is quite tricky, and will not be given until Chapter VI. But for $n = 1$ the proof is both easy and illuminating, so we deal with this case now.

For any limit ordinal $\alpha > \omega$ and any $n \in \omega$, $\models_{L_\alpha}^{\Sigma_n}$ denotes the restriction of $\models_{L_\alpha}$ to the Σ_n sentences of $\mathscr{L}_{L_\alpha}$.

6.3 Lemma. *Let* $\alpha > \omega$ *be a limit ordinal. Then the relation* $\models_{L_\alpha}^{\Sigma_0}$ *is (uniformly for all such* α) $\Delta_1^{L_\alpha}$.

Proof. For the purposes of this proof, we shall regard the Σ_1 formula $\mathrm{Sat}(u, \varphi)$ as being expressed in the language $\mathscr{L}$, rather than in LST as defined previously.

If φ is a Σ_0 sentence of $\mathscr{L}_{L_\alpha}$, then by Σ_0-absoluteness, if $\gamma < \alpha$ is such that $\varphi \in L_\gamma$,

$$\models_{L_\alpha} \varphi \quad \text{iff } \models_{L_\gamma} \varphi.$$

Moreover, absoluteness considerations also tell us that for any $u \in L_\alpha$ and any formula ψ of $\mathscr{L}_u$,

$$\models_u \psi \quad \text{iff } \models_{L_\alpha} \mathrm{Sat}(\mathring{u}, \mathring{\psi}).$$

Hence, for φ as above,

$$\models_{L_\alpha} \varphi \quad \text{iff } \models_{L_\alpha} (\exists \gamma)[(\mathring{\varphi} \in L_\gamma) \wedge \mathrm{Sat}(L_\gamma, \mathring{\varphi})],$$

and,

$$\models_{L_\alpha} \varphi \quad \text{iff } \models_{L_\alpha} (\forall \gamma) [(\mathring{\varphi} \in L_\gamma) \rightarrow \mathrm{Sat}(L_\gamma, \mathring{\varphi})].$$

By I.9.12, the class Fml^{Σ_0} is $\Delta_1^{L_\alpha}$, and by I.9.10 the class Sat is $\Delta_1^{L_\alpha}$. Also, by 2.8, the relation $(x = L_\gamma)$ is $\Delta_1^{L_\alpha}$. Thus the relation $\models_{L_\alpha}^{\Sigma_0}$ is $\Delta_1^{L_\alpha}$ by virtue of the two definitions above (which are uniform for α). □

6.4 Lemma. *Let $\alpha > \omega$ be a limit ordinal, and let $n \geqslant 1$. Then the relation $\models_{L_\alpha}^{\Sigma_0}$ is (uniformly in α) $\Sigma_n^{L_\alpha}$.*

Proof. Let φ be a Σ_n sentence of $\mathscr{L}_{L_\alpha}$. In case n is odd, the following is clearly equivalent to $\models_{L_\alpha} \varphi$:

$$\begin{aligned} \models_{L_\alpha} [&\exists x_1 \forall x_2 \exists x_3 \ldots \exists x_n \exists \psi \exists u \exists f \exists \theta [\mathrm{Fml}(\psi) \wedge \mathrm{Fr}(\psi, u) \wedge (f: n \leftrightarrow u) \\ &\wedge (\mathring{\varphi} = \exists f(0) \forall f(1) \exists f(2) \ldots \exists f(n-1) \psi) \wedge \mathrm{Finseq}(\theta) \\ &\wedge (\mathrm{dom}(\theta) = n+1) \wedge (\theta_0 = \psi) \wedge (\forall i \in n)(\mathrm{Sub}(\theta_{i+1}, \theta_i, f(i), \mathring{x}_{i+1})) \\ &\wedge (\models_{L_\alpha}^{\Sigma_0} \theta_n)]]. \end{aligned}$$

In case n is even, then equivalent to $\models_{L_\alpha} \varphi$ we have:

$$\begin{aligned} \models_{L_\alpha} [&\exists x_1 \forall x_2 \exists x_3 \ldots \forall x_n \forall \psi \forall u \forall f \forall \theta [[\mathrm{Fml}(\psi) \wedge \mathrm{Fr}(\psi, u) \wedge (f: n \leftrightarrow u) \\ &\wedge (\mathring{\varphi} = \exists f(0) \forall f(1) \exists f(2) \ldots \forall f(n-1) \psi) \wedge \mathrm{Finseq}(\theta) \\ &\wedge (\mathrm{dom}(\theta) = n+1) \wedge (\theta_0 = \psi) \wedge (\forall i \in n)(\mathrm{Sub}(\theta_{i+1}, \theta_i, f(i), \mathring{x}_{i+1})] \\ &\rightarrow [\models_{L_\alpha}^{\Sigma_0} \theta_n]]]. \end{aligned}$$

In either case, the $\mathscr{L}$-formula which says that

$$\mathring{\varphi} = \exists f(0) \forall f(1) \exists f(2) \ldots - f(n-1) \psi$$

is easily seen to be Σ_0 (given the values of $f(0), \ldots, f(n-1)$), being simply a long sequence of conjuncts concerning the values of the sequence φ. So by 6.3, the above expressions give a (uniformly in α) Σ_n definition of $\models_{L_\alpha}^{\Sigma_n}$. □

Using 6.4, we can now show that for limit $\alpha > \omega$, L_α has a Σ_1 skolem function.

6.5 Lemma. *Let $\alpha > \omega$ be a limit ordinal. Then L_α has a Σ_1 skolem function. Indeed, there is a Σ_0 formula $\Theta(v_0, v_1, v_2, v_3)$ of $\mathscr{L}$ such that for any limit ordinal $\alpha > \omega$, h_α is Σ_1 skolem function for L_α, where*

$$y = h_\alpha(i, x) \leftrightarrow \models_{L_\alpha} \exists z\, \Theta(z, \mathring{y}, \mathring{i}, \mathring{x}).$$

Proof. By an argument similar to the proof of I.9.6 (see also I.9.13), the relation "φ is an $\mathscr{L}$-formula of the form $\exists v_2 \bar{\varphi}(v_0, v_1, v_2)$ where $\bar{\varphi}$ is Σ_0" is uniformly $\Delta_1^{L_\alpha}$. So, if we define an enumeration $(\varphi_i | i < \omega)$ of all formulas of the form $\varphi_i = \exists v_2 \bar{\varphi}_i(v_0, v_1, v_2)$ where $\bar{\varphi}_i$ is Σ_0, in the same way that we well-ordered the formulas of $\mathscr{L}$ in our definition of $<_L$ (see immediately following 3.1), then it is easily seen that this enumeration is uniformly $\Delta_1^{L_\alpha}$.

Define a partial function r_α on $\omega \times L_\alpha$ by:

$$r_\alpha(i, x) \simeq \text{the } <_L\text{-least } w \in L_\alpha \text{ such that}$$
$$\models_{L_\alpha} (\text{“}\mathring{w} \text{ is an ordered pair”}) \wedge \bar{\varphi}_i((\mathring{w})_0, \mathring{x}, (\mathring{w})_1).$$

By (the proof of) 5.9, r_α is $\Sigma_1^{L_\alpha}$. Hence the partial function h_α is $\Sigma_1^{L_\alpha}$, where we define h_α on $\omega \times L_\alpha$ by:

$$h_\alpha(i, x) \simeq (r_\alpha(i, x))_0 .$$

We show that h_α is a Σ_1 skolem function for L_α with good parameter $\emptyset$ (i.e. effectively with *no* parameter).

Let A be a non-empty $\Sigma_1^{L_\alpha}(\{x\})$ subset of L_α. For some $i < \omega$,

$$z \in A \leftrightarrow \models_{L_\alpha} \varphi_i(\mathring{z}, \mathring{x}) .$$

Since $A \neq \emptyset$,

$$\models_{L_\alpha} \exists z\, \varphi_i(z, \mathring{x}) .$$

Hence,

$$\models_{L_\alpha} \exists w\, [\text{“}w \text{ is an ordered pair”} \wedge \bar{\varphi}_i((w)_0, \mathring{x}, (w)_1)].$$

Thus $r_\alpha(i, x)$ is defined, say $w = r_\alpha(i, x)$. Hence $h_\alpha(i, x)$ is defined and $h_\alpha(i, x) = (w)_0$. Clearly,

$$\models_{L_\alpha} \varphi_i((\mathring{w})_0, \mathring{x}) .$$

Hence $h_\alpha(i, x) \in A$, as required.

Notice that the above proof did not depend upon α. Hence there is a single Σ_0 formula Θ as described in the lemma. □

Notice that in general, the above procedure will not produce a Σ_n skolem function if $n > 1$. For if the formulas φ_i are Σ_n, then the formulas $\bar{\varphi}_i$ will be Π_{n-1}, which means that the function r_α will be Π_n, as is easily seen by writing out the definition of r_α more fully. (The procedure works in the case $n = 1$ because a bounded universal quantifier prefixing a Σ_0 formula results in another Σ_0 formula, whereas if $n > 1$, a bounded universal quantifier prefixing a Σ_{n-1} formula gives a Π_n formula.)

The function h_α defined in 6.5 is called the *canonical* Σ_1 skolem function for L_α. We illustrate its use in 6.8 below, for which application we require two lemmas.

6.6 Lemma (Gödel's Pairing Function). *There is a* Δ_1^{KP} *formula* $\Phi(v_0, v_1, v_2)$ *of* LST *such that, if*

$$G = \{(\gamma, (\alpha, \beta)) \mid \Phi(\alpha, \beta, \gamma)\},$$

then:

(i) G *is uniformly* $\Sigma_1^{L_\alpha}$ *for limit* $\alpha > \omega$;
(ii) G: $\mathrm{On} \times \mathrm{On} \leftrightarrow \mathrm{On}$;
(iii) $G(\alpha, \beta) \geqslant \alpha, \beta$ *for all* a, β.

Proof. Define a well-ordering $<^*$ of On × On by setting $(\alpha, \beta) <^* (\gamma, \delta)$ iff:

(i) $\max(\alpha, \beta) < \max(\gamma, \delta)$; or
(ii) $\max(\alpha, \beta) = \max(\gamma, \delta)$ and $\alpha < \gamma$; or
(iii) $\max(\alpha, \beta) = \max(\gamma, \delta)$ and $\alpha = \gamma$ and $\beta < \delta$.

Let $G(\alpha, \beta)$ be the order-type of the set of predecessors of (α, β) under $<^*$. Thus

$$G\colon (\mathrm{On} \times \mathrm{On}, <^*) \cong (\mathrm{On}, <).$$

It is not hard to see that:

$$\text{(I)} \qquad G(0, \beta) = \sup_{\nu<\beta} G(\nu, \nu);$$

$$\text{(II)} \qquad G(\alpha, \beta) = \begin{cases} G(0, \beta) + \alpha, & \text{if } \alpha < \beta, \\ G(0, \alpha) + \alpha + \beta, & \text{if } \alpha \geqslant \beta. \end{cases}$$

By (I) and (II) we can define the unary function $G(0, \beta)$ by means of the recursion

$$\text{(III)} \qquad G(0, \beta) = \sup_{\nu<\beta} (G(0, \nu) + \nu + \nu).$$

Thus (c.f. the proof of the KP Recursion Theorem, I.11.8), the function $G(0, \beta)$ can be defined by means of a Σ_1^{KP} formula of LST, and is (by checking that the relevant existential quantifiers can be restricted to L_α) uniformly $\Sigma_1^{L_\alpha}$ for limit $\alpha > \omega$. But using (II) we can define the binary function $G(\alpha, \beta)$ from $G(0, \beta)$ in Σ_1 fashion. Hence G is definable by means of a Σ_1^{KP} formula of LST and is uniformly $\Sigma_1^{L_\alpha}$ for limit $\alpha > \omega$. (In connection with the definability results for L_α in the above, it should perhaps be emphasised that there is no suggestion that L_α should be closed under the function G; rather that for each limit $\alpha > \omega$ the class $G \cap (L_\alpha)^3$ is (uniformly) $\Sigma_1^{L_\alpha}$.) Since G is a total function on On × On, it is in fact definable by a Δ_1^{KP} formula of LST. (But since $G \cap (L_\alpha)^3$ is not total on $\alpha \times \alpha$, it is not the case that G is $\Delta_1^{L_\alpha}$.) □

6.7 Lemma. *Let $\alpha > \omega$ be a limit ordinal. Then there is a $\Sigma_1(L_\alpha)$ map of α onto $\alpha \times \alpha$.*

Proof. Before we start, we remark that there is no suggestion of any uniformity here, and indeed for many ordinals α we shall make use of parameters in order to define the mapping of α onto $\alpha \times \alpha$.

Set

$$Q = \{\alpha \mid G\colon \alpha \times \alpha \leftrightarrow \alpha\}.$$

It is easily seen that Q is closed and unbounded in On, and that

$$Q = \{\alpha \mid G(0, \alpha) = \alpha\}.$$

We prove the lemma by induction on α. Assume that it holds for all $\beta < \alpha$. To prove it for α we consider three cases.

Case 1. $\alpha \in Q$.

In this case $G^{-1} \restriction \alpha$ is sufficient.

Case 2. $\alpha = \gamma + \omega$ for some limit ordinal γ.
Define $j\colon \alpha \leftrightarrow \gamma$ by

$$j(\xi) = \begin{cases} 2\xi, & \text{if } \xi < \omega, \\ \xi, & \text{if } \omega \leqslant \xi < \gamma, \\ 2n+1, & \text{if } \xi = \gamma + n. \end{cases}$$

Clearly, j is $\Sigma_1^{L_\alpha}(\{\omega, \gamma\})$.
By induction hypothesis there is a $\Sigma_1(L_\gamma)$ map

$$g\colon \gamma \xrightarrow{\text{onto}} \gamma \times \gamma .$$

Define a map

$$f\colon \alpha \to \alpha \times \alpha$$

by

$$f(\xi) = (j^{-1}((g \circ j(\xi))_0), j^{-1}((g \circ j(\xi))_1)).$$

Clearly, f is $\Sigma_1(L_\alpha)$ and maps α onto $\alpha \times \alpha$. (The function g is an element of L_α, and is thus a parameter in the definition of f.)

Case 3. Otherwise.
Set $(\nu, \tau) = G^{-1}(\alpha)$. Since $\alpha \notin Q$, $\nu, \tau < \alpha$. Let

$$C = \{z \mid z <^* (\nu, \tau)\},$$

where $<^*$ is the well-ordering of $\mathrm{On} \times \mathrm{On}$ defined in the proof of 6.6. Notice that $C \in L_\alpha$. Now, $g = G \restriction C$ maps C one-one onto α (by definition of G from $<^*$). So by 6.6 (i), g is $\Sigma_1^{L_\alpha}(\{C\})$.

Let $\gamma > \omega$ be a limit ordinal such that $\nu, \tau < \gamma < \alpha$. By the induction hypothesis there is a $\Sigma_1(L_\gamma)$ map

$$k\colon \gamma \xrightarrow{\text{onto}} \gamma \times \gamma .$$

Define

$$\bar{k}\colon \gamma \times \gamma \xrightarrow{\text{one-one}} \gamma$$

by setting

$$\bar{k}(\xi, \zeta) = \text{the least } \iota \text{ such that } k(\iota) = (\xi, \zeta).$$

Then l is one-one from α to γ where we set $l = \bar{k} \circ g^{-1}$. (Since $\nu, \tau < \gamma$, $C \subseteq \gamma \times \gamma$, so $\operatorname{ran}(g^{-1}) = C \subseteq \gamma \times \gamma = \operatorname{dom}(\bar{k})$.) Now define

$$h\colon \alpha \times \alpha \xrightarrow{\text{one-one}} \gamma \times \gamma$$

by setting $h(\xi, \zeta) = (l(\xi), l(\zeta))$, and define

$$p\colon \alpha \times \alpha \xrightarrow{\text{one-one}} \gamma$$

by $p = \bar{k} \circ h$. Now, $\operatorname{ran}(l) = \bar{k}''(g^{-1}{}''\alpha) = \bar{k}''C$. Hence $\operatorname{ran}(h) = (\bar{k}''C) \times (\bar{k}''C)$. Hence $\operatorname{ran}(p) = \bar{k}''\operatorname{ran}(h) = \bar{k}''[(\bar{k}''C) \times (\bar{k}''C)]$. But k is L_γ-definable, so $\bar{k} \in L_\alpha$. Since we know that $C \in L_\alpha$, it follows that $D \in L_\alpha$, where $D = \operatorname{ran}(p)$. Thus the function

$$f\colon \alpha \to \alpha \times \alpha$$

is $\Sigma_1(L_\alpha)$, where we set

$$f(\xi) = \begin{cases} p^{-1}(\xi), & \text{if } \xi \in D, \\ (0, 0), & \text{if } \xi \notin D. \end{cases}$$

Since f is clearly onto $\alpha \times \alpha$, we are done. $\square$

We are now able to give our promised use of the Σ_1 skolem function. It is a "localised" version of 3.6.

6.8 Lemma. *Let $\alpha > \omega$ be a limit ordinal. Then there is a $\Sigma_1(L_\alpha)$ map of α onto L_α. (The map is not uniformly definable, and may involve parameters in its definition.)*

Proof. By 6.7, let f be a $\Sigma_1^{L_\alpha}(\{p\})$ *map of α onto* $\alpha \times \alpha$, chosen so that p is the $<_L$-least element of L_α for which such an f exists. Define "inverse functions" f^0, f^1 to f by the requirement

$$f(v) = (f^0(v), f^1(v)) \quad (v \in \alpha).$$

For each n, define a $\Sigma_1^{L_\alpha}(\{p\})$ function f_n from α onto α^n so that the following conditions are satisfied:

$$f_1 = \operatorname{id} \upharpoonright \alpha;$$
$$f_{n+1}(v) = (f^0(v), f_n \circ f^1(v)).$$

(For each n, the precise definition of f_n is obtained by unravelling the above "recursion".)

Let $h = h_\alpha$, the canonical Σ_1 skolem function for L_α, and let Θ be the canonical Σ_0 formula of $\mathscr{L}$ which defines h (see 6.5). Set

$$X = h''(\omega \times (\alpha \times \{p\})).$$

Claim 1. X is closed under ordered pairs.

To see this, let $x_1, x_2 \in X$. Pick $j_1, j_2 \in \omega$ and $v_1, v_2 \in \alpha$ so that

$$x_1 = h(j_1, (v_1, p)), \qquad x_2 = h(j_2, (v_2, p)).$$

Let $\tau \in \alpha$ be such that

$$(v_1, v_2) = f_2(\tau).$$

Clearly, $\{(x_1, x_2)\}$ is a $\Sigma_1^{L_\alpha}(\{(\tau, p)\})$ predicate on L_α. Hence by the properties of h,

$$(x_1, x_2) \in X,$$

which proves the claim.

By claim 1 and 6.2,

$$X \prec_1 L_\alpha .$$

By the Condensation Lemma, let

$$\pi : X \cong L_\beta .$$

Since $\alpha \subseteq X$, we must have $\beta = \alpha$, so in fact

$$\pi : X \cong L_\alpha .$$

Claim 2. For all $i \in \omega$, $x \in X$,

$$\pi(h(i, x)) \simeq h(i, \pi(x)) .$$

To see this, suppose first that $y = h(i, x)$ is defined. Since h is $\Sigma_1^{L_\alpha}$ and $x \in X \prec_1 L_\alpha$, we have $y \in X$. Since $y = h(i, x)$, we have

$$\models_{L_\alpha} \exists z \, \Theta(z, \mathring{y}, \mathring{i}, \mathring{x}) .$$

So, as $x, y, i \in X \prec_1 L_\alpha$,

$$\models_X \exists z \, \Theta(z, \mathring{y}, \mathring{i}, \mathring{x}) .$$

So for some $z \in X$,

$$\models_X \Theta(\mathring{z}, \mathring{y}, \mathring{i}, \mathring{x}) .$$

Applying π gives

$$\models_{L_\alpha} \Theta(\pi(\mathring{z}), \pi(\mathring{y}), \mathring{i}, \pi(\mathring{x})) .$$

Hence,

$$\models_{L_\alpha} \exists z \, \Theta(z, \pi(\mathring{y}), \mathring{i}, \pi(\mathring{x})) .$$

This means that

$$\pi(y) = h(i, \pi(x))$$

(and in particular that $h(i, \pi(x))$ is defined).

Now suppose that $h(i, \pi(x))$ is defined. Then $h(i, \pi(x)) \in L_\alpha = \pi'' X$, so for some $y \in X$,

$$h(i, \pi(x)) = \pi(y) .$$

By reversing the argument above we obtain

$$\pi^{-1}(h(i, \pi(x))) = h(i, x) ,$$

and the claim is proved.

Now, $f \subseteq \alpha \times \alpha \times \alpha$, and $\pi \restriction \alpha = \text{id} \restriction \alpha$, so

$$\pi'' f = f.$$

Moreover, since $p \in X \prec_1 L_\alpha$ and $\pi: X \cong L_\alpha$, $\pi'' f$ is $\Sigma_1^{L_\alpha}(\{\pi(p)\})$. So by choice of p we must have $p \leqslant_L \pi(p)$. But by the properties of π (see 5.2), $\pi(p) \leqslant_L p$. Hence $\pi(p) = p$. So by claim 2, if $i \in \omega$ and $v \in \alpha$,

$$\pi(h(i, (v, p))) \simeq h(i, (v, p)).$$

Thus $\pi = \text{id} \restriction X$, which means that $X = L_\alpha$. It follows at once that the function r defined on a subset of α by

$$r(v) \simeq h(((f(v))_0, ((f(v))_1, p))$$

is a $\Sigma_1(L_\alpha)$ map such that $r'' \alpha = L_\alpha$. But this does not prove the lemma, since we are looking for a *total* function from α onto L_α. However, a simple modification to the function r will suffice. Define g from α^3 to L_α by:

$$g(i, v, \tau) = \begin{cases} y, & \text{if } \models_{L_\alpha} \exists w [w = L_\tau \wedge (\exists z \in w)\, \Theta(z, \mathring{y}, \mathring{i}, (\mathring{v}, \mathring{p}))], \\ \emptyset, & \text{if} \cdot \models_{L_\alpha} \exists w [w = L_\tau \wedge (\forall z \in w) \neg\, \Theta(z, \mathring{y}, \mathring{i}, (\mathring{v}, \mathring{p}))]. \end{cases}$$

It is easily seen that g is $\Sigma_1(L_\alpha)$. And clearly,

$$g''(\alpha \times \alpha \times \alpha) = h''(\omega \times (\alpha \times \{p\})) = L_\alpha.$$

Thus $g \circ f_3$ is a required for the lemma. □

We give further applications of the Σ_1 skolem function in the next section.

7. Admissible Ordinals

The notion of an admissible set has already been introduced in I.11. An ordinal α is said to be *admissible* iff there is an admissible set M such that $M \cap \text{On} = \alpha$.

By I.11.2, every uncountable cardinal is an admissible ordinal. The converse is not true, and indeed it is a simple exercise involving the Condensation Lemma to show that if κ is an uncountable cardinal, there are κ many admissible ordinals below κ. The starting point for this proof is the following lemma.

7.1 Lemma. *An ordinal α is admissible iff L_α is an admissible set.*

Proof. If L_α is an admissible set, then α is an admissible ordinal since $L_\alpha \cap \text{On} = \alpha$.

Conversely, suppose that α is an admissible ordinal. Let M be an admissible set such that $M \cap \text{On} = \alpha$. Clearly, α is a limit ordinal greater than ω. Hence by 2.1, L_α is amenable. Thus we must prove that for any $\Sigma_0(L_\alpha)$ relation $R \subseteq L_\alpha \times L_\alpha$, if

$$(\forall x \in L_\alpha)(\exists y \in L_\alpha) R(y, x),$$

then for any $u \in L_\alpha$ there is a $v \in L_\alpha$ such that

$$(\forall x \in u)(\exists y \in v) R(y, x).$$

Let $\varphi(v_0, v_1)$ be a Σ_0-formula of $\mathscr{L}_{L_\alpha}$ such that for all $x, y \in L_\alpha$,

$$R(y, x) \leftrightarrow \models_{L_\alpha} \varphi(\mathring{y}, \mathring{x}).$$

Thus

$$\models_{L_\alpha} \forall x \exists y \varphi(y, x).$$

Let $u \in L_\alpha$ be given. We seek a $v \in L_\alpha$ such that

$$\models_{L_\alpha} (\forall x \in \mathring{u})(\exists y \in \mathring{v}) \varphi(y, x).$$

Define a function g from u to α by

$$g(x) = \text{the least } \gamma \text{ such that } (\exists y \in L_\gamma)[\models_{L_\alpha} \varphi(\mathring{y}, \mathring{x})].$$

Since $L_\alpha = \bigcup_{\gamma<\alpha} L_\gamma$, g is well-defined. Now, since φ is Σ_0, for $x, y \in L_\alpha$, we have , by I.9.14,

$$\models_{L_\alpha} \varphi(\mathring{y}, \mathring{x}) \quad \text{iff } \models_M \varphi(\mathring{y}, \mathring{x}).$$

Moreover, by 2.10, $(L_\gamma)^M = L_\gamma$ for all $\gamma < \alpha$. Hence for any $x, \gamma \in M$,

$$\begin{aligned}\gamma = g(x) \leftrightarrow \models_M [(\mathring{x} \in \mathring{u}) \wedge (\exists w)[(w = L_{\mathring{\gamma}}) \wedge (\exists y \in w) \varphi(y, \mathring{x}) \\ \wedge (\forall v \in w) \neg (\exists y \in v) \varphi(y, \mathring{x})]].\end{aligned}$$

Thus g is $\Sigma_1(M)$. So by Localised Σ_1 Collection (I.11.5) for the admissible set M there is a $v \in M$ such that

$$(\forall x \in u)(\exists \gamma \in v)(\gamma = g(x)).$$

Since M is amenable,

$$\delta = \bigcup (v \cap \mathrm{On}) \in M.$$

Then by definition of g,

$$(\forall x \in u)(\exists y \in L_\delta)[\models_{L_\alpha} \varphi(\mathring{y}, \mathring{x})].$$

Since $\delta \in M$ we have $\delta < \alpha$, so $L_\delta \in L_\alpha$ and we are done. □

Using 7.1 we may prove:

7.2 Lemma. *Let $\alpha > \omega$ be a limit ordinal. Then α is admissible iff there is no $\Sigma_1(L_\alpha)$ mapping from an ordinal $\delta < \alpha$ cofinally into α.*

Proof. Suppose first that α is admissible. Let $\delta < \alpha$, and let

$$f\colon \delta \to \alpha$$

be $\Sigma_1(L_\alpha)$. Then

$$(\forall \xi \in \delta)(\exists \zeta \in \alpha)(\zeta = f(\xi)).$$

By Σ_1 Collection for L_α there is a $\gamma < \alpha$ such that

$$(\forall \xi \in \delta)(\exists \zeta \in \gamma)(\zeta = f(\xi)).$$

Thus $f''\delta \subseteq \gamma < \alpha$, showing that f cannot be cofinal in α.

Conversely, suppose there is no $\Sigma_1(L_\alpha)$ function from an ordinal $\delta < \alpha$ cofinally into α. Then certainly α cannot be of the form $\gamma + \omega$ for any γ, so α is a limit of limit ordinals. So by 6.8 there can be no $\Sigma_1(L_\alpha)$ function from any L_δ, $\delta < \alpha$, into α whose range is unbounded in α. We show that this implies that L_α is an admissible set. By 2.1, L_α is amenable. So, given a $\Sigma_0(L_\alpha)$ relation $R(y, x)$ on L_α such that

$$(\forall x \in L_\alpha)(\exists y \in L_\alpha)\, R(y, x),$$

and given a $u \in L_\alpha$, we must find a $v \in L_\alpha$ such that

$$(\forall x \in u)(\exists y \in v)\, R(y, x).$$

Pick $\delta < \alpha$ so that $u \in L_\delta$, and define a function f from L_δ *to* α by:

$$f(x) = \begin{cases} \text{the least } \gamma \text{ such that } (\exists y \in L_\gamma)\, R(y, x), & \text{if } x \in u, \\ 0, & \text{otherwise.} \end{cases}$$

It is easily seen that f is $\Sigma_1(L_\alpha)$. By the above remarks, we know that f cannot be cofinal in α, so there is a $\varrho < \alpha$ such that

$$f''L_\delta \subseteq \varrho.$$

By definition of f,

$$(\forall x \in u)(\exists y \in L_\varrho)\, R(y, x),$$

so we are done. □

Our next result strengthens 7.2 considerably. To state the result, it is convenient to introduce the following extension of the concept of amenability, an extension which we shall make frequent use of during our later development.

A structure

$$\mathbf{M} = \langle M, \in, A_1, \ldots, A_k \rangle$$

is said to be *amenable* if M is an amenable set and for each $i = 1, \ldots, k$,

$$u \in M \quad \text{implies } A_i \cap u \in M.$$

(This condition can be regarded as an extension of the "Σ_0 Comprehension" axiom for amenable sets, that if $R \subseteq M$ is $\Sigma_0(M)$, then $R \cap u \in M$ for all $u \in M$.)

7.3 Theorem. *Let $\alpha > \omega$ be a limit ordinal. Then the following are equivalent:*

(i) *α is admissible;*
(ii) *the structure $\langle L_\alpha, A \rangle$ is amenable for any $\Delta_1(L_\alpha)$ set $A \subseteq L_\alpha$;*
(iii) *there is no $\Delta_1(L_\alpha)$ function from an ordinal $\delta < \alpha$ onto α.*

Proof. (i) → (ii). This is an immediate consequence of the Δ_1 Comprehension Principle (I.11.1).

(ii) → (iii). We assume that (ii) holds and (iii) fails and use a diagonalisation argument to obtain a contradiction. By 6.8 and the failure of (iii) there is a $\delta < \alpha$ and a $\Sigma_1(L_\alpha)$ map f from δ onto L_α. Being total, f is in fact $\Delta_1(L_\alpha)$. Hence D is $\Delta_1(L_\alpha)$, where we set

$$D = \{\nu \in \delta \mid \nu \notin f(\nu)\}.$$

By (iii),

$$D = D \cap \delta \in L_\alpha.$$

Hence $D = f(\nu)$ for some $\nu < \delta$. But then

$$\nu \in f(\nu) \leftrightarrow \nu \in D \leftrightarrow \nu \notin f(\nu),$$

a contradiction.

(iii) → (i). Suppose (iii) holds but (i) fails. By 7.2 and the failure of (i) there is a $\delta < \alpha$ and a $\Sigma_1(L_\alpha)$ map f from δ cofinally into α. Let f be $\Sigma_1^{L_\alpha}(\{p\})$. By (iii), α cannot be of the form $\gamma + \omega$ for any γ, so we can pick a limit ordinal $\gamma < \alpha$ such that $\delta, p \in L_\gamma$. Set

$$X = h_\alpha''(\omega \times L_\gamma).$$

Since L_γ is closed under ordered pairs, 6.1 (ii) tells us that

$$L_\gamma \subseteq X \prec_1 L_\alpha.$$

By the Condensation Lemma, let

$$\pi: X \cong L_\beta.$$

Notice that $\pi \restriction L_\gamma = \mathrm{id} \restriction L_\gamma$.

Claim. $\pi \upharpoonright X = \mathrm{id} \upharpoonright X$.

To see this, let Θ be as in 6.5 (namely, the canonical Σ_0 formula of $\mathcal{L}$ which defines h_α over L_α). Let $i \in \omega$, $x \in L_\gamma$, $y \in L_\alpha$ be such that

$$y = h_\alpha(i, x).$$

Then

$$\models_{L_\alpha} \exists z \, \Theta(z, \mathring{y}, \mathring{i}, \mathring{x}).$$

Since $y, x \in X \prec_1 L_\alpha$, this gives

$$\models_X \exists z \, \Theta(z, \mathring{y}, \mathring{i}, \mathring{x}).$$

Applying π,

$$\models_{L_\beta} \exists z \, \Theta(z, \mathring{\pi(y)}, \mathring{i}, \mathring{x}).$$

By U-absoluteness (I.9.14), it follows that

$$\models_{L_\alpha} \exists z \, \Theta(z, \mathring{\pi(y)}, \mathring{i}, \mathring{x}).$$

In other words,

$$\pi(y) = h_\alpha(i, x) = y.$$

This proves the claim.

By the claim, $X = L_\beta$. Now, f is $\Sigma_1^{L_\alpha}(\{p\})$, and $p \in X \prec_1 L_\alpha$, so X is closed under f. But $\delta \subseteq X$ and f is cofinal in α. Thus as $X = L_\beta$, which is transitive, we must have $\alpha \subseteq X$. Thus $\beta = \alpha$ and $X = L_\alpha$.

Define a function g from $\omega \times \delta \times L_\gamma$ into L_α by:

$$g(i, v, x) = \begin{cases} y, & \text{if } (\exists z \in L_{f(v)})[\models_{L_\alpha} \Theta(\mathring{z}, \mathring{y}, \mathring{i}, \mathring{x})], \\ \emptyset, & \text{otherwise.} \end{cases}$$

It is easily seen that g is $\Sigma_1^{L_\alpha}(\{p\})$. (We leave this to the reader. A similar argument was used towards the end of the proof of 6.8.) Also,

$$g''(\omega \times \delta \times L_\gamma) = h_\alpha''(\omega \times L_\gamma) = X = L_\alpha.$$

(Because f is cofinal in α.) But it follows easily from 6.8 that there is a $\Sigma_1(L_\gamma)$ map, j, from γ onto $\omega \times \delta \times L_\gamma$. Then $g \circ j$ is a $\Sigma_1(L_\alpha)$ map from γ onto L_α, contradicting (iii). □

It is perhaps woth noting the following fact, used implicitly in the proof of the above lemma.

7.4 Lemma. *Let α, β be limit ordinals, $\omega < \alpha < \beta$. Then $h_\alpha \subseteq h_\beta$.*

Proof. Suppose that

$$y = h_\alpha(i, x).$$

Then with Θ as in 6.5,

$$\models_{L_\alpha} \exists z\, \Theta(z, \mathring{y}, \mathring{i}, \mathring{x}).$$

By U-absoluteness,

$$\models_{L_\beta} \exists z\, \Theta(z, \mathring{y}, \mathring{i}, \mathring{x}).$$

Thus

$$y = h_\beta(i, x). \quad \square$$

Clearly, it is the uniformity of the Σ_1 skolem function which lies behind 7.4. We often use 7.4 without mention.

Exercises

1. *Primitive Recursive Set Functions* (Section 2)

A function $f\colon V^n \to V$ is said to be *primitive recursive* (p.r.) iff it is geneated by the following schemas:

(i) $f(x_1, \ldots, x_n) = x_i \quad (1 \leqslant i \leqslant n)$;
(ii) $f(x_1, \ldots, x_n) = \{x_i, x_j\} \quad (1 \leqslant i, j \leqslant n)$;
(iii) $f(x_1, \ldots, x_n) = x_i - x_j \quad (1 \leqslant i, j \leqslant n)$;
(iv) $f(x_1, \ldots, x_n) = h(g_1(x_1, \ldots, x_n), \ldots, g_k(x_1, \ldots, x_n))$, where h, $g_1, \ldots, g_k$ are all p.r.;
(v) $f(y, x_1, \ldots, x_n) = \bigcup_{z \in y} g(z, x_1, \ldots, x_n)$, where g is p.r.;
(vi) $f(x_1, \ldots, x_n) = \omega$;
(vii) $f(y, x_1, \ldots, x_n) = g(y, x_1, \ldots, x_n, (f(z, x_1, \ldots, x_n) \mid z \in h(y)))$, where g and h are p.r. and where

$$z \in h(y) \to \operatorname{rank}(z) < \operatorname{rank}(y).$$

(Functions generated by schemas (i) through (v) are said to be *rudimentary*, and play a basic role in our later work on constructibility theory.)

1 A. Show that the following functions are p.r.:

$$f(x_1, \ldots, x_n) = \bigcup x_i \quad (1 \leqslant i \leqslant n);$$
$$f(x_1, \ldots, x_n) = x_i \cup x_j \quad (1 \leqslant i, j \leqslant n);$$
$$f(x_1, \ldots, x_n) = \{x_1, \ldots, x_n\};$$
$$f(x_1, \ldots, x_n) = (x_1, \ldots, x_n);$$
$$f(x_1, \ldots, x_n) = \emptyset.$$

1 B. Show that if the function $f(y, x_1, \ldots, x_n)$ is p.r., so too is the function $g(y, x_1, \ldots, x_n) = (f(z, x_1, \ldots, x_n) \mid z \in y)$.

A relation $R \subseteq V^n$ is said to be *primitive recursive* (p.r.) iff there is a p.r. function $f\colon V^n \to V$ such that

$$R = \{(x_1, \ldots, x_n) \mid f(x_1, \ldots, x_n) \neq \emptyset\}.$$

1 C. Prove the following:

(i) If f and R are p.r., so is

$$g(x_1, \ldots, x_n) = \begin{cases} f(x_1, \ldots, x_n), & \text{if } R(x_1, \ldots, x_n) \\ \emptyset, & \text{if } \neg R(x_1, \ldots, x_n). \end{cases}$$

(ii) R is p.r. iff χ_R (the characteristic function of R) is p.r.

(iii) R is p.r. iff $\neg R$ is p.r.

(iv) Let $f_i\colon V^n \to V$ be p.r. for $i = 1, \ldots, m$. Let $R_i \subseteq V^n$ be p.r. for $i = 1, \ldots, m$, such that $R_i \cap R_j = \emptyset$ for $i \neq j$ and $\bigcup_{i=1}^{m} R_i = V^n$. Define $f\colon V^n \to V$ by

$$f(x_1, \ldots, x_n) = f_i(x_1, \ldots, x_n) \quad \text{iff } R_i(x_1, \ldots, x_n).$$

Then f is p.r.

(v) If $R(y, x_1, \ldots, x_n)$ is p.r., so too is

$$f(y, x_1, \ldots, x_n) = \{z \in y \mid R(z, x_1, \ldots, x_n)\}.$$

(vi) Let $R(y, x_1, \ldots, x_n)$ be p.r. and such that

$$(\forall x_1 \ldots x_n)(\exists!\, y)\, R(y, x_1, \ldots, x_n).$$

Define f by

$$f(y, x_1, \ldots, x_n) = \begin{cases} \text{that } z \in y \text{ such that } R(z, x_1, \ldots, x_n), \text{ if such} \\ \text{a } z \text{ exists,} \\ \emptyset, \text{ if no such } z \text{ exists.} \end{cases}$$

Then f is p.r.

(vii) If $R(y, x_1, \ldots, x_n)$ is p.r., so too is $(\exists z \in y)\, R(z, x_1, \ldots, x_n)$.

(viii) If $R_i \subseteq V^n$ are p.r. for $i = 1, \ldots, m$, so too are $\bigcup_{i=1}^{m} R_i$ and $\bigcap_{i=1}^{m} R_i$.

(ix) The functions $(x)_0$, $(x)_1$, $\operatorname{dom}(x)$, $\operatorname{ran}(x)$ are p.r.

(x) the relations $x = y$ and $x \in y$ are p.r.

1 D. Show that if $f\colon V^n \to V$ is p.r., then there is a Σ_1 formula Φ of LST such that

$$y = f(x_1, \ldots, x_n) \leftrightarrow \Phi(y, x_1, \ldots, x_n).$$

1 E. Show that the ordinal functions $\alpha + 1$, $\alpha + \beta$, $\alpha \cdot \beta$, α^β are p.r.

1 F. Let $f(y, x_1, \ldots, x_n)$ be p.r. By recursion, define functions f^ν, $\nu \in \mathrm{On}$, by:

$$f^0(y, x_1, \ldots, x_n) = y;$$
$$f^{\nu+1}(y, x_1, \ldots, x_n) = f(f^\nu(y, x_1, \ldots, x_n), x_1, \ldots, x_n);$$
$$f^\lambda(y, x_1, \ldots, x_n) = \bigcup_{\nu<\lambda} f^\nu(y, x_1, \ldots, x_n), \quad \text{if } \lim(\lambda).$$

Let g be defined by

$$g(\nu, y, x_1, \ldots, x_n) = f^\nu(y, x_1, \ldots, x_n).$$

Show that g is p.r.

1 G. Show that the transitive closure function, TC, is p.r.

1 H. Show that any predicate defined by a Σ_0 formula of LST is p.r. (Hint: By induction on formulas, using 1 C (iii), (vii), and (viii).)

1 I. Show that the following functions are p.r.:

(i) $f(u) = \{x \mid \mathrm{Const}(x, u)\}$;

(ii) $f(u) = \{x\ \mathrm{Vbl}(x)\}$;

(iii) $f(u) = \{x \mid \mathrm{PFml}(x, u)\}$;

(iv) $f(u) = \{x \mid \mathrm{Fml}(x, u)\}$;

(v) $f(x) = \begin{cases} \text{the set of free variables of } x, & \text{if } \mathrm{Fml}(x), \\ \emptyset, & \text{if } \neg\, \mathrm{Fml}(x); \end{cases}$

(vi) $f(x, y, z) = \begin{cases} \text{that } x' \text{ such that } \mathrm{Sub}(x', x, y, z), \\ \qquad \text{if } \mathrm{Fml}(x) \wedge \mathrm{Vbl}(y) \wedge \mathrm{Const}(z), \\ \emptyset, \quad \text{otherwise}; \end{cases}$

(vii) $f(u) = \{x \mid \mathrm{Sat}(u, x)\}$;

(viii) $f(u) = \mathrm{Def}(u)$.

1 J. Show that the function $(L_\nu \mid \nu \in \mathrm{On})$ is p.r.

2. *Relative Constructibility* (Section 2)

Given some set A, we define a class $L[A]$ which has many of the nice properties of L, but in which the set A is, to some extent, available. The class $L[A]$ is called the *universe of sets constructible relative to* A, and is defined by analogy with the definition of L.

If X is a set, $\mathrm{Def}^A(X)$ denotes the set of all subsets of X which are definable in the structure $\langle X, \in, A \cap X\rangle$ by means of a formula of $\mathscr{L}_X(\mathring{A})$ having one free variable. (The language $\mathscr{L}_V(\mathring{A})$ was discussed briefly at the end of I.9 and the beginning of I.10.) The *hierarchy of sets constructible relative to A* is defined by the following recursion:

$$L_0[A] = \emptyset; \quad L_{\alpha+1}[A] = \mathrm{Def}^A(L_\alpha[A]);$$
$$L_\lambda[A] = \bigcup_{\alpha<\lambda} L_\alpha[A], \quad \text{if } \lim(\lambda).$$

The class $L[A]$ is then defined thus:

$$L[A] = \bigcup_{\alpha \in \mathrm{On}} L_\alpha[A].$$

2A. Prove the following analogues of Lemma 1.1:

(i) $\alpha \leqslant \beta$ implies $L_\alpha[A] \subseteq L_\beta[A]$;
(ii) $L_\alpha[A] \subseteq V_\alpha$ for all α;
(iii) Each $L_\alpha[A]$ is transitive, and (hence) $L[A]$ is transitive;
(iv) $\alpha < \beta$ implies α, $L_\alpha[A] \in L_\beta[A]$;
(v) $L[A] \cap \alpha = L_\alpha[A] \cap \alpha = L_\alpha[A] \cap \mathrm{On} = \alpha$;
(vi) for $\alpha \leqslant \omega$, $L_\alpha[A] = V_\alpha$;
(vii) for $\alpha \geqslant \omega$, $|L_\alpha[A]| = |\alpha|$.

2B. Prove that $L[A]$ is an inner model of ZF (in the sense of 1.2).

2C. A structure of the form $\langle M, \in, A\rangle$ is said to be *amenable* iff M is an amenable set and $A \cap u \in M$ for all $u \in M$. (This notion was introduced in Section 7). Prove that for any limit ordinal $\alpha > \omega$, the structure $\langle L_\alpha[A], \in, A \cap L_\alpha[A]\rangle$ is amenable.

Now, the intuition behind the construction of $L[A]$ is that the predicate "$x \in A$" should be available. Consequently, it is common practice to abbrebriate by $L_\alpha[A]$ the structure $\langle L_\alpha[A], \in, A \cap L_\alpha[A]\rangle$, just as we used L_α to mean $\langle L_\alpha, \in\rangle$. In particular, to say that $L_\alpha[A]$ is amenable means that the structure $\langle L_\alpha[A], \in, A \cap L_\alpha[A]\rangle$ is amenable, as defined above.

2D. Show that there is a Δ_1^{KP} formula $D(v, u, a)$ of LST such that

$$D(v, u, a) \quad \text{iff } v = \mathrm{Def}^a(u).$$

2E. Show that the function Def^A is uniformly $\Delta_1^{L_\alpha[A]}$ for all limit $\alpha > \omega$.

2F. Show that there is a Δ_1^{KP} formula $H(x, \alpha, a)$ of LST such that

$$H(x, \alpha, a) \quad \text{iff } x - L_\alpha[a].$$

2G. Show that the function $v \mapsto L_v[A]$ is uniformly $\Delta_1^{L_\alpha[A]}$ for limit $\alpha > \omega$.

2H. Show that if M is an admissible set or else an inner model of KP, and if $a \in M$, then for any $\alpha \in M$, $L_\alpha[a] \in M$ and $(L_\alpha[a])^M = L_\alpha[a]$.

2I. Show that if $\alpha > \omega$ is a limit ordinal, then for any $v < \alpha$,

$$L_\gamma[A] = (L_v[B])^{L_\alpha[A]},$$

where $B = A \cap L_v[A]$. (By 2C, $B \in L_\alpha[A]$.)

2J. Prove that if $\alpha > \omega$ is a limit ordinal and $B = A \cap L_\alpha[A]$, then

$$L_\alpha[A] = L_\alpha[B].$$

2 K. Prove that if $B = A \cap L[A]$, then

$$L[A] = L[B] = (L[B])^{L[A]}.$$

Deduce that

$$(V = L[B])^{L[A]}.$$

2 L. Show that there is a Σ_1 formula $\mathrm{WO}(x, y, a)$ of LST such that

$$\mathrm{KP} \vdash \text{“}\{(x, y) \mid \mathrm{WO}(x, y, a)\} \text{ is a well-ordering of } L[a]\text{”},$$

and such that if $<_{L[A]}$ denotes the well-ordering of $L[A]$ determined by WO, then for any limit ordinal $\alpha > \omega$, $<_{L[A]} \cap (L_\alpha[A])^2$ is $\Sigma_1^{L_\alpha[A]}$.

2 M. Prove $(\mathrm{AC})^{L[A]}$.

2 N. Prove that $L[A]$ is the smallest inner model of ZF which contains the set $A \cap L[A]$. (i.e. $L[A]$ is the smallest inner model M of ZF such that $A \cap M \in M$.)

3. *Use of the Condensation Lemma* (Section 5)

We investigate the question: as α varies over all limit ordinals, how many different sets of $\mathscr{L}$-sentences are theories of some L_α?

3 A. Let Σ be the set of all sets of $\mathscr{L}$-sentences of the form

$$\{\varphi \mid \models_{L_\alpha} \varphi\}$$

for some limit ordinal α. Show that

$$|\Sigma| \leqslant |\omega_1^L|.$$

(Is it also the case that

$$|\Sigma|^L \leqslant |\omega_1^L|^L?).$$

3 B. Let $(\varphi_n \mid n < \omega)$ be the “lexicographic” enumeration of the sentences of $\mathscr{L}$, as described in Section 3. Show that there is no formula $\varphi(v_0)$ of $\mathscr{L}$ such that

$$\models_{L_\alpha} \varphi_n \quad \text{iff } \models_{L_\alpha} \varphi(\mathring{n}).$$

(Hint: Diagonalisation. Let $(\psi_n \mid n < \omega)$ be the lexicographic enumeration of the formulas of $\mathscr{L}$ with free variable at most v_0. Consider the formula

$$\text{“}v_0 \text{ is a natural number”} \wedge \exists k [\text{“}k \text{ is a natural number”} \wedge (\varphi_k = {}^*\psi_{v_0}(v_0)^*) \wedge \neg\varphi(k)],$$

where ${}^*\psi_m(n)^*$ denotes the formula obtained from $\psi_m(v_0)$ by replacing every free occurrence of v_0 by the term denoting the integer n.)

3 C. Show that $|\Sigma| = |\omega_1^L|$. (Hint: First reduce this to proving that $|\Sigma|^L > \omega$. Then suppose that $|\Sigma|^L = \omega$ and let T be the $<_L$-least subset of $\omega \times \omega$ such that $(T''\{n\} \mid n < \omega)$ enumerates all members of Σ via the enumeration $(\varphi_n \mid n < \omega)$ of the $\mathscr{L}$-sentences in 3 B above. Now look at the $\mathscr{L}$-theory of $L_{\omega_1^L}$ and work for a contradiction with 3 B.)

An alternative solution to the original question can be obtained by exhibiting an unbounded set $A \subseteq \omega_1^L$ such that whenever $\alpha, \beta \in A$ and $\alpha \neq \beta$, then the theories of L_α and L_β are different. This can be done as follows.

3 D. Set

$$A = \{\alpha \in \omega_1^L \mid \lim(\alpha) \wedge \text{every element of } L_\alpha \text{ is definable (without the use of parameters) in } L_\alpha\}.$$

Show that A is unbounded in ω_1^L. (Hint: To show that A is non-empty, use 5.3 and 5.10. To show that A is unbounded in ω_1^L, suppose otherwise and consider $\lambda = \sup(A)$.)

3 E. Show that if $\alpha, \beta \in A$ and $\alpha \neq \beta$, then L_α and L_β have different theories. (Hint: Use 3 B again.)

4. *The Condensation Lemma and the* GCH *in* $L[A]$ (Section 5)

We continue the investigation of $L[A]$ commenced in Exercises 2 above.

4 A. Prove that if $\alpha > \omega$ is a limit ordinal and $X \prec_1 L_\alpha[A]$, there are unique π, β such that

$$\pi: X \cong L_\beta[B],$$

where $B = \pi''(A \cap X)$.

4 B. Show that if $A \in L_\varrho[A]$, $\alpha > \varrho$, $\alpha > \omega$, α a limit ordinal, and if

$$L_\varrho[A] \subseteq X \prec_1 L_\alpha[A],$$

then there are unique π, β such that

$$\pi: X \cong L_\beta[A].$$

As we saw in 4 A above, the "condensation lemma" for $L[A]$ does not in general lead to a structure in the $L[A]$ hierarchy. Thus we cannot prove GCH in $L[A]$ as we did for L. Indeed, if κ were a cardinal such that $2^\kappa = \kappa^{++}$, we could let $A \subseteq \kappa^{++}$ code all subsets of κ, so $2^\kappa \geqslant \kappa^{++}$ would hold in $L[A]$. However, 4 B enables us to obtain a partial GCH result.

4 C. Prove that if $V = L[A]$, where A is a subset of an infinite cardinal κ, then $2^\lambda = \lambda^+$ for all cardinals $\lambda \geqslant \kappa$.

A strengthening of 4 C is possible. We require a preliminary result.

4 D. Let κ be an uncountable regular cardinal, and let $\mathbf{M} = \langle M, \in, \dots \rangle$ be a structure such that $\kappa \subseteq M$. Let $X \subseteq M$, $|X| < \kappa$. Prove that there is a structure $\mathbf{N} \prec \mathbf{M}$ such that $X \subseteq N$ and $N \cap \kappa \in \kappa$. (Hint: Construct $\mathbf{N}$ as the union of a suitably chosen ω-sequence of submodels of $\mathbf{M}$.)

4 E. Let $V = L[A]$, where $A \subseteq \kappa^+$. Then $2^\kappa = \kappa^+$. (Hint: Use 4 D to prove a special case of 4 A, and note that if $\gamma, \delta < \kappa^+$, then

$$L_\gamma[A \cap \delta] \in L_{\kappa^+}[A].)$$

4 F. Show that if $V = L[A]$, where $A \subseteq \omega_1$, then GCH is valid.

5. Σ_n *Skolem Functions* (Section 6)

We show that there is no uniform Σ_2 skolem function for L_α, where $\lim(\alpha)$, $\alpha > \omega$. (It can be shown that each limit L_α does possess a Σ_2 skolem function, and indeed a Σ_n skolem function for any n, but the Σ_n definitions are not uniform for $n \geqslant 2$. See Chapter VI for details.)

It is convenient to assume $V = L$ throughout. We use α to denote an arbitrary countable limit ordinal.

5 A. Show that the predicate

$$x \leqslant \alpha$$

is uniformly $\Pi_2^{L_{\omega_1+\alpha}}(\{\omega_1\})$.

5 B. Show that the predicate

$$x = \omega_1$$

is uniformly $\Pi_1^{L_{\omega_1+\alpha}}$.

5 C. Show that the predicate

$$x > \alpha$$

is uniformly $\Sigma_2^{L_{\omega_1+\alpha}}$.

5 D. Show that the predicate

$$P_\alpha(x)\colon \lim(x) \wedge (\alpha < x < \omega_1)$$

is uniformly $\Sigma_2^{L_{\omega_1+\alpha}}$.

5 E. Suppose that there were a uniform Σ_2 skolem function h_γ for L_γ, where $\gamma > \omega$ is a limit ordinal. Let φ be a Σ_0 formula of $\mathscr{L}$ such that for any limit ordinal $\gamma > \omega$,

$$y = h_\gamma(i, x) \quad \text{iff } \models_{L_\gamma} \exists u \forall v \varphi(\mathring{y}, \mathring{i}, \mathring{x}, u, v).$$

Show that for each α there is an integer i_α such that

$$P_\alpha(h_{\omega_1+\alpha}(i_\alpha, \emptyset)),$$

and deduce that for a stationary set $A \subseteq \omega_1$ there is an integer i such that for any $\alpha \in A$,

$$P_\alpha(h_{\omega_1+\alpha}(i, \emptyset)).$$

5 F. Define a sequence $(\alpha_\nu \mid \nu < \omega_1)$ thus:

$$\alpha_0 = 0;$$
$$\alpha_{\nu+1} = h_{\omega_1+\alpha_\nu}(i, \emptyset);$$
$$\alpha_\nu = \sup_{\tau<\nu} \alpha_\tau, \quad \text{if } \lim(\nu).$$

Show that $(\alpha_\nu \mid \nu < \omega_1)$ is a strictly increasing, continuous sequence of countable limit ordinals.

5 G. Pick a limit ordinal ν such that $\alpha_\nu \in A$ and for arbitrarily large $\tau < \nu$, $\alpha_\tau \in A$. Let $\gamma = \alpha_{\nu+1}$. By considering φ, show that there is a $\tau < \nu$ such that $\alpha_\tau \in A$ and $\gamma = \alpha_{\tau+1}$, and deduce that there can be no uniform Σ_2 skolem function for limit L_α.

Chapter III
ω_1-Trees in L

Tree theory forms a rich and interesting part of combinatorial set theory, having applications in other parts of set theory as well as in other areas of mathematics (in particular, in general topology). We study trees here because tree theory is greatly enhanced by the assumption $V = L$, and affords a good example of the application of the methods of constructibility theory. In this chapter we concentrate on ω_1-trees, and as we shall demonstrate, these arise out of some very basic questions in mathematics. Later chapters deal with generalisations to higher cardinals.

1. The Souslin Problem. ω_1-Trees. Aronszajn Trees

The Souslin Problem has its origin in a classical theorem of Cantor concerning the real line. In order to consider this theorem we need some definitions.

A *densely ordered set* is a linearly ordered set $\langle X, \leqslant \rangle$ such that whenever $x, y \in X$ and $x < y$, there is a $z \in X$ such that $x < z < y$.

An *interval* in a linearly ordered set $\langle X, \leqslant \rangle$ is a subset of X of the form

$$(x, y) = \{z \in X \mid x < z < y\}$$

for some $x, y \in X$, $x < y$. (We call this set the interval *determined by* x and y.)

An *ordered continuum* is a densely ordered set $\langle X, \leqslant \rangle$ such that whenever Y is a subset of an interval of X, there is a least $z \in X$ such that $(\forall y \in Y)(y \leqslant z)$ and a greatest $x \in X$ such that $(\forall y \in Y)(x \leqslant y)$. (We call z the *supremum* of Y, x the *infimum* of Y.)

A linearly ordered set is said to be *open* if it has no end-points.

A subset Y of a densely ordered set $\langle X, \leqslant \rangle$ is said to be *dense* in X if, whenever $x, z \in X$ are such that $x < z$, there is a $y \in Y$ such that $x < y < z$.

Cantor proved that, considered as a linearly ordered set, the real line ($\mathbb{R}$) is characterised, up to isomorphism, by being an open, ordered continuum having a countable dense subset (the rationals). In 1920, M. Souslin asked whether a natural weakening of these conditions still suffices to characterise $\mathbb{R}$.

Let us say that a linearly ordered set X has the *Souslin Property* if every set of pairwise disjoint, non-empty intervals of X is countable. (This condition is often referred to as the "countable chain condition".) Clearly, if a densely ordered set X has a countable dense subset Y, it must have the Souslin Property, since any non-empty interval of X must contain an element of Y. The question Souslin raised was this: Is it the case that $\mathbb{R}$ is characterised by being an open, ordered continuum having the Souslin Property? Although Souslin did not publish any indication that he thought a positive answer was likely, it has become common to refer to a positive answer as *The Souslin Hypothesis.*

We now know that the Souslin Problem cannot be solved in ZFC set theory, even if we assume GCH. We shall show that if we assume $V = L$, however, then the problem can be solved, with Souslin's Hypothesis being false.

We shall solve the Souslin Problem (assuming $V = L$) by first reformulating it in terms of trees. But before we do that, let us notice that the Souslin Hypothesis is equivalent (in ZFC) to the following assertion:

> *Every densely ordered set with the Souslin Property has a countable dense subset.*

(We shall denote this last assertion by *SH*.) The proof (of equivalence) in one direction is immediate. Assuming *SH*, if we are given an open, ordered continuum having the Souslin Property, then by *SH* it will have a countable dense subset, and so by Cantor's theorem it will be isomorphic to $\mathbb{R}$. For the proof in the other direction, suppose we are given a densely ordered set, X, with the Souslin Property. Let X' be obtained from X by introducing a copy of the rationals at each end (to obtain an *open* ordered set). Let X'' be the Dedekind completion of X'. It is easily seen that X'' is an open, ordered continuum with the Souslin Property. By the Souslin Hypothesis (as formulated by Souslin), X'' is isomorphic to $\mathbb{R}$. Hence X is isomorphic to a dense subset of an interval of $\mathbb{R}$. Thus X has a countable dense subset.

We shall prove that if $V = L$ then *SH* is false, by using $V = L$ to construct a densely ordered set having the Souslin Property but no countable dense subset. We achieve this by way of trees.

A *tree* is a partially ordered set $\mathbf{T} = \langle T, \leqslant_T \rangle$ such that for every $x \in T$, the set

$$\hat{x} = \{ y \in T \mid y <_T x \}$$

is well-ordered by $\leqslant_T$.

The order-type of the set $\hat{x}$ under $<_T$ is called the *height* of x in $\mathbf{T}$, denoted by $ht_{\mathbf{T}}(x)$.

If α is an ordinal, the α-th *level* of $\mathbf{T}$ is the set

$$T_\alpha = \{ x \in T \mid ht_{\mathbf{T}}(x) = \alpha \}.$$

We often write $T \restriction \alpha$ to denote the set $\bigcup_{\beta < \alpha} T_\beta$, and $\mathbf{T} \restriction \alpha$ for the restriction of the structure $\mathbf{T}$ to this set.

Sometimes we blur the distinction between a tree and its underlying set, writing T instead of **T**, etc.

In a tree **T**, if we are at any point x, there is only one path "downwards", namely $\hat{x}$, though there may be several (or none) paths "upwards" from x. It is customary to represent trees pictorally as in Figure 1, using vertical connecting lines to denote the ordering $<_T$ in the upward direction, drawing the levels of the tree on a horizontal line.

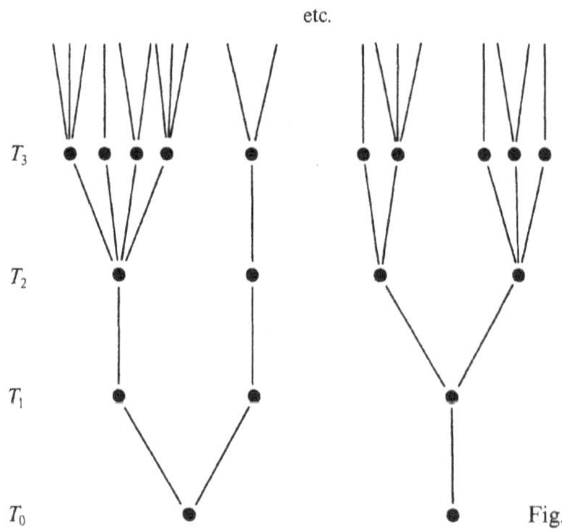

Fig. 1

Let **T** be a tree. A linearly ordered subset b of T with the property that whenever $x \in b$, then $y <_T x$ implies $y \in b$, is called a *branch* of **T**. If α is the order-type of b under $<_T$, we say that b is an *α-branch*. A branch is *maximal* if it is not properly contained in any other branch of **T**. By the Axiom of Choice, every branch can be extended to a maximal branch. Every set $\hat{x}$ is a branch of **T**. If x has no *successors* in **T** (i.e. there are no points $y \in T$ such that $x <_T y$), then $\hat{x} \cup \{x\}$ is a maximal branch of **T**.

An *antichain* of **T** is a subset of T, no two elements of which are comparable under the ordering $<_T$. An antichain is *maximal* if it is not properly contained in any other antichain of **T** (or, equivalently, iff every point of **T** is comparable with some member of the antichain under $<_T$). By the Axiom of Choice, every antichain of **T** can be extended to a maximal antichain. If $T_\alpha \neq \emptyset$, then T_α is a maximal antichain of **T**.

Let θ be an ordinal, λ a cardinal. A tree **T** is said to be a (θ, λ)-*tree* iff:

(i) $(\forall \alpha < \theta)(T_\alpha \neq \emptyset)$;
(ii) $T_\theta = \emptyset$;
(iii) $(\forall \alpha < \theta)(|T_\alpha| < \lambda)$.

In words, a (θ, λ)-tree is one of "height" θ and "width" less than λ. (We demand $|T_\alpha| < \lambda$ rather than $|T_\alpha| \leqslant \lambda$ in (iii) to allow for the case where λ is a limit cardinal.)

A tree $\mathbf{T}$ is said to have *unique limits* if, whenever α is a limit ordinal and $x, y \in T_\alpha$, if $\hat{x} = \hat{y}$ then $x = y$.

A (θ, λ)-tree $\mathbf{T}$ is said to be *normal* if $\mathbf{T}$ has unique limits and each of the following conditions is satisfied:

(i) $|T_0| = 1$;

(ii) if $\alpha, \alpha + 1 < \theta$ and $x \in T_\alpha$, then there are distinct $y_1, y_2 \in T_{\alpha+1}$ such that $x <_T y_1$ and $x <_T y_2$;

(iii) if $\alpha < \beta < \theta$ and $x \in T_\alpha$, there is a $y \in T_\beta$ such that $x <_T y$.

Let κ be an infinite cardinal. A *κ-tree* is a normal (κ, κ)-tree.

It is trivial to show that every ω-tree has an ω-branch. (By recursion, pick $x_n \in T_n$ so that $x_n <_T x_{n+1}$.) And it is tempting to imagine that this simple result generalises to ω_1-trees. However, as was first demonstrated by N. Aronszajn, there are ω_1-trees having no ω_1-branch. Such trees are now known as *Aronszajn trees.*

1.1 Theorem. *There is an Aronszajn tree (i.e. an ω_1-tree with no ω_1-branch).*

Proof. By recursion on the levels, we construct an ω_1-tree $\mathbf{T}$. The elements of T_α will be strictly increasing α-sequences of rational numbers, and the tree ordering will be $x <_T y$ iff x is an initial segment of y (i.e. iff $x \subset y$). Notice that if b were an ω_1-branch of such a tree, $\bigcup b$ would be a strictly increasing ω_1-sequence of rationals, which is impossible. Hence our tree certainly can have no ω_1-branches, and our problem is simply to construct the tree. In order to do this, we ensure that at each stage in the construction, $\mathbf{T} \restriction \alpha$ satisfies the following condition:

> $P(\alpha)$: $\mathbf{T} \restriction \alpha$ is a normal (α, ω_1)-tree, and for every $\beta < \gamma < \alpha$ and every $x \in T_\beta$ and every rational $q > \sup(x)$, there is a $y \in T_\gamma$ such that $x \subset y$ and $q > \sup(y)$,

where $\sup(x)$ here denotes the supremum (in the reals) of the range of values of the rational sequence x.

To commence the construction, we set

$$T_0 = \{\emptyset\}.$$

If $\mathbf{T} \restriction (\alpha + 1)$ is defined and satisfies $P(\alpha + 1)$, we define

$$T_{\alpha+1} = \{x \frown \langle q \rangle \mid x \in T_\alpha \wedge q \in \mathbb{Q} \wedge q > \sup(x)\}.$$

Clearly, $\mathbf{T} \restriction (\alpha + 2)$ then satisfies $P(\alpha + 2)$.

Finally, suppose α is a limit ordinal and $\mathbf{T} \restriction \alpha$ has been defined and satisfies $P(\alpha)$. (Notice that if $P(\beta)$ is valid for $\mathbf{T} \restriction \beta$ for all $\beta < \alpha$ then $P(\alpha)$ is automatically valid for $\mathbf{T} \restriction \alpha$.) The construction of T_α depends upon the following claim.

Claim. For each $x \in T \restriction \alpha$ and each rational $q > \sup(x)$, there is an α-branch b of $\mathbf{T} \restriction \alpha$ such that $x \in b$ and $\sup(\bigcup b) \leqslant q$.

To prove the claim, given x, q as above, pick a strictly increasing ω-sequence $(\alpha_n \mid n < \omega)$ of ordinals, cofinal in α, so that $x \in T \restriction \alpha_0$. Since $P(\alpha)$ is valid, we can inductively pick elements $y_n \in T_{\alpha_n}$ so that $x \subset y_0 \subset y_1 \subset y_2 \subset \ldots$ and $\sup(y_n) < q$. Set

$$b = \{y \in T \restriction \alpha \mid (\exists n < \omega)(y \subset y_n)\}.$$

Celarly, b is an α-branch of $\mathbf{T} \restriction \alpha$ which contains x and is such that $\sup(\bigcup b) \leqslant q$, proving the claim.

Using the claim, we construct T_α as follows. For each $x \in T \restriction \alpha$ and each rational $q > \sup(x)$, pick one α-branch $b(x, q)$ of $\mathbf{T} \restriction \alpha$ as in the claim, and set

$$T_\alpha = \{\bigcup b(x, q) \mid x \in T \restriction \alpha \wedge q \in \mathbb{Q} \wedge q > \sup(x)\}.$$

It is easily seen that $\mathbf{T} \restriction (\alpha + 1)$ satisfies $P(\alpha + 1)$. In particular, T_α is countable because both $T \restriction \alpha$ and $\mathbb{Q}$ are countable.

That completes the construction of $\mathbf{T}$. Since $\mathbf{T} \restriction \alpha$ satisfies $P(\alpha)$ for all $\alpha < \omega_1$, $\mathbf{T}$ is an ω_1-tree, and so we are done. □

Related to the notion of an Aronszajn tree is that of a *Souslin tree*. This is defined to be an ω_1-tree having no uncountable antichain. (We shall see later that Souslin trees are closely connected with the Souslin Problem.) As the following result shows, Souslin trees are just special kinds of Aronszajn trees.

1.2 Theorem. *Every Souslin tree is an Aronszajn tree.*

Proof. Let $\mathbf{T}$ be a Souslin tree. Let b be any branch of $\mathbf{T}$. We show that b must be countable. Since $\mathbf{T}$ is normal, for each $x \in b$ we can pick an element $x^* \in T$ such that $x <_T x^*$, $ht(x^*) = ht(x) + 1$, and $x^* \notin b$. It is easily seen that $\{x^* \mid x \in b\}$ is an antichain of $\mathbf{T}$. But if $x, y \in b$ are such that $x \neq y$, then $x^* \neq y^*$. So as $\mathbf{T}$ has no uncountable antichain, b must be countable. □

The above proof made use of the normality requirements on a Souslin tree. These are rather strong conditions, since they tend to point in the opposite direction to the Aronszajn and Souslin requirements of no uncountable branches or antichains. In the case of Aronszajn trees, the somewhat "paradoxical" situation arose (in 1.1) that essential use was made of normality requirements in order to construct an Aronszajn tree. But in the case of Souslin trees, the full normality requirements turn out to be a burden as far as construction of such trees in connection with the Souslin Problem is concerned. The next lemma shows that this burden is easily shed.

1.3 Lemma. (i) *Let* $\mathbf{T}$ *be an* (ω_1, ω_1)*-tree with unique limits, having no uncountable branch. Then there is a subset* T^* *of* T *such that, under the induced ordering,* T^* *is an Aronszajn tree.*

(ii) *Let* $\mathbf{T}$ *be an* (ω_1, ω_1)*-tree with unique limits, having no uncountable branch and no uncountable antichain. Then there is a subset* T^* *of* T *such that, under the induced ordering,* T^* *is a Souslin tree.*

Proof. (i) Since T_0 is countable, we can find an element x_0 of T_0 such that T' is uncountable, where we set

$$T' = \{x \in T \mid x_0 \leqslant_T x\}.$$

Let T'' be the set of all members of T' which have extensions on all higher levels of T'. It is easily seen that each member of T'' has extensions on all higher levels of T'' itself. It follows that for every point $x \in T''$ there are points $y, z \in T''$ such that $x <_T y$, $x <_T z$, and y and z are incomparable in **T**. (Otherwise the extensions of x would form an uncountable branch of **T**.) Hence we can define a function $f\colon \omega_1 \to \omega_1$ by the following recursion:

$$\begin{aligned} f(0) &= 0; \\ f(\alpha+1) &= \text{the least } \beta > f(\alpha) \text{ such that for all } x \in T''_{f(\alpha)} \text{ there are } y, z \in T''_\beta \\ &\quad\ \text{such that } x <_T y,\ x <_T z, \text{ and } y \neq z; \\ f(\lambda) &= \sup\nolimits_{\nu<\lambda} f(\nu), \quad \text{if } \lim(\lambda). \end{aligned}$$

Set

$$T^* = \bigcup_{\alpha<\omega_1} T''_{f(\alpha)}.$$

It is easily checked that T^* is as required.

(ii) The above proof works in this case also. □

Notice that unique limits played no role in the above proof. We could have omitted this requirement from all definitions and results, but it is common to include it, and we shall always do so.

Our next result indicates our usage of the phrase "Souslin tree".

1.4 Theorem. *Souslin's Hypothesis is equivalent to the non-existence of a Souslin tree.*

Proof. Assume first that there is a Souslin tree. We construct a counterexample to *SH*, i.e. a densely ordered set having the Souslin Property but no countable dense subset.

Let **T** be a Souslin tree. By replacing **T** by its restriction to the limit levels of **T**, if necessary, we may assume that each member of **T** has infinitely many successors on the next level of **T**. For each non-zero $\alpha < \omega_1$, let $<_\alpha$ be a linear ordering of T_α, isomorphic to the rationals, so that the set of all successors on $T_{\alpha+1}$ of any element of T_α is ordered as the rationals by $<_{\alpha+1}$. Let X be the set of all maximal branches of **T**, and define a linear ordering on X by setting $b <_X d$ iff $b(\alpha) <_\alpha d(\alpha)$, where α is the least ordinal such that $b \cap T_\alpha \neq d \cap T_\alpha$ and $b(\alpha)$ denotes the unique element of $b \cap T_\alpha$, $d(\alpha)$ the unique element of $d \cap T_\alpha$. Clearly, $\langle X, \leqslant_X\rangle$ is a densely ordered set of cardinality 2^ω.

We show that X has the Souslin Property. Let I be any interval of X, say $I = (b, d)$. Choose α minimal so that $b(\alpha) \neq d(\alpha)$. Pick $x_I \in T_\alpha$ so that $b(\alpha) <_\alpha x_I <_\alpha d(\alpha)$. Let $e(I)$ be a maximal branch of **T** containing x_I. Thus $e(I) \in I$. Suppose now that I and J are disjoint intervals of X. Then $e(I) \notin J$ and $e(J) \notin I$, so x_I and x_J must be incomparable in **T**. Since **T** has no uncountable antichains, it follows that any pairwise disjoint collection of intervals of X must be countable.

We complete the proof of this half of the theorem by showing that X has no countable dense subset. Let A be any countable subset of X. For each pair b, d of distinct elements of A, let $\alpha(b, d)$ be the least ordinal α such that $b(\alpha) \neq d(\alpha)$. Let

$$\gamma = \sup\{\alpha(b, d) \mid b, d \in A \ \& \ b \neq d\}.$$

Since A is countable, $\gamma < \omega_1$. Let $w \in T_\gamma$ and choose $x, y, z \in T_{\gamma+1}$ so that $w <_T x, y, z$ and $x <_{\gamma+1} y <_{\gamma+1} z$. Let b_x be a maximal branch of **T** containing x, and choose b_y, b_z similarly. If A were dense in X, we could find $d, d' \in A$ such that $b_x <_X d <_X b_y$ and $b_y <_X d' <_X b_z$. But since b_x, b_y, b_z all contain w, we would have $\alpha(d, d') > \gamma$, contrary to the choice of γ. Hence A cannot be dense in X.

Thus X is a counterexample to SH.

We now assume that SH is false and construct a tree satisfying the hypotheses of 1.3 (ii), which by virtue of 1.3 (ii) at once implies the existence of a Souslin tree.

By the failure of SH, let X be a densely ordered set with the Souslin Property but no countable dense subset. By recursion on the levels we define a *partition tree* $\mathbf{T} = \langle T, \supseteq \rangle$ of X, elements of which are non-empty "intervals" of X. To commence, we set $T_0 = \{X\}$.

Suppose we have defined T_α. For every $I \in T_\alpha$ of cardinality greater than 1, choose an interior point $x(I)$ of I. (Since X is densely ordered, if I has at least two elements, such a point always exists.) Let

$$I_0 = \{y \in I \mid y <_X x(I)\}$$
$$I_1 = \{y \in I \mid x(I) \leqslant_X y\}.$$

Set

$$T_{\alpha+1} = \{I_0 \mid I \in T_\alpha \wedge |I| > 1\} \cup \{I_1 \mid I \in T_\alpha \wedge |I| > 1\}.$$

Now suppose that $\lim(\alpha)$ and T_β has been defined for all $\beta < \alpha$. In this case, set

$$T_\alpha = \{\bigcap b \mid b \text{ is an } \alpha\text{-branch of } \mathbf{T} \restriction \alpha \text{ such that } |\bigcap b| > 1\}.$$

That defines **T**. Let θ be the least ordinal such that $T_\theta = \emptyset$. We shall show that $\theta = \omega_1$ and that **T** satisfies the hypotheses of 1.3 (ii). It is clear that T has unique limits. We show first that **T** has no uncountable branch (so that, in particular, $\theta \leqslant \omega_1$).

Suppose that B were an uncountable branch of T. Let $(I_\alpha \mid \alpha < \omega_1)$ be the canonical enumeration of the first ω_1 elements of B. Set

$$A_0 = \{\alpha < \omega_1 \mid (\forall y \in I_{\alpha+1})(y <_X x(I_\alpha))\},$$
$$A_1 = \{\alpha < \omega_1 \mid (\forall y \in I_{\alpha+1})(x(I_\alpha) \leqslant_X y)\}.$$

Thus A_0 and A_1 constitute a disjoint partition of ω_1. Hence at least one of A_0, A_1 is uncountable. Suppose, for the sake of argument, that A_0 were uncountable.

(The other case is handled similarly.) For $\alpha \in A_0$, let J_α be the X-interval

$$J_\alpha = (x(I_\beta), x(I_\alpha)),$$

where β is the least element of A_0 above α. Now, if $\alpha \in A_0$ and $\alpha < \beta$, we have $x(I_\beta) <_X x(I_\alpha)$. Hence $\{J_\alpha \mid \alpha \in A_0\}$ is an uncountable set of pairwise disjoint intervals of X, which is impossible. Thus **T** has no uncountable branch.

Moreover, **T** has no uncountable antichain. Essentially this is because incomparability in **T** means disjointness as "intervals" in X. For suppose $\{I_\alpha \mid \alpha < \omega_1\}$ were an uncountable antichain of **T**. Then for each $\alpha < \omega_1$ we could choose $x_\alpha, y_\alpha \in I_\alpha$, $x_\alpha <_X y_\alpha$, whence $\{(x_\alpha, y_\alpha) \mid \alpha < \omega_1\}$ would be an uncountable set of pairwise disjoint intervals of X, which is impossible.

Since **T** has no uncountable antichains, each level of **T** must be countable. If we can show that T is uncountable, we shall thus be able to conclude that **T** is an (ω_1, ω_1)-tree and be done. But it follows easily from the construction of **T** that the set $\{x(I) \mid I \in T\}$ is dense in X. (Roughly speaking, this is because we keep on "splitting" intervals of X until it is not possible to go any further.) So, as X has no countable dense subset, we see that T is indeed uncountable. □

1.4 enables us to prove that SH fails if we assume $V = L$.

1.5 Theorem. *Assume $V = L$. Then there is a Souslin tree.*

Proof. We construct an ω_1-tree, **T**, by recursion on the levels. The elements of T_α will be sequences from ${}^\alpha 2$, and the ordering of **T** will be sequence extension ($=$ set-theoretic inclusion). We carry out the construction so that at each stage $\alpha < \omega_1$, $\mathbf{T} \restriction \alpha$ is a normal (α, ω_1)-tree. This will ensure that **T** is an ω_1-tree, so the only problem will be to ensure that **T** has no uncountable antichains.

To commence, set

$$T_0 = \{\emptyset\}.$$

The definition of $T_{\alpha+1}$ is dictated by the normality requirements. If $\mathbf{T} \restriction \alpha + 1$ is defined, we set

$$T_{\alpha+1} = \{s^\frown\langle i\rangle \mid s \in T_\alpha \wedge i = 0, 1\}.$$

If $\mathbf{T} \restriction \alpha + 1$ is a normal $(\alpha + 1, \omega_1)$-tree, then $\mathbf{T} \restriction \alpha + 2$ is clearly a normal $(\alpha + 2, \omega_1)$-tree.

There remains the definition of T_α when α is a limit ordinal and $\mathbf{T} \restriction \alpha$ has been defined. Notice first that if $\mathbf{T} \restriction \beta$ is a normal (β, ω_1)-tree for all $\beta < \alpha$, $\mathbf{T} \restriction \alpha$ will be a normal (α, ω_1)-tree. Now, if $s \in {}^\alpha 2$ is to be a member of T_α, $\{s \restriction \beta \mid \beta < \alpha\}$ will have to be an α-branch of $\mathbf{T} \restriction \alpha$. Hence for some collection, B_α, of α-branches of $\mathbf{T} \restriction \alpha$ we shall have

$$T_\alpha = \{\bigcup b \mid b \in B_\alpha\}.$$

What properties must the set B_α have? Certainly it must be countable. And to preserve normality requirements, each element of $T \restriction \alpha$ must be a member of some branch in B_α. Since trees consisting of sequences as in this case necessarily have unique limits, these two conditions on B_α suffice to ensure that $\mathbf{T} \upharpoonright \alpha + 1$ will be a normal $(\alpha + 1, \omega_1)$-tree. So we are left with choosing B_α to ensure that $\mathbf{T}$ will be a Souslin tree. How can we do this? Well, the final tree, $\mathbf{T}$, will be a subset of $\bigcup_{\alpha<\omega_1} {}^{\alpha}2$ of cardinality ω_1. By GCH, the set T will have ω_2 many uncountable subsets. We must choose the collections B_α so that none of these uncountable subsets of T is an antichain of $\mathbf{T}$. To see how this might be achieved, suppose that in fact there were an uncountable antichain in $\mathbf{T}$. Then there would be a maximal uncountable antichain, A. For each $\alpha < \omega_1$, $A \cap (T \upharpoonright \alpha)$ is an antichain in $\mathbf{T} \upharpoonright \alpha$. Let

$$C = C_A = \{\alpha \in \omega_1 \mid \lim(\alpha) \wedge A \cap (T \upharpoonright \alpha) \text{ is a } \textit{maximal} \text{ antichain in } \mathbf{T} \upharpoonright \alpha\}.$$

The set C is club in ω_1. Closure is immediate, of course. To prove the unboundedness of C in ω_1, given $\alpha_0 < \omega_1$, define $\alpha_n < \omega_1$ recursively by setting α_{n+1} to be the least ordinal $\gamma > \alpha_n$ such that each element of $T \upharpoonright \alpha_n$ is comparable with some member of $A \cap (T \upharpoonright \gamma)$, in which case it is easily seen that $\alpha = \bigcup_{n<\omega} \alpha_n \in C$. Suppose now that we can somehow choose the sets B_α so that for each maximal uncountable antichain A of $\mathbf{T}$, there is an $\alpha \in C_A$ for which the definition of T_α prevents the addition of any elements to $\mathbf{T}$ which are incomparable with all of the elements of $A \cap (T \upharpoonright \alpha)$. This would then ensure that in fact there are *no* uncountable antichains in $\mathbf{T}$. (The above discussion would be a proof by contradiction of this fact.) Now, constructing T_α so that some specific maximal antichain $A \cap (T \upharpoonright \alpha)$ of $\mathbf{T} \upharpoonright \alpha$ does not "grow" in $\mathbf{T}$ (at any subsequent stage) is easy. Define B_α so that each element of B_α contains a member of $A \cap (T \upharpoonright \alpha)$. Since $A \cap (T \upharpoonright \alpha)$ is a *maximal* antichain in $\mathbf{T} \upharpoonright \alpha$, each element of $T \upharpoonright \alpha$ is comparable with some member of $A \cap (T \upharpoonright \alpha)$, so constructing B_α with this property causes no difficulties, and will ensure that every element of T_α extends a member of $A \cap (T \upharpoonright \alpha)$, and hence that any element of $\mathbf{T}$ of height greater than α will have to extend an element of $A \cap (T \upharpoonright \alpha)$. Our problem now reduces to one of cardinalities. In constructing $\mathbf{T}$ there are ω_1 limit stages α where we can "kill off" maximal antichains $A \cap (T \upharpoonright \alpha)$ of $\mathbf{T} \upharpoonright \alpha$ in the above sense. But there are ω_2 many potential sets A. So we must somehow deal with ω_2 possibilities in ω_1 steps. This is where we use $V = L$.

Suppose then that we are at stage α, where $\lim(\alpha)$ and $\mathbf{T} \upharpoonright \alpha$ has been defined. Let A_α be the $<_L$-least maximal antichain of $\mathbf{T} \upharpoonright \alpha$ with the property that the set

$$\{\gamma < \alpha \mid A_\alpha \cap T_\gamma \neq \emptyset\}$$

is unbounded in α. (Such a set always exists, as is easily seen.) For each $x \in T \upharpoonright \alpha$, let b_x be the $<_L$-least α-branch of $\mathbf{T} \upharpoonright \alpha$ such that $x \in b_x$ and $b_x \cap A_\alpha \neq \emptyset$. Since A_α is a maximal antichain of $\mathbf{T} \upharpoonright \alpha$, b_x is always defined. Let

$$B_\alpha = \{b_x \mid x \in T \upharpoonright \alpha\}.$$

Set

$$T_\alpha = \{\textstyle\bigcup b \mid b \in B_\alpha\}.$$

That completes the definition of **T**. We must check that **T** has no uncountable antichain. Suppose, on the contrary, that it did. Let A be the $<_L$-least maximal uncountable antichain of **T**. Now, all of the sets involved in the above definition of **T** are members of L_{ω_2}, so we could in fact carry out the construction of **T** within the set L_{ω_2}. Thus **T** is a definable element of L_{ω_2}. Hence A is also a definable element of L_{ω_2}. Moreover, we clearly have (by a trivial absoluteness observation):

$$\models_{L_{\omega_2}} \text{“}A \text{ is the } <_L\text{-least maximal antichain of } \mathbf{T} \text{ such that the set } \{\gamma \in \omega_1 \mid A \cap T_\gamma \neq \emptyset\} \text{ is unbounded in } \omega_1\text{”.}$$

Let M be the smallest elementary submodel of L_{ω_2}. By II.5.11, $M \cap L_{\omega_1}$ is transitive and of the form L_α for some $\alpha < \omega_1$. (M is, of course, countable.) Since **T** and A are definable in L_{ω_2}, they are elements of M. We have

$$T \cap M = T \restriction \alpha.$$

To see this, suppose first that $\beta < \alpha$. Then there is a surjection $f: \omega \to T_\beta$. Hence there is such a surjection in M. But $\omega \subseteq M$, so it follows that $T_\beta = f''\omega \subseteq M$. Thus $T \restriction \alpha \subseteq M$. Again, if $x \in T \cap M$, then (again because $M \prec L_{\omega_2}$) $ht(x) \in M$, so $ht(x) < \alpha$, so $x \in T \restriction \alpha$. We also have

$$A \cap M = A \cap (T \restriction \alpha).$$

(This is an immediate consequence of the previous equality.) So, if we let

$$\pi: M \cong L_\beta$$

(by the Condensation Lemma) we have:

$$\pi \restriction L_\alpha = \mathrm{id} \restriction L_\alpha, \quad \pi(\omega_1) = \alpha, \quad \pi(\mathbf{T}) = \mathbf{T} \restriction \alpha, \quad \pi(A) = A \cap (T \restriction \alpha).$$

(These are all easy consequences of the properties of the collapsing isomorphism. Such considerations will occur often in our later development.) Thus, by elementary substructure and isomorphism, we have:

$$\models_{L_\beta} \text{“}(A \cap T \restriction \alpha) \text{ is the } <_L\text{-least maximal antichain of } \mathbf{T} \restriction \alpha \text{ such that the set } \{\gamma \in \alpha \mid (A \cap T \restriction \alpha) \cap (\mathbf{T} \restriction \alpha)_\gamma \neq \emptyset\} \text{ is unbounded in } \alpha\text{”.}$$

By elementary absoluteness considerations, this clearly implies that $A \cap T \restriction \alpha$ really is the $<_L$-least maximal antichain of $\mathbf{T} \restriction \alpha$ such that the set $\{\gamma < \alpha \mid (A \cap T \restriction \alpha) \cap T_\gamma \neq \emptyset\}$ is unbounded in α. Hence

$$A \cap T \restriction \alpha = A_\alpha.$$

But then by the construction of T_α, every element of **T** of height greater than or equal to α is comparable with some element of $A \cap T \restriction \alpha$. This contradicts the fact that A is an uncountable antichain of **T**. Hence **T** must be a Souslin tree, and we are done. □

In section 3 we shall analyse the use of $V = L$ in the above proof.

2. The Kurepa Hypothesis

We have seen that there are ω_1-trees with no ω_1-branches. And by making simple modifications to an Aronszajn tree it is possible to construct ω_1-trees with exactly κ many ω_1-branches, where κ is any of the cardinals $1, 2, 3, \ldots, n, \ldots, \omega, \omega_1$. Now, any ω_1-tree is a set of cardinality ω_1, so the maximum possible number of branches is 2^{ω_1}. Thus, if we assume GCH, no ω_1-tree can have more than ω_2 many ω_1-branches. A natural question is whether in fact there are *any* ω_1-trees which have ω_2 many ω_1-branches. This turns out to be related to an old question of D. Kurepa concerning the Generalised Continuum Hypothesis (see later), and as a result, an ω_1-tree with ω_2 (or more) ω_1-branches is called a *Kurepa tree.*

In ZFC, or even in ZFC + GCH, it is not possible to decide whether or not Kurepa trees exist. The sharpest results are these:

(I) If ZF is consistent, so too is the theory

ZFC + GCH + "*there is a Kurepa tree*".

(II) If the theory

ZFC + "*there is an inaccessible cardinal*"

is consistent, so too is the theory

ZFC + GCH + "*there are no Kurepa trees*".

(III) If the theory

ZFC + "*there are no Kurepa trees*"

is consistent, so too is the theory

ZFC + "*there is an inaccessible cardinal*".

Hence the non-existence of Kurepa trees is closely bound up with the notion of inaccessible cardinals. We shall prove that if $V = L$, there is a Kurepa tree. But before we do this, we relate the notion of Kurepa trees to the problem of Kurepa, mentioned earlier.

The *Kurepa Hypothesis* (*KH*) is the assertion that there is a family $\mathscr{F} \subseteq \mathscr{P}(\omega_1)$ of cardinality ω_2 such that for all $\alpha < \omega_1$, the set

$$\mathscr{F} \upharpoonright \alpha = \{x \cap \alpha \mid x \in \mathscr{F}\}$$

is countable. The following lemma is due to Kurepa himself.

2.1 Lemma. *The Kurepa Hypothesis is equivalent to the existence of a Kurepa tree.*

Proof. Suppose first that there is a Kurepa tree, **T**. We may clearly assume that $\mathbf{T} = \langle \omega_1, \leqslant_T \rangle$ and that $\alpha <_T \beta$ implies $\alpha < \beta$. Let $\mathscr{F}$ be the set of all ω_1-branches of **T**. It is immediately clear that $\mathscr{F}$ satisfies KH.

Conversely, let $\mathscr{F} \subseteq \mathscr{P}(\omega_1)$ satisfy KH. For each $x \in \mathscr{F}$, define a function $f_x \colon \omega_1 \to \mathscr{P}(\omega_1)$ by setting

$$f_x(\alpha) = x \cap \alpha .$$

Let

$$T = \{ f_x \upharpoonright \alpha \mid x \in \mathscr{F} \wedge \alpha < \omega_1 \} .$$

For $g_1, g_2 \in T$, say $g_1 <_T g_2$ iff $g_1 \subset g_2$. It is clear that $\mathbf{T} = \langle T, \leqslant_T \rangle$ is a tree such that $T_\alpha \subseteq {}^\alpha\mathscr{P}(\omega_1)$. Since $\mathscr{F} \upharpoonright \alpha$ is countable for each $\alpha < \omega_1$, each level of **T** is countable. Hence **T** is an (ω_1, ω_1)-tree. For each $x \in \mathscr{F}$, the set

$$b_x = \{ f_x \upharpoonright \alpha \mid \alpha < \omega_1 \}$$

is an ω_1-branch of **T**, and if $x \neq y$ then $b_x \neq b_y$. Hence **T** has ω_2 many ω_1-branches. Hence we shall be done if we can show that **T** is normal. Well, it is easily seen that **T** satisfies all of the normality requirements except possibly the requirement that each element of **T** has at least two immediate successors. But this is easily achieved: simply add two copies of an Aronszajn tree above each point of **T**. The resulting tree will then be a Kurepa tree. □

2.2 Theorem. *Assume $V = L$. Then there is a Kurepa tree.*

Proof. We verify KH, rather than construct a Kurepa tree directly, as this turns out to be marginally simpler (because there is less to check).

Using II.5.4 and II.5.10, we can define a function $f \colon \omega_1 \to \omega_1$ by letting $f(\alpha)$ be the least ordinal $\gamma > \alpha$ such that $L_\gamma \prec L_{\omega_1}$. Notice that $L_{f(\alpha)}$ will be a model of the theory ZF^- (= ZF minus the Power Set Axiom). (As is often the case in such situations, we are being a little sloppy here. As formulated, ZF^- will be a theory in LST, and we have no concept of a model for an LST-theory. We can avoid this sloppiness either by formulating a "copy" of the theory ZF^- in the language $\mathscr{L}$, or else defining within set theory the notion of "a model of ZF^-" in an entirely semantic fashion, just as we defined the notions of amenable sets and admissible sets to provide us with the notions of "models" of the theories BS and KP, respectively (Chapter I). What matters to us is that, working inside $L_{f(\alpha)}$, we can carry out any construction which can be carried out in ZF without use being made of the Power Set Axiom.)

Define $\mathscr{F} \subseteq \mathscr{P}(\omega_1)$ by:

$$\mathscr{F} = \{ x \subseteq \omega_1 \mid (\forall \alpha < \omega_1)(x \cap \alpha \in L_{f(\alpha)}) \} .$$

For any $\alpha < \omega_1$, $\mathscr{F} \upharpoonright \alpha \subseteq L_{f(\alpha)}$, so certainly $|\mathscr{F} \upharpoonright \alpha| \leqslant \omega$. What we must show, in order to prove that $\mathscr{F}$ satisfies KH, is that $|\mathscr{F}| = \omega_2$. Intuitively, this is because, although countable, $f(\alpha)$ is "much larger" than α (in the sense that $L_{f(\alpha)}$ is a "partial universe" as far as the theory ZF^- is concerned).

We shall assume that $|\mathscr{F}| \neq \omega_2$ and work for a contradiction. By this assumption, $\mathscr{F}$ has an ω_1-enumeration (not necessarily one-one). Let $X = (x_\alpha \mid \alpha < \omega_1)$ be the $<_L$-least ω_1-enumeration of $\mathscr{F}$. Notice that the function f is definable in L_{ω_2} (because the definition of f given above only involves sets in L_{ω_2}), whence both $\mathscr{F}$ and X are definable in L_{ω_2}.

By recursion, we define elementary submodels $N_\nu \prec L_{\omega_2}$ for $\nu < \omega_1$ as follows:

$$N_0 = \text{the smallest } N \prec L_{\omega_2};$$

$$N_{\nu+1} = \text{the smallest } N \prec L_{\omega_2} \text{ such that } N_\nu \cup \{N_\nu\} \subseteq N;$$

$$N_\delta = \bigcup_{\nu<\delta} N_\nu, \qquad \text{if } \lim(\delta).$$

By II.5.11, $N_\nu \cap \omega_1$ is transitive for each $\nu < \omega_1$. Let $\alpha_\nu = N_\nu \cap \omega_1$. Now, by a simple induction, we see that each N_ν is countable, so each α_ν is a countable ordinal. Moreover, since $N_\nu \in N_{\nu+1} \prec L_{\omega_2}$, we have $\alpha_\nu = N_\nu \cap \omega_1 \in N_{\nu+1}$, so $\alpha_\nu < \alpha_{\nu+1}$. Hence $(\alpha_\nu \mid \nu < \omega_1)$ is a normal sequence in ω_1. (Continuity follows from the continuity of the sequence $(N_\nu \mid \nu < \omega_1)$, of course.) Set

$$x = \{\alpha_\nu \mid \nu < \omega_1 \wedge \alpha_\nu \notin x_\nu\}.$$

For each $\nu < \omega_1$, $x \neq x_\nu$, so $x \notin \mathscr{F}$. We obtain our contradiction by showing that $x \cap \alpha \in L_{f(\alpha)}$ for all $\alpha < \omega_1$.

Fix $\alpha < \omega_1$ arbitrarily. We prove that $x \cap \alpha \in L_{f(\alpha)}$. Let η be the largest limit ordinal such that $\alpha_\eta \leqslant \alpha$. (If no such η exists, then $x \cap \alpha$ is finite and hence $x \cap \alpha \in L_{f(\alpha)}$.) Since $x \cap \alpha$ differs from $x \cap \alpha_\eta$ by only a finite amount, and since $L_{f(\alpha)}$ is amenable, it clearly suffices to prove that $x \cap \alpha_\eta \in L_{f(\alpha)}$. But $\alpha_\eta \leqslant \alpha$ and f is clearly monotone, so it suffices to prove that $x \cap \alpha_\eta \in L_{f(\alpha_\eta)}$. Hence we may assume that $\alpha = \alpha_\eta$, where $\lim(\eta)$.

Now, we have

$$x \cap \alpha = \{\alpha_\nu \mid \nu < \eta \wedge \alpha_\nu \notin x_\nu \cap \alpha\},$$

so as $L_{f(\alpha)}$ is a model of ZF^- we shall be done if we can show that

$$(\alpha_\nu \mid \nu < \eta), \qquad (x_\nu \cap \alpha \mid \nu < \eta) \in L_{f(\alpha)}.$$

Let

$$\pi\colon N_\eta \cong L_\beta.$$

Clearly,

$$\pi \upharpoonright L_\alpha = \mathrm{id} \upharpoonright L_\alpha, \qquad \pi(\omega_1) = \alpha, \qquad \pi(X) = (x_\nu \cap \alpha \mid \nu < \alpha).$$

In particular,

$$(x_\nu \cap \alpha \mid \nu < \alpha) \in L_\beta.$$

So as $\eta \leqslant \alpha$,

$$(x_\nu \cap \alpha \mid \nu < \eta) \in L_\beta.$$

Now, $\alpha \in L_{f(\alpha)} \prec L_{\omega_1}$, so

$$\models_{L_{f(\alpha)}} \text{"}\alpha \text{ is countable"}.$$

But since $\pi(\omega_1) = \alpha$,

$$\alpha = \omega_1^{L_1}.$$

Hence,

$$\beta < f(\alpha).$$

Thus

$$(x_\nu \cap \alpha \mid \nu < \eta) \in L_{f(\alpha)}.$$

It remains only to prove that $(\alpha_\nu \mid \nu < \eta) \in L_{f(\alpha)}$.
For each $\nu < \eta$, let

$$\pi_\nu \colon N_\nu \cong L_{\beta(\nu)}.$$

Then,

$$\pi_\nu \upharpoonright L_{\alpha_\nu} = \mathrm{id} \upharpoonright L_{\alpha_\nu}, \qquad \pi_\nu(\omega_1) = \alpha_\nu.$$

Since $\alpha_\nu = \omega_1^{L_{\beta(\nu)}}$, the sequence $(\alpha_\nu \mid \nu < \eta)$ is definable from the sequence $(\beta(\nu) \mid \nu < \eta)$ in ZF^-, so we shall be done if we can prove that

$$(\beta(\nu) \mid \nu < \eta) \in L_{f(\alpha)}.$$

Well, we proved above that $\beta < f(\alpha)$, so certainly $\beta \in L_{f(\alpha)}$. Moreover, $L_{f(\alpha)}$ is a model of ZF^-. So, working inside $L_{f(\alpha)}$ we can define a sequence $(N'_\nu \mid \nu < \eta')$ of elementary submodels of L_β (for some η') as follows:

$$N'_0 = \text{the smallest } N \prec L_\beta;$$
$$N'_{\nu+1} = \text{the smallest } N \prec L_\beta \text{ such that } N'_\nu \cup \{N'_\nu\} \subseteq N;$$
$$N'_\delta = \bigcup_{\nu<\delta} N'_\nu, \qquad \text{if } \lim(\delta).$$

(The ordinal η' is the largest $\eta' \leqslant \eta$ for which the above construction is possible: in a moment we shall see that in fact $\eta' = \eta$.) Still inside $L_{f(\alpha)}$, let

$$\pi'_\nu \colon N'_\nu \cong L_{\beta'(\nu)} \qquad (\nu < \eta').$$

Thus

$$(\beta'(\nu) \mid \nu < \eta') \in L_{f(\alpha)}.$$

Now recall the dfinition of the original sequence $(N_\nu \mid \nu < \omega_1)$. Since $\nu < \eta$ implies $N_\nu \prec N_\eta \prec L_{\omega_2}$, in the definition of the initial part $(N_\nu \mid \nu < \eta)$ of this sequence we could equally well use N_η in place of L_{ω_2}. That is to say, for $\nu < \eta$ we have:

$$N_0 = \text{the smallest } N \prec N_\eta;$$
$$N_{\nu+1} = \text{the smallest } N \prec N_\eta \text{ such that } N_\nu \cup \{N_\nu\} \subseteq N;$$
$$N_\delta = \bigcup_{\nu<\delta} N_\nu, \quad \text{if } \lim(\delta).$$

Now.

$$\pi\colon N_\eta \cong L_\beta,$$

so an easy induction argument shows that for each $\nu < \eta$,

$$(\pi \restriction N_\nu)\colon N_\nu \cong N'_\nu.$$

(The successor step uses II.5.3.) Hence $\eta' = \eta$ and for each $\nu < \eta$, the structures N_ν and N'_ν have the same transitive collapse, i.e.

$$\nu < \eta \to \beta(\nu) = \beta'(\nu).$$

Thus

$$(\beta(\nu) \mid \nu < \eta) \in L_{f(\alpha)},$$

and we are done. □

3. Some Combinatorial Principles Related to the Previous Constructions

Both for later use and for independent interest, we shall analyse the use of the condensation lemma in the two previous constructions using $V = L$. We begin with the construction of a Souslin tree (1.5). If we try to eliminate the use of the elementary substructure argument of 1.5, we see that what we need is the following:

> *There should be a sequence $(A_\alpha \mid \alpha < \omega_1)$ such that $A_\alpha \subseteq T \restriction \alpha$, with the property that whenever $A \subseteq T$, then for any club set $C \subseteq \omega_1$ there is an $\alpha \in C$ such that $A \cap (T \restriction \alpha) = A_\alpha$.*

For then, given an uncountable maximal antichain $A \subseteq \mathbf{T}$, we take

$$C = \{\alpha \in \omega_1 \mid A \cap (T \restriction \alpha) \text{ is a maximal antichain of } \mathbf{T} \restriction \alpha\}$$

and find an $\alpha \in C$ for which $A \cap (T \restriction \alpha) = A_\alpha$.

The problem with the above approach is that the sequence $(A_\alpha \mid \alpha < \omega_1)$ is too closely bound up with the tree, **T**, which we are trying to define. And until T_α has been defined, we do not know which members of ${}^\alpha 2$ will lie in **T**, of course. However, this problem is easily overcome. By taking the elements of **T** to be countable binary sequences as we did, we fixed in advance the *ordering* of **T** (namely $\subseteq$), and concentrated all our efforts upon choosing the correct subset of $\bigcup_{\alpha < \omega_1} {}^\alpha 2$ for the *domain* of **T**. An alternative approach is to fix in advance the *domain* of **T**, say the set ω_1, and to define the *ordering*, $<_T$, by recursion. Thus we can commence by setting $T_0 = \{0\}$, and if $\mathbf{T} \restriction (\alpha + 1)$ is defined, then for each $x \in T_\alpha$ we can pick the first two unused ordinals in ω_1 and appoint them as successors to x in $T_{\alpha+1}$ (subject to some well-ordering of T_α). For limit ordinals α, if $\mathbf{T} \restriction \alpha$ is defined, we use the next ω unused ordinals to provide extensions in T_α of each member of a suitably chosen countable collection, B_α, of α-branches of $\mathbf{T} \restriction \alpha$. Analysis of the proof in this form leads to the following combinatorial principle:

There should be a sequence $(S_\alpha \mid \alpha < \omega_1)$ such that $S_\alpha \subseteq \alpha$ and for each $X \subseteq \omega_1$ and each club $C \subseteq \omega_1$ there is an $\alpha \in C$ such that $X \cap \alpha = S_\alpha$.

(See 3.2 below for a construction of a Souslin tree using this principle.)

The above principle implicitly involves the classical set-theoretic concept of a stationary set, which we now consider briefly.

A subset, E, of a limit ordinal λ is said to be *stationary* in λ iff E has a non-empty intersection with every club subset of λ.

It is immediate that stationary sets are unbounded. They need not be club, since the result of removing one (limit) point from any stationary set is a stationary set, of course. If κ is an uncountable, regular cardinal, every club set $C \subseteq \kappa$ is stationary (by I.6.1), and in this case the property of being stationary lies strictly between the properties of being club and of being unbounded. For example, in the case of ω_2, the set $\{\alpha + 1 \mid \alpha \in \omega_2\}$ is unbounded in ω_2 but not stationary, whilst the set $\{\alpha \in \omega_2 \mid \mathrm{cf}(\alpha) = \omega\}$ is stationary in ω_2 but not club. A classical result of Ulam (which we do not prove here) states that if $E \subseteq \omega_1$ is stationary, there are disjoint stationary sets $E_\nu \subseteq \omega_1$, for $\nu < \omega_1$, such that $E = \bigcup_{\nu < \omega_1} E_\nu$.

Stationary sets are closely connected with "regressive functions". If λ is an ordinal and $E \subseteq \lambda$, a function $f\colon E \to \lambda$ is said to be *regressive* iff, for each non-zero $\alpha \in E$, $f(\alpha) < \alpha$.

3.1 Theorem (Fodor's Theorem). *Let κ be an uncountable regular cardinal, and let $E \subseteq \kappa$ be stationary. If $f\colon E \to \kappa$ is regressive, then for some $\beta \in \kappa$, the set*

$$\{\alpha \in E \mid f(\alpha) = \beta\}$$

is stationary in κ.

Proof. Suppose that, on the contrary, for each $\beta \in \kappa$ the set

$$\{\alpha \in E \mid f(\alpha) = \beta\}$$

is not stationary in κ. Then for each $\beta \in \kappa$ we can find a club set $C_\beta \subseteq \kappa$ such that

$$\alpha \in C_\beta \cap E \to f(\alpha) \neq \beta .$$

Let

$$C = \{\alpha \in \kappa \mid \alpha \in \bigcap_{\beta < \alpha} C_\beta\} .$$

This set, C, is called the *diagonal intersection* of the sets C_β, $\beta < \kappa$. It is not hard to see that C is club in κ. Hence, as E is stationary in κ we can find a non-zero ordinal $\alpha \in C \cap E$. For $\beta < \alpha$, we have $\alpha \in C_\beta$, so $f(\alpha) \neq \beta$. (Since $\alpha \in E$, $f(\alpha)$ is defined, of course.) Thus $f(\alpha) \geqslant \alpha$. But this is absurd, since f is regressive. The theorem is proved. □

As an easy exercise, the reader might like to prove that if $E \subseteq \kappa$ is not stationary, there is a regressive function on E which is not constant on any unbounded set. Thus stationary sets may be characterised as those unbounded sets E such that all regressive functions on E are constant on an unbounded subset of E.

In terms of stationary sets, our previous combinatorial principle can be expressed as follows:

> *There is a sequence $(S_\alpha \mid \alpha < \omega_1)$ such that $S_\alpha \subseteq \alpha$, with the property that whenever $X \subseteq \omega_1$, the set $\{\alpha \in \omega_1 \mid X \cap \alpha = S_\alpha\}$ is stationary in ω_1.*

In turns out that this combinatorial principle has many applications, and thus deserves a name. Following Jensen, who discovered it, we call it $\diamondsuit$ (i.e. "diamond").

By amending the argument of 1.5 we prove:

3.2 Theorem. *$\diamondsuit$ implies the existence of a Souslin tree.*

Proof. Assume $\diamondsuit$, and let $(S_\alpha \mid \alpha < \omega_1)$ be a $\diamondsuit$-sequence as described above. By recursion on the levels we construct a Souslin tree, $\mathbf{T}$, with domain ω_1. The elements of $T \upharpoonright \omega$ will be the finite ordinals, and for infinite α the elements of T_α will be the oridinals in the set

$$\{\xi \mid \omega\alpha \leqslant \xi < \omega\alpha + \omega\} .$$

We shall carry out the construction so that for each $\alpha < \omega_1$, $\mathbf{T} \upharpoonright \alpha$ is a normal (α, ω_1)-tree.

Set $T_0 = \{0\}$. If $n \in \omega$ and $\mathbf{T} \upharpoonright n + 1$ is defined, define $\mathbf{T} \upharpoonright n + 2$ by taking the elements of T_n in turn, for each one picking the next two unused finite ordinals to be its successors in T_{n+1}. If $\alpha \geqslant \omega$ and $\mathbf{T} \upharpoonright \alpha + 1$ is defined, define $\mathbf{T} \upharpoonright \alpha + 2$ by using the ordinals in the set $\{\xi \mid \omega\alpha \leqslant \xi < \omega\alpha + \omega\}$ to provide each element of T_α with two successors on $T_{\alpha+1}$. Since T_α is countable, this is easily arranged. There remains the case where $\lim(\alpha)$ and $\mathbf{T} \upharpoonright \alpha$ is defined. By the normality of $\mathbf{T} \upharpoonright \alpha$, for each $x \in T \upharpoonright \alpha$ we can pick an α-branch b_x of $\mathbf{T} \upharpoonright \alpha$ containing x. The exact choice of b_x is unimportant except when S_α is a maximal antichain of $\mathbf{T} \upharpoonright \alpha$, in which case we ensure that $b_x \cap S_\alpha \neq \emptyset$, which is easy to do by virtue of the maximality of the antichain S_α in $\mathbf{T} \upharpoonright \alpha$. The ordinals in the set $\{\xi \mid \omega\alpha \leqslant \xi < \omega a + \omega\}$ are then used to provide one-point extensions in T_α of each of the (countably many) branches b_x, $x \in T \upharpoonright \alpha$.

The above construction clearly provides us with an ω_1-tree, $\mathbf{T}$. We need to check that $\mathbf{T}$ is Souslin. It suffices to show that every *maximal* antichain of $\mathbf{T}$ is countable. Let $X \subseteq \omega_1$ be a maximal antichain of $\mathbf{T}$. Set

$$C = \{\alpha \in \omega_1 \mid \omega\alpha = \alpha \wedge X \cap \alpha \text{ is a maximal antichain of } \mathbf{T} \upharpoonright \alpha\}.$$

Now, if $\omega\alpha = \alpha$, then $T \upharpoonright \alpha = T \cap \alpha$, so $X \cap \alpha$ is certainly an antichain of $\mathbf{T} \upharpoonright \alpha$. It is easy to see that C is club in ω_1 now. (The argument was given in 1.5.) So by $\diamond$, we can pick an $\alpha \in C$ so that $X \cap \alpha = S_\alpha$. By the construction of T_α, every element of T_α lies above an element of $X \cap \alpha$. Hence $X \cap \alpha$ is a maximal antichain in $\mathbf{T}$. Thus $X = X \cap \alpha$, which means that X is countable, as required. □

Notice that $\diamond$ implies CH: for if $(S_\alpha \mid \alpha < \omega_1)$ is a $\diamond$-sequence, then for each set $X \subseteq \omega$ there is an ordinal α such that $X = X \cap \alpha = S_\alpha$. In fact $\diamond$ can be regarded as a sort of "super-CH". This is highlighted by the following fact, whose proof is left as an exercise (see Exercise 3). $\diamond$ is equivalent to the existence of a sequence $(S_\alpha \mid \alpha < \omega_1)$ such that $S_\alpha \subseteq \alpha$ for each α and, whenever $X \subseteq \omega_1$ there is *at least one* infinite ordinal α such that $X \cap \alpha = S_\alpha$. CH, on the other hand, is equivalent to the existence of a sequence $(S_\alpha \mid \alpha < \omega_1)$ such that $S_\alpha \subseteq \alpha$ for each α and, whenever $X \subseteq \omega_1$, then for all $\alpha < \omega_1$ there is a $\beta < \omega_1$ such that $X \cap \alpha = S_\beta$.

The following result completes our analysis of the proof of 1.5.

3.3 Theorem. *Assume $V = L$. Then $\diamond$ is valid.*

Proof. By recursion on α we define sets $S_\alpha \subseteq \alpha$, $C_\alpha \subseteq \alpha$ for each $\alpha < \omega_1$.

To commence, we set

$$S_0 = C_0 = \emptyset.$$

If S_α, C_α are defined, set

$$S_{\alpha+1} = C_{\alpha+1} = \alpha + 1.$$

Finally, suppose $\lim(\alpha)$ and S_γ, C_γ are defined for all $\gamma < \alpha$. Let (S_α, C_α) be the $<_L$-least pair of subsets of α such that:

(i) C_α is club in α;
(ii) $(\forall \gamma \in C_\alpha)(S_\alpha \cap \gamma \neq S_\gamma)$,

providing that such sets exist, and set

$$S_\alpha = C_\alpha = \alpha,$$

otherwise.

Notice that, by the above definition, the sequence $((S_\alpha, C_\alpha) \mid \alpha < \omega_1)$ is definable in L_{ω_2}. We show that the sequence $(S_\alpha \mid \alpha < \omega_1)$ satisfies $\diamond$.

Suppose that $(S_\alpha | \alpha < \omega_1)$ were not a $\diamondsuit$-sequence. Then for some set $S \subseteq \omega_1$, the set

$$\{\alpha \in \omega_1 | S \cap \alpha = S_\alpha\}$$

would fail to be stationary in ω_1, so there would be a club set $C \subseteq \omega_1$ such that

$$(\forall \gamma \in C)(S \cap \gamma \neq S_\gamma).$$

Let (S, C) be the $<_L$-least pair of such sets S, C. Notice that this definition will define (S, C) in L_{ω_2}.

Let $X \prec L_{\omega_2}$ be countable, and let $\pi: X \cong L_\beta$. By II.5.11, $X \cap L_{\omega_1}$ is transitive. Let $\alpha = X \cap \omega_1$. Then

$$\pi \upharpoonright L_\alpha = \mathrm{id} \upharpoonright L_\alpha \quad \text{and} \quad \pi(\omega_1) = \alpha.$$

Moreover, as is easily checked (cf. similar arguments in 1.5)

$$\pi(S) = S \cap \alpha, \quad \pi(C) = C \cap \alpha, \quad \pi((S_\gamma | \gamma < \omega_1)) = (S_\gamma | \gamma < \alpha),$$
$$\pi((C_\gamma | \gamma < \omega_1)) = (C_\gamma | \gamma < \alpha).$$

Now, by elementary absoluteness considerations, we have

$\models_{L_{\omega_2}}$ "(S, C) is the $<_L$-least pair of subset of ω_1 such that C is club in ω_1 and $(\forall \gamma \in C)(S \cap \gamma \neq S_\gamma)$".

So, as $\pi^{-1}: L_\beta \prec L_{\omega_2}$,

$\models_{L_\beta}$ "$(S \cap \alpha, C \cap \alpha)$ is the $<_L$-least pair of subsets of α such that $C \cap \alpha$ is club in α and $(\forall \gamma \in C \cap \alpha)((S \cap \alpha) \cap \gamma \neq S_\gamma)$".

Thus, by another simple absoluteness observation (together with II.3.4(i)), we see that $(S \cap \alpha, C \cap \alpha)$ really is the $<_L$-least pair of such subsets of α. But by definition, this means that $S_\alpha = S \cap \alpha$ and $C_\alpha = C \cap \alpha$.

Now, as we saw above,

$\models_{L_\beta}$ "$C \cap \alpha$ is unbounded in α".

Thus $C \cap \alpha$ really must be unbounded in α. But C is closed in ω_1. Hence $\alpha \in C$. But this implies that $S \cap \alpha \neq S_\alpha$, so we have a contradiction. The proof is complete. □

A natural strenghtening of $\diamondsuit$ would be the following: there is a sequence $(S_\alpha | \alpha < \omega_1)$ such that $S_\alpha \subseteq \alpha$ for each α and whenever $X \subseteq \omega_1$ there is a *club* set $C \subseteq \omega_1$ such that $X \cap \alpha = S_\alpha$ for all $\alpha \in C$. However, it is an easy exercise to show that this is impossible. But by modifying the formulation of $\diamondsuit$ a little, we can obtain an equivalent statement which can be strengthened in the above fashion.

Let $\diamondsuit'$ mean the following assertion:

> *There is a sequence $(T_\alpha \mid \alpha < \omega_1)$ such that for each α, T_α is a countable subset of $\mathscr{P}(\alpha)$, with the property that whenever $X \subseteq \omega_1$, the set $\{\alpha \in \omega_1 \mid X \cap \alpha \in T_\alpha\}$ is stationary in ω_1.*

Clearly, $\diamondsuit'$ is a consequence of $\diamondsuit$: if $(S_\alpha \mid \alpha < \omega_1)$ is a $\diamondsuit$-sequence, then $(T_\alpha \mid \alpha < \omega_1)$ is a $\diamondsuit'$-sequence, where we set $T_\alpha = \{S_\alpha\}$ for all $\alpha < \omega_1$. In fact, $\diamondsuit'$ and $\diamondsuit$ are equivalent, as we now show.

3.4 Lemma. *$\diamondsuit'$ and $\diamondsuit$ are equivalent.*

Proof. Let $(T_\alpha \mid \alpha < \omega_1)$ be a $\diamondsuit'$-sequence. We first of all use $(T_\alpha \mid \alpha < \omega_1)$ in order to construct a "$\diamondsuit'$-sequence" on $\omega_1 \times \omega$. That is, we define a sequence $(U_\alpha \mid \alpha < \omega_1)$ such that U_α is a countable subset of $\mathscr{P}(\alpha \times \omega)$ and for each set $X \subseteq \omega_1 \times \omega$, the set

$$\{\alpha \in \omega_1 \mid X \cap (\alpha \times \omega) \in U_\alpha\}$$

is stationary in ω_1.

To this end, choose a bijection

$$j\colon \omega_1 \leftrightarrow \omega_1 \times \omega$$

so that for all limit $\alpha < \omega_1$,

$$(j \upharpoonright \alpha)\colon \alpha \leftrightarrow \alpha \times \omega .$$

For instance, using the fact that any ordinal in ω_1 has a unique expression of the form

$$\delta + 2^m \cdot (2n + 1) - 1 ,$$

where δ is either 0 or else a limit ordinal, and where $m, n \in \omega$, we can set

$$j(\delta + 2^m \cdot (2n + 1) - 1) = (\delta + m, n) .$$

For each $\alpha < \omega_1$, now, set

$$U_\alpha = \begin{cases} \{j'' U \mid U \in T_\alpha\}, & \text{if } \lim(\alpha), \\ \emptyset, & \text{otherwise}. \end{cases}$$

It is easily checked that $(U_\alpha \mid \alpha < \omega_1)$ has the desired properties.

Now let $(U_\alpha^n \mid n < \omega)$ enumerate U_α, for each $\alpha < \omega_1$. Thus $U_\alpha^n \subseteq \alpha \times \omega$ and whenever $X \subseteq \omega_1 \times \omega$ there is a stationary set $E \subseteq \omega_1$ such that for every $\alpha \in E$ there is an $n \in \omega$ such that $X \cap (\alpha \times \omega) = U_\alpha^n$. Now, in general, the n here, for which $X \cap (\alpha \times \omega) = U_\alpha^n$, will depend upon α. But as we shall show below, this is not always the case.

Claim. If $X \subseteq \omega_1 \times \omega$, there is a stationary set $F \subseteq \omega_1$ such that for some *fixed* $n \in \omega$, $X \cap (\alpha \times \omega) = U_\alpha^n$ for all $\alpha \in F$.

To see this, let $X \subseteq \omega_1 \times \omega$ be given. Choose $E \subseteq \omega_1$ stationary so that

$$\alpha \in E \rightarrow (\exists n \in \omega)[X \cap (\alpha \times \omega) = U_\alpha^n].$$

Define f: $E \rightarrow \omega$ by setting $f(n) = 0$ for $n \in E \cap \omega$, and letting $f(\alpha)$ be the least n such that $X \cap (\alpha \times \omega) = U_\alpha^n$, otherwise. Since f is regressive, Fodor's Theorem (3.1) tells us that for some $n \in \omega$, the set

$$F = \{\alpha \in E \mid f(\alpha) = n\}$$

is stationary in ω_1. Clearly, F is a claimed.

For each $n < \omega$ and each $\alpha < \omega_1$, now, set

$$S_\alpha^n = \{\nu \in \alpha \mid (\nu, n) \in U_\alpha^n\}.$$

We show that for some $n \in \omega$, $(S_\alpha^n \mid \alpha < \omega_1)$ is a $\diamondsuit$-sequence. Well, suppose otherwise. Thus for each $n \in \omega$ we can find a set $X_n \subseteq \omega_1$ and a club set $C_n \subseteq \omega_1$ such that

$$\alpha \in C_n \rightarrow X_n \cap \alpha \neq S_\alpha^n.$$

Set

$$X = \bigcup_{n<\omega} (X_n \times \{n\}),$$

$$C = \bigcap_{n<\omega} C_n.$$

Then C is club in ω_1 and for all $n < \omega$,

$$\alpha \in C \rightarrow X \cap (\alpha \times \omega) \neq U_\alpha^n.$$

This contradicts our earlier claim, and completes the proof. □

The following principle, known as $\diamondsuit^*$ ("diamond-star") is an obvious strengthening of $\diamondsuit'$.

> $\diamondsuit^*$: *there is a sequence $(S_\alpha \mid \alpha < \omega_1)$ such that S_α is a countable subset of $\mathscr{P}(\alpha)$ and for any $X \subseteq \omega_1$ there is a club set $C \subseteq \omega_1$ such that $X \cap \alpha \in S_\alpha$ for all $\alpha \in C$.*

It is clear that $\diamondsuit^*$ implies $\diamondsuit'$ (hence $\diamondsuit$). And it is known that $\diamondsuit^*$ does not follow from $\diamondsuit$. The next theorem provides us with an alternative proof of $\diamondsuit$ from $V = L$.

3.5 Theorem. *Assume $V = L$. Then $\diamondsuit^*$ is true.*

Proof. Define a function f: $\omega \rightarrow \omega_1$ by setting

$$f(\alpha) = \text{the least } \gamma > \alpha \text{ such that } \vDash_{L_\gamma} \text{"}\alpha \text{ is countable"}.$$

Let

$$S_\alpha = \mathscr{P}(\alpha) \cap L_{f(\alpha)}, \quad \alpha < \omega_1.$$

We show that $(S_\alpha|\alpha<\omega_1)$ is a $\diamond^*$-sequence. Since each S_α is clearly a countable subset of $\mathscr{P}(\alpha)$, what we must prove is that if $X \subseteq \omega_1$ is given, there is a club set $C \subseteq \omega_1$ such that $X \cap \alpha \in S_\alpha$ for all $\alpha \in C$.

By recursion, we define a sequence of elementary submodels

$$N_\nu \prec L_{\omega_2}, \quad \nu < \omega_1.$$

Let

$$N_0 = \text{the smallest } N \prec L_{\omega_2} \text{ such that } X \in N;$$

$$N_{\nu+1} = \text{the smallest } N \prec L_{\omega_2} \text{ such that } N_\nu \cup \{N_\nu\} \subseteq N;$$

$$N_\delta = \bigcup_{\nu<\delta} N_\nu, \quad \text{if } \lim(\delta).$$

By II.5.11 we can define $\alpha_\nu \in \omega_1$ by

$$\alpha_\nu = N_\nu \cap \omega_1.$$

Clearly, the set $C = \{\alpha_\nu | \nu < \omega_1\}$ is club in ω_1. We show that $X \cap \alpha \in S_\alpha$ for each $\alpha \in C$. Let $\nu < \omega_1$ be given. Let

$$\pi\colon N_\nu \cong L_\beta.$$

Then,

$$\pi \restriction \alpha_\nu = \text{id} \restriction \alpha_\nu, \quad \pi(\omega_1) = \alpha_\nu, \quad \pi(X) = X \cap \alpha_\nu.$$

In particular,

$$X \cap \alpha_\nu \in L_\beta.$$

But

$$\models_{L_{f(\alpha_\nu)}} \text{“}\alpha_\nu \text{ is countable”},$$

whereas

$$\alpha_\nu = \omega_1^{L_\beta}.$$

Hence $\beta < f(\alpha_\nu)$ and we see that

$$X \cap \alpha_\nu \in L_{f(\alpha_\nu)}.$$

Thus $X \cap \alpha_\nu \in S_{\alpha_\nu}$ and we are done. □

We turn now to our analysis of the construction of a Kurepa tree from $V = L$ (2.2). The essential combinatorial property of L used here is the following generalisation of $\diamond^*$ known as $\diamond^+$ (“diamond-plus”):

> $\diamond^+$: *there is a sequence $(S_\alpha|\alpha<\omega_1)$ such that S_α is a countable subset of $\mathscr{P}(\alpha)$ and whenever $X \subseteq \omega_1$ there is a club set $C \subseteq \omega_1$ such that for all $\alpha \in C$, both $X \cap \alpha \in S_a$ and $C \cap \alpha \in S_\alpha$.*

It is clear that $\diamondsuit^*$ is just an apparently weaker version of $\diamondsuit^+$. In fact $\diamondsuit^+$ is a real strengthening of $\diamondsuit^*$. In particular, $\diamondsuit^*$ does not imply the existence of a Kurepa tree, whereas $\diamondsuit^+$ does, as we show below.

3.6 Theorem. *Assume $\diamondsuit^+$. Then there is a Kurepa tree.*

Proof. As in 2.2, we choose to establish the existence of a family $\mathscr{F} \subseteq \mathscr{P}(\omega_1)$ such that $|\mathscr{F}| = \omega_2$ and $|\mathscr{F} \upharpoonright \alpha| \leqslant \omega$ for all $\alpha < \omega_1$, rather than construct a Kurepa tree outright.

Let $(S_\alpha | \alpha < \omega_1)$ be a $\diamondsuit^+$-sequence. Recalling that H_{ω_1} is the set of all hereditarily countable sets, for each $\alpha < \omega_1$, let $M_\alpha \prec H_{\omega_1}$ be countable and such that

$$(\alpha + 1) \cup (\bigcup_{\beta \leqslant \alpha} S_\beta) \subseteq M_\alpha .$$

Set

$$\mathscr{F} = \{x \subseteq \omega_1 | (\forall \alpha < \omega_1)(x \cap \alpha \in M_\alpha)\} .$$

If we can prove that $|\mathscr{F}| = \omega_2$ we shall clearly be done. Suppose that, on the contrary, $|\mathscr{F}| = \omega_1$. (It is clear that $\mathscr{F}$ is at least uncountable, since $\{\alpha\} \in \mathscr{F}$ for all $\alpha < \omega_1$.) Let $(x_\nu | \nu < \omega_1)$ enumerate all unbounded members of $\mathscr{F}$. (This sequence does not have to be one-one. Hence, as we clearly have $\omega_1 \in \mathscr{F}$, the sequence does exist.) For each $\nu < \omega_1$, let

$$B_\nu = \{\alpha \in \omega_1 | \lim(\alpha) \wedge x_\nu \cap \alpha \text{ is unbounded in } \alpha\} .$$

It is easily seen that B_ν is club in ω_1. Set

$$B = \{\alpha \in \omega_1 | \lim(\alpha) \wedge (\forall \nu < \alpha)(\alpha \in B_\nu)\} .$$

It is easily seen that B is club in ω_1. (B is essentially the *diagonal intersection* of the sequence $(B_\nu | \nu < \omega_1)$, already mentioned in 3.1.) Applying $\diamondsuit^+$ to the set $B \subseteq \omega_1$ we obtain a club set $C \subseteq \omega_1$ such that

$$\alpha \in C \to B \cap \alpha, \quad C \cap \alpha \in S_\alpha .$$

Let $(\alpha_\nu | \nu < \omega_1)$ enumerate, monotonically, the club set

$$\{\alpha \in B | \alpha = \sup(C \cap \alpha)) .$$

For $\nu < \omega_1$, set

$$\beta_\nu = \min(C - (\alpha_\nu + 1)) .$$

Thus

$$\alpha_\nu < \beta_\nu < \alpha_{\nu+1} .$$

Set

$$x = \{\beta_\nu | \nu < \omega_1\} .$$

For any $\nu < \omega_1$,

$$\nu \leqslant \alpha_\nu < \alpha_{\nu+1} \in B,$$

so $x_\nu \cap \alpha_{\nu+1}$ is unbounded in $\alpha_{\nu+1}$. But

$$x \cap \alpha_{\nu+1} = \{\beta_\tau \mid \tau \leqslant \nu\} \subseteq \beta_\nu + 1 < \alpha_{\nu+1}.$$

Hence $x \neq x_\nu$. We obtain our desired contradiction now by proving that $x \in \mathscr{F}$, i.e. that $x \cap \alpha \in M_\alpha$ for all $\alpha < \omega_1$.

If $x \cap \alpha$ is finite, then it is immediate that $x \cap \alpha \in M_\alpha$, since

$$\alpha \subseteq M_\alpha \prec H_{\omega_1}.$$

So assume $x \cap \alpha$ is infinite. Let $\beta \leqslant \alpha$ be the greatest limit point of $x \cap \alpha$. Since $x \cap \alpha$ differs from $x \cap \beta$ by at most finitely many points, and since M_α is a model of ZF$^-$, it suffices to prove that $x \cap \beta \in M_\alpha$. Now, β is a limit point of x and $x \subseteq C$, so as C is closed in ω_1, $\beta \in C$. Thus

$$B \cap \beta, \quad C \cap \beta \in S_\beta.$$

But $\beta \leqslant \alpha$. Hence

$$B \cap \beta, \quad C \cap \beta \in M_\alpha.$$

Let λ be such that

$$\beta = \sup_{\nu<\lambda} \beta_\nu.$$

Then

$$\{\alpha_\nu \mid \nu < \lambda\} = \{\alpha \in B \cap \beta \mid \alpha = \sup[(C \cap \beta) \cap \alpha]\}.$$

So, as

$$B \cap \beta, C \cap \beta \in M_\alpha \prec H_{\omega_1},$$

we conclude that

$$\{\alpha_\nu \mid \nu < \lambda\} \in M_\alpha.$$

But for $\nu < \lambda$,

$$\beta_\nu = \min[(C \cap \beta) - (\alpha_\nu + 1)].$$

Hence

$$x \cap \beta = \{\beta_\nu \mid \nu < \lambda\} \in M_\alpha,$$

and we are done. □

To complete our analysis of 2.2 now, we prove:

3.7 Theorem. *Assume $V = L$. Then $\diamondsuit^+$ is valid.*

Proof. As in 2.2 we may define a function $f\colon \omega_1 \to \omega_1$ by letting $f(\alpha)$ be the least ordinal such that

$$\alpha \in L_{f(\alpha)} \prec L_{\omega_1}.$$

Set

$$S_\alpha = \mathscr{P}(\alpha) \cap L_{f(\alpha)}.$$

Notice that f and $(S_\alpha \mid \alpha < \omega_1)$ are definable in L_{ω_2} (using the above definitions). We prove that $(S_\alpha \mid \alpha < \omega_1)$ satisfies $\diamondsuit^+$.

Suppose that $(S_\alpha \mid \alpha < \omega_1)$ did not satisfy $\diamondsuit^+$, and let X be the $<_L$-least subset of ω_1 such that for all club sets $C \subseteq \omega_1$ there is an $\alpha \in C$ such that it is not the case that both $X \cap \alpha$ and $C \cap \alpha$ lie in S_α. Notice that X is definable in L_{ω_2} by means of this definition.

By recursion, define a sequence of elementary submodels $N_\nu \prec L_{\omega_2}$, $\nu < \omega_1$, as follows:

$$N_0 = \text{the smallest } N \prec L_{\omega_2};$$

$$N_{\nu+1} = \text{the smallest } N < L_{\omega_2} \text{ such that } N_\nu \cup \{N_\nu\} \subseteq N;$$

$$N_\delta = \bigcup_{\nu<\delta} N_\nu, \quad \text{if } \lim(\delta).$$

By II.5.11, $N \cap L_{\omega_1}$ is transitive. Set

$$\alpha_\nu = N_\nu \cap \omega_1.$$

Clearly, $(\alpha_\nu \mid \nu < \omega_1)$ is a normal sequence in ω_1. Let

$$\pi_\nu\colon N_\nu \cong L_{\beta(\nu)}.$$

Clearly,

$$\pi_\nu \restriction L_{\alpha_\nu} = \mathrm{id} \restriction L_{\alpha_\nu}, \qquad \pi_\nu(\omega_1) = \alpha_\nu, \qquad \pi_\nu(X) = X \cap \alpha_\nu.$$

Let C be the set of all limit points of the set $\{\beta(\nu) \mid \nu < \omega_1\}$. C is club in ω_1. We obtain our contradiction by showing that for all $\alpha \in C$,

$$X \cap \alpha, C \cap \alpha \in S_\alpha.$$

Let $\alpha \in C$ be given. For some limit ordinal $\lambda < \omega_1$,

$$\alpha = \sup\nolimits_{\nu<\lambda} \beta(\nu).$$

Claim 1. $\alpha = \alpha_\lambda$.

To see this, it suffices to prove that for all $\nu < \omega_1$,

$$\alpha_\nu < \beta(\nu) < \alpha_{\nu+1}.$$

Well, clearly, $\alpha_\nu < \beta(\nu)$. But $\beta(\nu)$ is definable from N_ν since $L_{\beta(\nu)}$ is the transitive collapse of N_ν, and moreover this definition relativises to L_{ω_2} (i.e. is absolute for L_{ω_2}). So, as

$$N_\nu \in N_{\nu+1} \prec L_{\omega_2},$$

we have $\beta(\nu) \in N_{\nu+1}$. Hence $\beta(\nu) \in \alpha_{\nu+1}$, and the claim is established.

Claim 2. $\beta(\lambda) < f(\alpha)$.

To see this, note first that by definition of f,

$$\models_{L_{f(\alpha)}} \text{"}\alpha \text{ is countable"}.$$

But,

$$\alpha = \alpha_\lambda = \omega_1^{L_{\beta(\lambda)}}.$$

Hence $\beta(\lambda) < f(\alpha)$, as claimed.

Now, by claim 1,

$$X \cap \alpha = \pi_\lambda(X) \in L_{\beta(\lambda)},$$

so by claim 2,

$$X \cap \alpha \in L_{f(\alpha)}.$$

Thus

$$X \cap \alpha \in S_\alpha,$$

and it remains to prove that $C \cap \alpha \in S_\alpha$. It clearly suffices to prove that

$$\{\beta(\nu) \mid \nu < \lambda\} \in L_{f(\alpha)}.$$

This is proved exactly as in 2.2, so we do not repeat the details here. Our proof is complete. □

Exercises

1. ω_1-*Trees and Souslin Trees* (Section 1)

Let **T** be an ω_1-tree, **P** a totally ordered set. **T** is said to be **P**-*embeddable* iff there is an order-preserving map f: **T** → **P**. Our interest concerns the cases when **P** is either the rationals, $\mathbb{Q}$, or else the reals, $\mathbb{R}$.

1 A. Show that an ω_1-tree, **T**, is $\mathbb{Q}$-embeddable iff there are antichains A_n, $n < \omega$, of **T** such that

$$T = \bigcup_{n<\omega} A_n.$$

1 B. Show that if an ω_1-tree, $\mathbf{T}$, is $\mathbb{R}$-embeddable, it is an Aronszajn tree but not a Souslin tree. (Hint: It is possible to utilise 1 A here.)

1 C. Construct a $\mathbb{Q}$-embeddable ω_1-tree. (The tree constructed in 1.1 *almost* suffices.) Such trees are sometimes referred to as *special* Aronszajn trees, though we shall use this name for a different notion (see Exercise IV.1).

1 D. It is known to be consistent with ZFC that every Aronszajn tree is $\mathbb{Q}$-embeddable. (See *Devlin and Johnsbråten* (1974).) Show that if $V = L$ there is a $\mathbb{R}$-embeddable Aronszajn tree which is not $\mathbb{Q}$-embeddable. (Hint: Take the elements of $\mathbf{T}$ to be countable one-one sequences of integers whose ranges are co-infinite in ω, ordered by inclusion. Construct $\mathbf{T}$ by recursion on the levels to satisfy the following condition:

if $\alpha < \beta < \omega_1$ and $s \in T_\alpha$ and σ is a finite set of integers, disjoint from $\mathrm{ran}(s)$, *there is a $t \in T_\beta$ such that $s \subset t$ and $\sigma \cap$* $\mathrm{ran}(t) = \emptyset$.

Use $V = L$ to ensure that if $f\colon T \to \mathbb{Q}$ were an embedding, there would be a limit ordinal $\alpha < \omega_1$ such that for each $x \in T_\alpha$ there is a $y \in T$, $y <_T x$, such that $f(y) = f(x)$.)

2. *Kurepa Trees* (Section 2)

2 A. Assume $V = L$. Define $f\colon \omega_1 \to \omega_1$ by setting

$$f(\alpha) = \text{the least } \gamma \text{ such that } \alpha \in L_\gamma \prec L_{\omega_1}.$$

Construct an ω_1-tree as follows. The elements of T_α will be members of ${}^\alpha 2$. The ordering of $\mathbf{T}$ will be $\subset$. Let $T_0 = \{\emptyset\}$. If T_α is defined, let

$$T_{\alpha+1} = \{s ^\frown \langle i \rangle \mid s \in T_\alpha \wedge i = 0, 1\}.$$

If $\lim(\alpha)$ and $T \restriction \alpha$ is defined, let

$$T_\alpha = \{\bigcup b \mid b \text{ is an } \alpha\text{-branch of } \mathbf{T} \restriction \alpha \text{ lying in } L_{f(\alpha)}\}.$$

Prove that $\mathbf{T}$ is a Kurepa tree.

2 B. Let $\mathbf{T}$ be the Kurepa tree constructed in 2 A. Show that there is a set $U \subseteq T$ which is a Souslin tree under the induced ordering.

3. *The Combinatorial Principle* $\diamondsuit$ (Section 3)

3 A. Let $\diamondsuit^-$ be the following principle: there is a sequence $(S_\alpha \mid \alpha < \omega_1)$ such that S_α is a countable subset of $\mathscr{P}(\alpha)$ for each α and whenever $X \subseteq \omega_1$ there is an infinite ordinal α such that $X \cap \alpha \in S_\alpha$. Prove that $\diamondsuit^-$ is equivalent to $\diamondsuit$. (Hint: First show that $\diamondsuit^-$ implies $\diamondsuit^{-+}$, where $\diamondsuit^{-+}$ is the same as $\diamondsuit^-$ except that the α which is asserted to exist is required to be a limit ordinal. Now let $(S_\alpha \mid \alpha < \omega_1)$ be as in $\diamondsuit^{-+}$. Define $j\colon \omega_1 \to \omega_1$ by $j(v) = 2v$. Set $T_\alpha = \{j^{-1}{}''x \mid x \in S_\alpha\}$. Then $(T_\alpha \mid \alpha < \omega_1)$ is a $\diamondsuit'$-sequence. The idea is that, given a club set $C \subseteq \omega_1$ from which we must find an α

with $X \cap \alpha \in T_\alpha$, for a given $X \subseteq \omega_1$, we construct a set $Y \subseteq \omega_1$ whose intersection with the even ordinals is $j''X$ and whose intersection with the odd ordinals is a diagonalisation set ensuring that if $Y \cap \alpha \in S_\alpha$, then $\alpha \in C$.)

3 B. Show that CH is equivalent to the existence of a sequence $(S_\alpha \mid \alpha < \omega_1)$ such that S_α is a countable subset of $\mathscr{P}(\alpha)$ $(\alpha < \omega_1)$ and whenever $X \subseteq \omega_1$, then

$$(\forall \alpha)(\exists \beta)(X \cap \alpha \in S_\beta).$$

3 C. Show that $\diamondsuit$ is equivalent to the existence of a sequence $(S_\alpha \mid \alpha < \omega_1)$ and a function $f\colon \omega_1 \to \omega_1$ such that S_α is a countable subset of $\mathscr{P}(\alpha)$ for each α and whenever $X \subseteq \omega_1$ then for uncountably many $\alpha < \omega_1$.

$$(\exists \beta < f(\alpha))(X \cap \alpha \in S_\beta).$$

3 D. Let P assert the existence of a sequence $(U_\alpha \mid \alpha < \omega_1 \wedge \lim(\alpha))$ such that U_α is an increasing ω-sequence, cofinal in α, with the property that whenever $X \subseteq \omega_1$ is uncountable there is an α such that $U_\alpha \subseteq X$. Show that in the presence of CH, P is equivalent to $\diamondsuit$. (It is known that in the absence of CH, P does not necessarily imply $\diamondsuit$.) (Hint: Let $(X_\alpha \mid \alpha < \omega_1)$ enumerate all bounded subsets of ω_1 so that each set appears cofinally often. Define $S_\alpha = \bigcup \{X_\beta \cap \alpha \mid \beta \in U_\alpha\}$ to obtain a $\diamondsuit^-$-sequence.)

3 E. Show that $\diamondsuit$ implies the existence of two non-isomorphic Souslin trees.

3 F. Show that $\diamondsuit$ implies the existence of an $\mathbb{R}$-embeddable tree which is not $\mathbb{Q}$-embeddable.

3 G. Show that $\diamondsuit$ implies the existence of a family $\{A_\nu \mid \nu < \omega_2\}$ of stationary subsets of ω_1 such that the intersection of any two of them is countable.

4. *$\diamondsuit$ and $\diamondsuit^+$ in* $L[A]$ (Section 3)

Using the same kind of ideas employed in Exercises II.2 and II.4, we prove that $\diamondsuit$ and $\diamondsuit^+$ hold in $L[A]$, where $A \subseteq \omega_1^{L[A]}$.

4 A. Assume $V = L[A]$, where $A \subseteq \omega_1$. Prove that $\diamondsuit$ is valid. (Hint: For each limit ordinal α, let (S_α, C_α) be the $<_{L[A\cap\alpha]}$-least pair of subsets of α lying in $L[A \cap \alpha]$ such that C_α is club in α and $S_\alpha \cap \gamma \neq S_\gamma$ for all $\gamma \in C_\alpha$, whenever possible. Now argue analogously to 3.3.)

4 B. Suppose $V = L[A]$, where $A \subseteq \omega_1$. Prove that if $\omega_1^{L[A\cap\alpha]} < \omega_1$ for all $\alpha < \omega_1$, then ω_1 is inaccessible in $L[A \cap \alpha]$ for all $\alpha < \omega_1$. (Hint: If ω_1 were not inaccessible in $L[A \cap \alpha]$ for all $\alpha < \omega_1$, then for some α we would have $\omega_1 = (\theta^+)^{L[A\cap\alpha]}$. By a condensation argument, θ can be shown to be countable in some $L[A \cap \gamma]$. Then $\omega_1 = \omega_1^{L[A\cap\delta]}$ for $\delta = \max(\alpha, \gamma)$.)

4 C. Assume $V = L[A]$, where $A \subseteq \omega_1$. Prove that $\diamondsuit^+$ is valid. (Hint: Define $\delta\colon \omega_1 \to \omega_1$ by cases, depending on A. If $\omega_1 = \omega_1^{L[A\cap\alpha]}$ for some $\alpha < \omega_1$, let α_0 be the least such, and let $\delta(\alpha) = \omega_1 \cap M_\alpha$, where M_α is the smallest $M \prec L_{\omega_1}[A]$ such

that $\alpha_0, \alpha \in M$. Otherwise let $\delta(\alpha) = \omega_3^{L[A \cap \alpha]}$ (which is countable by virtue of 4 B), and set $\alpha_0 = \omega$. For $\alpha < \omega_1$ now, let $\hat{\alpha} = \max(\alpha, \alpha_0)$. Set $S_\alpha = \mathscr{P}(\alpha) \cap L_{\delta(\alpha)}[A \cap \hat{\alpha}]$. Now argue as in 3.7, except for the fact that there are now the two cases to consider instead of one.)

4 D. Prove that if there is no Kurepa tree, then ω_2 is inaccessible in L. (Hint: Use 4 C, together with an absoluteness argument concerning Kurepa trees.)

Chapter IV
κ^+-Trees in L and the Fine Structure Theory

In this chapter we shall investigate natural generalisations of the Souslin and Kurepa hypotheses to cardinals above ω_1. In the case of the Souslin hypothesis this will require some combinatorial properties of L which we shall only be able to prove by developing the theory of the constructible hierarchy more thoroughly than hitherto. (This is the so-called "fine-structure theory".)

1. κ^+-Trees

Let κ be an infinite cardinal. The concept of a *κ-tree* was defined in Chapter III. By a *κ-Aronszajn tree* we mean a κ-tree with no κ-branch. A *κ-Souslin tree* is a κ-tree with no antichain of cardinality κ. Just as in III.1.2, every κ-Souslin tree is κ-Aronszajn. And by arguments as in III.1.3, if κ is regular, then any (κ, κ)-tree with unique limits which has no κ-branch has a subtree which is κ-Aronszajn; if in addition the original tree has no antichain of cardinality κ, it has a subtree which is κ-Souslin. The regularity of κ is essential here. Indeed, for singular κ, the notion of a κ-tree is somewhat pathalogical. For example, if κ is singular there is a (κ, κ)-tree with no κ-branch and no antichain of cardinality κ (namely the disjoint union of the well-ordered sets $(\kappa_\nu, \varepsilon)$, $\nu < \mathrm{cf}(\kappa)$, where $(\kappa_\nu \mid \nu < \mathrm{cf}(\kappa))$ is cofinal in κ), but every κ-tree has an antichain of cardinality κ (an easy exercise). We therefore restrict our attention to κ-trees for regular κ only. Since we shall be assuming $V = L$ for our main results, GCH will hold, and hence the only regular limit cardinals are the (strongly) inaccessible cardinals. In this context we may therefore expect the notion of a κ-tree for κ a regular limit cardinal to be bound up with the notion of large cardinals. As we shall see in Chapter VII, this is in fact the case. In this chapter we concentrate only upon the successor cardinals.

By a *κ^+-Kurepa tree* we mean a κ^+-tree with κ^{++} many κ^+-branches. (Adopting a similar definition of a "κ-Kurepa tree" for inaccessible κ does not lead to any interesting notions, as we see in Exercise 3. More care is required in order to define a reasonable notion of a κ-Kurepa tree in this case.) As in III.2.1, the existence of a κ^+-Kurepa tree can be shown to be equivalent to the existence of a certain kind of family of subsets of κ^+. Moreover, the proof that such a family exists in L is a straightforward generalisation of the proof for the ω_1 case, given in III.2.2.

However, when we try to construct a κ^+-Souslin tree in L we run into some difficulties. It turns out to be slightly easier to try to generalise the proof using $\diamondsuit$ (III.3.2 and III.3.3) rather than the original construction (III.1.5). Now, the proof of $\diamondsuit$ generalises from ω_1 to any uncountable regular cardinal in a straightforward manner. Hence the generalised construction of the tree hinges upon a generalisation of the proof of III.3.3. This is not so easy. For suppose we try to construct a κ^+-Souslin tree by recursion on the levels. Consideration of the proof of III.3.3 tells us that on a stationary set of levels we must be very restrictive in the choice of branches to extend, in order that all antichains be eventually "killed-off". But consider now some limit stage α "late" in the construction. We have defined the tree $\mathbf{T} \upharpoonright \alpha$ and wish to define T_α. Each point of T_α must extend some α-branch of $\mathbf{T} \upharpoonright \alpha$. But unless $\mathrm{cf}(\alpha) = \omega$, how can we be sure that $\mathbf{T} \upharpoonright \alpha$ has *any* α-branches? Our attempts to kill off antichains at earlier limit stages may have resulted in $\mathbf{T} \upharpoonright \alpha$ having a sort of "Aronszajn property". To overcome this problem we introduce a combinatorial principle, $\square_\kappa$ ("square κ"), which enables us to split the construction of the limit levels of the tree into two cases. At some limit stages we kill off antichains, using the generalised $\diamondsuit$ principle. At the remaining limit stages we ensure that enough branches are extended in order that the construction will never break down. The penalty we must pay in order to be able to do this lies in the proof of $\square_\kappa$. This requires a detailed analysis of the levels of the constructible hierachy (the "fine-structure theory"). This will occupy the later parts of this chapter.

2. κ^+-Souslin Trees

We prove that if $V = L$, then for all infinite cardinals κ there is a κ^+-Souslin tree. Our first step is to formulate and prove a generalisation of the combinatorial principle $\diamondsuit$.

Let κ be any uncountable regular cardinal, E a stationary subset of κ. By $\diamondsuit_\kappa(E)$ we mean the following assertion:

> *There is a sequence $(S_\alpha \mid \alpha \in E)$ such that $S_\alpha \subseteq \alpha$ for all α and whenever $X \subseteq \kappa$, the set $\{\alpha \in E \mid X \cap \alpha = S_\alpha\}$ is stationary in κ.*

We denote $\diamondsuit_\kappa(\kappa)$ by $\diamondsuit_\kappa$. Thus $\diamondsuit_{\omega_1}$ is the same as our original principle $\diamondsuit$.

In order to prove that $\diamondsuit_\kappa(E)$ is valid in L we require the following simple lemma.

2.1 Lemma. *Let κ be an uncountable regular cardinal, λ a limit ordinal greater than κ. Let $X \subseteq L_\lambda$, $|X| < \kappa$. Then there is an $N \prec L_\lambda$ such that $X \subseteq N$, $|N| < \kappa$, and $N \cap \kappa \in \kappa$.*

Proof. Let N_0 be the smallest $N \prec L_\lambda$ such that $X \subseteq N$, and set

$$\alpha_0 = \sup(N_0 \cap \kappa).$$

Since $|N_0| = \max(|X|, \omega) < \kappa$, and κ is regular, we have $\alpha_0 < \kappa$. Proceeding recursively now, let N_{n+1} be the smallest $N \prec L_\lambda$ such that $N_n \cup \alpha_n \subseteq N$, and set

$$\alpha_{n+1} = \sup(N_{n+1} \cap \kappa).$$

If $|N_n| < \kappa$ and $\alpha_n < \kappa$, then

$$|N_{n+1}| = \max(|N_n|, |\alpha_n|) < \kappa,$$

so as κ is regular, $\alpha_{n+1} < \kappa$.

Let

$$N = \bigcup_{n<\omega} N_n.$$

Then

$$X \subseteq N \prec L_\lambda,$$

and

$$N \cap \kappa = (\bigcup_{n<\omega} N_n) \cap \kappa = \bigcup_{n<\omega} (N_n \cap \kappa).$$

But for $n > 0$,

$$\alpha_{n-1} \subseteq N_n \cap \kappa \subseteq \alpha_n.$$

Hence

$$N \cap \kappa = \bigcup_{n<\omega} \alpha_n.$$

So, if we set

$$\alpha = \sup\nolimits_{n<\omega} \alpha_n,$$

we have $N \cap \kappa = \alpha$. But κ is regular. Thus $|N| < \kappa$ and $\alpha < \kappa$, so we are done. □

2.2 Theorem. *Assume $V = L$. Let κ be any uncountable regular cardinal, E a stationary subset of κ. Then $\diamondsuit_\kappa(E)$ is valid.*

Proof. By recursion on $\alpha \in E$, define (S_α, C_α) to be the $<_L$-least pair of subsets of α such that C_α is club in α and

$$\gamma \in C_\alpha \cap E \to S_\alpha \cap \gamma \neq S_\gamma,$$

provided $\lim(\alpha)$ and such a pair exists, and define $S_\alpha = C_\alpha = \emptyset$ in all other cases. We show that $(S_\alpha \mid \alpha \in E)$ satisfies $\diamondsuit_\kappa(E)$.

Suppose that $(S_\alpha \mid \alpha \in E)$ is not a $\diamondsuit_\kappa(E)$-sequence. Let (S, C) be the $<_L$-least pair of subsets of κ such that C is club in κ and

$$\gamma \in C \cap E \to S \cap \gamma \neq S_\gamma.$$

Now, the sequence $((S_\alpha, C_\alpha) | \alpha \in E)$ is clearly definable from E in L_{κ^+}. (The definition given above is absolute for L_{κ^+}.) Hence (S, C) is also definable from E in L_{κ^+}. Using 2.1, we now define a sequence of submodels $N_\nu \prec L_{\kappa^+}$ $(\nu < \kappa)$, by the following recursion:

$$N_0 = \text{the smallest } N \prec L_{\kappa^+} \text{ such that } |N| < \kappa,\ N \cap \kappa \in \kappa, \text{ and } E \in N;$$

$$N_{\nu+1} = \text{the smallest } N \prec L_{\kappa^+} \text{ such that } |N| < \kappa,\ N \cap \kappa \in \kappa, \text{ and } N_\nu \cup \{N_\nu\} \subseteq N;$$

$$N_\lambda = \bigcup_{\nu<\lambda} N_\nu, \text{ if } \lim(\lambda). \text{ (Clearly, } |N_\lambda| < \kappa \text{ and } N_\lambda \cap \kappa \in \kappa \text{ here also.)}$$

Set

$$\alpha_\nu = N_\nu \cap \kappa \quad (\nu < \kappa).$$

Then $(\alpha_\nu | \nu < \kappa)$ is a normal sequence in κ, so the set

$$Z = \{\alpha_\nu | \alpha_\nu = \nu\}$$

is club in κ. Hence

$$E \cap Z \cap C \neq \emptyset.$$

Let $\alpha_\nu \in E \cap Z \cap C$. Let

$$\pi: N_\nu \cong L_\beta.$$

Then,

$$\pi \upharpoonright L_\nu = \mathrm{id} \upharpoonright L_\nu, \quad \pi(\kappa) = \nu, \quad \pi(E) = E \cap \nu,$$
$$\pi((S_\alpha, C_\alpha) | \alpha \in E)) = ((S_\alpha, C_\alpha) | \alpha \in E \cap \nu), \quad \pi((S, C)) = (S \cap \nu, C \cap \nu).$$

Since $\pi^{-1}: L_\beta \prec L_{\kappa^+}$, $(S \cap \nu, C \cap \nu)$ is the $<_L$-least pair of subsets of ν such that $C \cap \nu$ is club in ν and

$$\gamma \in (C \cap \nu) \cap (E \cap \nu) \rightarrow (S \cap \nu) \cap \gamma \neq S_\gamma.$$

Hence $(S \cap \nu, C \cap \nu) = (S_\nu, C_\nu)$, and in particular $S \cap \nu = S_\nu$. But $\nu \in C \cap E$, so this contradicts the choice of (S, C), and we are done. □

Using $\diamondsuit_{\kappa^+}(E)$ for a suitable set E, in the case where κ is regular it is quite easy to construct a κ^+-Souslin tree in L. We take

$$E = \{\alpha \in \kappa^+ | \mathrm{cf}(\alpha) = \kappa\},$$

and construct the tree by recursion on the levels, following the pattern of III.3.3. At limit stages $\alpha \in E$ we extend branches to "kill off" S_α, if S_α happens to be a maximal antichain of $\mathbf{T} \upharpoonright \alpha$. At all other limit stages α we extend *all* α-branches of $\mathbf{T} \upharpoonright \alpha$, noting that as $\mathrm{cf}(\alpha) < \kappa$ in such cases, there are at most $\kappa^{\mathrm{cf}(\alpha)} = \kappa$ (by GCH) such branches, so that T_α will not be too big. We leave the details to the reader (see Exercise 2).

If κ is singular, however, the above idea will not work. It is in order to handle this case that we need to introduce the combinatorial principle $\square_\kappa(E)$. Using $\square_\kappa(E)$, we shall give a construction of a κ^+-Souslin tree which works in all cases.

Let κ be an infinite cardinal, E a subset of κ^+. By $\square_\kappa(E)$ we mean the following assertion:

There is a sequence $(C_\alpha \mid \alpha < \kappa^+ \wedge \lim(\alpha))$ *such that*:

(i) C_α *is club in* α;
(ii) $\mathrm{cf}(\alpha) < \kappa \to \mathrm{otp}(C_\alpha) < \kappa$;
(iii) *if* $\bar{\alpha} < \alpha$ *is a limit point of* C_α, *then* $\bar{\alpha} \notin E$ *and* $C_{\bar{\alpha}} = \bar{\alpha} \cap C_\alpha$.

Notice that by virtue of condition (iii), condition (ii) can be extended to give the implication

$$\text{(ii)}' \qquad \mathrm{cf}(\alpha) = \kappa \to \mathrm{otp}(C_\alpha) = \kappa .$$

Notice also that if κ is singular, we shall have $\mathrm{cf}(\alpha) < \kappa$ for all relevant α, so $\mathrm{otp}(C_\alpha) < \kappa$ for all α.

For any set $E \subseteq \omega_1$, $\square_\omega(E)$ is a theorem of ZFC, since for each limit ordinal $\alpha < \omega_1$ we can take C_α to be any ω-sequence cofinal in α. But already $\square_{\omega_1}(E)$ is a significant proposition, not provable in ZFC alone.

We shall write $\square_\kappa$ in place of $\square_\kappa(\emptyset)$.

In 2.10 we shall prove that if $\square_\kappa$, then there is a stationary set $E \subseteq \kappa^+$ such that $\square_\kappa(E)$. And then in section 5 we shall prove the following theorem.

2.3 Theorem. *Assume* $V = L$. *Let* κ *be an infinite cardinal. Then* $\square_\kappa$ *is valid.* □

We are now ready to construct a κ^+-Souslin tree in L.

2.4 Theorem. *Assume* $V = L$. *Let* κ *be an infinite cardinal. Then there is a* κ^+-*Souslin tree.*

Proof. By 2.3 and 2.10, let $E \subseteq \kappa^+$ be stationary and let $(C_\alpha \mid \alpha < \kappa^+ \wedge \lim(\alpha))$ satisfy $\square_\kappa(E)$. By 2.2, let $(S_\alpha \mid \alpha \in E)$ satisfy $\diamondsuit_{\kappa^+}(E)$. We shall construct a κ^+-Souslin tree, $\mathbf{T}$, by recursion on the levels, ensuring as we proceed that for each infinite $\alpha < \kappa^+$, $\mathbf{T} \upharpoonright \alpha$ is a normal (α, α^+)-tree. The elements of $\mathbf{T}$ will be the ordinals in κ^+, and we shall ensure that

$$\alpha <_T \beta \to \alpha < \beta .$$

To commence, set

$$T_0 = \{0\} .$$

If $\mathbf{T} \upharpoonright \alpha + 1$ is defined, $T_{\alpha+1}$ is obtained by using new ordinals from κ^+ to provide each element of T_α with two successors in $T_{\alpha+1}$. There remains the case where $\lim(\alpha)$ and $\mathbf{T} \upharpoonright \alpha$ is defined. This is where we must proceed carefully.

For each $x \in T \upharpoonright \alpha$ we attempt to define an α-branch b_α^x of $\mathbf{T} \upharpoonright \alpha$ such that $x \in b_\alpha^x$. Let $(\gamma_\alpha(\nu) \mid \nu < \lambda_\alpha)$ be the monotone enumeration of C_α. Given $x \in T \upharpoonright \alpha$, let

$\nu_\alpha(x)$ be the least ν such that $x \in T \upharpoonright \gamma_\alpha(\nu)$. Define a sequence $(p_\alpha^x(\nu) \mid \nu_\alpha(x) \leqslant \nu < \lambda_\alpha)$ of elements of $T \upharpoonright \alpha$ as follows:

$$p_\alpha^x(\nu_\alpha(x)) = \text{the least (as an ordinal) } y \in T_{\gamma_\alpha(\nu_\alpha(x))} \text{ such that } x <_T y;$$

$$p_\alpha^x(\nu + 1) = \text{the least } y \in T_{\gamma_\alpha(\nu+1)} \text{ such that } p_\alpha^x(\nu) <_T y;$$

$$p_\alpha^x(\eta) = \text{the unique } y \in T_{\gamma_\alpha(\eta)} \text{ such that } (\forall \nu < \eta)(\nu \geqslant \nu_\alpha(x) \rightarrow p_\alpha^x(\nu) <_T y), \text{ provided such a } y \text{ exists (otherwise undefined), if } \lim(\eta).$$

Should the above construction prove impossible (because for some limit ordinal $\eta < \lambda_\alpha$, $p_\alpha^x(\eta)$ is not defined), the entire construction of **T** breaks down. But for the time being, let us assume that this is not the case and see how b_α^x is defined. Later on we shall prove (by induction on α) that the construction never breaks down. Set

$$b_\alpha^x = \{y \in T \upharpoonright \alpha \mid (\exists \nu < \lambda_\alpha)(y \leqslant_T p_\alpha^x(\nu))\}.$$

Clearly, b_α^x is an α-branch of $\mathbf{T} \upharpoonright \alpha$ which contains x and each point $p_\alpha^x(\nu)$ for $\nu_\alpha(x) \leqslant \nu < \lambda_\alpha$. We now define T_α as follows.

Suppose first that $\alpha \notin E$. In this case we use new ordinals from κ^+ to provide each branch b_α^x, $x \in T \upharpoonright \alpha$, with an extension in T_α.

Now suppose that $\alpha \in E$, but that S_α is not a maximal antichain of $\mathbf{T} \upharpoonright \alpha$. In this case construct T_α just as in the last case.

Finally, suppose that $\alpha \in E$ and that S_α is a maximal antichain of $\mathbf{T} \upharpoonright \alpha$. Then use new ordinals from κ^+ to provide an extension in T_α of each branch b_α^x such that $x \in T \upharpoonright \alpha$ lies above an element of S_α. (Since S_α is assumed to be a maximal antichain here, T_α will still contain a point above each member of $T \upharpoonright \alpha$, so normality will be preserved.)

The definition is complete. We show that **T** is a κ^+-Souslin tree. It is clearly a κ^+-tree. So, given a maximal antichain, A, of **T**, we must show that $|A| \leqslant \kappa$. Set

$$C = \{\alpha \in \kappa^+ \mid T \upharpoonright \alpha \subseteq \alpha \wedge A \cap \alpha \text{ is a maximal antichain of } \mathbf{T} \upharpoonright \alpha\}.$$

It is easily seen that C is club in κ^+. So by $\diamondsuit_{\kappa^+}(E)$ there is a limit ordinal $\alpha \in C \cap E$ such that $A \cap \alpha = S_\alpha$. Thus, in particular, S_α is a maximal antichain of $\mathbf{T} \upharpoonright \alpha$. But $\alpha \in E$, so by construction every element of T_α lies above a member of $A \cap \alpha$. Thus $A \cap \alpha$ is a maximal antichain of **T**. Hence $A = A \cap \alpha$, and we are done.

It remains to check that the construction of **T** never broke down. Suppose, on the contrary, that it did. Let α be the least limit ordinal for which we cannot define all the α-branches b_α^x, $x \in T \upharpoonright \alpha$. Pick $x \in T \upharpoonright \alpha$ so that b_α^x cannot be defined. Thus for some limit ordinal η, $\nu_\alpha(x) < \eta < \lambda_\alpha$, there is no point in $T_{\gamma_\alpha(\eta)}$ which extends all the points $p_\alpha^x(\nu)$ for $\nu_\alpha(x) \leqslant \nu < \eta$. Since $\lim(\eta)$, $\gamma_\alpha(\eta)$ is a limit point of C_α. Hence by the $\square_\kappa(E)$ properties, $\gamma_\alpha(\eta) \notin E$ and

$$C_{\gamma_\alpha(\eta)} = \gamma_\alpha(\eta) \cap C_\alpha = \{\gamma_\alpha(\nu) \mid \nu < \eta\}.$$

By this last equality, $b^x_{\gamma_\alpha(\eta)}$ contains all the points $p^x_\alpha(v)$ for $v_\alpha(x) \leqslant v < \eta$. But since $\gamma_\alpha(\eta) \notin E$, $b^x_{\gamma_\alpha(\eta)}$ has an extension in $T_{\gamma_\alpha(\eta)}$. But this extension is precisely what we assumed did not exist: an extension of each point $p^x_\alpha(v)$, $v_\alpha(x) \leqslant v < \eta$. This contradiction shows that the construction of **T** does not, in fact, break down, and thereby completes the proof. □

Notice that what we have in fact just proved is the following result.

2.5 Theorem. *Let κ be an infinite cardinal. If there is a stationary set $E \subseteq \kappa^+$ such that both $\square_\kappa(E)$ and $\diamondsuit_{\kappa^+}(E)$ hold, then there is a κ^+-Souslin tree.* □

Using 2.5, we shall show that κ^+-Souslin trees exist under much weaker assumptions than $V = L$. We need some preliminary combinatorial results.

By an argument as in III.3.4 we have:

2.6 Lemma. *Let κ be any infinite cardinal, and let $E \subseteq \kappa^+$ be stationary. Then $\diamondsuit_{\kappa^+}(E)$ is equivalent to the principle $\diamondsuit'_{\kappa^+}(E)$, which asserts the existence of a sequence $(S_\alpha \mid \alpha \in E)$ such that $S_\alpha \subseteq \mathscr{P}(\alpha)$, $|S_\alpha| \leqslant \kappa$, and whenever $X \subseteq \kappa^+$, the set $\{\alpha \in E \mid X \cap \alpha \in S_\alpha\}$ is stationary in κ^+.* □

Using 2.6, we now prove (see also Exercise 7):

2.7 Lemma. *Assume* GCH. *Let κ be an infinite cardinal such that* $\mathrm{cf}(\kappa) > \omega$. *Let $W \subseteq \kappa^+$ be the stationary set*

$$W = \{\alpha \in \kappa^+ \mid \mathrm{cf}(\alpha) = \omega\}.$$

Then $\diamondsuit_{\kappa^+}(W)$ is valid.

Proof. By GCH there are exactly κ^+ many subsets of κ^+ of cardinality at most κ. Let $(X_v \mid v < \kappa^+)$ enumerate them in such a way that $X_v \subseteq v$ for each $v < \kappa^+$. For each $\alpha < \kappa^+$, set

$$\Gamma_\alpha = \{X_v \mid v < \alpha\}.$$

For each $\alpha \in W$, let

$$S_\alpha = \{\bigcup \mathrm{ran}(f) \mid f\colon \omega \to \Gamma_\alpha\}.$$

Since $|\Gamma_\alpha| \leqslant \kappa$ and $\mathrm{cf}(\kappa) > \omega$,

$$|S_\alpha| \leqslant |\Gamma_\alpha|^\omega \leqslant \kappa^\omega = \kappa.$$

And of course

$$S_\alpha \subseteq \mathscr{P}(\alpha).$$

We show that $(S_\alpha \mid \alpha \in W)$ is a $\diamondsuit'_{\kappa^+}(W)$-sequence (see 2.6).

Let $X \subseteq \kappa^+$ be given. Let $C \subseteq \kappa^+$ be club. We must find an $\alpha \in C \cap W$ such that $X \cap \alpha \in S_\alpha$. To this end, define a strictly increasing sequence $(\alpha_n \mid n < \omega)$ of elements of C as follows, by recursion. Let α_0 be the smallest infinite ordinal in C. If $\alpha_n \in C$ is defined, let α_{n+1} be the least element of C such that $\alpha_{n+1} > \alpha_n$ and $X \cap \alpha_n \in \Gamma_{\alpha_{n+1}}$. Let

$$\alpha = \sup_{n<\omega} \alpha_n.$$

Since C is closed in κ^+, $\alpha \in C$. Moreover, $\operatorname{cf}(\alpha) = \omega$, so $\alpha \in W$. Define $f\colon \omega \to \Gamma_\alpha$ by

$$f(n) = X \cap \alpha_n \quad (n < \omega).$$

Clearly,

$$X \cap \alpha = \bigcup \operatorname{ran}(f) \in S_\alpha,$$

so we are done. $\square$

In the above proof, we used the assumption $\operatorname{cf}(\kappa) > \omega$ in order to ensure that the sets S_α had cardinality at most κ. But what about the status of $\diamondsuit_{\kappa^+}(W)$ when $\operatorname{cf}(\kappa) = \omega$? Well, if we assume $\square_\kappa$ in addition to GCH, we can modify the proof of 2.7 to cover this case also, as we show next. (See also Exercise 8.)

2.8 Lemma. *Assume* GCH. *Let κ be an uncountable cardinal such that* $\operatorname{cf}(\kappa) = \omega$, *and let $W \subseteq \kappa^+$ be the stationary set*

$$W = \{\alpha \in \kappa^+ \mid \operatorname{cf}(\alpha) = \omega\}.$$

If $\square_\kappa$ holds, then $\diamondsuit_{\kappa^+}(W)$ is valid.

Proof. Define Γ_α, $\alpha < \kappa^+$ as in 2.7. Let $(C_\lambda \mid \lambda < \kappa^+ \wedge \lim(\lambda))$ be a $\square_\kappa$-sequence, and for each λ let $(c_\nu^\lambda \mid \nu < \theta_\lambda)$ be the canonical enumeration of C_λ. (Thus $\theta_\lambda = \operatorname{otp}(C_\lambda)$.)

Let A_ν, $\nu < \kappa$, be disjoint subsets of κ of cardinality κ such that $\kappa = \bigcup_{\nu<\kappa} A_\nu$. For each $\delta < \kappa^+$ and each $\nu < \kappa$, let

$$f_\nu^\delta\colon \Gamma_\delta \xrightarrow{1-1} A_\nu.$$

Then for each limit $\lambda < \kappa^+$ we can define

$$f_\lambda\colon \Gamma_\lambda \xrightarrow{1-1} \kappa$$

by setting $f_\lambda(x) = f_\nu^{c_\nu^\lambda}(x)$ where $\nu < \theta_\lambda$ is least such that $x \in \Gamma_{c_\nu^\lambda}$. The important point to notice here is the following:

(*) If $\alpha < \lambda$ is a limit point of C_λ, then $f_\lambda \upharpoonright \Gamma_\alpha = f_\alpha$.

(This is immediate from the fact that $C_\alpha = \alpha \cap C_\lambda$ in this case.)

For $\alpha \in W$ now, set

$$S_\alpha = \{\bigcup f_\alpha^{-1}[x] \mid x \text{ is a countable, bounded subset of } \kappa\}.$$

Then $S_\alpha \subseteq \mathscr{P}(\alpha)$, and, since the number of countable, *bounded* subsets of κ is κ, $|S_\alpha| \leqslant \kappa$. We show that $(S_\alpha \mid \alpha \in W)$ is a $\diamondsuit'_{\kappa^+}(W)$-sequence (as in 2.6).

Let $X \subseteq \kappa^+$ be given. Let $C \subseteq \kappa^+$ be club. We seek an $\alpha \in C \cap W$ such that $X \cap \alpha \in S_\alpha$. Define

$$A = \{\lambda \in \kappa^+ \mid (\forall \nu < \lambda)(X \cap \nu \in \Gamma_\lambda)\}.$$

Clearly, A is club in κ^+. Let λ be a limit point of $A \cap C$ such that $\mathrm{cf}(\lambda) = \omega_1$. Since C_λ is club in λ we can pick a strictly increasing, continuous sequence $(b_\nu \mid \nu < \omega_1)$ of elements of $A \cap C \cap C_\lambda$, cofinal in λ. Notice that

$$X \cap b_\nu \in \Gamma_{b_{\nu+1}}$$

for all $\nu < \omega_1$.

Let $(\kappa_n \mid n < \omega)$ be a strictly increasing sequence of cardinals, cofinal in κ. Define $h\colon \omega_1 \to \omega$ by:

$$h(\nu) = \text{the least } n \text{ such that } f_\lambda(X \cap b_\nu) < \kappa_n.$$

By Fodor's Theorem there is a stationary set $E \subseteq \omega_1$ such that for some fixed $n < \omega$, $h(\nu) = n$ for all $\nu \in E$. Let $(\gamma(i) \mid i < \omega)$ enumerate (in order) the first ω elements of E, and set $\gamma = \sup_{i<\omega} \gamma(i)$. Let $\alpha = b_\gamma$. Notice that $\mathrm{cf}(\gamma) = \omega$, so $\alpha \in W$. Moreover, by choice of the elements b_ν, α is a limit point of $C \cap C_\lambda$, and in particular $\alpha \in C$.

Now,

$$\alpha = b_\gamma = \sup_{i<\omega} b_{\gamma(i)},$$

so

$$X \cap \alpha = \bigcup_{i<\omega} (X \cap b_{\gamma(i)}).$$

Thus

$$X \cap \alpha = \bigcup f_\lambda^{-1}[x],$$

where $x \subseteq \kappa$ is defined by

$$x = \{f_\lambda(X \cap b_{\gamma(i)}) \mid i < \omega\}.$$

But by choice of E,

$$x \subseteq \kappa_n < \kappa,$$

so x is a countable, bounded subset of κ. Moreover, by (*),

$$f_\lambda \upharpoonright \Gamma_\alpha = f_\alpha.$$

Hence

$$X \cap \alpha \in S_\alpha,$$

and we are done. □

2.9 Lemma. *Let κ be any infinite cardinal, and let $E \subseteq \kappa^+$ be stationary. Suppose that $\diamondsuit_{\kappa^+}(E)$ is valid. Let*

$$E = \bigcup_{\nu<\kappa} E_\nu$$

be a disjoint partition of E. Then for some $\nu < \kappa$, E_ν is stationary and $\diamondsuit_{\kappa^+}(E_\nu)$ is valid.

Proof. Much as in III.3.4, by $\diamondsuit_{\kappa^+}(E)$ we can find a sequence $(T_\alpha \mid \alpha \in E)$ such that $T_\alpha \subseteq \alpha \times \kappa$ and for each $X \subseteq \kappa^+ \times \kappa$, the set

$$\{\alpha \in E \mid X \cap (\alpha \times \kappa) = T_\alpha\}$$

is stationary in κ^+. For each $\nu < \kappa$, define $(S_\alpha^\nu \mid \alpha \in E_\nu)$ by

$$S_\alpha^\nu = T_\alpha'' \{\nu\}.$$

We show that for some $\nu < \kappa$, $(S_\alpha^\nu \mid \alpha \in E_\nu)$ is a $\diamondsuit_{\kappa^+}(E_\nu)$-sequence. (This will automatically entail that E_ν is stationary, of course.) Suppose that, on the contrary, no sequence $(S_\alpha^\nu \mid \alpha \in E_\nu)$ is a $\diamondsuit_{\kappa^+}(E_\nu)$-sequence. Then for each $\nu < \kappa$ we can find a set $X_\nu \subseteq \kappa^+$ and a club set $C_\nu \subseteq \kappa^+$ such that

$$\alpha \in C_\nu \cap E_\nu \to X_\nu \cap \alpha \neq S_\alpha^\nu.$$

Set

$$X = \bigcup_{\nu<\kappa} (X_\nu \times \{\nu\}),$$

$$C = \bigcap_{\nu<\kappa} C_\nu.$$

Then C is club in κ^+, and, since $X'' \{\nu\} = X_\nu$ for each $\nu < \kappa$,

$$\alpha \in C \cap E \to X \cap (\alpha \times \kappa) \neq T_\alpha,$$

which is a contradiction. The lemma is proved. □

2.10 Lemma. *Let κ be any uncountable cardinal for which $\square_\kappa$ is valid. Let $W \subseteq \kappa^+$ be the stationary set*

$$W = \{\alpha \in \kappa^+ \mid \mathrm{cf}(\alpha) = \omega\}.$$

Then there is a stationary set $E \subseteq W$ such that:

(i) *$\square_\kappa(E)$ is valid;*
(ii) *if $\diamondsuit_{\kappa^+}(W)$, then $\diamondsuit_{\kappa^+}(E)$.*

(*Thus, by* 2.7 *and* 2.8, *if* GCH *holds, then $\diamondsuit_{\kappa^+}(E)$ follows from* (ii).)

Proof. Let $(A_\lambda | \lambda < \kappa^+ \wedge \lim(\lambda))$ be a $\Box_\kappa$-sequence. For each λ, let B_λ be the set of limit points of A_λ below λ. The sequence $(B_\lambda | \lambda < \kappa^+ \wedge \lim(\lambda))$ has the following properties:

(i) B_λ is a closed subset of λ;
(ii) if $\mathrm{cf}(\lambda) > \omega$, then B_λ is unbounded in λ;
(iii) $\gamma \in B_\lambda \to B_\gamma = \gamma \cap B_\lambda$;
(iv) $\mathrm{cf}(\lambda) < \kappa \to |B_\lambda| < \kappa$.

By (iii) and (iv), $\mathrm{otp}(B_\lambda) \leqslant \kappa$ for all λ, so we can define a partition

$$W = \bigcup_{\nu \leqslant \kappa} W_\nu$$

by setting

$$W_\nu = \{\lambda \in W \,|\, \mathrm{otp}(B_\lambda) = \nu\}.$$

Now, W is stationary, so for at least one $\nu \leqslant \kappa$, W_ν must be stationary. Indeed, by 2.9 we can pick a $\nu \leqslant \kappa$ such that W_ν is stationary and

$$\diamondsuit_{\kappa^+}(W) \to \diamondsuit_{\kappa^+}(W_\nu).$$

Let $E = W_\nu$ for such a ν. We prove that $\Box_\kappa(E)$ holds. For each limit ordinal $\lambda < \kappa^+$, define D_λ as follows. If $\mathrm{otp}(B_\lambda) \leqslant \nu$, let $D_\lambda = B_\lambda$.Otherwise, let D_λ consist of all members of B_λ beyond the $(1 + \nu)$-th element, i.e.

$$D_\lambda = B_\lambda - \{\alpha \in B_\lambda \,|\, \mathrm{otp}(B_\alpha) \leqslant \nu\}.$$

It is easily checked that the sequence $(D_\lambda \,|\, \lambda < \kappa^+ \wedge \lim(\lambda))$ has properties (i)–(iv) above. And clearly, $D_\lambda \cap E = \emptyset$ for all λ. Define C_λ for limit $\lambda < \kappa^+$ by recursion on λ as follows:

$$C_\lambda = \begin{cases} \bigcup \{C_\gamma | \gamma \in D_\lambda\}, & \text{if } \sup(D_\lambda) = \lambda, \\ \bigcup \{C_\gamma | \gamma \in D_\lambda\} \cup \{\theta_n^\lambda \,|\, n < \omega\}, & \text{otherwise, where } (\theta_n^\lambda \,|\, n < \omega) \end{cases}$$

is any strictly increasing ω-sequence cofinal in λ such that $\theta_0^\lambda = \bigcup D_\lambda$. (By (ii) for D_λ, we have $\mathrm{cf}(\lambda) = \omega$ in case $\sup(D_\lambda) < \lambda$.)

We shall prove that $(C_\lambda | \lambda < \kappa^+ \wedge \lim(\lambda))$ is a $\Box_\kappa$-sequence and that D_λ is the set of all limit points of C_λ below λ for each λ (which implies at once that $(C_\lambda | \lambda < \kappa^+ \wedge \lim(\lambda))$ is in fact a $\Box_\kappa(E)$-sequence, since $D_\lambda \cap E = \emptyset$ for all λ).

A trivial induction on λ shows that C_λ is unbounded in λ for each λ. Now, by induction on λ, we prove:

(a) if $\gamma \in D_\lambda$, then $C_\gamma = \gamma \cap C_\lambda$.

Assume (a) holds below λ. Let $\gamma \in D_\lambda$. Then by definition of C_λ, $C_\gamma \subseteq C_\lambda$. So $C_\gamma \subseteq \gamma \cap C_\lambda$. To prove the reverse inclusion, let $\xi \in \gamma \cap C_\lambda$. We show that $\xi \in C_\gamma$. By the definition of C_λ, for some $\delta \in D_\lambda$ we have $\xi \in \gamma \cap C_\delta$. If $\delta = \gamma$ then $\xi \in C_\gamma$

is immediate. Suppose that $\delta < \gamma$. Since $\gamma \in D_\lambda$, we have $D_\gamma = \gamma \cap D_\lambda$. Thus $\delta \in D_\gamma$. So by definition of C_γ, $C_\delta \subseteq C_\gamma$. Thus $\xi \in C_\gamma$. Finally, suppose that $\delta > \gamma$. Then $\gamma \in \delta \cap D_\lambda$. But $\delta \in D_\lambda$, so $D_\delta = \delta \cap D_\lambda$. Thus $\gamma \in D_\delta$. So by induction hypothesis at δ, $C_\gamma = \gamma \cap C_\delta$. Thus $\xi \in C_\gamma$, and we are done.

The next step is to prove:

(b) D_λ is the set of all limit points of C_λ below λ.

Again we proceed by induction on λ. Assume that (b) holds below λ. Let $\xi \in D_\lambda$. Then by definition of C_λ, $C_\xi \subseteq C_\lambda$. But C_ξ is unbounded in ξ. Thus ξ is a limit point of C_λ. Conversely, let $\xi < \lambda$ be a limit point of C_λ. We consider first the case where $\sup(D_\lambda) < \lambda$. Then

$$C_\lambda = \bigcup \{C_\gamma \mid \gamma \in D_\lambda\} \cup \{\theta_n^\lambda \mid n < \omega\},$$

and so ξ must be a limit point of $\bigcup \{C_\gamma \mid \gamma \in D_\lambda\}$. Now, D_λ is closed in λ, so $\delta = \bigcup D_\lambda \in D_\lambda$. Thus $D_\delta = \delta \cap D_\lambda$ and

$$\bigcup \{C_\gamma \mid \gamma \in D_\lambda\} = (\bigcup \{C_\gamma \mid \gamma \in D_\delta\}) \cup C_\delta = C_\delta \cup C_\delta = C_\delta .$$

Thus ξ is a limit point of C_δ. Then by induction hypothesis at δ, $\xi \in D_\delta$. But $D_\delta = \delta \cap D_\lambda$. Thus $\xi \in D_\lambda$, as required. We turn to the other case, where $\sup(D_\lambda) = \lambda$. Let $\gamma \in D_\lambda$, $\gamma > \xi$. Thus ξ is a limit point of $\gamma \cap C_\lambda$. But by (a), $\gamma \cap C_\lambda = C_\gamma$. Thus by induction hypothesis at γ, $\xi \in D_\gamma$. But $\gamma \in D_\lambda$, so $D_\gamma = \gamma \cap D_\lambda$. Thus $\xi \in D_\lambda$, and again we are done.

By virtue of (a) and (b) we shall be done if we prove that each C_λ is closed in λ and that if $\operatorname{cf}(\lambda) < \kappa$ then $\operatorname{otp}(C_\lambda) < \kappa$. Well, we prove that C_λ is closed in λ by induction on λ. Assume it is true below λ. Let $\gamma < \lambda$ be a limit point of C_λ. We prove that $\gamma \in C_\lambda$. By (b), $\gamma \in D_\lambda$. If $\gamma = \bigcup D_\lambda$, then $\gamma = \theta_0^\lambda \in C_\lambda$ and we are done. Otherwise, there is an $\alpha \in D_\lambda$ such that $\alpha > \gamma$. By (a), $C_\alpha = \alpha \cap C_\lambda$. Thus γ is a limit point of C_α. So by induction hypothesis, $\gamma \in C_\alpha$. Thus $\gamma \in \quad \cap C_\lambda \subseteq C_\lambda$, and again we are done. Finally now, if $\operatorname{otp}(C_\lambda) \geqslant \kappa$, then C_λ must have at least κ limit points, so by (b), $|D_\lambda| \geqslant \kappa$. But if $\operatorname{cf}(\lambda) < \kappa$, this is not the case. The proof is complete. □

Notice that in proving the above result, we have demonstrated that $\Box_\kappa$ is equivalent to the existence of a sequence $(B_\lambda \mid \lambda < \kappa^+ \wedge \lim(\lambda))$ which satisfies (i)–(iv) as stated in that proof. A stronger result of this nature will be proved in section 5.

We are now ready to say a little more concerning the existence of κ^+-Souslin trees.

2.11 Theorem. *Assume* GCH. *Let κ be an uncountable cardinal for which $\Box_\kappa$ holds. Then there exists a κ^+-Souslin tree.*

Proof. If $\operatorname{cf}(\kappa) > \omega$, then by 2.7, $\diamondsuit_{\kappa^+}(W)$ is valid, where

$$W = \{\alpha \in \kappa^+ \mid \operatorname{cf}(\alpha) = \omega\}.$$

If $\operatorname{cf}(\kappa) = \omega$, then by 2.8, $\diamondsuit_{\kappa^+}(W)$ is valid. Thus in all cases, $\diamondsuit_{\kappa^+}(W)$ holds. Hence by 2.10 there is a stationary set $E \subseteq \kappa^+$ such that both $\diamondsuit_{\kappa^+}(E)$ and $\Box_\kappa(E)$ are valid. So by 2.5 there is a κ^+-Souslin tree. □

3. κ^+-Kurepa Trees

A κ^+-*Kurepa tree*, it may be recalled, is a κ^+-tree with κ^{++} many κ^+-branches. A κ^+-*Kurepa family* is a family $\mathscr{F} \subseteq \mathscr{P}(\kappa^+)$ such that $|\mathscr{F}| = \kappa^{++}$ and for all $\alpha < \kappa^+$, $|\mathscr{F} \upharpoonright \alpha| \leqslant \kappa$, where

$$\mathscr{F} \upharpoonright \alpha = \{x \cap \alpha \mid x \in \mathscr{F}\}.$$

Exactly as in III.2.1, we can show that the existence of a κ^+-Kurepa tree is equivalent to the existence of a κ^+-Kurepa family. By generalising the proof of III.2.2 we shall show that if $V = L$, there is a κ^+-Kurepa family for every infinite cardinal κ. We require two lemmas, generalisations of II.5.10 and II.5.11, respectively.

3.1 Lemma. *Assume $V = L$. Let κ be an infinite cardinal. If*

$$\kappa \subseteq X \prec L_{\kappa^+},$$

then $X = L_\alpha$ for some $\alpha \leqslant \kappa^+$, $\alpha > \kappa$.

Proof. It suffices to prove that X is transitive, since the lemma then follows at once from the condensation lemma. But

$$\models_{L_{\kappa^+}} \forall x (|x| \leqslant \kappa),$$

so this is proved just as in II.5.10. □

3.2 Lemma. *Assume $V = L$. Let κ be an infinite cardinal. If*

$$\kappa \subseteq X \prec L_{\kappa^{++}},$$

then $X \cap L_{\kappa^+} = L_\alpha$ for some $\alpha \leqslant \kappa^+$, $\alpha > \kappa$.

Proof. This follows from 3.1 in the same way that II.5.11 follows from II.5.10. □

We can now prove:

3.3 Theorem. *Assume $V = L$. Let κ be any infinite cardinal. Then there is a κ^+-Kurepa tree.*

Proof. It suffices to construct a κ^+-Kurepa family. We proceed much as in III.2.2.

By 3.1 we can define a function $f\colon \kappa^+ \to \kappa^+$ by letting $f(\alpha)$ be the least ordinal such that

$$\kappa \cup \{\alpha\} \subseteq L_{f(\alpha)} \prec L_{\kappa^+}.$$

Set

$$\mathscr{F} = \{x \subseteq \kappa^+ \mid (\forall \alpha < \kappa^+)(x \cap \alpha \in L_{f(\alpha)})\}.$$

For each $\alpha < \kappa^+$, $|\mathscr{F} \upharpoonright \alpha| \leqslant \kappa$, so in order to show that $\mathscr{F}$ is a κ^+-Kurepa family we need only prove that $|\mathscr{F}| = \kappa^{++}$.

Suppose, on the contrary, that $|\mathscr{F}| \leqslant \kappa^+$, and let

$$X = (x_\alpha \mid \alpha < \kappa^+)$$

be the $<_L$-least κ^+-enumeration of $\mathscr{F}$. Since the function f is clearly definable in $L_{\kappa^{++}}$, so too are $\mathscr{F}$ and X.

By recursion, we define a sequence $(N_\nu \mid \nu < \kappa^+)$ of elementary submodels of $L_{\kappa^{++}}$ as follows:

$$N_0 = \text{the smallest } N \prec L_{\kappa^{++}} \text{ such that } \kappa \subseteq N;$$
$$N_{\nu+1} = \text{the smallest } N \prec L_{\kappa^{++}} \text{ such that } N_\nu \cup \{N_\nu\} \subseteq N;$$
$$N_\delta = \bigcup_{\nu<\delta} N_\nu, \quad \text{if } \lim(\delta).$$

By 3.2,

$$\alpha_\nu = N_\nu \cap \kappa^+ \in \kappa^+,$$

for each $\nu < \kappa^+$. Clearly, $(\alpha_\nu \mid \nu < \kappa^+)$ is a normal sequence in κ^+.

Set

$$x = \{\alpha_\nu \mid \nu < \kappa^+ \wedge \alpha_\nu \notin x_\nu\}.$$

Since $x \neq x_\nu$ for each $\nu < \kappa^+$, $x \notin \mathscr{F}$, and we obtain our contradiction by proving that $x \cap \alpha \in L_{f(\alpha)}$ for all $\alpha < \kappa^+$.

Let $\alpha < \kappa^+$ be given. Let η be the largest limit ordinal such that $\alpha_\eta \leqslant \alpha$. (If no such η exists, then $x \cap \alpha$ is finite and we are done.) Since $x \cap \alpha$ differs from $x \cap \alpha_\eta$ by at most finitely many points, in order to show that $x \cap \alpha \in L_{f(\alpha)}$ it suffices to show that $x \cap \alpha_\eta \in L_{f(\alpha)}$. In fact we show that $x \cap \alpha_\eta \in L_{f(\alpha_\eta)}$, which is if anything a stronger result. Since we shall have no further recourse to the original α, let us write α for α_η from now on.

Now,

$$x \cap \alpha = \{\alpha_\nu \mid \nu < \eta \wedge \alpha_\nu \notin x_\nu\},$$

so if $(\alpha_\nu \mid \nu < \eta)$ and $(x_\nu \cap \alpha \mid \nu < \eta)$ are elements of $L_{f(\alpha)}$ we shall be done. (Recall that $L_{f(\alpha)}$ is a model of ZF^-, though nothing like the full power of ZF^- is required in order to define $x \cap \alpha$ from the above two sequences, of course.)

Let

$$\pi\colon N_\eta \cong L_\beta.$$

Clearly,

$$\pi \restriction L_\alpha = \mathrm{id} \restriction L_\alpha, \quad \pi(\kappa^+) = \alpha, \quad \pi(X) = (x_\nu \cap \alpha \mid \nu < \alpha).$$

Now,

$$\alpha \in L_{f(\alpha)} \prec L_{\kappa^+},$$

so

$$\models_{L_{f(\alpha)}} [|\alpha| \leqslant \kappa].$$

But

$$\alpha = (\kappa^+)^{L_\beta}.$$

Hence

$$\beta < f(\alpha).$$

So, as $\pi(X) = (x_\nu \cap \alpha \mid \nu < \alpha)$, we have

$$(x_\nu \cap \alpha \mid \nu < \alpha) \in L_{f(\alpha)}.$$

In particular,

$$(x_\nu \cap \alpha \mid \nu < \eta) \in L_{f(\alpha)}.$$

It remains to show that $(\alpha_\nu \mid \nu < \eta) \in L_{f(\alpha)}$. To this end, for $\nu < \eta$, let

$$\pi_\nu \colon N_\nu \cong L_{\beta(\nu)}.$$

For each ν,

$$\pi_\nu(\kappa^+) = \alpha_\nu,$$

so

$$\alpha_\nu = [\text{the largest cardinal}]^{L_{\beta(\nu)}}.$$

So, as $L_{f(\alpha)}$ is a model of ZF$^-$, it is sufficient to prove that

$$(\beta(\nu) \mid \nu < \eta) \in L_{f(\alpha)}.$$

We define, by recursion on ν, a sequence of elementary submodels $N'_\nu \prec L_\beta$, for $\nu < \eta' \leqslant \eta$, as follows (see below concerning η'):

$$N'_0 = \text{the smallest } N \prec L_\beta \text{ such that } \kappa \subseteq N;$$
$$N'_{\nu+1} = \text{the smallest } N \prec L_\beta \text{ such that } N'_\nu \cup \{N'_\nu\} \subseteq N;$$
$$N'_\delta = \bigcup_{\nu<\delta} N'_\nu, \quad \text{if } \lim(\delta).$$

The ordinal η' is the largest $\eta' \leqslant \eta$ for which the above recursion is possible. (We shall prove that $\eta' = \eta$.)

Clearly,

$$(N'_\nu \mid \nu < \eta') \in L_{f(\alpha)}.$$

Hence

$$(\beta'(\nu) \mid \nu < \eta') \in L_{f(\alpha)},$$

where we define

$$\pi'_\nu: N'_\nu \cong L_{\beta'(\nu)}$$

for each $\nu < \eta'$.

But

$$\nu < \eta \to N_\nu \prec N_\eta \prec L_{\kappa^{++}},$$

so in the definition of N_ν for $\nu < \eta$ we can replace $L_{\kappa^{++}}$ by N_η. That is:

$$\begin{aligned} N_0 &= \text{the smallest } N \prec N_\eta \text{ such that } \kappa \subseteq N; \\ N_{\nu+1} &= \text{the smallest } N \prec N_\eta \text{ such that } N_\nu \cup \{N_\nu\} \subseteq N; \\ N_\delta &= \bigcup_{\nu<\delta} N_\nu, \quad \text{if } \lim(\delta). \end{aligned}$$

But

$$\pi: N_\eta \cong L_\beta,$$

so an easy induction on ν now yields the result

$$\nu < \eta \to (\pi \upharpoonright N_\nu): N_\nu \cong N'_\nu.$$

Hence $\eta' = \eta$ and $\beta(\nu) = \beta'(\nu)$ for all $\nu < \eta$. In particular, we have $(\beta(\nu) \mid \nu < \eta) \in L_{f(\alpha)}$, so we are done. □

By modifying the above proof along the lines of III.3.5 we may prove that $V = L$ implies $\diamondsuit^+_{\kappa^+}$ for all infinite cardinals κ, where $\diamondsuit^+_{\kappa^+}$ is obtained from $\diamondsuit^+$ by replacing ω_1 by κ^+ throughout (so $\diamondsuit^+$ is $\diamondsuit^+_{\omega_1}$). And an argument as in III.3.6 shows that $\diamondsuit^+_{\kappa^+}$ implies the existence of a κ^+-Kurepa family. (See Exercise 4.)

The notion of a κ-Kurepa tree and the principle $\diamondsuit^+_\kappa$ in the case of κ an innaccessible cardinal will be dealt with in Chapter VII.

4. The Fine Structure Theory

The deeper results concerning the constructible universe, including the proof that $\square_\kappa$ is valid in L, require a detailed study of the individual levels of the constructible hierarchy. (Actually, there is an alternative approach as far as $\square_\kappa$ is concerned: the so-called "Silver machine" method. This is described in Chapter IX.)

The detailed study of the individual levels of the constructible hierarchy needed to prove $\square_\kappa$ and related results was begun by Jensen in the late 1960's, and is known as the "fine structure theory". Initially this really was a study of the properties of the individual sets L_α as defined in Chapter II. However, it soon became clear that the sets L_α do not lend themselves easily to such a study. If one tries to carry out simple set theoretic arguments within an arbitrary L_α, then unless α is a limit ordinal one meets a host of minor, but troublesome difficulties. For instance, unless α is a limit ordinal, L_α is not closed under the formation of ordered pairs. Since the ordered pair function is one which is used all the time in even the most elementary set-theoretical arguments, this is an annoying problem. Certainly, it is possible to overcome this, and similar difficulties, but in so doing a great deal of cumbersome apparatus needs to be introduced, and much of the naturalness of set theory is lost. The difficulty is the more annoying because it arises for an essentially irrelevant reason. The very simple functions which we would like our levels to be closed under (ordered pairs, etc.) are all highly "constructible", and we only fail to achieve closure because they increase rank. And there lies the root of the problem. The trouble is, when we defined the constructible hierarchy, we mimicked the definition of the cumulative hierarchy, insisting that at each stage only *subsets* of the stage could appear at the next stage. But for constructibility the crucial point lies in our other requirement, that at each stage we allow only those new sets which are *constructible* from the sets already available. And there are many set-theoretic operations which are, under any definition, "constructible", but which increase rank by more than one level, and hence violate the "subsets only" requirement. The way out of this dilemma is easy. We modify the definition of the constructible hierarchy so that each level of the hierarchy is an amenable set. This was first done by Jensen, and we thus refer to the modified hierarchy as the *Jensen hierarchy*. It is this hierarchy whose "fine structure" is usually investigated. The α-th level of the Jensen hierarchy is denoted by J_α. Roughly speaking, J_α possesses all of the properties of the limit levels of the usual L_α-hierarchy of constructible sets. And we can think of J_α as being a "constructibly inessential" extension of the structure L_α. (By virtue of the closure properties we obtain for the sets J_α, this picture is not totally accurate, but by and large is the way in which the beginner should view matters: when you read "J_α", think "L_α, $\lim(\alpha)$"!)

In this section we outline the fine structure theory, developed to the stage where we can prove $\Box_\kappa$ (assuming $V = L$). However, by its very nature, the fine structure theory is very intricate, and some of the proofs tend to be long (though except for the early development they are rarely boring). Consequently, we omit practically all proofs in our outline. For applications of the fine structure theory of the type we shall consider, however, it is not at all necessary to know anything about these proofs, a knowledge of a few, readily appreciated key results being sufficient. So we do not lose a great deal by our approach. Then, in section 5, we use the fine structure theory outlined in order to give a rigorous proof of $\Box_\kappa$ in L. The interested reader may then investigate the fine structure theory itself in Chapter VI, where we develop the entire theory rigorously.

Now to our outline of the fine structure theory. Our first step is to define a new "constructible hierarchy". Since we are interested in functional closure of the levels of the hierarchy, rather than pure definability, our approach will be functional. We shall define the hierarchy by iteratively closing up under various set theoretical functions. All of these functions will be "constructible" in some sense. Moreover, they will be sufficient to ensure that at the very least we obtain all of the usual constructible sets at each stage, i.e. $L_\alpha \subseteq J_\alpha$. The collection we use is described below.

A function $f: V^n \to V$ is said to be *rudimentary* (*rud* for short) iff it is generated by the following schemas:

(i) $f(x_1, \ldots, x_n) = x_i \qquad (1 \leqslant i \leqslant n)$;
(ii) $f(x_1, \ldots, x_n) = \{x_i, x_j\} \qquad (1 \leqslant i, j \leqslant n)$;
(iii) $f(x_1, \ldots, x_n) = x_i - x_j \qquad (1 \leqslant i, j \leqslant n)$;
(iv) $f(x_1, \ldots, x_n) = h(g_1(x_1, \ldots, x_n), \ldots, g_k(x_1, \ldots, x_n))$, where $h, g_1, \ldots, g_k$ are rudimentary;
(v) $f(y, x_1, \ldots, x_{n-1}) = \bigcup_{z \in y} g(z, x_1, \ldots, x_{n-1})$, where g is rudimentary.

It is clear that rudimentary functions are "constructible", so that any hierarchy we define using them can reasonably be called a "constructible hierarchy". Indeed, it can be shown that all rudimentary functions are Σ_0^{ZF}. The converse to this is false, but if we define a relation $A \subseteq V^n$ to be *rudimentary* iff its characteristic function is rudimentary, then for relations the notions of being rudimentary and of being Σ_0^{ZF} do coincide. Another point which should perhaps be mentioned here is that although rudimentary functions increase rank they do so by a finite amount only.

If X is a set, the *rudimentary closure* of X is the smallest set $Y \supseteq X$ such that Y is closed under all rudimentary functions. If X is transitive, so is its rudimentary closure. For transitive sets X we set

$$\mathrm{rud}(X) = \text{the rudimentary closure of the transitive set } X \cup \{X\}.$$

That the rudimentary functions will constitute an ideal class for defining a constructible hierarchy which is only an "inessential" extension of the usual one follows from the fact that for any transitive set X,

$$\mathrm{rud}(X) \cap \mathscr{P}(X) = \mathrm{Def}(X).$$

(In fact,

$$\Sigma_0(\mathrm{rud}(X)) \cap \mathscr{P}(X) = \mathrm{Def}(X).)$$

The *Jensen hierarchy* is defined as follows:

$$J_0 = \emptyset;$$
$$J_{\alpha+1} = \mathrm{rud}(J_\alpha);$$
$$J_\lambda = \bigcup_{\alpha<\lambda} J_\alpha, \quad \text{if } \lim(\lambda).$$

Thus each J_α is transitive, the hierarchy is cumulative, and for each α, the rank of J_α is $\omega\alpha$, and

$$J_\alpha \cap \mathrm{On} = \omega\alpha.$$

(This last fact has the effect that in arguments involving the Jensen hierarchy, ordinals of the type $\omega\alpha$ appear all the time.) For $\alpha > 1$, J_α is amenable; the canonical LST formula which says

$$x = J_\alpha$$

is Σ_1^{ZF}; and $(J_\nu \mid \nu < \alpha)$ is uniformly $\Sigma_1^{J_\alpha}$ for $\alpha > 1$. For each α,

$$J_{\alpha+1} \cap \mathscr{P}(J_\alpha) = \mathrm{Def}(J_\alpha).$$

The relationship between the Jensen hierarchy and the usual constructible hierarchy is:

(i) $(\forall \alpha)(L_\alpha \subseteq J_\alpha \subseteq L_{\omega\alpha})$;
(ii) $L_\alpha = J_\alpha \quad$ iff $\omega\alpha = \alpha$.

(It is possible to say a little more, but the exact relationship between L_α and J_α is rather complicated, and in any case is of no use to us.) In particular,

$$L = \bigcup_{\alpha \in \mathrm{On}} J_\alpha .$$

There is a well-ordering $<_J$ of L which is definable by means of a Σ_1^{ZF} formula of LST which is absolute for L and for any set J_α, $\alpha > 1$, and which is such that $<_J \cap (J_\alpha \times J_\alpha)$ is uniformly $\Sigma_1^{J_\alpha}$ for all $\alpha > 1$.

There is a $\Sigma_1(J_\alpha)$ map of $\omega\alpha$ onto J_α for each $\alpha > 1$.

The concept of a Σ_n skolem function has been met in II.6, and in II.6.5 we proved that each limit L_α has a (uniformly Σ_1) Σ_1 skolem function. Essentially the same proof shows that each J_α, $\alpha > 1$, has a (uniformly Σ_1) Σ_1 skolem function. A rather more complicated proof shows that for $\alpha > 1$, J_α has a Σ_n skolem function for any $n \geqslant 1$. But there is no uniform Σ_n skolem function for J_α except for the case $n = 1$. (The proof developed in Exercise II.5 can be used to show that the Jensen hierarchy has no uniform Σ_2 skolem function.) This is a serious drawback. Even the rather simple result proved in II.6.8 shows how useful uniformity properties are in skolem function applications. And in order to prove results such as $\square_\kappa$, we need to be able to carry out Σ_n condensation arguments of a type generalising II.6.8 for any $n \geqslant 1$. In order to facilitate this, we proceed as follows.

Recall that a structure of the form $\langle M, A\rangle$ (i.e. $\langle M, \in, A\rangle$), where $A \subseteq M$, is said to be *amenable* iff M is an amenable set and

$$u \in M \rightarrow A \cap u \in M .$$

It is easily seen that most of the results about limit levels of the constructible hierarchy given in Chapter II are in fact valid (by almost the same proof in each case) for amenable structures of the form $\langle L_\alpha, A\rangle$. Moreover, each of these results has a valid analogue for amenable structures $\langle J_\alpha, A\rangle$. (In this connection, remember that J_α is an amenable set for all $\alpha > 1$.) In particular, there is a uniformly Σ_1 Σ_1-skolem function for the amenable structures $\langle J_\alpha, A\rangle$. The main idea behind the fine structure theory is to capitalise on this fact, by reducing Σ_n predicates over a J_α to Σ_1 predicates over some amenable structure $\langle J_\varrho, A\rangle$, and then working with $\langle J_\varrho, A\rangle$ instead of J_α.

Let $h_{\alpha, A}$ denote the canonical, uniform Σ_1 skolem function for any amenable structure $\langle J_\alpha, A\rangle$, and let $H_{\alpha, A}$ be the uniform $\Sigma_0^{\langle J_\alpha, A\rangle}$ preidcate on J_α such that

$$y \simeq h_{\alpha, A}(i, x) \leftrightarrow (\exists z \in J_\alpha) H_{\alpha, A}(z, y, i, x).$$

(The function $h_{\alpha, A}$ is defined in precisely the same manner as the canonical Σ_1 skolem function h_α for limit L_α in II.6. Thus,

$$h_{\alpha, A}(i, x) \simeq (r_{\alpha, A}(i, x))_0,$$

where

$$r_{\alpha, A}(i, x) \simeq \text{the } <_J\text{-least } w \in J_\alpha \text{ such that}$$
$$\models_{\langle J_\alpha, A\rangle} (\text{“}\mathring{w} \text{ is an ordered pair”}) \wedge \bar{\varphi}_i((\mathring{w})_0, \mathring{x}, (\mathring{w})_1),$$

where $(\bar{\varphi}_i(v_0, v_1, v_2) \mid i < \omega)$ enumerates (in a uniformly $\Delta_1^{J_\alpha}$ fashion) all Σ_0 formulas of $\mathscr{L}(\mathring{A})$ having free variables amongst v_0, v_1, v_2.)

We now describe the means by which Σ_n predicates on a J_α can be coded as Σ_1 predicates on an amenable $\langle J_\varrho, A\rangle$.

Let $\alpha > 1$, $n > 0$. The *Σ_n-projectum* of α, denoted by ϱ_α^n, is the smallest $\varrho \leqslant \alpha$ such that there is a $\Sigma_n(J_\alpha)$ map f for which $f'' J_\varrho = J_\alpha$. It can be shown that ϱ_α^n is the largest $\varrho \leqslant \alpha$ such that $\langle J_\varrho, A\rangle$ is amenable for any $\Sigma_n(J_\alpha)$ subset A of J_ϱ. Moreover, ϱ_α^n equals the smallest ϱ such that $\Sigma_n(J_\alpha) \cap \mathscr{P}(\omega\varrho) \nsubseteq J_\alpha$.

It is easily seen that

$$m < n \rightarrow \varrho_\alpha^n \leqslant \varrho_\alpha^m.$$

For later convenience, we set

$$\varrho_\alpha^0 = \alpha.$$

For each $\alpha > 1$, $n \geqslant 0$, we can associate with α a *standard code*, A_α^n, and a *standard parameter* , p_α^n, with the following properties:

1. $A_\alpha^n \subseteq J_{\varrho_\alpha^n}$, $A_\alpha^n \in \Sigma_n(J_\alpha)$;
2. $\langle J_{\varrho_\alpha^n}, A_\alpha^n\rangle$ is amenable;
3. $A_\alpha^0 = p_\alpha^0 = \emptyset$;
4. For all $m > 0$,

$$\Sigma_m(\langle J_{\varrho_\alpha^n}, A_\alpha^n\rangle) = \mathscr{P}(J_{\varrho_\alpha^n}) \cap \Sigma_{n+m}(J_\alpha);$$

5. p_α^{n+1} is the $<_J$-least $p \in J_{\varrho_\alpha^n}$ such that

$$J_{\varrho_\alpha^n} = h''_{\varrho_\alpha^n, A_\alpha^n}(\omega \times (J_{\varrho_\alpha^{n+1}} \times \{p\})).$$

By definition of the Σ_n-projectum, ϱ_α^n, there is a $\Sigma_n(J_\alpha)$ map f such that $f'' J_{\varrho_\alpha^n} = J_\alpha$. Suppose now that P is a $\Sigma_n(J_\alpha)$ predicate on J_α. Set

$$Q = \{x \in J_{\varrho_\alpha^n} \mid f(x) \in P\}.$$

Then Q is a $\Sigma_n(J_\alpha)$ subset of $J_{\varrho_\alpha^n}$. By Fact 4 in the above list, Q is $\Sigma_1(\langle J_{\varrho_\alpha^n}, A_\alpha^n\rangle)$. In this way, instead of working with Σ_n predicates on J_α, we may work with "equivalent" (in the sense that Q and P are "equivalent" in the above discussion) Σ_1 predicates on $\langle J_{\varrho_\alpha^n}, A_\alpha^n\rangle$, thereby being able to utilise the uniform Σ_1 skolem function possessed by the structures $\langle J_{\varrho_\alpha^n}, A_\alpha^n\rangle$. (Actually, from the above account it would appear that the coding of a Σ_n predicate on J_α by a Σ_1 predicate on $\langle J_{\varrho_\alpha^n}, A_\alpha^n\rangle$ is via an arbitrary $\Sigma_n(J_\alpha)$ function f. In practice we use, in effect, a canonical such function constructed from the standard parameters and the canonical Σ_1 skolem functions. See the definition of the standard parameter p_α^{n+1} above.) What is now needed in order to make this procedure work is a suitable condensation lemma. For suppose that

$$\langle X, A_\alpha^n \cap X\rangle \prec_1 \langle J_{\varrho_\alpha^n}, A_\alpha^n\rangle.$$

By the standard condensation lemma, there are unique $\bar\varrho$, $\bar A$ such that

$$\langle X, A_\alpha^n \cap X\rangle \cong \langle J_{\bar\varrho}, \bar A\rangle.$$

But if we are to be able to work with the structures $\langle J_{\varrho_\alpha^n}, A_\alpha^n\rangle$ instead of the original J_α, we shall require that the $\bar\varrho$, $\bar A$ obtained in this manner are of the form $\bar\varrho = \varrho_{\bar\alpha}^n$, $\bar A = A_{\bar\alpha}^n$ for some unique $\bar\alpha$. In other words, what we need is a condensation lemma for the hierarchy of structures

$$\langle J_{\varrho_\alpha^n}, A_\alpha^n\rangle \quad (\alpha \in \mathrm{On}).$$

This is provided by the following property of the standard codes:

6. Let $\alpha > 1$, $m \geqslant 0$, $n \geqslant 1$. Let $\langle J_{\bar\varrho}, \bar A\rangle$ be amenable, and let

$$\pi: \langle J_{\bar\varrho}, \bar A\rangle \prec_m \langle J_{\varrho_\alpha^n}, A_\alpha^n\rangle.$$

Then there is a unique $\bar\alpha \geqslant \bar\varrho$ such that $\bar\varrho = \varrho_{\bar\alpha}^n$, $\bar A = A_{\bar\alpha}^n$. Moreover, there is a unique $\tilde\pi \supseteq \pi$ such that

$$\tilde\pi: J_{\bar\alpha} \prec_{m+n} J_\alpha,$$

and such that for all $i = 1, \ldots, n$:

(a) $\quad \tilde\pi(p_{\bar\alpha}^i) = p_\alpha^i$;

(b) $\quad (\tilde\pi \restriction J_{\varrho_{\bar\alpha}^i}): \langle J_{\varrho_{\bar\alpha}^i}, A_{\bar\alpha}^i\rangle \prec_{m+n-i} \langle J_{\varrho_\alpha^i}, A_\alpha^i\rangle.$

The assertions concerning the extension $\tilde\pi$ here should not be too surprising, since A_α^n codes all the Σ_n information about J_α. The heart of assertion 6 is the fact that the standard codes are preserved under condensation arguments, indeed even under "Σ_0 condensation arguments". This is essentially the case because of the canonical manner in which the standard codes are defined. We fix some simple (hence uniformly Δ_1) enumeration $(\varphi_i | i < \omega)$ of the Σ_1 formulas of $\mathscr{L}(\mathring{A})$ with free variables v_0 and v_1, and define, by recursion on n:

$$A_\alpha^{n+1} = \{(i, x) | i \in \omega \wedge x \in J_{\varrho_\alpha^n} \wedge \models_{\langle J_{\varrho_\alpha^n}, A_\alpha^n \rangle} \varphi_i(\mathring{x}, \mathring{p}_\alpha^{n+1})\}.$$

5. The Combinatorial Principle $\Box_\kappa$

Using the fine structure theory outlined above, we shall prove that if $V = L$, then $\Box_\kappa$ is valid for all infinite cardinals κ. We begin by recalling the statement of $\Box_\kappa$.

$\Box_\kappa$: There is a sequence $(C_\alpha | \alpha < \kappa^+ \wedge \lim(\alpha))$ such that:

(i) C_α is a club subset of α;
(ii) $\mathrm{cf}(\alpha) < \kappa \to |C_\alpha| < \kappa$;
(iii) if $\bar{\alpha}$ is a limit point of C_α, then $C_{\bar{\alpha}} = \bar{\alpha} \cap C_\alpha$.

Let $\Box'_\kappa$ assert the existence of a sequence $(B_\alpha | \alpha < \kappa^+ \wedge \lim(\alpha))$ such that:

(i) B_α is a closed subset of α such that $(\forall \gamma \in B_\alpha) \lim(\gamma)$;
(ii) $\mathrm{cf}(\alpha) > \omega \to B_\alpha$ is unbounded in α;
(iii) $\mathrm{otp}(B_\alpha) \leqslant \kappa$;
(iv) $\bar{\alpha} \in B_\alpha \to B_{\bar{\alpha}} = \bar{\alpha} \cap B_\alpha$.

5.1 Lemma. *Let κ be any uncountable cardinal. Then $\Box_\kappa$ and $\Box'_\kappa$ are equivalent.*

Proof. Before we commence, notice that a weaker version of this result was proved during the course of 2.10. The present proof is a refinement of the argument used there.

First of all, suppose $(C_\alpha | \alpha < \kappa^+ \wedge \lim(\alpha))$ is as in $\Box_\kappa$. For each α, let B_α be the set of limit points of C_α below α. It is clear that the sequence $(B_\alpha | \alpha < \kappa^+ \wedge \lim(\alpha))$ satisfies $\Box'_\kappa$.

Conversely, let $(B_\alpha | \alpha < \kappa^+ \wedge \lim(\alpha))$ be as in $\Box'_\kappa$. By recursion on α we define sets C_α as follows. If $\bigcup B_\alpha = \alpha$, set

$$C_\alpha = \bigcup \{C_\gamma | \gamma \in B_\alpha\}.$$

Otherwise, if B_α is not cofinal in α, then by (ii) of $\Box'_\kappa$, $\mathrm{cf}(\alpha) = \omega$, so we may fix some strictly increasing ω-sequence $(\theta_n^\alpha | n < \omega)$, cofinal in α, with $\theta_0^\alpha = \bigcup B_\alpha$, and set

$$C_\alpha = (\bigcup \{C_\gamma | \gamma \in B_\alpha\}) \cup \{\theta_n^\alpha | n < \omega\}.$$

The following are proved exactly as in 2.10:

(a) If $\gamma \in B_\alpha$, then $C_\gamma = \gamma \cap C_\alpha$.

(b) B_α is the set of all limit points of C_α below α.

(c) C_α is a club subset of α.

Moreover, we have

(d) $\mathrm{otp}(C_\alpha) \leqslant \kappa$.

For suppose, on the contrary, that $\mathrm{otp}(C_\alpha) > \kappa$. Then, since κ is an uncountable cardinal, it follows from (b) that $\mathrm{otp}(B_\alpha) > \kappa$. This is not the case, by choice of B_α.

Now, if κ is regular, then since C_α is cofinal in α, (d) implies that

$$\mathrm{cf}(\alpha) < \kappa \rightarrow |C_\alpha| < \kappa,$$

and hence $(C_\alpha \mid \alpha < \kappa^+ \wedge \lim(\alpha))$ satisfies $\Box_\kappa$. On the other hand, if κ is singular, we must modify the sets C_α in order to obtain a $\Box_\kappa$-sequence, as follows. Let $\bar{\kappa} = \mathrm{cf}(\kappa)$, and let $(\theta_\nu \mid \nu < \bar{\kappa})$ be a strictly increasing, continuous sequence of limit ordinals, cofinal in κ, with $\theta_0 = 0$. Set $\theta_{\bar{\kappa}} = \kappa$. Define sets C_α as follows. If there is a $\nu < \bar{\kappa}$ such that

$$\theta_\nu < \mathrm{otp}(C_\alpha) \leqslant \theta_{\nu+1},$$

set

$$C'_\alpha = \{\gamma \in C_\alpha \mid \mathrm{otp}(\gamma \cap C_\alpha) \geqslant \theta_\nu\}.$$

If no such ν exists, then we must have $\mathrm{otp}(C_\alpha) = \theta_\nu$ for some *limit* ordinal $\nu \leqslant \bar{\kappa}$, in which case we set

$$C'_\alpha = \{\gamma \in C_\alpha \mid (\exists \tau < \nu)(\mathrm{otp}(\gamma \cap C_\alpha) = \theta_\tau)\}.$$

It is routine to verify that $(C'_\alpha \mid \alpha < \kappa^+ \wedge \lim(\alpha))$ is a $\Box_\kappa$-sequence. That completes the proof. $\Box$

Assume $V = L$ from now on. We shall prove that $\Box_\kappa$ holds for all infinite cardinals κ. Since $\Box_\omega$ is trivially valid (in ZFC), we may ignore the case $\kappa = \omega$. By 5.1, given some uncountable cardinal κ, it suffices to prove $\Box'_\kappa$. The basic idea is to construct sets B_α to satisfy (i)–(iii) of $\Box'_\kappa$ by means of a construction which is sufficiently uniform to enable (iv) to be proved by a condensation argument. In order to do this we must set up some machinery.

Let α be a limit ordinal, and let $\omega\beta \geqslant \alpha$. We say that α is *singular over* J_β iff there is a J_β-definable map of a bounded subset of α cofinally into α; otherwise we say that α is *regular over* J_β. Let $n \geqslant 1$. We say that α is Σ_n-*singular over* J_β iff there is a $\Sigma_n(J_\beta)$ map of a bounded subset of α cofinally into α; otherwise we say that α is Σ_n-*regular over* J_β.

Clearly, α is regular over J_β iff it is Σ_n-regular over J_β for all n. If α is singular over J_β, then α is singular over J_γ for all $\gamma \geqslant \beta$. And if α is Σ_n-singular over J_β, then α is Σ_m-singular over J_β for all $m \geqslant n$. Moreover, by $V = L$, if α is singular, then there are β, n such that α is Σ_n-singular over J_β.

Let

$$S = \{\alpha \in \kappa^+ \mid (\alpha > \kappa) \wedge (\omega\alpha = \alpha) \wedge (\forall \gamma < \alpha)(|\gamma|^{J_\alpha} \leqslant \kappa)\}.$$

It is easily seen that S is a club subset of κ^+. We shall construct a sequence $(C_\alpha \mid \alpha \in S)$ such that:

(i) C_α is a closed subset of $S \cap \alpha$;
(ii) $\mathrm{cf}(\alpha) > \omega \rightarrow C_\alpha$ is unbounded in α;
(iii) $\mathrm{otp}(C_\alpha) \leqslant \kappa$;
(iv) $\bar{\alpha} \in C_\alpha \rightarrow C_{\bar{\alpha}} = \bar{\alpha} \cap C_\alpha$.

If we then identify S with $\{\alpha \in \kappa^+ \mid \lim(\alpha)\}$ in the obvious manner, we obtain a $\square'_\kappa$-sequence, as required.

Let $\alpha \in S$. Then α is a limit ordinal between κ and κ^+. So in particular, α is a
$\beta(\alpha)$ singular limit ordinal. Let $\beta(\alpha)$ be the least ordinal β such that α is singular over
$n(\alpha)$ J_β. Let $n(\alpha)$ be the least integer $n \geqslant 1$ such that α is Σ_n-singular over $J_{\beta(\alpha)}$. The definition of C_α splits into two cases, depending upon the nature of $\beta(\alpha)$ and $n(\alpha)$.

Define

Q $$Q = \{\alpha \in S \mid \beta(\alpha) \text{ is a successor ordinal and } n(\alpha) = 1\};$$

R $$R = S - Q.$$

5.2 Lemma. $\alpha \in Q \rightarrow \mathrm{cf}(\alpha) = \omega$.

Proof. Let $\beta = \beta(\alpha) = \gamma + 1$. Notice that as $\alpha \in S$, we have $\lim(\alpha)$, so we must have $\gamma \geqslant \alpha$ here. Let f be a $\Sigma(J_\beta)$ map of a subset, u, of an ordinal $\delta < \alpha$ cofinally into α. Let P be a $\Sigma_0(J_\beta)$ predicate such that

$$f(v) = \tau \leftrightarrow (\exists z \in J_\beta)\, P(z, \tau, v).$$

Now, $J_\beta = \mathrm{rud}(J_\gamma)$, so every element of J_β can be obtained by the successive application of finitely many rud functions to finitely many elements of the set $J_\gamma \cup \{J_\gamma\}$. But amongst the rud functions are the identity function, the pairing function, and the inverses to the pairing function. Moreover, the rud functions are closed under composition. Thus, given any $x \in J_\beta$ we can in fact find a single rud function g and a single element y of J_γ such that $x = g(y, J_\gamma)$. Hence, if $(g_i \mid i < \omega)$ is an enumeration of all the binary rud functions, we have:

(*) $$J_\beta = \{g_i(x, J_\gamma) \mid x \in J_\gamma \wedge i \in \omega\}.$$

For each $i < \omega$, define a partial function f_i on u by:

$$f_i(v) = \tau \leftrightarrow (\exists x \in J_\gamma)\, P(g_i(x, J_\gamma), \tau, v).$$

By (*),

$$f = \bigcup_{i<\omega} f_i,$$

so

$$\sup_{i<\omega} \bigcup (f_i{}'' \delta) = \alpha.$$

Thus the lemma will be proved if we can show that $\bigcup(f_i''\delta) < \alpha$ for each $i < \omega$. Since α is regular over J_γ, it suffices to prove that for each $i < \omega, f_i$ is J_γ-definable.

Now, the predicate (of x, τ, ν)

$$P(g_i(x, J_\gamma), \tau, \nu)$$

is $\Sigma_0(J_\beta)$ on J_γ. But one of the properties of the rudimentary functions that we mentioned in section 4 was that for any transitive set X,

$$\Sigma_0(\mathrm{rud}(X)) \cap \mathscr{P}(X) = \mathrm{Def}(X),$$

so in particular we have

$$\Sigma_0(J_\beta) \cap \mathscr{P}(J_\gamma) = \mathrm{Def}(J_\gamma).$$

Thus the predicate (of x, τ, ν) $P(g_i(x, J_\gamma), \tau, \nu)$ is J_γ-definable. It follows at once that f_i is J_γ-definable, and we are done. $\square$

By virtue of 5.2, we may define

$$C_\alpha = \emptyset$$

for the case $\alpha \in Q$, and there is nothing further to check.

We consider now the case $\alpha \in R$.

5.3 Lemma. *If $\alpha \in R$, then $\varrho_{\beta(\alpha)}^{n(\alpha)} = \kappa$.*

Proof. Let $\beta = \beta(\alpha)$, $n = n(\alpha)$. Since κ is a cardinal, II.5.5 (for the Jensen hierar- β, n
chy) implies that $\mathscr{P}(\omega\gamma) \subseteq J_\kappa \subseteq J_\beta$ for all $\gamma < \kappa$. Thus we certainly have $\mathscr{P}(\omega\gamma) \cap \Sigma_n(J_\beta) \subseteq J_\beta$ for all $\gamma < \kappa$. Thus $\varrho_\beta^n \geqslant \kappa$.

Now, by choice of β, n there is a $\Sigma_n(J_\beta)$ map, f, of a bounded subset of α f
cofinally into α. We may code f as a subset of α in a simple fashion (e.g. using a $\Sigma_1(J_\alpha)$ map of α onto J_α). But as f is cofinal in α, $f \notin J_\alpha$, and if $\alpha < \beta$ then by definition of β, α is regular within J_β, so again as f is cofinal in α, we have $f \notin J_\beta$. Thus, in all cases, $\mathscr{P}(\omega\alpha) \cap \Sigma_n(J_\beta) \nsubseteq J_\beta$, and so $\varrho_\beta^n \leqslant \alpha$.

By definition of S, if $\mathrm{dom}(f) \subseteq \gamma < \alpha$, then $|\gamma|^{J_\alpha} \leqslant \kappa$, so J_α contains a map from κ onto γ. Consequently, by composing f with such a map if necessary, we may assume that $\mathrm{dom}(f) \subseteq \kappa$. We may also assume (again by making trivial alter- f
ations to f if necessary) that $f(\nu) > \kappa$ for all $\nu \in \mathrm{dom}(f)$. f

Again, since $\alpha \in S$, for each γ such that $\kappa < \gamma < \alpha$ there is a function $g_\gamma \in J_\alpha$ such that $g_\gamma\colon \kappa \leftrightarrow \gamma$. In fact we may take g_γ to be the $<_J$-least such map, and then the g_γ
sequence $(g_\gamma \mid \kappa < \gamma < \alpha)$ is $\Sigma_1(J_\alpha)$.

Let $(U_\nu \mid \nu < \kappa)$ be a J_κ-definable partition of κ into κ many disjoint sets of size κ, and let $(j_\nu \mid \nu < \kappa)$ be a J_κ-definable sequence of maps $j_\nu\colon U_\nu \leftrightarrow \kappa$. (Practically any $(U_\nu \mid \nu < \kappa)$ and $(j_\nu \mid \nu < \kappa)$ which are explicitly defined will be J_κ-definable.)

Set

$$k = \bigcup\{g_{f(\nu)} \circ j_\nu \mid \nu \in \mathrm{dom}(f)\}.$$

Clearly, k is a $\Sigma_n(J_\beta)$ map of κ onto α. Since $\varrho_\beta^n \leqslant \alpha$, it follows at once that $\varrho_\beta^n \leqslant \kappa$, and we are done. $\square$

For $\alpha \in R$, now, we set

$\varrho(\alpha)$, $A(\alpha)$
$$\varrho(\alpha) = \varrho_{\beta(\alpha)}^{n(\alpha)-1}, \qquad A(\alpha) = A_{\beta(\alpha)}^{n(\alpha)-1}.$$

If $n(\alpha) = 1$, then $\varrho(\alpha) = \beta(\alpha)$, so as $\alpha \in R$ we shall have $\lim(\varrho(\alpha))$. And if $n(\alpha) > 1$, then $\varrho(\alpha)$ will be admissible, so again $\lim(\varrho(\alpha))$. Thus $\varrho(\alpha)$ is a limit ordinal. Moreover $\varrho(\alpha) \geqslant \alpha$. For if $n(\alpha) = 1$, then $\varrho(\alpha) = \beta(\alpha) \geqslant \alpha$. On the other hand, suppose $n(\alpha) > 1$. Now, α is $\Sigma_{n(\alpha)-1}$-regular over $J_{\beta(\alpha)}$, so there is certainly no $\Sigma_{n(a)-1}(J_{\beta(\alpha)})$ map of a bounded subset of α cofinally into α. But by definition of $\varrho(\alpha)$, there is a $\Sigma_{n(\alpha)-1}(J_{\beta(\alpha)})$ map from a subset of $J_{\varrho(\alpha)}$ onto J_α, and hence there is a $\Sigma_{n(\alpha)-1}(J_{\beta(\alpha)})$ map from a subset of $\omega \cdot \varrho(\alpha)$ onto α. Thus $\omega \cdot \varrho(\alpha) \geqslant \alpha$. But $\omega\alpha = \alpha$. Hence $\varrho(\alpha) \geqslant \alpha$, as stated.

Fix $\alpha \in R$ now, and set

$\beta, n, \varrho, A, h, H$
$$\beta = \beta(\alpha), \quad n = n(\alpha), \quad \varrho = \varrho(\alpha), \quad A = A(\alpha), \quad h = h_{\varrho, A}, \quad H = H_{\varrho, A}.$$

So, in particular, h is the canonical Σ_1 skolem function for $\langle J_\varrho, A\rangle$ and H is a $\Sigma_0^{\langle J_\rho, A\rangle}$ predicate with the property that

$$y = h(i, x) \leftrightarrow (\exists z \in J_\varrho)\, H(z, y, i, x).$$

h_τ For $\tau < \varrho$, define a partial function h_τ from $\omega \times J_\tau$ into J_τ by

$$y = h_\tau(i, x) \leftrightarrow (\exists z \in J_\tau)\, H(z, y, i, x).$$

Now, the canonical Σ_1 skolem function $h_{\xi, U}$ is uniform for all amenable $\langle J_\xi, U\rangle$. In particular, whenever $\tau < \varrho$ is such that $\langle J_\tau, A \cap J_\tau\rangle$ is amenable, then the function h_τ defined above is its canonical Σ_1 skolem function, i.e. $h_\tau = h_{\tau, A \cap J_\tau}$.

Now, by definition of ϱ_β^n, together with 5.3 (and the J_γ-analogue of II.6.8) there

$f, \bar f$ is a $\Sigma_n(J_\beta)$ map f such that $f''\kappa = J_\beta$. Let $\bar f = f \cap (J_\varrho \times \kappa)$. Then $\bar f$ is a $\Sigma_n(J_\beta)$ subset of J_ϱ. So by the properties of the standard codes given in section 4, $\bar f$ is $\Sigma_1(\langle J_\varrho, A\rangle)$. Moreover, $\bar f''\kappa = J_\varrho$. By the properties of the Σ_1 skolem function, if $\bar f$ is $\Sigma_1^{\langle J_\rho, A\rangle}(\{p\})$, we will have

$$h''(\omega \times (\kappa \times \{p\})) = J_\varrho.$$

So we may define

p
$$p = p(\alpha) = \text{the } <_J\text{-least } p \in J_\varrho \text{ such that } J_\varrho = h''(\omega \times (\kappa \times \{p\})).$$

Define a map g from a subset of κ into J_ϱ by setting

g
$$g(\omega\nu + i) \simeq h(i, (\nu, p)).$$

G By choice of p, $g''\kappa = J_\varrho$. Moreover, g is $\Sigma_1^{\langle J_\rho, A\rangle}(\{p\})$. Let G be the canonical $\Sigma_0^{\langle J_\rho, A\rangle}(\{p\})$ predicate (obtained from H) such that

$$g(\nu) = x \leftrightarrow (\exists z \in J_\varrho)\, G(z, x, \nu).$$

By recursion we shall define functions $k: \theta \to \kappa$, $m: \theta \to \varrho$, $(X_\nu \mid \nu < \theta)$, $(\alpha_\nu \mid \nu < \theta)$, for some θ which is to be determined during the course of the definition. The exact order of this definition will be examined as soon as the definition has been given in full.

$k(\nu) =$ the least $\tau \in \mathrm{dom}(g)$ such that $\alpha_\nu < g(\tau) < \alpha$ and $|\alpha_\nu|^{J_{g(\tau)}} \leqslant \kappa$. $k(\nu)$

$m(0) =$ the least $\gamma \geqslant \kappa$ such that $p \in J_\gamma$; $m(\nu)$

$m(\nu + 1) =$ the least $\gamma > m(\nu), \alpha_\nu, g \circ k(\nu)$ such that:

(i) $A \cap J_{m(\nu)} \in J_\gamma$;

(ii) $m(\nu), \alpha_\nu, g \circ k(\nu) \in h''_\gamma(\omega \times (\kappa \times \{p\}))$;

(iii) $(\exists z \in J_\gamma)(G(z, g \circ k(\nu), k(\nu))$;

$m(\lambda) = \sup_{\nu < \lambda} m(\nu)$, if $\lim(\lambda)$ and this supremum is less than ϱ (undefined if the supremum equals ϱ).

$X_\nu = h''_{m(\nu)}(\omega \times (\kappa \times \{p\}))$. X_ν

$\alpha_\nu = \sup(X_\nu \cap \alpha)$. α_ν

Our $\square'_\kappa$ set C_α will be the set

$$C_\alpha = \{\alpha_\nu \mid \nu < \theta \wedge \lim(\nu)\},$$

where θ is the first ordinal for which the above definition breaks down. We shall θ
show that $\lim(\theta)$ and that θ is the least ordinal such that $\sup_{\nu<\theta} m(\nu) = \varrho$. We shall also show that the function k is order-preserving, so $\theta \leqslant \kappa$ and $\mathrm{otp}(C_\alpha) \leqslant \kappa$. (The function g is used precisely in order to obtain this result.) A condensation argument will be used to show that $C_{\bar\alpha} = \bar\alpha \cap C_\alpha$ whenever $\bar\alpha \in C_\alpha$. The rather complicated definition of the function m is designed to facilitate this part of the proof. And now down to business.

Let us examine the way in which the above definition proceeds, and how it may break down. The definition of $m(0)$ comes first, and is unproblematical. Suppose now that $m(\nu)$ is defined for some ν. Then we may define X_ν and α_ν. We show that $\alpha_\nu < \alpha$. Suppose not. In other words, suppose that $h''_{m(\nu)}(\omega \times (\kappa \times \{p\})) \cap \alpha$ is cofinal in α. Now, $\langle J_\varrho, A\rangle$ is amenable, so $A \cap J_{m(\nu)} \in J_\varrho$. Thus $\langle J_{m(\nu)}, A \cap J_{m(\nu)}\rangle \in J_\varrho$. Thus $h_{m(\nu)} \in J_\varrho$. Now, there is a J_κ-definable map of κ onto $\omega \times (\kappa \times \{p\})$. Thus J_ϱ contains a map from a subset of κ cofinally into α. If $\varrho = \alpha$ this is already a contradiction. What if $\varrho > \alpha$? Well, in this case, since $\varrho \leqslant \beta(\alpha)$, α is a regular cardinal inside J_ϱ, and again we have a contradiction. Thus $\alpha_\nu < \alpha$. Since $\alpha \in S$, it follows that $k(\nu)$ is defined. And now we may define $m(\nu + 1)$ without any difficulty. Thus the only way in which the construction can break down is when a reach a limit ordinal θ such that $\sup_{\nu<\theta} m(\nu) = \varrho$.

For each $\nu < \theta$, by definition of $m(\nu + 1)$ we have $\alpha_\nu \in h''_{m(\nu+1)}(\omega \times (\kappa \times \{p\}))$, so $\alpha_\nu \in X_{\nu+1} \cap \alpha$. Thus $\alpha_\nu < \alpha_{\nu+1}$. Moreover, since the function m is continuous (by definition), for any limit ordinal $\lambda < \theta$ we have, by virtue of the manner in which the functions h_τ were defined, $X_\lambda = \bigcup_{\nu<\lambda} X_\nu$, and hence $\alpha_\lambda = \sup_{\nu<\lambda} \alpha_\nu$. Thus the sequence $(\alpha_\nu \mid \nu < \theta)$ is strictly increasing and continuous at limits. Again, since

$\sup_{\nu<\theta} m(\nu) = \varrho$, we have

$$\bigcup_{\nu<\theta} X_\nu = h''_\varrho(\omega \times (\kappa \times \{p\})) = J_\varrho,$$

so $\sup_{\nu<\theta} \alpha_\nu = \alpha$.

We show next that the function k is strictly increasing. Let $\nu < \tau < \theta$. By the definition of $k(\tau)$, $g \circ k(\tau) > \alpha_\tau$ and $|\alpha_\tau|^{J_{g \circ k(\tau)}} \leqslant \kappa$. So as $\alpha_\nu < \alpha_\tau$, we have $g \circ k(\tau) > \alpha_\nu$ and $|\alpha_\nu|^{J_{g \circ k(\tau)}} \leqslant \kappa$. So by the minimality of $k(\nu)$ in its definition, $k(\nu) \leqslant k(\tau)$. But by definition of $m(\nu+1)$, $g \circ k(\nu) \in X_{\nu+1} \cap \alpha$, so $g \circ k(\nu) < \alpha_{\nu+1} \leqslant \alpha_\tau \leqslant g \circ k(\tau)$, and in particular $k(\nu) \neq k(\tau)$. Thus $k(\nu) < k(\tau)$.

Since k is strictly increasing from θ into κ, we must have $\theta \leqslant \kappa$.

We set

C_α $$C_\alpha = \{\alpha_\nu \mid \nu < \theta \wedge \lim(\nu)\}.$$

By the above results, C_α is closed in α, has order-type at most κ, and if $\mathrm{cf}(\alpha) > \omega$ then C_α is unbounded in α. Moreover, by the definition of k and the inequality $\alpha_\nu < g \circ k(\nu) < \alpha_{\nu+1}$ (noted during the proof that k is increasing), we have $C_\alpha \subseteq S$. Thus, all that remains to be proved now is that if $\bar\alpha \in C_\alpha$, then $C_{\bar\alpha} = \bar\alpha \cap C_\alpha$.

$\bar\alpha, \lambda$ Let $\bar\alpha \in C_\alpha$ be given. For some limit ordinal $\lambda < \theta$, $\bar\alpha = \alpha_\lambda$. Note that by the definition of m, $\lim(\lambda)$ implies that $\langle J_{m(\lambda)}, A \cap J_{m(\lambda)}\rangle$ is amenable.

5.4 Lemma. $\bar\alpha \subseteq X_\lambda$.

Proof. Since $\bar\alpha = \sup_{\nu<\lambda} \alpha_\nu$, it suffices to show that $\alpha_\nu \subseteq X_\lambda$ for all $\nu < \lambda$. So let $\nu < \lambda$. Then by definition of $m(\nu+1)$, $\alpha_\nu \in X_{\nu+1} \subseteq X_\lambda$. But $\bar\alpha \in S$, so $|\alpha_\nu|^{J_{\bar\alpha}} \leqslant \kappa$. Hence $|\alpha_\nu|^{J_{m(\lambda)}} \leqslant \kappa$. Since

$$\{\alpha_\nu\} \cup \kappa \subseteq X_\lambda \prec_1 J_{m(\lambda)},$$

we have $\alpha_\nu \subseteq X_\lambda$, as required. □

By the condensation lemma, let

$\pi, \bar\varrho, \bar A$ $$\pi\colon \langle J_{\bar\varrho}, \bar A\rangle \cong \langle X_\lambda, A \cap X_\lambda\rangle.$$

Thus

$$\pi\colon \langle J_{\bar\varrho}, \bar A\rangle \prec_1 \langle J_{m(\lambda)}, A \cap J_{m(\lambda)}\rangle.$$

But by transitivity,

$$\langle J_{m(\lambda)}, A \cap J_{m(\lambda)}\rangle \prec_0 \langle J_\varrho, A\rangle.$$

Hence

$$\pi\colon \langle J_{\bar\varrho}, \bar A\rangle \prec_0 \langle J_\varrho, A\rangle.$$

It follows from the fine-structure theory (section 4) that there is a unique $\bar\beta$ such

$\bar\beta, \tilde\pi$ that $\bar\varrho = \varrho_{\bar\beta}^{n-1}$ and $\bar A = A_{\bar\beta}^{n-1}$, and a unique $\tilde\pi \supseteq \pi$ such that, in particular,

$$\tilde\pi\colon J_{\bar\beta} \prec_{n-1} J_\beta.$$

$\bar h, \bar p$ Let $\bar h = h_{\bar\varrho, \bar A}$, and set $\bar p = \pi^{-1}(p)$.

By 5.4, $\pi \upharpoonright \bar\alpha = \mathrm{id} \upharpoonright \bar\alpha$. If $\bar\alpha < \bar\beta$, then since $\bar\alpha = \alpha_\lambda = X_\lambda \cap \alpha = \mathrm{ran}(\pi) \cap \alpha$ and $\pi \subseteq \tilde\pi$, we have $\tilde\pi(\bar\alpha) \geqslant \alpha$. (If $\alpha \in \mathrm{ran}(\tilde\pi)$, then in fact $\tilde\pi(\bar\alpha) = \alpha$, but we have no reason to suppose that this is the case.)

Define a map $\bar g_0$ from a subset of κ into $J_{m(\lambda)}$ by $\bar g_0$

$$\bar g_0(\nu) = x \leftrightarrow (\exists z \in J_{m(\lambda)})\, G(z, x, \nu)\,.$$

Now, $\bar g_0$ is $\Sigma_1^{\langle J_{m(\lambda)},\, A \cap J_{m(\lambda)}\rangle}(\{p\})$. Hence $\bar g_1 = \bar g_0 \cap (\bar\alpha \times \kappa)$ is $\Sigma_1^{\langle J_{m(\lambda)},\, A \cap J_{m(\lambda)}\rangle}(\{p\})$. $\bar g_1$
But $\pi \upharpoonright \bar\alpha = \mathrm{id} \upharpoonright \bar\alpha$. Thus $\pi^{-1}{}''\bar g_1 = \bar g_1$, and so $\bar g_1$ is $\Sigma_1^{\langle J_{\bar\rho},\, \bar A\rangle}(\{\bar p\})$. So, as $\bar\varrho = \varrho_{\bar\beta}^{n-1}$, $\bar A = A_{\bar\beta}^{n-1}$, $\bar g_1$ is $\Sigma_n(J_{\bar\beta})$.

By definition of m, $k''\lambda \subseteq \mathrm{dom}(\bar g_1)$ and $\bar g_1 \upharpoonright (k''\lambda) = g \upharpoonright (k''\lambda)$. But $\alpha_\nu < g \circ k(\nu) < \alpha_{\nu+1}$ for all ν. Thus $\bar g_1$ is cofinal in $\bar\alpha$.

5.5 Lemma. $\varrho_{\bar\beta}^n = \kappa$.

Proof. Since $\bar\alpha \in S$, we can use the function $\bar g_1$ to prove this by the same argument we used in 5.3. $\square$

5.6 Lemma. $\bar\beta = \beta(\bar\alpha)$.

Proof. The existence of $\bar g_1$ shows that $\beta(\bar\alpha) \leqslant \bar\beta$. If $\bar\beta = \bar\alpha$ we are done. So assume $\bar\beta > \bar\alpha$. Suppose that $\beta(\bar\alpha) < \bar\beta$. Then $J_{\bar\beta}$ will contain a map f from a subset of some $\gamma < \bar\alpha$ cofinally into $\bar\alpha$. Since $\tilde\pi$: $J_{\bar\beta} \prec_{n-1} J_\beta$, $\tilde\pi(f)$ will be a map from a subset of $\tilde\pi(\gamma) < \tilde\pi(\bar\alpha)$ cofinally into $\tilde\pi(\bar\alpha)$. Now, $\gamma \in \bar\alpha$, so $\tilde\pi(\gamma) = \gamma$. Thus $\tilde\pi(f)$ maps a subset of γ cofinally into $\tilde\pi(\bar\alpha)$. But since $f \subseteq \bar\alpha \times \bar\alpha$, $f \subseteq \tilde\pi(f)$. So as $\mathrm{dom}(f) \subseteq \gamma$ and $\mathrm{dom}(\tilde\pi(f)) \subseteq \gamma$, we must have $\tilde\pi(f) = f$. Thus f maps a subset of γ cofinally into $\tilde\pi(\bar\alpha)$. But $\mathrm{ran}(f) \subseteq \bar\alpha < \alpha \leqslant \tilde\pi(\bar\alpha)$, so this is impossible. This proves the lemma. $\square$

5.7 Lemma. $n = n(\bar\alpha)$.

Proof. By 5.6 and the properties of $\bar g_1$, $n(\bar\alpha) \leqslant n$. If $n = 1$ we are done. So assume $n > 1$. We must prove that $\bar\alpha$ is Σ_{n-1} regular over $J_{\bar\beta}$. Suppose not. Then there is a $\Sigma_{n-1}(J_{\bar\beta})$ map of a bounded subset of $\bar\alpha$ cofinally into $\bar\alpha$. Since $\bar\alpha \in S$, an argument as in 5.3 now shows that $\varrho_{\bar\beta}^{n-1} = \kappa$. But $\varrho_{\bar\beta}^{n-1} = \bar\varrho \geqslant \bar\alpha > \kappa$, a contradiction. $\square$

5.8 Lemma. $\bar p = p(\bar\alpha)$.

Proof. By 5.5, 5.6, and 5.7, $p(\bar\alpha)$ is (by definition) the $<_J$-least element of $J_{\bar\varrho}$ such that $J_\varrho = \bar h''(\omega \times (\kappa \times \{p(\bar\alpha)\}))$. Now,

$$\pi\colon \langle J_{\bar\varrho}, \bar A\rangle \prec_1 \langle J_{m(\lambda)}, A \cap J_{m(\lambda)}\rangle,$$

so

$$\pi''[\bar h''(\omega \times (\kappa \times \{\bar p\}))] = h''_{m(\lambda)}(\omega \times (\kappa \times \{p\})) = X_\lambda\,.$$

Thus,

$$\bar h''(\omega \times (\kappa \times \{\bar p\})) = \pi^{-1}{}''X_\lambda = J_{\bar\varrho}\,.$$

This proves that $p(\bar\alpha) \leqslant_J \bar p$. Let $p' = \pi(p(\bar\alpha))$. Pick $i \in \omega$, $\nu \in \kappa$ so that $\bar p = \bar h(i, (\nu, p(\bar\alpha)))$. Applying π, we get $p = h(i, (\nu, p'))$. Thus

$$h''(\omega \times (\kappa \times \{p\})) \subseteq h''(\omega \times (\kappa \times \{p'\}))\,.$$

Thus $J_\varrho = h''(\omega \times (\kappa \times \{p'\}))$. So by choice of $p, p \leqslant_J p'$. Applying π^{-1}, $\bar{p} \leqslant_J p(\bar{\alpha})$. The lemma is proved. □

$\bar{g}$ Now define $\bar{g}$ from $\bar{\alpha}$ exactly as g was defined from α.

5.9 Lemma. $\bar{g} \cap (\bar{\alpha} \times k''\lambda) = g \cap (\bar{\alpha} \times k''\lambda)$.

Proof. By virtue of 5.5 through 5.8, for $\nu < \kappa$, $\tau < \bar{\alpha}$, we have

$$\bar{g}(\omega\nu + i) = \tau \leftrightarrow \bar{h}(i, (\nu, \bar{p})) = \tau.$$

But $\pi\colon \langle J_\varrho, \bar{A}\rangle \prec_1 \langle J_{m(\lambda)}, A \cap J_{m(\lambda)}\rangle$ and $\pi \restriction \bar{\alpha} = \mathrm{id} \restriction \bar{\alpha}$. Hence

$$\bar{h}(i, (\nu, \bar{p})) = \tau \leftrightarrow h_{m(\lambda)}(i, (\nu, p)) = \tau.$$

Thus:

$$(*) \qquad \bar{g}(\omega\nu + i) = \tau \leftrightarrow h_{m(\lambda)}(i, (\nu, p)) = \tau.$$

Now, by the uniformity of the Σ_1 skolem function,

$$h_{m(\lambda)}(i, (\nu, p)) = \tau \quad \text{implies} \quad h(i, (\nu, p)) = \tau.$$

Thus by (*),

$$\bar{g}(\omega\nu + i) = \tau \quad \text{implies} \quad g(\omega\nu + i) = \tau.$$

Suppose that, in addition, $\omega\nu + i \in k''\lambda$. Assume that $g(\omega\nu + i) = \tau$. Then by the definition of the function m,

$$(\exists z \in J_{m(\lambda)})\, G(z, \tau, \omega\nu + i).$$

So by the canonical, uniform nature of the Σ_0 predicate G,

$$h_{m(\lambda)}(i, (\nu, p)) = \tau.$$

Thus by (*), $\bar{g}(\omega\nu + i) = \tau$, and we are done. □

Now define $\bar{k}, \bar{m}, (\bar{X}_\nu \mid \nu < \bar{\theta})$, $(\bar{\alpha}_\nu \mid \nu < \bar{\theta})$ from $\bar{\alpha}$ exactly as we defined $k, m, (X_\nu \mid \nu < \theta), (\alpha_\nu \mid \nu < \theta)$ from α. Thus, in particular, provided that $\bar{\alpha} \in R$, we will have

$$C_{\bar{\alpha}} = \{\bar{\alpha}_\nu \mid \nu < \bar{\theta} \wedge \lim(\nu)\}.$$

5.10 Lemma. *For all* $\nu < \lambda$, $\bar{k}(\nu) = \pi(\bar{k}(\nu)) = k(\nu)$, $\pi(\bar{m}(\nu)) = m(\nu)$, $\pi''\bar{X}_\nu = X_\nu$, $\bar{\alpha}_\nu = \pi(\bar{\alpha}_\nu) = \alpha_\nu$.

Proof. Since $\pi\colon \langle J_{\bar{\varrho}}, \bar{A}\rangle \prec_1 \langle J_{m(\lambda)}, A \cap J_{m(\lambda)}\rangle$ and $\pi \restriction \bar{\alpha} = \mathrm{id} \restriction \bar{\alpha}$, this follows from 5.5 through 5.9 by a straightforward induction on ν. □

Since $(\alpha_\nu \mid \nu < \lambda)$ is cofinal in $\alpha_\lambda = \bar{\alpha}$, it follows from the above lemma that $\bar{\theta} = \lambda$. Hence,

$$\{\bar{\alpha}_\nu \mid \nu < \bar{\theta} \wedge \lim(\nu)\} = \{\alpha_\nu \mid \nu < \lambda \wedge \lim(\nu)\} = \bar{\alpha} \cap C_\alpha .$$

We shall be done provided that $\bar{\alpha} \in R$. Suppose that $n = 1$. Then $\bar{\varrho} = \varrho^0_{\bar{\beta}} = \bar{\beta}$. But $\pi\colon J_{\bar{\varrho}} \prec J_{m(\lambda)}$ and $\lim(m(\lambda))$. Hence $\lim(\bar{\varrho})$. Thus $n = 1$ implies $\lim(\bar{\beta})$. Thus $\bar{\alpha} \in R$, and we are done. $\square'_\kappa$ is proved.

Exercises

1. κ^+-*Aronszajn trees* (Section 2)

Let κ be an infinite cardinal. By a *special* κ^+-Aronszajn tree we mean a κ^+-tree T such that for each $\alpha < \kappa^+$, $T_\alpha \subseteq \{f \mid f\colon \alpha \xrightarrow{1-1} \kappa\}$, with the ordering $f <_{\mathbf{T}} g$ iff $f \subseteq g$. It is immediate that any such tree must be κ^+-Aronszajn, of course.

1 A. Prove that there is a special ω_1-Aronszajn tree, first of all by making a simple modification to the tree constucted in III.1.1, and then by means of a direct recursion on the levels, much as in the proof of III.1.1. (As then, the problem is to ensure that the construction does not break down at some stage.)

1 B. Prove that if κ is a regular cardinal such that $2^{<\kappa} = \kappa$, then there is a special κ^+-Aronszajn tree. (Generalise the direct proof of 1 A above. The hypotheses on κ are used to ensure that the construction does not break down.)

1 C. Prove that if $V = L$ (or more generally if $2^{<\kappa} = \kappa$ and $\square_\kappa$ holds), then for any infinite cardinal κ there is a special κ^+-Aronszajn tree. (The $\square_\kappa$-sequence is used to ensure that the construction does not break down. See the proof of 2.4.)

2. κ^+-*Souslin trees* (Section 2)

Let κ be an uncountable regular cardinal. Assume GCH together with $\diamondsuit_{\kappa^+}(E)$, where $E = \{\alpha \in \kappa^+ \mid \mathrm{cf}(\alpha) = \kappa\}$. Prove that there is a κ^+-Souslin tree.

3. κ-*Kurepa trees* (Section 1)

Show that if κ is inaccessible, there is a κ-tree with 2^κ many κ-branches. Suggest a definition of a κ-Kurepa tree which avoids this example (and indeed any other example one can construct in ZFC alone).

4. *The combinatorial principle* $\diamondsuit^+_{\kappa^+}$ (Section 3)

Formulate the principle $\diamondsuit^+_{\kappa^+}$ by analogy with $\diamondsuit^+$ for ω_1. Prove that $V = L$ implies $\diamondsuit^+_{\kappa^+}$ and that $\diamondsuit^+_{\kappa^+}$ implies the existence of a κ^+-Kurepa tree.

5. $\square_\kappa$ *in* $L[A]$ (Section 5)

Prove that $\square_\kappa$ holds if $V = L[A]$, where $A \subseteq \kappa^+$ is such that

$$(\forall \alpha < \kappa^+)[|\alpha|^{L[A \cap \alpha]} \leqslant \kappa].$$

(This requires some reworking of the fine-structure theory, and is quite a demanding exercise.)

6. *On the failure of* $\Box_\kappa$ (Section 5)

From the result of exercise 5 above, deduce that if $\Box_\kappa$ fails, then κ^+ is Mahlo in L. (It can be proved that if it is consistent with ZFC that a Mahlo cardinal exists, then it is consistent with ZFC that $\Box_{\omega_1}$ is false.)

7. *GCH and the principles* $\Diamond_{\kappa^+}$ (Section 2)

Prove the following generalisation of lemma 2.7: Assume $2^\kappa = \kappa^+$ and that $\lambda < \kappa$ is a regular cardinal such that either $\kappa^\lambda = \kappa$ or else $[\lambda \neq \mathrm{cf}(\kappa)$ and $(\forall \theta < \kappa) \cdot (\theta^\lambda \leqslant \kappa)]$. Then $\Diamond_{\kappa^+}(\{\delta < \kappa^+ \mid \mathrm{cf}(\delta) = \lambda\})$ holds. (Even better, conclude that $\Diamond^+_{\kappa^+}(\{\delta < \kappa^+ \mid \mathrm{cf}(\delta) = \lambda\})$ holds.)

8. $\Box_\kappa$ *and the principles* $\Diamond_{\kappa^+}$ (Section 2)

Prove the following generalisation of lemma 2.8: Assume $\Box_\kappa$ and that $(\forall \theta < \kappa) \cdot (\theta^{\mathrm{cf}(\kappa)} \leqslant \kappa)$ & $2^\kappa = \kappa^+$. Then $\Diamond_{\kappa^+}(\{\delta < \kappa^+ \mid \mathrm{cf}(\delta) = \mathrm{cf}(\kappa)\})$ holds. (Can the above be strengthened to get $\Diamond^*_{\kappa^+}(\{\delta < \kappa^+ \mid \mathrm{cf}(\delta) = \mathrm{cf}(\kappa)\})$?)

Chapter V
The Story of $0^{\#}$

In this chapter we investigate the effect upon $V = L$ of the postulated existence of various large cardinals in the universe. This represents a different approach to constructibility from that adopted hitherto. Previously we have been looking at the *internal* structure of the constructible universe. We now step back and regard L *from the outside* as it were.

It is assumed that the reader has a prior acquaintance with large cardinal theory. Admittedly, our account is self-contained (except for the omission of some proofs); but the results we shall obtain cannot really be appreciated without some familiarity with the standard theory of the cardinal properties concerned. The relevant material can be found in *Drake* (1974) and *Jech* (1978).

We shall make considerable use of model-theoretic techniques, usually for models of the languages $\mathscr{L}(A_1, \ldots, A_n)$. It will be convenient to use some of the standard notation of model theory. In particular, we shall write the satisfaction relation as

$$\langle M, \in, A_1, \ldots, A_n \rangle \models \varphi$$

rather than

$$\models_{\langle M, \in, A_1, \ldots, A_n \rangle} \varphi .$$

We shall also not bother to distinguish between an element, x, of a structure and the constant, $\mathring{x}$, of $\mathscr{L}_V$ which denotes it. If $t(\mathring{x}_0, \ldots, \mathring{x}_m)$ is a term of $\mathscr{L}_M(A_1, \ldots, A_n)$ (so $x_0, \ldots, x_m \in M$), we write $t^{\mathscr{A}}(x_0, \ldots, x_m)$ for the interpretation of the term $t(\mathring{x}_0, \ldots, \mathring{x}_m)$ in the structure $\mathscr{A} = \langle M, \ , A_1, \ldots, A_n \rangle$.

We shall also speak of models of ZFC, BS, etc. In each such case we mean these theories formulated in the language $\mathscr{L}$, and not in LST as was originally the case.

1. A Brief Review of Large Cardinals

A cardinal κ is said to be *weakly inaccessible* iff it is an uncountable, regular limit cardinal, and (*strongly*) *inaccessible* iff it is uncountable and regular and has the property that $(\forall \lambda < \kappa)(2^\lambda < \kappa)$. It is clear that all inaccessible cardinals are weakly inaccessible, and that if the GCH be assumed then the two notions of inaccessibility coincide.

If κ is inaccessible, then V_κ (and L_κ) is a model of ZFC. Hence by Gödel's Second Incompleteness Theorem, the existence of inaccessible cardinals is not provable in ZFC.

1.1 Theorem.
(i) *If κ is a cardinal, then $[\kappa$ is a cardinal$]^L$.*
(ii) *If κ is a limit cardinal, then $[\kappa$ is a limit cardinal$]^L$.*
(iii) *If κ is a regular cardinal, then $[\kappa$ is a regular cardinal$]^L$.*
(iv) *If κ is a weakly inaccessible cardinal, then $[\kappa$ is an inaccessible cardinal$]^L$.*

Proof. (i)–(iii). Each of the properties is easily seen to be Π_1, and hence D-absolute.
(iv) By (i)–(iii) and the fact that $[\mathrm{GCH}]^L$. □

A cardinal κ is said to be *Mahlo* iff it is inaccessible and the set

$$\{\lambda \in \kappa \mid \lambda \text{ is inaccessible}\}$$

is stationary in κ.

1.2 Theorem. *If κ is Mahlo, then $[\kappa$ is Mahlo$]^L$.*

Proof. Again, this property is easily seen to be D-absolute. □

A cardinal κ is said to be *weakly compact* iff it is uncountable and satisfies the partition property

$$\kappa \to (\kappa)^2_2 .$$

What does this mean? In order to explain we need some notation. If X is a set of ordinals and α is an ordinal, $[X]^\alpha$ denotes the set of all strictly increasing α-sequences of members of X. We set

$$[X]^{<\alpha} = \bigcup_{\beta<\alpha} [X]^\beta$$

$$[X]^{\leqslant\alpha} = \bigcup_{\beta\leqslant\alpha} [X]^\beta .$$

Let X be a set of ordinals, α an ordinal, μ a cardinal. By a *μ-partition* of $[X]^\alpha$ we mean a function

$$f\colon [X]^\alpha \to \mu ,$$

which we regard as partitioning $[X]^\alpha$ into μ disjoint classes. A subset Y of X is said to be *homogeneous* for the partition f iff

$$|f''[Y]^\alpha| = 1 ,$$

i.e. iff all strictly increasing α-sequences of members of Y lie in the same partition class. We write

$$\kappa \to (\lambda)^\alpha_\mu$$

iff every μ-partition of $[\kappa]^\alpha$ has a homogeneous set of cardinality λ. This notation is due to Erdös and Rado. The idea behind it is the easily observed fact that a valid partition relation remains valid if the parameter on the left of the arrow is increased or if any parameter on the right of the arrow is decreased.

The well known *Ramsey's Theorem* states that

$$\omega \to (\omega)^m_n$$

for all $m, n \in \omega$. Generalising this to some extent is the *Erdös-Rado Theorem* that for any cardinal κ and any $n \in \omega$,

$$\mathcal{J}_n(\kappa)^+ \to (\kappa^+)^{n+1}_\kappa,$$

where $\mathcal{J}_n(\kappa)$ is the n-th iterate of the exponential function 2^λ, starting from κ (i.e. $\mathcal{J}_0(\kappa) = \kappa$, $\mathcal{J}_1(\kappa) = 2^\kappa$, $\mathcal{J}_2(\kappa) = 2^{\mathcal{J}_1(\kappa)}$, etc.).

A weakly compact cardinal, then, is one for which the generalised Ramsey's Theorem

$$\kappa \to (\kappa)^2_2$$

holds. All weakly compact cardinals are Mahlo. The name "weakly compact" stems from the equivalent definition that a weakly compact cardinal is a cardinal κ for which the "κ-compactness property" is valid for any "κ-language". By a *κ-language* we mean a first-order language having κ many basic symbols, whose syntax allows conjunctions and disjunctions of any length less than κ and quantification over any sequence of variables of length less than κ. (In this context, an ordinary first-order language would be called an "ω-language".) The *κ-compactness property* for such a language says that if a set of at most κ sentences of the language is *κ-satisfiable* (i.e. any subset of cardinality less than κ has a model), then the entire set has a model. The whole idea is to generalise to an uncountable cardinal κ, everything connected with the compactness theorem of ordinary logic.

The following theorem lists several standard, equivalent formulations of the notion of weak compactness. Proofs of the various equivalences may be found in *Drake* (1974) or *Jech* (1978).

1.3 Theorem. *Let κ be an uncountable cardinal. The following are equivalent:*

(i) *κ is weakly compact (i.e. $\kappa \to (\kappa)^2_2$);*
(ii) $(\forall n \in \omega)(\forall \lambda < \kappa)[\kappa \to (\kappa)^n_\lambda]$;
(iii) *the κ-compactness property holds for any κ-language;*
(iv) *κ is Π^1_1-indescribable: i.e. if φ is a sentence of $\mathcal{L}(U, A_1, \ldots, A_n)$ such that*

$$(\forall U \subseteq V_\kappa)[\langle V_\kappa, \in, U, A_1, \ldots, A_n\rangle \models \varphi]$$

for $A_1, \ldots, A_n \subseteq V_\kappa$, then for some $\alpha < \kappa$,

$$(\forall U \subseteq V_\alpha)[\langle V_\alpha, \in, U, A_1 \cap V_\alpha, \ldots, A_n \cap V_\alpha\rangle \models \varphi];$$

(v) *κ has Keisler's Extension Property: every structure of the form $\langle V_\kappa, \varepsilon, U\rangle$ has a transitive elementary extension which contains κ;*

(vi) *If $\mathscr{F}$ is a κ-complete field of sets which is κ-generated by a set of cardinality at most κ, then $\mathscr{F}$ has a κ-complete ultrafilter;*

(vii) *κ is inaccessible and if $\mathscr{F}$ is a κ-complete field of sets of cardinality κ, then $\mathscr{F}$ has a κ-complete ultrafilter;*

(viii) *κ is inaccessible and there is no κ-Aronszajn tree.* □

The following lemma is relevant to our present purposes.

1.4 Lemma. *Let κ be a weakly compact cardinal, and let $A \subseteq \kappa$. If $A \cap \alpha \in L$ for all $\alpha < \kappa$, then $A \in L$.*

Proof. By assumption,

$$\langle V_\kappa, \in, A \rangle \models (\forall \alpha)(A \cap \alpha \in L).$$

By Keisler's Extension Property, let M be a transitive set such that $V_\kappa \cup \{\kappa\} \subseteq M$ and for some set $A' \subseteq M$,

$$\langle V_\kappa, \in, A \rangle \prec \langle M, \in, A' \rangle.$$

We have

$$\langle M, \in, A' \rangle \models (\forall \alpha)(A' \cap \alpha \in L).$$

In particular, since $\kappa \in M$,

$$\langle M, \in, A' \rangle \models A' \cap \kappa \in L.$$

But $A' \cap \kappa = A$. So, noting that set-membership is absolute for M,

$$A \in (L)^M.$$

Now, as κ is inaccessible, V_κ is a model of ZFC. Thus M is a model of ZFC. So by II.2.10, $(L)^M \subseteq L$. Thus $A \in L$, and we are done. □

Utilising 1.4 we have (see also Exercise 1.):

1.5 Theorem. *If κ is weakly compact, then [κ is weakly compact]^L.*

Proof. By 1.1 we know that [κ is inaccessible]^L. So by 1.3 it suffices to prove that [there are no κ-Aronszajn trees]^L.

Let $T \in L$ be, in L, a κ-tree. We may assume that T has domain κ and that $\alpha <_T \beta$ implies $\alpha < \beta$. It is clear that T is a κ-tree in the real world. Hence as κ is weakly compact, there is (in V) a κ-branch, b, of T. For any $\alpha < \kappa$, let γ be the least ordinal in $b - \alpha$. Then

$$b \cap \alpha = \{\xi \in T \mid \xi <_T \gamma\} \in L.$$

So by 1.4, $b \in L$. But clearly,

$$[b \text{ is a } \kappa\text{-branch of } T]^L.$$

Thus T is not a κ-Aronszajn tree in the sense of L. The proof is complete. □

We shall return to the notion of weak compactness, and a strengthening of it (ineffability) in Chapter VII. In the meantime we consider a much more powerful large cardinal notion.

We write

$$\kappa \to (\lambda)_{\mu}^{<\omega}$$

if, whenever f is a μ-partition of $[\kappa]^{<\omega}$ (i.e. $f\colon [\kappa]^{<\omega} \to \mu$) there is a set $X \subseteq \kappa$ of cardinality λ such that

$$(\forall n \in \omega)[|f''[X]^n| = 1].$$

Thus, X is simultaneously homogeneous for each of the partitions $f \restriction [X]^n$, $n \in \omega$. We cannot expect X to be "homogeneous" for all of f in the sense that $|f''[X]^{<\omega}| = 1$, since the value of f could depend upon the length of the argument. There is thus no real danger of confusion if we agree to say that a set X for which $(\forall n \in \omega)[|f''[X]^n| = 1]$ is *homogeneous* for f.

Already the existence of a cardinal κ such that

$$\kappa \to (\omega)_2^{<\omega}$$

is a powerful assumption, implying the existence of (many) weakly compact cardinals.

For any cardinal λ, if there is a cardinal κ such that

$$\kappa \to (\lambda)_2^{<\omega},$$

then the least such κ is denoted by $\kappa(\lambda)$. The cardinals $\kappa(\lambda)$ are called the *Erdös cardinals.* They are all inaccessible, and $\kappa(\omega)$ exceeds the first weakly compact cardinal (and the first ineffable cardinal). If $\lambda < \mu$, then $\kappa(\lambda) < \kappa(\mu)$. (See *Drake* (1974) or *Jech* (1978) for all details.)

A cardinal κ such that $\kappa(\kappa) = \kappa$ is called a *Ramsey cardinal.* Thus κ is a Ramsey cardinal iff

$$\kappa \to (\kappa)_1^{<\omega}.$$

Since this is not a simple generalisation of Ramsey's Theorem (the obvious generalisation being provided by the weakly compact cardinals), the name "Ramsey cardinal" is slightly misleading, but is now well established.

The Erdös cardinals have powerful model-theoretic properties, as we show next. We consider structures of the form

$$\mathscr{A} = \langle A, <_A, \ldots \rangle,$$

where $<_A$ linearly orders some subset of A, called the *field* of $<_A$. By the *length* of $\mathscr{A}$ we mean the cardinality of the set of all functions, relations and constants of $\mathscr{A}$. If the length of $\mathscr{A}$ is infinite, then this is just the cardinality of the language of $\mathscr{A}$.

An infinite subset, H, of the field of $<_A$ is said to be *$\mathscr{A}$-indiscernible* iff for each $n \in \omega$ and each pair $(a_0, \ldots, a_n)$, $(b_0, \ldots, b_n) \in [H]^{n+1}$ it is the case that for all formulas $\varphi(v_0, \ldots, v_n)$ in the language of $\mathscr{A}$:

$$\mathscr{A} \models \varphi(a_0, \ldots, a_n) \quad \text{iff } \mathscr{A} \models \varphi(b_0, \ldots, b_n).$$

In other words, as far as first-order properties are concerned, for each $n \in \omega$, all increasing n-tuples from H look the same to $\mathscr{A}$.

1.6 Theorem. *Let λ be an infinite cardinal. The following conditions on κ are equivalent*:

(i) $\kappa \to (\lambda)_2^{<\omega}$;

(ii) $\kappa \to (\lambda)_{2^\omega}^{<\omega}$;

(iii) *for all* $\mu < \kappa(\lambda)$, $\kappa \to (\lambda)_\mu^{<\omega}$;

(iv) *every structure of the form*

$$\mathscr{A} = \langle A, <_A, \ldots \rangle,$$

of countable length, such that $\kappa \subseteq \text{field}(<_A)$ *and* $<_A \restriction \kappa$ *is the usual order on* κ, has an *$\mathscr{A}$-indiscernible subset of cardinality* λ;

(v) *as in* (iv) *except that* $\mathscr{A}$ *may have any length less than* $\kappa(\lambda)$.

Proof. The proofs of the equivalence of (i), (ii) and (iii) can be found in *Drake* (1974) and *Jech* (1978), but, since they are not really relevant to us here we shall not give them. We prove the equivalence of (iii) and (v), this being the result that we require. (A similar argument yields the equivalence of (iv) and (ii), as is easily seen.)

Assume (iii). Let $\mathscr{A}$ be a stated in (v). Define a function f on $[\kappa]^{<\omega}$ by letting $f(a_0, \ldots, a_n)$ be the set of all formulas $\varphi(v_0, \ldots, v_n)$ in the language of $\mathscr{A}$ such that

$$\mathscr{A} \models \varphi(a_0, \ldots, a_n).$$

Since *length* $(\mathscr{A}) < \kappa(\lambda)$ and $\kappa(\lambda)$ is inaccessible, the range of f has cardinality less than $\kappa(\lambda)$. So, by (iii), f has a homogeneous set, H, of cardinality λ. Clearly, H is $\mathscr{A}$-indiscernible.

Now assume (v). Given a partition

$$f\colon [\kappa]^{<\omega} \to \mu < \kappa(\lambda),$$

consider the structure

$$\mathscr{A} = \langle \kappa, <, (f \restriction [\kappa]^n)_{n<\omega}, (\xi)_{\xi<\mu} \rangle.$$

The length of $\mathscr{A}$ is less than $\kappa(\lambda)$, so by (v), $\mathscr{A}$ has an $\mathscr{A}$-indiscernible subset, H, of cardinality λ. Clearly, H is homogeneous for f. $\square$

A *measurable* cardinal is an uncountable cardinal κ such that there is a function

$$\mu\colon \mathscr{P}(\kappa) \to 2$$

with the properties:

(i) $\mu(\{\alpha\}) = 0$ for all $\alpha \in \kappa$;
(ii) $\mu(\kappa) = 1$;
(iii) if X_α, $\alpha < \lambda$, are disjoint subsets of κ, where $\lambda < \kappa$, then

$$\mu(\bigcup_{\alpha<\lambda} X_\alpha) = \sum_{\alpha<\lambda} \mu(X_\alpha).$$

(Such a function μ is called a *two-valued measure* on κ.) Measurable cardinals are extremely large. In particular, if κ is measurable, then κ is Ramsey, and indeed is the κ-th Ramsey cardinal. However, as far as L is concerned, cardinals well below the first measurable cardinal (if it exists) already have a highly significant effect. The critical point is the jump from $\kappa(\omega)$ to $\kappa(\omega_1)$, as we show next.

1.7 Theorem. *If $\kappa \to (\omega)_2^{<\omega}$, then $[\kappa \to (\omega)_2^{<\omega}]^L$.*

Proof. Let $f \in L$ be such that, in the sense of L, $f\colon [\kappa]^{<\omega} \to 2$. Then, clearly, f is such a function in the real world. Define

$$H = \{\sigma \in [\kappa]^{<\omega} \mid (\forall n \leqslant |\sigma|)[|f \upharpoonright [\sigma]^n| = 1]\}.$$

Notice that the definition of H is absolute for L. We regard H as a poset under the ordering $\supseteq$. Since $\kappa \to (\omega)_2^{<\omega}$, f has an infinite homogeneous set X. Let σ_n consist of the first n elements of X, for each $n \in \omega$. Then $\sigma_n \in H$, and $(\sigma_\nu \mid n < \omega)$ is a $\supseteq$-decreasing chain in the poset H. Thus H is not well-founded. So by I.8.7 and I.8.3,

$$[H \text{ is not well-founded}]^L.$$

So let $(\tau_n \mid n < \omega)$ be a $\supseteq$-decreasing chain from H in L. Then $Y = \bigcup_{n<\omega} \tau_n \in L$ is an infinite homogeneous set for f. This proves the theorem. □

1.8 Theorem. *If there exists a κ such that $\kappa \to (\omega_1)_2^{<\omega}$, then $\mathscr{P}^L(\omega)$ is countable (so in particular, $V \neq L$).*

Proof. Let $\kappa = \kappa(\omega_1)$, and consider the structure

$$\mathscr{A} = \langle L_\kappa, \in, \mathscr{P}^L(\omega)\rangle.$$

Let $X \subseteq \kappa$ be an uncountable, $\mathscr{A}$-indiscernible set, and let

$$\mathscr{B} = \langle B, \in, \mathscr{P}^L(\omega) \cap B\rangle$$

be the smallest $\mathscr{B} \prec \mathscr{A}$ such that $X \subseteq B$ (see II.5.3). Then every element of B is of the form $t^{\mathscr{A}}(\vec{x})$ for some term t of set theory and some $(\vec{x}) \in [X]^{<\omega}$. Suppose that $t^{\mathscr{A}}(\vec{x}) \subseteq \omega$. Now, each $n \in \omega$ is definable in $\mathscr{A}$, and the validity of the sentence

$$n \in t(\vec{x})$$

in $\mathscr{A}$ is independent of the exact choice of $(\vec{x})$ from $[X]^{<\omega}$. Hence $t^{\mathscr{A}}(\vec{x})$ does not depend upon $\vec{x}$. But there are only countably many terms t. Thus $\mathscr{P}^L(\omega) \cap B$ must be countable.

Now let

$$\pi: \mathscr{B} \cong \mathscr{C} = \langle L_\lambda, \in, U \rangle .$$

Since $\lambda \geqslant \omega_1$, we have $\mathscr{P}^L(\omega) \subseteq L_\lambda$. But

$$\mathscr{A} \models (\forall x \subseteq \omega)(x \in \mathscr{P}^L(\omega)),$$

so

$$\mathscr{C} \models (\forall x \subseteq \omega)(x \in U).$$

Hence $\mathscr{P}^L(\omega) \subseteq U$. But $|U| = |\mathscr{P}^L(\omega) \cap B| = \omega$, so we are done. □

2. L-Indiscernibles and $0^\#$

In this section we shall obtain a considerable strengthening of 1.8, by proving that if $\kappa(\omega_1)$ exists, then the class of all uncountable cardinals is L-indiscernible. (In particular, this will imply that every uncountable cardinal is inaccessible in the sense of L, giving the conclusion of 1.8 at once.) The existence of $\kappa(\omega_1)$ will also be shown to imply the existence of a truth definition for L, so we may consider the set of all formulas $\varphi(v_0, \dots, v_n)$ of $\mathscr{L}$ such that $\models_L \varphi(\kappa_0, \dots, \kappa_n)$ for any strictly increasing sequence $\kappa_0, \dots, \kappa_n$ of uncountable cardinals. Denoting the set of all Gödel numbers of formulas in this set by the symbol $0^\#$, we shall go on to show that the set $0^\#$ has an alternative definition, which does not depend upon the existence of L-indiscernibles and a truth definition for L, and that the mere existence of a set of integers satisfying this definition is itself sufficient to ensure that the uncountable cardinals are L-indiscernible.

The techniques which we shall employ are essentially model-theoretic, and originate with some work of Ehrenfeucht and Mostowski concerning models with indiscernibles.

By examining the proof of II.2.9, we see that there is an extension of the $\mathscr{L}$-theory BS, let us call it BSL, which consists of BS together with finitely many instances of the Σ_0-Collection Schema of KP, such that:

(i) $L_\lambda \models$ BSL for any limit ordinal $\lambda > \omega$; and
(ii) if M is a transitive model of BSL, then for any $\alpha \in M$, $(L_\alpha)^M = L_\alpha$.

(This relates to the proof of II.2.12. We simply require enough instances of Σ_0-Collection to enable us to define the constructible hierarchy.)

Suppose $\mathscr{A} = \langle A, E \rangle$ is a model of the $\mathscr{L}$-theory BSL + $(V = L)$. If $X \subseteq A$, we denote by $\mathscr{A} \upharpoonright X$ the set

$$\{t^{\mathscr{A}}(x_0, \dots, x_n) \mid t \text{ is a term of } \mathscr{L} \text{ and } x_0, \dots, x_n \in X\}.$$

It follows from the fact that L has a definable well-ordering that $\mathscr{A} \upharpoonright X \prec \mathscr{A}$. (See II.5.3 in this connection.) We say that $\mathscr{A} \upharpoonright X$ is the elementary substructure of $\mathscr{A}$ *generated by* X.

We shall call a set, Σ, of formulas of $\mathscr{B}$ an *Ehrenfeucht-Mostowski set* (or *E–M set* for short) iff there is a model $\mathscr{A} = \langle A, E\rangle$ of BSL + $(V = L)$ and an infinite set $H \subseteq \mathrm{On}^{\mathscr{A}}$ which is $\mathscr{A}$-indiscernible, such that Σ is the set of all $\mathscr{L}$-formulas which are valid in $\mathscr{A}$ on increasing sequences of indiscernibles from H.

Let Σ be an E–M set, and let α be an infinite ordinal. By a *(Σ, α)-model* we mean a pair $(\mathscr{A}, H)$ such that:

(i) $\mathscr{A} = \langle A, E\rangle$ is a model of BSL + $(V = L)$;
(ii) $H \subseteq \mathrm{On}^{\mathscr{A}}$ is an $\mathscr{A}$-indiscernible set of order-type α (under $<^{\mathscr{A}}_{\mathrm{On}}$);
(iii) $\mathscr{A} = \mathscr{A} \upharpoonright H$ (i.e. H generates $\mathscr{A}$);
(iv) Σ is the set of all $\mathscr{L}$-formulas which are valid in $\mathscr{A}$ on increasing tuples from H.

2.1 Lemma. *Let Σ be an E–M set, and let α, β be infinite ordinals, $\alpha \leqslant \beta$. Let $(\mathscr{A}_\alpha, H_\alpha)$ be a (Σ, α)-model, $(\mathscr{A}_\beta, H_\beta)$ a (Σ, β)-model, and let $h\colon H_\alpha \to H_\beta$ be order-preserving. Then there is an embedding $\tilde{h}\colon \mathscr{A}_\alpha \prec \mathscr{A}_\beta$ such that $h \subseteq \tilde{h}$. Moreover, if $\beta = \alpha$ and h is onto H_β, then $\tilde{h}$ is an isomorphism of $\mathscr{A}_\alpha$ onto $\mathscr{A}_\beta$; so in particular, the (Σ, α)-model is unique up to isomorphism.*

Proof. Since H_α generates $\mathscr{A}_\alpha$, for any $a \in \mathscr{A}_\alpha$ there is a term t and elements $\vec{x}$ of H_α such that $a = t^{\mathscr{A}_\alpha}(\vec{x})$. Set $\tilde{h}(a) = t^{\mathscr{A}_\beta}(\overrightarrow{hx})$.

We must first of all check that $\tilde{h}$ is well-defined. Suppose that there are terms t_1, t_2 and elements $x_1, \ldots, x_n, y_1, \ldots, y_m$ of H_α such that

$$a = t_1^{\mathscr{A}_\alpha}(x_1, \ldots, x_n) = t_2^{\mathscr{A}_\alpha}(y_1, \ldots, y_m).$$

Let $z_1, \ldots, z_k$ enumerate the set $\{x_1, \ldots, x_n, y_1, \ldots, y_m\}$ in increasing order, and let $\varphi(z_1, \ldots, z_k)$ be the formula

$$t_1(x_1, \ldots, x_n) = t_2(y_1, \ldots, y_m).$$

Then $\varphi(z_1, \ldots, z_k) \in \Sigma$, since φ is true in $\mathscr{A}_\alpha$ on the increasing sequence $z_1, \ldots, z_k$ from H_α. Hence φ is true in $\mathscr{A}_\beta$ on any increasing sequence from H_β. But $h(z_1), \ldots, h(z_k)$ is an increasing sequence from H_β. Thus

$$\mathscr{A}_\beta \models \varphi(h(z_1), \ldots, h(z_k)).$$

In other words,

$$t_1^{\mathscr{A}_\beta}(h(x_1), \ldots, h(x_n)) = t_2^{\mathscr{A}_\beta}(h(y_1), \ldots, h(y_m)),$$

so h is well-defined.

Similarly, we can show that $\tilde{h}$ is one-one and preserves the $\in$-relations of the two models. To show that $\tilde{h}$ is elementary, it suffices to show that $\tilde{h}$ preserves the validity of formulas on tuples from H_α only (since H_α generates $\mathscr{A}_\alpha$), which again can be done by passing through Σ as above. The rest of the lemma follows easily now. □

2.2 Lemma. *Let Σ be an $E-M$ set. For each infinite ordinal α there is a unique (up to isomorphism) (Σ, α)-model.*

Proof. Uniqueness was established in 2.1, so we need only concentrate on existence. We introduce new individual constants c_ν, $\nu < \alpha$, to the language $\mathscr{L}$. Let $\mathscr{A}$ be a model of BSL + $(V = L)$, and let $H \subseteq \mathrm{On}^{\mathscr{A}}$ be an $\mathscr{A}$-indiscernible set such that Σ is the set of all $\mathscr{L}$-formulas true in $\mathscr{A}$ on increasing sequences from H. (Such $\mathscr{A}, H$ exist because Σ is an $E-M$ set.) Consider the following theory in the language $\mathscr{L} \cup \{c_\nu \mid \nu < \alpha\}$:

$$T = \{\varphi \mid \varphi \text{ is a sentence of } \mathscr{L} \text{ and } \mathscr{A} \models \varphi\} \cup \{\mathrm{On}(c_\nu) \mid \nu < \alpha\}$$
$$\cup \{c_\nu < c_\tau \mid \nu < \tau < \alpha\} \cup \{\varphi(c_{\nu_0}, \ldots, c_{\nu_n}) \mid \varphi \in \Sigma \ \&\ \nu_0 < \ldots < \nu_n < \alpha\}.$$

It is clear that T is finitely satisfiable in $\mathscr{A}$. So by the compactness theorem, T has a model, say $\mathscr{B}$. Let

$$K = \{c_\nu^{\mathscr{B}} \mid \nu < \alpha\}.$$

Clearly, $\mathscr{B}$ is a model of BSL + $(V = L)$, $K \subseteq \mathrm{On}^{\mathscr{B}}$ is a $\mathscr{B}$-indiscernible set of order-type α, and Σ is the set of formulas of $\mathscr{L}$ which are valid in $\mathscr{B}$ on increasing tuples from K. Thus $(\mathscr{B} \upharpoonright K, K)$ is a (Σ, α)-model. $\square$

So far we have said nothing regarding the existence of an $E-M$ set. In fact the results of this section will depend not just upon the existence of an $E-M$ set, but of an $E-M$ set with some very special properties. We shall describe these properties and their implications for the (Σ, α)-models next, before turning out attention to the construction of an $E-M$ set of the type desired (which will require the existence of large cardinals).

An $E-M$ set Σ is said to be *cofinal* if it contains all formulas of the form:

$$\mathrm{On}(t(v_0, \ldots, v_{n-1})) \to t(v_0, \ldots, v_{n-1}) < v_n$$

for any $\mathscr{L}$-term t.

2.3 Lemma. *Let Σ be an $E-M$ set. The following are equivalent:*

(i) *Σ is cofinal;*
(ii) *for every limit ordinal α, if $(\mathscr{A}, H)$ is the (Σ, α)-model, then H is cofinal in $\mathrm{On}^{\mathscr{A}}$;*
(iii) *for some limit ordinal α, if $(\mathscr{A}, H)$ is the (Σ, α)-model, then H is cofinal in $\mathrm{On}^{\mathscr{A}}$.*

Proof. (i) $\to$ (ii). If $(\mathscr{A}, H)$ is the (Σ, α)-model and $x \in \mathrm{On}^{\mathscr{A}}$, then there is a term t and elements $\vec{h}$ of H such that $x = t^{\mathscr{A}}(\vec{h})$. But then if $k \in H$, $k > \vec{h}$, we have $x < k$ by the requirements on Σ.

(ii) $\to$ (iii). Trivial.

(iii) $\to$ (i). Let t be any term, and let $\varphi(v_0, \ldots, v_n)$ be the formula

$$\mathrm{On}(t(v_0, \ldots, v_{n-1})) \to t(v_0, \ldots, v_{n-1}) < v_n.$$

We must show that $\varphi \in \Sigma$. It suffices to show that for some increasing sequence $h_0, \ldots, h_n$ from H,

$$\mathscr{A} \models \varphi(h_0, \ldots, h_n).$$

Choose $h_0, \ldots, h_{n-1}$ arbitrary increasing from H. If $t^{\mathscr{A}}(h_0, \ldots, h_{n-1}) \notin \mathrm{On}^{\mathscr{A}}$, we are done already. Otherwise, by our assumption on H we can find $h_n \in H$, $h_n > h_{n-1}$, such that $h_n > t^{\mathscr{A}}(h_0, \ldots, h_{n-1})$, and again we are done. □

An $E-M$ set Σ is said to be *remarkable* if, for every term t of $\mathscr{L}$, if the formula

$$t(v_0, \ldots, v_{n-1}, v_n, \ldots, v_{n+m}) < v_n$$

is in Σ, then so too is the formula

$$t(v_0, \ldots, v_{n-1}, v_n, \ldots, v_{n+m}) = t(v_0, \ldots, v_{n-1}, v_{n+m+1}, \ldots, v_{n+2m+1}).$$

2.4 Lemma. *Let Σ be a remarkable, cofinal, $E-M$ set. Let λ be a limit ordinal, and let $(\mathscr{A}, H)$ be the (Σ, λ)-model. Let $(h_\gamma \mid \gamma < \lambda)$ be the monotone enumeration of H. Let $\alpha < \lambda$ be a limit ordinal, and set $K = \{h_\gamma \mid \gamma < \alpha\}$. Let $\mathscr{B} = \mathscr{A} \restriction K$. Then $(\mathscr{B}, K)$ is the (Σ, α)-model and*

$$\mathrm{On}^{\mathscr{B}} = \{x \in \mathrm{On}^{\mathscr{A}} \mid x < h_\alpha\}.$$

Proof. It is immediate (by uniqueness) that $(\mathscr{B}, K)$ is the (Σ, α)-model. And since Σ is cofinal, 2.3 (ii) tells us that K is cofinal in $\mathrm{On}^{\mathscr{B}}$, so

$$\mathrm{On}^{\mathscr{B}} \subseteq \{x \in \mathrm{On}^{\mathscr{A}} \mid x < h_\alpha\}.$$

Hence the lemma boils down to proving that if $x \in \mathrm{On}^{\mathscr{A}}$ and $x < h_\alpha$, then in fact $x \in \mathrm{On}^{\mathscr{B}}$.

Well, since H generates $\mathscr{A}$, there is a term t and elements $k_0, \ldots, k_{n-1}$ of K, $l_0, \ldots, l_m$ of $H - K$, such that $k_0 < \ldots < k_{n-1} < l_0 < \ldots < l_m$ and $x = t^{\mathscr{A}}(\vec{k}, \vec{l})$. By virtue of our convention concerning the indication of variables present in terms, we may assume that $l_0 = h_\alpha$ here. Now, $x < h_\alpha$, so $t^{\mathscr{A}}(\vec{k}, \vec{l}) < h_\alpha$. Thus the formula

$$t(v_0, \ldots, v_{n-1}, v_n, \ldots, v_{n+m}) < v_n$$

is in Σ. So, by remarkability, the formula

$$t(v_0, \ldots, v_{n-1}, v_n, \ldots, v_{n+m}) = t(v_0, \ldots, v_{n-1}, v_{n+m+1}, \ldots, v_{n+2m+1})$$

is in Σ. Thus for any increasing sequence $l'_0, \ldots, l'_m$ from H with $k_{n-1} < l'_0$, we have $t^{\mathscr{A}}(\vec{k}, \vec{l}) = t^{\mathscr{A}}(\vec{k}, \vec{l}')$. But α is a limit ordinal, so we can find such $l'_0, \ldots, l'_m$ with $l'_m < h_\alpha$. Then, since $\vec{k}, \vec{l}' \in K$, we have

$$x = t^{\mathscr{A}}(\vec{k}, \vec{l}) = t^{\mathscr{A}}(\vec{k}, \vec{l}') \in \mathscr{A} \restriction K = \mathscr{B},$$

as required. □

Thus, if Σ is a remarkable, cofinal, $E-M$ set and $(\mathscr{A}, H)$ is the (Σ, λ)-model for some limit ordinal λ, then if we pick any limit ordinal $\alpha < \lambda$ and let K consist of the first α elements of H, the ordinals of the (Σ, α)-model $(\mathscr{A} \upharpoonright K, K)$ form an initial segment of the ordinals of $\mathscr{A}$. Another consequence of remarkability is that the indiscernibles form a club subset of the ordinals of the model:

2.5 Lemma. *Let Σ be a remarkable, cofinal $E-M$ set. Let λ be a limit ordinal, and let $(\mathscr{A}, H)$ be the (Σ, λ)-model. Then H is closed and unbounded in* $\mathrm{On}^{\mathscr{A}}$.

Proof. Unboundedness was proved in 2.3. We verify closure. Let $(h_\gamma \mid \gamma < \lambda)$ be the monotone enumeration of H. Let $\alpha < \lambda$ be a limit ordinal. We must prove that h_α is the least upper bound of the set $K = \{h_\gamma \mid \gamma < \alpha\}$ in $\mathrm{On}^{\mathscr{A}}$. Well, we know from 2.4 that $(\mathscr{A} \upharpoonright K, K)$ is the (Σ, α)-model. But Σ is cofinal, so K is a cofinal subset of $\mathrm{On}^{\mathscr{A} \upharpoonright K}$. It thus suffices to show that h_α is the least upper bound of $\mathrm{On}^{\mathscr{A} \upharpoonright K}$ in $\mathrm{On}^{\mathscr{A}}$. But by 2.4 again,

$$\mathrm{On}^{\mathscr{A} \upharpoonright K} = \{x \in \mathrm{On}^{\mathscr{A}} \mid x < h_\alpha\},$$

so in particular, h_α is the least upper bound of $\mathrm{On}^{\mathscr{A} \upharpoonright K}$ in $\mathrm{On}^{\mathscr{A}}$. $\square$

We shall be particularly interested in well-founded (Σ, α)-models. For suppose $\mathscr{A}$ is a well-founded (Σ, α)-model. Then $\mathscr{A}$ is a well-founded model of the Axiom of Extensionality, in particular, so by the collapsing lemma there is an isomorphism

$$\pi \colon \mathscr{A} \cong \langle M, \in \rangle,$$

where M is a transitive set, Now, M is a transitive model of the theory BSL + $(V = L)$. So by virtue of our choice of this theory (see earlier)

$$M = (V)^M = (L)^M = L_\lambda,$$

where $\lambda = \sup(M \cap \mathrm{On})$. Hence $\mathscr{A} \cong L_\lambda$.

The well-foundedness of the (Σ, α)-model will depend upon the $E-M$ set Σ. We shall call an $E-M$ set Σ *well-founded* if, for all infinite ordinals α, the (Σ, α)-model is well-founded.

2.6 Lemma. *Let Σ be an $E-M$ set. The following are equivalent:*

(i) *Σ is well-founded;*

(ii) *for some $\alpha \geqslant \omega_1$, the (Σ, α)-model is well-founded;*

(iii) *for all infinite $\alpha < \omega_1$, the (Σ, α)-model is well-founded.*

Proof. (i) $\to$ (ii). Immediate.

(ii) $\to$ (iii). Choose $\alpha \geqslant \omega_1$ so that the (Σ, α)-model is well-founded. As we observed earlier, up to isomorphism the (Σ, β)-model is a submodel of the (Σ, α)-model for any infinite $\beta < \omega_1$, which proves (iii).

(iii) $\to$ (i). Suppose Σ were not well-founded. Then for some infinite α, the (Σ, α)-model, $(\mathscr{A}, H)$ say, is not well-founded. Let $a_n \in A$, $n < \omega$, be such that

$a_{n+1} E a_n$, where $\mathscr{A} = \langle A, E \rangle$. Each a_n is of the form $t_n^{\mathscr{A}}(\vec{h}_n)$ for some $\mathscr{L}$-term t and some $\vec{h}_n \in H$. Let K be a countably infinite subset of H which contains all $\vec{h}_n$, $n < \omega$. Let $\mathscr{B} = \mathscr{A} \upharpoonright K$. Then $(\mathscr{B}, K)$ is the (Σ, β)-model, where $\beta = \operatorname{otp}(K) < \omega_1$. But $\alpha_n \in B$ for all n, where $\mathscr{B} = \langle B, E \rangle$, so $\mathscr{B}$ is not well-founded. This contradicts (iii). □

If Σ is a well-founded $E-M$ set, then for any infinite ordinal α, there is a unique transitive (Σ, α)-model. We denote this model by $M(\Sigma, \alpha)$. We observed above that $M(\Sigma, \alpha)$ has the form $(\langle L_\lambda, \in \rangle, H)$, where λ is a limit ordinal greater than ω, and where $H \subseteq \lambda$. In case α is an uncountable cardinal, we can say even more, namely:

2.7 Lemma. *Let Σ be a well-founded, remarkable, cofinal, $E-M$ set. If κ is an uncountable cardinal, then the universe of $M(\Sigma, \kappa)$ is L_κ.*

Proof. Let $M(\Sigma, \kappa)$ be (L_γ, H). Since $H \subseteq \gamma$ and $|H| = \kappa$, we know that $\gamma \geqslant \kappa$. Suppose that $\gamma > \kappa$. Since $H = \{h_\alpha \mid \alpha < \kappa\}$ is cofinal in γ, we can find a limit ordinal $\alpha < \kappa$ such that $h_\alpha > \kappa$. Let $K = \{h_\beta \mid \beta < \alpha\}$ and set $N = L_\gamma \upharpoonright K$. By 2.4,

$$\mathrm{On}^N = \{x \in \gamma \mid x < h_\alpha\} = h_\alpha.$$

Thus $\kappa \subseteq \mathrm{On}^N$. But $|N| = |K| = |\alpha| < \kappa$, so this is absurd. Hence $\gamma = \kappa$ and we are done. □

For each uncountable cardinal κ, let H_κ denote the unique subset of κ (if it exists) such that (L_κ, H_κ) is the (Σ, κ)-model $M(\Sigma, \kappa)$. By 2.5, we know that H_κ is a club subset of κ.

2.8 Lemma. *If $\kappa < \lambda$ are uncountable cardinals, then $H_\kappa = H_\lambda \cap \kappa$ and $L_\kappa = L_\lambda \upharpoonright H_\kappa$.*

Proof. Let $(h_\nu \mid \nu < \lambda)$ enumerate H_λ in increasing order. Set $K = \{h_\nu \mid \nu < \kappa\}$, and let $N = L_\lambda \upharpoonright K$. Then (N, K) is a (Σ, k)-model, so $N \cong L_\kappa$. But On^N is an initial segment of λ. Hence N must be transitive. But then we must have $N = L_\kappa$, and moreover $K = H_\kappa$, $h_\kappa = \kappa$, and $H_\kappa = K = H_\lambda \cap \kappa$. □

2.9 Corollary. *If λ is an uncountable cardinal, then H_λ contains all uncountable cardinals below λ.*

Proof. Let $\kappa < \lambda$ be an uncountable cardinal. Then, as we saw above,

$$\kappa = h_\kappa \in H_\lambda. \quad \square$$

Of course, we have still said nothing concerning the existence of $E-M$ sets. We are now about to rectify this omission. We show first that if there is a well-founded, remarkable, cofinal $E-M$ set, then it must be unique.

2.10 Lemma. *If there is a well-founded, remarkable, cofinal, $E-M$ set, then it is unique.*

Proof. Let Σ be a well-founded, remarkable, cofinal $E-M$ set. Now, $(L_{\omega_\omega}, H_{\omega_\omega})$ is the transitive (Σ, ω_ω)-model, and by 2.9, $\omega_n \in H_{\omega_\omega}$ for all $n < \omega$. Thus for any $\mathscr{L}$-formula φ,

$$\varphi(v_1, \ldots, v_n) \in \Sigma \quad \text{iff } L_{\omega_\omega} \models \varphi(\omega_1, \ldots, \omega_n).$$

This determines Σ uniquely. □

The unique well-founded, remarkable, cofinal $E-M$ set, if it exists, is denoted by the symbol $0^{\#}$ ("zero sharp"). It is possible to carry out a similar development for the relativised universe $L[a]$ for any set $a \subseteq \omega$, in which case the corresponding $E-M$ set is denoted by $a^{\#}$. (This is considered in Exercise 2.) Summarising our previous results, we have:

2.11 Theorem. *Assume $0^{\#}$ exists. Then there is a club class H of ordinals such that:*

(i) *H contains all uncountable cardinals;*

and for any uncountable cardinal κ, if we set $H_\kappa = H \cap \kappa$, then:

(ii) *H_κ has order-type κ and is club in κ;*
(iii) *H_κ is L_κ-indiscernible;*
(iv) *$L_\kappa = L_\kappa \upharpoonright H_\kappa$.*

Proof. We set $H = \bigcup_\kappa H_\kappa$, where H_κ is a described earlier. The theorem is immediate now. □

2.12 Theorem. *Assume $0^{\#}$ exists. If $\kappa < \lambda$ are uncountable cardinals, then $L_\kappa \prec L_\lambda$.*

Proof. We know that

$$M(0^{\#}, \kappa) = (L_\kappa, H_\kappa), \qquad M(0^{\#}, \lambda) = (L_\lambda, H_\lambda).$$

So, by 2.8, we have

$$L_\kappa = L_\lambda \upharpoonright H_\kappa \prec L_\lambda. \quad \square$$

The existence of $0^{\#}$ also provides us with a truth definition for L:

2.13 Theorem (Metatheorem). *There is a formula $\Theta(x)$ of* LST *such that, for any* LST *formula $\Phi(v_0, \ldots, v_n)$, if φ is the $\mathscr{L}$-formula corresponding to Φ* (as in I.9.11), ZF $\vdash$ "*if $0^{\#}$ exists, then* $(\forall a_0, \ldots, a_n \in \mathrm{L})[\Phi^L(a_0, \ldots, a_n) \leftrightarrow \Theta(\varphi(\mathring{a}_0, \ldots, \mathring{a}_n))]$".

Proof. Given any formula $\Phi(v_0, \ldots, v_n)$ of LST, the reflection principle (I.8.2) provides us with an uncountable cardinal κ such that

$$(\forall \vec{a} \in L_\kappa)[\Phi^L(\vec{a}) \leftrightarrow \Phi^{L_\kappa}(\vec{a})].$$

But by 2.12, together with I.9.11, the actual choice of κ here is irrelevant in the case that $0^{\#}$ exists. Thus, given any $\vec{a} \in L$, if κ is any uncountable cardinal such that $\vec{a} \in L_\kappa$, then providing that $0^{\#}$ exists, we have (using I.9.11)

$$\Phi^L(\vec{a}) \leftrightarrow \Phi^{L_\kappa}(\vec{\alpha}) \leftrightarrow \models_{L_\kappa} \varphi(\vec{\mathring{a}}).$$

Thus $\Theta(x)$ is the LST formula which says:

> "x is a sentence of $\mathscr{L}_L$, and if κ is the least uncountable cardinal such that $x \in L_\kappa$, then $\models_{L_\kappa} x$".

By virtue of 2.13, we may speak about "elementary submodels of L" quite openly, and indeed may state the following theorem:

2.14 Theorem. *If $0^\#$ exists, then for any uncountable cardinal κ, $L_\kappa \prec L$.*

Proof. Since

$$L = \bigcup \{L_\kappa \mid \kappa \text{ is an uncountable cardinal}\},$$

this follows easily from 2.12. □

Before we turn to an existence proof for $0^\#$, we give one more consequence of its existence.

2.15 Theorem. *Assume that $0^\#$ exists, and let κ be any uncountable cardinal. Then:*

(i) $[\kappa \text{ is inaccessible}]^L$;

(ii) $[\kappa \to (\omega)_2^{<\omega}]^L$;

(iii) $|\mathscr{P}^L(\kappa)| = \kappa$.

Proof. (i) If $\lambda = \omega_1$, then

$$[\lambda \text{ is regular}]^L,$$

and if $\mu = \omega_\omega$, then

$$[\mu \text{ is a limit cardinal}]^L.$$

So by the L-indiscernibility of the cardinals,

$$[\kappa \text{ is a regular limit cardinal}]^L.$$

Since $[\mathrm{GCH}]^L$, this proves (i).

(ii) Suppose not, and let $f \in L$, $f\colon [\kappa]^{<\omega} \to 2$ be the $<_L$-least partition with no infinite homogeneous set (in the sense of L). In the real world, $f\colon [\kappa]^{<\omega} \to 2$, of course. Now, f is definable from κ in L_{κ^+} (by the above definition, which is clearly absolute for L_{κ^+}). It follows that, in the real world, H_κ is homogeneous for f. For if t is a term such that

$$f(\sigma) = t^{L_{\kappa^+}}(\sigma, \kappa)$$

for all $\sigma \in [\kappa]^{<\omega}$, then for any $\alpha_1 < \ldots < \alpha_n$, $\beta_1 < \ldots < \beta_n$ from H_κ, if $i = 0, 1$, then

$$\begin{aligned} f(\alpha_1, \ldots, \alpha_n) = i \quad &\text{iff } L_{\kappa^+} \models t(\alpha_1, \ldots, \alpha_n, \kappa) = i \\ &\text{iff } L_{\kappa^+} \models t(\beta_1, \ldots, \beta_n, \kappa) = i \\ &\text{iff } f(\beta_1, \ldots, \beta_n) = i. \end{aligned}$$

But then, exactly as in 1.7 it follows that there is, in L, an infinite set which is homogeneous for f (in the sense of L), contradicting the choice of f. This proves (ii).

(iii) By (i) $[\lambda \text{ is inaccessible}]^L$, where $\lambda = \kappa^+$. This implies (iii) at once. □

Note that a particular consequence of the above (and previous) results is that the existence of $0^\#$ cannot be established in ZFC alone. We shall now look into this question of existence of $0^\#$.

2.16 Theorem. *The following are equivalent:*

(i) $0^\#$ *exists;*

(ii) *for every uncountable cardinal* κ, L_κ *has an uncountable set of indiscernibles;*

(iii) *for some uncountable cardinal* κ, L_κ *has an uncountable set of indiscernibles.*

Proof. (i) → (ii). By 2.11 (iii).

(ii) → (iii). Trivial.

(iii) → (i). Let λ be the least limit ordinal such that L_λ has an uncountable set of indiscernibles. Let $H \subseteq \lambda$ be an L_λ-indiscernible set of order-type ω_1, chosen so that h_ω is as small as possible, where $(h_\nu \mid \nu < \omega_1)$ is the monotone enumeration of H. Let Σ be the E–M set determined by the indiscernible set H in L_λ. We show that Σ is well-founded, remarkable, and cofinal.

(a) Σ is well-founded. Well, clearly, $L_\lambda \upharpoonright H$ is well-founded. But $(L_\lambda \upharpoonright H, H)$ is the (Σ, ω_1)-model. So by 2.6, Σ is well-founded.

(b) Σ is cofinal. For suppose not. Then by 2.3, H is not cofional in (the ordinals of) $L_\lambda \upharpoonright H$. So for some $\mathscr{L}$-term t and some $\nu_1 < \ldots < \nu_n < \omega_1$,

$$\gamma = t^{L_\lambda}(h_{\nu_1}, \ldots, h_{\nu_n}) \geqslant \sup(H).$$

We may assume that γ is a limit ordinal here. (For otherwise, if $\gamma = \delta + m$, we may replace t by the term

$$t'(h_{\nu_1}, \ldots, h_{\nu_n}) = t(h_{\nu_1}, \ldots, h_{\nu_n}) - m.)$$

Let

$$K = \{h_\nu \mid \nu_n < \nu < \omega_1\}.$$

Clearly, K is a set of indiscernibles for L_γ. But $\gamma < \lambda$, so this contradicts the choice of λ. Hence Σ must be cofinal.

(c) Σ is remarkable. To see this, suppose that the formula

$$t(v_0, \ldots, v_{n-1}, v_n, \ldots, v_{n+m}) < v_n$$

is in Σ, for some $\mathscr{L}$-term t. Partition H into increasing, finite pieces

$$\vec{c}, \vec{d}_0, \vec{d}_1, \ldots, \vec{d}_\nu, \ldots \quad (\nu < \omega_1),$$

where $\vec{c}$ has length n and each $\vec{d}_\nu$ has length $m + 1$, and where

$$\max(\vec{c}) < \min(\vec{d}_0) < \max(\vec{d}_0) < \min(\vec{d}_1) < \max(\vec{d}_1) < \min(\vec{d}_2) < \ldots.$$

Notice that, in particular,

$$\vec{d}_\omega = h_\omega, h_{\omega+1}, \ldots, h_{\omega+m}.$$

By indiscernibility, one of the following must occur:

(A) $t^{L_\lambda}(\vec{c}, \vec{d}_\nu) = t^{L_\lambda}(\vec{c}, \vec{d}_\tau)$ for all $\nu < \tau < \omega_1$;

(B) $t^{L_\lambda}(\vec{c}, \vec{d}_\nu) < t^{L_\lambda}(\vec{c}, \vec{d}_\tau)$ for all $\nu < \tau < \omega_1$;

(C) $t^{L_\lambda}(\vec{c}, \vec{d}_\nu) > t^{L_\lambda}(\vec{c}, \vec{d}_\tau)$ for all $\nu < \tau < \omega_1$.

Since Σ is determined by H in L_λ, if we can prove that (A) must occur, we shall be done, since this will imply that Σ contains the formula

$$t(v_0, \ldots, v_{n-1}, v_n, \ldots, v_{n+m}) = t(v_0, \ldots, v_{n-1}, v_{n+m+1}, \ldots, v_{n+2m+1}).$$

Well, (C) is clearly impossible, since that would give us a decreasing ω_1-sequence of ordinals. So let us assume (B) and work for a contradiction. Set

$$h'_\nu = t^{L_\lambda}(\vec{c}, \vec{d}_\nu), \qquad \nu < \omega_1.$$

By (B), the sequence $(h'_\nu \mid \nu < \omega_1)$ is strictly increasing. And it is easily checked that $\{h_\nu \mid \nu < \omega_1\}$ is L_λ-indiscernible. But by choice of t, $h'_\omega < h_\omega$, so this contradicts our choice of H, h_ω, and we are done. □

2.17 Corollary. *If $\kappa(\omega_1)$ exists, then $0^{\#}$ exists. Hence if there is a measurable cardinal, then $0^{\#}$ exists.*

Proof. By 1.6. □

3. *Definability of* $0^{\#}$

We have already seen that the existence of $0^{\#}$ has a profound effect upon the constructible universe. In this section we investigate the logical complexity of the set $0^{\#}$ as a subset of the set of all formulas of $\mathscr{L}$. In particular we shall show that $0^{\#}$ has strong absoluteness properties.

3.1 Lemma. *There is a Π_1 formula $\Phi(x)$ of* LST *such that*

$$\Phi(x) \leftrightarrow x = 0^{\#}.$$

Proof. By 2.10, $0^{\#}$ is unique, if it exists, and what we must show is that the predicate

"x is a well-founded, remarkable, cofinal $E-M$ set"

can be expressed in a Π_1 fashion.

We commence by examining the predicate

"Σ is an $E-M$ set".

Let $\mathscr{L}^+$ be the language $\mathscr{L}$ together with the extra constant symbols c_n, $n < \omega$. For any set, Σ, of $\mathscr{L}$-formulas, let Σ^+ be the set of $\mathscr{L}^+$-sentences which consists of:

(i) the axioms of BSL + $(V = L)$;

(ii) $\varphi(c_0, \ldots, c_n)$, for each $\varphi(v_0, \ldots, v_n) \in \Sigma$;

(iii) $\mathrm{On}(c_n)$, for all $n < \omega$;

(iv) $(c_n < c_m)$, for all $n < m < \omega$;

(v) $\varphi(c_{i_1}, \ldots, c_{i_n}) \leftrightarrow \varphi(c_{j_1}, \ldots, c_{j_n})$, for each $\varphi(v_0, \ldots, v_{n-1}) \in \Sigma$ and each $i_1 < \ldots < i_n < \omega$, $j_1 < \ldots < j_n < \omega$.

Claim. Σ is an $E-M$ set iff Σ^+ is consistent.

Proof of claim: Suppose Σ is an $E-M$ set. Then Σ is the set of all $\mathscr{L}$-formulas which are true on increasing tuples from an $\mathscr{A}$-indiscernible set $\{a_n | n < \omega\}$ in some model $\mathscr{A}$ of BSL + $(V = L)$. Clearly, $\langle \mathscr{A}, (a_n)_{n<\omega} \rangle$ is a model of Σ^+, so Σ^+ is consistent.

Conversely, suppose Σ^+ is consistent, and let $\langle \mathscr{A}, (a_n)_{n<\omega} \rangle$ be a model of Σ^+. Clearly, $\{a_n | n < \omega\}$ is $\mathscr{A}$-indiscernible and $(\mathscr{A}, \{a_n | n < \omega\})$ is a (Σ, ω)-model, so Σ is an $E-M$ set. The claim is proved.

By the claim we have:

Σ is an $E-M$ set iff there does not exist a proof of the sentence $(0 = 1)$ from the sentences in Σ^+.

More precisely:

Σ is an $E-M$ set iff there does not exist a finite sequence of $\mathscr{L}^+$-formulas such that the last formula in the sequence is $(0 = 1)$ and each formula of the sequence is either a consequence of previous formulas by modus ponens or else is an axiom of logic or else an axiom of BSL + $(V = L)$ or else is of the form $\varphi(c_0, \ldots, c_n)$ for some $\varphi(v_0, \ldots, v_n) \in \Sigma$ or else of the form $\mathrm{On}(c_n)$ for some $n < \omega$ or else of the form $(c_n < c_m)$ for some $n < m < \omega$ or else of the form $(\varphi(c_{i_1}, \ldots, c_{i_n}) \leftrightarrow \varphi(c_{j_1}, \ldots, c_{j_n}))$ for some $\varphi(v_0, \ldots, v_{n-1})$ in Σ and some $i_1 < \ldots < i_n < \omega$, $j_1 < \ldots < j_n < \omega$.

Now, provided that the constants c_n are suitably chosen (e.g. take $\mathscr{L}_\omega$ as the language $\mathscr{L}^+$ and use the constant symbol $\mathring{n}$ for c_n), all quantifiers in the above definition can be bound (without loss of generality) by V_ω. Thus the above characterisation of the predicate "Σ is an $E-M$ set" is Σ_0 in the parameter V_ω.

It is easily seen that the predicates "Σ is cofinal" and "Σ is remarkable" are also Σ_0 in the parameter V_ω. Thus there is a Σ_0 formula $\Psi(x, y)$ of LST such that

$$\Sigma \text{ is a remarkable, cofinal } E-M \text{ set} \leftrightarrow \Psi(\Sigma, V_\omega).$$

But $V_\omega = L_\omega$. Hence

$$\begin{aligned}&\Sigma \text{ is a remarkable, cofinal } E{-}M \text{ set}\\ &\leftrightarrow \forall\alpha\,\forall u\,[[\mathrm{On}(\alpha) \wedge \lim(\alpha) \wedge (\forall\beta\in\alpha)(\beta = 0 \vee \mathrm{succ}(\beta)) \wedge u = L_\alpha]\\ &\qquad \rightarrow \Psi(\Sigma, u)].\end{aligned}$$

The formula on the right here is Π_1. (In fact there is an equivalent Σ_1 formula, as the reader may readily verify, but we do not require this fact.) So we are left with proving that the predicate

"Σ is well-founded"

(for a remarkable, cofinal $E{-}M$ set Σ) is Π_1.

By definition,

Σ is well-founded iff for all α, the (Σ, α)-model is well-founded.

Now, if Σ is a remarkable, cofinal $E{-}M$ set, then for every limit ordinal α there is a unique (up to isomorphism) (Σ, α)-model, and we can find one of the form $(\langle A, E\rangle, \alpha)$, where $E \cap (\alpha\times\alpha) = \in \cap (\alpha\times\alpha)$. Let us call such a model a *standardised* (Σ, α)-model. Then:

$(\langle A, E\rangle, \alpha)$ is a standardised (Σ, α)-model iff

(i) $\langle A, E\rangle$ is a model of BSL $+ (V = L) \wedge$

(ii) $\alpha \subseteq \mathrm{On}^{\langle A, E\rangle} \wedge E \cap (\alpha\times\alpha) = \in \cap (\alpha\times\alpha) \wedge$

(iii) α is $\langle A, E\rangle$-indiscernible $\wedge$

(iv) α generates $\langle A, E\rangle \wedge$

(v) Σ is the set of all $\mathscr{L}$-formulas valid in $\langle A, E\rangle$ on increasing tuples from α.

Now, in each of the clauses (i)–(v) above, all necessary quantifiers may be bound either by A or by α or by V_ω. (This is a routine matter which we leave to the reader to check.) Thus there is a Σ_0 formula $\Theta(w, x, y, z)$ of LST such that

$(\langle A, E\rangle, \alpha)$ is a standardised (Σ, α)-model iff $\Theta(\Sigma, \langle A, E\rangle, \alpha, V_\omega)$.

But, clearly,

$$\begin{aligned}\Theta(\Sigma, \langle A, E\rangle, \alpha, V_\omega) \quad &\text{iff } \exists\gamma\,\exists u\,[\mathrm{On}(\gamma) \wedge \lim(\gamma)\\ &\wedge (\forall\beta\in\gamma)(\beta = 0 \vee \mathrm{succ}(\beta)) \wedge (u = L_\gamma)\\ &\wedge \Theta(\Sigma, \langle A, E\rangle, \alpha, u)].\end{aligned}$$

Thus the predicate "$(\langle A, E\rangle, \alpha)$ is a standardised (Σ, α)-model" (as a predicate on $\langle A, E\rangle, \alpha, \Sigma$) is Σ_1. But (for a remarkable, cofinal $E{-}M$ set Σ):

$$\begin{aligned}\Sigma \text{ is well-founded} \leftrightarrow \forall\alpha\,\forall\langle A, E\rangle\,[&\text{if } (\langle A, E\rangle, \alpha) \text{ is a standardised}\\ &(\Sigma, \alpha)\text{-model, then } E \text{ is well-founded on } A].\end{aligned}$$

This is easily seen to be Π_1, so we are done. □

3.2 Corollary. $0^\# \notin L$.

Proof. If $0^\# \in L$, then since Π_1 properties are D-absolute, $\Phi(0^\#)$ implies $\Phi^L(0^\#)$, where Φ is the Π_1 formula from 3.1. But then we can prove all of the results of section 2 *inside* L, which is absurd. □

4. $0^\#$ and Elementary Embeddings

The existence of $0^\#$ is closely connected with the existence of elementary embeddings of the form $j: L_\kappa \prec L_\kappa$, where κ is a cardinal. The simplest such result is the following:

4.1 Theorem. *If $0^\#$ exists, then for any uncountable cardinal κ there is a non-trivial embedding $j: L_\kappa \prec L_\kappa$.*

Proof. Let $(h_\alpha \mid \alpha < \kappa)$ be the monotone enumeration of H_κ. Define $j: H_\kappa \to H_\kappa$ by $j(h_\alpha) = h_{\alpha+1}$. By 2.1, j extends to an embedding $\tilde{j}: L_\kappa \prec L_\kappa$. □

The main effort in this section is directed towards proving the converse to 4.1. In fact we shall prove a stronger result. In order to imply the existence of $0^\#$ it is enough to have an embedding $j: L_\alpha \prec L_\beta$ for some limit ordinals α, β such that $j(\gamma) \neq \gamma$ for some $\gamma < |\alpha|$. In order to do this we shall first of all prove a converse to 4.1 under some additional assumptions. We require some prior definitions.

Say that a cardinal κ is of *γ-type* 0 if it is a limit cardinal and $\mathrm{cf}(\kappa) > \gamma$. Notice that there are arbitrarily large cardinals of γ-type 0, for any given ordinal γ. Moreover, if $(\kappa_\nu \mid \nu < \theta)$ is an increasing sequence of cardinals of γ-type 0 such that $\mathrm{cf}(\theta) > \gamma$, then $\sup_{\nu<\theta} \kappa_\nu$ is of γ-type 0.

A cardinal κ is said to be of *γ-type* 1 if it is of γ-type 0 and

$$|\{\lambda \in \kappa \mid \lambda \text{ is of } \gamma\text{-type } 0\}| = \kappa.$$

Since the γ-type 0 cardinals are closed under limits of η-sequences whenever $\mathrm{cf}(\eta) > \gamma$, it is easily proved that there are arbitrarily large cardinals of γ-type 1. Moreover, it is clear that the γ-type 1 cardinals are closed under limits of η-sequences whenever $\mathrm{cf}(\eta) > \gamma$.

Proceeding in a recursive fashion now, say that a cardinal κ is of *γ-type* $\nu + 1$ if it is of γ-type ν and

$$|\{\lambda \in \kappa \mid \lambda \text{ is of } \gamma\text{-type } \nu\}| = \kappa.$$

Provided the γ-type ν cardinals are unbounded and closed under limits of η-sequences whenever $\mathrm{cf}(\eta) > \delta$ for some $\delta \geqslant \gamma$, the same will be true of the γ-type $\nu + 1$ cardinals.

If τ is a limit ordinal, we say that a cardinal κ is of *γ-type τ* iff it is of γ-type ν for every $\nu < \tau$. If, for each $\nu < \tau$, the γ-type ν cardinals are unbounded and closed under limits of η-sequences whenever $\mathrm{cf}(\eta) > \delta$ for some $\delta \geqslant \gamma$, then, provided $\tau < \delta$, the same will be true of the γ-type τ cardinals.

4.2 Theorem. *Let κ be a cardinal. Suppose that there is an embedding*

$$e: L_\kappa \prec L_\kappa$$

such that for some ordinal $\gamma < \kappa$:

(i) $e \restriction \gamma = \mathrm{id} \restriction \gamma$;
(ii) $e(\gamma) > \gamma$;
(iii) *if $\lambda > \kappa$ is of γ-type 0, then $e(\lambda) = \lambda$.*

Suppose further that κ is of γ-type ω_1. Then $0^{\#}$ exists.

Proof. By 2.16, it suffices to show that L_κ has an uncountable, indiscernible subset.

For each $\nu < \omega_1$, let

$$U_\nu = \{\lambda \in \kappa \mid \lambda \text{ is a } \gamma\text{-type } \nu \text{ cardinal}\}.$$

Since κ has γ-type ω_1, $|U_\nu| = \kappa$ for each $\nu < \omega_1$. Moreover,

$$U_0 \supseteq U_1 \supseteq \ldots \supseteq U_\nu \supseteq \ldots \quad (\nu < \omega_1),$$

and for each $\nu < \omega_1$,

$$U_{\nu+1} = \{\lambda \in U_\nu \mid |U_\nu \cap \lambda| = \lambda\},$$

with

$$U_\delta = \bigcap_{\nu<\delta} U_\nu, \quad \text{if } \lim(\delta),\ \delta < \omega_1.$$

For each $\nu < \omega_1$, let

$$M_\nu = L_\kappa \restriction (\gamma \cup U_\nu).$$

Thus,

$$M_\nu \prec L_\kappa \quad \text{and} \quad |M_\nu| = \kappa.$$

In particular, the transitive collapse of M_ν is L_κ. Let

$$i_\nu: L_\kappa \cong M_\nu.$$

Thus

$$i_\nu: L_\kappa \prec L_\kappa.$$

Set

$$\gamma_\nu = i_\nu(\gamma).$$

Claim 1. Let $\nu, \tau < \omega_1$. Then:

(i) γ_ν is the least ordinal in $M_\nu - (\gamma + 1)$;
(ii) if $\nu < \tau$ and $x \in M_\tau$, then $i_\nu(x) = x$;
(iii) if $\nu < \tau$, then $i_\nu(\gamma_\tau) = \gamma_\tau$;
(iv) if $\nu < \tau$, then $\gamma_\nu < \gamma_\tau$.

Proof. (i) Since $\gamma \in M_\nu$, $i_\nu \restriction \gamma = \mathrm{id} \restriction \gamma$ and $i_\nu(\gamma)$ is the least element of $M_\nu - \gamma$. So it suffices to prove that $\gamma \notin M$. Since $M_0 \supseteq M_1 \supseteq \ldots \supseteq M_\nu \supseteq \ldots$ $(\nu < \omega_1)$, it is enough to prove that $\gamma \notin M_0$. Consider any $x \in M_0$. Then $x = t^{L_\kappa}(\eta_1, \ldots, \eta_n)$ for some $\mathscr{L}$-term t and some $\eta_1, \ldots, \eta_n \in \gamma \cup U_0$. By the assumptions (i) and (iii) of the lemma, $e(\eta_1) = \eta_1, \ldots, e(\eta_n) = \eta_n$. Thus

$$e(x) = e(t^{L_\kappa}(\eta_1, \ldots, \eta_n)) = t^{L_\kappa}(e(\eta_1), \ldots, e(\eta_n)) = t^{L_\kappa}(\eta_1, \ldots, \eta_n) = x.$$

So by assumption (ii) of the lemma, $x \neq \gamma$. Thus $\gamma \notin M_0$.

(ii) Let $x \in M_\tau$. Then for some $\mathscr{L}$-term t and some $\eta_1, \ldots, \eta_n \in \gamma \cup U_\tau$, $x = t^{L_\kappa}(\eta_1, \ldots, \eta_n)$. If $\eta \in \gamma$, then since $\gamma \subseteq M_\nu$, $i_\nu(\eta) = \eta$. If $\eta \in U_\tau$, then since $\nu < \tau$, $|U_\nu \cap \eta| = \eta$, so $i_\nu^{-1}(\eta) = \eta$, so $i_\nu(\eta) = \eta$. Thus

$$i_\nu(x) = i_\nu(t^{L_\kappa}(\eta_1, \ldots, \eta_n)) = t^{L_\kappa}(i_\nu(\eta_1), \ldots, i_\nu(\eta_n)) = t^{L_\kappa}(\eta_1, \ldots, \eta_n) = x.$$

(iii) An immediate consequence of (ii).

(iv) If $\nu < \tau$, then $M_\nu \subseteq M_\tau$, so $\gamma_\nu \leqslant \gamma_\tau$. Now by result (i) of this claim, $\gamma_\nu > \gamma$, so applying i_ν, we get $i_\nu(\gamma_\nu) > i_\nu(\gamma) = \gamma_\nu$. But by result (iii), $i_\nu(\gamma_\tau) = \gamma_\tau$. Hence $\gamma_\nu \neq \gamma_\tau$. Thus $\gamma_\nu < \gamma_\tau$.

The claim is proved.

For $\nu < \tau < \omega_1$, set

$$M_{\nu\tau} = L_\kappa \restriction (\gamma_\nu \cup U_\tau).$$

Let

$$i_{\nu\tau}\colon L_\kappa \cong M_{\nu\tau}.$$

Thus

$$i_{\nu\tau}\colon L_\kappa \prec L_\kappa.$$

Claim 2. Let $\nu < \tau$. Then:

(i) if $\xi < \nu$, then $i_{\nu\tau}(\gamma_\xi) = \gamma_\xi$;
(ii) $i_{\nu\tau}(\gamma_\nu) = \gamma_\tau$;
(iii) if $\xi > \tau$, then $i_{\nu\tau}(\gamma_\xi) = \gamma_\xi$.

Proof. (i) Since $\gamma_\nu \subseteq M_{\nu\tau}$, we have $i_{\nu\tau} \restriction \gamma_\nu = \mathrm{id} \restriction \gamma_\nu$, so this is immediate.

(ii) Since $\gamma_\nu > \gamma$, we have $M_\tau \subseteq M_{\nu\tau}$, so $\gamma_\tau \in M_{\nu\tau}$. But $i_{\nu\tau} \restriction \gamma_\nu = \mathrm{id} \restriction \gamma_\nu$, so $i_{\nu\tau}(\gamma_\nu)$ is the least ordinal in $M_{\nu\tau}$ greater than or equal to γ_ν. Hence $\gamma_\nu \leqslant i_{\nu\tau}(\gamma_\nu) \leqslant \gamma_\tau$. It therefore suffices to show that there is no ordinal $\delta \in M_{\nu\tau}$ such that $\gamma_\nu \leqslant \delta < \gamma_\tau$. Suppose that there were such a δ. Then for some $\mathscr{L}$-term t, $\delta = t^{L_\kappa}(\xi_1, \ldots, \xi_n, \eta_1, \ldots, \eta_k)$, where $\xi_1, \ldots, \xi_n \in \gamma_\nu$ and $\eta_1, \ldots, \eta_k \in U_\tau$. Thus

$$L_\kappa \models (\exists \xi_1, \ldots, \xi_n < \gamma_\nu)[\gamma_\nu \leqslant t(\xi_1, \ldots, \xi_n, \eta_1, \ldots, \eta_k) < \gamma_\tau].$$

Applying i_ν^{-1}, we get (since $i_\nu(\gamma_\tau) = \gamma_\tau$, $i_\nu \upharpoonright U_\tau = \mathrm{id} \upharpoonright U_\tau$, and $i_\nu(\gamma) = \gamma_\nu$)

$$L_\kappa \models (\exists \xi_1, \ldots, \xi_n < \gamma)[\gamma \leqslant t(\xi_1, \ldots, \xi_n, \eta_1, \ldots, \eta_k) < \gamma_\tau].$$

So for some $\xi_1, \ldots, \xi_n < \gamma$ we have

$$\gamma \leqslant t^{L_\kappa}(\xi_1, \ldots, \xi_n, \eta_1, \ldots, \eta_k) < \gamma_\tau.$$

But $t^{L_\kappa}(\xi_1, \ldots, \xi_n, \eta_1, \ldots, \eta_k) \in M_\tau$, so this contradicts (i) of Claim 1.

(iii) If $x \in M_{\tau+1}$, then $x = t^{L_\kappa}(\eta_1, \ldots, \eta_n)$ for some $\mathscr{L}$-term t and some $\eta_1, \ldots, \eta_n \in \gamma \cup U_{\tau+1}$. Now, $i_{\nu\tau} \upharpoonright \gamma_\nu = \mathrm{id} \upharpoonright \gamma_\nu$, so $i_{\nu\tau} \upharpoonright \gamma = \mathrm{id} \upharpoonright \gamma$. And if $\eta \in U_{\tau+1}$, then $|U_\tau \cap \eta| = \eta$, so $i_{\nu\tau}^{-1}(\eta) = \eta$, giving $i_{\nu\tau}(\eta) = \eta$. Thus $i_{\nu\tau}(x) = x$. In particular, $i_{\nu\tau}(\gamma_\xi) = \gamma_\xi$ for all $\xi > \tau$.

The claim is proved.

Claim 3. The set $\{\gamma_\nu \mid \nu < \omega_1\}$ is L_κ-indiscernible.

Proof. Let $\varphi(v_1, \ldots, v_n)$ be any $\mathscr{L}$-formula, and let $\nu_1 < \ldots < \nu_n < \omega_1$, $\tau_1 < \ldots < \tau_n < \omega_1$. We show that

$$L_\kappa \models \varphi(\gamma_{\nu_1}, \ldots, \gamma_{\nu_1}) \quad \text{iff } L_\kappa \models \varphi(\gamma_{\tau_1}, \ldots, \gamma_{\tau_n}).$$

Pick $\delta_1 < \ldots < \delta_n < \omega_1$ so that $\nu_n, \tau_n < \delta_1$. Applying $i_{\nu_n \delta_n}$ we get, using Claim 2,

$$L_\kappa \models \varphi(\gamma_{\nu_1}, \ldots, \gamma_{\nu_{n-1}}, \gamma_{\nu_n}) \quad \text{iff } L_\kappa \models \varphi(\gamma_{\nu_1}, \ldots, \gamma_{\nu_{n-1}}, \gamma_{\delta_n}).$$

Applying $i_{\nu_{n-1} \delta_{n-1}}$ now gives

$$L_\kappa \models \varphi(\gamma_{\nu_1}, \ldots, \gamma_{\nu_{n-2}}, \gamma_{\nu_{n-1}}, \gamma_{\delta_n}) \quad \text{iff } L_\kappa \models \varphi(\gamma_{\nu_1}, \ldots, \gamma_{\nu_{n-2}}, \gamma_{\delta_{n-1}}, \gamma_{\delta_n}).$$

Successively applying $i_{\nu_{n-2} \delta_{n-2}}, \ldots, i_{\nu_1 \delta_1}$ now gives, in the end, the equivalence

$$L_\kappa \models \varphi(\gamma_{\nu_1}, \ldots, \gamma_{\nu_n}) \quad \text{iff } L_\kappa \models \varphi(\gamma_{\delta_1}, \ldots, \gamma_{\delta_n}).$$

Repeating the above procedure with $\tau_1, \ldots, \tau_n$ in place of $\nu_1, \ldots, \nu_n$, we get

$$L_\kappa \models \varphi(\gamma_{\tau_1}, \ldots, \gamma_{\tau_n}) \quad \text{iff } L_\kappa \models \varphi(\gamma_{\delta_1}, \ldots, \gamma_{\delta_n}).$$

The above two equivalences combine to give the desired result. That proves the claim, and with it the theorem. □

We shall make use of 4.2 in our proof of the next result, the (strong) converse to 4.1.

4.3 Theorem. *Assume there is an embedding $j: L_\alpha \prec L_\beta$, where α, β are limit ordinals, and that $j(\gamma) \neq \gamma$ for some $\gamma < |\alpha|$. Then $0^{\#}$ exists.* □

The proof of 4.3 will take some time. Fix j, α, β as above, and let γ be the least ordinal such that $j(\gamma) \neq \gamma$. Thus $j \upharpoonright \gamma = \mathrm{id} \upharpoonright \gamma$ and $j(\gamma) > \gamma$. Let κ be a cardinal of γ-type ω_1. We prove 4.3 by using j to construct an embedding $e: L_\kappa \prec L_\kappa$ to satisfy the hypotheses of 4.2.

Let $\lambda = \kappa^+$. Notice that since $\gamma < |\alpha|$, $\mathscr{P}(\gamma) \cap L \subseteq L_\alpha$. Hence we may define

$$D = \{X \subseteq \gamma \mid X \in L \;\&\; \gamma \in j(X)\}.$$

Since $j: L_\alpha \prec L_\beta$ is elementary, the following lemma is easily proved.

4.4 Lemma.

(i) *$\gamma \in D$ and $\emptyset \notin D$;*
(ii) *if $X \in D$ and $Y \in D$, then $X \cap Y \in D$;*
(iii) *if $X \in D$ and $X \subseteq Y \subseteq \gamma$, where $Y \in L$, then $Y \in D$;*
(iv) *if $X \subseteq \gamma$, $X \in L$, then either $X \in D$ or else $\gamma - X \in D$;*
(v) *if $\bar{\gamma} < \gamma$ and $\{X_\xi \mid \xi < \bar{\gamma}\} \subseteq D$ and $(X_\xi \mid \xi < \bar{\gamma}) \in L$, then $\bigcap_{\xi < \bar{\gamma}} X_\xi \in D$.* □

Thus D is an ultrafilter in the field of sets $\mathscr{P}^L(\gamma)$ which is γ-complete with regards to families of sets in L. We do not necessarily have $D \in L$; indeed, it is a consequence of our ensuing results that $D \notin L$.

We use D to construct a kind of "ultrapower" of L_λ. Set

$$F = \{f \in L \mid f: \gamma \to L_\lambda\}.$$

Notice that as $\mathrm{cf}(\lambda) = \lambda > \gamma$, if $f \in F$ then in fact $f \in L_\lambda$. This fact will be relevant later on. Define an equivalence relation on F by

$$f \sim g \quad \text{iff } \{\nu \in \gamma \mid f(\nu) = g(\nu)\} \in D.$$

(Since $\{\nu \in \gamma \mid f(\nu) = g(\nu)\} \in L$ whenever $f, g \in F$, this definition makes sense. And, using the results of 4.4, it is easily checked that $\sim$ is an equivalence relation.) Let $[f]$ denote the equivalence class of f, and set

$$M = \{[f] \mid f \in F\}.$$

Define a binary relation, E, on M by

$$[f]\,E\,[g] \quad \text{iff } \{\nu \in \gamma \mid f(\nu) \in g(\nu)\} \in D.$$

(Again, for $f, g \in F$, $\{\nu \in \gamma \mid f(\nu) \in g(\nu)\} \in L$. And using 4.4 it is easily seen that E is well-defined on M.)

4.5 Lemma. *Let $\varphi(v_0, \ldots, v_n)$ be any $\mathscr{L}$-formula, and let $[f_0], \ldots, [f_n] \in M$. Then*

$$\langle M, E \rangle \models \varphi([f_0], \ldots, [f_0]) \quad \textit{iff } \{v \in \gamma \mid L_\lambda \models \varphi(f_0(v), \ldots, f_n(v))\} \in D.$$

Proof. Notice first that, since $f_0, \ldots, f_n \in L_\lambda$, $\{v \in \gamma \mid L_\lambda \models \varphi(f_0(v), \ldots, f_n(v))\} \in L$. The lemma is proved by induction on the length of φ.

If φ is primitive, the result is true by definition of $\langle M, E \rangle$.

If φ is of the form $\neg\psi$ or else $\psi_1 \wedge \psi_2$, the induction step is trivial, using the results of 4.4.

Suppose finally that φ has the form $\exists y \psi(y, v_0, \ldots, v_n)$ and that the result holds for ψ. If

$$\langle M, E \rangle \models \exists y \psi(y, [f_0], \ldots, [f_n]),$$

then for some $[g] \in M$,

$$\langle M, E \rangle \models \psi([g], [f_0], \ldots, [f_n]),$$

so by induction hypothesis

$$X = \{v \in \gamma \mid L_\lambda \models \psi(g(v), f_0(v), \ldots, f_n(v))\} \in D.$$

But clearly,

$$X \subseteq \{v \in \gamma \mid L_\lambda \models \exists y \psi(y, f_0(v), \ldots, f_n(v))\} \in L.$$

Hence

$$\{v \in \gamma \mid L_\lambda \models \exists y \psi(y, f_0(v), \ldots, f_n(v))\} \in D.$$

Conversely, suppose that

$$Y = \{v \in \gamma \mid L_\lambda \models \exists y \psi(y, f_0(v), \ldots, f_n(v))\} \in D.$$

In particular, $Y \in L$. Define $g: \gamma \to L_\lambda$ by

$$g(v) = \begin{cases} \text{the } <_L\text{-least } y \text{ such that } L_\lambda \models \psi(y, f_0(v), \ldots, f_n(v)), & \text{if } v \in Y, \\ \emptyset, \quad \text{if } v \notin Y. \end{cases}$$

Clearly, $g \in L$. Hence $[g] \in M$. But

$$\{v \in \gamma \mid L_\lambda \models \psi(g(v), f_0(v), \ldots, f_n(v))\} = Y \in D.$$

So by induction hypothesis,

$$\langle M, E \rangle \models \psi([g], [f_0], \ldots, [f_n]).$$

Hence

$$\langle M, E \rangle \models \exists y \psi(y, [f_0], \ldots, [f_n]).$$

The proof is complete. □

4.6 Lemma. *$\langle M, E\rangle$ is well-founded.*

Proof. Suppose not, and let $[g_{n+1}] \ni [g_n]$ for all $n < \omega$. Now, $g_n \in L_\lambda$ for all $n < \omega$, so pick $X \prec L_\lambda$ such that $(\gamma + 1) \cup \{g_n | n < \omega\} \subseteq X$ and $|X| = |\gamma|$. Let σ: $X \cong L_\delta$. Then $|\delta| = |\gamma| < |\alpha|$, and hence $\delta < \alpha$. Thus $\bar{g}_n \in L_\alpha$, where we set $\bar{g}_n = \sigma(g_n)$. Now, σ^{-1}: $L_\delta \prec L_\lambda$ and $\sigma \upharpoonright \gamma = \mathrm{id} \upharpoonright \gamma$, so for each $n < \omega$,

$$\{\nu \in \gamma \,|\, \bar{g}_{n+1}(\nu) \in \bar{g}_n(\nu)\} = \{\nu \in \gamma \,|\, g_{n+1}(\nu) \in g_n(\nu)\} \in D.$$

Thus for each $n < \omega$,

$$\gamma \in j(\{\nu \in \gamma \,|\, \bar{g}_{n+1}(\nu) \in \bar{g}_n(\nu)\}) = \{\nu \in j(\gamma) \,|\, [j(\bar{g}_{n+1})](\nu) \in [j(\bar{g}_n)](\nu)\}.$$

In other words, for all $n < \omega$, we have

$$[j(\bar{g}_{n+1})](\gamma) \in [j(\bar{g}_n)](\gamma).$$

But this is absurd. The lemma is proved. □

We can define a map k: $L_\lambda \to M$ by

$$k(x) = [c_x],$$

where c_x is the constant function $(x \,|\, \nu < \gamma)$. Using 4.5, it is easily seen that

$$k\colon \langle L_\lambda, \in\rangle \prec \langle M, E\rangle,$$

so by 4.6 there is an isomorphism

$$\varrho\colon \langle M, E\rangle \cong \langle L_\mu, \in\rangle,$$

for some $\mu \geqslant \lambda$. Let $\pi = \varrho \circ k$. Thus

$$\pi\colon L_\lambda \prec L_\mu.$$

4.7 Lemma.

(i) $\pi \upharpoonright \gamma = \mathrm{id} \upharpoonright \gamma$;
(ii) $\pi(\gamma) > \gamma$;
(iii) *if $\theta < \lambda$ is a cardinal of γ-type* 0, *then* $\pi(\theta) = \theta$.

Proof. (i) Let $\nu < \gamma$. Then

$$\pi(\nu) = \varrho \circ k(\nu) = \varrho([c_\nu]).$$

So, as ϱ is the collapsing isomorphism for $\langle M, E\rangle$,

$$\pi(\nu) = \mathrm{otp}(\langle A, E\rangle),$$

where

$$A = \{[f] \in M \,|\, [f]\, E\, [c_\nu]\}.$$

Now,

$$\begin{aligned}[f]\,E\,[c_\nu] \quad &\text{iff } \{\xi \in \gamma \mid f(\xi) \in c_\nu(\xi)\} \in D \\ &\text{iff } \{\xi \in \gamma \mid f(\xi) \in \nu\} \in D \\ &\text{iff } \bigcup_{\zeta<\nu} \{\xi \in \gamma \mid f(\xi) = \zeta\} \in D .\end{aligned}$$

Using 4.4 (v), we get

$$\begin{aligned}[f]\,E\,[c_\nu] \quad &\text{iff } (\exists \zeta < \nu)[\{\xi \in \gamma \mid f(\xi) = \zeta\} \in D] \\ &\text{iff } (\exists \zeta < \nu)[[f] = [c_\zeta]].\end{aligned}$$

Thus

$$A = \{[c_\zeta] \mid \zeta < \nu\}.$$

But

$$\xi < \zeta \rightarrow [c_\xi]\,E\,[c_\zeta].$$

Thus

$$\pi(\nu) = \mathrm{otp}(\langle A, E\rangle) = \nu .$$

(ii) For all $\nu < \gamma$, $\gamma - \nu \in D$, so

$$\nu < \gamma \rightarrow [c_\nu]\,E\,[\mathrm{id} \restriction \gamma]\,E\,[c_\gamma].$$

Thus $\pi(\gamma) \geqslant \gamma + 1$.

(iii) Suppose that $[g]\,E\,[c_\theta]$. Thus

$$\{\nu \in \gamma \mid g(\nu) \in \theta\} \in D .$$

Define $f\colon \gamma \rightarrow \theta$ by

$$f(\nu) = \begin{cases} g(\nu), & \text{if } g(\nu) \in \theta, \\ 0 \quad , & \text{if } g(\nu) \notin \theta . \end{cases}$$

Then $f \in L$, so $f \in F$, and $[f] = [g]$. But $\mathrm{cf}(\theta) > \gamma$, so $f''\gamma \subset \nu$ for some $\nu < \theta$. Thus $[f]\,E\,[c_\nu]$, i.e. $[g]\,E\,[c_\nu]$. We have therefore shown that the set $\{[c_\nu] \mid \nu < \theta\}$ is E-cofinal in $[c_\theta]$, i.e. that $\{k(\nu) \mid \nu < \theta\}$ is E-cofinal in $k(\theta)$. But ϱ is the collapsing isomorphism for $\langle M, E\rangle$. Thus

$$\varrho(k(\theta)) = \sup\nolimits_{\nu<\theta} \varrho(k(\nu)),$$

i.e.

$$\pi(\theta) = \sup\nolimits_{\nu<\theta} \pi(\nu).$$

But for $\nu < \theta$, if $[g]\,E\,[c_\nu]$, then as above we have $[g] = [f]$ for some $f \in ({}^\gamma\nu) \cap L$, so, noting that θ is a limit cardinal and that $[\mathrm{GCH}]^L$, we have

$$|\pi(\nu)| = |\varrho \circ k(\nu)| = |\varrho([c_\nu])| = |\{[g] \mid [g]\,E\,[c_\nu]\}| \leqslant |({}^\gamma\nu) \cap L| < \theta .$$

Thus $\pi(\theta) \leqslant \theta$, and so, in fact, $\pi(\theta) = \theta$. □

Since $\kappa < \lambda$ is of γ-type ω_1, it follows from 4.7 (iii) that $\pi(\kappa) = \kappa$. Hence

$$(\pi \upharpoonright L_\kappa)\colon L_\kappa \prec L_\kappa .$$

Setting $e = \pi \upharpoonright L_\kappa$, 4.7 implies that e is as in 4.2. That completes the proof of 4.3.

5. The Covering Lemma

It is the very essence of $0^{\#}$ that its existence implies that V is very different from L. In this section we show that if $0^{\#}$ does not exist, then V is very similar to L. More precisely, we shall prove the following result.

5.1 Theorem (The Covering Lemma). *Assume* $0^{\#}$ *does not exist. If* X *is an uncountable set of ordinals, then there is a constructible set,* Y, *of ordinals such that* $X \subseteq Y$ *and* $|Y| = |X|$. (Thus, every uncountable set of ordinals is *covered* by a constructible set of ordinals of the same (real) cardinality.)

The proof will take some time. Before we commence, let us notice that if $0^{\#}$ does exist, then the conclusion of 5.1 fails badly. For example, if $0^{\#}$ exists, then ω_ω is inaccessible in L, so the countable set $\{\omega_n \mid n < \omega\}$, being cofinal in ω_ω, can only be covered by a constructible set of cardinality at least ω_ω.

It is also instructive to give some examples of how the covering lemma effects the set theory of V, making it resemble L to some extent.

5.2 Theorem. *Assume* $0^{\#}$ *does not exist. Let* κ *be a singular cardinal. If* $2^{\mathrm{cf}(\kappa)+\omega_1} \leqslant \kappa^+$, *then* $\kappa^{\mathrm{cf}(\kappa)} = \kappa^+$. *In particular, if* κ *is a singular cardinal such that* $(\forall \lambda < \kappa)(2^\lambda < \kappa)$, *then* $2^\kappa = \kappa^+$.

Proof. Let κ be a singular cardinal such that $2^{\mathrm{cf}(\kappa)+\omega_1} \leqslant \kappa^+$. Let A be the set of all subsets of κ of cardinality $\mathrm{cf}(\kappa)$. We know (see I.5.8) that $|A| > \kappa$, so we must prove here that $|A| \leqslant \kappa^+$ in order to obtain the first part of the theorem. (The second part follows easily by cardinal arithmetic.)

Let $X \in A$. By the covering lemma there is a set $Y \in L$, $Y \subseteq \kappa$, such that $X \subseteq Y$ and $|Y| = \mathrm{cf}(\kappa) + \omega_1$. Given such a set Y, how many subsets can it have (in V)? It has $2^{|Y|}$ many, of course. So, by hypothesis, Y has at most κ^+ subsets (in V). Now we ask ourselves how many such sets Y there are? Clearly, there are at most $|(2^\kappa)^L|$. But GCH is valid in L. So the number of possible sets Y is at most $|(\kappa^+)^L| \leqslant \kappa^+$. So the set X is one of at most κ^+ subsets of one of at most κ^+ constructible sets. There are thus at most κ^+ sets $X \in A$. □

Further consequences of the covering lemma for cardinal arithmetic are considered in Exercise 3.

5.3 Theorem. *Assume* $0^{\#}$ *does not exist. Let* κ *be a singular cardinal. Then* $[\kappa$ *is singular*$]^L$.

Proof. Let $X \subseteq \kappa$ be cofinal in κ, $|X| = \mathrm{cf}(\kappa)$. Let $X \subseteq Y \subseteq \kappa$, $Y \in L$, $|Y| = |X| + \omega_1$. Since $Y \in L$, $|Y| < \kappa$, and $\sup(Y) = \kappa$, we must have $[\kappa$ is singular$]^L$. □

Notice that as an immediate consequence of 5.3 we have:

$$0^\# \text{ exists iff } \omega_\omega \text{ is regular in } L.$$

5.4 Theorem. *Assume* $0^\#$ *does not exist. Let* κ *be a singular cardinal. If* $(\forall \alpha < \kappa)[\mathscr{P}(\alpha) \subseteq L]$, *then* $\mathscr{P}(\kappa) \subseteq L$.

Proof. Let $A \subseteq \kappa$. We show that $A \in L$. Let $\lambda = \mathrm{cf}^L(\kappa)$, and let $(\kappa_\nu \mid \nu < \lambda) \in L$ be cofinal in κ. By 5.3, $\lambda < \kappa$. Let $f \in L$, $f\colon \kappa \leftrightarrow L_\kappa$. For each $\nu < \lambda$, $A \cap \kappa_\nu \in L$, so $A \cap \kappa_\nu \in L_\kappa$, and we can find an $\alpha_\nu < \kappa$ so that $A \cap \kappa_\nu = f(\alpha_\nu)$. Let $X = \{\alpha_\nu \mid \nu < \lambda\}$. Pick $Y \in L$, $Y \subseteq \kappa$, so that $X \subseteq Y$ and $|Y| = |X| + \omega_1 < \kappa$. Then $\mu = |Y|^L < \kappa$. Let $j \in L, j\colon \mu \leftrightarrow Y$. Since $j^{-1}{}''X \subseteq \mu < \kappa$, we have $j^{-1}{}''X \in L$. So, as $j \in L$, we have $X \in L$. But $f \in L$, so it follows that $A = \bigcup \{f(\alpha) \mid \alpha \in X\} \in L$. □

5.5 Theorem. *Assume* $0^\#$ *does not exist. If* κ *is a singular cardinal, then* $(\kappa^+)^L = \kappa^+$.

Proof. Let $\lambda = (\kappa^+)^L$. Suppose that $\lambda < \kappa^+$. Thus $|\lambda| = \kappa$, and so $\mathrm{cf}(\lambda) < \kappa$. Let $X \subseteq \lambda$ be cofinal in λ, $|X| = \mathrm{cf}(\lambda)$. Let $Y \in L$, $X \subseteq Y \subseteq \lambda$, $|Y| = |X| + \omega_1 < \kappa$. Then $|Y|^L < \kappa$. So as Y is cofinal in λ, $[\mathrm{cf}(\lambda)]^L < \kappa < \lambda$. But $[\lambda \text{ is regular}]^L$. Contradiction. □

5.6 Theorem. *Assume* $0^\#$ *does not exist. Let* κ *be a singular cardinal. Then* $\square_\kappa$ *holds.*

Proof. In L, $\square_\kappa$ is valid, so let $(C_\alpha \mid \alpha < (\kappa^+)^L \wedge \lim(\alpha)) \in L$ be a $\square_\kappa$-sequence *in the sense of* L. By 5.5, this sequence is clearly a $\square_\kappa$-sequence in the real world. □

5.7 Theorem. *Assume* $0^\#$ *does not exist. If* GCH *holds, then for every singular cardinal* κ *there is a* κ^+*-Souslin tree.*

Proof. By 5.6 and IV.2.11. □

A slight strengthening of 5.7 is considered in Exercise ID.

We turn now to the proof of the Covering Lemma. It turns out to be a little more convenient to work with the Jensen hierarchy of constructible sets, $(J_\alpha \mid \alpha \in \mathrm{On})$, rather than the hierarchy $(L_\alpha \mid \alpha \in \mathrm{On})$. The Jensen hierarchy was introduced briefly in IV.4, and is studied in detail in Chapter VI. In the meantime, we summarise the facts we need concerning this hierarchy. Note that although the Covering Lemma can be proved using the Fine Structure Theory outlined in IV.4, we shall give here a proof which is free of Fine Structure. Consequently, this section may be read independently of IV.4.

The *rudimentary functions* were defined in IV.4, so, even though you are not require to have read IV.4, there seems little point in repeating the definition here. For any set U, $\mathrm{rud}(U)$ denotes the closure of $U \cup \{U\}$ under the rudimentary functions. If U is transitive, so is $\mathrm{rud}(U)$. The Jensen hierarchy is defined by the recursion

$$J_0 = \emptyset;$$
$$J_{\alpha+1} = \mathrm{rud}(J_\alpha);$$
$$J_\lambda = \bigcup_{\alpha<\lambda} J_\alpha, \quad \text{if } \lim(\lambda).$$

Each J_α is transitive, $\alpha < \beta$ implies $J_\alpha \cup \{J_\alpha\} \subseteq J_\beta$, and $J_\alpha \cap \mathrm{On} = \omega\alpha$. We have $L_\alpha \subseteq J_\alpha \subseteq L_{\omega\alpha}$, so $J_\alpha = L_\alpha$ iff $\omega\alpha = \alpha$. Each J_α is an amenable set, and for all α,

$$J_{\alpha+1} \cap \mathscr{P}(J_\alpha) = \mathrm{Def}(J_\alpha).$$

The Jensen hierarchy thus resembles the usual L_α-hierarchy to a great extent, the main difference being that the slightly more rapid growth of the Jensen hierarchy makes *each level* amenable, not just the limit levels as is the case with the L_α-hierarchy.

There is a single rudimentary function $\mathbf{S}$ such that $U \cup \{U\} \subseteq \mathbf{S}(U)$, and in case U is transitive, $\mathrm{rud}(U) = \bigcup_{n<\omega} \mathbf{S}^n(U)$, where $\mathbf{S}^n$ denotes the n'th iterate of $\mathbf{S}$. We define a refinement of the Jensen hierarchy by the recursion

$$\begin{aligned} S_0 &= \emptyset; \\ S_{\alpha+1} &= \mathbf{S}(S_\alpha); \\ S_\lambda &= \bigcup_{\alpha<\lambda} S_\alpha, \quad \text{if } \lim(\lambda). \end{aligned}$$

Then $\alpha < \beta$ implies $S_\alpha \cup \{S_\alpha\} \subseteq S_\beta$, $S_\alpha \cap \mathrm{On} = \alpha$, and $J_\alpha = S_{\omega\alpha}$.

Every rudimentary function is Σ_0 and uniformly $\Sigma_0^{J_\alpha}$ for all $\alpha > 0$. Consequently, both $(J_\alpha \mid \alpha \in \mathrm{On})$ and $(S_\alpha \mid \alpha \in \mathrm{On})$ are Σ_1, and if $\alpha > 0$, then $(J_\nu \mid \nu < \alpha)$ and $(S_\nu \mid \nu < \omega\alpha)$ are uniformly $\Sigma_1^{J_\alpha}$.

There is a well-ordering $<_J$ of L, which is Σ_1, such that $<_J \cap (J_\alpha \times J_\alpha)$ is an initial segment of $<_J \cap (J_\beta \times J_\beta)$ whenever $\alpha < \beta$. If $\alpha > 0$, $x <_J y \in J_\alpha$ implies $x \in J_\alpha$. Moreover, $<_J \cap (J_\alpha \times J_\alpha)$ is uniformly $\Sigma_1^{J_\alpha}$ for $\alpha > 0$.

The Condensation Lemma is valid for the Jensen hierarchy: if $\alpha > 0$ and $X \prec_1 J_\alpha$, then $X \cong J_\beta$ for some unique $\beta \leqslant \alpha$.

We have already mentioned that we shall give here a proof of the Covering Lemma which does not require any of the Fine Structure Theory. For those familiar with that theory (from IV.4, perhaps) we mention that it is the following, relatively crude notion which suffices here in place of the full Fine Structure apparatus.

Let $\varphi(v_0, v_1, \ldots, v_m)$ be any $\mathscr{L}$-formula. Let $\alpha > 0$. The *J_α-skolem function for*
h_α^φ φ is the function $h_\alpha^\varphi\colon (J_\alpha)^m \to J_\alpha$ defined by

$$h_\alpha^\varphi(x_1, \ldots, x_m) = \begin{cases} \text{the } <_J\text{-least } y \in J_\alpha \text{ such that } \models_{J_\alpha} \varphi(y, x_1, \ldots, x_m), \\ \qquad \text{if such a } y \text{ exists}, \\ \emptyset, \quad \text{if no such } y \text{ exists}. \end{cases}$$

$H_\alpha^n(A)$ Let $\alpha > 0$, $n < \omega$, $A \subseteq J_\alpha$. We denote by $H_\alpha^n(A)$ the closure of A under all J_α-skolem functions h_α^φ for which φ is Σ_n. It is easily seen that if $n > 0$,
$H_\alpha^\omega(A)$ $A \subseteq H_\alpha^n(A) \prec_n J_\alpha$. Similarly, if we denote by $H_\alpha^\omega(A)$ the closure of A under all J_α-skolem functions, then $A \subseteq H_\alpha^\omega(A) \prec J_\alpha$. (We sometimes write $\prec_\omega$ to mean $\prec$ in such contexts.) We also have $A \subseteq H_\alpha^0(A) \prec_1 J_\alpha$. To see this, suppose $\psi(v_0, \ldots, v_n)$ is Σ_1 and that $x_1, \ldots, x_n \in H_\alpha^0(A)$ are such that

$$\models_{J_\alpha} \exists y \varphi(y, x_1, \ldots, x_n)$$

Let φ be a Σ_0 formula such that $\psi(y, \vec{x})$ is equivalent to $\exists z\varphi(z, y, \vec{x})$. Then

$$\models_{J_\alpha} \exists w\varphi((w)_0, (w)_1, \vec{x}).$$

By definition,

$$\models_{J_\alpha} \varphi((h_\alpha^\varphi(\vec{x}))_0, (h_\alpha^\varphi(\vec{x}))_1, \vec{x}).$$

So

$$\models_{J_\alpha} \exists z\varphi(z, (h_\alpha^\varphi(\vec{x}))_1, \vec{x}),$$

i.e.

$$\models_{J_\alpha} \varphi((h_\alpha^\varphi(\vec{x}))_1, \vec{x}).$$

So we shall be done if we can show that $(h_\alpha^\varphi(\vec{x}))_1 \in H_\alpha^0(A)$. Well, for any ordered pair $z \in H_\alpha^0(A)$,

$$(z)_1 = \text{the } <_J\text{-least } y \in J_\alpha \text{ such that } \models_{J_\alpha} \theta(y, z),$$

where θ is the Σ_0-formula $(\exists x \in z)[z = (x, y)]$. Hence $(z)_1 = h_\alpha^\theta(z) \in H_\alpha^0(A)$, and we are done.

5.8 Lemma. *Let* $\alpha > 0$, $1 \leqslant n \leqslant \omega$. *Let* $j\colon J_\alpha \prec_n J_\beta$. *Let* $\varphi(v_0, \ldots, v_k)$ *be any* Σ_n*-formula of* $\mathscr{L}$. *Then for all* $x_1, \ldots, x_k \in J_\alpha$,

$$j(h_\alpha^\varphi(x_1, \ldots, x_k)) = h_\beta^\varphi(j(x_1), \ldots, j(x_k)).$$

Proof. Suppose first that there is no $y \in J_\alpha$ such that $\models_{J_\alpha} \varphi(y, x_1, \ldots, x_k)$. Thus

$$\models_{J_\alpha} \neg \exists y\varphi(y, x_1, \ldots, x_k).$$

Applying j, we get

$$\models_{J_\beta} \neg \exists y\varphi(y, j(x_1), \ldots, j(x_k)).$$

Hence in this case, we have

$$h_\alpha^\varphi(x_1, \ldots, x_k) = \emptyset \quad \text{and} \quad h_\beta^\varphi(j(x_1), \ldots, j(x_k)) = \emptyset,$$

and the lemma is immediate.

Now suppose there is a $y \in J_\alpha$ such that $\models_{J_\alpha} \varphi(y, x_1, \ldots, x_k)$. Set

$$y_0 = h_\alpha^\varphi(x_1, \ldots, x_k).$$

Then

$$\models_{J_\alpha} \varphi(y_0, x_1, \ldots, x_k).$$

So, applying j,

$$\models_{J_\beta} \varphi(j(y_0), j(x_1), \ldots, j(x_k)).$$

Thus if $j(y_0) \neq h_\beta^\varphi(j(x_1), \ldots, j(x_k))$, we must have

$$\models_{J_\beta} \exists z [z <_J j(y_0) \wedge \varphi(z, j(x_1), \ldots, j(x_k))].$$

Applying j^{-1}, we get

$$\models_{J_\alpha} \exists z [z <_J y_0 \wedge \varphi(z, x_1, \ldots, x_k)],$$

contrary to the choice of y_0. □

The main technique involved in the proof of the Covering Lemma is that of constructing limits of directed systems of embeddings. For the benefit of readers not familiar with this technique, we give here a brief outline of what is involved.

A *directed set* is a poset $(I, \leqslant)$ such that whenever $i, j \in I$ there is a $k \in I$ such that $i, j \leqslant k$. A simple example of such is the set $([X]^{<\omega}, \subseteq)$ of all finite subsets of a set X, ordered by inclusion.

Let $n \leqslant \omega$. A *directed Σ_n-elementary system* consists of a family $(\mathscr{A}_i \mid i \in I)$ of structures (of the same kind), indexed by members of a directed set I, together with embeddings $\sigma_{ij}: \mathscr{A}_i \prec_n \mathscr{A}_j$ for each $i, j \in I$, $i \leqslant j$, satisfying the *commutativity* condition $\sigma_{ik} = \sigma_{jk} \circ \sigma_{ij}$ for $i \leqslant j \leqslant k$. In case $n = \omega$, here, we speak simply of a *directed elementary system.*

A *direct limit* of a directed Σ_n-elementary system $\langle (\mathscr{A}_i)_{i \in I}, (\sigma_{ij})_{i \leqslant j} \rangle$ consists of a structure $\mathscr{A}$ (of the same kind as all the $\mathscr{A}_i$), together with embeddings σ_i: $\mathscr{A}_i \prec_n \mathscr{A}$ such that $\sigma_{ij} = \sigma_j^{-1} \circ \sigma_i$ for $i \leqslant j$, satisfying the condition that if $x \in \mathscr{A}$ then $x \in \operatorname{ran}(\sigma_i)$ for some $i \in I$. If $\langle \mathscr{A}, (\sigma_i)_{i \in I} \rangle$, $\langle \mathscr{B}, (\tau_i)_{i \in I} \rangle$ are direct limits of the same system $\langle (\mathscr{A}_i)_{i \in I}, (\sigma_{ij})_{i \leqslant j} \rangle$ we may define an isomorphism π: $\mathscr{A} \cong \mathscr{B}$ as follows: let $x \in \mathscr{A}$. Pick $i \in I$ so that $x = \sigma_i(\bar{x})$ for some $\bar{x} \in \mathscr{A}_i$. Let $\pi(x) = \tau_i(\bar{x})$. It is easily checked (using the commutativity condition) that the choice of i is unimportant here, and that π is a well-defined isomorphism. Since any two direct limits are isomorphic, we often speak of *the* direct limit of a directed elementary system. That there always is a direct limit may be demonstrated as follows.

Let $\langle (\mathscr{A}_i)_{i \in I}, (\sigma_{ij})_{i \leqslant j} \rangle$ be a directed Σ_n-elementary system. For simplicity, suppose that $\mathscr{A}_i = \langle A_i, R_i \rangle$, where $R_i \subseteq A_i^n$. Set

$$U = \bigcup_{i \in I} A_i.$$

Define an equivalence relation $\sim$ on U as follows. Let $x, y \in U$. Pick $i, j \in I$ so that $x \in A_i$, $y \in A_j$. Say $x \sim y$ iff there is a $k \geqslant i, j$ such that $\sigma_{ik}(x) = \sigma_{jk}(y)$. (We leave it to the reader to check that this is an equivalence relation.) Let A be the set of equivalence classes of elements of U under $\sim$. Define a relation $R \subseteq A^n$ as follows. Let $X_1, \ldots, X_n \in A$. Since I is directed we can find an $i \in I$ such that there are elements $x_1 \in X_1 \cap A_i, \ldots, x_n \in X_n \cap A_i$. Set $R(X_1, \ldots, X_n)$ iff $R_i(x_1, \ldots, x_n)$. (We leave it to the reader to check that R is well-defined here.) Let $\mathscr{A} = \langle A, R \rangle$. For $i \in I$, define σ_i: $A_i \to A$ by letting $\sigma_i(x)$ be the equivalence class of x. It is routine to show that σ_i: $\mathscr{A}_i \prec_n \mathscr{A}$.

In cases where the direct limit of a system is well-founded, we usually take the transitive collapse of the limit as "the direct limit" to work with. In this connection, the following result is sometimes useful.

5.9 Lemma. *If $e: J_\alpha \prec_0 M$, where M is transitive, then $e(S_\nu) = S_{e(\nu)}$ for all $\nu < \omega\alpha$.*

Proof. Let φ be the canonical Σ_0 formula which defines the S_ν-hierarchy: that is, for any γ and any $\nu < \omega\gamma$, if N is a transitive set such that $J_\gamma \subseteq N$, then

$$x = S_\nu \quad \text{iff } \models_N \exists y \varphi(y, x, \nu).$$

Let $\nu < \omega\alpha$ be given. Set $x = S_\nu$. Then

$$\models_{J_\alpha} \exists y \varphi(y, x, \nu).$$

So for some $y \in J_\alpha$,

$$\models_{J_\alpha} \varphi(y, x, \nu).$$

Applying $e: J_\alpha \prec_0 M$,

$$\models_M \varphi(e(y), e(x), e(\nu)).$$

Thus

$$\models_M \exists y \varphi(y, e(x), e(\nu)).$$

Thus $e(x) = S_{e(\nu)}$, as required. □

There are various ways of obtaining a structure J_δ as a direct limit of a directed system of smaller structues. We describe below three methods that will be of use to us.

Let $\delta > \omega$ be given. For each integer $n > 0$ and each infinite ordinal $\eta < \omega\delta$, we define a directed Σ_n-elementary system $S_\delta^n(\eta)$ whose limit is J_δ, as follows. Let $I = I_\delta^n(\eta)$ be the set of pairs (α, p) such that $\alpha < \eta$ and p is a finite subset of J_δ. Partially order I by setting $S_\delta^n(\eta)$ $I_\delta^n(\eta)$

$$(\alpha, p) \leqslant (\beta, q) \quad \text{iff } \alpha \leqslant \beta \text{ and } p \subseteq q.$$

Under this ordering, I is a directed set. We use I to index the system $S_\delta^n(\eta)$.

Let $i = (\alpha, p) \in I$. Then

$$\alpha \cup p \subseteq H_\delta^n(\alpha \cup p) \prec_n J_\delta.$$

By the Condensation Lemma, let

$$\sigma_i: J_{\varrho(i)} \cong H_\delta^n(\alpha \cup p).$$

For $i, j \in I$, $i \leqslant j$, set

$$\sigma_{ij} = \sigma_j^{-1} \circ \sigma_i.$$

(This is clearly well-defined.) Thus

$$\sigma_{ij}: J_{\varrho(i)} \prec_n J_{\varrho(j)},$$

and

$$\sigma_i: J_{\varrho(i)} \prec_n J_\delta.$$

Then $S_\delta^n(\eta) = \langle (J_{\varrho(i)})_{i \in I}, (\sigma_{ij})_{i \leqslant j} \rangle$ is a directed Σ_n-elementary system whose direct limit is $\langle J_\delta, (\sigma_i)_{i \in I} \rangle$.

$S_\delta^\omega(\eta)$ $I_\delta^\omega(\eta)$ We may also represent J_δ as the limit of a directed Σ_1-elementary system $S_\delta^\omega(\eta)$, described next. As above, let $\omega \leqslant \eta < \omega\delta$. Let $I = I_\delta^\omega(\eta)$ be the set of all triples (k, α, p) such that $0 < k < \omega$, $\alpha < \eta$, and p is a finite subset of J_δ. Partially order I by setting

$$(k, \alpha, p) \leqslant (l, \beta, q) \quad \text{iff } k \leqslant l \text{ and } \alpha \leqslant \beta \text{ and } p \subseteq q.$$

The directed set I is used to index the system $S_\delta^\omega(\eta)$.

Let $i = (k, \alpha, p) \in I$. Then

$$\alpha \cup p \subseteq H_\delta^k(\alpha \cup p) \prec_k J_\delta.$$

Since $k > 0$, by the Condensation Lemma we may let

$$\sigma_i \colon J_{\varrho(i)} \cong H_\delta^k(\alpha \cup p).$$

Then $\sigma_i \colon J_{\varrho(i)} \prec_1 J_\delta$, and in fact $\sigma_i \colon J_{\varrho(i)} \prec_k J_\delta$. For $i \leqslant j$, we may clearly define

$$\sigma_{ij} = \sigma_j^{-1} \circ \sigma_i.$$

Then

$$\sigma_{ij} \colon J_{\varrho(i)} \prec_1 J_{\varrho(j)}.$$

$S_\delta^\omega(\eta)$ is the directed Σ_1-elementary system

$$\langle (J_{\varrho(i)})_{i \in I}, (\sigma_{ij})_{i \leqslant j} \rangle.$$

Its direct limit is, of course,

$$\langle J_\delta, (\sigma_i)_{i \in I} \rangle.$$

Our third directed system to give J_δ as its limit will apply only in the case when δ is a limit ordinal. For $\omega \leqslant \eta < \omega\delta$, we define the directed Σ_1-elementary system
$S_\delta^0(\eta)$, $S_\delta^0(\eta)$ as follows. Let $I = I_\delta^0(\eta)$ be the set of all triples (ν, α, p) such that $0 < \nu < \delta$,
$I_\delta^0(\eta)$ $\alpha < \eta$, $\alpha \leqslant \nu$, and p is a finite subset of J_ν. Partially order I by setting

$$(\nu, \alpha, p) \leqslant (\mu, \beta, q) \quad \text{iff } [\nu = \mu \text{ or } J_\nu \in q] \text{ and } \alpha \leqslant \beta \text{ and } p \subseteq q.$$

The directed set I will be used to index the system $S_\delta^0(\eta)$.

Let $i = (\nu, \alpha, p) \in I$. Then

$$\alpha \cup p \subseteq H_\nu^0(\alpha \cup p) \prec_1 J_\nu.$$

By the Condensation Lemma,

$$\sigma_i \colon J_{\varrho(i)} \cong H_\nu^0(\alpha \cup p).$$

Thus

$$\sigma_i \colon J_{\varrho(i)} \prec_1 J_\nu.$$

Suppose that $i = (\nu, \alpha, p)$, $j = (\mu, \beta, q)$ are elements of I, $i \leqslant j$. Let $x \in H_\nu^0(\alpha \cup p)$. If $\nu = \mu$, then $x \in H_\mu^0(\alpha \cup p) \subseteq H_\mu^0(\beta \cup q)$. If $\nu \neq \mu$, then since $i \leqslant j$ we must have $J_\nu \in q$. Now, x is Σ_1-definable from elements of $\alpha \cup p$ in J_ν. So there is a Σ_0-formula φ of $\mathscr{L}$ such that x is the unique element of J_ν such that

$$\models_{J_\mu} \varphi(x, \eta_1, \ldots, \eta_n, y_1, \ldots, y_m, J_\nu),$$

where $\eta_1, \ldots, \eta_n \in \alpha$ and $y_1, \ldots, y_m \in p$. [To obtain φ, take a formula which defines x from elements of $\alpha \cup p$ in J_ν and bind all quantifiers by J_ν.] Thus $x \in H_\mu^0(\beta \cup q)$. So we have proved that $H_\nu^0(\alpha \cup p) \subseteq H_\mu^0(\beta \cup q)$. Hence we may define

$$\sigma_{ij} = \sigma_j^{-1} \circ \sigma_i .$$

Since Σ_0 formulas are absolute for transitive sets (I.9.14), $J_\nu \prec_0 J_\mu$. Hence

$$\sigma_{ij}\colon J_{\varrho(i)} \prec_0 J_{\varrho(j)} .$$

$S_\delta^0(\eta)$ is the directed Σ_0-elementary system

$$\langle (J_{\varrho(i)})_{i \in I}, (\sigma_{ij})_{i \leqslant j} \rangle .$$

Its direct limit is

$$\langle J_\delta, (\sigma_i)_{i \in I} \rangle .$$

(Again, by Σ_0-absoluteness, $J_\nu \prec_0 J_\delta$ for all $\nu < \delta$, so $\sigma_i\colon J_{\varrho(i)} \prec_0 J_\delta$.)

The relevance of the above directed system "representations" of structures J_δ lies in the fact that they enable us to represent a possibly large J_δ in terms of small structures $J_{\varrho(i)}$. For, although the directed system will have to be large, in the sense that the index set I must be large, the individual structures $J_{\varrho(i)}$ may all be relatively small. We investigate this phenomenon next.

Consider any of the systems $S_\delta^n(\eta)$ just defined, where $\delta > \omega$, $0 \leqslant n \leqslant \omega$, $\omega \leqslant \eta < \delta$, with $\lim(\delta)$ in case $n = 0$. Let γ be any admissible ordinal. We shall say that $S_\delta^n(\eta)$ is *below* γ if $\varrho(i) < \gamma$ for all $i \in I_\delta^n(\eta)$.

5.10 Lemma. *If $S_\delta^n(\eta)$ is below γ, then $\sigma_{ij} \in J_\gamma$ for all $i, j \in I_\delta^n(\eta)$, $i \leqslant j$.*

Proof. Consider first the case $0 < n < \omega$. Let $i, j \in I_\delta^n(\eta)$, $i \leqslant j$, $i = (\alpha, p)$, $j = (\beta, q)$. Then $\sigma_{ij} = \sigma_j^{-1} \circ \sigma_i$, where

$$\sigma_i\colon J_{\varrho(i)} \cong H_\delta^n(\alpha \cup p), \qquad \sigma_j\colon J_{\varrho(j)} \cong H_\delta^n(\beta \cup q) .$$

Now, $H_\delta^n(\alpha \cup p)$ is the closure of $\alpha \cup p$ under the Σ_n skolem functions h_δ^φ. Since $\alpha \leqslant \beta$ and $p \subseteq q$, $H_\delta^n(\alpha \cup p) \subseteq H_\delta^n(\beta \cup q)$. Using 5.8 and applying

$$\sigma_j\colon J_{\varrho(j)} \cong H_\delta^n(\beta \cup q) \prec_n J_\delta$$

"backwards", we see that the set

$$\operatorname{ran}(\sigma_j^{-1} \circ \sigma_i) = \sigma_j^{-1} {}'' H_\delta^n(\alpha \cup p)$$

is the closure of $\alpha \cup \sigma_j^{-1}(p)$ under the Σ_n skolem functions $h^{\varphi}_{\varrho(j)}$, i.e.

$$\operatorname{ran}(\sigma_{ij}) = H^n_{\varrho(j)}(\alpha \cup \sigma_j^{-1}(p)).$$

Consider the definition of $H^n_{\varrho(j)}(\alpha \cup \sigma_j^{-1}(p))$ from $\varrho(j)$, n, α, $\sigma_j^{-1}(p)$. It is the subset of $J_{\varrho(j)}$ consisting of all those elements x of $J_{\varrho(j)}$ which may be obtained from elements of $\alpha \cup \sigma_j^{-1}(p)$ by finitely many applications of functions of the form $h^{\varphi}_{\varrho(j)}$ where φ is a Σ_n formula of $\mathscr{L}$. Thus, it is easily seen that $H^n_{\varrho(j)}(\alpha \cup \sigma_j^{-1}(p))$ is a $\Delta_1(J_\gamma)$ subset of $J_{\varrho(j)}$. But J_γ is admissible. Hence by Δ_1-Comprehension (I.11.1), $H^n_{\varrho(j)}(\alpha \cup \sigma_j^{-1}(p)) \in J_\gamma$.

Again, σ_{ij}^{-1} is the collapsing isomorphism for the set $H^n_{\varrho(j)}(\alpha \cup \sigma_j^{-1}(p))$, so σ_{ij}^{-1} is a $\Delta_1(J_\gamma)$ subset of $J_{\varrho(j)}$, whence $\sigma_{ij}^{-1} \in J_\gamma$. Thus $\sigma_{ij} \in J_\gamma$, as required.

Consider next the case $n = \omega$. If $i, j \in I^{\omega}_{\delta}(\eta)$, $i \leqslant j$, $i = (k, \alpha, p)$, $j = (l, \beta, q)$, then, much as above, we have

$$\sigma_{ij}\colon J_{\varrho(i)} \cong H^k_{\varrho(j)}(\alpha \cup \sigma_j^{-1}(p)),$$

and again as before this implies that $\sigma_{ij} \in J_\gamma$.

Finally suppose $n = 0$. Let $i, j \in I^0_\delta(\eta)$, $i \leqslant j$, $i = (\nu, \alpha, p)$, $j = (\mu, \beta, q)$. If $\nu = \mu$, then

$$\sigma_i\colon J_{\varrho(i)} \cong H^0_\nu(\alpha \cup p), \qquad \sigma_j\colon J_{\varrho(j)} \cong H^0_\nu(\beta \cup q),$$

so as in the above cases

$$\sigma_{ij}\colon J_{\varrho(i)} \cong H^0_{\varrho(j)}(\alpha \cup \sigma_j^{-1}(p)),$$

and as before we can conclude that $\sigma_{ij} \in J_\gamma$.

Now suppose $J_\nu \in q$. Then

$$\sigma_i\colon J_{\varrho(i)} \cong H^0_\nu(\alpha \cup p), \qquad \sigma_j\colon J_{\varrho(j)} \cong H^0_\mu(\beta \cup q).$$

Now, if φ is Σ_0 and $x_1, \ldots, x_m, y \in J_\nu$, then, as is easily seen,

$$y = h^{\varphi}_{\nu}(x_1, \ldots, x_m) \quad \text{iff } y = h^{\varphi}_{\mu}(x_1, \ldots, x_m).$$

Thus we may in fact apply the same argument as before to obtain

$$\sigma_{ij}\colon J_{\varrho(i)} \cong H^0_{\varrho(j)}(\alpha \cup \sigma_j^{-1}(p)).$$

Again this implies that $\sigma_{ij} \in J_\gamma$, so we are done in all cases. □

We are now ready to begin our proof of the Covering Lemma. We shall
τ assume from now on that the Covering Lemma is false. We fix τ the least ordinal
X such that there is an uncountable set $X \subseteq \tau$ which is not a subset of any constructible set of cardinality $|X|$. This choice of τ has two immediate consequences.

5.11 Lemma. $[\tau$ *is a cardinal*$]^L$.

Proof. Suppose otherwise. Let $\lambda = |\tau|^L$, and let $j \in L$, $j: \lambda \leftrightarrow \tau$. Let $\bar{X} = j^{-1}{}''X$. Since $\bar{X} \subseteq \lambda < \tau$, the minimality of τ guarantees the existence of a set $\bar{Y} \in L$, $\bar{X} \subseteq \bar{Y} \subseteq \lambda$, $|\bar{Y}| = |\bar{X}|$. Let $Y = j''\bar{Y}$. Then $Y \in L$, $X \subseteq Y \subseteq \tau$, and $|Y| = |X|$. Contradiction. □

5.12 Lemma. *If $Y \in L$ and $|Y|^L < \tau$, then Y cannot cover X.*

Proof. Let $\lambda = |Y|^L$, and let $j \in L$, $j: \lambda \leftrightarrow Y$. Suppose $X \subseteq Y$. Let $\bar{X} = j^{-1}{}''X$. Then $\bar{X} \subseteq \lambda < \tau$, so by choice of τ there is a set $\bar{Z} \in L$, $\bar{X} \subseteq \bar{Z} \subseteq \lambda$, $|\bar{Z}| = |\bar{X}|$. Let $Z = j''\bar{Z}$. Then $Z \in L$, $X \subseteq Z \subseteq \tau$, $|Z| = |X|$, contrary to the choice of X. □

The overall strategy behind our proof of the Covering Lemma is as follows. Let $M \prec J_\tau$, $|M| = |X|$, $X \subseteq M$, and let $\pi: J_\gamma \cong M$. Since $X \subseteq \tau$, we know that (by choice of X) $|X| < |\tau|$ (for otherwise $\tau \in L$ is a cover of X of cardinality $|X|$). Thus $|J_\gamma| < |\tau|$. But X is cofinal in τ and $X \subseteq \operatorname{ran}(\pi)$. Hence $\pi: J_\gamma \prec J_\tau$ is non-trivial. Let β be least such that $\pi(\beta) > \beta$. If $\beta < |\gamma|$, then by 4.3, $0^\#$ exists, and we have our sought-after contradiction. What if $\beta \geqslant |\gamma|$? Then we try to find a $\delta \geqslant \gamma$, $|\delta| > \beta$, such that it is possible to find an embedding $\tilde{\pi}: J_\delta \prec J_\nu$ which extends π (so $\tilde{\pi}(\beta) > \beta$), in which case 4.3 may again be applied. The question is, how might we extend π as desired? Well, by choosing M carefully in the first place, we find a δ such that J_δ is the direct limit of a system which is below γ. (Note that as $J_\gamma \equiv J_\tau$, γ is admissible, by virtue of 5.11.) Thus the map π sends the members of this system to a directed system inside J_τ. If the direct limit of this system is well-founded, and thus of the form J_ν for some ν, then it will be easy to construct an embedding $\tilde{\pi}$: $J_\delta \prec J_\nu$ which extends π, as we shall see. However, as we shall discover, the choice of M, in particular, must be made very carefully indeed, making use of the special properties of τ and X.

We defer until later the actual choice of the submodel $M \prec J_\tau$. We assume simply that we have found some embedding $\pi: J_\gamma \prec J_\tau$. Note that by 5.11, γ will be an admissible ordinal. Let us further assume that δ, n, η are such that $\delta > \omega$, $0 \leqslant n \leqslant \omega$, $\omega \leqslant \eta < \delta$ and that $S^n_\delta(\eta)$ is below γ. Then we may define a directed system $\pi^* S^n_\delta(\eta)$, of the same degree of elementarity as $S^n_\delta(\eta)$, as follows. As index set we take the set $I^n_\delta(\eta)$. Associated with $i \in I^n_\delta(\eta)$ will be the structure $J_{\pi(\varrho(i))}$. For $i, j \in I^n_\delta(\eta)$, $i \leqslant j$, the embedding associated with i, j will be $\pi(\sigma_{ij})$. Since π is elementary, if $\sigma_{ij}: J_{\varrho(i)} \prec_k J_{\varrho(j)}$, then $\pi(\sigma_{ij}): J_{\pi(\varrho(i))} \prec_k J_{\pi(\varrho(j))}$, so this makes sense, and moreover, $\pi^* S^n_\delta(\eta)$ so defined is a directed system. The lemma below shows how, under these circumstances, it is possible to extend π from J_γ to J_δ. (Actually, in the form stated, all that we get is that $\tilde{\pi}$ extends $\pi \upharpoonright \eta$, but in our main application of the lemma we shall have $\eta = \gamma$, in which case we really will have $\pi \subseteq \tilde{\pi}$.)

Margin notes beside this paragraph: π, γ — δ, n, η — $\pi^* S^n_\delta(\eta)$

5.13 Lemma. *Let $\langle\langle U, E\rangle, (\theta_i)_{i\in I}\rangle$ be the direct limit of the system $\pi^* S^n_\delta(\eta)$. Then there is an embedding $\tilde{\pi}$: $\langle J_\delta, \in\rangle \prec_{1+n} \langle U, E\rangle$. Moreover, if $\langle U, E\rangle$ is well-founded, we may take $\langle U, E\rangle$ to be of the form $\langle J_\mu, \in\rangle$ for some μ, in which case $\tilde{\pi} \upharpoonright \eta = \pi \upharpoonright \eta$.*

Proof. Let $x \in J_\delta$. For some $i \in I^n_\delta(\eta)$, $x \in \operatorname{ran}(\sigma_i)$, say $x = \sigma_i(\bar{x})$, where $\bar{x} \in J_{\varrho(i)}$. Let $\bar{y} = \pi(\bar{x})$, and set $\tilde{\pi}(x) = \theta_i(\bar{y})$. (Thus $\tilde{\pi}(x) = \theta_i \circ \pi \circ \sigma_i^{-1}(x)$.) It is routine to verify that $\tilde{\pi}$ is well-defined. And in the cases $n < \omega$, it is immediate that $\tilde{\pi}$ is Σ_n-elementary. To show that in these cases $\tilde{\pi}$ is in fact Σ_{n+1}-elementary, we argue as follows.

Suppose that φ is Π_n and that

$$\langle U, E\rangle \models \exists y \varphi(y, \tilde{\pi}(x)).$$

Then for some $y \in U$,

$$\langle U, E\rangle \models \varphi(y, \tilde{\pi}(x)).$$

Pick i so that $\tilde{\pi}(x)$, $y \in \operatorname{ran}(\theta_i)$, say $\tilde{\pi}(x) = \theta_i(\bar{x})$, $y = \theta_i(\bar{y})$. Since

$$\theta_i\colon \langle J_{\pi(\varrho(i))}, \in\rangle \prec_n \langle U, E\rangle,$$

we have

$$J_{\pi(\varrho(i))} \models \varphi(\bar{y}, \bar{x}).$$

But $\bar{x} = \theta_i^{-1} \circ \tilde{\pi}(x) = \pi \circ \sigma_i^{-1}(x)$. So we may rewrite the above as

$$J_{\pi(\varrho(i))} \models \varphi(\bar{y}, \pi \circ \sigma_i^{-1}(x)).$$

Thus

$$J_{\pi(\varrho(i))} \models \exists z \varphi(z, \pi \circ \sigma_i^{-1}(x)).$$

Then, since $\pi\colon J_\gamma \prec J_\tau$, we deduce that

$$J_{\varrho(i)} \models \exists z \varphi(z, \sigma_i^{-1}(x)).$$

So for some $z \in J_{\varrho(i)}$,

$$J_{\varrho(i)} \models \varphi(z, \sigma_i^{-1}(x)).$$

But $\sigma_i\colon J_{\varrho(i)} \prec_n J_\delta$. So

$$J_\delta \models \varphi(\sigma_i(z), x).$$

Thus

$$J_\delta \models \exists y \varphi(y, x).$$

The argument in the other direction is similar, and we leave it to the reader to supply.

In the case $n = \omega$, to prove that $\tilde{\pi}$ is elementary, we argue as follows. Let φ be a formula which we wish to show is preserved by $\tilde{\pi}$. Suppose that φ is Σ_m. Pick $i = (k, \alpha, p)$ in $I_\delta^\omega(\eta)$ "large" enough so that ran (σ_i) contains all parameters involved and so that $k \geqslant m$. Then use the fact that σ_i and θ_i are Σ_k-elementary. (We leave the details to the reader.)

Now suppose that $\langle U, E\rangle$ is well-founded. Then we may assume that U is transitive and that $E = \in \cap U^2$. Let $U \cap \mathrm{On} = \omega\mu$. (It is clear that $U \cap \mathrm{On}$ must be a limit ordinal.) We prove that $U = J_\mu$. First of all set $x \in U$. Pick i so that

$x = \theta_i(\bar{x})$ for some $\bar{x} \in J_{\pi(\varrho(i))}$. For some $\nu < \omega \cdot \pi(\varrho(i))$, $\bar{x} \in S_\nu$. Applying $\theta_i\colon J_{\pi(\varrho(i))} \prec_0 U$ and using 5.9, we have

$$x = \theta_i(\bar{x}) \in \theta_i(S_\nu) = S_{\theta_i(\nu)} \subseteq \bigcup_{\xi < \omega\mu} S_\xi = J_\mu.$$

Now let $x \in J_\mu$. For some $\nu < \omega\mu$, $x \in S_\nu$. Since $\nu \in U$ we can find an i such that $\nu = \theta_i(\bar{\nu})$, where $\bar{\nu} < \pi(\varrho(i))$. Then by 5.9 again, $\theta_i(S_{\bar{\nu}}) = S_{\theta_i(\bar{\nu})} = S_\nu$, so $S_\nu \in \operatorname{ran}(\theta_i)$. Thus $S_\nu \in U$. But U is transitive. Hence $x \in U$.

Finally, assume now that $U = J_\mu$. We show that $\tilde{\pi} \upharpoonright \eta = \pi \upharpoonright \eta$. Let $\xi < \eta$ be given. Pick $i \in I^n_\delta(\eta)$ so that $\xi \in \alpha$, where $i = (\alpha, p)$ if $0 < n < \omega$, $i = (\nu, \alpha, p)$ if $n = 0$, and $i = (k, \alpha, p)$ if $n = \omega$. Then $\sigma_{ij}(\xi) = \xi$ for all $j \geqslant i$, so $\sigma_i(\xi) = \xi$. Again, since $\sigma_{ij}(\xi) = \xi$ for all $j \geqq i$, applying $\pi\colon J_\gamma \prec J_\tau$ we have $[\pi(\sigma_{ij})](\pi(\xi)) = \pi(\xi)$ for all $j \geqslant i$. Hence as U is transitive, $\theta_i(\pi(\xi)) = \pi(\xi)$. Thus $\tilde{\pi}(\xi) = \pi(\xi)$. □

The proof of the following lemma is very complicated, and is deferred until later.

5.14 Lemma. *There is an admissible ordinal γ and an embedding $\pi\colon J_\gamma \prec J_\tau$ such that $|\gamma| = |X|$, $X \subseteq \operatorname{ran}(\pi)$, and whenever $\delta \geqslant \gamma$, then $S^\omega_\delta(\gamma)$ is below γ and the direct limit of $\pi^* S^\omega_\delta(\gamma)$ is well-founded.* □

Using 5.14, it is very easy to obtain the contradiction which proves the Covering Lemma. Namely:

5.15 Lemma. $0^\#$ *exists.*

Proof. Since $X \subseteq \operatorname{ran}(\pi)$ and $|\gamma| = |X| < |\tau|$ we can find a β such that $\pi(\beta) \neq \beta$. Pick $\delta \geqslant \gamma$, $|\delta| > \beta$. By 5.14, $S^\omega_\delta(\gamma)$ is below γ and the direct limit of $\pi^* S^\omega_\delta(\gamma)$ is well-founded. So by 5.13 we may take this limit to be J_μ for some μ, and there is an embedding $\tilde{\pi}\colon J_\delta \prec J_\mu$ such that $\tilde{\pi} \upharpoonright \gamma = \pi \upharpoonright \gamma$. But $\tilde{\pi}(\beta) \neq \beta$ and $\beta < |\delta|$. So by 4.3, $0^\#$ exists. □

Now let us begin our attack on 5.14. The part that makes use of the fact that X cannot be covered by a constructible set Y such that $|Y|^L < \tau$ (see 5.12) is the proof that $S^\omega_\delta(\gamma)$ is below γ for any $\delta \geqslant \gamma$. In fact, we shall prove, by induction on δ, n, that if $\delta \geqslant \gamma$ and $0 < n \leqq \omega$, then $S^n_\delta(\gamma)$ is below γ. (This is why we need to consider three types of directed system, not just $S^\omega_\delta(\gamma)$.) This in turn means that we must be even more careful in our original choice of γ, π. More precisely, instead of simply proving 5.14 as stated, we prove the following two results, which together imply 5.14 at once.

5.16 Lemma. *There is an admissible ordinal γ and an embedding $\pi\colon J_\gamma \prec J_\tau$ such that:*

(i) *$|\gamma| = |X|$ and $X \subseteq \operatorname{ran}(\pi)$;*

(ii) *if $\delta \geqslant \gamma$, $n \leqslant \omega$, and if* lim (δ) *in case $n = 0$, then,* IF *$S^n_\delta(\gamma)$ is below γ, then the direct limit of $\pi^* S^n_\delta(\gamma)$ is well-founded.* □

5.17 Lemma. *Let $\delta \geqslant \gamma$, $0 < n \leqslant \omega$. Then $S^n_\delta(\gamma)$ is below γ.* □

We prove 5.17 first, since 5.16 is the more complex of the two. It is clear that 5.17 follows directly from the following lemma (which is in fact only a reformulation of 5.17 in the cases $n < \omega$, being stronger only in the case $n = \omega$).

5.18 Lemma. *Let $\delta \geqslant \gamma$, $0 < n \leqslant \omega$. For every $\alpha < \gamma$ and every finite set $p \subseteq J_\delta$,*

$$\mathrm{otp}\,[H_\delta^n(\alpha \cup p) \cap \mathrm{On}] < \gamma.$$

Proof. Suppose the lemma is false. Let $\delta \geqslant \gamma$ be the least ordinal for which it fails (for some n), and let n be least such that $0 < n \leqslant \omega$ and the lemma fails for δ, n. We wish to apply 5.16 (ii) to $\delta, n-1$. In order to do this we must know that if $n = 1$, then $\lim(\delta)$. This is in fact the case, but we shall defer the proof for a moment, and simply assume it.

Claim 1. $S_\delta^{n-1}(\gamma)$ is below γ. (If $n = \omega$, then of course $n - 1 = \omega$.)

Proof. Suppose first that $n > 1$. Let $i \in I_\delta^{n-1}(\gamma)$. Then

$$J_{\varrho(i)} \cong H_\delta^k(\alpha \cup p)$$

for some k, $0 < k < n$ and some finite $p \subseteq J_\delta$. (If $n < \omega$, then in fact $k = n - 1$.) By the minimality of n,

$$\mathrm{otp}\,[H_\delta^k(\alpha \cup p) \cap \mathrm{On}] < \gamma.$$

Hence $\varrho(i) < \gamma$.

Now consider the case $n = 1$. Let $i \in I_\delta^0(\gamma)$. Then

$$J_{\varrho(i)} \cong H_\mu^0(\alpha \cup p)$$

for some $\mu < \delta$, $\alpha < \gamma$, and some finite $p \subseteq J_\mu$. If $\mu \geqslant \gamma$, then by the minimality of δ,

$$\mathrm{otp}\,[H_\mu^0(\alpha \cup p) \cap \mathrm{On}] < \gamma,$$

whilst if $\mu < \gamma$, then trivially

$$\mathrm{otp}\,[H_\mu^0(\alpha \cup p) \cap \mathrm{On}] \leqslant \omega\mu < \gamma,$$

so again we have $\varrho(i) < \gamma$. The claim is proved.

Claim 2. There are $\alpha_0 < \gamma$, $p_0 \subseteq J_\delta$, p_0 finite, such that

$$J_\delta = H_\delta^n(\alpha_0 \cup p_0).$$

Proof. Pick $\alpha < \gamma$, $p \subseteq J_\delta$, p finite, such that

$$\mathrm{otp}\,[H_\delta^n(\alpha \cup p) \cap \mathrm{On}] \geqslant \gamma.$$

Let

$$j\colon J_{\bar\delta} \cong H_\delta^n(\alpha \cup p).$$

Set $\alpha_0 = \alpha$, $p_0 = j^{-1}(p)$. By 5.8,

$$J_{\bar\delta} = H_{\bar\delta}^n(\alpha_0 \cup p_0).$$

But $\bar\delta \geqslant \gamma$. So by the minimality of δ we have $\bar\delta = \delta$. The claim is proved.

By Claim 1 and 5.16 (ii), the direct limit of $\pi^* S_\delta^{n-1}(\gamma)$ is well-founded. Thus by 5.13 we may take this limit to be of the form J_ν, and there is an embedding $\tilde{\pi}$: $J_\delta \prec_n J_\nu$ such that $\tilde{\pi} \upharpoonright \gamma = \pi \upharpoonright \gamma$. Let $\beta_0 = \tilde{\pi}(\alpha_0)$, $q_0 = \tilde{\pi}(p_0)$. By Claim 2 and 5.8, we have, applying $\tilde{\pi}$,

$$\operatorname{ran}(\tilde{\pi}) = \tilde{\pi}'' J_\delta = \tilde{\pi}'' H_\delta^n(\alpha_0 \cup p_0) \subseteq H_\nu^n(\beta_0 \cup q_0).$$

But $X \subseteq \operatorname{ran}(\pi) \subseteq \operatorname{ran}(\tilde{\pi})$. Hence

$$X \subseteq H_\nu^n(\beta_0 \cup q_0).$$

Now, clearly, $Y = H_\nu^n(\beta_0 \cup q_0) \in L$. Moreover $|Y|^L = |\beta_0|^L + \omega$. But

$$\beta_0 = \tilde{\pi}(\alpha_0) = \pi(\alpha_0) \in J_\tau,$$

so $\beta_0 < \tau$. Thus by 5.11, $|\beta_0|^L < \tau$. Thus Y contradicts 5.12, and we are done.

We are left with the proof that if $n = 1$, then $\lim(\delta)$.[5] Suppose, on the contrary, that we had $n = 1$ and $\delta = \beta + 1$. Note that as $\delta \geqslant \gamma$ and γ is a limit ordinal, we must have $\beta \geqslant \gamma$. Choose $\alpha < \gamma$, $p \subseteq J_\delta$ finite, so that

$$\operatorname{otp}[H_\delta^1(\alpha \cup p) \cap \mathrm{On}] \geqslant \gamma.$$

Now,

$$H_\delta^1(\alpha \cup p) \cap \mathrm{On} = H_\delta^1(\alpha \cup p) \cap \omega\delta.$$

Since $\delta = \beta + 1$, if we intersect $H_\delta^1(\alpha \cup p) \cap \omega\delta$ with $\omega\beta$ we lose at most ω elements. But

$$\operatorname{otp}[H_\delta^1(\alpha \cup p) \cap \omega\delta] \geqslant \gamma = \omega\gamma.$$

Thus we must have

$$(*) \qquad \operatorname{otp}[H_\delta^1(\alpha \cup p) \cap \omega\beta] \geqslant \gamma.$$

Let $p = \{a_1, \ldots, a_l\}$. Since $a_1, \ldots, a_l \in J_\delta - \operatorname{rud}(J_\beta)$, there are rudimentary functions $f_1, \ldots, f_l$ and elements $b_1, \ldots, b_l$ of J_β such that

$$a_1 = f_1(J_\beta, b_1), \ldots, a_1 = f_l(J_\beta, b_l).$$

Let $q = \{b_1, \ldots, b_l\}$. We prove that

$$(**) \qquad H_\delta^1(\alpha \cup p) \cap \omega\beta \subseteq H_\beta^\omega(\alpha \cup q) \cap \omega\beta.$$

5 This part of the proof makes use of some technical facts concerning the Jensen hierarchy of constructible sets and the properties of rudimentary functions. These facts are proved in Chapter VI, and we simply quote them in the present account. Consequently, the reader not already familiar with the Jensen hierarchy may prefer to simply take the result $n = 1 \rightarrow \lim(\delta)$ on trust, or else to merely skip through the account given. In any event, it hardly seems worth postponing a proof of the Covering Lemma until after Chapter VI, when this one technical detail in the proof is the only point where such knowledge is required.

By (*), this implies that $\operatorname{otp}[H^\omega_\beta(\alpha\cup q)\cap \mathrm{On}]\geqslant\gamma$, contrary to the choice of δ, which completes the proof.

So let $\xi\in H^1_\delta(\alpha\cup p)\cap\omega\beta$. We must prove that $\xi\in H^\omega_\beta(\alpha\cup q)$. Let φ be a Σ_0-formula, and let $\xi_1,\dots,\xi_k<\alpha$ be such that ξ is the least (which for ordinals is the same as the $<_J$-least) ordinal in $\omega\delta$ such that

$$(1)\qquad \models_{J_\delta}\exists x\,\varphi(x,\xi,\xi_1,\dots,\xi_k,a_1,\dots,a_l).$$

Pick $x\in J_\delta$ such that

$$(2)\qquad \models_{J_\delta}\varphi(x,\xi,\xi_1,\dots,\xi_k,a_1,\dots,a_l).$$

Then we can find a rudimentary function f and an element y of J_β such that $x=f(J_\beta,y)$. So,

$$(3)\qquad \models_{J_\delta}\varphi(f(J_\beta,y),\xi,\xi_1,\dots,\xi_k,f_1(J_\beta,b_1),\dots,f_l(J_\beta,b_l)).$$

Since φ is Σ_0 and $f,f_1,\dots,f_l$ are rudimentary, the formula

$$\varphi(f(x,y),\xi,\xi_1,\dots,\xi_k,f_1(x,b_1),\dots,f_l(x,b_l))$$

is Σ_0 in the variables $x,y,\xi,\xi_1,\dots,\xi_k,b_1,\dots,b_l$. This depends upon a property of rudimentary functions that we have not mentioned before, that if $R(x)$ is a Σ_0 predicate and f is rudimentary, then $R(f(\vec{x}))$ is a Σ_0 predicate. For a proof of this fact we refer the reader to VI.1.3. It follows that there is a formula ψ of $\mathscr{L}$ such that (3) is equivalent to

$$(4)\qquad \models_{J_\beta}\psi(y,\xi,\xi_1,\dots,\xi_k,b_1,\dots,b_l).$$

This requires another result not yet proved, which says that Σ_0-definability over $\operatorname{rud}(U)$ for elements of a transitive rud closed set U, using parameters $U,\vec{a}$, where $\vec{a}\in U$, is equivalent to definability over U using parameters $\vec{a}$. This is proved in VI.1.18. By (4) we have

$$(5)\qquad \models_{J_\beta}\exists y\,\psi(y,\xi,\xi_1,\dots,\xi_k,b_1,\dots,b_l).$$

Moreover, ξ is the least such. For suppose, on the contrary, that $\xi'<\xi$ is such that

$$(6)\qquad \models_{J_\beta}\exists y\,\psi(y,\xi',\xi_1,\dots,\xi_k,b_1,\dots,b_l).$$

Then, using the equivalence of (3) and (4) we can find a $y'\in J_\beta$ such that

$$(7)\qquad \models_{J_\delta}\varphi(f(J_\beta,y'),\xi',\xi_1,\dots,\xi_k,f_1(J_\beta,b_1),\dots,f_l(J_\beta,b_l)).$$

So, setting $x'=f(J_\beta,y')$, we have

$$(8)\qquad \models_{J_\delta}\varphi(x',\xi',\xi_1,\dots,\xi_k,a_1,\dots,a_l).$$

This contradicts the choice of ξ. Since ξ is the least ordinal satisfying (5), we have $\xi\in H^\omega_\beta(\alpha\cup q)$. This proves (**), and completes our proof of the lemma. □

This leaves us with the proof of 5.16. We shall define, by recursion, a chain of submodels

$$M_0 \prec M_1 \prec \ldots \prec M_\theta \prec \ldots \prec J_\tau \qquad (\theta \leqslant \omega_1)$$

such that $X \subseteq M_0$. Setting

$$\pi_\theta \colon J_{\gamma(\theta)} \cong M_\theta$$ $\pi_\theta, \gamma(\theta)$

for each $\theta \leqslant \omega_1$, we shall let $\gamma = \gamma(\omega_1)$, $\pi = \pi_{\omega_1}$ to obtain the lemma. That is, we shall have $|\gamma| = |X|$, and whenever $\delta \geqslant \gamma$ and $n \leqslant \omega$, with $\lim(\delta)$ in case $n = 0$, if $S_\delta^n(\gamma)$ is below γ, then the direct limit of $\pi^* S_\delta^n(\gamma)$ is well-founded. The idea is to include in the models M_θ, $\theta < \omega_1$, withnesses to any possible failure of well-foundedness of any eventual $\pi^* S_\delta^n(\gamma)$, so that the well-foundedness can be established by a proof by contradiction. γ, π

To commence, we set

$$M_0 = H_\tau^\omega(X).$$ M_θ

And a limit stages $\theta < \omega_1$, we set

$$M_\theta = \bigcup_{\psi < \theta} M_\varphi.$$

This leaves us with the case where $\theta < \omega_1$ and M_θ has been defined.

Consider a pair (n, η) such that $n \leqslant \omega$ and $\omega < \eta \leqslant \gamma(\theta)$. Suppose that there is a $\delta \geqslant \eta$ such that $S_\delta^n(\eta)$ is below $\gamma(\theta)$ and the direct limit of $\pi_\theta^* S_\delta^n(\eta)$ is not well-founded. Let δ_0 be the least such δ. Since the limit of $\pi_\theta^* S_{\delta_0}^n(\eta)$ is not well-founded, we can find a sequence $(a_k \mid k < \omega)$ and elements $j_k \in I_{\delta_0}^n(\eta)$, $j_k \leqslant j_{k+1}$, such that $a_k \in J_{\pi_\theta(\varrho(j_k))}$ and $a_{k+1} \in [\pi_\theta(\sigma_{j_k, j_{k+1}})](a_k)$, where $\varrho(i)$, σ_{ij} relate to the system $S_{\delta_0}^n(\eta)$ here.[6]

For each pair (n, η) as above, we pick one such sequence $(a_k \mid k < \omega)$. We let N_θ be the set of all elements a_k, $k < \omega$, chosen in this way. (Of course, it is possible that $N_\theta = \emptyset$.) We set

$$M_{\theta+1} = H_\tau^\omega(M_\theta \cup N_\theta).$$

By induction on $\theta < \omega_1$ we easily see that $|N_\theta| \leqslant |X|$ for all $\theta < \omega_1$. Thus $|M_\theta| = |X|$ for all $\theta \leqslant \omega_1$. In particular, if $\gamma = \gamma(\omega_1)$, then $|\gamma| = |X|$. Setting $\pi = \pi_{\omega_1}$, we have $\pi \colon J_\gamma \prec J_\tau$ and $X \subseteq \operatorname{ran}(\pi)$. So what we must show is that if

6 To avoid the necessity of extra notation in a situation where the notational complexity is already at the limit of human tolerance, we shall frequently use the symbols $\varrho(i)$, σ_{ij}, etc. to refer to various dirceted systems of any of the three basic types described earlier, and merely observe which system is referred to each time. In each case, $\varrho(i)$, σ_{ij}, etc. will have the meaning originally defined, *but for the system under consideration at the time.* This desire for notational "simplicity" is also the reason why we made no notational distinction between the three types of directed system which we introduced; with $\varrho(i)$, ϱ_{ij}, etc. having the same meaning in each case.

δ, n $\delta \geqslant \gamma$ and $n \leqslant \omega$, with $\lim(\delta)$ in case $n = 0$, and if $S^n_\delta(\gamma)$ is below γ, then the limit of $\pi^* S^n_\delta(\gamma)$ is well-founded. We assume otherwise and work for a contradiction.

We consider first the case $0 < n < \omega$. Pick sequences $(b_m | m < \omega)$, $(i_m | m < \omega)$
b_m, i_m so that $i_m \in I^n_\delta(\gamma)$, $i_m \leqslant i_{m+1}$, $i_m = (\alpha_m, p_m)$, $\alpha_m < \alpha_{m+1}$, $b_m \in J_{\pi(\varrho(i_m))}$, $b_{m+1} \in$
α_m, p_m $[\pi(\sigma_{i_m, i_{m+1}})](b_m)$, where $\varrho(i)$, σ_{ij} refer to the system $S^n_\delta(\gamma)$.

Now, in order to obtain a contradiction with the construction of M, what we require is that such a sequence $(b_m | m < \omega)$ exists for a system which is below $\gamma(\theta)$ for some $\theta < \omega_1$. But all that we know about $S^n_\delta(\gamma)$ is that it is below γ. (Indeed, for the system $S^n_\delta(\gamma)$ itself, the ordinal γ is clearly the least ordinal such that the system is below γ.) However, the subsystem $\langle (J_{\varrho(i_m)})_{m<\omega}, (\sigma_{i_m, i_s})_{m \leqslant s} \rangle$ is such that the limit of $\langle (J_{\pi(\varrho(i_m))})_{m<\omega}, (\pi(\sigma_{i_m, i_s}))_{m \leqslant s} \rangle$ is not well-founded. The idea now is to use this countable system to construct a system $S^n_{\bar\delta}(\bar\eta)$ which is below $\gamma(\theta)$ for some $\theta < \omega_1$ and for which the direct limit of $\pi^*_\theta S^n_{\bar\delta}(\bar\eta)$ is not well-founded.

Now, $S^n_\delta(\gamma)$ is below γ, so for each m, $\varrho(i_m)$, α_m, $\sigma^{-1}_{i_m}(p_m) \in J_\gamma$. Thus for each m,
θ $\pi(\varrho(i_m))$, $\pi(\alpha_m)$, $\pi(\sigma^{-1}_{i_m}(p_m)) \in M_{\omega_1}$. Since $M_{\omega_1} = \bigcup_{\theta<\omega_1} M_\theta$, we can find a $\theta < \omega_1$
j such that $\pi(\varrho(i_m))$, $\pi(\alpha_m)$, $\pi(\sigma^{-1}_{i_m}(p_m)) \in M_\theta$ for all $m < \omega$. Let $j = \pi^{-1} \circ \pi_\theta$. Thus

$$j\colon J_{\gamma(\theta)} \prec J_\gamma .$$

Our next move is to use j in order to "pull back" from J_γ to $J_{\gamma(\theta)}$ the system $\langle (J_{\varrho(i_m)})_{m<\omega}, (\sigma_{i_m, i_k})_{m \leqslant k} \rangle$. Since $\operatorname{ran}(j) = \operatorname{ran}(\pi^{-1} \restriction M_\theta)$, we have $\varrho(i_m)$, α_m, $\sigma^{-1}_{i_m}(p_m) \in \operatorname{ran}(j)$ for all $m < \omega$. For each $m < \omega$, let $\bar\varrho_m < \gamma(\theta)$ be such that
$\bar\varrho_m, \bar\alpha_m$ $j(\bar\varrho_m) = \varrho(i_m)$, let $\bar\alpha_m \leqslant \omega\varrho_m$ be such that $j(\bar\alpha_m) = \alpha_m$, and let $\bar p_m \subseteq J_{\bar\varrho_m}$ be such that
$\bar p_m$ $j(\bar p_m) = \sigma^{-1}_{i_m}(p_m)$.

Now, by definition,

$$\sigma_{i_m}\colon J_{\varrho(i_m)} \cong H^n_\delta(\alpha_m \cup p_m) \prec_n J_\delta .$$

So, using 5.8,

$$J_{\varrho(i_m)} = H^n_{\varrho(i_m)}(\alpha_m \cup \sigma^{-1}_{i_m}(p_m)) .$$

But $j\colon J_{\gamma(\theta)} \prec J_\gamma$, so applying j^{-1} we have

$$J_{\bar\varrho_m} = H^n_{\bar\varrho_m}(\bar\alpha_m \cup \bar p_m) .$$

Let $m \leqslant s$. Now, if $x \in \bar\alpha_m$, then $j(x) \in j(\bar\alpha_m) = \alpha_m$. So as $\sigma_{i_m, i_s} \restriction \alpha_m = \mathrm{id} \restriction \alpha_m$, we have $\sigma_{i_m, i_s}(j(x)) = j(x)$, and hence $j^{-1}(\sigma_{i_m, i_s}(j(x))) = x$ is defined. Again, suppose $x \in \bar p_m$. Then $j(x) \in j(\bar p_m) = \sigma^{-1}_{i_m}(p_m)$. So $\sigma_{i_m, i_s}(j(x)) \in \sigma^{-1}_{i_s}(p_m)$. But $\sigma^{-1}_{i_s}(p_s) = j(\bar p_s) \in \operatorname{ran}(j) \prec J_\gamma$. So as p_m is a finite subset of the finite set p_s, $\sigma^{-1}_{i_s}(p_m) \in \operatorname{ran}(j)$. Thus $j^{-1}(\sigma_{i_m, i_s}(j(x))) = j^{-1}(\sigma^{-1}_{i_s}(p_m))$ is defined.

Thus we can define an embedding

$\bar\sigma_{ms}$
$$\bar\sigma_{ms}\colon J_{\bar\varrho_m} \prec_n J_{\bar\varrho_s}$$

by setting

$$\bar\sigma_{ms}(h^\varphi_{\bar\varrho_m}(x_1, \ldots, x_k)) = h^\varphi_{\bar\varrho_s}(j^{-1}(\sigma_{i_m, i_s}(j(x_1))), \ldots, j^{-1}(\sigma_{i_m, i_s}(j(x_k)))) ,$$

for any Σ_n formula $\varphi(v_1, \ldots, v_k)$ and any $x_1, \ldots, x_k \in \bar{\alpha}_m \cup \bar{p}_m$. (By 5.8, this does define a Σ_n-elementary embedding.) Then

$$\langle (J_{\bar{\varrho}_m})_{m<\omega}, (\bar{\sigma}_{ms})_{m \leqslant s} \rangle$$

is a directed Σ_n-elementary system, and moreover the following infinite diagram commutes:

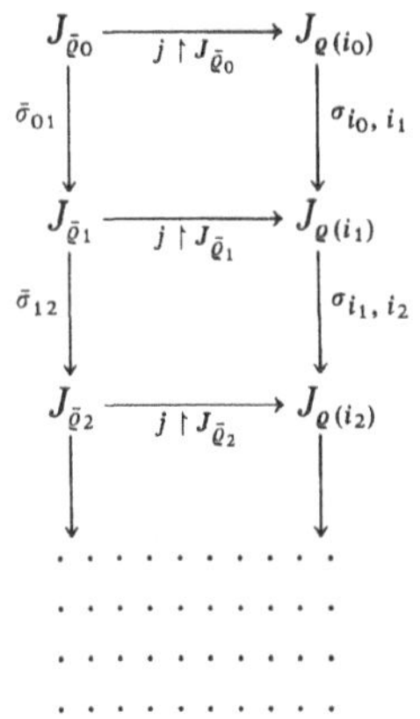

Let $\langle \langle U, E \rangle, (\bar{\sigma}_m)_{m<\omega} \rangle$ be the direct limit of the system $\bar{\sigma}_m$

$$\langle (J_{\bar{\varrho}_m})_{m<\omega}, (\bar{\sigma}_{ms})_{m \leqslant s} \rangle .$$

We have $\bar{\sigma}_m \colon J_{\bar{\varrho}_m} \prec_n \langle U, E \rangle$, $\sigma_{i_m} \colon J_{\varrho(i_m)} \prec_n J_\delta$, and $\sigma_{i_m, i_s} = \sigma_{i_s}^{-1} \circ \sigma_{i_m}$ for $m \leqslant s$. So we can define an embedding

$$e \colon \langle U, E \rangle \prec_n \langle J_\delta, \in \rangle$$

as follows. Let $u \in U$. Pick $m < \omega$ so that $u = \bar{\sigma}_m(x)$ for some $x \in J_{\bar{\varrho}_m}$. Set $e(u) = \sigma_{i_m}(j(x))$. (It is routine to check that e is well-defined and Σ_n-elementary.) In particular, $\langle U, E \rangle$ is well-founded, and we may assume that $\langle U, E \rangle = \langle J_{\bar{\delta}}, \in \rangle$ for some $\bar{\delta}$. $\bar{\delta}$

So, starting with a system

$$\langle (J_{\pi(\varrho(i_m))})_{m<\omega}, (\pi(\sigma_{i_m, i_s}))_{m \leqslant s} \rangle$$

which has a non-well-founded limit (witnessed by the elements b_m), we picked a $\theta < \omega_1$ sufficiently large for us to be able to use $j = \pi^{-1} \circ \pi_\theta$ (so $j \colon J_{\gamma(\theta)} \prec J_\gamma$) in order to "pull-back" from J_γ the system

$$\langle (J_{\varrho(i_m)})_{m<\omega}, (\sigma_{i_m, i_s})_{m \leqslant s} \rangle$$

to a system

$$\langle (J_{\bar{\varrho}_m})_{m<\omega}, (\bar{\sigma}_{ms})_{m \leqslant s} \rangle$$

with limit

$$\langle J_{\bar\delta}, (\bar\sigma_m)_{m<\omega}\rangle .$$

$\bar\eta$ We shall show that for $\bar\eta = \sup_{m<\omega}\bar\alpha_m$, the direct limit of $\pi_\theta^* S_{\bar\delta}^n(\bar\eta)$ is not well-founded. Indeed, we shall show that $\pi_\theta(\bar\sigma_{ms}) = \pi(\sigma_{i_m, i_s})$ for $m \leqslant s$, so that the same elements b_m witness this non-well-foundedness, just as they did for the original system.

Let $\bar\eta = \sup_{m<\omega}\bar\alpha_m$. Since $(\alpha_m \mid m<\omega)$ is strictly increasing and $j(\bar\alpha_m) = \alpha_m$, $(\bar\alpha_m \mid m<\omega)$ is strictly increasing. Hence $\bar\alpha_m < \bar\eta$ for all $m<\omega$. Since $\bar\sigma_m\colon J_{\bar\varrho_m} \prec_n J_{\bar\delta}$, we have $\bar\alpha_m \leqslant \omega . \bar\varrho_m \leqslant \omega . \bar\delta$ for all $m<\omega$. Hence $\bar\eta \leqslant \omega\bar\delta$. So we may consider the directed Σ_n-elementary system $S_{\bar\delta}^n(\bar\eta)$.

$q_m, \bar i_m$ Set $q_m = \bar\sigma_m(\bar p_m)$, $\bar i_m = (\bar\alpha_m, \bar q_m)$. Then $\bar i_m \in I_{\bar\delta}^n(\bar\eta)$ and $m \leqslant s$ implies $\bar i_m \leqslant \bar i_s$. And for the system $S_{\bar\delta}^n(\bar\eta)$ we have $\bar\varrho_m = \varrho(\bar i_m)$, $\bar\sigma_m = \sigma_{\bar i_m}$, $\bar\sigma_{ms} = \sigma_{\bar i_m, \bar i_s}$. (This is not a fact that requires any *proof*. We have simply started with a system $\langle (J_{\bar\varrho_m})_{m<\omega}, (\bar\sigma_{ms})_{m\leqslant s}\rangle$ and then defined $\bar\delta$, $\bar\eta$, $\bar i_m$ so that the above equalities are true by definition.)

5.19 Lemma. *In the system $S_{\bar\delta}^n(\bar\eta)$, for each $i \in I_{\bar\delta}^n(\bar\eta)$ there is an $m<\omega$ such that*

$$\sigma_i{}''J_{\varrho(i)} \subseteq \sigma_{\bar i_m}{}''J_{\varrho(\bar i_m)} .$$

Proof. Let $i = (\alpha, p) \in I_{\bar\delta}^n(\bar\eta)$. Since $p \subseteq J_{\bar\delta}$ is finite and $J_{\bar\delta}$ is the direct limit of the system

$$\langle (J_{\bar\varrho_m})_{m<\omega}, (\bar\sigma_{ms})_{m\leqslant s}\rangle,$$

there is an $m<\omega$ such that $p \subseteq \bar\sigma_m{}''J_{\bar\varrho_m}$. Moreover, since $\alpha < \bar\eta$ we can choose m here so that $\bar\alpha_m \geqslant \alpha$. But $\bar\sigma_m = \sigma_{\bar i_m}$ and $\sigma_{\bar i_m}\colon J_{\varrho(\bar i_m)} \cong H_{\bar\delta}^n(\bar\alpha_m \cup \bar q_m)$, so $\bar\sigma_m \restriction \bar\alpha_m = \mathrm{id} \restriction \bar\alpha_m$. Thus

$$\alpha \cup p \subseteq \bar\sigma_m{}''J_{\bar\varrho_m} \prec_n J_{\bar\delta} .$$

It follows that $H_{\bar\delta}^n(\alpha \cup p) \subseteq \bar\sigma_m{}''J_{\bar\varrho_m}$. But in $S_{\bar\delta}^n(\bar\eta)$, by definition,

$$\sigma_i\colon J_{\varrho(i)} \cong H_{\bar\delta}^n(\alpha \cup p) .$$

Thus $\sigma_i{}''J_{\varrho(i)} \subseteq \bar\sigma_m{}''J_{\bar\varrho_m}$. Since $\bar\sigma_m = \sigma_{\bar i_m}$, $\varrho_m = \varrho(\bar i_m)$ we are done. □

It follows from 5.19 that $S_{\bar\delta}^n(\bar\eta)$ is below $\gamma(\theta)$. To see this, let $i \in I_{\bar\delta}^n(\bar\eta)$ be given. Pick $m<\omega$ so that $\sigma_i{}''J_{\varrho(i)} \subseteq \sigma_{\bar i_m}{}''J_{\varrho(\bar i_m)}$. Since $\sigma_i, \sigma_{\bar i_m}$ are one-one and $\in$-preserving, it follows that $\varrho(i) \leqslant \varrho(\bar i_m) = \bar\varrho_m < \gamma(\theta)$, as required.

Since $S_{\bar\delta}^n(\bar\eta)$ is below $\gamma(\theta)$, $\pi_\theta^* S_{\bar\delta}^n(\bar\eta)$ is defined. Now,

$$\sigma_{i_m}\colon J_{\varrho(i_m)} \cong H_\delta^n(\alpha_m \cup p_m),$$
$$\sigma_{i_s}\colon J_{\varrho(i_s)} \cong H_\delta^n(\alpha_s \cup p_s).$$

Thus

$$(*) \qquad \sigma_{i_m, i_s} = \sigma_{i_s}^{-1} \circ \sigma_{i_m}\colon J_{\varrho(i_m)} \cong H_{\varrho(i_s)}^n(\alpha_m \cup \sigma_{i_s}^{-1}(p_m)) .$$

Now, by choice of θ, $\varrho(i_m)$, $\varrho(i_s)$, $\alpha_m \in \operatorname{ran}(j)$. Moreover, the choice of θ ensures that $\sigma_{i_s}^{-1}(p_s) \in \operatorname{ran}(j)$, so as p_m is a finite subset of the finite set p_s, we have $\sigma_{i_s}^{-1}(p_m) \in \operatorname{ran}(j)$. Thus as $j: J_{\gamma(\theta)} \prec J_\gamma$, we have $\sigma_{i_m, i_s} \in \operatorname{ran}(j)$. But $j(\bar{\varrho}_m) = \varrho(i_m)$, $j(\bar{\varrho}_s) = \varrho(i_s)$, $j(\bar{\alpha}_m) = \alpha_m$. Thus from (*), applying j^{-1}, we get

$$(**) \qquad j^{-1}(\sigma_{i_m, i_s}): J_{\bar{\varrho}_m} \cong H^n_{\bar{\varrho}_s}(\bar{\alpha}_m \cup j^{-1} \circ \sigma_{i_s}^{-1}(p_m)).$$

Now,

$$\begin{aligned}
j^{-1} \circ \sigma_{i_s}^{-1}(p_m) &= j^{-1} \circ \sigma_{i_s}^{-1} \circ \sigma_{i_m} \circ j(\bar{p}_m) && \text{(by choice of } \bar{p}_m) \\
&= j^{-1} \circ \sigma_{i_m, i_s} \circ j(\bar{p}_m) && \text{(by definition of } \sigma_{i_m, i_s}) \\
&= \bar{\sigma}_{ms}(\bar{p}_m) && \text{(by commutativity of the diagram above)} \\
&= \bar{\sigma}_s^{-1} \circ \bar{\sigma}_m(\bar{p}_m) && \text{(by definition of } \bar{\sigma}_m, \bar{\sigma}_s) \\
&= \bar{\sigma}_s^{-1}(\bar{q}_m) && \text{(by definition of } \bar{q}_m).
\end{aligned}$$

Thus by (**),

$$j^{-1}(\sigma_{i_m, i_s}): J_{\bar{\varrho}_m} \cong H^n_{\bar{\varrho}_s}(\bar{\alpha}_m \cup \bar{\sigma}_s^{-1}(\bar{q}_m)).$$

In other words, since $\bar{\varrho}_m = \varrho(\bar{i}_m)$, $\bar{\varrho}_s = \varrho(\bar{i}_s)$, $\bar{\sigma}_s = \sigma_{\bar{i}_s}$,

$$j^{-1}(\sigma_{i_m, i_s}): J_{\varrho(\bar{i}_m)} \cong H^n_{\varrho(\bar{i}_s)}(\bar{\alpha}_m \cup \sigma_{\bar{i}_s}^{-1}(\bar{q}_m)).$$

But $\bar{i}_m = (\bar{\alpha}_m, \bar{q}_m)$, $\bar{i}_s = (\bar{\alpha}_s, \bar{q}_s)$. Thus, in the same way that we deduced (*), we may obtain

$$\sigma_{\bar{i}_m, \bar{i}_s}: J_{\varrho(\bar{i}_m)} \cong H^n_{\varrho(\bar{i}_s)}(\bar{\alpha}_m \cup \sigma_{\bar{i}_s}^{-1}(\bar{q}_m)).$$

Hence

$$j^{-1}(\sigma_{i_m, i_s}) = \sigma_{\bar{i}_m, \bar{i}_s},$$

i.e.

$$j(\sigma_{\bar{i}_m, \bar{i}_s}) = \sigma_{i_m, i_s}.$$

Therefore, applying π,

$$\pi \circ j(\sigma_{\bar{i}_m, \bar{i}_s}) = \pi(\sigma_{i_m, i_s}).$$

But $j = \pi^{-1} \circ \pi_\theta$. So,

$$\pi_\theta(\sigma_{\bar{i}_m, \bar{i}_s}) = \pi(\sigma_{i_m, i_s}).$$

Thus by choice of the elements b_m,

$$b_{m+1} \in [\pi_\theta(\sigma_{\bar{i}_m, \bar{i}_{m+1}})](b_m)$$

for all $m < \omega$. Hence the direct limit of the system $\pi_\theta^* S_\delta^n(\bar{\eta})$ is not well-founded.

We have now arrived at the following situation. We started with a $\delta \geqslant \gamma$ and a $0 < n < \omega$, such that $S^n_\delta(\gamma)$ is below γ and the limit of $\pi^* S^n_\delta(\gamma)$ is not well-founded. By choosing a suitable embedding $j\colon J_{\gamma(\theta)} \prec J_\gamma$, we were able to "pull back" $S^n_\delta(\gamma)$ (or at least a subsystem of this large enough to give a non-well-founded π-image) to a system $S^n_{\bar\delta}(\bar\eta)$ which is below $\gamma(\theta)$, such that the direct limit of $\pi^*_\theta S^n_{\bar\delta}(\bar\eta)$ is not well-founded. (Remember also that $\pi = \pi_{\omega_1}$.)

Consider now the definition of $M_{\theta+1}$. When the pair $(n, \bar\eta)$ was considered, $\bar\delta$ was, by the above, a candidate in the choice of what we then called δ_0. Hence as δ_0 was chosen minimally, $\delta_0 \leqslant \bar\delta$. Let $(a_k \mid k < \omega)$, $(j_k \mid k < \omega)$ be the sequences
a_k, j_k chosen for δ_0, n as described: that is, $j_k \in I^n_{\delta_0}(\bar\eta)$, $j_k \leqslant j_{k+1}$, $a_k \in J_{\pi_\theta(\varrho(j_k))}$, a_{k+1}
β_k, q_k $\in [\pi_\theta(\sigma_{j_k, j_{k+1}})](a_k)$. Let $j_k = (\beta_k, q_k)$.

It is easy to construct an increasing sequence $(m_k \mid k < \omega)$ of integers such that $\beta_k \leqslant \bar\alpha_{m_k}$, $q_k \subseteq \sigma_{\bar i_{m_k}}{}''J_{\varrho(\bar i_{m_k})}$, and in case $\delta_0 < \bar\delta$, such that $J_{\delta_0} \in \sigma_{\bar i_{m_k}}{}''J_{\varrho(\bar i_{m_k})}$, where, as before, these relate to the system $S^n_{\bar\delta}(\bar\eta)$. (To get $\beta_k \leqslant \bar\alpha_{m_k}$ we use the fact that $\beta_k < \bar\eta = \sup_{i<\omega} \bar\alpha_i$. To get $q_k \subseteq \sigma_{\bar i_{m_k}}{}''J_{\varrho(\bar i_{m_k})}$ we use the facts that $\delta_0 \leqslant \bar\delta$ and $J_{\bar\delta}$ is the direct limit of $S^n_{\bar\delta}(\bar\eta)$, together with 5.19. Likewise to obtain $J_{\delta_0} \in \sigma_{\bar i_{m_k}}{}''J_{\varrho(\bar i_{m_k})}$ in case $\delta_0 < \bar\delta$.)

For each $k < \omega$,

$$\beta_k \cup q_k \subseteq [\sigma_{\bar i_{m_k}}{}''J_{\varrho(\bar i_{m_k})}] \cap J_{\delta_0} \prec_n J_{\delta_0}.$$

(For if $\delta_0 = \bar\delta$, this just says that

$$\sigma_{\bar i_{m_k}}{}''J_{\varrho(\bar i_{m_k})} \prec_n J_{\bar\delta},$$

which we know already. Whilst if $\delta_0 < \bar\delta$, then from the fact that

$$\sigma_{\bar i_{m_k}}{}''J_{\varrho(\bar i_{m_k})} \prec_n J_{\bar\delta}$$

and

$$J_{\delta_0} \in \sigma_{\bar i_{m_k}}{}''J_{\varrho(\bar i_{m_k})},$$

we deduce easily that

$$[\sigma_{\bar i_{m_k}}{}''J_{\varrho(\bar i_{m_k})}] \cap J_{\delta_0} \prec J_{\delta_0},$$

i.e. full elementarity.) It follows that, if σ_{j_k}, $\varrho(j_k)$ refer to the system $S^n_{\delta_0}(\bar\eta)$,

$$\sigma_{j_k}{}''J_{\varrho(j_k)} = H^n_{\delta_0}(\beta_k \cup q_k) \subseteq [\sigma_{\bar i_{m_k}}{}''J_{\varrho(\bar i_{m_k})}] \cap J_{\delta_0}.$$

Thus we can define embeddings

$$e_k\colon J_{\varrho(i_k)} \prec_0 J_{\varrho(\bar i_{m_k})}$$

by

e_k $$e_k\colon \sigma^{-1}_{\bar i_{m_k}} \circ \sigma_{j_k}.$$

Now,

$$\sigma_{j_k}: J_{\varrho(j_k)} \cong H^n_{\delta_0}(b_k \cup q_k),$$

and

$$\sigma_{\bar{\imath}_{m_k}}: J_{\bar{\varrho}_{m_k}} \cong H^n_{\bar{\delta}}(\bar{\alpha}_{m_k} \cup \bar{q}_{m_k}),$$

so

$$e_k: J_{\varrho(j_k)} \cong H^n_{\psi}(\beta_k \cup \sigma^{-1}_{\bar{\imath}_{m_k}}(q_k)),$$

where

$$\psi = \begin{cases} \sigma^{-1}_{\bar{\imath}_{m_k}}, & \text{if } \delta_0 < \bar{\delta} \\ \bar{\varrho}_{m_k}, & \text{if } \delta_0 = \bar{\delta}. \end{cases}$$

Thus $e_k \in J_{\gamma(\theta)}$, and so $\pi_\theta(e_k)$ is defined. Moreover, the following diagram clearly commutes:

$$\begin{array}{ccc}
J_{\varrho(j_0)} & \xrightarrow{\quad e_0 \quad} & J_{\varrho(\bar{\imath}_{m_0})} \\
\sigma_{j_0,j_1}\Big\downarrow & & \Big\downarrow\sigma_{\bar{\imath}_{m_0},\bar{\imath}_{m_1}} \\
J_{\varrho(j_1)} & \xrightarrow{\quad e_1 \quad} & J_{\varrho(\bar{\imath}_{m_1})} \\
\sigma_{j_1,j_2}\Big\downarrow & & \Big\downarrow\sigma_{\bar{\imath}_{m_1},\bar{\imath}_{m_2}} \\
J_{\varrho(j_2)} & \xrightarrow{\quad e_2 \quad} & J_{\varrho(\bar{\imath}_{m_2})} \\
\Big\downarrow & & \Big\downarrow \\
\cdots & \cdots & \cdots \\
\cdots & \cdots & \cdots \\
\cdots & \cdots & \cdots
\end{array}$$

Applying π_θ we obtain the commutative diagram:

$$\begin{array}{ccc}
J_{\pi_\theta(\varrho(j_0))} & \xrightarrow{\quad \pi_\theta(e_0) \quad} & J_{\pi_\theta(\varrho(\bar{\imath}_{m_0}))} \\
\pi_\theta(\sigma_{j_0,j_1})\Big\downarrow & & \Big\downarrow\pi_\theta(\sigma_{\bar{\imath}_{m_0},\bar{\imath}_{m_1}}) \\
J_{\pi_\theta(\varrho(j_1))} & \xrightarrow{\quad \pi_\theta(e_1) \quad} & J_{\pi_\theta(\varrho(\bar{\imath}_{m_1}))} \\
\pi_\theta(\sigma_{j_1,j_2})\Big\downarrow & & \Big\downarrow\pi_\theta(\sigma_{\bar{\imath}_{m_1},\bar{\imath}_{m_2}}) \\
J_{\pi_\theta(\varrho(j_2))} & \xrightarrow{\quad \pi_\theta(e_2) \quad} & J_{\pi_\theta(\varrho(\bar{\imath}_{m_2}))} \\
\Big\downarrow & & \Big\downarrow \\
\cdots & \cdots & \cdots \\
\cdots & \cdots & \cdots \\
\cdots & \cdots & \cdots
\end{array}$$

c_k Let $c_k = [\pi_\theta(e_k)](a_k)$. We know that $a_k \in M_{\omega_1}$ and that $\pi_\theta(e_k) \in M_{\omega_1}$. So, as
$\bar{c}_k$ $M_{\omega_1} \prec J_\tau$, we have $c_k \in M_{\omega_1}$. Let $\bar{c}_k = \pi^{-1}(c_k)$.

By its definition, $c_k \in J_{\pi_\theta(\varrho(\bar{\imath}_{m_k}))}$. Also, $a_{k+1} \in [\pi_\theta(\sigma_{j_k, j_{k+1}})](a_k)$. So, referring to the above diagram, we have (by commutativity)

$$c_{k+1} = [\pi_\theta(e_k)](a_{k+1}) \in [\pi_\theta(e_k) \circ \pi_\theta(\sigma_{j_k, j_{k+1}})](a_k)$$
$$= [\pi_\theta(\sigma_{\bar{\imath}_{m_k}, \bar{\imath}_{m_{k+1}}}) \circ \pi_\theta(e_k)](a_k) = [\pi_\theta(\sigma_{\bar{\imath}_{m_k}, \bar{\imath}_{m_{k+1}}})](c_k).$$

But $\pi_\theta(\varrho(\bar{\imath}_m)) = \pi(\varrho(i_m))$ and $\pi_\theta(\sigma_{\bar{\imath}_m, \bar{\imath}_s}) = \pi(\sigma_{i_m, i_s})$. (The first of these equalities is easily seen, the second was proved earlier.) So, applying π^{-1} to the above results, we get

$$\bar{c}_k \in J_{\varrho(i_k)} \quad \text{and} \quad \bar{c}_{k+1} \in \sigma_{i_{m_k}, i_{m_{k+1}}}(\bar{c}_k).$$

But then

$$\sigma_{i_{m_{k+1}}}(\bar{c}_{k+1}) \in \sigma_{i_{m_k}}(\bar{c}_k)$$

for all $k < \omega$, which is absurd. That completes the proof in the case $0 < n < \omega$.

The case $n = \omega$ is handled in an entirely similar fashion. The only difference is that we must ensure that the sequences $(b_m \mid m < \omega)$, $(i_m \mid m < \omega)$ are chosen so that, if $i_m = (k_m, \alpha_m, p_m)$, then $k_m < k_{m+1}$. We may then proceed as for $0 < n < \omega$. (We leave it to the reader to check all the details. Note that we dealt with the proof of 5.10 in this fashion, giving full details for the case $0 < n < \omega$ and simply indicating the modifications required for the case $n = \omega$. With this as a model, there should be no difficulty for the reader in handling the case $n = \omega$ here as well.)

The case $n = 0$ is also similar. We start with sequences $(b_m \mid m < \omega)$, $(i_m \mid m < \omega)$ chosen so that $\mu_m < \mu_{m+1}$, where $i_m = (\mu_m, \alpha_m, p_m)$, so that, in particular, $J_{\mu_m} \in p_{m+1}$ for all m. It is then easy to modify the proof for the case $0 < n < \omega$ to work in this case. At various points we need to rely upon Σ_0-absoluteness between the structures J_δ, J_η involved. Again, the proof of 5.10 indicates the type of modification required, so once again we leave it to the reader to supply the missing details.

That completes our proof of 5.16, and with it the Covering Lemma.

Exercises

1. *The Tree Property* (Section 1)

An uncountable regular cardinal κ is said to have the *tree property* iff there is no κ-Aronszajn tree. By Theorem 1.3 (viii), if κ is weakly compact then κ has the tree property. It follows from Theorems IV.2.4 and VII.1.3 that if $V = L$, the tree property is equivalent to weak compactness. On the other hand, Silver has proved (see *Mitchell* (1972)) that if ZFC + "there is a weakly compact cardinal" is consistent, so too is ZFC + "ω_2 has the tree property". The results below show that the assumption concerning weak compactness here is essential. It is shown that if κ has the tree property, then κ is a weakly compact cardinal in the sense of L.

1 A. Show that if κ has the tree property and [κ is inaccessible]^L, then [κ is weakly compact]^L.

(Outline: Use Theorem 1.3 (vii). Let $\mathscr{F} \in L$ be, in the sense of L, a κ-complete field of subsets of $\mathscr{F}$ of cardinality κ. Pick $\lambda < (\kappa^+)^L$ admissible such that $|\mathscr{F}|^{L_\lambda} = \kappa$. Let $c \in L$, $c\colon \kappa \leftrightarrow \mathscr{P}(\kappa) \cap L_\lambda$. For each $\alpha < \kappa$, let

$$T_\alpha = \{f \in {}^\alpha 2 \cap L \mid |\bigcap \{c(v) \mid f(v) = 1\} \cap \bigcap \{\kappa - c(v) \mid f(v) = 0\}| = \kappa\},$$

and set

$$T = \bigcup_{\alpha < \kappa} T_\alpha.$$

Show that, under inclusion, T is a tree of height κ and width κ. By the tree property, let $f\colon \kappa \to 2$ define a κ-branch of T, and set

$$D = \{x \in \mathscr{P}(\kappa) \cap L \mid f(c^{-1}(x)) = 1\}.$$

Show that D is a κ-complete non-principal ultrafilter on $\mathscr{P}(\kappa) \cap L_\lambda$. Let

$$M = \{f \in L_\lambda \mid f\colon \kappa \to L_\lambda\},$$

and form the ultrapower M/D. Let $i\colon L \prec M/D$ be the canonical embedding. M/D is well-founded, so let $J\colon M/D \cong L_\gamma$ be the collapsing isomorphism. Let $g \in L_\lambda$, $g\colon \kappa \leftrightarrow \mathscr{F}$. Then $j \circ i(g) \in L$ and $j \circ i(g) \restriction \kappa = (j \circ i(g(v)) \mid v < \kappa)$, so $U = \{g(v) \mid \kappa \in j \circ i(g(v))\} \in L$. Since U is, in the sense of L, a κ-complete ultrafilter on $\mathscr{F}$, the proof is complete.)

1 B. Show that if κ has the tree property, then [κ is weakly compact]^L.

(Hint: By 1 A it suffices to show that [κ is inaccessible]^L. Suppose not. Then for some $\mu < \kappa$, $\kappa = (\mu^+)^L$. By Exercise IV.1, in L let T be a special μ^+-Aronszajn tree. In V, T is a κ-tree. Since $T_\alpha \subseteq \{f \mid f\colon \alpha \xrightarrow{1-1} \mu\}$ for all $\alpha < \kappa$ and the ordering on T is inclusion, T is κ-Aronszajn. Contradiction.)

The following exercise provides an alternative solution to 1 B.

1 C. Let κ be a regular cardinal in L, not weakly compact in L. Show that there is a tree T on κ in L such that if, in the real world, there is a κ-branch through T, then $\mathrm{cf}(\kappa) = \omega$.

(Hint: Let T_0 be, in L, a κ-Aronszajn tree. We may assume that T_0 is an initial part of $2^{<\kappa}$. Define T by putting a triple (α, M, b) into T iff $\alpha < \kappa$, $M = L_\beta$ for some limit ordinal β, $b \in M$, $\alpha \subseteq M$, M is the smallest $M \prec M$ such that $\alpha \cup \{b\} \subseteq M$, b is a function with domain containing α as a subset, and $b \restriction \alpha \in T_0$. We have $(\alpha, M, b) <_T (\alpha', M', b')$ iff $\alpha \leqslant \alpha'$, M is the transitive collapse of the skolem hull of $\alpha \cup \{b'\}$ in M', and b' collapses to b. Show that T is a tree, $T \in L$, and that (α, M, b) has height α in T. Show further that if $((\alpha, M_\alpha, b_\alpha) \mid \alpha < \kappa)$ is a branch through T, and $\langle M, E, b\rangle$ is the limit of the elementary system $\langle M_\alpha, \in, b_\alpha\rangle$, $\alpha < \kappa$, then $\langle M, E\rangle$ is a model of BS + $V = L$ + "b is a function", $\kappa \subseteq M$, and for each $\alpha < \kappa$, $b \restriction \alpha \in T_0$. Thus $b \restriction \kappa$ is a branch through T_0. Thus $b \notin L$, which implies that $\langle M, E\rangle$ cannot be well-founded. This implies that $\mathrm{cf}(\kappa) = \omega$.)

1 D. The following result extends Theorem 5.7. Assume $0^{\#}$ does not exist. Then for every strong limit cardinal κ, there is a Souslin κ^{+}-tree.

(Hint: Use Exercise IV.8 to strengthen IV.2.11 appropriately, and combine this with 5.2 and 5.6.)

2. *The Sharp Operation* (Section 2)

Show that for any set $a \subseteq \omega$ there is a set $a^{\#} \subseteq \omega$ which has the same effect upon $L[a]$ as does $0^{\#}$ upon L. (i.e. Show that the development of section 2 goes through for $L[a]$ whenever $a \subseteq \omega$.) Is it the case that if $a, b \subseteq \omega$ are such that $a \in L[b]$, then $a^{\#} \in L[b^{\#}]$? Investigate the relationship between the various sets $a^{\#}$, $a \subseteq \omega$.

3. *On the Existence of* $0^{\#}$ (Section 2)

Show that $0^{\#}$ exists iff for some (all) uncountable regular cardinal κ, every constructible set $X \subseteq \kappa$ either contains or is disjoint from a club subset of κ.

(Hint: If $0^{\#}$ exists, show that if $X \subseteq \kappa$, $X \in L$, then either X or else $\kappa - X$ contains $H_\kappa - \gamma$ for some $\gamma < \kappa$. For the converse, let

$$D = \{X \in \mathscr{P}(\kappa) \cap L \mid X \text{ contains a club}\},$$

show that D is an ultrafilter on $\mathscr{P}(\kappa) \cap L$ which is κ-complete for families in L, and use D to construct an ultrapower which allows the use of Theorem 4.3.).

4. *The Covering Lemma and Cardinal Arithmetic* (Section 5)

By the *Singular Cardinals Hypothesis* (SCH) we mean the assertion that for all singular cardinals κ,

$$2^{\mathrm{cf}(\kappa)} < \kappa \quad \text{implies} \quad \kappa^{\mathrm{cf}(\kappa)} = \kappa^{+}.$$

Clearly, GCH implies SCH. As is shown in the following exercises, SCH completely determines the cardinal exponentiation of singular cardinals.

4 A. Show that (in ZFC) if κ is a singular cardinal, then

$$2^{\kappa} = (2^{<\kappa})^{\mathrm{cf}(\kappa)}.$$

4 B. Show that SCH implies that for any singular cardinal κ,

$$2^{\kappa} = \begin{cases} 2^{<\kappa}, & \text{if } (\exists \lambda < \kappa)(2^{<\kappa} = 2^{\lambda}) \\ (2^{<\kappa})^{+}, & \text{otherwise.} \end{cases}$$

4 C. Show that SCH implies that for any cardinals κ, λ, singular or regular,

$$\kappa^{\lambda} = \begin{cases} 2^{\lambda}, & \text{if } 2^{\lambda} \geqslant \kappa \\ \kappa, & \text{if } \lambda < \mathrm{cf}(\kappa) \text{ and } 2^{\lambda} < \kappa \\ \kappa^{+}, & \text{otherwise.} \end{cases}$$

4D. Use the Covering Lemma to show that if $0^{\#}$ does not exist, then SCH is valid.

5. *An Application of the Covering Lemma*

Prove that if $0^{\#}$ does not exist, and $\kappa \geqslant \omega_2$ is any cardinal such that $2^{<\kappa} = \kappa$, then there is a set $A \subseteq \kappa$ such that $X \in L[A]$ for every set $X \subseteq \mathrm{On}$ such that $|X| < \kappa$.

Part B

Advanced Theory

Chapter VI

The Fine Structure Theory

The basic ideas of the fine structure theory have already been outlined in IV.4. In this chapter we develop rigorously the material sketched there. We commence with a certain class of set functions – the rudimentary functions – and then, with the aid of these functions we shall define a new hierarchy of constructible sets, namely the Jensen hierarchy, $(J_\alpha \mid \alpha \in \mathrm{On})$. This hierarchy has all of the important properties of the usual L_α-hierarchy, with the difference that each level in the Jensen hierarchy has many of the properties of the limit levels of the L_α-hierarchy (notably amenability). The Jensen hierarchy is thus a more convenient hierarchy as far as a detailed examination of individual levels is concerned. Certainly it is possible to carry out a comparable study of the sets L_α, but only at the cost of some considerable (though in a sense "trivial") technical difficulties. Intuitively, we may regard J_α as a slightly expanded version of L_α which is closed under simple set functions such as ordered pairs, etc. This is not totally accurate (as we shall see), but it should serve the reader well enough until a more complete understanding is achieved.

1. Rudimentary Functions

The definition of rudimentary functions has already been given in Chapter IV, but is repeated here for convenience.

A function $f: V^n \to V$ is said to be *rudimentary* (*rud* for short) iff it is generated by the following schemas:

(i) $f(x_1, \ldots, x_n) = x_i \qquad (1 \leqslant i \leqslant n)$;

(ii) $f(x_1, \ldots, x_n) = \{x_i, x_j\} \qquad (1 \leqslant i, j \leqslant n)$;

(iii) $f(x_1, \ldots, x_n) = x_i - x_j \qquad (1 \leqslant i, j \leqslant n)$;

(iv) $f(x_1, \ldots, x_n) = h(g_1(x_1, \ldots, x_n), \ldots, g_k(x_1, \ldots, x_n))$, where $h, g_1, \ldots, g_k$ are rudimentary;

(v) $f(y, x_2, \ldots, x_n) = \bigcup_{z \in y} g(z, x_2, \ldots, x_n)$, where g is rudimentary.

Notice that in the above definition we have made use of proper classes, which is not strictly allowable in ZF set theory. There are two ways of avoiding this, both of relevance to our later development. Firstly, since any "rudimentary function" from V^n to V will be built up from functions of types (i)–(iii) in the above list by means of finitely many applications of the composition rules (iv) and (v), we could replace any mention of the "function" by the LST formula which is implicit in the construction of the function via these schemas. In other words, we are just making use of our usual (and we hope familiar by now) conventions concerning proper classes in ZF set theory (Chapter I). An alternative approach is to regard the "V" in the above definition as being some set (e.g. a V_α) which is large enough to contain all of the sets which we are interested in at any one time, in which case the rudimentary functions defined are genuine functions (i.e. they are *sets*). Since the untimate goal in set theory is to study the properties of *sets*, this second approach is clearly adequate. Nevertheless we choose to take the "class function" approach as basic for one important reason: it emphasises the *uniformity* of the rudimentary functions; how their construction is quite independent of any particular set domain under consideration.

A similar situation has already arisen in Chapter II. When we studied the L_α-hierarchy, we proved "global" results concerning the logical complexity of the LST formulas which define the L_α-hierarchy, as well as "local" results concerning the definability (using the language $\mathscr{L}_V$) of the hierarchy within given levels of the hierarchy (I.2.6 and I.2.7 provide good examples of this parallel development). Here, the rudimentary functions are used (instead of the language $\mathscr{L}_V$) to define the Jensen hierarchy of constructible sets: global results will be proved using *class* rudimentary "functions" (which correspond to LST formulas), and local results will be proved using *set* rudimentary functions (which are genuine sets, as are the formulas of $\mathscr{L}_V$).

From now on, except for occasional remarks, we leave it to the reader to supply the relevant "rigorisation" of our development in the appropriate fashion.

To continue with our definition then, if A is a class we say that a function $f: V^n \to V$ is *rudimentary relative to A* (*A-rud* for short) iff it is generated by schemas (i)–(v) above and the schema:

(vi) $\quad f(x_1, \ldots, x_n) = A \cap x_i \quad (1 \leqslant i \leqslant n).$

If p is a set, we say that a function $f: V^n \to V$ is *rudimentary in parameter p* (or simply *rudimentary in p*) iff it is generated by schemas (i)–(v) and the schema:

(vii) $\quad f(x_1, \ldots, x_n) = p.$

By a *rudimentary definition* of a rudimentary function f we mean a sequence $f_0, \ldots, f_n$ of functions such that $f_n = f$ and for each $i \leqslant n$, f_i is obtained from $f_0, \ldots, f_{i-1}$ by means of a single application of one of the schemas (i)–(v) above. Similarly for an *A-rud definition* of an A-rud function and a *rud in p definition* of a function rud in p.

A class $R \subseteq V^n$ is said to be *rudimentary* iff there is a rudimentary function $f: V^n \to V$ such that

$$R = \{(x_1, \ldots, x_n) \mid f(x_1, \ldots, x_n) \neq \emptyset\}.$$

Similarly for an *A-rud* class and a *rud in p* class.

The following lemma lists some of the basic properties of rudimentary functions. In each case, the simple proof is given in parentheses alongside the statement of the result.

1.1 Lemma.

(1) *The function* id (*the identity function*) *is rud.* (By schema (i).)

(2) *The function* $f(x) = \bigcup x$ *is rud.* (By schema (v) together with (1) above.)

(3) *The function* $f(x, y) = x \cup y$ *is rud.* ($f(x, y) = \bigcup\{x, y\}$, so use schema (ii) and (2) above.)

(4) *The function* $f(x_1, \ldots, x_n) = \{x_1, \ldots, x_n\}$ *is rud.* (By schema (ii), the function $g(x_1, \ldots, x_n) = \{x_n\}$ is rud for each n. But,

$$\{x_1, \ldots, x_{n+1}\} = \{x_1, \ldots, x_n\} \cup \{x_{n+1}\}.$$

So argue by induction on n, using schema (iv), together with (3).)

(5) *The function* $f(x_1, \ldots, x_n) = (x_1, \ldots, x_n)$ *is rud.* (By definition,

$$(x_1, \ldots, x_n) = \{\{x_1\}, \{x_1, (x_2, \ldots, x_n)\}\}.$$

So argue by induction on n, using schemas (ii) and (iv).)

(6) *The function* $f_m(x) = m$ *is rud for each* $m \in \omega$. (We have:

$$f_0(x) = 0 = x - x; \quad f_1(x) = 1 = \{0\}; \quad f_2(x) = 2 = \{0, 1\}; \quad \text{etc.}$$

So use schemas (iii) and (iv), together with (4), and proceed by induction on m.)

(7) *The relations* $(x \notin y)$ *and* $(x \neq y)$ *are rud.* (We have:

$$(x \notin y) \leftrightarrow \{x\} - y \neq \emptyset;$$
$$(x \neq y) \leftrightarrow (x - y) \cup (y - x) \neq \emptyset.$$

The result is clear now in view of earlier results.)

(8) *If* $f(y, \vec{x})$ *is rud, so is the function* $g(y, \vec{x}) = (f(z, \vec{x}) \mid z \in y)$. (Use schema (v), together with previous results and the identity

$$g(y, \vec{x}) = \bigcup_{z \in y} \{(f(z, \vec{x}), z)\}.)$$

(9) *If* $f\colon V^n \to V$ *is rud and* $R \subseteq V^n$ *is rud, then* $g\colon V^n \to V$ *is rud, where we set*

$$g(\vec{x}) = \begin{cases} f(\vec{x}), & \text{if } \quad R(\vec{x}), \\ \emptyset, & \text{if } \neg R(\vec{x}). \end{cases}$$

(Let r be a rud function such that

$$R(\vec{x}) \leftrightarrow r(\vec{x}) \neq \emptyset.$$

Then

$$g(\vec{x}) = \bigcup_{y \in r(\vec{x})} f(\vec{x}).)$$

(10) *Let χ_R be the characteristic function of R. Then R is rud iff χ_R is rud.* (If χ_R is rud, then since

$$R(\vec{x}) \leftrightarrow \chi_R(\vec{x}) \neq \emptyset,$$

R is rud, by definition. Conversely, if R is rud, then χ_R is rud by (6) and (9).)

(11) *R is rud iff $\neg R$ is rud.* (By (10), since $\chi_R(\vec{x}) = 1 - \chi_{\neg R}(\vec{x})$.)

(12) *The relations $(x \in y)$ and $(x = y)$ are rud.* (By (7) and (11).)

(13) (*Definition by Cases*) *Let f_i: $V^n \to V$ be rud for $i = 1,, \ldots, m$. Let $R_i \subseteq V^n$ be rud for $i = 1, \ldots, m$, and such that $R_i \cap R_j = \emptyset$ for $i \neq j$ and $R_1 \cup \ldots \cup R_m = V^n$. Define f: $V^n \to V$ by*

$$f(\vec{x}) = f_i(\vec{x}) \leftrightarrow R_i(\vec{x}).$$

Then f is rud. (For each $i = 1, \ldots, m$, set

$$f_i'(\vec{x}) = \begin{cases} f_i(\vec{x}), & \text{if } \quad R_i(\vec{x}) \\ \emptyset, & \text{if } \neg R_i(\vec{x}). \end{cases}$$

By (9), each f_i' is rud. But then f is rud, since

$$f(\vec{x}) = f_1'(\vec{x}) \cup \ldots \cup f_m'(\vec{x}).)$$

(14) *If $R(z, \vec{x})$ is rud, so is the function*

$$f(y, \vec{x}) = y \cap \{z \mid R(z, \vec{x})\}.$$

(Set

$$h(z, \vec{x}) = \begin{cases} \{z\}, & \text{if } \quad R(z, \vec{x}) \\ \emptyset, & \text{if } \neg R(z, \vec{x}). \end{cases}$$

By (9), h is rud. Hence f is rud, since

$$f(y, \vec{x}) = \bigcup_{z \in y} h(z, \vec{x}).)$$

(15) *Let $R(z, \vec{x})$ be rud and such that for any $\vec{x}$ there is at most one z such that $R(z, \vec{x})$. Then f is rud, where we define*

$$f(y, \vec{x}) = \begin{cases} \text{that } z \in y \text{ such that } R(z, \vec{x}), \text{ if such a } z \text{ exists,} \\ \emptyset, \text{ if no such } z \text{ exists.} \end{cases}$$

(By (14) and the identity

$$f(y, \vec{x}) = \bigcup(y \cap \{z \mid R(z, \vec{x})\}).)$$

(16) *If* $R(y, \vec{x})$ *is rud, so are* $(\exists z \in y)\, R(z, \vec{x})$ *and* $(\forall z \in y)\, R(z, \vec{x})$. (Let r be a rud function such that

$$R(y, \vec{x}) \leftrightarrow r(y, \vec{x}) \neq \emptyset.$$

Then

$$(\exists z \in y)\, R(z, \vec{x}) \leftrightarrow \bigcup_{z \in y} r(z, \vec{x}) \neq \emptyset,$$

so $(\exists z \in y)\, R(z, \vec{x})$ is rud. The second part now follows using (11).)

(17) *The function* $f(x) = \bigcap x$ *is rud.* (Use (12), (16) and (14) and the identity

$$f(x) = (\bigcup x) \cap \{z \mid (\forall y \in x)\,(z \in y)\}.)$$

(18) *The function* $f(x, y) = x \cap y$ *is rud.* (Because $f(x, y) = \bigcap \{x, y\}$.)

(19) *If* $R_i \subseteq V^n$ *are rud for* $i = 1, \ldots, m$, *then* $S = R_1 \cup \ldots \cup R_m$ *and* $T = R_1 \cap \ldots \cap R_m$ *are rud.* (Let $r_i = \chi_{R_i}$ for each i. Then

$$S(\vec{x}) \leftrightarrow r_1(\vec{x}) \cup \ldots \cup r_m(\vec{x}) \neq \emptyset,$$
$$T(\vec{x}) \leftrightarrow r_1(\vec{x}) \cup \ldots \cup r_m(\vec{x}) \neq \emptyset.$$

The result follows easily now.)

(20) *The functions* $(x)_0$ *and* $(x)_1$ *are rud.* (For example,

$$(x)_0 = \begin{cases} \text{that } z \in \bigcup x \text{ such that } (\exists v \in \bigcup x)\,(x = (z, v)), \text{ if such a } z \text{ exists,} \\ \emptyset, \text{ if no such } z \text{ exists.} \end{cases}$$

Now use (15).)

(21) *Define*

$$x(y) = \begin{cases} \textit{that } z \in \bigcup\bigcup x \textit{ such that } (z, y) \in x, \textit{ if there is a unique such } z, \\ \emptyset, \textit{ if there is no unique such } z. \end{cases}$$

Then the function $f(x, y) = x(y)$ *is rud.* (By Definition by cases.)

(22) *The functions* $\mathrm{dom}(x)$ *and* $\mathrm{ran}(x)$ *are rud.* (We have:

$$\mathrm{dom}(x) = \{z \in \bigcup\bigcup x \mid (\exists w \in \bigcup\bigcup x)\,((w, z) \in x)\};$$
$$\mathrm{ran}(x) = \{z \in \bigcup\bigcup x \mid (\exists w \in \bigcup\bigcup x)\,((z, w) \in x)\}.)$$

(23) *The function* $f(x, y) = x \times y$ *is rud.* (By the identity

$$x \times y = \bigcup_{u \in x} \bigcup_{v \in y} \{(u, v)\}.)$$

(24) *The function* $f(x, y) = x \restriction y$ *is rud.* (By the identity

$$x \restriction y = x \cap (\mathrm{ran}(x) \times y).)$$

(25) *The function* $f(x, y) = x''y$ *is rud.* (Since $x''y = \mathrm{ran}(x \restriction y)$.)

(26) *The function* x^{-1} *is rud.* (By definition,

$$x^{-1} = u''(x \cap (\mathrm{ran}(x) \times \mathrm{dom}(x))), \quad \text{where}$$

$$u(z) = ((z)_1, (z)_0).) \quad \square$$

By now, the reader may well have observed that all of the results in the above lemma are valid if we replace "rud" by "Σ_0". (In class terms, a function is said to be Σ_0 iff it is of the form

$$\{(y, \vec{x}) \mid \Phi(y, \vec{x})\},$$

where Φ is a Σ_0 formula of LST. In set theoretic terms, a function f is said to be Σ_0 iff there is a Σ_0 formula φ of $\mathscr{L}$ such that for any $\vec{x}$, y, if M is a transitive set such that $\vec{x}, y \in M$, then

$$f(\vec{x}) = y \leftrightarrow \models_M \varphi(\mathring{y}, \vec{\mathring{x}}).$$

By I.9.15, these notions are, in a sense, "equivalent".) However, it is not the case that the class of rud functions is the same as the class of Σ_0 functions. As we shall show presently, the rud functions form a proper subcollection of the Σ_0 functions. Strange as it may at first seem, in the case of relations, the notions of being rud and of being Σ_0 do coincide (as we prove later). The reason why there is no paradox here is that, whereas a function f is Σ_0 just in case it is of the form $\{(y, \vec{x}) \mid \Phi(y, \vec{x})\}$, where Φ is a Σ_0 formula of LST (so the fundamental concept is that of a relation, functions being treated as simply special kinds of relation), a function f is rud iff it can be built up using the schemas for rud functions (i.e. the fundamental concept is that of a function, and relations are effectively identified with their characteristic function).

In order to show that every rud function is Σ_0, it is convenient to introduce the following auxiliary notion.

Say a function $f\colon V^n \to V$ is *simple* iff, whenever $R(z, \vec{y})$ is a Σ_0 relation, the relation $R(f(\vec{x}), \vec{y})$ is also Σ_0.

The following lemma shows that simplicity is characterised by two special cases of the simplicity requirement.

1.2 Lemma. *A function* $f\colon V^n \to V$ *is simple iff:*

(i) *the predicate* $z \in f(\vec{x})$ *is* Σ_0; *and*

(ii) *whenever* $A(z)$ *is a* Σ_0 *predicate, so too is* $(\exists z \in f(\vec{x}))\, A(z)$.

Proof. ($\rightarrow$) Trivial, since (i) and (ii) are special cases of the simplicity requirements.

($\leftarrow$). Using (i) and (ii) we shall prove by induction on the logical complexity of R that if $R(z, \vec{y})$ is Σ_0, so too is $R(f(\vec{x}), \vec{y})$.

(a) Suppose first that $R(z, \vec{y})$ has the form $(z = y_i)$. Then

$$R(f(\vec{x}), \vec{y}) \leftrightarrow f(\vec{x}) = y_i$$
$$(\forall z \in f(\vec{x}))(z \in y_i) \wedge (\forall z \in y_i)(z \in f(\vec{x})).$$

By (i), the clause $(\forall z \in y_i)(z \in f(\vec{x}))$ is Σ_0, and by (ii) the clause $(\forall z \in f(\vec{x}))(z \in y_i)$ is Σ_0. Hence $R(f(\vec{x}), \vec{y})$ is Σ_0.

(b) Now suppose $R(z, \vec{y})$ has the form $(z \in y_i)$. Then

$$R(f(\vec{x}), \vec{y}) \leftrightarrow f(\vec{x}) \in y_i$$
$$\leftrightarrow (\exists z \in y_i)(f(\vec{x}) = z).$$

By part (a) above, the clause $(f(\vec{x}) = z)$ is Σ_0. Hence $R(f(\vec{x}), \vec{y})$ is Σ_0.

(c) Suppose that $R(z, \vec{y})$ has the form $(y_i \in z)$. Then

$$R(f(\vec{x}), \vec{y}) \leftrightarrow y_i \in f(\vec{x}).$$

This is Σ_0 by (i).

That takes care of all the primitive (i.e. atomic) cases.

(d) If $R(z, \vec{y})$ has the form $S(z, \vec{y}) \wedge T(z, \vec{y})$ the induction step is immediate.

(e) If $R(z, \vec{y})$ has the form $\neg S(z, \vec{y})$ the induction step is also immediate.

(f) Suppose that $R(z, \vec{y})$ has the form $(\exists u \in y_i)\, S(u, z, \vec{y})$. Then

$$R(f(\vec{x}), \vec{y}) \leftrightarrow (\exists u \in y_i)\, S(u, f(\vec{x}), \vec{y}),$$

and the induction step follows at once.

(g) Finally, suppose that $R(z, \vec{y})$ has the form $(\exists u \in z)\, S(u, z, \vec{y})$. Then

$$R(f(\vec{x}), \vec{y}) \leftrightarrow (\exists u \in f(\vec{x}))\, S(u, f(\vec{x}), \vec{y}),$$

and the induction step follows from (ii). □

1.3 Lemma. *If f is rud, then f is simple. Hence all rud functions are Σ_0.*

Proof. Let f be rud, and let $f_0, \ldots, f_n$ be a rud definition of f. Using 1.2, we shall prove by induction on $i \leqslant n$ that f_i is simple. (Such a proof is said to be "by induction on a rud definition of f".)

It is clear that schemas (i), (ii) and (iii) for rud functions all give simple functions. (In each case it is trivial to check conditions (i) and (ii) of 1.2.)

To handle schema (iv) we use the definition of simplicity. Let

$$f(\vec{x}) = h(g_1(\vec{x}), \ldots, g_k(\vec{x})),$$

where $h, g_1, \ldots, g_k$ are already known to be simple. Let $R(z, \vec{y})$ be Σ_0. Define S by

$$S(z_1, \ldots, z_k, \vec{y}) \leftrightarrow R(h(z_1, \ldots, z_k), \vec{y}).$$

Since h is simple, S is Σ_0. But

$$R(f(\vec{x}), \vec{y}) \leftrightarrow S(g_1(\vec{x}), \ldots, g_k(\vec{x}), \vec{y}).$$

So, as $g_1, \ldots, g_k$ are simple it follows (in k steps) that $R(f(\vec{x}), \vec{y})$ is Σ_0.

Finally, for schema (v) we use 1.2 again. Suppose that

$$f(y, x_2, \ldots, x_n) = \bigcup_{u \in y} g(u, x_2, \ldots, x_n),$$

where g is known to be simple. Then

$$z \in f(y, \vec{x}) \leftrightarrow (\exists u \in y)(z \in g(u, \vec{x})).$$

Since g is simple, by 1.2(i) the clause $(z \in g(u, \vec{x}))$ is Σ_0. Hence $(z \in f(y, \vec{x}))$ is Σ_0. Again, if $A(z)$ is Σ_0, then

$$(\exists z \in f(y, \vec{x}))\, A(z) \leftrightarrow (\exists u \in y)(\exists z \in g(u, \vec{x}))\, A(z).$$

Since g is simple, by 1.2(ii) the clause $(\exists z \in g(u, \vec{x}))\, A(z)$ is Σ_0. Hence $(\exists z \in f(y, \vec{x}))\, A(z)$ is Σ_0. The proof is complete. □

That the converse to 1.3 is false will follow from the following result.

1.4 Lemma (Finite Rank Property). *Let $f: V^n \to V$ be rud. Then there is a $p \in \omega$ such that for all $x_1, \ldots, x_n$,*

$$\text{rank}(f(x_1, \ldots, x_n)) < \max(\text{rank}(x_1), \ldots, \text{rank}(x_n)) + p.$$

Proof. By induction on a rud definition of f. The details are trivial. □

Consider now the constant function $f: V \to V$ defined by

$$f(x) = \omega \quad (\text{all } x).$$

By 1.4, f cannot be rud. But f has the Σ_0 definition

$$y = f(x) \leftrightarrow \text{On}(y) \wedge \lim(y) \wedge (\forall z \in y)(\text{succ}(z) \vee z = \emptyset).$$

But as the next lemma shows, the *graph* of f (i.e. the set $\{(y, x) \mid y = f(x)\}$) is rud.

1.5 Lemma. *Let $R \subseteq V^n$. Then R is rud iff it is Σ_0.*

Proof. If R is rud, then χ_R is rud, so by 1.4 χ_R is Σ_0, so R is Σ_0. Conversely, by 1.1, parts (11), (12), (16), (19), the class of all Σ_0 relations is a subclass of the class of all rud relations. □

A useful consequence of 1.5 is that, because of 1.1 (14), if $R(y, \vec{x})$ is a Σ_0 relation, then the function

$$f(y, \vec{x}) = \{z \in y \mid R(z, \vec{x})\}$$

is rud. We utilise this fact in our next, highly relevant result.

A class M is said to be *rudimentary closed* iff $f''M^n \subseteq M$ for all rud functions $f: V^n \to V$ (all n).

1.6 Lemma. *Let M be a transitive set containing ω. If M is rud closed then it is amenable.*

Proof. Recall that a transitive set M is amenable iff:

(i) $(\forall x, y \in M)(\{x, y\} \in M)$;
(ii) $(\forall x \in M)(\bigcup x \in M)$;
(iii) $\omega \in M$;
(iv) $(\forall x, y \in M)(x \times y \in M)$;
(v) if $R \subseteq M$ is $\Sigma_0(M)$, then $(\forall u \in M)(R \cap u \in M)$.

Assume that M is rud closed. By definition, the functions $f(x, y) = \{x, y\}$ and $f(x) = \bigcup x$ are rud, so M satisfies (i) and (ii) above. And by the hypotheses of the lemma, M satisfies (iii). By 1.1 (23), the function $f(x, y) = x \times y$ is rud, so (iv) is valid. That leaves us with (v). Let $R \subseteq M$ be $\Sigma_0(M)$. Suppose R is $\Sigma_0^M((p_1, \ldots, p_n))$. Let S be Σ_0^M such that

$$(\forall x \in M)\,[R(x) \leftrightarrow S(x, p_1, \ldots, p_n)].$$

Since S is Σ_0, it is rud (by 1.5). So by 1.1 (14), the function

$$f(u, x_1, \ldots, x_n) = u \cap \{x \mid S(x, x_1, \ldots, x_n)\}$$

is rud. Hence as $p_1, \ldots, p_n \in M$ and M is rud closed,

$$u \in M \to f(u, p_1, \ldots, p_n) \in M.$$

In other words,

$$u \in M \to u \cap R \in M,$$

as required. (Notice that we have here made use of "localised" versions of 1.5 and 1.1.) □

The converse to 1.6 is false. But by strenthening amenability clause (v) a little, it is possible to obtain a complete characterisation of rud closure in amenability like terms. We leave this as an exercise for the reader. (Hint: See what is required in order to prove the "converse" to 1.6.)

The *rudimentary closure* of a set X is the smallest rudimentary closed set Y such that $X \subseteq Y$. It is immediate that the rudimentary closure of X is of the form

$$\{f(\vec{x}) \mid f \text{ is rud and } \vec{x} \in X\}.$$

1.7 Lemma. *If U is transitive, then the rud closure of U is transitive.*

Proof. Let W be the rud closure of U. We prove by induction on a rud definition of f that for any rud function $f\colon V^n \to V$ and any $x_1, \ldots, x_n \in W$,

$$(*) \qquad TC(x_1) \subseteq W \wedge \ldots \wedge TC(x_n) \subseteq W \rightarrow TC(f(x_1, \ldots, x_n)) \subseteq W.$$

Since U is transitive and, as noted above,

$$W = \{f(\vec{x}) \mid f \text{ is rud and } \vec{x} \in U\}.$$

this proves the lemma.

If $f(x_1, \ldots, x_n) = x_i$, (*) is a propositional tautology.

If $f(x_1, \ldots, x_n) = \{x_i, x_j\}$, then

$$TC(f(x_1, \ldots, x_n)) = TC(\{x_i, x_j\}) = \{x_i, x_j\} \cup TC(x_i) \cup TC(x_j),$$

and (*) is immediate.

If $f(x_1, \ldots, x_n) = x_i - x_j$, then

$$TC(f(x_1, \ldots, x_n)) = TC(x_i - x_j) \subseteq TC(x_i),$$

and again (*) is immediate.

If $f(x_1, \ldots, x_n) = h(g_1(x_1, \ldots, x_n), \ldots, g_k(x_1, \ldots, x_n))$, where $h, g_1, \ldots, g_k$ are rudimentary and where (*) holds for $h, g_1, \ldots, g_k$, then (*) for f follows from the application of (*) first to each of $g_1, \ldots, g_k$ and then to h.

Finally, suppose $f(y, x_2, \ldots, x_n) = \bigcup_{z \in y} g(z, x_2, \ldots, x_n)$, where g is rudimentary and where (*) holds for g. If $TC(y) \subseteq W$, then $TC(z) \subseteq W$ for all $z \in y$, so by applying (*) to $g(z, x_2, \ldots, x_n)$ for each $z \in y$ we get (*) for f by taking the union according to the definition of f.

The proof is complete. □

We consider now the notion of relatively rudimentary functions. We show that these reduce, in a natural way, to combinations of rud functions and the function $f(x) = A \cap x$.

1.8 Lemma. *Let $A \subseteq V$. If $f\colon V^n \to V$ is A-rud, then f is expressible, in a uniform way with respect to any given A-rud definition of f, as a combination of rud functions and the function $a(x) = A \cap x$.*

Proof. Let $P(f)$ mean that f is expressible as a composition of rud functions and the function a defined above. We shall show that if f is A-rud, then $P(f)$. The proof is by induction on a rud definition of f. (The uniformity will be an immediate consequence of the proof.)

Clauses (i), (ii), (iii), and (vi) in the definition of A-rud functions cause no difficulties in the induction. And clause (iv) is taken care of by virtue of the fact that a composition of compositions is itself a composition. The only trickly step is the proof that if $P(g)$ holds and f is defined by

$$f(y, \vec{x}) = \bigcup_{z \in y} g(z, \vec{x}),$$

then $P(f)$ holds. We do this by induction on the "complexity" of g. More precisely, let $P_0(h)$ mean that h is rud, and, inductively, let $P_{n+1}(h)$ mean that

$$h(\vec{x}) = h_0(\vec{x}, A \cap h_1(\vec{x}), \ldots, A \cap h_m(\vec{x}))$$

for some $h_0, h_1, \ldots, h_m$ such that $P_0(h_0)$ and $P_n(h_1), \ldots, P_n(h_m)$ are all valid. By the definition of P, it is clear that

$$P(h) \leftrightarrow \exists n P_n(h).$$

So it suffices to prove that $R(n)$ holds for all n, where $R(n)$ means:

$$\text{if } P_n(g) \text{ and } f(y, \vec{x}) = \bigcup_{z \in y} g(z, \vec{x}), \text{ then } P(f).$$

We do this by induction on n.

For $n = 0$ there is nothing to prove, since in this case f is itself rud. So suppose that $n > 0$ and that $R(n-1)$ holds. Let g be given such that $P_n(g)$. Thus

$$g(z, \vec{x}) = h_0(z, \vec{x}, A \cap h_1(z, \vec{x}), \ldots, A \cap h_m(z, \vec{x})),$$

where $P_0(h_0)$ and $P_{n-1}(h_1), \ldots, P_{n-1}(h_m)$. Set

$$\tilde{g}(z, \vec{x}, u) = h_0(z, \vec{x}, u \cap h_1(z, \vec{x}), \ldots, u \cap h_m(z, \vec{x})).$$

Clearly, $P_{n-1}(\tilde{g})$. Set

$$\tilde{f}(y, \vec{x}, u) = \bigcup_{z \in y} \tilde{g}(z, \vec{x}, u),$$

$$\tilde{h}(y, \vec{x}) = [\bigcup_{z \in y} h_1(z, \vec{x})] \cup \ldots \cup [\bigcup_{z \in y} h_m(z, \vec{x})].$$

By $R(n-1)$, both $P(\tilde{f})$ and $P(\tilde{h})$. But

$$f(y, \vec{x}) = \bigcup_{z \in y} g(z, \vec{x}) = \tilde{f}(y, \vec{x}, A \cap \tilde{h}(y, \vec{x})).$$

This proves $R(n)$. □

A structure of the form $\mathbf{M} = \langle M, A \rangle$, where $A \subseteq M$, is said to be *rud closed* iff $f''M^n \subseteq M$ for all A-rud functions $f\colon V^n \to V$ (all n).

1.9 Lemma. *Let* $\ = \langle M, A \rangle$, *where M is transitive and $A \subseteq M$. Then* $\mathbf{M}$ *is a rud closed structure iff M is a rud closed set and the structure* $\mathbf{M}$ *is amenable.*

Proof. A direct consequence of 1.6 and 1.8. □

1.10 Lemma. *Let $A \subseteq V$. If $f\colon V^n \to V$ is A-rud, then $f \upharpoonright M^n$ is uniformly $\Sigma_1^{\langle M, A \cap M\rangle}$ for all transitive, rud closed structures $\langle M, A \cap M\rangle$.*

Proof. By 1.3 and 1.8. □

The following lemma shows that, in a certain, obvious sense, the rud functions have a finite "basis". In the statement of the lemma, we allow the use of "dummy variables" so that, for later convenience, all of the "basis" functions are binary.

1.11 Lemma (The Basis Lemma). *Every rudimentary function is a composition of some or all of the following rudimentary functions*:

$$F_0(x, y) = \{x, y\};$$

$$F_1(x, y) = x - y;$$

$$F_2(x, y) = x \times y;$$

$$F_3(x, y) = \{(u, z, v) \mid z \in x \wedge (u, v) \in y\};$$

$$F_4(x, y) = \{(u, v, z) \mid z \in x \wedge (u, v) \in y\};$$

$$F_5(x, y) = \bigcup x;$$

$$F_6(x, y) = \mathrm{dom}(x);$$

$$F_7(x, y) = \in \cap (x \times x);$$

$$F_8(x, y) = \{x''\{z\} \mid z \in y\}.$$

Proof. It is easily seen that each of the above functions is rudimentary. Hence if $\mathscr{C}$ denotes the class of all functions obtainable from $F_0, \ldots, F_8$ by composition, then every function in $\mathscr{C}$ is rudimentary. We prove the converse, that every rudimentary function is a member of $\mathscr{C}$.

If φ is an $\mathscr{L}$-formula and $x_0, \ldots, x_n$ are variables of $\mathscr{L}$, say $x_0 = v_{i(0)}, \ldots, x_n = v_{i(n)}$, we usually write $\varphi(x_0, \ldots, x_n)$ to indicate that the free variables of φ are all amongst $x_0, \ldots, x_n$. Let us call the expression "$\varphi(x_0, \ldots, x_n)$" a *representation* of φ. Thus, any $\mathscr{L}$-formula has infinitely many representations: if the free variables of φ are all amongst $v_0, \ldots, v_n$, then

$$\varphi(v_0, \ldots, v_n), \qquad \varphi(v_0, \ldots, v_n, v_{n+1}), \qquad \varphi(v_0, \ldots, v_n, v_{n+10}, v_{n+3})$$

are all representations of φ.

For each representation $\varphi(x_0, \ldots, x_n)$ of an $\mathscr{L}$-formula φ we define a function $t_{\varphi(x_0, \ldots, x_n)}$ as follows:

$$t_{\varphi(x_0, \ldots, x_n)}(u) = \{(a_0, \ldots, a_n) \mid a_0, \ldots, a_n \in u \wedge \models_u \varphi(\mathring{a}_0, \ldots, \mathring{a}_n)\}.$$

As a first step towards proving the lemma, we show that $t_{\varphi(x_0, \ldots, x_n)} \in \mathscr{C}$ for any $\varphi(x_0, \ldots, x_n)$. The proof is by induction on the construction of φ.

(a) Suppose that $\varphi(x_0, \ldots, x_n)$ is the formula $(x_i \in x_j)$, where $0 \leqslant i < j \leqslant n$. Thus

$$t_{\varphi(x_0,\ldots,x_n)}(u) = \{(a_0, \ldots, a_n) \mid a_0, \ldots, a_n \in u \wedge \models_u (\mathring{a}_i \in \mathring{a}_j)\}$$
$$= \{(a_0, \ldots, a_n) \mid a_0, \ldots, a_n \in u \wedge a_i \in a_j\}.$$

The main complicating factor is the presence of the "superfluous" variables $x_0, \ldots, x_{i-1}, x_{i+1}, \ldots, x_{j-1}, x_{j+1}, \ldots, x_n$. This is where we use the functions F_3 and F_4. (Remember that, by definition,

$$(x_0, \ldots, x_n) = (x_0, (x_1, \ldots, x_n)) = (x_0, (x_1, (x_2, \ldots, x_n))) = \ldots\ldots)$$

We shall assume that $0 < i, i + 1 < j, j < n$. This is the most complicated case, with "superfluous" variables in all possible locations. All other cases are degenerate versions of this one. Let us write $G^0(x, y)$ for $F_2(x, y)$ and, inductively, $G^{m+1}(x, y)$ for $F_2(x, G^m(x, y))$. Thus $G^m \in \mathscr{C}$ for all m. Note that, in particular, $G^m(u, u) = u^{m+2}$ for all m. Let

$$H(u) = F_4(G^{n-j-2}(u, u), F_7(u, u)).$$

Then $H \in \mathscr{C}$. But we have

$$H(u) = F_4(u^{n-j}, \in \cap\, u^2)$$
$$= \{(a, b, c) \mid c \in u^{n-j} \wedge (a, b) \in (\in \cap\, u^2)\}$$
$$= \{(a, b, c) \mid c \in u^{n-j} \wedge a, b \in u \wedge a \in b\}.$$

Thus,

$$F_3(u, H(u)) = \{(d, e, f) \mid e \in u \wedge (d, f) \in H(u)\}$$
$$= \{(a, e, (b, c)) \mid e \in u \wedge (a, b, c) \in H(u)\}$$
$$= \{(a, e, b, c) \mid e \in u \wedge (a, b, c) \in H(u)\}.$$

Similarly,

$$F_3(u, F_3(u, H(u))) = \{(a, e, f, b, c) \mid e, f \in u \wedge (a, b, c) \in H(u)\}.$$

So if we write $F_x(y)$ for $F_3(x, y)$ we have

$$F_u^{j-i-1}(H(u)) = \{(a, e_1, \ldots, e_{j-i-1}, b, c) \mid e_1, \ldots, e_{j-i-1} \in u \wedge (a, b, c) \in H(u)\}.$$

Then

$$G^i(u, F_u^{j-i-1}(H(u))) = \{(f_1, \ldots, f_{i-1}, a, e_1, \ldots, e_{j-i-1}, b, c) \mid f_1, \ldots, f_{i-1}, e_1, \ldots, e_{j-i-1} \in u \wedge (a, b, c) \in H(u)\}$$
$$= t_{\varphi(x_0,\ldots,x_n)}(u).$$

Thus $t_{\varphi(x_0,\ldots,x_n)} \in \mathscr{C}$.

(b) Suppose that $\varphi = \psi \vee \theta$ and that $t_{\psi(x_0,\ldots,x_n)}, t_{\theta(x_0,\ldots,x_n)} \in \mathscr{C}$. Then $t_{\varphi(x_0,\ldots,x_n)} \in \mathscr{C}$, because,

$$\begin{aligned} t_{\varphi(x_0,\ldots,x_n)}(u) &= t_{\psi(x_0,\ldots,x_n)}(u) \cup t_{\theta(x_0,\ldots,x_n)}(u) \\ &= F_5(F_0(t_{\psi(x_0,\ldots,x_n)}(u), t_{\theta(x_0,\ldots,x_n)}(u)), u). \end{aligned}$$

(c) Suppose that $\varphi = \neg\psi$ and that $t_{\psi(x_0,\ldots,x_n)} \in \mathscr{C}$. Then $t_{\varphi(x_0,\ldots,x_n)} \in \mathscr{C}$ because

$$t_{\varphi(x_0,\ldots,x_n)}(u) = u - t_{\psi(x_0,\ldots,x_n)}(u) = F_1(u, t_{\psi(x_0,\ldots,x_n)}(u)).$$

(d) If $\varphi = \psi \wedge \theta$ and $t_{\psi(x_0,\ldots,x_n)}, t_{\theta(x_0,\ldots,x_n)} \in \mathscr{C}$, then $t_{\varphi(x_0,\ldots,x_n)} \in \mathscr{C}$ by (b) and (c).

(e) If $\varphi = \exists y \psi$ and $t_{\psi(y,x_0,\ldots,x_n)} \in \mathscr{C}$, then $t_{\varphi(x_0,\ldots,x_n)} \in \mathscr{C}$, because

$$t_{\varphi(x_0,\ldots,x_n)}(u) = \operatorname{dom}(t_{\psi(y,x_0,\ldots,x_n)}(u) = F_6(t_{\psi(y,x_0,\ldots,x_n)}(u), u).$$

(f) If $\varphi = \forall y \psi$ and $t_{\psi(y,x_0,\ldots,x_n)} \in \mathscr{C}$, then $t_{\varphi(x_0,\ldots,x_n)} \in \mathscr{C}$ by (e) and (c).

(g) If $\varphi(x_0, \ldots, x_n)$ is the formula $(x_i = x_j)$, where $0 \leqslant i, j \leqslant n$, then $t_{\varphi(x_0,\ldots,x_n)} \in \mathscr{C}$. To see this, let $\theta(y, x_0, \ldots, x_n)$ be the formula

$$(y \in x_i) \leftrightarrow (y \in x_j).$$

By (a), together with (b), (c), (d), $t_{\theta(y,x_0,\ldots,x_n)} \in \mathscr{C}$. Let $\psi(x_0, \ldots, x_n)$ be the formula $\forall y \theta(y, x_0, \ldots, x_n)$. By (f), $t_{\psi(x_0,\ldots,x_n)} \in \mathscr{C}$. But clearly,

$$\models_u \varphi(\mathring{a}_0, \ldots, \mathring{a}_n) \quad \text{iff } \models_{u \cup (\bigcup u)} \psi(\mathring{a}_0, \ldots, \mathring{a}_n).$$

Thus,

$$\begin{aligned} t_{\varphi(x_0,\ldots,x_n)}(u) &= \{(a_0, \ldots, a_n) \mid a_0, \ldots, a_n \in u \wedge \models_{u \cup (\bigcup u)} \psi(\mathring{a}_0, \ldots, \mathring{a}_n)\} \\ &= u^{n+1} \cap t_{\psi(x_0,\ldots,x_n)}(u \cup (\bigcup u)). \end{aligned}$$

But we saw in (a) that the function $F(u) = u^{n+1}$ is in $\mathscr{C}$ (if $n = 0$, use $F(u) = u - (u - u)$ instead), and by F_5, F_0 the function $F(u) = u \cup (\bigcup u)$ is in $\mathscr{C}$. Thus $t_{\varphi(x_0,\ldots,x_n)} \in \mathscr{C}$.

(h) Now suppose that $\varphi(x_0, \ldots, x_n)$ is the formula $(x_i \in x_j)$ where $0 \leqslant j < i \leqslant n$. To see that $t_{\varphi(x_0,\ldots,x_n)} \in \mathscr{C}$, argue as follows. Let $\psi(y, z, x_0, \ldots, x_n)$ be the formula

$$(y \in z) \wedge (y = x_i) \wedge (z = x_j).$$

By (a), (g), (d), $t_{\psi(y,z,x_0,\ldots,x_n)} \in \mathscr{C}$. But clearly,

$$\models_u \varphi(\mathring{a}_0, \ldots, \mathring{a}_n) \quad \text{iff } \models_u \exists y \exists z \psi(y, z, \mathring{a}_0, \ldots, \mathring{a}_n).$$

So by (e), $t_{\varphi(x_0,\ldots,x_n)} \in \mathscr{C}$.

By (a), (h), (g), (b), (c), (d), we see that $t_{\varphi(x_0,\ldots,x_n)} \in \mathscr{C}$ whenever φ is a quantifier free formula of $\mathscr{L}$. Hence by (e), (f), $t_{\varphi(x_0,\ldots,x_n)} \in \mathscr{C}$ for any $\mathscr{L}$-formula φ.

As the next step towards proving the lemma, for any $f\colon V^n \to V$ we define $f^*\colon V \to V$ by

$$f^*(u) = f''u^n.$$

We prove that if f is rudimentary, then $f^* \in \mathscr{C}$. The proof is by induction on a rudimentary definition of f.

(a) Suppose that $f(x_1, \ldots, x_n) = x_i$. Then

$$f^*(u) = f''u^n = u - (u - u)$$

and so $f^* \in \mathscr{C}$.

(b) Suppose $f(x_1, \ldots, x_n) = x_i - x_j$. Then

$$f^*(u) = f''u^n = \{x - y \mid x, y \in u\}.$$

Let $\varphi(z, y, x)$ be the formula $z \in (x - y)$. Let

$$\begin{aligned} F(u) &= t_{\varphi(z,x,y)}(u \cup (\bigcup u)) \cap (\bigcup u \times u^2) \\ &= \{(z, x, y) \mid x, y \in u \wedge z = x - y\}. \end{aligned}$$

Since $t_{\varphi(z,x,y)} \in \mathscr{C}$ we have $F \in \mathscr{C}$. But then $f^* \in \mathscr{C}$, since

$$\begin{aligned} F_8(F(u), u^2) &= \{F(u)''\{a\} \mid a \in u^2\} \\ &= \{F(u)''\{(x, y)\} \mid x, y \in u\} \\ &= \{\{z\} \mid x, y \in u \wedge z = x - y\} \\ &= \{\{x - y\} \mid x, y \in u\} \\ &= f^*(u). \end{aligned}$$

(c) Let $f(x_1, \ldots, x_n) = \{x_i, x_j\}$. Then

$$f^*(u) = \{\{x, y\} \mid x, y \in u\} = \bigcup(u^2),$$

so $f^* \subset \mathscr{C}$.

(d) Let $f(x_1, \ldots, x_n) = h(g_1(x_1, \ldots, x_n), \ldots, g_k(x_1, \ldots, x_n))$, where $h, g_1, \ldots, g_k$ are rudimentary and $h^*, g_1^*, \ldots, g_k^* \in \mathscr{C}$. Let

$$G(u) = g_1^*(u) \cup \ldots \cup g_k^*(u), \quad H(u) = h^*(G(u)), \quad K(u) = u^n \cup G(u) \cup H(u).$$

By our assumptions, $G, H, K \in \mathscr{C}$.

By 1.3 there is a Σ_0-formula $\Phi(z_1, \ldots, z_k, x_1, \ldots, x_n)$ of LST such that

$$\begin{aligned} &\Theta(z_1, \ldots, z_k, x_1, \ldots, x_n) \quad \text{iff} \\ &z_1 = g_1(x_1, \ldots, x_n) \wedge \ldots \wedge z_k = g_k(x_1, \ldots, x_n) \end{aligned}$$

and a Σ_0-formula $\Psi(y, z_1, \ldots, z_k)$ of LST such that

$$\Psi(y, z_1, \ldots, z_k) \quad \text{iff } y = h(z_1, \ldots, z_k).$$

Let r exceed the number of quantifiers which occur in Θ and Ψ, and define D by

$$D(u) = u \cup (\bigcup u) \cup (\bigcup \bigcup u) \cup \ldots \cup (\bigcup^r u).$$

Then $D \in \mathscr{C}$, and moreover, by I.9.15, if θ, ψ are the $\mathscr{L}$-analogues of Θ, Ψ, then for any set u and any $y, z_1, \ldots, z_k, x_1, \ldots, x_n \in u$,

$$\Theta(z_1, \ldots, z_k, x_1, \ldots, x_n) \quad \text{iff } \models_{D(u)} \theta(\mathring{z}_1, \ldots, \mathring{z}_k, \mathring{x}_1, \ldots, \mathring{x}_n)$$

and

$$\Psi(y, z_1, \ldots, z_k) \quad \text{iff } \models_{D(u)} \psi(\mathring{y}, \mathring{z}_1, \ldots, \mathring{z}_k).$$

(Strictly speaking, I.9.15 is not adequate for the above, since this would require $D(u)$ to be transitive. However, as is easily seen, the choice of the integer r above makes $D(u)$ resemble a transitive set sufficiently for the proof of I.9.15 to go through for the formulas concerned here.) Let $\varphi(y, x_1, \ldots, x_n)$ be the $\mathscr{L}$-formula

$$\exists z_1 \ldots z_k[\theta(z_1, \ldots, z_k, x_1, \ldots, x_n) \wedge \psi(y, z_1, \ldots, z_k)].$$

Now, $K(u)$ consists of u^n, together with all values of $g_1, \ldots, g_k$ on u and all values of f on u. Thus by the definition of φ,

$$\begin{aligned} &t_{\varphi(y, x_1, \ldots, x_n)}(D \circ K(u)) \cap (f''u^n \times u^n) \\ &\quad = \{(f(x_1, \ldots, x_n), x_1, \ldots, x_n) \mid x_1, \ldots, x_n \in u\}. \end{aligned}$$

Thus

$$f^*(u) = \bigcup F_8(t_{\varphi(y, x_1, \ldots, x_n)}(D \circ K(u)) \cap (H(u) \times u^n), u^n).$$

This shows that $f^* \in \mathscr{C}$.

(e) Suppose that $f(y, x_1, \ldots, x_n) = \bigcup_{v \in y} g(v, x_1, \ldots, x_n)$, where g is rudimentary and $g^* \in \mathscr{C}$. By 1.3 there is a Σ_0-formula $\Phi(z, y, x_1, \ldots, x_n)$ of LST such that

$$\Phi(z, y, x_1, \ldots, x_n) \quad \text{iff } (\exists v \in y)\,[z \in g(v, x_1, \ldots, x_n)].$$

Suppose that Φ has fewer than r quantifiers, and define D as in the above case (d). Then, if φ is the $\mathscr{L}$-analogue of Φ, we have, as above, for any $z, y, x_1, \ldots, x_n \in u$,

$$\Phi(z, y, x_1, \ldots, x_n) \quad \text{iff } \models_{D(u)} \varphi(\mathring{z}, \mathring{y}, \mathring{x}_1, \ldots, \mathring{x}_n).$$

Then

$$\begin{aligned} t_{\varphi(z, y, x_1, \ldots, x_n)}(D(u)) &= \{(z, y, x_1, \ldots, x_n) \mid z, y, x_1, \ldots, x_n \in D(u) \\ &\qquad \wedge (\exists v \in y)\,(z \in g(v, x_1, \ldots, x_n))\} \\ &= \{(z, y, x_1, \ldots, x_n) \mid z, y, x_1, \ldots, x_n \in D(u) \\ &\qquad \wedge z \in f(y, x_1, \ldots, x_n)\}. \end{aligned}$$

So

$$F_8(t_{\varphi(z,y,x_1,\ldots,x_n)}(D(u)), u^{n+1})$$
$$= \{\{z\} \mid y, x_1, \ldots, x_n \in u \wedge z \in f(y, x_1, \ldots, x_n)\}.$$

Thus

$$f^*(u) = \bigcup F_8(t_{\varphi(z,y,x_1,\ldots,x_n)}(D(u)), u^{n+1}),$$

which shows that $f^* \in \mathscr{C}$.

We have proved that $f^* \in \mathscr{C}$ for any rudimentary function f. We are now able to complete the proof of the lemma. Let $f: V^n \to V$ be a given rudimentary function. We prove that $f \in \mathscr{C}$. Define $\tilde{f}: V \to V$ by

$$\tilde{f}(x) = \begin{cases} f(z_1, \ldots, z_n), & \text{if } x = (z_1, \ldots, z_n) \\ \emptyset, & \text{in all other cases.} \end{cases}$$

By 1.1(9), $\tilde{f}$ is rudimentary. So by the above, $\tilde{f}^* \in \mathscr{C}$. Moreover, $g \in \mathscr{C}$, where we define $g: V^n \to V$ by

$$g(x_1, \ldots, x_n) = \{(x_1, \ldots, x_n)\}.$$

(By repeated use of F_0.) But

$$\begin{aligned} f(x_1, \ldots, x_n) &= \bigcup \{f(x_1, \ldots, x_n)\} = \bigcup \{\tilde{f}((x_1, \ldots, x_n))\} \\ &= \bigcup \tilde{f}''\{(x_1, \ldots, x_n)\} = \bigcup \tilde{f}^*((x_1, \ldots, x_n)) \\ &= \bigcup \tilde{f}^*(g(x_1, \ldots, x_n)). \end{aligned}$$

Thus $f \in \mathscr{C}$, and we are done. □

As an immediate corollary of 1.8 and 1.11, we have:

1.12 Lemma (Extended Basis Lemma). *Let $A \subseteq V$, and define F_9 by*

$$F_9(x, y) = A \cap x.$$

Then every A-rudimentary function may be expressed as a composition of some or all of the A-rud functions $F_0, \ldots, F_9$. □

Lemma 1.14 below provides an immediate application of the above basis result. It concerns the semantics of the languages $\mathscr{L}_V(A)$. These languages (or rather more general languages $\mathscr{L}_V(A_1, \ldots, A_k)$) were defined in I.9. As was mentioned there, the basic syntactics and semantics of these languages differs only in a trivial way from that of the language $\mathscr{L}_V$, and so there is no need to spend any time on such a development. Suffice it to say that, what comes out of it is the following. There is a Σ_1 formula $Sat^A(u, a, \varphi)$ of LST (in three variables, u, a, φ) which says that:

> "u is a non-empty set" $\wedge$ "$a \subseteq u$" $\wedge$ "φ is a sentence of $\mathscr{L}_u(A)$ which is true in the structure $\langle u, a \rangle$ under the canonical interpretation".

Just as in I.9.10, we get:

1.13 Lemma. *The* LST *formula* $Sat^A(u, a, \varphi)$ *is* Δ_1^{BS}. $\square$

As usual, we usually write $\models_{\langle u,a\rangle} \varphi$ rather than $Sat^A(u, a, \varphi)$. For any $n \in \omega$, we denote by $\models^{\Sigma_n}_{\langle u,a\rangle}$ the restriction of the relation $\models_{\langle u,a\rangle}$ to the Σ_n sentences of $\mathscr{L}_u(A)$.

The following lemma will provide us with an analogue to II.6.3 for the Jensen hierarchy of constructible sets, defined in the next section.

1.14 Lemma. $\models^{\Sigma_0}_{[M,A\rangle}$ *is uniformly* $\Sigma_1^{\langle M,A\rangle}$ *for transitive, rud closed structures* $\langle M, A\rangle$.

Proof. Consider the language Γ_M which consists of the variables v_n, $n \in \omega$, of $\mathscr{L}$ (i.e. $v_n = (2, n)$), the constant symbols $\mathring{x}(= (3, x))$, for each $x \in M$, and the binary function symbols $\mathring{F}_0, \ldots, \mathring{F}_9$. (More formally, for each $i = 0, \ldots, 9$, $\mathring{F}_i(x, y)$ denotes the set $(0, i, x, y)$.) The syntax of Γ_M is particularly simple. Each variable and each constant of $\mathscr{L}_M$ is a *term* of Γ_M, and if t_1, t_2 are terms of Γ_M, then $\mathring{F}_0(t_1, t_2), \ldots, \mathring{F}_9(t_1, t_2)$ are all terms of Γ_M. Note that each term of Γ_M is an element of M. A *constant* term is one which contains no variables. Each constant term, t, of Γ_M has an obvious interpretation in $\langle M, A\rangle$, where we let x interpret $\mathring{x}$ for each $x \in M$ and F_i interpret $\mathring{F}_i$ for each $i = 0, \ldots, 9$. Since $\langle M, A\rangle$ is rud closed, the interpretation, $t^{\langle M,A\rangle}$ of each constant term t is an element of M. Clearly, for each constant term t and each $x \in M$, we have:

$$\begin{aligned}
&x = t^{\langle M,A\rangle} \quad \text{iff} \\
&\exists f \exists g[\text{Finseq}(f) \wedge \text{Finseq}(g) \wedge \text{dom}(f) = \text{dom}(g) \wedge g(\text{dom}(g) - 1) \\
&\quad = x \wedge (\forall i \in \text{dom}(f))\,[\text{Const}_M(f(i)) \vee (\exists j, k \in i)\,[f(i) \\
&\quad = \mathring{F}_0(f(j), f(k)) \vee \ldots \vee f(i) = \mathring{F}_9(f(j), f(k))] \wedge (\forall i \in \text{dom}(f)) \\
&\qquad [\text{Const}_M(f(i)) \rightarrow g(i) = (f(i))_1] \wedge (\forall i \in \text{dom}(f))\,(\forall j, k \in i) \\
&\qquad [(f(i) = \mathring{F}_0(f(j), f(k)) \rightarrow g(i) = F_0(g(j), g(k))) \wedge \ldots \wedge (f(i) \\
&\quad = \mathring{F}_9(f(j), f(k)) \rightarrow g(i) = F_9(g(j), g(k)))]]].
\end{aligned}$$

Now, if such f, g as above exist, they will certainly be elements of M. Moreover, by 1.10, each of the functions $F_0, \ldots, F_9$ is (uniformly) $\Sigma_1^{\langle M,A\rangle}$. Hence the above equivalence shows that the relation $x = t^{\langle M,A\rangle}$ (as a relation of x, t) is (uniformly) $\Sigma_1^{\langle M,A\rangle}$. The idea now is to utilise this fact by associating with each Σ_0 sentence φ of $\mathscr{L}_M(A)$ a constant term t_φ of Γ_M so that:

(i) the map $\varphi \mapsto t^\varphi$ is $\Sigma_1^{\langle M,A\rangle}$ (uniformly);

(ii) $\models_{\langle M,A\rangle} \varphi$ iff $t_\varphi^{\langle M,A\rangle} = 1$.

In fact, in order to do this, we need to define t_φ for any formula φ, not just sentences. (This is why we allow variables in the language Γ_M.)

As our starting point we take 1.5. This tells us that if $R(\vec{x})$ is a Σ_0 relation, there is a rud function $f(\vec{x})$ such that

$$R(\vec{x}) \leftrightarrow f(\vec{x}) = 1.$$

By 1.11, we know that the function f here may be expressed as a composition of the basic functions $F_0, \ldots, F_9$. Now, the existence of the function f is established by proceeding inductively on the logical structure of R, using 1.1(11), (12), (16), (19), and the proof of 1.11 is (essentially) by induction on a rudimentary definition of f. And by virtue of 1.8, we can extend all of this to allow for the unary predicate A, introducing the extra basic function F_9. So by examining the inductive proofs of 1.5, 1.11, and 1.8, we obtain the required map $\varphi \mapsto t_\varphi$.

We proceed inductively following the logical construction of the Σ_0 formula φ, using the techniques of 1.1, 1.11, and 1.8. Now, if you have spent any time on the proofs of these results, particularly 1.11, you will appreciate that it would be pointless trying to write out explicitly the definition of the function $\varphi \mapsto t_\varphi$. But it should be clear that the following is the case.

From I.9 (extended to the language $\mathscr{L}_M(A)$) we know that there are Σ_0 formulas $F_\in, F_=, F_A, F_\wedge, F_\neg, F_\exists$ of LST such that (see, in particular, I.9.3):

$$F_\in(\theta, x, y) \leftrightarrow \theta \text{ is the } \mathscr{L}_M(A) \text{ formula } (x \in y);$$

$$F_=(\theta, x, y) \leftrightarrow \theta \text{ is the } \mathscr{L}_M(A) \text{ formula } (x = y);$$

$$F_A(\theta, x) \leftrightarrow \theta \text{ is the } \mathscr{L}_M(A) \text{ formula } \mathring{A}(x);$$

$$F_\wedge(\theta, \varphi, \psi) \leftrightarrow \theta \text{ is the } \mathscr{L}_M(A) \text{ formula } (\varphi \wedge \psi);$$

$$F_\neg(\theta, \varphi) \leftrightarrow \theta \text{ is the } \mathscr{L}_M(A) \text{ formula } (\neg \varphi);$$

$$F_\exists(\theta, u, \varphi) \leftrightarrow \theta \text{ is the } \mathscr{L}_M(A) \text{ formula } (\exists u \varphi).$$

These LST formulas simply describe the way in which the formulas of $\mathscr{L}_M(A)$ are constructed. Implicit in the proofs of 1.1, 1.11, and 1.8 is the fact that there are Σ_0 formulas $G_\in, G_=, G_A, G_\wedge, G_\neg, G_\exists$ of LST such that:

$$G_\in(t, x, y) \leftrightarrow t = t_{(x \in y)};$$

$$G_=(t, x, y) \leftrightarrow t = t_{(x = y)};$$

$$G_A(t, x) \leftrightarrow t = t_{A(x)};$$

$$G_\wedge(t, t_\varphi, t_\psi) \leftrightarrow t = t_{(\varphi \wedge \psi)};$$

$$G_\neg(t, t_\varphi) \leftrightarrow t = t_{(\neg \varphi)};$$

$$G_\exists(t, t_\varphi) \leftrightarrow t = t_{(\exists y \in x)\varphi},$$

where for each φ, t_φ is a term of Γ_M which satisfies (ii) above. These G-formulas describe the way in which the terms t_φ must be combined (together with specific of the function symbols $\mathring{F}_0, \ldots, \mathring{F}_9$) to make (ii) valid, and thus correspond to the induction steps of the proofs of the 1.1, 1.11, and 1.7 (all rolled into one).

It follows that there is a Σ_1 formula H of LST such that

$$H(t, \varphi) \leftrightarrow \text{“}\varphi \text{ is a } \Sigma_0 \text{ formula of } \mathscr{L}_M(A)\text{”} \wedge t = t_\varphi.$$

In essence (though not totally accurate), $H(t, \varphi)$ is as follows (see I.9, in particular I.9.6):

$$\exists f \exists g[\text{Build}(f, \varphi) \wedge \text{Finseq}(g) \wedge \text{dom}(g) = \text{dom}(f)$$
$$\wedge (\forall i \in \text{dom}(f))((F_\in(f(i), x, y) \to G_\in(g(i), x, y)))$$
$$\wedge \ldots \wedge (F_\exists(f(i), u, f(j)) \to G_\exists(g(i), u, g(j)))].$$

Notice that if $H(t, \varphi)$ is true, it is always possible to find such f, g as above in M. Consequently, if $h(t, \varphi)$ denotes the $\mathscr{L}$-analogue of the LST formula $H(t, \varphi)$, we have, by I.9.15, for any $x, y \in M$,

$$H(y, x) \leftrightarrow \models_M h(\mathring{y}, \mathring{x}).$$

This proves (i) and (ii) and thus completes the proof of the lemma. □

The following result, which will provide us with an analogue of II.6.4 for the Jensen hierarchy, is deduced from 1.14 in exactly the same way that II.6.4 was deduced from II.6.3:

1.15 Lemma. *For any $n \geqslant 1$, the relation $\models^{\Sigma_n}_{\langle M, A\rangle}$ is uniformly $\Sigma_n^{\langle M, A\rangle}$ for all transitive rud closed structures $\langle M, A\rangle$.* □

For any set U, we define the set $\text{rud}(U)$ to be the rudimentary closure of the set $U \cup \{U\}$, i.e. the smallest rud closed set that contains U as a subset and as an element. Notice that by 1.7, we have:

1.16 Lemma. *If U is transitive, then* $\text{rud}(U)$ *is transitive.*

Proof. Immediate, since if U is transitive, then $U \cup \{U\}$ is transitive. □

We shall use the function $\text{rud}(U)$ in order to define the Jensen hierarchy. The following lemmas will be of use to us in this connection. The first of them will enable us to compare the rates of growth of the two constructible hierarchies. The other two will help us to define well-orderings of the levels of the Jensen hierarchy.

1.17 Lemma[7]**.** *Let U be a transitive set. Then*

$$\text{rud}(U) \cap \mathscr{P}(U) = \text{Def}(U).$$

In fact

$$\Sigma_0(\text{rud}(U)) \cap \mathscr{P}(U) = \text{Def}(U).$$

Proof. We commence by proving that

$$(*) \qquad \Sigma_0(U \cup \{U\}) \cap \mathscr{P}(U) = \text{Def}(U).$$

First of all let $A \in \text{Def}(U)$. Thus for some formula $\varphi(x)$ of $\mathscr{L}_U$,

$$A = \{x \in U \mid \models_U \varphi(\mathring{x})\}.$$

7 In the statement of this lemma we extend our notation a little by using $\Sigma_n(M)$ to mean the set of all $\Sigma_n(M)$ subsets of M. This notational extension will be used several times from now on.

Let $\psi(x)$ be the formula of $\mathscr{L}_{U \cup \{U\}}$ obtained from $\varphi(x)$ by binding all unbounded quantifiers by $\mathring{U}$. Clearly, for any $x \in U$.

$$\models_U \varphi(\mathring{x}) \quad \text{iff } \models_{U \cup \{U\}} \psi(\mathring{x}).$$

Thus

$$A = \{x \in U \cup \{U\} \mid \models_{U \cup \{U\}} \mathring{x} \in \mathring{U} \wedge \psi(\mathring{x})\} \in \Sigma_0(U \cup \{U\}) \cap \mathscr{P}(U).$$

Conversely, let $A \in \Sigma_0(U \cup \{U\}) \cap \mathscr{P}(U)$. Thus for some Σ_0 formula $\varphi(x)$ of $\mathscr{L}_{U \cup \{U\}}$,

$$A = \{x \in U \mid \models_{U \cup \{U\}} \varphi(\mathring{x})\}.$$

To show that $A \in \mathrm{Def}(U)$, it suffices to show that for any Σ_0 formula $\varphi(\vec{x})$ of $\mathscr{L}_{U \cup \{U\}}$, there is a formula $\varphi^*(\vec{x})$ of $\mathscr{L}_U$ such that for any $\vec{x} \in U$

$$\models_{U \cup \{U\}} \varphi(\mathring{\vec{x}}) \quad \text{iff } \models_U \varphi^*(\mathring{\vec{x}}).$$

The proof of the above is by induction on φ. Suppose first that φ is primitive. If φ does not involve $\mathring{U}$, take $\varphi^* = \varphi$, in which case the result is clear. Suppose that φ involves $\mathring{U}$. If φ is of the form $(a \in \mathring{U})$ where $a \in Vbl \cup \mathrm{Const}_U$, take φ^* to be $(a = a)$. If φ is of the form $(\mathring{U} = \mathring{U})$, take φ^* to be $\forall x(x = x)$. In all other cases, take φ^* to be $\exists x(x \neq x)$. It is easily seen that φ^* is as required. In case $\varphi = \psi \wedge \theta$ now, we take $\varphi^* = (\psi^*) \wedge (\theta^*)$, and in case $\varphi = \neg \psi$, we take $\varphi^* = \neg (\psi^*)$. Suppose next that φ is of the form $(\exists x \in a)\, \psi$, where $a \in Vbl \cup \mathrm{Const}_U$. In this case take φ^* to be $(\exists x \in a)(\psi^*)$. Finally, suppose that φ is of the form $(\exists x \in \mathring{U})\, \psi$. Then we take φ^* to be $\exists x(\psi^*)$. The result is clear now.

By (*), in order to prove the first part of the lemma, it suffices to show that

$$\Sigma_0(U \cup \{U\}) \cap \mathscr{P}(U) = \mathrm{rud}(U) \cap \mathscr{P}(U).$$

First of all, let $A \in \Sigma_0(U \cup \{U\}) \cap \mathscr{P}(U)$. Thus for some Σ_0 formula $\varphi(x)$ of $\mathscr{L}_{U \cup \{U\}}$,

$$A - \{x \in U \mid \vdash_{U \cup \{U\}} \varphi(\mathring{x})\}.$$

By Σ_0-absoluteness,

$$A = \{x \in U \mid \models_{\mathrm{rud}(U)} \varphi(\mathring{x})\}.$$

But $\mathrm{rud}(U)$ is amenable (by 1.6). Thus by definition of amenability, $A \in \mathrm{rud}(U)$. For the converse, let $A \in \mathrm{rud}(U) \cap \mathscr{P}(U)$. Then for some rudimentary function f and some $a \in U$, $A = f(a, U)$. Now by 1.3 and 1.2 (or rather by localised versions of them where V is taken to be the transitive, rudimentary closed set $\mathrm{rud}(U)$), there is a Σ_0 formula φ of $\mathscr{L}$ such that for any $x \in \mathrm{rud}(U)$,

$$x \in f(a, U) \quad \text{iff } \models_{\mathrm{rud}(U)} \varphi(\mathring{x}, \mathring{a}, \mathring{U}).$$

Thus

$$A = \{x \in U \mid \models_{\mathrm{rud}(U)} \varphi(\mathring{x}, \mathring{a}, \mathring{U})\}.$$

By Σ_0-absoluteness it follows that

$$A = \{x \in U \mid \models_{U \cup \{U\}} \varphi(\mathring{x}, \mathring{a}, \mathring{U})\}.$$

Hence $A \in \Sigma_0(U \cup \{U\})$.

For the second part of the lemma it suffices to prove that

$$\Sigma_0(\text{rud}(U)) \cap \mathscr{P}(U) \subseteq \text{rud}(U).$$

Let $A \in \Sigma_0(\text{rud}(U)) \cap \mathscr{P}(U)$. Then for some Σ_0 formula $\varphi(x)$ of $\mathscr{L}_{\text{rud}(U)}$,

$$A = \{x \in U \mid \models_{\text{rud}(U)} \varphi(\mathring{x})\}.$$

So as $\text{rud}(U)$ is amenable, $A \in \text{rud}(U)$. The proof is complete. □

By tracing through the proof of the above lemma, we see that we have in fact proved the following result:

1.18 Lemma. *Let $\varphi(y, \vec{x})$ be a Σ_0 formula of $\mathscr{L}$. Then there is a formula $\psi(\vec{x})$ of $\mathscr{L}$ such that for any transitive, rudimentary closed set U,*

$$(\forall \vec{x} \in U)\,[\models_{\text{rud}(U)} \varphi(U, \vec{x}) \quad \text{iff } \models_U \psi(\vec{x})]. \quad \square$$

This lemma will be of use to us in dealing with successor levels of the Jensen hierarchy of constructible sets. (See, for example, the proof of V.5.18.) The following consequence of 1.18 will be required in Chapter VIII.

1.19 Lemma. *Let M, N be transitive, rud closed sets, and let*

$$\sigma: M \prec N.$$

Then there is a unique embedding

$$\tilde{\sigma}: \text{rud}(M) \prec_1 \text{rud}(N)$$

such that $\sigma \subseteq \tilde{\sigma}$.

Proof. We show first that if f, g are rudimentary functions and $x, y \in M$ are such that $f(M, x) = g(M, y)$, then $f(N, \sigma(x)) = g(N, \sigma(y))$.

By 1.3 (and Σ_0 absoluteness), let φ be a Σ_0 formula of $\mathscr{L}$ such that for any transitive, rud closed set U and any $a, b, c \in U$.

$$f(a, b) = g(a, c) \leftrightarrow \models_U \exists z\, \varphi(z, \mathring{a}, \mathring{b}, \mathring{c}).$$

Then we have

$$\models_{\text{rud}(M)} \exists z\, \varphi(z, \mathring{M}, \mathring{x}, \mathring{y}).$$

So for some $z \in \text{rud}(M)$,

$$\models_{\text{rud}(M)} \varphi(\mathring{z}, \mathring{M}, \mathring{x}, \mathring{y}).$$

For some rudimentary function h and some $w \in M$, we have $z = h(M, w)$. So

(1) $$\models_{\text{rud}(M)} \varphi(h(M, w)^{\circ}, \mathring{M}, \mathring{x}, \mathring{y}).$$

Since h is rudimentary, hence simple, the formula $\varphi(h(M, w), M, x, y)$ is in fact Σ_0 in the variables M, w, x, y. So by 1.18 there is an $\mathscr{L}$-formula ψ, which depends upon φ but not upon M, such that (1) is equivalent to

(2) $$\models_M \psi(\mathring{w}, \mathring{x}, \mathring{y})$$

(for any such M). Applying σ to (2) we get

(3) $$\models_N \psi(\sigma\,(\mathring{w}), \sigma\,(\mathring{x}), \sigma\,(\mathring{y})).$$

But the equivalence of (1) and (2) holds for N as well as M. Hence by (3), we get

(4) $$\models_{\text{rud}(N)} \varphi(h(N, \sigma(w))^{\circ}, \mathring{N}, \sigma\,(\mathring{x}), \sigma\,(\mathring{y})).$$

Hence

(5) $$\models_{\text{rud}(N)} \exists z\, \varphi(z, \mathring{N}, \sigma\,(\mathring{x}), \sigma\,(\mathring{y})).$$

So by choice of φ, we conclude that $f(N, \sigma(x)) = g(N, \sigma(y))$, as required.

By the above result we may define a function $\tilde{\sigma}$: $\text{rud}(M) \to \text{rud}(N)$ by setting

$$\tilde{\sigma}(f(M, x)) = f(N, \sigma(x))$$

for all rudimentary functions f and all $x \in M$. We show that $\tilde{\sigma}$ is Σ_1 elementary. (Uniqueness of $\tilde{\sigma}$ will then be immediate, of course, since any Σ_1 elementary embedding which extends σ must satisfy the above defining equation.)

Let $\varphi(x, y)$ be a Σ_0 formula of $\mathscr{L}$. Suppose first that for some $x \in \text{rud}(M)$,

$$\models_{\text{rud}(M)} \exists y\, \varphi(\mathring{x}, y).$$

Pick $y \in \text{rud}(M)$ such that

$$\models_{\text{rud}(M)} \varphi(\mathring{x}, \mathring{y}).$$

There are rudimentary functions f, g and elements $\bar{x}, \bar{y} \in M$ such that $x = f(M, \bar{x})$, $y = g(M, \bar{y})$. Thus

(*) $$\models_{\text{rud}(M)} \varphi(f(M, \bar{x})^{\circ}, g(M, \bar{y})^{\circ}).$$

Since f, g are simple the formula $\varphi(f(M, \bar{x}), g(M, \bar{y}))$ is Σ_0 in variables $M, \bar{x}, \bar{y}$. So there is an $\mathscr{L}$-formula ψ, independent of M, such that (*) is equivalent to

$$(**) \qquad \models_M \psi(\mathring{\bar{x}}, \mathring{\bar{y}}).$$

Applying σ to (**), we get

$$\models_N \psi(\sigma(\mathring{\bar{x}}), \sigma(\mathring{\bar{y}})).$$

Since the equivalence of (*) and (**) is valid for N in place of M, we get

$$\models_{\mathrm{rud}(N)} \varphi(f(N, \sigma(\bar{x}))^{\circ}, g(N, \sigma(\bar{y}))^{\circ}),$$

i.e.

$$\models_{\mathrm{rud}(N)} \varphi(\tilde{\sigma}(\mathring{x}), \tilde{\sigma}(\mathring{y})).$$

Thus

$$\models_{\mathrm{rud}(N)} \exists y\, \varphi(\tilde{\sigma}(\mathring{x}), y).$$

This is what we set out to prove.

Conversely, suppose that $x \in \mathrm{rud}(M)$ is such that

$$\models_{\mathrm{rud}(N)} \exists y\, \varphi(\tilde{\sigma}(\mathring{x}), y).$$

Let $x = f(M, \bar{x})$, where f is rudimentary and $\bar{x} \in M$. Pick $y \in \mathrm{rud}(N)$ so that

$$\models_{\mathrm{rud}(N)} \varphi(\tilde{\sigma}(\mathring{x}), \mathring{y}).$$

Let $y = g(N, \bar{y})$ where g is rudimentary and $\bar{y} \in N$. Then

$$(+) \qquad \models_{\mathrm{rud}(N)} \varphi(f(N, \sigma(\bar{x}))^{\circ}, g(N, \bar{y})^{\circ}).$$

As above, let ψ be an $\mathscr{L}$-formula such that (+) is equivalent to

$$(++) \quad \models_N \psi(\sigma(\mathring{\bar{x}}), \mathring{\bar{y}})$$

for any such N. We have (since (+) is valid)

$$\models_N \exists y'\, \psi(\sigma(\mathring{\bar{x}}), y').$$

So, as $\sigma: M \prec N$,

$$\models_M \exists y'\, \psi(\mathring{\bar{x}}, y').$$

So for some $y' \in M$,

$$\models_M \psi(\mathring{\bar{x}}, \mathring{y}').$$

By the equivalence of (+) and (+ +) applied to M, we get

$$\models_{\mathrm{rud}(M)} \varphi(\mathring{f(M, \vec{x})}, \mathring{g(M, y')}).$$

Thus

$$\models_{\mathrm{rud}(M)} \exists y \varphi(\mathring{\vec{x}}, y),$$

and the proof is complete. □

The following lemma provides us with a useful hierarchy for the construction of rud(U) from U.

1.20 Lemma. *There is a rudimentary function* **S** *such that whenever U is transitive,*

$$U \cup \{U\} \subseteq \mathbf{S}(U) \quad \text{and} \quad \mathrm{rud}(U) = \bigcup_{n<\omega} \mathbf{S}^n(U).$$

Proof. Set

$$\mathbf{S}(U) = [U \cup \{U\}] \cup \left[\bigcup_{i=0}^{8} F_i''(U \cup \{U\})^2\right].$$

The result follows from 1.11. □

1.21 Lemma. *There is a rudimentary function* **Wo** *such that whenever u is transitive and r is a well-ordering of u,* **Wo**(u, r) *is an end-extension of r which well-orders* **S**(u).

Proof. The idea is roughly the same as in II.4.4. Since

$$\mathbf{S}(u) = [u \cup \{u\}] \cup \left[\bigcup_{i=0}^{8} F_i''(u \cup \{u\})^2\right],$$

r induces, via the functions $F_0, \ldots, F_8$, a well-ordering of $\mathbf{S}(u)$. The function **Wo** will be rudimentary because of 1.1(14) and 1.5, since we shall obtain **Wo** by the definition

$$\mathbf{Wo}(u, r) = \mathbf{S}(u)^2 \cap \{(x, y) \mid \Phi(u, r, x, y)\},$$

where Φ is a Σ_0 formula of LST (see below).

Before we formulate Φ precisely, let us indicate what this formula is intended to say. Let $\tilde{r}$ denote the ordering r with u added as a greatest element. To see if $\Phi(u, r, x, y)$, we first check if $x, y \in u \cup \{u\}$, in which case we order x, y according to r, i.e. $\Phi(u, r, x, y)$ iff $x \tilde{r} y$. If $x \in u \cup \{u\}$ and $y \notin u \cup \{u\}$, then $\Phi(u, r, x, y)$ unconditionally holds. If $x \notin u \cup \{u\}$ and $y \in u \cup \{u\}$, then $\neg\, \Phi(u, r, x, y)$. Now suppose that $x, y \notin u \cup \{u\}$. First we see if the least i for which $x \in F_i''(u \cup \{u\})^2$ is smaller than the least i for which $y \in F_i''(u \cup \{u\})^2$, in which case $\Phi(u, r, x, y)$. If the two indices here are ordered in the opposite way, then $\neg\, \Phi(u, r, x, y)$. Otherwise, let i be the common least index here, and proceed as follows. Let x_1 be the $\tilde{r}$-least element of $u \cup \{u\}$ for which $x \in F_i''(\{x_1\} \times (u \cup \{u\}))$, and let y_1 be defined analogously for y. If $x_1 \tilde{r} y_1$, then $\Phi(u, r, x, y)$, and if $y_1 \tilde{r} x_1$, then $\neg\, \Phi(u, r, x, y)$. Otherwise, $x_1 = y_1$, and we define x_2 to be the $\tilde{r}$-least member of $u \cup \{u\}$ such that $x = F_i(x_1, x_2)$ and define y_2 for y, y_1 analogously, and set $\Phi(u, r, x, y)$ iff $x_2 \tilde{r} y_2$.

Precisely, $\Phi(u, r, x, y)$ is the following Σ_0 formula of LST (which we write in an abbreviated form for clarity):

$$\begin{aligned}
&[(x \in u) \wedge (y \in u) \wedge (x r y)] \vee [(x \in u) \wedge (y \notin u)] \\
&\quad \vee [(x = u) \wedge (y \notin u) \wedge (y \neq u)] \vee \bigvee_{i=0}^{8} [(x \notin u) \wedge (x \neq u) \wedge (y \notin u) \\
&\quad \wedge (y \neq u) \wedge \bigwedge_{j<i} (x \notin F_j''(u \cup \{u\})^2 \wedge y \notin F_j''(u \cup \{u\})^2) \\
&\quad \wedge [(x \in F_i''(u \cup \{u\})^2 \wedge y \notin F_i''(u \cup \{u\})^2) \vee (\exists x_1, x_2 \in u \cup \{u\}) \\
&\qquad [x = F_i(x_1, x_2) \wedge (\forall y_1, y_2 \in u \cup \{u\}) (y_1 r x_1 \vee y_1 = x_1 \\
&\quad \rightarrow y \neq F_i(y_1, y_2))] \vee (\exists x_1 \in u \cup \{u\}) (\exists y_1, y_2 \in u \cup \{u\}) \\
&\qquad [x = F_i(x_1, y_1) \wedge y = F_i(x_1, y_2) \wedge (\forall z_1, z_2 \in u \cup \{u\}) (z_1 r x_1 \\
&\quad \vee (z_1 \in u \wedge x_1 = u) \rightarrow x \neq F_i(z_1, z_2) \wedge y \neq F_i(z_1, z_2)) \\
&\quad \wedge (y_1 r y_2 \vee (y_1 \in u \wedge y_2 \notin u))]].
\end{aligned}$$

In connection with the above formula, the following points should be noted. $\bigvee_{i=0}^{8}$ denotes the disjunction of nine formulas for $i = 0, \ldots, 8$, and $\bigwedge_{j<i}$ is the conjunction of i formulas for $j = 0, \ldots, i-1$. In the case $i = 0$, the conjunction $\bigwedge_{j<i}$ should be dropped, whereas for $i = 1$ the conjunction is a degenerate one consisting of a single formula only. Expressions such as $x \in F_i''(u \cup \{u\})^2$ should be written as

$$(\exists y \in u \cup \{u\}) (\exists z \in u \cup \{u\}) (x = F_i(y, z)).$$

Since the function $u \cup \{u\}$ is simple, quantifiers of the forms $(\exists x \in u \cup \{u\})$ and $(\forall x \in u \cup \{u\})$ are allowed in a Σ_0 formula of course.

An examination of the above formula $\Phi(u, r, x, y)$ should complete the proof of the lemma now. □

To complete this section we prove a result which we shall need in order to prove the Condensation Lemma for the Jensen hierarchy.

1.22 Lemma. *Let M be a transitive, rudimentary closed set, and let $X \prec_1 M$. Then X is rudimentary closed and $\langle X, \in \rangle$ satisfies the Axiom of Extensionality. Let $\pi: \langle X, \in \rangle \cong \langle W, \in \rangle$, where W is transitive. If $f: M^n \rightarrow M$ is rudimentary, then for all $\vec{x} \in X$, $\pi(f(\vec{x})) = f(\pi(\vec{x}))$.*

Proof. Since M is transitive, $\langle M, \in \rangle$ satisfies the Axiom of Extensionality. So for any $x, y \in X$,

$$\models_M [x \neq y \leftrightarrow \exists z (z \in x \leftrightarrow z \notin y)].$$

Thus if $x \neq y$, then since $X \prec_1 M$, we have

$$\models_X \exists z (z \in x \leftrightarrow z \notin y).$$

Hence

$$\models_X [x \neq y \leftrightarrow \exists z(z \in x \leftrightarrow z \notin y)].$$

and so $\langle X, \in \rangle$ satisfies the Axiom of Extensionality. And by 1.3, X is, of course, rudimentary closed, so in particular, if $f\colon M^n \to M$ is rudimentary, then $f(\vec{x}) \in X$ whenever $\vec{x} \in X$. We shall prove by means of an induction on a rudimentary definition of f that $\pi(f(\vec{x})) = f(\pi(\vec{x}))$ for all $\vec{x} \in X$. Cases (i) through (iv) of the rudimentary function schemata cause no problems in this induction, as is easily seen. For case (v), suppose that $f(y, \vec{x}) = \bigcup_{z \in y} g(z, \vec{x})$, where g is rudimentary and for $z, \vec{x} \in X$, it is the case that $\pi(g(z, \vec{x})) = g(\pi(z), \pi(\vec{x}))$. Let $y, \vec{x} \in X$. We show that $\pi(f(y, \vec{x})) = f(\pi(y), \pi(\vec{x}))$.

By definition of π,

$$\pi(f(y, \vec{x})) = \pi''[f(y, \vec{x}) \cap X].$$

And by definition of f,

$$\begin{aligned} f(\pi(y), \pi(\vec{x})) &= \bigcup\{g(z, \pi(\vec{x})) \mid z \in \pi(y)\} \\ &= \bigcup\{g(z, \pi(\vec{x})) \mid z \in \pi''(y \cap X)\} \\ &= \bigcup\{g(\pi(z), \pi(\vec{x})) \mid z \in y \cap X\} \\ &= \bigcup\{\pi(g(z, \vec{x})) \mid z \in y \cap X\}. \end{aligned}$$

So it suffices to show that

$$\pi''[f(y, \vec{x}) \cap X] = \bigcup\{\pi(g(z, \vec{x})) \mid z \in y \cap X\}.$$

Suppose first that $v \in \pi''[f(y, \vec{x}) \cap X]$, say $v = \pi(u)$ where $u \in f(y, \vec{x}) \cap X$. Since $u \in f(y, \vec{x})$, we have $(\exists z \in y)\,(u \in g(z, \vec{x}))$. But this sentence is $\Sigma_1^M(\{u, y, \vec{x}\})$ and $u, y, \vec{x} \in X \prec_1 M$, so $(\exists z \in y \cap X)\,(u \in g(z, \vec{x}))$. Hence $v = \pi(u) \in \bigcup\{\pi(g(z, \vec{x})) \mid z \in y \cap X\}$.

Now suppose that $v \in \bigcup\{\pi(g(z, \vec{x})) \mid z \in y \cap X\}$. Pick $z \in y \cap X$ such that $v \in \pi(g(z, \vec{x}))$. Then $v \in \pi''[g(z, \vec{x}) \cap X]$, so for some $u \in g(z, \vec{x}) \cap X$, we have $v = \pi(u)$. But then $u \in \bigcup\{g(z, \vec{x}) \mid z \in y\}$ and $u \in X$, so $u \in f(y, \vec{x}) \cap X$, which gives $v = \pi(u) \in \pi''[f(y, \vec{x}) \cap X]$. The proof is complete. □

2. *The Jensen Hierarchy of Constructible Sets*

The *Jensen hierarchy*, $(J_\alpha \mid \alpha \in \mathrm{On})$, is defined by the following recursion:

$$\begin{aligned} J_0 &= \emptyset; \\ J_{\alpha+1} &= \mathrm{rud}(J_\alpha); \\ J_\lambda &= \bigcup_{\alpha<\lambda} J_\alpha, \quad \text{if } \lim(\lambda). \end{aligned}$$

2.1 Lemma.

(i) *Each J_α is transitive.*
(ii) $\alpha \leqslant \beta$ *implies* $J_\alpha \subseteq J_\beta$.
(iii) $rank(J_\alpha) = J_\alpha \cap \mathrm{On} = \omega\alpha$.

Proof. (i) By 1.16.
(ii) Immediate.
(iii) By induction on α. For $\alpha = 0$ the result is trivial. Limit stages in the induction are immediate. For successor steps, we use the finite rank property of rudimentary functions (1.4) to show that

$$\mathrm{rank}(J_{\alpha+1}) = \mathrm{rank}(\mathrm{rud}(J_\alpha)) = \mathrm{rank}(J_\alpha) + \omega.$$

The details are left to the reader. □

Note in particular that in passing from J_α to $J_{\alpha+1}$, exactly ω new ordinals appear: $\omega\alpha, \omega\alpha + 1, \omega\alpha + 2, \ldots, \omega\alpha + n, \ldots, (n \in \omega)$, whereas by 1.17,

$$J_{\alpha+1} \cap \mathscr{P}(J_\alpha) = \mathrm{Def}(J_\alpha).$$

Thus, although $J_{\alpha+1}$ only contains those subsets of J_α which are J_α-definable, these sets appear in a hierarchy which is "stretched" from one level of rank, as is the case with the usual constructible hierarchy, to ω levels of rank. Moreover, this stretched hierarchy is closed under many simple set-theoretic functions such as ordered pairs, union, cartesian product, etc.

To facilitate our handling of the Jensen hierarchy, we define a sub-hierarchy as follows.

$$S_0 = \emptyset;$$
$$S_{\alpha+1} = \mathbf{S}(S_\alpha);$$
$$S_\lambda = \bigcup_{\alpha<\lambda} S_\alpha, \quad \text{if } \lim(\lambda).$$

Clearly, the sets J_α are just the limit levels of this new hierarchy. In fact:

2.2 Lemma.

(i) $\alpha \leqslant \beta$ *implies* $S_\alpha \subseteq S_\beta$;
(ii) $J_\alpha = \bigcup_{\nu<\omega\alpha} S_\nu = S_{\omega\alpha}$.

Proof. (i) Immediate.
(ii) By induction. The only non-trivial step is the successor step. Here we have:

$$J_{\alpha+1} = \mathrm{rud}(J_\alpha) = \bigcup_{n\in\omega} \mathbf{S}^n(J_\alpha) = \bigcup_{n\in\omega} \mathbf{S}^n(S_{\omega\alpha}) = \bigcup_{n\in\omega} S_{\omega\alpha+n} = S_{\omega\alpha+\omega}$$
$$= S_{\omega(\alpha+1)}. \quad □$$

We shall use the S-hierarchy in order to assist our detailed study of the Jensen hierarchy. But before we commence this study, let us digress for a moment to examine the relationship between the Jensen hierarchy and the usual constructible hierarchy. (In particular, we have not yet proved that the Jensen hierarchy does consist only of constructible sets, and that all constructible sets do appear in the Jensen hierarchy.)

Will, we have $J_0 = L_0 = \emptyset$, of course. And it is easily seen that $J_1 = H_\omega = L_\omega$. In view of these two facts, and our knowledge that $J_\alpha \cap \mathrm{On} = \omega\alpha$ and $L_\alpha \cap \mathrm{On} = \alpha$ for all α, one might be tempted into thinking that $J_\alpha = L_{\omega\alpha}$ for all α. This is not the case, however. (The proof that the above equality is false makes a good little exercise for the reader.) Nevertheless, we do have $J_\alpha = L_\alpha$ whenever $\omega\alpha = \alpha$. As a first step towards proving this, we have:

2.3 Lemma. *For all* α, $L_\alpha \subseteq J_\alpha$ *and* L_α, $(L_\beta \mid \beta \leqslant \alpha) \in J_{\alpha+1}$.

Proof. We first of all prove that:

$$(*) \qquad u \in J_\alpha \to \mathrm{Def}(u) \subseteq J_\alpha.$$

For $\alpha = 0$ there is nothing to prove, and for $\alpha = 1$ the result is trivial, since $J_1 = H_\omega$, so we shall assume that $\alpha > 1$ from now on. During the proof of 1.11, we showed that for any representation $\varphi(x_0, \ldots, x_n)$ of an $\mathscr{L}$-formula φ, the function $t_{\varphi(x_0,\ldots,x_n)}$ is rudimentary, where

$$t_{\varphi(x_0,\ldots,x_n)}(u) = \{(x_0, \ldots, x_n) \mid x_0, \ldots, x_n \in u \wedge \models_u \varphi(\mathring{x}_0, \ldots, \mathring{x}_n)\}.$$

It follows that the functions $d_{\varphi(x_0,\ldots,x_n)}$ are rudimentary, where we define

$$d_{\varphi(x_0,\ldots,x_n)}(u, x_1, \ldots, x_n) = \begin{cases} \{x_0 \in u \mid \models_u \varphi(\mathring{x}_0, \ldots, \mathring{x}_n)\}, & \text{if } x_1, \ldots, x_n \in u \\ \emptyset, & \text{otherwise.} \end{cases}$$

Since J_α is rudimentary closed, for each $\varphi(x_0, \ldots, x_n)$ we have

$$u, x_1, \ldots, x_n \in J_\alpha \to d_{\varphi(x_0,\ldots,x_n)}(u, x_1, \ldots, x_n) \in J_\alpha.$$

But for any set u,

$$\mathrm{Def}(u) = \{d_{\varphi(x_0,\ldots,x_n)}(u, x_1, \ldots, x_n) \mid \varphi(x_0, \ldots, x_n) \text{ is a representation of an } \mathscr{L}\text{-formula } \varphi \text{ and } x_1, \ldots, x_n \in u\}.$$

Thus $u \in J_\alpha$ implies $\mathrm{Def}(u) \subseteq J_\alpha$, which proves (*).

We prove the lemma by induction now. For $\alpha = 0$ there is nothing to prove. For the successor case, suppose we know that $L_\alpha \subseteq J_\alpha$, $L_\alpha \in J_{\alpha+1}$, and $(L_\beta \mid \beta \leqslant \alpha) \in J_{\alpha+1}$. Since $L_\alpha \in J_{\alpha+1}$, (*) tells us at once that $L_{\alpha+1} = \mathrm{Def}(L_\alpha) \subseteq J_{\alpha+1}$. We show next that $L_{\alpha+1} \in \mathrm{Def}(J_{\alpha+1})$, whence $L_{\alpha+1} \in J_{\alpha+2}$, of course. Well, we have

$$\begin{aligned} L_{\alpha+1} &= \{x \subseteq L_\alpha \mid (\exists \varphi)(\exists(\vec{a}))\,[\mathrm{Fml}_\emptyset(\varphi) \wedge \vec{a} \in L_\alpha \\ &\quad \wedge (\forall z \in L_\alpha)(z \in x \leftrightarrow \models_{L_\alpha} \varphi(\mathring{z}, \vec{\mathring{a}}))]\} \\ &= \{x \in J_{\alpha+1} \mid x \in L_\alpha \wedge (\exists \varphi)(\exists(\vec{a}))\,[\mathrm{Fml}_\emptyset(\varphi) \wedge \vec{a} \in L_\alpha \\ &\quad \wedge (\forall z \in L_\alpha)(z \in x \leftrightarrow \models_{L_\alpha} \varphi(\mathring{z}, \vec{\mathring{a}}))]\} \\ &= \{x \in L_{\alpha+1} \mid x \subseteq L_\alpha \wedge (\exists \varphi)(\exists(\vec{a}))\,[\mathrm{Fml}_\emptyset(\varphi) \wedge \vec{a} \in L_\alpha \\ &\quad \wedge (\forall z \in L_\alpha)(z \in x \leftrightarrow \mathrm{Sat}(L_\alpha, \mathrm{Sub}(\varphi, \vec{v}, \mathring{z}, \vec{\mathring{a}})))]\}, \end{aligned}$$

where for clarity we have abused slightly the notation developed in II.2, using Sub as a function rather than as a relation. Now, for amenable sets M, the predicate $\mathrm{Fml}_\emptyset(-)$ is Δ_1^M (by II.2.4), the function Sub is Δ_1^M (by II.2.7), and the predicate Sat is Δ_1^M (by II.2.8). But $J_{\alpha+1}$ is rudimentary closed, and hence amenable. Moreover, the set $\mathrm{Fml}_\emptyset$ is a subset of $J_{\alpha+1}$. Hence by Δ_1-absoluteness,

$$\begin{aligned} L_{\alpha+1} = \{x \in J_{\alpha+1} \models_{J_{\alpha+1}} \text{“}(x \subseteq L_\alpha) \wedge (\exists \varphi)(\exists(\vec{a}))\,[\mathrm{Fml}_\emptyset(\varphi) \wedge \vec{a} \in L_\alpha \\ \wedge (\forall z \in L_\alpha)(z \in x \leftrightarrow \mathrm{Sat}(L_\alpha, \mathrm{Sub}(\varphi, \vec{v}, \mathring{z}, \vec{\mathring{a}})))]\text{”}\}. \end{aligned}$$

Hence $L_{\alpha+1} \in \mathrm{Def}(J_{\alpha+1})$, giving $L_{\alpha+1} \in J_{\alpha+2}$, as required. Finally, we have

$$(L_\beta \mid \beta \leqslant \alpha + 1) = (L_\beta \mid \beta < \alpha) \cup \{(L_{\alpha+1}, \alpha + 1)\},$$

so by induction hypothesis and the fact that $L_{\alpha+1} \in J_{\alpha+2}$, since $J_{\alpha+2}$ is rudimentary closed, we see that $(L_\beta \mid \beta \leqslant \alpha + 1) \in J_{\alpha+2}$.

There remains the limit case of the induction. Suppose that $\alpha > 0$ is a limit ordinal, and that for all $\beta < \alpha$, $L_\beta \subseteq J_\beta$ and L_β, $(L_\gamma \mid \gamma \leqslant \beta) \in J_{\beta+1}$. So, as J_α is transitive, $L_\beta \subseteq J_\alpha$ for all $\beta < \alpha$. Hence $L_\alpha = \bigcup_{\beta<\alpha} L_\beta \subseteq J_\alpha$. Again,

$$L_\alpha = \{x \in J_\alpha \mid (\exists \nu < \alpha)(x \in L_\nu)\},$$

so we have

$$\begin{aligned} L_\alpha = \{x \in J_\alpha \mid (\exists f)\,[f \text{ is a function } \wedge \mathrm{dom}(f) \in \alpha \wedge f(0) = \emptyset \\ \wedge (\forall \nu \in \mathrm{dom}(f))\,[(\lim(\nu) \rightarrow f(\nu) = \bigcup_{\tau<\nu} f(\tau)) \wedge (\mathrm{succ}(\nu) \\ \rightarrow f(\nu) = \mathrm{Def}(f(\nu - 1)))] \wedge x \in \mathrm{ran}(f)]\}. \end{aligned}$$

But by induction hypothesis, $(L_\gamma \mid \gamma \leqslant \beta) \in J_\alpha$ for all $\beta < \alpha$, so the quantifier $(\exists f)$ in the above can be restricted to J_α (without affecting the meaning). Moreover, the unbounded quantifiers involved in the definition of the function Def can also be restricted to J_α, since they only refer to elements of $\bigcup \mathrm{ran}(f)$ (see the proof of II.2.12). Hence, if φ is the $\mathscr{L}$-formula which we have just been (implicitly) discussing, we have

$$L_\alpha = \{x \in J_\alpha \mid \models_{J_\alpha} \varphi(\mathring{x})\}.$$

Thus $L_\alpha \in \mathrm{Def}(J_\alpha) \subseteq J_{\alpha+1}$. Similar considerations lead to the conclusion that $(L_\beta \mid \beta < \alpha) \in \mathrm{Def}(J_\alpha)$, and so

$$(L_\beta \mid \beta \leqslant \alpha) = (L_\beta \mid \beta < \alpha) \cup \{(L_\alpha, \alpha)\} \in J_{\alpha+1}.$$

The lemma is proved. □

Using 2.3, we may now show that

$$L = \bigcup_{\alpha \in \mathrm{On}} J_\alpha.$$

In fact we show that the sets J_α and L_α are equal for many ordinals α.

2.4 Lemma.

(i) $L_\alpha \subseteq J_\alpha \subseteq L_{\omega\alpha}$.

(ii) $J_\alpha = L_\alpha$ *iff* $\omega\alpha = \alpha$.

(iii) $L = \bigcup_{\alpha \in \mathrm{On}} J_\alpha$.

Proof. Clearly, (i) → (ii) → (iii). We prove (i). By 2.3, we know already that $L_\alpha \subseteq J_\alpha$. We show that $J_\alpha \subseteq L_{\omega\alpha}$. As a first step we prove that

(∗) for all α: $u \in L_\alpha \rightarrow \mathbf{S}(u) \in L_{\alpha+5}$.

It is easily seen that for each $i = 0, \ldots, 8$,

$$x, y \in L_\alpha \rightarrow F_i(x, y) \in L_{\alpha+4}.$$

Thus if $u \in L_\alpha$, we have $\mathbf{S}(u) \subseteq L_{\alpha+4}$. So, by Σ_0-absoluteness,

$$\mathbf{S}(u) = \{x \in L_{\alpha+4} \mid \models_{L_{\alpha+4}} \text{“}(\mathring{x} \in \mathring{u}) \wedge (\exists v, w \in \mathring{u} \cup \{\mathring{u}\}) \\ [\mathring{x} = F_0(v, w) \vee \ldots \vee \mathring{x} = F_8(v, w)]\text{”}\}.$$

Hence $\mathbf{S}(u) \in \mathrm{Def}(L_{\alpha+4}) = L_{\alpha+5}$, which proves (∗).

In order to prove that $J_\alpha \subseteq L_{\omega\alpha}$, since $L_{\omega\alpha}$ is transitive and $J_\alpha = \bigcup_{\nu<\omega\alpha} S_\nu$, it suffices to show that $S_\nu \in L_{\omega\alpha}$ for all $\nu < \omega\alpha$. By (∗), $\mathbf{S}'' L_{\omega\alpha} \subseteq L_{\omega\alpha}$. In particular, $L_{\omega\alpha}$ is rudimentary closed and (by 1.3) there is a Σ_0 formula $\varphi(v_0, v_1)$ of $\mathscr{L}$, independent of α, such that for $x, y \in L_{\omega\alpha}$,

$$y = \mathbf{S}(x) \quad \text{iff} \models_{L_{\omega\alpha}} \varphi(\mathring{y}, \mathring{x}).$$

By induction on α we prove the following result:

$P(\alpha)$: if $\nu < \omega\alpha$, then S_ν, $(S_\tau \mid \tau \leqslant \nu) \in L_{\omega\alpha}$ and the sequence $(S_\nu \mid \nu < \omega\alpha)$ is uniformly $\Sigma_1^{L_{\omega\alpha}}$.

This, of course, will complete the proof of the lemma.

Let $\theta(f)$ be the following Σ_0 formula of $\mathscr{L}$ (to define the hierarchy $(S_\nu \mid \nu \in \mathrm{On})$):

$$\text{“}f \text{ is a function”} \wedge \text{“dom}(f) \text{ is an ordinal”} \wedge f(0) = \emptyset$$
$$\wedge (\forall \nu \in \mathrm{dom}(f)) [(\mathrm{succ}(\nu) \rightarrow \varphi(f(\nu), f(\nu - 1)))$$
$$\wedge (\lim(\nu) \rightarrow f(\nu) = \bigcup_{\tau \in \nu} f(\tau)))].$$

By our above remarks, it is clear that for any α and any $\nu < \omega\alpha$, if

$$\models_{L_{\omega\alpha}} \exists f [\theta(f) \wedge y = f(\nu)],$$

then $y = S_\nu$. We prove the part of $P(\alpha)$ concerning Σ_1 definability by showing that, in fact, for any α and any $\nu < \omega\alpha$,

$$y = S_\nu \quad \text{iff } \models_{L_{\omega\alpha}} \exists f [\theta(f) \wedge y = f(\nu)].$$

Now for the proof of $P(\alpha)$. For $\alpha = 0$ there is nothing to prove. Now assume $P(\alpha)$. Then, in particular, $(S_\tau \mid \tau < \omega\alpha)$ is $\Sigma_1^{L_{\omega\alpha}}$, and hence is an element of $L_{\omega\alpha+1}$. Thus $J_\alpha = \bigcup_{\tau < \omega\alpha} S_\tau \in L_{\omega\alpha+2} \subseteq L_{\omega(\alpha+1)}$. For any $n < \omega$, since $L_{\omega\alpha}$ is rudimentary closed, we have $S_{\omega\alpha+n} = \mathbf{S}^n(J_\alpha) \in L_{\omega(\alpha+1)}$. Thus $S_\nu \in L_{\omega(\alpha+1)}$ for all $\nu < \omega(\alpha+1)$. Again, for any $n < \omega$, $(S_\tau \mid \tau \leqslant \omega\alpha + n) = (S_\tau \mid \tau < \omega\alpha) \cup \{(S_{\omega\alpha+m}, \omega\alpha + m) \mid m \leqslant n\}$, so as $L_{\omega(\alpha+1)}$ is rudimentary closed, $(S_\tau \mid \tau \leqslant \omega\alpha + n) \in L_{\omega(\alpha+1)}$, and so $(S_\tau \mid \tau \leqslant \nu) \in L_{\omega(\alpha+1)}$ for all $\nu < \omega(\alpha+1)$. Finally, to show that for any $\nu < \omega(\alpha+1)$,

$$y = S_\nu \quad \text{iff } \models_{L_{\omega(\alpha+1)}} \exists f [\theta(f) \wedge y = f(\nu)],$$

it clearly suffices to show that whenever $\nu < \omega(\alpha+1)$ and $y = S_\nu$, then there is an $f \in L_{\omega(\alpha+1)}$ such that

$$\models_{L_{(\alpha+1)}} \theta(f) \wedge y = f(\nu).$$

But $(S_\tau \mid \tau \leqslant \nu) \in L_{\omega(\alpha+1)}$ is such an f, so we are done.

Finally, assume $\delta > 0$ is a limit ordinal and that $P(\alpha)$ holds for all $\alpha < \delta$. It is then trivial that $S_\nu, (S_\tau \mid \tau \leqslant \nu) \in L_{\omega\delta}$ for all $\nu < \omega\delta$. And since $(S_\tau \mid \tau \leqslant \nu) \in L_{\omega\delta}$ for all $\nu < \omega\delta$, the same argument as above shows that for $\nu < \omega\delta$,

$$x = S_\nu \quad \text{iff } \models_{L_{\omega\delta}} \exists f [\theta(f) \wedge y = f(\nu)].$$

The proof is complete. □

Returning now to our study of the Jensen hierarchy itself, the same argument as in 2.4 above shows that

2.5 Lemma. *The sequence $(S_\nu \mid \nu < \omega\alpha)$ is uniformly $\Sigma_1^{J_\alpha}$ for all α.* □

2.6 Corollary. *The sequence $(J_\nu \mid \nu < \alpha)$ is uniformly $\Sigma_1^{J_\alpha}$ for all α.*

Proof. Clearly, the sequence $(\omega\nu \mid \nu < \alpha)$ is uniformly $\Sigma_1^{J_\alpha}$ for all α, so the result follows easily from 2.5. □

2.7 Lemma. *There are well-orderings $<_\nu^S$ of the sets S_ν such that:*

(i) *$\nu_1 < \nu_2$ implies $<_{\nu_1}^S \subseteq <_{\nu_2}^S$;*
(ii) *$<_{\nu+1}^S$ is an end-extension of $<_\nu^S$;*
(iii) *the sequence $(<_\nu^S \mid \nu < \omega\alpha)$ is uniformly $\Sigma_1^{J_\alpha}$ for all α.*

Proof. We use 1.21. Set $<_0^S = \emptyset$, and, by recursion, let

$$<_{\nu+1}^S = \mathbf{Wo}(S_\nu, <_\nu^S),$$
$$<_\lambda^S = \bigcup_{\nu<\lambda} <_\nu^S, \quad \text{if } \lim(\lambda).$$

Then (i) and (ii) are immediate, whilst (iii) is proved by an argument as in 2.4 and 2.5. □

2.8 Lemma. *There are well-orderings $<_\alpha$ of the sets J_α such that:*

(i) *$\alpha_1 < \alpha_2$ implies $<_{\alpha_1} \subseteq <_{\alpha_2}$;*
(ii) *$<_{\alpha+1}$ is an end-extension of $<_\alpha$;*
(iii) *the sequence $(<_\beta \mid \beta < \alpha)$ is uniformly $\Sigma_1^{J_\alpha}$ for all α;*
(iv) *$<_\alpha$ is uniformly $\Sigma_1^{J_\alpha}$ for all α;*
(v) *the function $\mathrm{pr}_\alpha(x) = \{z \mid z <_\alpha x\}$ is uniformly $\Sigma_1^{J_\alpha}$ for all α.*

Proof. Set $<_\alpha = <_{\omega\alpha}^S$ for all α. Then (i)–(iii) are immediate by 2.7. For (iv), note simply that $x <_\alpha y$ iff $(\exists \nu \in J_\alpha)(x <_\nu^S y)$. For (v), note that

$$y = \mathrm{pr}_\alpha(x) \quad \text{iff } (\exists \nu \in J_\alpha)(x \in S_\nu \wedge y = \{z \mid z <_\nu^S x\})$$

and that $\nu < \omega\alpha$ implies $<_\nu^S \in J_\alpha$, and use 2.5 and 2.7. □

By 2.4 we can define a well-ordering $<_J$ of L by setting

$$<_J = \bigcup_{\alpha\in\mathrm{On}} <_\alpha.$$

Then, as was the case with the well-ordering $<_L$, $<_J$ is a Σ_1 well-ordering of L.

2.9 Lemma (Condensation Lemma). *Let α be any ordinal. Let $X \prec_1 J_\alpha$. Then there is a unique ordinal β and a unique isomorphism π such that:*

(i) *$\pi: X \cong J_\beta$;*
(ii) *$\pi(\nu) \leqslant \nu$ for all $\nu \in X \cap \omega\alpha$;*
(iii) *$\pi(x) \leqslant_J x$ for all $x \in X$;*
(iv) *if $Y \subseteq X$ is transitive, then $\pi \restriction Y = \mathrm{id} \restriction Y$.*

Proof. By 1.22 there are unique π, W such that $\pi: X \cong W$, where W is transitive. Let $\beta = \pi''(X \cap \omega\alpha)$. We show that $W = J_\beta$, which proves (i). First we establish a simple claim.

Claim. $\gamma \in X \cap \omega\alpha \to [S_\gamma \in X \wedge \pi(S_\gamma) = S_{\pi(\gamma)}]$.

We prove the claim by induction on γ. Clearly, $0 \in X \cap \omega\alpha$, $S_0 = \emptyset \in X$, and $\pi(S_0) = \pi(\emptyset) = \emptyset = S_0 = S_{\pi(0)}$. Suppose now that $\gamma = \delta + 1$ and we have proved the claim below γ. Since $\gamma \in X$, we have $\delta \in X$ also. And by 2.5, we have $S_\gamma, S_\delta \in X$. Using 1.22 now, together with the induction hypothesis,

$$\pi(S_\gamma) = \pi(S_{\delta+1}) = \pi(\mathbf{S}(S_\delta)) = \mathbf{S}(\pi(S_\delta)) = \mathbf{S}(S_{\pi(\delta)}) = S_{\pi(\delta)+1} = S_{\pi(\gamma)}.$$

Finally, suppose that $\gamma > 0$ is a limit ordinal and we have proved the claim below γ. Notice that $\models_{J_\alpha} \lim(\gamma)$, so $\lim(\mathrm{otp}(X \cap \gamma))$, so $\lim(\pi(\gamma))$. Now, $S_\gamma = \bigcup_{\delta<\gamma} S_\delta$, so $\pi(S_\gamma) = \pi''(S_\gamma \cap X) = \pi''(\bigcup_{\delta<\gamma} (S_\delta \cap X))$, so it suffices to show that $S_{\pi(\gamma)} = \pi''(\bigcup_{\delta<\gamma} (S_\delta \cap X))$. First of all, let $x \in S_{\pi(\gamma)}$. Thus for some $\xi < \pi(\gamma)$, $x \in S_\xi$. But $\mathrm{ran}(\pi)$ is transitive, so $\xi = \pi(\zeta)$ for some $\zeta \in X \cap \gamma$. Thus by induction hypothesis, $x \in S_{\pi(\zeta)} = \pi(S_\zeta) = \pi''(S_\zeta \cap X) \subseteq \pi''(\bigcup_{\delta<\gamma} (S_\delta \cap X))$. Conversely, let $x \in \pi''(\bigcup_{\delta<\gamma} (S_\delta \cap X))$. Thus $x = \pi(y)$, where $y \in \bigcup_{\delta<\gamma} (S_\delta \cap X)$. Now, $\models_{J_\alpha} (\exists \delta < \gamma)(y \in S_\delta)$, so as $y, \gamma \in X \prec_1 J_\alpha$, we have $\models_X (\exists \delta < \gamma)(y \in S_\delta)$, so we can pick $\delta \in X \cap \gamma$ with $y \in S_\delta$. Then by induction hypothesis, $x = \pi(y) \in \pi(S_\delta) = S_{\pi(\delta)}$. But $\pi(\delta) < \pi(\gamma)$. Hence $x \in S_{\pi(\gamma)}$. This proves the claim.

Using the claim, it is now easy to prove that $W = S_{\omega\beta} = J_\beta$. Suppose first that $w \in W$. Thus $w = \pi(x)$ for some $x \in X$. Now $\models_{J_\alpha} (\exists \gamma)(x \in S_\gamma)$, so as $x \in X \prec_1 J_\alpha$, we have $\models_X (\exists \gamma)(x \in S_\gamma)$. So pick $\gamma \in X \cap \omega\alpha$ with $x \in S_\gamma$. Then $w = \pi(x) \in \pi(S_\gamma) = S_{\pi(\gamma)} \subseteq S_{\omega\beta} = J_\beta$. Conversely, suppose that $y \in J_\beta$. Then $y \in S_\gamma$ for some $\gamma < \omega\beta$. But $\gamma = \pi(\delta)$ for some $\delta \in X \cap \omega\alpha$. Thus $y \in S_{\pi(\delta)} = \pi(S_\delta) = \pi''(S_\delta \cap X)$, whence $y \in \mathrm{ran}(\pi) = W$.

That proves part (i) of the lemma. Part (iv) holds by definition of π. And (ii) follows from (iii). So we need to prove (iii). Notice that as $<_\alpha$ is uniformly $\Sigma_1^{J_\alpha}$, we have

$$x <_\alpha y \quad \text{iff } \pi(x) <_\beta \pi(y).$$

Suppose that $x <_J \pi(x)$ for some $x \in X$. Let x be the $<_J$-least such. Since $\pi(x) \in J_\beta$, we must have $x \in J_\beta$ here, so $x = \pi(x')$ for some $x' \in X$. But $\pi(x') = x <_J \pi(x)$ so $x' <_J x$. Thus by choice of x, $\pi(x') \leqslant_J x'$. But this means that $x \leqslant_J x'$, which is absurd. The lemma is proved. □

3. The Σ_1 Skolem Function

The general notion of a Σ_n skolem function was already introduced in II.6. Recall that if $\mathbf{M} = \langle M, (A_i)_{i<\omega}\rangle$, where M is an amenable set and $A_i \subseteq M$, then by a Σ_n-*skolem function* for $\mathbf{M}$ we mean a $\Sigma_n^M(\{p\})$ function h (for some $p \in M$) with

$\mathrm{dom}(h) \subseteq \omega \times M$, such that whenever $p \in \Sigma_n^M(\{x, p\})$ for some $x \in M$, then $\exists y P(y) \to (\exists i \in \omega) P(h(i, x))$. (In which case we say that p is a *good* parameter for h.)

In this section we shall be concerned with structures of the form $\langle M, A\rangle$, where $A \subseteq M$. Notice that if M is rudimentary closed, it is amenable. Hence we may reformulate II.6.1 through II.6.3 as follows.

If h is a function with $\mathrm{dom}(h) \subseteq \omega \times M$, and if $X \subseteq M$, then we shall denote by $h^*(X)$ the set $h''(\omega \times X)$. In what follows we assume $n \geqslant 1$.

3.1 Lemma. *Let $\langle M, A\rangle$ be transitive and rudimentary closed. Let h be a Σ_n skolem function for $\langle M, A\rangle$. If $x \in M$, then $x \in h^*(\{x\}) \prec_n \langle M, A\rangle$. (More precisely, $\langle h^*(\{x\}), A \cap h^*(\{x\})\rangle \prec_n \langle M, A\rangle$.)* $\square$

3.2 Lemma. *Let $\langle M, A\rangle, h$ be as above. Let $q \in M$, and let $X \subseteq M$ be closed under ordered pairs. Then $X \cup \{q\} \subseteq h^*(X \times \{q\}) \prec_n \langle M, A\rangle$.* $\square$

3.3 Lemma. *Let $\langle M, A\rangle$, h be as above. Let $X \subseteq M$, and suppose that $h^*(X)$ is closed under ordered pairs. Then $X \subseteq h^*(X) \prec_n \langle M, A\rangle$.* $\square$

Now, in II.6.6 we showed that each limit $L_\alpha (\alpha > \omega)$ has a Σ_1 skolem function. And an entirely parallel proof will show that each $J_\alpha (\alpha > 1)$ has a Σ_1 skolem function. But as our discussion in section 1 indicate, we require slightly more than this. We need to know that each amenable structure $\langle J_\alpha, A\rangle$ has a (uniform) Σ_1 skolem function, and that even in the absence of amenability, the definition of this skolem function still defines a function having "skolem-like" properties. This is where 1.15 comes in. By 1.15 (together with 1.9) we have:

3.4 Lemma. *Let $n \geqslant 1$. If $\alpha > 1$ and $\langle J_\alpha, A\rangle$ is amenable, then $\models^{\Sigma_n}_{\langle J_\alpha, A\rangle}$ is (uniformly) $\Sigma_n^{\langle J_\alpha, A\rangle}$.* $\square$

We now fix, once and for all, some simple enumeration $(\varphi_i \mid i < \omega)$ of all the formulas of $\mathscr{L}(A)$ of the form

$$\varphi_i = \varphi_i(v_0, v_1) = \exists v_2 \bar{\varphi}_i(v_0, v_1, v_2),$$

where $\bar{\varphi}_i$ is Σ_0. The exact definition of this enumeration is not important. All we need to know is that it is $\Delta_1^{J_1}$, which will be the case for any "effective" enumeration. We leave it to the reader to supply any details felt neccessary.

Fix $\langle J_\alpha, A\rangle$ now. For $i \in \omega$ and $x \in J_\alpha$, set:

$$r_{\alpha,A}(i, x) \simeq \text{the } <_J\text{-least } z \in J_\alpha \text{ such that } \models_{\langle J_\alpha, A\rangle} \bar{\varphi}_i((\mathring{z})_0, \mathring{x}, (\mathring{z})_1)$$

$$h_{\alpha,A}(i, x) \simeq (r_{\alpha,A}(i, x))_0.$$

Thus, for $i \in \omega$ and $x, y \in J_\alpha$:

$$y = h_{\alpha,A}(i, x) \leftrightarrow \text{there is a } z \in J_\alpha \text{ such that } (z)_0 = y \text{ and } z \text{ is the } <_J\text{-least } z \text{ in } J_\alpha \text{ such that } \models_{\langle J_\alpha, A\rangle} \bar{\varphi}_i((\mathring{z})_0, \mathring{x}, (\mathring{z})_1).$$

In other words:

$$y = h_{\alpha, A}(i, x) \leftrightarrow \exists z \exists w \ [(z)_0 = y \wedge w = \{v \mid v <_J z\} \\ \wedge \models_{\langle J_\alpha, A\rangle}[\bar{\varphi}_i((\mathring{z})_0, \mathring{x}, (\mathring{z})_1) \wedge (\forall v \in \mathring{w}) \neg \bar{\varphi}_i((v)_0, \mathring{x}, (v)_1)]].$$

Let θ be the canonical Σ_0 formula such that for all $\alpha > 1$ and all $z \in J_\alpha$,

$$w = \{v \mid v <_J z\} \leftrightarrow \models_{J_\alpha} \exists t \theta(\mathring{w}, \mathring{z}, t).$$

(See 2.8 (v).)

Then we have:

$$y = h_{\alpha, A}(i, x) \leftrightarrow \exists z \exists w \exists t [(z)_0 = y \wedge \models_{\langle J_\alpha, a\rangle}[\theta(\mathring{w}, \mathring{z}, \mathring{t}) \wedge \bar{\varphi}_i((\mathring{z})_0, \mathring{x}, (\mathring{z})_1) \\ \wedge (\forall v \in \mathring{w}) \neg \bar{\varphi}_i((v)_0, \mathring{x}, (v)_1)]].$$

Let $\theta_i(u, y, x)$ be the Σ_0 $\mathscr{L}$-formula:

$$[((u)_0)_0 = y \wedge \theta((u)_1, (u)_0, (u)_2) \wedge \bar{\varphi}_i(((u)_0)_0, x, ((u)_0)_1) \\ \wedge (\forall v \in (u)_1) \neg \bar{\varphi}_i((v)_0, x, (v)_1)].$$

(More precisely, let θ_i be the canonical rendering of this formula in true Σ_0 form.)

Then θ_i is independent of the choice of α, A. But clearly, for any $\langle J_\alpha, A\rangle$,

$$y = h_{\alpha, A}(i, x) \leftrightarrow (\exists u \in J_\alpha) [\models_{\langle J_\alpha, A\rangle} \theta_i(\mathring{u}, \mathring{y}, \mathring{x})].$$

We establish several important facts concerning the functions $h_{\alpha, A}$.

3.5 Lemma. *The sequence* $(\theta_i \mid i < \omega)$ *is* $\Delta_1^{J_1}$.

Proof. Since the sequence $(\bar{\varphi}_i \mid i < \omega)$ is $\Delta_1^{J_1}$. □

3.6 Lemma. *Let* $1 < \bar{\alpha} < \alpha$, $A \subseteq J_\alpha$. *If* $y = h_{\bar{\alpha}, A \cap J_{\bar{\alpha}}}(i, x)$, *then* $y = h_{\alpha, A}(i, x)$.

Proof. By Σ_0-absoluteness (I.9.14). □

Notice that we have not so far required that the structure $\langle J_\alpha, A\rangle$ is amenable. As we shall show presently, in the case where we do have amenability, the function $h_{\alpha, A}$ is Σ_1-definable over $\langle J_\alpha, A\rangle$. In such cases, it is possible to deduce our next three lemmas from II.6.1–II.6.3. We do not do it this way because we shall need these results in cases where amenability is not available.

3.7 Lemma. *Let* $A \subseteq J_\alpha$, $x \in J_\alpha$. *Then*

$$x \in h^*_{\alpha, A}(\{x\}) \prec_1 \langle J_\alpha, A\rangle.$$

Proof. Set $h = h_{\alpha, A}$, $N = h^*(\{x\})$. Let $P \in \Sigma_1^{\langle J_\alpha, A\rangle}(N) \cap \mathscr{P}(J_\alpha)$. We show that if $P \neq \emptyset$ then $P \cap N \neq \emptyset$.

Let P be $\Sigma_1^{\langle J_\alpha, A\rangle}(\{x_1, \ldots, x_n\})$, where $x_1, \ldots, x_n \in N$. Pick $i_1, \ldots, i_n \in \omega$ so that $x_1 = h(i_1, x), \ldots, x_n = h(i_n, x)$. For each $k = 1, \ldots, n$, x_k is the unique y in J_α such that $\models_{\langle J_\alpha, A\rangle} \exists z \theta_{i_k}(z, \mathring{y}, \mathring{x})$. Hence for each such k, x_k is Σ_1-definable from x in $\langle J_\alpha, A\rangle$. Hence P is in fact $\Sigma_1^{\langle J_\alpha, A\rangle}(\{x\})$. Thus for some $i \in \omega$,

$$P(y) \leftrightarrow \models_{\langle J_\alpha, A\rangle} \varphi_i(\mathring{y}, \mathring{x}).$$

Since $P \neq \emptyset$, let y be the $<_J$-least element of P. Then clearly, $y = h(i, x)$. Hence $y \in N$, proving that $P \cap N \neq \emptyset$. □

By modifying the proof of the above lemma along the lines of II.6.2 and II.6.3, we obtain:

3.8 Lemma. *Let $A \subseteq J_\alpha$, $p \in J_\alpha$, $X \subseteq J_\alpha$. If X is closed under ordered pairs, then*

$$X \cup \{p\} \subseteq h^*_{\alpha, A}(X \times \{p\}) \prec_1 \langle J_\alpha, A\rangle. \quad \square$$

3.9 Lemma. *Let $A \subseteq J_\alpha$, $X \subseteq J_\alpha$. If $h^*_{\alpha, A}(X)$ is closed under ordered pairs, then*

$$X \subseteq h^*_{\alpha, A}(x) \prec_1 \langle J_\alpha, A\rangle. \quad \square$$

3.10 Lemma. *If $\langle J_\alpha, A\rangle$ is amenable, the function $h_{\alpha, A}$ is (uniformly) $\Sigma_1^{\langle J_\alpha, A\rangle}$.*

Proof. We have

$$y = h_{\alpha, A}(i, x) \leftrightarrow \models_{\langle J_\alpha, A\rangle} \exists u \theta_i(u, \mathring{y}, \mathring{x}).$$

By 3.5 and 3.4, the result follows immediately. □

Let $H_{\alpha, A}$ denote the uniformly $\Sigma_0^{\langle J_\alpha, A\rangle}$ predicate such that for amenable $\langle J_\alpha, A\rangle$,

$$y = h_{\alpha, A}(i, x) \leftrightarrow (\exists z \in J_\alpha)\, H_{\alpha, A}(z, y, i, x).$$

As an immediate corollary to the above result we have:

3.11 Lemma.

(i) *The function $h_{\alpha, A}$ is a (uniformly Σ_1) Σ_1 skolem function for amenable $\langle J_\alpha, A\rangle$ with $\alpha > 1$.*

(ii) *The function $h_{\alpha, \emptyset}$ is a (uniformly Σ_1) Σ_1 skolem function for J_α for each $\alpha > 1$.* □

We often write h_α for $h_{\alpha, \emptyset}$. The notation $h_{\alpha, A}$, h_α, θ_i, $H_{\alpha, A}$ $H_\alpha (= H_{\alpha, \emptyset})$ is fixed for the rest of this book.

As an illustration of the use of the skolem functions h_α, we shall prove an analogue of II.6.8 for the Jensen hierarchy, showing that for any ordinal α there is a $\Sigma_1(J_\alpha)$ map from $\omega\alpha$ onto J_α. This will require some preliminary lemmas, but before we give them we introduce an important notion which should throw some light upon our construction of the Σ_1 skolem function.

A function r is said to *uniformise* a relation R iff $\mathrm{dom}(r) = \mathrm{dom}(R)$ and for all x,

$$\exists y R(y, \vec{x}) \leftrightarrow R(r(\vec{x}), \vec{x}).$$

We say a structure of the form $\mathbf{M} = \langle M, (A_i)_{i<\omega}\rangle$ is Σ_n*-uniformisable* iff every $\Sigma_n(\mathbf{M})$ relation on M is uniformised by a $\Sigma_n(\mathbf{M})$ function.

In general, Σ_n-uniformisability is a very strong condition to demand of a structure. Indeed, the existence of *any* uniformising function definable over the structure concerned is quite a strong property, let alone the existence of one whose definition is no more complex than that of the relation it is uniformising. It is thus perhaps rather surprising to learn that for all $\alpha > 1$ and all $n \geqslant 1$, J_α is Σ_n-uniformisable. In the general case the proof is rather tricky, and will be given in the next section, where Σ_n-uniformisation will play an important role in our study of the Σ_n-projectum. But the case $n = 1$ is quite straightforward, and we shall consider this case here, using it to obtain an analogue of II.6.7 for the Jensen hierarchy. (In the proof of II.6.7 we did in fact make implicit use of the fact that for limit $\alpha > \omega$, L_α is Σ_1-uniformisable, but we did not dwell upon this point there.)

First let us see how Σ_n-uniformisability affects the existence of Σ_n-skolem functions.

3.12 Lemma. *Let $n \geqslant 1$, $\alpha > 1$. If J_α is Σ_n-uniformisable, then it has a Σ_n-skolem function.*

Proof. Let $(\varphi_i \mid i < \omega)$ be a $\Delta_1^{J_1}$ enumeration of all Σ_n-formulas of $\mathscr{L}$ with free variables v_0, v_1. By 3.4, the relation

$$\{(y, i, x) \mid \models_{J_\alpha} \varphi_i(\mathring{y}, \mathring{x})\}$$

is $\Sigma_n^{J_\alpha}$. Let r be a $\Sigma_n(J_\alpha)$ function uniformising this relation. Pick $p \in J_\alpha$ so that r is $\Sigma_n^{J_\alpha}(\{p\})$. Set

$$h(i, x) \simeq r(i, (x, p)) \qquad (x \in J_\alpha).$$

It is easily seen that h is a Σ_n skolem function for J_α and that p is a good parameter for h. □

We note that the converse to the above lemma is trivially true.

For the case $n = 1$ we now prove:

3.13 Lemma. *Let $\alpha > 1$. Then J_α is Σ_1-uniformisable.*

Proof. Let $R(y, \vec{x})$ be a $\Sigma_1(J_\alpha)$ relation on J_α. Let S be a $\Sigma_0(J_\alpha)$ relation such that

$$R(y, \vec{x}) \leftrightarrow (\exists z \in J_\alpha)\, S(z, y, \vec{x}).$$

Define g on J_α by

$$g(\vec{x}) \simeq \text{the } <_J\text{-least } w \text{ such that } S((w)_0, (w)_1, \vec{x}).$$

The function g is $\Sigma_1(J_\alpha)$. For it has the definition

$$\begin{aligned} w = g(\vec{x}) \leftrightarrow\ & S((w)_0, (w)_1, \vec{x}) \\ & \wedge \exists u[u = \{w' \mid w' <_J w\} \wedge (\forall w' \in u) \neg S((w')_0, (w')_1, \vec{x})], \end{aligned}$$

which is $\Sigma_1(J_\alpha)$ by 2.8(v). Now set

$$r(\vec{x}) \simeq (g(\vec{x}))_1.$$

Then r is $\Sigma_1(J_\alpha)$, and r clearly uniformises R. □

At this point the reader might like to see what goes wrong when we try to generalise the above argument to the case $n > 1$. (As we shall see in the next section, proving Σ_n-uniformisability of J_α for $n > 1$ is by no means a simple matter, though it is achieved by somehow pushing through an argument such as the above.)

Now for our analogue of II.6.8. As in II.6.6, let

$$\Phi\colon \mathrm{On} \times \mathrm{On} \leftrightarrow \mathrm{On}$$

be Gödel's pairing function. By the same argument as in II.6.6, we have:

3.14 Lemma. *$\Phi^{-1} \upharpoonright \omega\alpha$ is uniformly $\Sigma_1^{J_\alpha}$ for all α.* □

Analogous to II.6.7 we have:

3.15 Lemma. *There is a $\Sigma_1(J_\alpha)$ map from $\omega\alpha$ onto $\omega\alpha \times \omega\alpha$.*

Proof. Set

$$Q = \{\alpha \mid \Phi\colon \alpha \times \alpha \leftrightarrow \alpha\},$$

a closed unbounded class of ordinals. It is easily seen that $\omega\alpha = \alpha$ for any ordinal α such that $\omega\alpha \in Q$. Moreover,

$$Q = \{\alpha \mid \Phi(0, \alpha) = \alpha\}.$$

We prove the lemma by induction on α. For $\alpha = 0$ the result is trivial, so we assume $\alpha > 0$ now and that the lemma holds for all $\beta < \alpha$. There are three cases to consider.

Case 1. $\omega\alpha \in Q$.

In this case, $\Phi^{-1} \upharpoonright \omega\alpha$ suffices.

Case 2. $\alpha = \beta + 1$.

If $\beta = 0$ here, then $\omega\alpha = \omega \in Q$ and we are done by Case 1. So we may assume that $\beta > 0$. Define $j\colon \omega\alpha \leftrightarrow \omega\beta$ by

$$j(\xi) = \begin{cases} 2\xi, & \text{if } \xi < \omega \\ \xi, & \text{if } \omega \leqslant \xi < \omega\beta \\ 2n+1, & \text{if } \omega\beta + n. \end{cases}$$

Clearly, j is $\Sigma_1^{J_\alpha}(\{\omega, \omega\beta\})$.

By induction hypothesis, there is a $\Sigma_1(J_\beta)$ map g from $\omega\beta$ onto $\omega\beta \times \omega\beta$. Let

$$G = \{(v, x) \mid g(v) = x\},$$

a $\Sigma_1(J_\beta)$ relation on J_β. Let $\bar{g}$ be a $\Sigma_1(J_\beta)$ function uniformising G. Clearly, $\bar{g}$ maps $\omega\beta \times \omega\beta$ one-one into $\omega\beta$. Now, $\bar{g} \in \mathrm{rud}(J_\beta) = J_\alpha$ (since $\mathrm{rud}(J_\beta) \cap \mathscr{P}(J_\beta) = Def(J_\beta)$), so f is a $\Sigma_1(J_\alpha)$ map from $\omega\alpha \times \omega\alpha$ one-one into $\omega\beta$, where we define f by

$$f((\nu, \tau)) = \bar{g}((j(\nu), j(\tau))).$$

Now, j is onto $\omega\beta$, so $\mathrm{ran}(f) = \mathrm{ran}(\bar{g}) \in J_\alpha$. Hence h is a $\Sigma_1(J_\alpha)$ map from $\omega\alpha$ onto $\omega\alpha \times \omega\alpha$, where we define h by

$$h(\nu) = \begin{cases} f^{-1}(\nu), & \text{if } \nu \in \mathrm{ran}(f) \\ (0, 0), & \text{otherwise.} \end{cases}$$

The map h is as required.

Case 3. $\omega\alpha \notin Q$ and $\lim(\alpha)$.

Set $(\nu, \tau) = \Phi^{-1}(\omega\alpha)$. Since $\omega\alpha \notin Q$, we have $\nu, \tau < \omega\alpha$. Let $<^*$ be the well-ordering of On $\times$ On used to define Φ (see II.6.6), and set

$$c = \{z \mid z <^* (\nu, \tau)\}.$$

Then $c \in J_\alpha$, and moreover, $\Phi \upharpoonright c$ is a $\Sigma_1(J_\alpha)$ bijection from c onto $\omega\alpha$. Pick $\gamma < \alpha$ such that $\nu, \tau < \omega\gamma$. (Possible since $\lim(\alpha)$.) Then $\Phi^{-1} \upharpoonright \omega\gamma$ is a $\Sigma_1(J_\alpha)$ map from $\omega\alpha$ one-one into $\omega\gamma$. Also, arguing as in Case 2, the induction hypothesis implies the existence of a map $\bar{g} \in J_\alpha$ one-one from $\omega\gamma \times \omega\gamma$ into $\omega\gamma$. Then f is a $\Sigma_1(J_\alpha)$ bijection from $\omega\alpha \times \omega\alpha$ onto $d = \bar{g}''[\bar{g}''c \times \bar{g}''c]$, where we define f by

$$f((\xi, \zeta)) = \bar{g}((\bar{g}(\Phi^{-1}(\xi)), \bar{g}(\Phi^{-1}(\zeta)))).$$

But $d \in J_\alpha$, so h is a $\Sigma_1(J_\alpha)$ map from $\omega\alpha$ onto $\omega\alpha \times \omega\alpha$, where we define h by

$$h(\xi) = \begin{cases} f^{-1}(\xi), & \text{if } \xi \in d, \\ (0, 0), & \text{otherwise.} \end{cases}$$

Then h is as required. The proof is complete. □

We may now prove our analogue of II.6.8.

3.16 Lemma. *Let $\alpha > 1$. There is a $\Sigma_1(J_\alpha)$ map from $\omega\alpha$ onto J_α.*

Proof. Let f be a $\Sigma_1^{J_\alpha}(\{p\})$ map from $\omega\alpha$ onto $\omega\alpha \times \omega\alpha$, where $p \in J_\alpha$ is the $<_J$-least for which such an f exists. Define f^0, f^1 by

$$f(\nu) = (f^0(\nu), f^1(\nu)) \quad (\nu \in \omega\alpha).$$

By induction, define f_n from $\omega\alpha$ onto $(\omega\alpha)^n$ by:

$$f_1 = id \upharpoonright \omega\alpha,$$
$$f_{n+1}(\nu) = (f^0(\nu), f_n \circ f^1(\nu)).$$

Notice that each f_n is $\Sigma_1^{J_\alpha}(\{p\})$.

Let $h = h_\alpha$, $H = H_\alpha$, and set $X = h^*(\omega\alpha \times \{p\})$.

Claim 1. X is closed under ordered pairs.

To see this, let $x_1, x_2 \in X$, say $x_i = h(j_i, (v_i, p))$. Let $(v_1, v_2) = f_2(\tau)$. Then $\{(x_1, x_2)\}$ is a $\Sigma_1^{J_\alpha}(\{\tau, p\})$ predicate on J_α. So by definition of h, $(x_1, x_2) \in X$, as claimed.

By claim 1 and 3.9, $X \prec_1 J_\alpha$. Let $\pi: X \cong J_\beta$, where $\beta \leqslant \alpha$, by the Condensation Lemma. Clearly, $\omega\alpha \subseteq X$, so we must have $\beta = \alpha$ here.

Claim 2. For all $i \in \omega$, $x \in X$,

$$\pi(h(i, x)) \simeq h(i, \pi(x)).$$

Let $i \in \omega$, $x \in X$. Suppose first that $y = h(i, x)$ is defined. Note that as $x \in X \prec_1 J_\alpha$, we have $y \in X$. Now (with $(\theta_i \mid i < \omega)$ as defined in the definition of the Σ_1 skolem function),

$$\models_{J_\alpha} \exists z\, \theta_i(z, \mathring{y}, \mathring{x}).$$

So as $x, y \in X \prec_1 J_\alpha$,

$$\models_X \exists z\, \theta_i(z, \mathring{y}, \mathring{x}).$$

Pick $z \in X$ such that

$$\models_X \theta_i(\mathring{z}, \mathring{y}, \mathring{x}).$$

Applying $\pi: X \cong J_\alpha$,

$$\models_{J_\alpha} \theta_i(\pi(z)^\circ, \pi(y)^\circ, \pi(x)^\circ).$$

Thus

$$\models_{J_\alpha} \exists z\, \theta_i(z, \pi(y)^\circ, \pi(x)^\circ).$$

Thus $\pi(y) = h(i, \pi(x))$.

Conversely, suppose $h(i, \pi(x))$ is defined. Then $h(i, \pi(x)) \in J_\alpha = \pi'' X$, so for some $y \in X$, $h(i, \pi(x)) = \pi(y)$, and we can reverse the above steps to obtain $y = h(i, x)$. This proves claim 2.

Now, $f: \omega\alpha \to \omega\alpha \times \omega\alpha$, so as $\pi \restriction \omega\alpha = id \restriction \omega\alpha$, $\pi'' f = f$. And by isomorphism, $\pi'' f$ is $\Sigma_1^{J_\alpha}(\{\pi(p)\})$. So by choice of p, $p \leqslant_J \pi(p)$. But by 2.9(iii), $\pi(p) \leqslant_J p$. Hence $\pi(p) = p$.

By claim 2 now, for $i \in \omega$, $v \in \omega\alpha$, we have

$$\pi(h(i, (v, p))) \simeq h(i, \pi((v, p))) \simeq h(i, (v, p)).$$

Thus $\pi \restriction X = id \restriction X$. Thus $X = J_\alpha$. It follows at once that if we set

$$r(v) \simeq h((f(v))_0, ((f(v))_1, p)),$$

then r is a $\Sigma_1(J_\alpha)$ map such that $r''\omega\alpha = J_\alpha$. However, we are not yet done, since the map r just defined is not total on $\omega\alpha$. To achieve this, define $g\colon \omega\alpha \times \omega\alpha \times \omega\alpha \to J_\alpha$ by:

$$g(i, v, \tau) = \begin{cases} y, & \text{if } (\exists z \in S_\tau)\, H(z, y, i, (v, p)) \\ \emptyset, & \text{otherwise.} \end{cases}$$

Then g is $\Sigma_1(J_\alpha)$. And clearly,

$$g''(\omega\alpha \times \omega\alpha \times \omega\alpha) = h^*(\omega\alpha \times \{p\}) = X = J_\alpha.$$

Thus $g \circ f_3$ satisfies the lemma. □

By examining the proofs of 3.15 and 3.16, we see that in the case where $\alpha \in Q$, no parameters are required in the functions we defined. Hence, noting that $\omega\alpha = \alpha$ whenever $\alpha \in Q$, we have:

3.17 Lemma. *If α is closed under the Gödel pairing function, there is a (uniform) $\Sigma_1^{J_\alpha}$ map from $\omega\alpha$ onto J_α.* □

4. The Σ_n-Projectum

As we indicated in IV.4, the Σ_n-projectum of an ordinal plays an important role in the reduction of Σ_n predicates to Σ_1 predicates, the main idea behind the fine structure theory. Indeed, if ϱ is the Σ_n-projectum of α, then it is as a Σ_1 predicate on $\langle J_\varrho, A\rangle$ for some set A that we shall code a given Σ_n predicate on J_α.

Let $n > 0$, $\alpha > 0$. The *Σ_n-projectum* of α, ϱ_α^n, is the least ordinal $\varrho \leqslant \alpha$ such that there is a $\Sigma_n(J_\alpha)$ function f over J_α such that $f''J_\varrho = J_\alpha$.

By 3.16, it is easily seen that ϱ_α^n is the least $\varrho \leqslant \alpha$ such that there is a $\Sigma_n(J_\alpha)$ map f for which $f''\omega\varrho = \omega\alpha$.

Clearly, $0 < m < n \to \varrho_\alpha^n \leqslant \varrho_\alpha^m$. So it is natural to define $\varrho_\alpha^0 = \alpha$ for each ordinal α.

4.1 Lemma. *If $\varrho_\alpha^n > 1$, then* $\lim(\varrho_\alpha^n)$.

Proof. Suppose that $\varrho = \varrho_\alpha^n = \gamma + 1$, where $\gamma > 0$. Let f be a $\Sigma_n(J_\alpha)$ function such that $f''\omega\varrho = \omega\alpha$. Define $g\colon \omega\gamma \to \omega\varrho$ by

$$g(v) = \begin{cases} m, & \text{if } v = 2m < \omega, \\ \omega\gamma + m, & \text{if } v = 2m + 1 < \omega, \\ v, & \text{if } \omega \leqslant v < \omega\gamma. \end{cases}$$

Clearly, g is $\Sigma_1(J_\alpha)$. Thus $f \circ g$ is $\Sigma_n(J_\alpha)$. But $(f \circ g)''\omega\gamma = \omega\alpha$, so this contradicts the choice of ϱ. □

In order to obtain more information about the Σ_n-projectum we shall prove that for all $\alpha > 1$ and all $n > 0$, J_α is Σ_n-uniformisable. The proof is fairly intricate, and requires several preliminary lemmas. Before we begin, we outline the general strategy.

We begin by examining the proof of Σ_1-uniformisation given in 3.13. This reduced to proving that every Σ_0 relation is uniformised by a Σ_1 function. (In 3.13, what we really did was to uniformise the Σ_0 relation S, obtaining the uniformisation of the Σ_1 relation R as a simple consequence.) This worked in the case $n = 1$ because, if $S(y, \vec{x})$ is Σ_0, then so too is $(\forall z \in y) \neg S(z, \vec{x})$. But consider now the analogous situation for $n > 1$. We seek a Σ_n uniformisation of a Π_{n-1} relation S. Now, if $S(y, \vec{x})$ is Π_{n-1}, then $(\forall z \in y) \neg S(z, x)$ is in general Σ_{n+1}, not Σ_n. Roughly speaking, we overcome this difficulty as follows. We reduce the predicate S on J_α to a predicate on $J_{\varrho_\alpha^{n-1}}$. The structure $\mathbf{J}_{\varrho_\alpha^{n-1}}$ is sufficiently suited to handling $\Sigma_{n-1}(J_\alpha)$ predicates on it that the canonical uniformisation procedure applied to the reduced predicate turns out to be $\Sigma_n(J_\alpha)$, thereby providing us with the desired Σ_n uniformisation of S. The precise property of the projectum which we need in order to make this work is described below.

Let $P(y, \vec{x})$ be any predicate on J_α. For $\varrho \leqslant \alpha$, we say that $P(y, \vec{x})$ is *$\Sigma_n(J_\alpha)$ on J_ϱ* iff there is a Σ_n formula $\varphi(y, \vec{x})$ of $\mathscr{L}_{J_\alpha}$ such that

$$(\forall y \in J_\varrho)(\forall \vec{x} \in J_\alpha)[P(y, \vec{x}) \leftrightarrow \models_{J_\alpha} \varphi(\mathring{y}, \mathring{\vec{x}})].$$

Similarly for *$\Pi_n(J_\alpha)$ on J_ϱ*.

For any predicate $R(y, \vec{x})$, we denote by $R^\forall(y, \vec{x})$ the predicate

$$\{(y, \vec{x}) \mid (\forall z \in y)\, R(z, \vec{x})\},$$

and by $R^\exists(y, \vec{x})$ the predicate

$$\{(y, \vec{x}) \mid (\exists z \in y)\, R(z, \vec{x})\}.$$

Let $\alpha > 1$, $n > 0$, $0 < \varrho \leqslant \alpha$. We denote by $\Gamma(\alpha, n, \varrho)$ the following property: whenever $R(y, \vec{x})$ is $\Sigma_n(J_\alpha)$, then $R^\forall(y, \vec{x})$ is $\Sigma_{n+1}(J_\alpha)$ on J_ϱ.

We shall prove that for any $\alpha > 1$, $n > 0$, $\Gamma(\alpha, n, \varrho_\alpha^n)$ is valid. Using $\Gamma(\alpha, n, \varrho_\alpha^n)$ we shall be able to prove that J_α is Σ_{n+1}-uniformisable, the proof being a variation of the proof for the Σ_1 case (3.13) as outlined above. (In fact the proof of $\Gamma(\alpha, n, \varrho_\alpha^n)$ and that of Σ_{n+1}-uniformisability proceeds by a simultaneous induction on n.) But first we need some preliminary results.

4.2 Lemma. *Let $\alpha > 1$, $n > 0$, $\varrho > 0$. Assume $\Gamma(\alpha, n, \varrho)$. Then:*

(i) *if $R(y, \vec{x})$ is $\Pi_n(J_\alpha)$, then $R^\exists(y, \vec{x})$ is $\Pi_{n+1}(J_\alpha)$ on J_ϱ;*

(ii) *if $R(y, \vec{x})$ is $\Sigma_n(J_\alpha)$, then $Q(y, \vec{x})$ is $\Sigma_{n+1}(J_\alpha)$ on J_ϱ, where $Q = \{(y, \vec{x}) \mid (\forall z <_J y)\, R(z, \vec{x})\}$.*

Proof. (i) This follows from $\Gamma(\alpha, n, \varrho)$ by taking negations.

(ii) For $y, \vec{x} \in J_\varrho$, we have

$$Q(y, \vec{x}) \leftrightarrow (\exists u, w, v \in J_\varrho)\,[y \in S_v \wedge w = {<}^{S}_{v} \wedge (\forall z)(z \in u \leftrightarrow (z, y) \in w) \wedge (\forall z \in u)\, R(z, \vec{x})].$$

Using $\Gamma(\alpha, n, \varrho)$, this is easily seen to be $\Sigma_{n+1}(J_\alpha)$. (In case $\varrho < \alpha$, we must use J_ϱ as a parameter to ensure that $u \in J_\varrho$. If $\varrho = \alpha$ there is no need to mention ϱ at all, of course.) □

4.3 Lemma. *Let $\alpha > 1$, $n > 0$, and set $\varrho = \varrho_\alpha^n$. Suppose that J_α is Σ_n-uniformisable. Then $\langle J_\varrho, A\rangle$ is amenable for all $A \in \Sigma_n(J_\alpha) \cap \mathscr{P}(J_\varrho)$.*

Proof. Let $A \in \Sigma_n(J_\alpha) \cap \mathscr{P}(J_\varrho)$. We show that $\langle J_\varrho, A\rangle$ is amenable. If $\varrho = 1$, then $J_\varrho = H_\omega$, so this is immediate. Now assume $\varrho > 1$. Thus by 4.1, $\lim(\varrho)$. So it suffices to show that $\gamma < \varrho$ implies $A \cap J_\gamma \in J_\varrho$.

Let $\gamma < \varrho$ be given. Set $B = A \cap J_\gamma$. Thus B is $\Sigma_n(J_\alpha)$. Let B be $\Sigma_n^{J_\alpha}(\{p\})$. Let $\varphi(v_0, v_1)$ be a Σ_n-formula such that

$$(*) \qquad x \in B \quad \text{iff } \models_{J_\alpha} \varphi(\mathring{x}, \mathring{p}).$$

By assumption, J_α is Σ_n-uniformisable, so by 3.12, J_α has a Σ_n skolem function, h. Set $X = h^*(J_\gamma \times \{p\})$. By 3.2, $X \prec_n J_\alpha$. Let $\pi: X \cong J_{\bar{\alpha}}$. Set $\bar{p} = \pi(p)$, $\bar{h} = \pi''(h \cap (X \times \omega \times X))$. Since $B \subseteq J_\gamma$, $\pi'' B = B$. So by $(*)$

$$(**) \qquad x \in B \quad \text{iff } \models_{J_\alpha} \varphi(\mathring{x}, \mathring{\bar{p}}).$$

Thus B is $\Sigma_n^{J_\alpha}(\{\bar{p}\})$. Hence $B \in J_{\bar{\alpha}+1}$. If $\bar{\alpha} < \varrho$ then this means that $B \in J_\varrho$ and we are done. So we are reduced to proving that $\bar{\alpha} < \varrho$.

Suppose, on the contary, that $\bar{\alpha} \geqslant \varrho$. By definition of X, $J_{\bar{\alpha}} = \bar{h}^*(J_\gamma \times \{\bar{p}\})$. So, as $\bar{h}$ is $\Sigma_n(J_{\bar{\alpha}})$, there is a $\Sigma_n(J_{\bar{\alpha}})$ function f such that $f'' J_\gamma = J_{\bar{\alpha}}$. Let g be a $\Sigma_n(J_{\bar{\alpha}})$ map such that $g'' J_\varrho = J_\alpha$. Since $\varrho \leqslant \bar{\alpha}$, $g \circ f$ is a $\Sigma_n(J_\alpha)$ map such that $g \circ f'' J_\gamma = J_\alpha$, contrary to $\gamma < \varrho$. The lemma is proved. □

Our proof of Σ_n uniformisability will be by induction on n. The key to the induction is provided by the following lemma.

4.4 Lemma. *Let $\alpha > 1$, $n > 0$, and assume $\Gamma(\alpha, n, \varrho_\alpha^n)$. If J_α is Σ_n-uniformisable, then it is Σ_{n+1}-uniformisable.*

Proof. The procedure is not unlike that adopted in proving Σ_1-uniformisability, except that we reduce the predicate to one on $J_{\varrho_\alpha^n}$ before we commence.

Let $R(y, \vec{x})$ be $\Sigma_{n+1}(J_\alpha)$, and let S be $\Pi_n(J_\alpha)$ such that

$$R(y, \vec{x}) \leftrightarrow (\exists z \in J_\alpha)\, S(z, y, \vec{x}).$$

Let $\varrho = \varrho_\alpha^n$, and let f be a $\Sigma_n(J_\alpha)$ function such that $f'' J_\varrho = J_\alpha$. We shall consider the case where $\varrho < \alpha$. The case where $\varrho = \alpha$ is a little simpler, since there is no need to mention ϱ at all. Set

$$r(\vec{x}) \simeq \text{the } <_J\text{-least } z \text{ such that } S((f(z))_0, (f(z))_1, \vec{x}),$$

$$\bar{r}(\vec{x}) \simeq (f \circ r(\vec{x}))_1.$$

Clearly, $\bar{r}$ uniformises R. If r is $\Sigma_{n+1}(J_\alpha)$, so too is $\bar{r}$, so what we must do is prove that r is indeed $\Sigma_{n+1}(J_\alpha)$. We have, by definition,

$$y = r(\vec{x}) \leftrightarrow [y \in \mathrm{dom}(f)] \wedge [\forall z(z = f(y) \to S((z)_0, (z)_1, \vec{x})]$$
$$\wedge [(\forall y' <_J y)(y' \in \mathrm{dom}(f) \to \neg S((f(y'))_0, (f(y'))_1, \vec{x})].$$

The first conjunct here is $\Sigma_n(J_\alpha)$ and the second is $\Pi_n(J_\alpha)$. Also, $\mathrm{dom}(f)$ is $\Sigma_n(J_\alpha)$, and for $y \in J_\varrho$, $\{y' \mid y' <_J y\} \in J_\varrho$, so by 4.3,

$$\mathrm{dom}(f) \cap \{y' \mid y' <_J y\} \in J_\varrho$$

for each $y \in J_\varrho$. Hence the third conjunct reduces to

$$(\exists u \in J_\varrho)\,[(\forall y' \in u)(y' <_J y \wedge y' \in \mathrm{dom}(f))$$
$$\wedge (\forall y')(y' <_J y \wedge y' \in \mathrm{dom}(f) \to y' \in u)$$
$$\wedge (\forall y' \in u)(\exists z)(z = f(y') \wedge \neg S((z)_0, (z)_1, \vec{x}))].$$

This is of the form

$$(\exists u \in J_\varrho)\,[(\forall y' \in u)(\Sigma_n(J_\alpha)) \wedge (\forall y')(\Pi_n(J_\alpha)) \wedge (\forall y' \in u)(\Sigma_n(J_\alpha))].$$

Using $\Gamma(\alpha, n, \varrho)$, we see that it is in fact of the form

$$(\exists u \in J_\varrho)\,[\Sigma_{n+1}(J_\alpha) \wedge \Pi_n(J_\alpha) \wedge \Sigma_{n+1}(J_\alpha)].$$

Hence r is $\Sigma_{n+1}(J_\alpha)$, as required. $\square$

4.5 Theorem (Uniformisation Theorem). *Let $\alpha > 1, n > 0$. Then J_α is Σ_n-uniformisable.*

Proof. By 3.13 we are done if $n = 1$. By 4.4, the result follows by induction if we can establish $\Gamma(\alpha, n, \varrho_\alpha^n)$ for all $n > 0$. We do this by induction on n as well.

Let $n \geqslant 1$, and in case $n > 1$ assume $\Gamma(\alpha, 1, \varrho_\alpha^1), \ldots, \Gamma(\alpha, n-1, \varrho_\alpha^{n-1})$. We prove that $\Gamma(\alpha, n, \varrho_\alpha^n)$. Note that by 4.4, J_α is Σ_m-uniformisable for all $m \leqslant n$, $m \geqslant 1$.

Set $\varrho = \varrho_\alpha^n$, $\eta = \varrho_\alpha^{n-1}$. Notice that $\varrho \leqslant \eta \leqslant \alpha$. There are two cases to consider.

Case 1. There is no $\Sigma_n(J_\alpha)$ map from any $\gamma < \omega\varrho$ cofinally into $\omega\eta$.

In this case we commence by proving a sort of Σ_n Collection Axiom.

Claim. If $R(y, x)$ is $\Sigma_n(J_\alpha)$ and $u \in J_\varrho$, then

$$(\forall x \in u)(\exists y \in J_\eta) R(y, x) \to (\exists v \in J_\eta)(\forall x \in u)(\exists y \in v) R(y, x).$$

Proof of claim. If $\varrho = 1$ the claim is trivial, so assume $\varrho > 1$. Hence $\lim(\varrho)$, and we can pick $\gamma < \varrho$ so that $u \in J_\gamma$. Let $j\colon \omega\gamma \xrightarrow{\text{onto}} J_\gamma$ be $\Sigma_1(J_\gamma)$. Let r be a $\Sigma_n(J_\alpha)$ function uniformising R. Define $f\colon \omega\gamma \to \omega\eta$ by

$$f(v) = \begin{cases} \text{the least } \tau < \omega\eta \text{ such that } r \circ j(v) \in S_\tau, & \text{if } j(v) \in u, \\ 0, & \text{otherwise}. \end{cases}$$

Thus:

$$\tau = f(v) \leftrightarrow \models_{J_\alpha} [(j(v) \in u) \wedge \exists z \exists f [z = r \circ j(v) \wedge f = (S_\xi \mid \xi \leqslant \tau) \wedge z \in f(\tau) \wedge (\forall \xi \in \tau)(z \notin f(\xi))]] \vee [(j(v) \notin u) \wedge (\tau = 0)].$$

Thus f is $\Sigma_n(J_\alpha)$. So, by assumption there is a $\delta < \omega\eta$ such that $f''\omega\gamma \subseteq \delta$. Then

$$(\forall x \in u)(\exists y \in S_\delta) R(y, x),$$

which proves the claim.

We must now consider two subcases.

Case 1.1. $n = 1$.

Let $R(y, \vec{x})$ be $\Sigma_1(J_\alpha)$. Let S be $\Sigma_0(J_\alpha)$ with

$$R(y, \vec{x}) \leftrightarrow (\exists t \in J_\alpha) S(t, y, \vec{x}).$$

Let $y \in J_\varrho$, $\vec{x} \in J_\alpha$. Since $\eta = \varrho_\alpha^0 = \alpha$, the claim gives

$$(\forall z \in y) R(z, \vec{x}) \leftrightarrow (\forall z \in y)(\exists t \in J_\eta) S(t, z, \vec{x}) \leftrightarrow (\exists v \in J_\eta)(\forall z \in y)(\exists t \in v) S(t, z, \vec{x}),$$

which is $\Sigma_1(J_\alpha)$. Thus $R^\forall$ is $\Sigma_1(J_\alpha)$ on J_ϱ, proving $\Gamma(\alpha, 1, \varrho)$.

Case 1.2. $n > 1$.

Let $R(y, \vec{x})$ be $\Sigma_n(J_\alpha)$, and let S be $\Pi_{n-1}(J_\alpha)$ with

$$R(y, \vec{x}) \leftrightarrow (\exists t \in J_\alpha) S(t, y, \vec{x}).$$

Let f be a $\Sigma_{n-1}(J_\alpha)$ function such that $f''J_\eta = J_\alpha$. Let $y \in J_\varrho$, $\vec{x} \in J_\alpha$. By the claim,

$$(\forall z \in y) R(z, \vec{x}) \leftrightarrow (\forall z \in y)(\exists t \in J_\eta) S(f(t), z, \vec{x}) \leftrightarrow (\exists v \in J_\eta)(\forall z \in y)(\exists t \in v) S(f(t), z, \vec{x}).$$

Now, J_α is Σ_{n-1}-uniformisable and $\mathrm{dom}(f)$ is $\Sigma_{n-1}(J_\alpha)$, so by 4.3,

$$v \in J_\eta \rightarrow \mathrm{dom}(f) \cap v \in J_\eta.$$

Hence

$$R^\forall(y, \vec{x}) \leftrightarrow (\exists v \in J_\eta) [(\forall x \in v)(x \in \mathrm{dom}(f)) \wedge (\forall z \in y)(\exists t \in v)(\forall w)[w = f(t) \rightarrow S(w, z, \vec{x})]].$$

This is of the form

$$(\exists v \in J_\eta) [\Pi_n(J_\alpha) \wedge (\forall z \in y)(\exists t \in v)(\Pi_{n-1}(J_\alpha))].$$

Using $\Gamma(\alpha, n-1, \eta)$, together with 4.2(i), this is in fact of the form

$$(\exists v \in J_\eta) [\Pi_n(J_\alpha) \wedge (\forall z \in y)(\Pi_n(J_\alpha))],$$

which is the same as

$$(\exists v \in J_\eta)\,(\Pi_n(J_\alpha)),$$

which is $\Sigma_{n+1}(J_\alpha)$, as required.

Case 2. Otherwise.

Let $\gamma < \omega\varrho$ be least such that there is a $\Sigma_n(J_\alpha)$ map g from γ cofinally into $\omega\eta$. Let $R(y, \vec{x})$ be $\Sigma_n(J_\alpha)$. We commence by proving:

Claim. There is a $\Delta_n(J_\alpha)$ predicate $Q(v, y, \vec{x})$ such that for any $y \in J_\varrho$, $\vec{x} \in J_\alpha$,

$$R(y, \vec{x}) \leftrightarrow (\exists v \in \gamma)\, Q(v, y, \vec{x}).$$

Proof of claim. Let f be a $\Sigma_{n-1}(J_\alpha)$ function such that $f''J_\eta = J_\alpha$. (If $n = 1$, then $\eta = \alpha$, so take $f = id \restriction J_\alpha$.) Let S be $\Pi_{n-1}(J_\alpha)$ with

$$R(y, \vec{x}) \leftrightarrow (\exists t \in J_\alpha)\, S(t, y, \vec{x}).$$

Define Q by

$$Q(v, y, \vec{x}) \leftrightarrow (v \in \gamma) \wedge (\exists t \in S_{g(v)})\, S(f(t), y, \vec{x}).$$

Since g is cofinal in $\omega\eta$ and $f''J_\eta = J_\alpha$, we have

$$R(y, \vec{x}) \leftrightarrow (\exists v \in \gamma)\, Q(v, y, \vec{x}).$$

We show that Q is $\Delta_n(J_\alpha)$. It is clearly $\Sigma_n(J_\alpha)$.

Define $\tilde{Q}$ by

$$\tilde{Q}(u, y, \vec{x}) \leftrightarrow (\exists t \in u)\, S(f(t), y, \vec{x}).$$

Thus:

$$\begin{aligned} Q(v, y, \vec{x}) &\leftrightarrow (v \in \gamma) \wedge \tilde{Q}(S_{g(v)}, y, \vec{x}) \\ &\leftrightarrow (v \in \gamma) \wedge \forall w \forall \tau [\tau = g(v) \wedge w = S_\tau \rightarrow \tilde{Q}(w, y, \vec{x})]. \end{aligned}$$

So it suffices to show that $\tilde{Q}$ is $\Pi_n(J_\alpha)$.

Well, if $n = 1$, then $f = \mathrm{id} \restriction J_\alpha$, so

$$\tilde{Q}(u, y, \vec{x}) \leftrightarrow (\exists t \in u)\, S(t, y, \vec{x}),$$

which is in fact $\Sigma_0(J_\alpha)$. So suppose $n > 1$. Then

$$\tilde{Q}(u, y, \vec{x}) \leftrightarrow (\exists t \in u \cap \mathrm{dom}(f))\,(\forall w)\,[w = f(t) \rightarrow S(w, y, \vec{x})].$$

Define T by

$$T(t, y, \vec{x}) \leftrightarrow (\forall w)\,[w = f(t) \rightarrow S(w, y, \vec{x})].$$

Then T is $\Pi_{n-1}(J_\alpha)$, and by the above

$$\tilde{Q}(u, y, \vec{x}) \leftrightarrow (\exists t \in u \cap \mathrm{dom}(f))\ T(t, y, \vec{x}).$$

Now, J_α is Σ_{n-1}-uniformisable and $\mathrm{dom}(f)$ is $\Sigma_{n-1}(J_\alpha)$, so by 4.3,

$$u \in J_\eta \rightarrow u \cap \mathrm{dom}(f) \in J_\eta.$$

Thus

$$\tilde{Q}(u, y, \vec{x}) \leftrightarrow (\forall v \in J_\eta)\ [v = u \cap \mathrm{dom}(f) \rightarrow (\exists t \in v)\ T(t, y, \vec{x})].$$

But we have

$$v = u \cap \mathrm{dom}(f) \leftrightarrow (\forall x \in v)\ (x \in u \wedge x \in \mathrm{dom}(f)) \wedge (\forall x \in u)\ (x \in \mathrm{dom}(f) \rightarrow x \in v).$$

This is of the form

$$(\forall x \in v)\ (\Sigma_{n-1}(J_\alpha)) \wedge (\forall x \in u)\ (\Pi_{n-1}(J_\alpha)).$$

Using $\Gamma(\alpha, n-1, \eta)$, as we may since $v \in J_\eta$, we see that this is of the form

$$\Sigma_n(J_\alpha) \wedge \Pi_{n-1}(J_\alpha),$$

and is thus $\Sigma_n(J_\alpha)$. Hence $\tilde{Q}(u, y, \vec{x})$ is of the form

$$\tilde{Q}(u, y, \vec{x}) \leftrightarrow (\forall v \in J_\eta)\ [\Sigma_n(J_\alpha) \rightarrow (\exists t \in v)\ (\Pi_{n-1}(J_\alpha))].$$

Using $\Gamma(\alpha, n-1, \eta)$ again, this is of the form

$$(\forall v \in J_\eta)\ [\Sigma_n(J_\alpha) \rightarrow \Pi_n(J_\alpha)],$$

which is $\Pi_n(J_\alpha)$. That completes the proof of the claim.

By the claim, we have, for $y \in J_\varrho$, $\vec{x} \in J_\alpha$,

$$R^\forall(y, \vec{x}) \leftrightarrow (\forall z \in y)\ (\exists v \in \gamma)\ Q(v, z, \vec{x}).$$

For each $\vec{x} \in J_\alpha$, we define

$$G(\vec{x}) = \{(v, z) \mid v \in \gamma \wedge z \in J_\varrho \wedge Q(v, z, \vec{x})\}.$$

Thus,

$$R^\forall(y, \vec{x}) \leftrightarrow (\forall z \in y)\ (\exists v \in \gamma)\ [(v, z) \in G(\vec{x})] \leftrightarrow \models_{\langle J_\rho, G(x)\rangle} \varphi(\mathring{y}, \mathring{\gamma}),$$

where φ is the Σ_0 formula

$$\varphi(v_0, v_1) : (\forall v_2 \in v_0)\ (\exists v_3 \in v_1)\ \mathring{A}((v_3, v_2)),$$

in the language $\mathscr{L}(A)$.

Now, $\lim(\varrho)$, so as φ is Σ_0, by Σ_0 absoluteness we have (cf. the proof of II.6.3)

$$\begin{aligned} R^{\forall}(y, \vec{x}) &\leftrightarrow (\exists w \in J_\varrho) [(w \text{ is transitive}) \wedge (y, \gamma \in w) \\ &\qquad \wedge (\models_{\langle w, G(\vec{x}) \cap w \rangle} \varphi(\mathring{y}, \mathring{\gamma}))] \\ &\leftrightarrow (\exists w \in J_\varrho) [(\forall u \in w)(\forall v \in u)(v \in w) \wedge (y, \gamma \in w) \\ &\qquad \wedge Sat^A(w, G(\vec{x}) \cap w, \varphi(\mathring{y}, \mathring{\gamma}))]. \end{aligned}$$

Now, Q is $\Sigma_n(J_\alpha)$, so for each $\vec{x} \in J_\alpha$, $G(\vec{x})$ is a $\Sigma_n(J_\alpha)$ subset of J_ϱ. Moreover, J_α is Σ_n-uniformisable. So by 4.3, for each $\vec{x} \in J_\alpha$ we have

$$w \in J_\varrho \to G(\vec{x}) \cap w \in J_\varrho.$$

Hence,

$$\begin{aligned} R^{\forall}(y, \vec{x}) \leftrightarrow (\exists w \in J_\varrho)(\exists a \in J_\varrho) &[(\forall u \in w)(\forall v \in u)(v \in w) \wedge (y, \gamma \in w) \\ &\wedge (a = G(\vec{x}) \cap w) \wedge Sat^A(w, a, \varphi(\mathring{y}, \mathring{\gamma}))]. \end{aligned}$$

So in order to show that $R^{\forall}(y, \vec{x})$ is $\Sigma_{n+1}(J_\alpha)$ on J_ϱ it suffices to show that the function $a(w, \vec{x}) = G(\vec{x}) \cap w$ is $\Sigma_{n+1}(J_\alpha)$.

Well, we have

$$a = a(w, \vec{x}) \leftrightarrow \forall z [z \in a \leftrightarrow z \in w \wedge (z)_0 \in \gamma \wedge (z)_1 \in J_\varrho \wedge Q((z)_0, (z)_1, \vec{x})].$$

So, as Q is $\Delta_n(J_\alpha)$, the function $a(w, \vec{x})$ is in fact $\Pi_n(J_\alpha)$. The proof is complete. □

With the aid of the Uniformisation Theorem, we are now able to provide some useful information about the Σ_n projectum.

4.6 Theorem. *Let $\alpha > 1$, $n > 0$. Then ϱ_α^n is equal to the largest ordinal δ such that $\langle J_\delta, A \rangle$ is amenable for all $A \in \Sigma_n(J_\alpha) \cap \mathscr{P}(J_\delta)$.*

Proof. By 4.5 and 4.3, $\langle J_\varrho, A \rangle$ is amenable for all $A \in \Sigma_n(J_\alpha) \cap \mathscr{P}(J_\varrho)$, where we have set $\varrho = \varrho_\alpha^n$ for convenience. Suppose δ were a larger ordinal with this property. Let f be a $\Sigma_n(J_\alpha)$ function such that $f'' J_\varrho = J_\alpha$. Set

$$A = \{u \in J_\varrho \mid u \notin f(u)\}.$$

A is $\Sigma_n(J_\alpha)$ and $A \subseteq J_\varrho$, so $\langle J_\delta, A \rangle$ is amenable. But then

$$A = A \cap J_\varrho \in J_\delta \subseteq J_\alpha,$$

so for some $u \in J_\varrho$, we have $A = f(u)$, which leads to the contradiction

$$u \in f(u) \leftrightarrow u \in A \leftrightarrow u \notin f(u).$$

This proves the theorem. □

4.7 Theorem. *Let $\alpha > 1$, $n > 0$. Then ϱ_α^n is equal to the smallest ordinal η such that $\mathscr{P}(\omega\eta) \cap \Sigma_n(J_\alpha) \nsubseteq J_\alpha$.*

Proof. Let $\varrho = \varrho_\alpha^n$, and let f be a $\Sigma_n(J_\alpha)$ function such that $f''J_\varrho = J_\alpha$. Let j be a $\Sigma_1(J_\varrho)$ map from $\omega\varrho$ onto J_ϱ. Set

$$A = \{v \in \omega\varrho \mid v \notin f \circ j(v)\}.$$

A is a $\Sigma_n(J_\alpha)$ subset of $\omega\varrho$. If $A \in J_\alpha$, then $A = f \circ j(v)$ for some $v < \omega\varrho$, and we get the contradiction

$$v \in A \leftrightarrow v \notin f \circ j(v) \leftrightarrow v \notin A.$$

Hence $\mathscr{P}(\omega\varrho) \cap \Sigma_n(J_\alpha) \nsubseteq J_\alpha$. But if $\eta < \varrho$ and $B \in \mathscr{P}(\omega\eta) \cap \Sigma_n(J_\alpha)$, then by 4.6, $\langle J_\varrho, B\rangle$ is amenable, so $B = B \cap J_\eta \in J_\varrho \subseteq J_\alpha$. Thus $\mathscr{P}(\omega\eta) \cap \Sigma_n(J_\alpha) \subseteq J_\alpha$. The theorem is proved. □

To complete this section, we state the following key fact that was used in our proof of the Uniformisation Theorem.

4.8 Lemma. *Let $\alpha > 1, n > 0, \varrho = \varrho_\alpha^n$. If $R(y, \vec{x})$ is $\Sigma_n(J_\alpha)$, then $R^\forall(y, \vec{x})$ is $\Sigma_{n+1}(J_\alpha)$ on J_ϱ. That is, there is a $\Sigma_{n+1}(J_\alpha)$ predicate $Q(y, \vec{x})$ such that*

$$(\forall y \in J_\varrho)\,(\forall \vec{x} \in J_\alpha)\,[(\forall z \in y)\, R(z, \vec{x}) \leftrightarrow Q(y, \vec{x})]. \quad \square$$

5. Standard Codes

Let $\alpha > 0$, $n > 0$. A Σ_n *code* for J_α is a set $A \subseteq J_{\varrho_\alpha^n}$. $A \in \Sigma_n(J_\alpha)$, such that for any $m \geqslant 1$,

$$\Sigma_{n+m}(J_\alpha) \cap \mathscr{P}(J_{\varrho_\alpha^n}) = \Sigma_m(\langle J_{\varrho_\alpha^n}, A\rangle).$$

In this section we show that not only does each J_α have a Σ_n code for each n, but there are particularly nice codes which are preserved under condensation arguments.

We begin by recalling the following result (V.5.9).

5.1 Lemma. *Let $\pi: J_{\bar{\alpha}} \prec_0 J_\alpha$. Then for any $v < \omega\bar{\alpha}$, $\pi(S_v) = S_{\pi(v)}$.* □

Using 5.1, we prove:

5.2 Lemma. *Let $\pi: \langle J_{\bar{\alpha}}, \bar{A}\rangle \prec_0 \langle J_\alpha, A\rangle$ and suppose that $\pi''\omega\bar{\alpha}$ is cofinal in $\omega\alpha$. Then in fact $\pi: \langle J_{\bar{\alpha}}, \bar{A}\rangle \prec_1 \langle J_\alpha, A\rangle$.*

Proof. Let φ be a Σ_0 formula of $\mathscr{L}$ such that

$$\models_{\langle J_\alpha, A\rangle} \exists z\, \varphi(z, \pi(\vec{x})).$$

Since $\pi''\omega\bar{\alpha}$ is cofinal in $\omega\alpha$, we can find a $v < \omega\bar{\alpha}$ such that

$$\models_{\langle J_\alpha, A\rangle} (\exists z \in S_{\pi(v)})\, \varphi(z, \pi(\vec{x})).$$

By 5.1, this can be written as

$$\models_{\langle J_\alpha, A\rangle} (\exists z \in \pi(S_\nu))\, \varphi(z, \pi(\vec{x})).$$

So as π is Σ_0-elementary, this gives,

$$\models_{\langle J_{\bar{\alpha}}, \bar{A}\rangle} (\exists z \in S_\nu)\, \varphi(z, \vec{x}).$$

So, as required,

$$\models_{\langle J_{\bar{\alpha}}, \bar{A}\rangle} \exists z \varphi(z, \vec{x}). \quad \square$$

Let $\alpha > 0$. The *standard codes*, A_α^n, and the *standard parameters*, p_α^n, are defined by recursion on n.

To commence, set

$$A_\alpha^0 = \emptyset, \quad p_\alpha^0 = \emptyset.$$

Now let $n \geqslant 0$ and assume that A_α^n and p_α^n are defined, and that if $n \geqslant 1$, A_α^n is a Σ_n code for J_α. We define A_α^{n+1} and p_α^{n+1}. By definition of ϱ_α^{n+1} there is a $\Sigma_{n+1}(J_\alpha)$ map $f \subseteq J_\alpha \times J_{\varrho_\alpha^{n+1}}$ such that $f''J_{\varrho_\alpha^{n+1}} = J_\alpha$. Let $\bar{f} = f \cap (J_{\varrho_\alpha^n} \times J_{\varrho_\alpha^{n+1}})$. Then $\bar{f}$ is also a $\Sigma_{n+1}(J_\alpha)$ map, and $\bar{f}''J_{\varrho_\alpha^{n+1}} = J_{\varrho_\alpha^n}$. But $\bar{f} \subseteq J_{\varrho_\alpha^n}$, so as A_α^n is a Σ_n code for J_α, $\bar{f}$ is in fact $\Sigma_1(\langle J_{\varrho_\alpha^n}, A_\alpha^n\rangle)$. Hence we may define

p_α^{n+1} = the $<_J$-least $p \in J_{\varrho_\alpha^n}$ such that every $x \in J_{\varrho_\alpha^n}$ is Σ_1-definable in $\langle J_{\varrho_\alpha^n}, A_\alpha^n\rangle$ from parameters in $J_{\varrho_\alpha^{n+1}} \cup \{p\}$.

As in section 3, $(\varphi_i \mid i < \omega)$ is a fixed $\Delta_1^{J_1}$ enumeration of all the Σ_1 formulas of $\mathscr{L}(A)$ of the form

$$\varphi_i(v_0, v_1) \equiv \exists v_2 \bar{\varphi}_i(v_0, v_1, v_2),$$

where $\bar{\varphi}_i$ is Σ_0. Set

$$A_\alpha^{n+1} = \{(i, x) \mid i \in \omega \wedge x \in J_{\varrho_\alpha^{n+1}} \wedge \models_{\langle J_{\varrho_\alpha^n}, A_\alpha^n\rangle} \varphi_i(\mathring{x}, \mathring{p}_\alpha^{n+1})\}.$$

5.3 Lemma. *A_α^{n+1} is a Σ_{n+1} code for J_α.*

Proof. By assumption, A_α^n is a $\Sigma_n(J_\alpha)$ set. So by 4.6, $\langle J_{\varrho_\alpha^n}, A_\alpha^n\rangle$ is amenable. So by 3.4, A_α^{n+1} is $\Sigma_1(\langle J_{\varrho_\alpha^n}, A_\alpha^n\rangle)$. Hence as A_α^n is a Σ_n code for J_α, A_α^{n+1} is $\Sigma_{n+1}(J_\alpha)$. We must show that for $m \geqslant 1$,

$$\Sigma_{n+1+m}(J_\alpha) \cap \mathscr{P}(J_{\varrho_\alpha^{n+1}}) = \Sigma_m(\langle J_{\varrho_\alpha^{n+1}}, A_\alpha^{n+1}\rangle).$$

Suppose first that $R \in \Sigma_0(\langle J_{\varrho_\alpha^{n+1}}, A_\alpha^{n+1}\rangle)$. Let φ be a Σ_0 formula of $\mathscr{L}$ and q an element from $J_{\varrho_\alpha^{n+1}}$ such that

$$R(x) \leftrightarrow \models_{\langle J_{\varrho_\alpha^{n+1}}, A_\alpha^{n+1}\rangle} \varphi(\mathring{x}, \mathring{q}).$$

Since $\langle J_{\varrho_\alpha^{n+1}}, A_\alpha^{n+1}\rangle$ is amenable, we have, by Σ_0-absoluteness:

$$R(x) \leftrightarrow (\exists u \in J_{\varrho_\alpha^{n+1}})(\exists a \in J_{\varrho_\alpha^{n+1}})[u \text{ is transitive} \wedge x \in u \wedge q \in u \wedge a = A_\alpha^{n+1} \cap u \wedge \models_{\langle u, a\rangle} \varphi(\mathring{x}, \mathring{q})].$$

Consider the function $a = A_\alpha^{n+1} \cap u$. Since A_α^{n+1} is J_α-definable, so is this function (as a function on $J_{\varrho_\alpha^{n+1}}$). Indeed, it has the definition

$$a = A_\alpha^{n+1} \cap u \leftrightarrow (\forall v \in a)(v \in u \wedge v \in A_\alpha^{n+1}) \wedge (\forall v \in u)(v \in A_\alpha^{n+1} \to v \in a).$$

This is of the form

$$a = A_\alpha^{n+1} \cap u \leftrightarrow (\forall v \in a)(\Sigma_{n+1}(J_\alpha)) \wedge (\forall v \in u)(\Pi_{n+1}(J_\alpha)).$$

By 4.8, for $a \in J_{\varrho_\alpha^{n+1}}$, this is of the form

$$\Sigma_{n+2}(J_\alpha) \wedge \Pi_{n+1}(J_\alpha),$$

and hence is $\Sigma_{n+2}(J_\alpha)$.

It follows at once from our above definition that R is $\Sigma_{n+2}(J_\alpha)$. Hence

$$\Sigma_0(\langle J_{\varrho_\alpha^{n+1}}, A_\alpha^{n+1}\rangle) \subseteq \Sigma_{n+2}(J_\alpha).$$

It follows immediately that

$$\Sigma_1(\langle J_{\varrho_\alpha^{n+1}}, A_\alpha^{n+1}\rangle) \subseteq \Sigma_{n+2}(J_\alpha).$$

By a simple induction on m, we get, for $m \geqslant 1$,

$$\Sigma_m(\langle J_{\varrho_\alpha^{n+1}}, A_\alpha^{n+1}\rangle) \subseteq \Sigma_{n+1+m}(J_\alpha).$$

It remains to prove that for every $m \geqslant 1$,

$$\Sigma_{n+1+m}(J_\alpha) \cap \mathscr{P}(J_{\varrho_\alpha^{n+1}}) \subseteq \Sigma_m(\langle J_{\varrho_\alpha^{n+1}}, A_\alpha^{n+1}\rangle).$$

Since A_α^n is a Σ_n code for J_α, it suffices to prove that

$$\Sigma_{m+1}(\langle J_{\varrho_\alpha^n}, A_\alpha^n\rangle) \cap \mathscr{P}(J_{\varrho_\alpha^{n+1}}) \subseteq \Sigma_m(\langle J_{\varrho_\alpha^{n+1}}, A_\alpha^{n+1}\rangle).$$

Let f be a $\Sigma_{n+1}(J_\alpha)$ function such that $f'' J_{\varrho_\alpha^{n+1}} = J_\alpha$. Set $\bar{f} = f \cap (J_{\varrho_\alpha^n} \times J_{\varrho_\alpha^{n+1}})$. Then $\bar{f}$ is $\Sigma_{n+1}(J_\alpha)$ and $\bar{f}'' J_{\varrho_\alpha^{n+1}} = J_{\varrho_\alpha^n}$. Moreover, $\bar{f} \subseteq J_{\varrho_\alpha^n}$, so $\bar{f}$ is $\Sigma_1(\langle J_{\varrho_\alpha^n}, A_\alpha^n\rangle)$.

Let $R \in \Sigma_{m+1}(\langle J_{\varrho_\alpha^n}, A_\alpha^n\rangle) \cap \mathscr{P}(J_{\varrho_\alpha^{n+1}})$. Assume for the sake of argument that m is even. Let P be a $\Sigma_1(\langle J_{\varrho_\alpha^n}, A_\alpha^n\rangle)$ relation such that for $x \in J_{\varrho_\alpha^{n+1}}$,

$$R(x) \leftrightarrow (\exists y_1 \in J_{\varrho_\alpha^n})(\forall y_2 \in J_{\varrho_\alpha^n}) \ldots (\exists y_{m-1} \in J_{\varrho_\alpha^n})(\forall y_m \in J_{\varrho_\alpha^n}) P(\vec{y}, x).$$

Define a relation $\tilde{P}$ by

$$\tilde{P}(\vec{z}, x) \leftrightarrow [(\vec{z}, x \in J_{\varrho_\alpha^{n+1}}) \wedge \exists y[y = \bar{f}(\vec{z}) \wedge P(y, x)]].$$

Now, there are $p, q \in J_{\varrho_\alpha^n}$ such that $\bar{f}$ is $\Sigma_1^{\langle J_{\varrho_\alpha^n}, A_\alpha^n\rangle}(\{p\})$ and P is $\Sigma_1^{\langle J_{\varrho_\alpha^n}, A_\alpha^n\rangle}(\{q\})$. By choice of p_α^{n+1}, the pair (p, q) is Σ_1-definable from elements of $J_{\varrho_\alpha^{n+1}} \cup \{p_\alpha^{n+1}\}$ in $\langle J_{\varrho_\alpha^n}, A_\alpha^n\rangle$. Hence both $\bar{f}$ and P are $\Sigma_1^{\langle J_{\varrho_\alpha^n}, A_\alpha^n\rangle}(\{u, p_\alpha^{n+1}\})$ for some $u \in J_{\varrho_\alpha^{n+1}}$. Thus $\tilde{P}$ is $\Sigma_1^{\langle J_{\varrho_\alpha^n}, A_\alpha^n\rangle}(\{u, p_\alpha^{n+1}\})$. (In case $\varrho_\alpha^{n+1} < \varrho_\alpha^n$, we may assume that ϱ_α^{n+1} is Σ_1-definable from u and p_α^{n+1} in $\langle J_{\varrho_\alpha^n}, A_\alpha^n\rangle$ as well.) So for some $i \in \omega$,

$$(*) \qquad \tilde{P}(\vec{z}, x) \leftrightarrow [(\vec{z}, x \in J_{\varrho_\alpha^n}) \wedge \models_{\langle J_{\varrho_\alpha^n}, A_\alpha^n\rangle} \varphi_i((\vec{z}, x, u)^\circ, \mathring{p}_\alpha^{n+1})]$$
$$\leftrightarrow [(\vec{z}, x \in J_{\varrho_\alpha^n}) \wedge (i, (\vec{z}, x, u)) \in A_\alpha^{n+1}].$$

Similarly, if we define D by

$$D(z) \leftrightarrow z \in \mathrm{dom}(\bar{f}),$$

then D is $\Sigma_1(\langle J_{\varrho_\alpha^n}, A_\alpha^n\rangle)$ and there is a $v \in J_{\varrho_\alpha^{n+1}}$ and a $j \in \omega$ such that

$$(**) \qquad D(z) \leftrightarrow [(z \in J_{\varrho_\alpha^n}) \wedge (j, (z, v)) \in A_\alpha^{n+1}].$$

Now, by definition of $\tilde{P}$ we have, for $x \in J_{\varrho_\alpha^{n+1}}$,

$$R(x) \leftrightarrow (\exists z_1 \in J_{\varrho_\alpha^{n+1}})(\forall z_2 \in J_{\varrho_\alpha^{n+1}}) \dots (\exists z_{m-1} \in J_{\varrho_\alpha^{n+1}})(\forall z_m \in J_{\varrho_\alpha^{n+1}})$$
$$\cdot [(D(z_1) \wedge D(z_3) \wedge \dots \wedge D(z_{m-1})) \wedge (D(z_2) \wedge D(z_4) \wedge \dots$$
$$\wedge D(z_m) \rightarrow \tilde{P}(\vec{z}, x))].$$

By (*) and (**), this is $\Sigma_m(\langle J_{\varrho_\alpha^{n+1}}, A_\alpha^{n+1}\rangle)$, as required. □

Let $\langle J_\alpha, A\rangle$ be amenable. The *Σ_n-projectum* of the structure $\langle J_\alpha, A\rangle$ is defined to be the largest ordinal $\varrho \leqslant \alpha$ such that $\langle J_\varrho, B\rangle$ is amenable for all $B \in \Sigma_n(\langle J_\alpha, A\rangle) \cap \mathscr{P}(J_\varrho)$, and is denoted by $\varrho_{\alpha,A}^n$. Note that this *definition* is not just a generalisation of the definition of the Σ_n-projectum of an ordinal. Though by 4.6, the *notion* is a generalisation of that of a Σ_n-projectum of an ordinal. Indeed, we can say more, as the next lemma indicates:

5.4 Lemma. *Let $\alpha > 1$, $n \geqslant 0$. Then $\varrho_\alpha^{n+1} = \varrho_{\varrho_\alpha^n, A_\alpha^n}^1$.*

Proof. By 4.6, ϱ_α^{n+1} is the largest $\varrho \leqslant \alpha$ such that $\langle J_\varrho, A\rangle$ is amenable for all $A \in \Sigma_{n+1}(J_\alpha) \cap \mathscr{P}(J_\varrho)$. Set $\eta = \varrho_{\varrho_\alpha^n, A_\alpha^n}^1$.

Suppose that $A \in \Sigma_1(\langle J_{\varrho_\alpha^n}, A_\alpha^n\rangle) \cap \mathscr{P}(J_{\varrho_\alpha^{n+1}})$. Then, as A_α^{n+1} is a Σ_{n+1} code for J_α, $A \in \Sigma_{n+1}(J_\alpha) \cap \mathscr{P}(J_{\varrho_\alpha^{n+1}})$. Thus by our above remark $\langle J_{\varrho_\alpha^{n+1}}, A\rangle$ is amenable. Thus by definition of η, $\varrho_\alpha^{n+1} \leqslant \eta$.

Now let $A \in \Sigma_{n+1}(J_\alpha) \cap \mathscr{P}(J_\eta)$. By choice of η, we have (trivially) $\eta \leqslant \varrho_\alpha^n$. Thus $A \in \Sigma_{n+1}(J_\alpha) \cap \mathscr{P}(J_{\varrho_\alpha^n})$. Hence $A \in \Sigma_1(\langle J_{\varrho_\alpha^n}, A_\alpha^n\rangle)$. Thus $\langle J_\eta, A\rangle$ is amenable. So, by definition, $\eta \leqslant \varrho_\alpha^{n+1}$. □

Again, let $\langle J_\alpha, A\rangle$ be amenable, and set $\varrho = \varrho^1_{\alpha,A}$. Suppose that every $x \in J_\alpha$ is Σ_1-definable in $\langle J_\alpha, A\rangle$ from parameters in $J_\varrho \cup \{p\}$ for some $p \in J_\alpha$. Then we define $p^1_{\alpha,A}$ to be the $<_J$-least such p, and set

$$A^1_{\alpha,A} = \{(i, x) \mid i \in \omega \wedge x \in J_\varrho \wedge \models_{\langle J_\alpha, A\rangle} \varphi_i(\mathring{x}, \mathring{p}^1_{\alpha,A})\}.$$

5.5 Lemma. *Let* $\alpha > 1$, $n \geqslant 0$. *Then*:

(i) $p^{n+1}_\alpha = p^1_{\varrho^n_\alpha, A^n_\alpha}$;
(ii) $A^{n+1}_\alpha = A^1_{\varrho^n_\alpha, A^n_\alpha}$.

Proof. (i) By definition,

p^{n+1}_α = the $<_J$-least $p \in J_{\varrho^n_\alpha}$ such that every $x \in J_{\varrho^n_\alpha}$ is Σ_1-definable in $\langle J_{\varrho^n_\alpha}, A^n_\alpha\rangle$ from parameters in $J_{\varrho^{n+1}_\alpha} \cup \{p\}$.

By 5.4, $\varrho^n_\alpha = \varrho^1_{\varrho^n_\alpha, A^n_\alpha}$. So by definition, $p^1_{\varrho^n_\alpha, A^n_\alpha} = p^{n+1}_\alpha$.

(ii) Likewise, by virtue of 5.4 and (i) above, the definitions of A^{n+1}_α and $A^1_{\varrho^n_\alpha, A^n_\alpha}$ coincide. □

It is the following result which will enable us to carry out condensation type arguments with structures of the form $\langle J_{\varrho^n_\alpha}, A^n_\alpha\rangle$, thereby enabling us to handle Σ_n predicates on the J_α's as coded Σ_1 predicates.

5.6 Theorem ("Condensation Lemma"). *Let* $\alpha > 1$, $n \geqslant 0$, $m \geqslant 0$. *Let* $\langle J_{\bar\varrho}, \bar A\rangle$ *be amenable, and let*

$$\pi\colon \langle J_{\bar\varrho}, \bar A\rangle \prec_m \langle J_{\varrho^n_\alpha}, A^n_\alpha\rangle.$$

Then:

(I) *There is a unique* $\bar\alpha \geqslant \bar\varrho$ *such that* $\bar\varrho = \varrho^n_{\bar\alpha}$, $A = A^n_{\bar\alpha}$.

(II) *There is a unique* $\tilde\pi \supseteq \pi$ *such that*:

(i) $\tilde\pi\colon J_{\bar\alpha} \prec_{m+n} J_\alpha$, *and*
(ii) *for* $i = 1, \ldots, n$, $\tilde\pi(p^i_{\bar\alpha}) = p^i_\alpha$.

(III) *For* $i = 1, \ldots, n$,

$$(\tilde\pi \restriction J_{\varrho^i_{\bar\alpha}})\colon \langle J_{\varrho^i_{\bar\alpha}}, A^i_{\bar\alpha}\rangle \prec_{m+n-i} \langle J_{\varrho^i_\alpha}, A^i_\alpha\rangle. \quad \square$$

The proof of 5.6 is quite long. Before we commence we make a few remarks. Firstly, notice that the result is indeed a condensation lemma. In many applications, the embedding π will simply be the inverse of the collapsing map obtained from some Σ_1 elementary submodel of $\langle J_{\varrho^n_\alpha}, A^n_\alpha\rangle$. Secondly, note that we allow for the case where $m = 0$. We will require this case in applications. Notice that this is the only case where we need to explicitly demand that $\langle J_{\bar\varrho}, \bar A\rangle$ be amenable. In all other cases this is automatic by the elementarity of π. Finally, some nomenclature. The embedding $\tilde\pi\colon J_{\bar\alpha} \to J_\alpha$ is called the *canonical extension* of $\pi\colon \langle J_{\bar\varrho}, \bar A\rangle \to \langle J_{\varrho^n_\alpha}, A^n_\alpha\rangle$.

Now for the proof of 5.6. This proceeds by induction on n. For $n = 0$ the theorem reduces to a triviality, so we are at once left with the proof that if the theorem holds for $n - 1$, then it holds for n, where $n > 0$. To simplify the notation, let us write $\varrho = \varrho_\alpha^n$, $A = A_\alpha^n$. So we are given an amenable structure $\langle J_{\bar\varrho}, \bar A\rangle$ and an embedding

$$\pi\colon \langle J_{\bar\varrho}, \bar A\rangle \prec_m \langle J_\varrho, A\rangle.$$

We shall show that there is a unique structure $\langle J_{\bar\beta}, \bar B\rangle$ such that $\bar\varrho = \varrho^1_{\bar\beta, \bar B}$ and $\bar A = A^1_{\bar\beta, \bar B}$, and a unique $\tilde\pi \supseteq \pi$ such that, setting $\beta = \varrho_\alpha^{n-1}$, $B = A_\alpha^{n-1}$, $p = p_\alpha^n$:

(i) $\tilde\pi\colon \langle J_{\bar\beta}, \bar B\rangle \prec_{m+1} \langle J_\beta, B\rangle$;

(ii) $\tilde\pi(p^1_{\bar\beta, \bar B}) = p$.

The induction step, and hence the theorem, follow directly from this. For by induction hypothesis there is a unique $\bar\alpha$ such that $\bar\beta = \varrho_{\bar\alpha}^{n-1}$, $\bar B = A_{\bar\alpha}^{n-1}$, etc., and we have, by 5.4 and 5.5, $p = p^1_{\beta, B}$, $\bar\varrho = \varrho^1_{\bar\beta, \bar B} = \varrho^n_{\bar\alpha}$, $\bar A = A^1_{\bar\beta, \bar B} = A^n_{\bar\alpha}$.

The function $\tilde\pi$ will be the inverse to a certain collapsing isomorphism. The set which $\tilde\pi^{-1}$ collapses is defined thus:

$$X = \{x \in J_\beta \mid x \text{ is } \Sigma_1\text{-definable in } \langle J_\beta, B\rangle \text{ from parameters in } \operatorname{ran}(\pi) \cup \{p\}\}.$$

Since $X \prec_1 \langle J_\beta, B\rangle$, there is an isomorphism

$$\tilde\pi\colon \langle J_{\bar\beta}, \bar B\rangle \cong \langle X, B \cap X\rangle,$$

for some unique $\bar\beta$, $\bar B$. Thus

$$\tilde\pi\colon \langle J_{\bar\beta}, \bar B\rangle \prec_1 \langle J_\beta, B\rangle.$$

Define $\tilde\varrho \leqslant \varrho$ by

$$\omega\tilde\varrho = \sup(\pi''\omega\bar\varrho).$$

Set

$$\tilde A = A \cap J_{\tilde\varrho}.$$

Then

$$\pi\colon \langle J_{\bar\varrho}, \bar A\rangle \prec_0 \langle J_{\tilde\varrho}, \tilde A\rangle.$$

But $\pi''\omega\bar\varrho$ is cofinal in $\omega\tilde\varrho$. So by 5.2,

$$\pi\colon \langle J_{\bar\varrho}, \bar A\rangle \prec_1 \langle J_{\tilde\varrho}, \tilde A\rangle.$$

5.7 Lemma. $\operatorname{ran}(\pi) = X \cap J_{\tilde\varrho}$.

Proof. Clearly, $\operatorname{ran}(\pi) \subseteq X \cap J_{\tilde\varrho}$. To prove the opposite inclusion, let $y \in X \cap J_{\tilde\varrho}$. Then for some $i \in \omega$ and some $x \in \operatorname{ran}(\pi)$,

$$y = \text{the unique } x \in J_\beta \text{ such that } \models_{\langle J_\beta, B\rangle} \varphi_i((\mathring{y}, \mathring{x}), \mathring{p}).$$

Thus by definition of $A = A^n_\alpha$,

$$y = \text{the unique } y \in J_{\bar\beta} \text{ such that } \tilde{A}(i, (y, x)).$$

But $x \in \operatorname{ran}(\pi)$ and $\pi\colon \langle J_{\bar\varrho}, \bar{A}\rangle \prec_1 \langle J_{\tilde\varrho}, \tilde{A}\rangle$, so we conclude that $y \in \operatorname{ran}(\pi)$. That proves the lemma. $\square$

By 5.7, π^{-1} is the unique collapsing isomorphism for $X \cap J_{\tilde\varrho}$. But $X \cap J_{\tilde\varrho}$ is an $\in$-initial segment of X and $\tilde\pi^{-1}$ is the unique collapsing isomorphism for X, so $\tilde\pi^{-1} \upharpoonright (X \cap J_{\tilde\varrho})$ is the unique collapsing isomorphism for $X \cap J_{\tilde\varrho}$. Thus $\tilde\pi^{-1} \upharpoonright (X \cap J_{\tilde\varrho}) = \pi^{-1}$. Thus $\pi = \tilde\pi \upharpoonright J_{\bar\varrho}$ and $\pi \subseteq \tilde\pi$. (Fig. 1 sums up the situation now.)

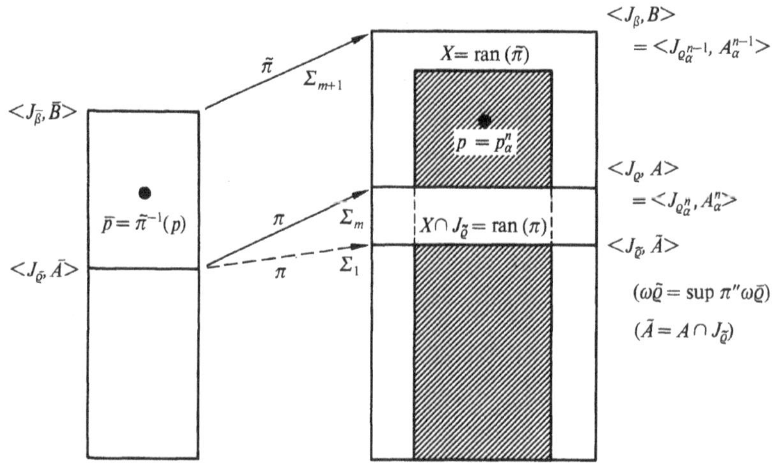

Shaded part $= X = \{x \in J_\beta \mid x$ is Σ_1-definable in $\langle J_\beta, B\rangle$ from parameters in $\operatorname{ran}(\pi) \cup \{p\}\}$.

Fig. 1

5.8 Lemma. $\tilde\pi\colon \langle J_{\bar\beta}, \bar{B}\rangle \prec_{m+1} \langle J_\beta, B\rangle$.

Proof. If $m = 0$ there is nothing to prove. So assume $m > 0$.

Let y be Σ_{m+1}-definable in $\langle J_\beta, B\rangle$ from parameters in $\operatorname{ran}(\tilde\pi)$. We must show that $y \in \operatorname{ran}(\tilde\pi)$. Now, by the definition of $\operatorname{ran}(\tilde\pi) = X$, y is Σ_{m+1}-definable in $\langle J_\beta, B\rangle$ from parameters in $\operatorname{ran}(\pi) \cup \{p\}$. Let φ be a Σ_{m+1}-formula of $\mathscr{L}(B)$ such that y is the unique $y \in J_\beta$ for which $\models_{\langle J_\beta, B\rangle} \varphi(\mathring{y}, \mathring{\vec{x}}, \mathring{p})$, where $\vec{x} \in \operatorname{ran}(\pi)$. Then we have

$$\varphi(u, \vec{v}, w) = \exists z_1 \forall z_2 \exists z_3 \ldots - z_m \psi(\vec{z}, u, \vec{v}, w),$$

where ψ is Σ_1 if m is even and Π_1 if m is odd.

Suppose first that $\varrho = \beta$. Now, y is the unique y such that

$$(*) \qquad (\exists z_1 \in J_\beta)(\forall z_2 \in J_\beta)\ldots(- z_m \in J_\beta)\,[\models_{\langle J_\beta, B\rangle} \psi(\mathring{\vec{z}}, \mathring{y}, \mathring{\vec{x}}, \mathring{p})].$$

But $\beta = \varrho = \varrho_\alpha^n = \varrho_{\beta,B}^1, p = p_\alpha^n = p_{\beta,B}^1$, and $A = A_\alpha^n = A_{\beta,B}^1$. So as ψ is Σ_1 or Π_1, (*) is a $\Sigma_m^{\langle J_\rho, A\rangle}(\{\vec{x}\})$ predicate of y. But $\vec{x} \in \mathrm{ran}(\pi) \prec_m \langle J_\varrho, A\rangle$. Thus $y \in \mathrm{ran}(\pi) \subseteq \mathrm{ran}(\tilde{\pi})$, and we are done.

Now suppose that $\varrho < \beta$. Let $h = h_{\beta,B}$, and set

$$\tilde{h}((i, x)) \simeq h(i, (x, p)).$$

Let $D = \mathrm{dom}(\tilde{h}) \cap J_\varrho$. For $u \in D$, $\tilde{h}(u)$ is Σ_1-definable in $\langle J_\beta, B\rangle$ from u, p, so if $u \in X$, then since $p \in X \prec_1 \langle J_\beta, B\rangle$, we have $\tilde{h}(u) \in X$. Thus in order to show that $y \in X$ it suffices to show that for some $u \in D \cap X$, we have

$$\models_{\langle J_\beta, B\rangle} \varphi(\tilde{h}(\mathring{u}), \mathring{\vec{x}}, \mathring{p}).$$

(For then by uniqueness, $y = \tilde{h}(u) \in X$.) Now, $\varrho = \varrho_\alpha^n$, so by definition of $p = p_\alpha^n$, every $x \in J_\beta$ is Σ_1-definable in $\langle J_\beta, B\rangle$ from parameters in $J_\varrho \cup \{p\}$. So in particular, $\tilde{h}'' J_\varrho = J_\beta$, i.e. $\tilde{h}'' D = J_\beta$. Thus it suffices to show that for some $u \in D \cap X$ we have

$$(**) \qquad (\exists z_1 \in D)(\forall z_2 \in D)\ldots(- z_m \in D)\,[\models_{\langle J_\beta, B\rangle} \psi(\mathring{\tilde{h}}(z_1), \ldots, \mathring{\tilde{h}}(z_m), \mathring{\tilde{h}}(u), \mathring{\vec{x}}, \mathring{p})].$$

If we can show that (**) is a $\Sigma_m^{\langle J_\rho, A\rangle}(\{\vec{x}\})$ predicate of u we shall be done, since $\vec{x} \in \mathrm{ran}(\pi) \prec_m \langle J_\varrho, A\rangle$ and $\mathrm{ran}(\pi) \subseteq X$.

Let us assume that m is even. (We deal with the similar case m odd later.) There is an $i_0 < \omega$ such that for any $z \in J_\varrho$,

$$\begin{aligned} z \in D &\leftrightarrow \exists y[y = \tilde{h}(z)] \\ &\leftrightarrow \exists y[y = h((z)_0, ((z)_1, p))] \\ &\leftrightarrow \models_{\langle J_\beta, B\rangle} \varphi_{i_0}(\mathring{z}, \mathring{p}) \\ &\leftrightarrow (i_0, z) \in A, \end{aligned}$$

where the last equivalence follows from the definition of $A = A_\alpha^n$. Similarly, as ψ is Σ_1 (for m even) there is a $j_0 < \omega$ such that for any $z_1, \ldots, z_m, u \in D$,

$$\begin{aligned} &\models_{\langle J_\beta, B\rangle} \psi(\mathring{\tilde{h}}(z_1), \ldots, \mathring{\tilde{h}}(z_m), \mathring{\tilde{h}}(u), \mathring{\vec{x}}, \mathring{p}) \\ &\leftrightarrow \models_{\langle J_\beta, B\rangle} \varphi_{j_0}((\mathring{z}_1, \ldots, \mathring{z}_m, \mathring{u}, \mathring{\vec{x}}), \mathring{p}) \\ &\leftrightarrow (j_0, (z_1, \ldots, z_m, u, \vec{x})) \in A. \end{aligned}$$

Hence (**) is equivalent to the following (for any $u \in J_\varrho$)

$$\begin{aligned} &[(i_0, u) \in A] \wedge (\exists z_1 \in J_\varrho)(\forall z_2 \in J_\varrho)(\exists z_3 \in J_\varrho)(\forall z_4 \in J_\varrho)\ldots \\ &\qquad (\exists z_{m-1} \in J_\varrho)(\forall z_m \in J_\varrho)[((i_0, z_1) \in A \\ &\qquad\quad \wedge (i_0, z_3) \in A \wedge \ldots \wedge (i_0, z_{m-1}) \in A) \wedge ((i_0, z_2) \in A \\ &\qquad\quad \wedge (i_0, z_4) \in A \wedge \ldots \wedge (i_0, z_m) \in A \rightarrow (j_0, (z_1, \ldots, z_m, u, \vec{x})) \in A)]. \end{aligned}$$

But this is $\Sigma_m^{\langle J_\varrho, A\rangle}(\{\mathring{x}\})$, so we are done.

The case m odd is fairly similar. The only difference is that we rewrite (**) as

$$(\exists z_1 \in D)(\forall z_2 \in D)\dots(\exists z_m \in D)\neg\,[\models_{\langle J_\beta, B\rangle}\neg\,\psi(\mathring{\tilde{h}}(z_1),\dots,\mathring{\tilde{h}}(z_m),\\ \mathring{\tilde{h}}(u), \vec{\mathring{x}}, \mathring{p})],$$

so that $(\neg\,\psi)$ is Σ_1. The rest of the proof is modified accordingly.

That completes the proof of the lemma. □

Now let $\bar{p} = \tilde{\pi}^{-1}(p)$. We must prove that $\bar{\varrho} = \varrho^1_{\bar{\beta},\bar{B}}$, $\bar{A} = A^1_{\bar{\beta},\bar{B}}$, $\bar{p} = p^1_{\bar{\beta},\bar{B}}$.

5.9 Lemma. $\bar{A} = \{(i, x) \mid i \in \omega \wedge x \in J_{\bar{\varrho}} \wedge \models_{\langle J_{\bar{\beta}}, \bar{B}\rangle} \varphi_i(\mathring{x}, \mathring{\bar{p}})\}$.

Proof. Since $\pi\colon \langle J_{\bar{\varrho}}, \bar{A}\rangle \prec_m \langle J_\varrho, A\rangle$, we have, for $x \in J_{\bar{\varrho}}$,

$$\bar{A}((i, x)) \leftrightarrow A((i, \pi(x))).$$

And since $\beta = \varrho_\alpha^{n-1}$, $B = A_\alpha^{n-1}$, $A = A_\alpha^n$, we have

$$A((i, \pi(x))) \leftrightarrow \models_{\langle J_\beta, B\rangle} \varphi_i(\mathring{\pi}(x), \mathring{p}).$$

Finally, since $\tilde{\pi}\colon \langle J_{\bar{\beta}}, \bar{B}\rangle \prec_1 \langle J_\beta, B\rangle$ and $\pi(x), p \subset \mathrm{ran}(\tilde{\pi})$, we have

$$\models_{\langle J_\beta, B\rangle} \varphi_i(\mathring{\pi}(x), \mathring{p}) \leftrightarrow \models_{\langle J_{\bar{\beta}}, \bar{B}\rangle} \varphi_i(\mathring{x}, \mathring{\bar{p}}).$$

The above three equivalences yield the lemma. □

5.10 Lemma. $\bar{\varrho} = \varrho^1_{\bar{\beta},\bar{B}}$.

Proof. Since $J_{\bar{\beta}}$ is the collapse of X, every $x \in J_{\bar{\beta}}$ is Σ_1-definiable in $\langle J_{\bar{\beta}}, \bar{B}\rangle$ from parameters in $J_{\bar{\varrho}} \cup \{\bar{p}\}$. Thus if $\bar{h} = h_{\bar{\beta},\bar{B}}$, we have

$$J_{\bar{\beta}} = \tilde{h}^*(J_{\bar{\varrho}} \times \{\bar{p}\}).$$

Hence there is a $\Sigma_1(\langle J_{\bar{\beta}}, \bar{B}\rangle)$ map f from a subset of $\omega\bar{\varrho}$ onto $J_{\bar{\beta}}$. It follows that $\varrho^1_{\bar{\beta},\bar{B}} \leqslant \bar{\varrho}$. For suppose, on the contary, that $\bar{\varrho} < \varrho^1_{\bar{\beta},\bar{B}}$. Let $E = \{\xi \in \omega\bar{\varrho} \mid \xi \notin f(\xi)\}$. Then E is a $\Sigma_1(\langle J_{\bar{\beta}}, \bar{B}\rangle)$ subset of $\omega\bar{\varrho}$. By definition of $\varrho^1_{\bar{\beta},\bar{B}}$, $\langle J_{\varrho^1_{\bar{\beta},\bar{B}}}, E\rangle$ must be amenable. Thus $E = E \cap \omega\bar{\varrho} \in J_{\varrho^1_{\bar{\beta},\bar{B}}} \subseteq J_{\bar{\beta}}$. So for some $\xi \in \omega\bar{\varrho}$, $E = f(\xi)$. But then we get

$$\xi \in f(\xi) \leftrightarrow \xi \in E \leftrightarrow \xi \notin f(\xi),$$

a contradiction. Thus, as claimed, $\varrho^1_{\bar{\beta},\bar{B}} \leqslant \bar{\varrho}$. We now prove the opposite inequality.

Let $C \in \Sigma_1(\langle J_{\bar{\beta}}, \bar{B}\rangle) \cap \mathscr{P}(J_{\bar{\varrho}})$. Since every member of $J_{\bar{\beta}}$ is Σ_1-definable from parameters in $J_{\bar{\varrho}} \cup \{\bar{p}\}$ in $\langle J_{\bar{\beta}}, \bar{B}\rangle$, $C \in \Sigma_1^{\langle J_{\bar{\beta}}, \bar{B}\rangle}(J_{\bar{\varrho}} \cup \{\bar{p}\})$. So for some $i \in \omega$ and some $y \in J_{\bar{\varrho}}$ we have, for $x \in J_{\bar{\varrho}}$,

$$x \in C \quad \text{iff } \models_{\langle J_{\bar{\beta}}, \bar{B}\rangle} \varphi_i((\mathring{x}, \mathring{y}), \mathring{\bar{p}}).$$

So by 5.9, we have, for $x \in J_{\bar{\varrho}}$,

$$x \in C \quad \text{iff } (i, (x, y)) \in \bar{A}.$$

Let $u \in J_{\bar{\varrho}}$, and set

$$v = \{(i, (x, y)) \mid x \in u\}.$$

Note that $v \in J_{\bar{\varrho}}$. Since $\langle J_{\bar{\varrho}}, \bar{A} \rangle$ is amenable, $\bar{A} \cap v \in J_{\bar{\varrho}}$. But look,

$$x \in C \cap u \quad \text{iff } (i, (x, y)) \in \bar{A} \cap v.$$

So as $J_{\bar{\varrho}}$ is rud closed, $C \cap u \in J_{\bar{\varrho}}$. Thus $\langle J_{\bar{\varrho}}, C \rangle$ is amenable. Thus, by definition, $\bar{\varrho} \leqslant \varrho^1_{\bar{\beta}, \bar{B}}$, and the lemma is proved. □

5.11 Lemma. $\bar{p} = p^1_{\bar{\beta}, \bar{B}}$.

Proof. Since every $x \in J_{\bar{\beta}}$ is Σ_1-definable from parameters in $J_{\bar{\varrho}} \cup \{\bar{p}\}$ in $\langle J_{\bar{\beta}}, \bar{B} \rangle$ and $\bar{\varrho} = \varrho^1_{\bar{\beta}, \bar{B}}$, it suffices to show that $\bar{p}$ is $<_J$-least with this property. Well suppose not, and let $\bar{p}' <_J \bar{p}$ have the same property. For some $i \in \omega$ and some $x \in J_{\bar{\varrho}}$, we have $\bar{p} = \bar{h}(i, \bar{p}'))$. Since $\tilde{\pi}: \langle J_{\bar{\beta}}, \bar{B} \rangle \prec_1 \langle J_\beta, B \rangle$ and $\bar{h} = h_{\bar{\beta}, \bar{B}}$, $h = h_{\beta, B}$, setting $p' = \tilde{\pi}(\bar{p}')$ and applying $\tilde{\pi}$ gives $p = h(i, (\tilde{\pi}(x), p'))$. Now, $\tilde{\pi}(x) = \pi(x) \in X \cap J_\varrho$. So, as every $y \in J_\beta$ is Σ_1-definable from parameters in $J_\varrho \cup \{p\}$ in $\langle J_\beta, B \rangle$, it follows that every $y \in J_\beta$ is Σ_1-definable from parameters in $J_\varrho \cup \{p'\}$ in $\langle J_\beta, B \rangle$. But $p' <_J p = p^n_\alpha$, so this contradicts the definition of p^n_α. The lemma is proved. □

Since $\bar{\varrho} = \varrho^1_{\bar{\beta}, \bar{B}}$ and $\bar{p} = p^1_{\bar{\beta}, \bar{B}}$, 5.9 implies immediately that $\bar{A} = A^1_{\bar{\beta}, \bar{B}}$. That proves the existence part of 5.6. We turn to the question of uniqueness.

Suppose that $\langle J_{\bar{\beta}_0}, \bar{B}_0 \rangle$ and $\langle J_{\bar{\beta}_1}, \bar{B}_1 \rangle$ are such that $\bar{\varrho} = \varrho^1_{\bar{\beta}_i, \bar{B}_i}$ and $\bar{A} = A^1_{\bar{\beta}_i, \bar{B}_i}$, $i = 0, 1$. Set $\bar{p}_i = p^1_{\bar{\beta}_i, \bar{B}_i}$. For each $j \in \omega$ and each $\vec{x} \in J_{\bar{\varrho}}$ we have

$$\models_{\langle J_{\bar{\beta}_0}, \bar{B}_0 \rangle} \varphi_j((\mathring{\vec{x}}), \mathring{\bar{p}}_0) \quad \text{iff } \bar{A}((j, (\vec{x}))) \quad \text{iff } \models_{\langle J_{\bar{\beta}_1}, \bar{B}_1 \rangle} \varphi_j((\mathring{\vec{x}}), \mathring{\bar{p}}_1).$$

Since $(\varphi_j \mid j < \omega)$ enumerates all the Σ_1 formulas of $\mathscr{L}(A)$ with free variables v_0, v_1, we have, for all x, y in $J_{\bar{\varrho}}$ and all $j, k \in \omega$,

(a) $\quad h_{\bar{\beta}_0, \bar{B}_0}(j, (x, p_0)) = h_{\bar{\beta}_0, \bar{B}_0}(k, (y, p_0)) \quad$ iff $h_{\bar{\beta}_1, \bar{B}_1}(j, (x, p_1)) = h_{\bar{\beta}_1, \bar{B}_1}(k, (y, p_1))$;

(b) $\quad h_{\bar{\beta}_0, \bar{B}_0}(j, (x, p_0)) \in h_{\bar{\beta}_0, \bar{B}_0}(k, (y, p_0)) \quad$ iff $h_{\bar{\beta}_1, \bar{B}_1}(j, (x, p_1)) \in h_{\bar{\beta}_1, \bar{B}_1}(k, (y, p_1))$;

(c) $\quad h_{\bar{\beta}_0, \bar{B}_0}(j, (x, p_0)) \in \bar{B}_0 \quad$ iff $h_{\bar{\beta}_1, \bar{B}_1}(j, (x, p_1)) \in \bar{B}_1$.

But

$$h^*_{\bar{\beta}_i, \bar{B}_i}(J_{\bar{\varrho}} \times \{p_i\}) = J_{\bar{\beta}_i}$$

for $i = 0, 1$, so by (a)–(c) we have

$$\sigma: \langle J_{\bar{\beta}_0}, \bar{B}_0 \rangle \cong \langle J_{\bar{\beta}_1}, \bar{B}_1 \rangle,$$

where for $x \in J_{\bar{\varrho}}$, $j \in \omega$, we set

$$\sigma(h_{\bar{\beta}_0, \bar{B}_0}(j, (x, p_0)) \simeq h_{\bar{\beta}_1, \bar{B}_1}(j, (x, p_1)).$$

This means that $\bar{\beta}_0 = \bar{\beta}_1$ and that $\sigma = \mathrm{id} \upharpoonright J_{\bar{\beta}_0}$, so $\bar{B}_0 = \bar{B}_1$ as well. Hence $\bar{\beta}, \bar{B}$ are unique and it remains only to show that $\tilde{\pi}$ is unique.

Let $\tilde{\pi}_i \supseteq \pi$, $\tilde{\pi}_i \colon \langle J_{\bar{\beta}}, \bar{B}\rangle \prec_{m+1} \langle J_\beta, B\rangle$, $\tilde{\pi}_i(\bar{p}) = p$, for $i = 0, 1$. Let $y \in J_{\bar{\beta}}$. For some $j \in \omega$ and some $x \in J_{\bar{\varrho}}$, we have $y = h_{\bar{\beta}, \bar{B}}(j, (x, \bar{p}))$. Then

$$\begin{aligned}\tilde{\pi}_0(x) &= \tilde{\pi}_0 \circ h_{\bar{\beta}, \bar{B}}(j, (x, \bar{p})) = h_{\beta, B}(j, (\tilde{\pi}_0(x), \tilde{\pi}_0(\bar{p}))) = h_{\beta, B}(j, \pi(x), p)) \\ &= h_{\beta, B}(j, (\tilde{\pi}_1(x), \tilde{\pi}_1(\bar{p}))) = \tilde{\pi}_1 \circ h_{\bar{\beta}, \bar{B}}(j, (x, \bar{p})) = \tilde{\pi}_1(y).\end{aligned}$$

Hence $\tilde{\pi}_0 = \tilde{\pi}_1$, and the proof of 5.6 is complete.

6. An Application: A Global $\square$-Principle

S Let S denote the class of all singular limit ordinals. Given any class E of limit ordinals, we shall denote the following principle by $\square(E)$: there is a sequence $(C_\alpha \mid \alpha \in S)$ such that:

(i) C_α is a club subset of α;
(ii) $\mathrm{otp}(C_\alpha) < \alpha$;
(iii) if $\bar{\alpha} < \alpha$ is a limit point of C_α, then $\bar{\alpha} \in S$, $\bar{\alpha} \notin E$, and $C_{\bar{\alpha}} = \bar{\alpha} \cap C_\alpha$.

Using our fine structure theory we shall prove the following theorem (which will be utilised in the next chapter):

6.1 Theorem. *Assume $V = L$. Then there is a class E of limit ordinals such that:*

(i) $\alpha \in E \to \mathrm{cf}(\alpha) = \omega$;
(ii) *if $\kappa > \omega$ is regular, then $E \cap \kappa$ is a stationary subset of κ;*
(iii) *$\square(E)$ is valid.* $\square$

In fact by a slightly different argument, it is possible to prove the following more general result.

6.1′ Theorem. *Assume $V = L$. Let A be a class of limit ordinals. Then there is a class $E \subseteq A$ such that:*

(i) *if $\kappa > \omega$ is regular and $A \cap \kappa$ is stationary in κ, then $E \cap \kappa$ is stationary in κ;*
(ii) *$\square(E)$.* $\square$

This more general result is proved in detail in Chapter IX, using Silver machines instead of the Fine Structure theory. It is also possible to adapt the proof given in this chapter using the fine structure (see Exercise 4), but in order to avoid making an already complicated proof look even worse, we prove the more specialised version (which in any case is enough for our needs here). As will be seen, the advantage with the specialised version is that the existence and behaviour of the set E can be relegated to a special case of the construction, and thus may be ignored for most of the proof. (This advantage does not arise with the machine proof, which does not involve a number of separate cases.)

Before we commence the proof, let us see how this relates to the principles $\Box_\kappa$ considered in Chapter IV. Let $\Box$ denote the principle $\Box(\emptyset)$. Clearly, if $F \subseteq E$, then $\Box(E)$ implies $\Box(F)$, so $\Box$ is the weakest of the global $\Box$-principles of the above kind.

6.2 Theorem. *Assume $\Box$. Then $\Box_\kappa$ holds for any uncountable cardinal κ.*

Proof. Recall that $\Box_\kappa$ asserts the existence of a sequence $(C_\alpha \mid \alpha < \kappa^+ \wedge \lim(\alpha))$ such that:

(i) C_α is a club subset of α;
(ii) $\mathrm{cf}(\alpha) < \kappa \to |C_\alpha| < \kappa$;
(iii) if $\bar{\alpha}$ is a limit point of C_α, then $C_{\bar{\alpha}} = \bar{\alpha} \cap C_\alpha$.

We shall denote by $\Box_\kappa^S$ the following principle: there is a sequence $(C_\alpha \mid \alpha \in S \cap \kappa^+)$ such that:

(i) C_α is a club subset of α;
(ii) $\mathrm{cf}(\alpha) < \kappa \to |C_\alpha| < \kappa$;
(iii) if $\bar{\alpha}$ is a limit point of C_α, then $\bar{\alpha} \in S \cap \kappa^+$ and $C_{\bar{\alpha}} = \bar{\alpha} \cap C_\alpha$.

We shall prove the implications $\Box \to \Box_\kappa^S \to \Box_\kappa$. We deal with the second implication first. Let $(C_\alpha \mid \alpha \in S \cap \kappa^+)$ be as in $\Box_\kappa^S$. Define a $\Box_\kappa$-sequence $(\tilde{C}_\alpha \mid \alpha < \kappa^+ \wedge \lim(\alpha))$ as follows.

Suppose first that κ is regular. Then we define $\tilde{C}_\alpha = C_\alpha - \kappa$ for $\kappa < \alpha < \kappa^+$, $\lim(\alpha)$, and $\tilde{C}_\alpha = \alpha$ for $\alpha \leqslant \kappa$, $\lim(\alpha)$. If $\kappa < \alpha < \kappa^+$, $\lim(\alpha)$, then $\alpha \in S \cap \kappa^+$, so C_α is defined. Hence $\tilde{C}_\alpha$ is defined for all limit ordinals $\alpha < \kappa^+$. Clearly, $(\tilde{C}_\alpha \mid \alpha < \kappa^+ \wedge \lim(\alpha))$ is a $\Box_\kappa$-sequence.

Now suppose κ is singular. In this case the above method will not work, since in order to satisfy $\Box_\kappa$ we shall require $|\tilde{C}_\kappa| < \kappa$, which prevents us from defining $\tilde{C}_\kappa = \kappa$. So we proceed as follows. Let $\theta = \mathrm{cf}(\kappa) < \kappa$. Let $\tilde{C}_\kappa$ be a club subset of κ of order-type θ with $\min(\tilde{C}_\kappa) = 0$. If $\alpha < \kappa$ is a limit point of $\tilde{C}_\kappa$, set $\tilde{C}_\alpha = \alpha \cap \tilde{C}_\kappa$. If $\alpha < \kappa$ is a limit ordinal but is not a limit point of $\tilde{C}_\kappa$, then there is a largest element $\nu \in \tilde{C}_\kappa$ such that $\nu < \alpha$, and we set $\tilde{C}_\alpha = \alpha - \nu$. Finally, in case $\kappa < \alpha < \kappa^+$, $\lim(\alpha)$, we set $\tilde{C}_\alpha = C_\alpha - \kappa$. It is easily seen that $(\tilde{C}_\alpha \mid \alpha < \kappa^+ \wedge \lim(\alpha))$ is a $\Box_\kappa$-sequence.

We turn now to the considerably less simple problem of deducing $\Box_\kappa^S$ from $\Box$. We start with a $\Box$-sequence $(C_\alpha^0 \mid \alpha \in S)$. For each $\alpha \in S \cap \kappa^+$, we set $C_\alpha^1 = C_\alpha^0 - \kappa$ in case $\alpha > \kappa$ and $C_\alpha^1 = C_\alpha^0$ in case $\alpha \leqslant \kappa$. It is clear that the sequence $(C_\alpha^1 \mid \alpha \in S \cap \kappa^+)$ satisfies the following conditions:

1 (i) C_α^1 is a club subset of α;
1 (ii) $\mathrm{otp}(C_\alpha^1) < \alpha$;
1 (iii) if $\bar{\alpha}$ is a limit point of C_α^1, then $\bar{\alpha} \in S \cap \kappa^+$ and $C_{\bar{\alpha}}^1 = \bar{\alpha} \cap C_\alpha^1$;
1 (iv) if $\alpha \in S \cap \kappa^+$, $\alpha > \kappa$, then $C_\alpha^1 \cap \kappa = \emptyset$.

We next define a sequence $(C_\alpha^2 \mid \alpha \in S \cap \kappa^+)$ such that:

2(i) $C_\alpha^2 \subseteq C_\alpha^1$;

2(ii) C_α^2 is a club subset of α;

2(iii) $\operatorname{otp}(C_\alpha^2) \leqslant \kappa$;

2(iv) if $\bar{\alpha}$ is a limit point of C_α^2, then $\bar{\alpha} \in S \cap \kappa^+$ and $C_{\bar{\alpha}}^2 = \bar{\alpha} \cap C_\alpha^2$.

For $\alpha \in S \cap \kappa^+$, let $\theta_\alpha = \operatorname{otp}(C_\alpha^1)$ and let $f_\alpha \colon \theta_\alpha \to C_\alpha^1$ be the monotone enumeration of C_α^1. We define C_α^2 by recursion on α.

For $\alpha \leqslant \kappa$, set $C_\alpha^2 = C_\alpha^1$. This part of the C^2-sequence clearly satisfies 2(i)–2(iv). And by 1(iv), the remaining case ($\alpha > \kappa$) will not affect the situation below κ, so we shall not beed to worry about any clashes when we come to check 2(iv) for the rest of the C^2-sequence.

Now suppose $\alpha > \kappa$ and we have defined $C_{\bar{\alpha}}^2$ for $\bar{\alpha} \in S \cap \alpha$. If $\theta_\alpha \leqslant \kappa$, we set $C_\alpha^2 = C_\alpha^1$. It is immediate that 2(i)–2(iii) are satisfied in this case. We check 2(iv). Let $\bar{\alpha}$ be a limit point of C_α^2. Then $\bar{\alpha}$ is a limit point of C_α^1, so by 1(iii), $\bar{\alpha} \in S$ and $C_{\bar{\alpha}}^1 = \bar{\alpha} \cap C_\alpha^1$. Thus $\theta_{\bar{\alpha}} = \operatorname{otp}(C_{\bar{\alpha}}^1) \leqslant \operatorname{otp}(C_\alpha^1) = \theta_\alpha \leqslant \kappa$. But by 1(iv), $\bar{\alpha} > \kappa$. Thus $C_{\bar{\alpha}}^2 = C_{\bar{\alpha}}^1 = \bar{\alpha} \cap C_\alpha^1 = \bar{\alpha} \cap C_\alpha^2$.

We are left with the case where $\theta_\alpha > \kappa$. In this case, θ_α is singular, since $\operatorname{cf}(\theta_\alpha) - \operatorname{cf}(\alpha) \leqslant \kappa < \theta_\alpha$. Hence $\theta_\alpha \in S \cap \kappa^+$. By 1(ii), $\theta_\alpha < \alpha$, so $C_{\theta_\alpha}^2$ is defined already. Set $C_\alpha^2 = f_\alpha'' C_{\theta_\alpha}^2$. Using the induction hypothesis, it is immediate that 2(i)–2(iii) are satisfied. We check 2(iv). Let $\bar{\alpha} < \alpha$ be a limit point of C_α^2. Then $\operatorname{cf}(\bar{\alpha}) < \operatorname{otp}(C_\alpha^2) \leqslant \kappa$. But by 1(iv), $\bar{\alpha} > \kappa$. Thus $\bar{\alpha} \in S$. Now, $\bar{\alpha}$ is a limit point of C_α^1, so $C_{\bar{\alpha}}^1 = \bar{\alpha} \cap C_\alpha^1$. Hence $\theta_{\bar{\alpha}} = \operatorname{otp}(C_{\bar{\alpha}}^1) < \operatorname{otp}(C_\alpha^1) = \theta_\alpha$ and $f_{\bar{\alpha}} = f_\alpha \restriction \theta_{\bar{\alpha}}$. Clearly, $f_\alpha(\theta_{\bar{\alpha}}) = \bar{\alpha}$. So as $\bar{\alpha}$ is a limit point of C_α^2 and $C_\alpha^2 = f_\alpha'' C_{\theta_\alpha}^2$, $\theta_{\bar{\alpha}}$ must be a limit point of $C_{\theta_\alpha}^2$. Thus $\theta_{\bar{\alpha}} \in S$ and $C_{\theta_{\bar{\alpha}}}^2 = \theta_{\bar{\alpha}} \cap C_{\theta_\alpha}^2$. But $C_{\theta_\alpha}^2 \subseteq C_{\theta_\alpha}^1$, so $\theta_{\bar{\alpha}}$ is a limit point of $C_{\theta_\alpha}^1$, so by 1(iv), $\theta_{\bar{\alpha}} > \kappa$. This means that $C_{\bar{\alpha}}^2 = f_{\bar{\alpha}}'' C_{\theta_{\bar{\alpha}}}^2$, and we have (since $f_\alpha(\theta_{\bar{\alpha}}) = \bar{\alpha}$ and $f_\alpha'' C_{\theta_\alpha}^2 = C_\alpha^2$ and $\bar{\alpha}$ is a limit point of C_α^2):

$$C_{\bar{\alpha}}^2 = f_{\bar{\alpha}}'' C_{\theta_{\bar{\alpha}}}^2 = f_{\bar{\alpha}}'' (\theta_{\bar{\alpha}} \cap C_{\theta_\alpha}^2) = f_\alpha''(\theta_{\bar{\alpha}} \cap C_{\theta_\alpha}^2) = \bar{\alpha} \cap C_\alpha^2.$$

That completes the definition of $(C_\alpha^2 \mid \alpha \in \kappa^+)$. If κ is regular, then $(C_\alpha^2 \mid \alpha \in S \cap \kappa^+)$ clearly satisfies $\Box_\kappa^S$, and we are done. If κ is singular, we extract from $(C_\alpha^2 \mid \alpha \in S \cap \kappa^+)$ a $\Box_\kappa^S$-sequence $(C_\alpha^3 \mid \alpha \in S \cap \kappa^+)$ in the same way as in the proof of IV.5.1 (at the very end). The proof of 6.2 is complete. □

Notice that in the above proof of 6.2 we commenced with a □-sequence $(C_\alpha \mid \alpha \in S)$ and constructed a $\Box_\kappa$-sequence $(\tilde{C}_\alpha \mid \alpha < \kappa^+ \wedge \lim(\alpha))$ such that, in particular, $\tilde{C}_\alpha \subseteq C_\alpha$ for $\kappa < \alpha < \kappa^+$. Thus the same argument establishes the following more general result:

6.2′ Theorem. *Assume $\Box(E)$. Then for any uncountable cardinal κ, $\Box_\kappa(F)$ holds, where $F = (E \cap \kappa^+) - (\kappa + 1)$. (So if $E \cap \kappa^+$ is stationary in κ^+, F is stationary in κ^+.)* □

This relates to 6.1′, of course.

We turn now to the proof of 6.1. We assume $V = L$ from now on.

Define a class E of limit ordinals as follows. E is the class of all limit ordinals E
α such that for some $\beta > \alpha$:

(i) α is regular over J_β; and
(ii) for some $p \in J_\beta$, if $p \in X \prec J_\beta$ and $X \cap \alpha$ is transitive, then $X = J_\beta$.

6.3 Lemma. *If $\kappa > \omega$ is regular, then $E \cap \kappa$ is stationary in κ.*

Proof. Let $C \subseteq \kappa$ be club in κ. We prove that $E \cap C \neq \emptyset$. Let N be the smallest $N \prec J_{\kappa^+}$ such that $C \in N$ and $N \cap \kappa$ is transitive. Since κ is regular, $N \cap \kappa \in \kappa$. Let $\alpha = N \cap \kappa$.

Let $\pi: J_\beta \cong N$. Then $\pi \upharpoonright \alpha = \mathrm{id} \upharpoonright \alpha$ and $\pi(\alpha) = \kappa$. Since $C \in N$, we have $C \cap \alpha \in J_\beta$ and $\pi(c \cap \alpha) = C$. Since C is club in κ, by absoluteness we have

$$\models_{J_{\kappa^+}} \text{“}C \text{ is club in } \kappa\text{”}.$$

So, as $\pi: J_\beta \prec J_{\kappa^+}$,

$$\models_{J_\beta} \text{“}C \cap \alpha \text{ is club in } \alpha\text{”}.$$

Thus by absoluteness again, $C \cap \alpha$ is indeed club in α. But C is closed in κ. Hence $\alpha \in C$. We show that $\alpha \in E$ as well.

Suppose that there were a J_β-definable map from a bounded subset of α cofinally into α. Then by applying $\pi: J_\beta \prec J_{\kappa^+}$ we would obtain a J_{κ^+}-definable map from a bounded subset of κ cofinally into κ, which is impossible. Hence α is regular over J_β.

Now suppose that $C \cap \alpha \in X \prec J_\beta$ and that $X \cap \alpha$ is transitive. Applying $\pi: J_\beta \cong N \prec J_{\kappa^+}$ we get $C \in (\pi''X) \prec N \prec J_{\kappa^+}$. But $\pi(\alpha) = \kappa$, so $(\pi''X) \cap \kappa = \pi''(X \cap \alpha) = X \cap \alpha$, which is transitive. So by the choice of N we must have $(\pi''X) = N$. Thus $X = J_\beta$.

Thus β and $C \cap \alpha$ testify that $\alpha \in E$. The proof is complete. □

6.4 Lemma. *Let $\alpha \in E$, and let $\beta > \alpha$ be as in the definition of E. Then* $\mathrm{cf}(\alpha) = \omega$ *and there is a $\Sigma_1(J_{\beta+1})$ map from ω cofinally into α.*

Proof. Let $p \in J_\beta$ be such that whenever $p \in X \prec J_\beta$ and $X \cap \alpha$ is transitive, then $X = J_\beta$. Let $h = h_{\beta+1}$, the canonical Σ_1 skolem function for $J_{\beta+1}$, and let $H = H_{\beta+1}$ be the uniformly $\Sigma_0^{J_{\beta+1}}$ predicate such that

$$y = h(i, x) \quad \text{iff } (\exists z \in J_{\beta+1})\, H(z, y, i, x).$$

For $n < \omega$, define partial functions h_n by

$$y = h_n(i, x) \quad \text{iff } x, y \in S_{\omega\beta+n} \wedge (\exists z \in S_{\omega\beta+n})\, H(z, y, i, x).$$

Since $J_{\beta+1}$ is amenable (and hence closed under Σ_0 subset formation), $h_n \in J_{\beta+1}$. And clearly, the sequence $(h_n \mid n < \omega)$ is $\Sigma_1(J_{\beta+1})$.

Define a sequence of sets $(X_n \mid n < \omega)$ and a sequence of ordinals $(\alpha_n \mid n < \omega)$ as follows.

$$\alpha_0 = 1;$$

$$X_n = h_n^*(J_{\alpha_n} \times \{(p, J_\beta)\});$$

$$\alpha_{n+1} = \sup(X_n \cap \alpha).$$

Let $X = \bigcup_{n<\omega} X_n$, and set $\alpha_\omega = \bigcup_{n<\omega} \alpha_n$. Then clearly, $X = h^*(J_{\alpha_\omega} \times \{(p, J_\beta)\})$ and $X \cap \alpha = \alpha_\omega$.

Let $Y = X \cap J_\beta$. Since $J_\beta \in X$ and $X \prec_1 J_{\beta+1}$, we clearly have $Y \prec J_\beta$. But $p \in Y$ and $Y \cap \alpha = X \cap \alpha = \alpha_\omega$. So by choice of p, $Y = J_\beta$. Thus $\alpha_\omega = Y \cap \alpha = \alpha$. This shows that $\alpha = \bigcup_{n<\omega} \alpha_n$. Since $(\alpha_n \mid n < \omega)$ is easily seen to be $\Sigma_1(J_{\beta+1})$, we shall be done if we can show that $\alpha_n < \alpha$ for all $n < \omega$.

For each $n < \omega$, let j_n be a J_{α_n}-definable map from $\omega\alpha_n$ onto J_{α_n}. For $\nu < \omega\alpha_n$, $i < \omega$, set

$$f_n(\nu, i) = \begin{cases} h_n(i, (j_n(\nu), (p, J_\beta))), & \text{if this is defined and is an element of } \alpha; \\ \text{undefined}, & \text{in all other cases.} \end{cases}$$

Since $h_n \in J_{\beta+1}$ and $J_{\beta+1}$ is closed under Σ_0 subset formation, $f_n \in J_{\beta+1}$. But $f_n \subseteq J_\beta$. So $f_n \in \mathrm{Def}(J_\beta)$, i.e. f_n is J_β-definable. Since α is regular over J_β, it follows that for each $\nu < \omega\alpha_n$, $\sup_{i<\omega} f_n(\nu, i) < \alpha$. Likewise, it then follows that if $\omega\alpha_n < \alpha$, then $\sup_{\nu<\omega\alpha_n} \sup_{i<\omega} f_n(\nu, i) < \alpha$. But clearly, $\sup_{\nu<\omega\alpha_n} \sup_{i<\omega} f_n(\nu, i) = \alpha_{n+1}$. Thus $\omega\alpha_n < \alpha$ implies $\alpha_{n+1} < \alpha$. But α is regular over J_β, so if $\alpha_{n+1} < \alpha$ then $\omega\alpha_{n+1} < \alpha$. Thus by induction on n we obtain $\alpha_n < \alpha$ for all $n < \omega$. The proof is complete. □

By 6.3 and 6.4, E is a class of limit ordinals, each cofinal with ω, such that $E \cap \kappa$ is stationary in κ for every regular $\kappa > \omega$. We complete the proof of 6.1 by showing that $\square(E)$ holds: that is, there is a sequence $(C_\alpha \mid \alpha \in S)$ such that:

(i) C_α is a club subset of α;

(ii) $\mathrm{otp}(C_\alpha) < \alpha$;

(iii) if $\bar{\alpha} < \alpha$ is a limit point of C_α, then $\bar{\alpha} \in S$, $\bar{\alpha} \notin E$, and $C_{\bar{\alpha}} = \bar{\alpha} \cap C_\alpha$.

In the definition of C_α there are several cases to consider.

Case 1. $\alpha < \omega_1$.

In this case, let C_α be any ω-sequence cofinal in α. There is nothing to check in this case.

In order to describe the next case we make use of the Gödel Pairing Function, Φ (see II.8.6). Set

Q $$Q = \{\alpha \mid \Phi''(\alpha \times \alpha) \subseteq \alpha\}.$$

By the properties of Φ,

$$Q = \{\alpha \mid (\Phi \restriction \alpha \times \alpha) : \alpha \times \alpha \leftrightarrow \alpha\}.$$

Q is clearly a club class. And it is an elementary exercise to verify that if $\alpha \in Q$, the next element of Q beyond α is α^ω.

Case 2. $\alpha > \omega_1$ and $\alpha \notin Q$.

Let β be the largest element of Q below α. Thus $\beta < \alpha < \beta^\omega$. Hence we can find a unique integer $n > 0$ and unique ordinals $\xi_0, \xi_1, \ldots, \xi_n$, $\xi_n \neq 0$, $0 \leqslant \xi_i < \beta$, such that

$$\alpha = \xi_n \beta^n + \xi_{n-1} \beta^{n-1} + \ldots + \xi_1 \beta + \xi_0.$$

Let m be the least integer such that $\xi_m \neq 0$.

Suppose first that $\xi_m = \zeta_m + 1$. Since $\lim(\alpha)$ we must have $m > 0$. Set

$$C_\alpha = \{(\xi_n \beta^n + \xi_{n-1} \beta^{n-1} + \ldots + \xi_{m+1} \beta^{m+1} + \zeta_m \beta^m + \xi \beta^{m-1}) \mid 1 \leqslant \xi < \beta\}.$$

It is easily seen that C_α is club in α and of order-type $\beta < \alpha$.

Now suppose that $\lim(\xi_m)$. Then set

$$C_\alpha = \{(\xi_n \beta^n + \xi_{n-1} \beta^{n-1} + \ldots + \xi_{m+1} \beta^{m+1} + \xi \beta^m) \mid 1 \leqslant \xi < \xi_m\}.$$

Again C_α is club in α. And C_α has order-type $\xi_m < \beta < \alpha$.

In either case now, if $\bar{\alpha} < \alpha$ is a limit point of C_α, then with β as above we have $\beta < \bar{\alpha} < \beta^\omega$ and $C_{\bar{\alpha}} = \bar{\alpha} \cap C_\alpha$. (This is elementary.) Moreover, it is clear that $E \subseteq Q$, so we have $\bar{\alpha} \notin E$.

Case 3. $\alpha > \omega_1$ and $\alpha \in Q$ and $\sup(Q \cap \alpha) < \alpha$.

Let $\beta = \sup(Q \cap \alpha)$. Then α is the successor of β in Q. Hence $\alpha = \beta^\omega$, and we may set

$$C_\alpha = \{\beta^n \mid n < \omega\}.$$

There is nothing to check in this case.

From now on we shall assume that α does not fall under Cases 1–3. Thus, $\alpha > \omega_1$ and α is a limit point of Q. Notice that, in particular, $\omega\alpha = \alpha$. Let

$$\beta = \beta(\alpha) = \text{the least } \beta \text{ such that } \alpha \text{ is singular over } J_\beta;$$ $\beta(\alpha), \beta$

$$n = n(\alpha) = \text{the least } n \text{ such that } \alpha \text{ is } \Sigma_n\text{-singular over } J_\beta.$$ $n(\alpha), n$

Case 4. $n = 1$ and β is a successor ordinal.

By IV.5.2, $\text{cf}(\alpha) = \omega$, so may let C_α be any ω-sequence cofinal in α. There is nothing to check in this case.

Notice that by 6.4, every element of E falls under Case 1 or Case 4.

Case 5. $n > 1$ or $\lim(\beta)$.

This is the only remaining case, and is by far the most difficult one. To commence, set

$$\varrho = \varrho(\alpha) = \varrho_\beta^{n-1}, \quad A = A(\alpha) = A_\beta^{n-1}.$$ $\varrho(\alpha), \varrho$ $A(\alpha), A$

Notice that we must have $\lim(\varrho)$ here.

By definition of ϱ_β^{n-1}, there is a $\Sigma_{n-1}(J_\beta)$ map from a subset of $\omega\varrho$ onto β. Hence there is a $\Sigma_{n-1}(J_\beta)$ map from a subset of $\omega\varrho$ onto α. But α is Σ_{n-1}-regular over J_β. Thus $\alpha \leqslant \omega\varrho$. Hence as $\omega\alpha = \alpha$, we have $\alpha \leqslant \varrho$. Again, there is a $\Sigma_n(J_\beta)$ map from a bounded subset of α cofinally into α. Since $\models_{J_\beta}$ "α is regular", this map cannot lie in J_β. Hence $\mathscr{P}(\alpha \times \alpha) \cap \Sigma_n(J_\beta) \nsubseteq J_\beta$. So, utilising Gödel's pairing function on $\alpha \times \alpha$, we see that $\mathscr{P}(\alpha) \cap \Sigma_n(J_\beta) \nsubseteq J_\beta$. Thus $\varrho_\beta^n \leqslant \alpha$. Hence we have proved that

$$\varrho_\beta^n \leqslant \alpha \leqslant \varrho .$$

$p(\alpha), p$ By virtue of the first of the above inequalities, we may define $p = p(\alpha) =$ the $<_J$-least $p \in J_\varrho$ such that every $x \in J_\varrho$ is Σ_1-definable from elements of $\alpha \cup \{p\}$ in $\langle J_\varrho, A\rangle$. (Thus $p \leqslant_J p_\beta^n$.)

h, H Let $h = h_{\varrho, A}$, the canonical Σ_1 skolem function for $\langle J_\varrho, A\rangle$, and let $H = H_{\varrho, A}$ be the uniformly $\Sigma_0^{\langle J_\rho, A\rangle}$ predicate such that

$$y = h(i, x) \leftrightarrow (\exists z \in J_\varrho)\, H(z, y, i, x).$$

6.5 Lemma. *There is a $\gamma < \alpha$ such that $h^*(\gamma \times \{p\}) \cap \alpha$ is unbounded in α.*

Proof. By choice of β there is a $\tau < \alpha$ and a $\Sigma_n(J_\beta)$ function f such that $f''\tau$ is cofinal in α. Since $\alpha \leqslant \varrho$, $f \subseteq J_\varrho$. But $\varrho = \varrho_\beta^{n-1}$, $A = A_\beta^{n-1}$. Thus f is $\Sigma_1(\langle J_\varrho, A\rangle)$. By choice of p, f will in fact be $\Sigma^{\langle J_\rho, A\rangle}(\{v, p\})$ for some $v < \alpha$. Since α is a limit point of Q, we can pick a $\gamma \in Q$ such that $v, \tau < \gamma < \alpha$. We show that $h^*(\gamma \times \{p\}) \cap \alpha$ is unbounded in α. It suffices to show that $f''\tau \subseteq h^*(\gamma \times \{p\})$.

Let $X = h^*(\gamma \times \{p\})$. We show that X is closed under the formation of ordered pairs. Let $x_0, x_1 \in X$, say $x_k = h(i_k, (\xi_k, p))$. Let $\xi = \Phi(\xi_0, \xi_1)$. Since $\gamma \in Q$, $\xi < \gamma$. Moreover, by the nature of Φ, ξ_0 and ξ_1 are Σ_1-definable from ξ in J_ϱ. Hence (x_0, x_1) is Σ_1-definable from ξ, p in $\langle J_\varrho, A\rangle$. So for some $i \in \omega$,

$$(x_0, x_1) = h(i, (\xi, p)) \in X .$$

Since X is closed under ordered pairs, 3.3 tells us that $X \prec_1 \langle J_\varrho, A\rangle$. But $\gamma \cup \{p\} \subseteq X$ and $\tau \subseteq \gamma$. So as f is $\Sigma_1^{\langle J_\rho, A\rangle}(\gamma \cup \{p\})$, we have $f''\tau \subseteq X$, as required. □

h_τ For $\tau < \varrho$, we shall write h_τ for $h_{\tau, A \cap J_\tau}$. Thus:

$$y = h_\tau(i, x) \quad \text{iff } (x, y \in J_\tau) \wedge (\exists z \in J_\tau)\, H(z, y, i, x).$$

$g^{(\alpha)}, g$ Define a map $g = g^{(\alpha)}$ from a subset of α onto J_ϱ by

$$g(\omega v + i) \simeq h(i, (v, p)).$$

G Thus g is $\Sigma_1^{\langle J_\rho, A\rangle}(\{p\})$. Let G be the canonical $\Sigma_0^{\langle J_\rho, A\rangle}(\{p\})$ predicate such that

$$g(v) = x \quad \text{iff } (\exists z \in J_\varrho)\, G(z, x, v).$$

Let γ be the smallest ordinal such that $\alpha \cap g''\gamma$ is unbounded in α. By 6.5, $\gamma < \alpha$. And it is clear that γ must be a limit ordinal. For $\gamma \leqslant \tau < \alpha$ we have $\bigcup(\alpha \cap g''\tau) = \alpha > \tau$. Hence there is a maximal $\kappa = \kappa^{(\alpha)} < \alpha$ such that $\bigcup(\alpha \cap g''\kappa) \leqslant \kappa$, and moreover $\kappa < \gamma$. We fix γ, κ for the rest of the proof. Note that $\bigcup(\alpha \cap g''\tau) > \tau$ whenever $\kappa < \tau < \gamma$. γ $\kappa^{(\alpha)}, \kappa$

6.6 Lemma. *If $(\kappa, p) \in X \prec_1 \langle J_\varrho, A\rangle$ and $X \cap \alpha$ is transitive, then $X \cap \alpha = \alpha$.*

Proof. Let X be as above, and set $\bar{\alpha} = X \cap \alpha$. Since $\kappa \in X$, $\bar{\alpha} > \kappa$. Thus if it were the case that $\bar{\alpha} < \alpha$, we should have $\sup(\alpha \cap g''\bar{\alpha}) > \bar{\alpha}$. So for some $\nu < \bar{\alpha}$, $\bar{\alpha} < g(\nu) < \alpha$. But $g(\nu) = h(i, (\tau, p))$, where $\nu = \omega\tau + i$, so as $p \in X$ and $\tau \leqslant \nu \in \bar{\alpha} \subseteq X$ and $X \prec_1 \langle J_\varrho, A\rangle$, we have $g(\nu) \in X$. Then $g(\nu) \in \bar{\alpha}$, a contradiction. Hence $\bar{\alpha} = \alpha$. □

We define, by recursion, functions $k: \theta \to \gamma$, $m: \theta \to \varrho$, and sequences $(X_\nu \mid \nu < \theta)$, $(\alpha_\nu \mid \nu < \theta)$, for some $\theta \leqslant \gamma$, as follows. (The exact order in which the definition proceeds is described after we have stated all of the clauses.)

$k(\nu) =$ the least $\tau \in \mathrm{dom}(g) - \kappa$ such that: $k(\nu)$

(i) $\tau \geqslant \bigcup(k''\nu)$;

(ii) $g(\tau) \in \alpha$ and $g(\tau) > \alpha_\nu$;

(iii) $m(\nu) \in h^*(g(\tau) \times \{p\})$.

$m(0) =$ the least $\eta > \kappa$ such that $p \in J_\eta$;

$m(\nu + 1) =$ the least $\eta > m(\nu)$ such that: $m(\nu)$

(i) $\eta > k(\nu), g \circ k(\nu)$;

(ii) $A \cap J_{m(\nu)} \in J_\eta$;

(iii) $m(\nu) \in h^*(g \circ k(\nu) \times \{p\})$;

(iv) $(\exists z \in J_\eta)\, G(z, g \circ k(\nu), k(\nu))$;

$m(\lambda) = \sup_{\nu<\lambda} m(\nu)$, if $\lim(\lambda)$ and $\sup_{\nu<\lambda} m(\nu) < \varrho$

(otherwise undefined).

$X_\nu = h^*_{m(\nu)}(J_\eta \times \{p\})$, where $\eta = \max(\bigcup[k''\nu], \bigcup[g \circ k''\nu])$. X_ν

$\alpha_\nu = \sup(X_\nu \cap \alpha)$. α_ν

We stop the construction when an ordinal θ is reached such that $k''\theta$ is cofinal in γ, unless the construction breaks down earlier. (We shall prove that this is not the case.) θ

Let us see how the construction proceeds. The definition of $m(0)$ is unproblematical. Now suppose that $m(\nu)$ is defined. This presupposes that we have not yet reached θ, so $\bigcup(k''\nu) < \gamma < \alpha$. Since $\bigcup(k''\nu) < \gamma$, the choice of γ implies that $\alpha \cap g''(\bigcup k''\nu)$ is bounded in α, so $\alpha \cap g \circ k''\nu$ will be bounded in α (because $g \circ k''\nu \subseteq g''(\bigcup k''\nu)$). Hence the η in the definition of X_ν satisfies $\eta < \alpha$. There is no difficulty in defining X_ν and α_ν of course. Since $m(\nu) < \varrho$ and $\langle J_\varrho, A\rangle$ is amenable, we have $h_{m(\nu)} \in J_\varrho \subseteq J_\beta$. So as α is a regular cardinal inside J_β and $\eta < \alpha$, we have $\alpha_\nu < \alpha$. By the choice of p, $h^*(\alpha \times \{p\}) = J_\varrho$, so there is now no problem in defining $k(\nu)$. Then we define $m(\nu+1)$. This causes no difficulty as far as clauses (i), (ii) and (iv) are concerned, but what about clause (iii)? Well, by definition of $k(\nu)$ we have $m(\nu) \in h^*(g \circ k(\nu) \times \{p\})$. So as $\lim(\varrho)$ there is an $\eta < \varrho$ such that $m(\nu) \in h^*_\eta(g \circ k(\nu) \times \{p\})$. Thus we can easily satisfy clause (iii) as well.

Now suppose that λ is a limit ordinal and that $k \upharpoonright \lambda$, $m \upharpoonright \lambda$, $(X_\nu \mid \nu < \lambda)$, $(\alpha_\nu \mid \nu < \lambda)$ are defined and that $\sup k''\lambda < \gamma$. Then by choice of γ, $\eta = \sup(g \circ k''\lambda) < \alpha$. Suppose it were not possible to define $m(\lambda)$. Thus it must be the case that $\sup_{\nu<\lambda} m(\nu) = \varrho$. Let $X = \bigcup_{\nu<\lambda} X_\nu$. Clearly, in this case, $X = h^*(J_\eta \times \{p\})$ and $X \cap \alpha = \sup_{\nu<\lambda}\alpha_\nu$. Now for all $\nu < \lambda$, by the definition of k we have $g \circ k(\nu) > \alpha_\nu$, and by the definition of $m(\nu+1)$ (clause (iv)), $g \circ k(\nu) \in X_{\nu+1}$, so $g \circ k(\nu) < \alpha_{\nu+1}$. Thus $\alpha_\nu < g \circ k(\nu) < \alpha_{\nu+1}$ for all $\nu < \lambda$. Hence $X \cap \alpha = \sup_{\nu<\lambda} g \circ k(\nu) = \eta < \alpha$. But $(\kappa, p) \in X_0 \subseteq X \prec_1 \langle J_\varrho, A\rangle$, so this contradicts 6.6. Hence $m(\lambda)$ can be defined. We may now define $X_\lambda, \alpha_\lambda, k(\lambda)$ without trouble, just as before.

Thus the construction proceeds until an ordinal θ is reached for which $\sup k''\theta = \gamma$. Clearly, θ must be a limit ordinal. Since k is monotone increasing from θ into γ, we have $\theta \leqslant \gamma$. Note also that, as we observed above, $\alpha_\nu < g \circ k(\nu) < \alpha_{\nu+1}$ for all $\nu < \theta$.

6.7 Lemma.

(i) $\sup_{\nu<\theta}\alpha_\nu = \alpha$.

(ii) $\sup_{\nu<\theta} m(\nu) = \varrho$.

(iii) $\bigcup_{\nu<\theta} X_\nu = J_\varrho$.

Proof. (i) By our last observation above,

$$\alpha_\nu < g \circ k(\nu) < \alpha_{\nu+1}$$

for all $\nu < \theta$. Hence

$$(*) \qquad \sup_{\nu<\theta}\alpha_\nu = \sup_{\nu<\theta} g \circ k(\nu).$$

Suppose now that (i) were false, and that

$$\eta = \sup_{\nu<\theta}\alpha_\nu = \sup_{\nu<\theta} g \circ k(\nu) < \alpha.$$

By choice of γ, $\alpha \cap g''\gamma$ is unbounded in α. So let $\tau_0 \in \mathrm{dom}(g)$ be least such that $\kappa, \eta < g(\tau_0) < \alpha$. By definition of κ, $\tau_0 \in \mathrm{dom}(g) - \kappa$. As τ_0 is minimal, by the choice of γ we must have $\tau_0 < \gamma$. So there is a least $\nu < \theta$ such that $k(\nu) > \tau_0$. Consider the definition of $k(\nu)$: namely, the least $\tau \in \mathrm{dom}(g) - \kappa$ such that $\tau \geqslant \bigcup(k''\nu)$, $\alpha_\nu < g(\tau) < \alpha$, and $m(\nu) \in h^*(g(\tau) \times \{p\})$. Now look at τ_0. We have already observed that $\tau_0 \in \mathrm{dom}(g) - \kappa$. By the minimality of ν, we have $k''\nu \subseteq \tau_0$, so $\tau_0 \geqslant \bigcup(k''\nu)$. By the choice of τ_0, we have $\alpha_\nu < \eta < g(\tau_0) < \alpha$. Finally, since $g \circ k(\nu) < \eta < g(\tau_0)$, we have (by the definition of $k(\nu)$) $m(\nu) \in h^*(g(\tau_0) \times \{p\})$. Thus τ_0 is a candidate in the choice of $k(\nu)$. Hence $k(\nu) \leqslant \tau_0$. But we chose ν so that $k(\nu) > \tau_0$. This contradiction proves (i).

(ii) Let $\bar{\varrho} = \sup_{\nu<\theta} m(\nu)$. Then for all $\nu < \theta$ we can find a $z \in J_{\bar{\varrho}}$ such that $G(z, g \circ k(\nu), k(\nu))$. Thus as $\sup_{\nu<\theta} g \circ k(\nu) = \sup_{\nu<\theta} \alpha_\nu = \alpha$ (by (i) and (*)), if we define f from a subset of γ into α by the $\Sigma_1(\langle J_{\bar{\varrho}}, A \cap J_{\bar{\varrho}}\rangle)$ definition

$$\zeta = f(\xi) \leftrightarrow (\exists z \in J_{\bar{\varrho}})\, G(z, \zeta, \xi),$$

then $f''\gamma$ is unbounded in α. But if $\bar{\varrho} < \varrho$, then as $\langle J_\varrho, A\rangle$ is amenable, $f \in J_\varrho \subseteq J_\beta$, so α is not regular inside J_β. Contradiction! Hence $\bar{\varrho} = \varrho$.

(iii) By (i), (ii) and (*), we have

$$\bigcup_{\nu<\theta} X_\nu = h^*(J_\alpha \times \{p\}).$$

So by choice of p,

$$\bigcup_{\nu<\theta} X_\nu = J_\varrho. \quad \square$$

For each $\tau < \theta$, define a map g_τ from a subset of α_τ into $J_{m(\tau)}$ by g_τ

$$g_\tau(\xi) = x \leftrightarrow (\exists z \in J_{m(\tau)})\, G(z, x, \xi).$$

By definition of m, if $\lim(\tau)$, then $\langle J_{m(\tau)}, A \cap J_{m(\tau)}\rangle$ is amenable, and in this case g_τ is $\Sigma_1^{\langle J_{m(\tau)}, A \cap J_{m(\tau)}\rangle}(\{p\})$.

We define κ_τ from g_τ in the same way that κ was defined from g: that is, we let κ_τ be the largest $\kappa_\tau \leqslant \alpha_\tau$ such that $\bigcup(\alpha_\tau \cap g''_\tau \kappa_\tau) \leqslant \kappa_\tau$. κ_τ

6.8 Lemma. *For sufficiently large ordinals $\tau < \theta$, $\kappa_\tau = \kappa$.*

Proof. Clearly, if $\nu < \tau < \theta$, then $g_\nu \subseteq g_\tau$. Moreover, $\bigcup_{\tau<\theta} g_\tau = g$. Thus for any $\tau < \theta$, $\bigcup(\alpha_\tau \cap g''_\tau \kappa) \leqslant \bigcup(\alpha \cap g''\kappa) \leqslant \kappa$. Thus $\kappa_\tau \geqslant \kappa$. Similarly, $\nu < \tau < \theta$ implies that $\kappa_\tau \leqslant \kappa_\nu$. So for some $\nu < \theta$ we must have $\kappa_\tau = \kappa_\nu \geqslant \kappa$ for all $\tau > \nu$. Suppose that $\kappa_\nu > \kappa$. Then $\bigcup(\alpha \cap g''\kappa_\nu) > \kappa_\nu$. So for some $\tau < \theta$, $\bigcup(\alpha \cap g''_\tau \kappa_\nu) > \kappa_\nu$. But we may assume that $\tau > \nu$ and that, in fact, $\bigcup(\alpha_\tau \cap g''\kappa_\nu) > \kappa_\nu$. Then $\kappa_\tau = \kappa_\nu$ and so $\bigcup(\alpha_\tau \cap g''_\tau \kappa_\tau) > \kappa_\tau$. Contradiction! That proves the lemma. □

By recursion, we define a strictly increasing, continuous function $t: \tilde{\theta} \to \theta$, for some $\tilde{\theta} \leqslant \theta$. First of all we let $t(0)$ be the least ν such that $(\nu \leqslant \tau < \theta) \to (\kappa_\tau = \kappa)$ and $\alpha_\nu > \omega_1$. t

In case $n = 1$, when $t(\iota)$ is defined we let $t(\iota + 1)$ be the least $\nu < t(\iota)$ such that $\Phi''(\alpha_{t(\iota)} \times \alpha_{t(\iota)}) \subseteq \alpha_\nu$. Since α is a limit point of Q, $t(\iota + 1) < \theta$ is always defined.

In case $n > 1$ and $t(\iota)$ is defined, we let $t(\iota + 1)$ be the least $\nu > t(\iota)$ such that $\Phi''(\alpha_{t(\iota)} \times \alpha_{t(\iota)}) \subseteq \alpha_\nu$ and

$$J_\alpha \cap h^*_{\varrho_\beta^{n-2}, A_\beta^{n-2}}(X_{t(\iota)} \times \{p_\beta^{n+1}\}) \subseteq X_\nu.$$

We must check that $t(\iota + 1) < \theta$ is well-defined.

Let

$$Y = J_\alpha \cap h^*_{\varrho^{n-2}_\beta, A^{n-2}_\beta}(X_{t(\iota)} \times \{p^{n-1}_\beta\}).$$

We must show that $Y \subseteq X_\xi$ for some $\xi < \theta$. Since $Y \subseteq J_\alpha$, it suffices to show that $Y \subseteq J_\tau$ for some $\tau < \alpha$; for if $\tau < \alpha$, then $\alpha_\xi > \tau$ for some $\xi < \theta$, and we have $g \circ k(\xi) > \alpha_\xi$, so by definition, $J_\tau \subseteq X_{\xi+1}$. Now, for some $\eta < \alpha$, we have $X_{t(\iota)} = h^*_{m(t(\iota))}(J_\eta \times \{p\})$. Since $\langle J_\varrho, A\rangle$ is amenable, $h_{m(t(\iota))} \in J_\varrho \subseteq J_\beta$. Thus J_β contains a function mapping $\omega\eta$ onto $\omega \times (X_{t(\iota)} \times \{p^{n-1}_\beta\})$. Again, by the definition of Y, Y is the image of a $\Sigma_1(\langle J_{\varrho^{n-2}_\beta}, A^{n-2}_\beta\rangle)$ function defined on a subset of $\omega \times (X_{t(\iota)} \times \{p^{n-1}_\beta\})$. By the properties of the standard code A^{n-2}_β, this function is $\Sigma_{n-1}(J_\beta)$. Combining these two functions gives us a $\Sigma_{n-1}(J_\beta)$ function f such that $f''\omega\eta = Y$. Since f is $\Sigma_{n-1}(J_\beta)$, so too is $\bar{f}\colon \omega\eta \to \alpha$, defined by letting $\bar{f}(v)$ be the least τ such that $f(v) \in J_\tau$. Since α is Σ_{n-1}-regular over J_β, $\bar{f}''\omega\eta \subseteq \tau$ for some $\tau < \alpha$. Then $Y \subseteq J_\tau$, as required.

Finally, if $\lim(\lambda)$ and $t \restriction \lambda$ is defined, we let $t(\lambda) = \sup_{\iota<\lambda} t(\iota)$, if this is less than θ, with $t(\lambda)$ undefined otherwise.

$\tilde{\theta}$ Thus for some limit ordinal $\tilde{\theta} \leqslant \theta$ we shall have $\sup_{\iota<\tilde{\theta}} t(\iota) = \theta$, at which point the definition of t is complete.

We define

C_α
$$C_\alpha = \{\alpha_{t(v)} \mid v < \tilde{\theta}\}.$$

Thus C_α is a club subset of α of order-type $\tilde{\theta} \leqslant \theta \leqslant \gamma < \alpha$. To complete the proof
$\bar{\alpha}$ of $\square$ we must show that if $\bar{\alpha} < \alpha$ is a limit point of C_α, then $\bar{\alpha} \in S$, $\bar{\alpha} \notin E$, and
λ $C_{\bar{\alpha}} = \bar{\alpha} \cap C_\alpha$. Let $\bar{\alpha} = \alpha_\lambda$, where $\lim(\lambda)$.

6.9 Lemma. *$\bar{\alpha} > \omega_1$ and $\bar{\alpha} \in Q$. Moreover, if $n > 1$ and f is a $\Sigma_1^{\langle J_{\varrho^{n-2}_\beta}, A^{n-2}_\beta\rangle}(X_\lambda \cup \{p^{n-1}_\beta\})$ function from a bounded subset of $\bar{\alpha}$ into $\bar{\alpha}$, then f is bounded in $\bar{\alpha}$.*

Proof. That $\bar{\alpha} > \omega_1$ and $\bar{\alpha} \in Q$ is an immediate consequence of the definition of t. Now let $n > 1$, and let f be as above. Since the function m is continuous, so too is the sequence $(X_v \mid v < \theta)$. Thus $X_\lambda = \bigcup_{v<\lambda} X_v$, and the finitely many parameters in the definition of f will all lie in X_v for some $v < \lambda$. We may choose v here so that $\mathrm{dom}(f) \subseteq \alpha_v$. Let $\iota < \tilde{\theta}$ be least such that $t(\iota) > v$. Since $\bar{\alpha} = \alpha_\lambda$ is a limit point of C_α, λ is a limit point of t and so $t(\iota)$, $t(\iota+1) < \lambda$. But f is $\Sigma_1^{\langle J_{\varrho^{n-2}_\beta}, A^{n-2}_\beta\rangle}(X_{t(\iota)} \times \{p^{n-1}_\beta\})$ and $\mathrm{dom}(f) \subseteq \alpha_{t(\iota)} \subseteq X_{t(\iota)}$. So by definition of $t(\iota+1)$, $\mathrm{ran}(f) \subseteq X_{t(\iota+1)}$. Thus $\mathrm{ran}(f) \subseteq \alpha_{t(\iota+1)} < \alpha_\lambda = \bar{\alpha}$. $\square$

Let

$\pi, \bar{\varrho}, \bar{A}$
$$\pi\colon \langle J_{\bar{\varrho}}, \bar{A}\rangle \cong \langle X_\lambda, A \cap X_\lambda\rangle.$$

Thus

$$\pi\colon \langle J_{\bar{\varrho}}, \bar{A}\rangle \prec_1 \langle J_{m(\lambda)}, A \cap J_{m(\lambda)}\rangle.$$

But by Σ_0-absoluteness,

$$\langle J_{m(\lambda)}, A \cap J_{m(\lambda)}\rangle \prec_0 \langle J_\varrho, A\rangle.$$

Thus

$$\pi\colon \langle J_{\bar{\varrho}}, \bar{A}\rangle \prec_0 \langle J_\varrho, A\rangle.$$

So by 5.6 there are unique $\bar{\beta}$, $\tilde{\pi}$ such that $\bar{\varrho} = \varrho_{\bar{\beta}}^{n-1}$, $A = A_{\bar{\beta}}^{n-1}$, $\tilde{\pi}: J_{\bar{\beta}} \prec_{n-1} J_{\beta}$, $\pi \subseteq \tilde{\pi}$, $\tilde{\pi}(p_{\bar{\beta}}^{n-1}) = p_{\beta}^{n-1}$. Note that by definition of k, $g \circ k(\nu) > \alpha_\nu$ for all $\nu < \lambda$, so by definition of $X_{\nu+1}$, $\alpha_\nu \subseteq X_{\nu+1}$ for all $\nu < \lambda$. Thus $\bar{\alpha} \subseteq X_\lambda$ and in fact $\bar{\alpha} = X_\lambda \cap \alpha$. So we have $\pi \upharpoonright \bar{\alpha} = \mathrm{id} \upharpoonright \bar{\alpha}$, and in case $\bar{\alpha} < \bar{\beta}$, $\tilde{\pi}(\bar{\alpha}) \geqslant \alpha$. $\bar{\beta}, \tilde{\pi}$

Let $\bar{h} = h_{\bar{\varrho}, \bar{A}}$, $\bar{H} = H_{\bar{\varrho}, \bar{A}}$. Set $\bar{p} = \pi^{-1}(p)$. $\bar{h}, \bar{H}, \bar{p}$

6.10 Lemma. *$\bar{p}$ = the $<_J$-least element of $J_{\bar{\varrho}}$ such that every $x \in J_{\bar{\varrho}}$ is Σ_1-definable from parameters in $\bar{\alpha} \cup \{\bar{p}\}$ in $\langle J_{\bar{\varrho}}, \bar{A}\rangle$.*

Proof. By definition,

$$X_\lambda = h_{m(\lambda)}^*(J_\eta \times \{p\}),$$

where $\eta = \max(\kappa + 1, \sup[g \circ k''\lambda])$. But $\alpha_\nu < g \circ k(\nu) < \alpha_{\nu+1}$ for all $\nu < \theta$. So as $\bar{\alpha} = \alpha_\lambda$ and $\lim(\lambda)$, $\eta = \bar{\alpha}$. Thus

$$X_\lambda = h_{m(h)}^*(J_{\bar{\alpha}} \times \{p\}).$$

Applying π^{-1}, we get

$$J_{\bar{\varrho}} = \bar{h}^*(J_{\bar{\alpha}} \times \{\bar{p}\}).$$

But by definition of t, we have $\bar{\alpha} \in Q$, so by 3.19 there is a $\Sigma_1^{J_\alpha}$ map from $\bar{\alpha}$ onto $J_{\bar{\alpha}}$. Hence

$$J_{\bar{\alpha}} = h_{\bar{\alpha}, \emptyset}^*(\bar{\alpha}) \subseteq \bar{h}^*(\bar{\alpha}).$$

Thus

$$J_{\bar{\varrho}} = \bar{h}^*(\bar{\alpha} \times \{\bar{p}\}).$$

This shows that every element of $J_{\bar{\varrho}}$ is Σ_1-definable from members of $\bar{\alpha} \cup \{\bar{p}\}$ in $\langle J_{\bar{\varrho}}, \bar{A}\rangle$. We must now show that $\bar{p}$ is the $<_J$-least such member of $J_{\bar{\varrho}}$. Suppose, on the contrary, that $\bar{p}' <_J \bar{p}$ also has this property. Then, in particular, for some $i < \omega$ and some $\nu < \bar{\alpha}$, we have $\bar{p} = \bar{h}(i, (\nu, \bar{p}'))$. Applying $\pi: \langle J_{\bar{\varrho}}, \bar{A}\rangle \prec_1 \langle J_{m(\lambda)}, A \cap J_{m(\lambda)}\rangle$, we get $p = h_{m(\lambda)}(i, (\nu, p'))$, where $p' = \pi(\bar{p}')$. Thus $p = h(i, (\nu, p'))$. Hence by choice of p, every element of J_ϱ will be Σ_1-definable from parameters in $\alpha \cup \{p'\}$ in $\langle J_\varrho, A\rangle$. But $\bar{p}' <_J \bar{p}$, so $p' <_J p$, and so we have contradicted the choice of p. $\Box$

Now define $\bar{g}$ from $\bar{h}, \bar{\alpha}, \bar{p}$ just as g was defined from h, α, p. Thus, we define $\bar{g}$ from a subset of $\bar{\alpha}$ into $J_{\bar{\varrho}}$ by $\bar{g}$

$$\bar{g}(\omega\nu + i) \simeq \bar{h}(i, (\nu, \bar{p})).$$

Let $\bar{G}$ be the canonical $\Sigma^{\langle J_{\bar{\varrho}}, A\rangle}(\{\bar{p}\})$ predicate such that $\bar{G}$

$$\bar{g}(\nu) = x \quad \text{iff } (\exists z \in J_{\bar{\varrho}})\, \bar{G}(z, x, \nu).$$

Note that the Σ_0 formula which defines $\bar{G}$ from $\bar{p}$ in $\langle J_{\bar{\varrho}}, \bar{A}\rangle$ will be the same as that which defines G from p in $\langle J_\varrho, A\rangle$. But

$$\pi: \langle J_{\bar{\varrho}}, \bar{A}\rangle \prec_1 \langle J_{m(\lambda)}, A \cap J_{m(\lambda)}\rangle,$$

$\pi \upharpoonright \bar\alpha = \mathrm{id} \upharpoonright \bar\alpha$, and $\pi(\bar p) = p$. Thus for $\nu, \tau \in \bar\alpha$,

$$\begin{aligned} \bar g(\nu) = \tau \quad & \text{iff } (\exists z \in J_{\bar\varrho})\, \bar G(z, \tau, \nu) \\ & \text{iff } (\exists z \in J_{m(\lambda)})\, G(z, \tau, \nu) \\ & \text{iff } g_\lambda(\nu) = \tau. \end{aligned}$$

Hence

(1) $$\bar g \cap (\bar\alpha \times \bar\alpha) = g_\lambda \cap (\bar\alpha \times \bar\alpha).$$

$\bar\kappa$ Next we define $\bar\kappa$ from $\bar g$, $\bar\alpha$ just as κ was defined from g, α. That is, let $\bar\kappa$ be the largest $\bar\kappa \leqslant \bar\alpha$ such that $\bigcup(\bar\alpha \cap \bar g''\bar\kappa) \leqslant \bar\kappa$. By (1) and the fact that $\bar\alpha = \alpha_\lambda$, this is the same as the definition of κ_λ, so $\bar\kappa = \kappa_\lambda$. But by the definition of $t(0)$, $\kappa_\lambda = \kappa$. Thus $\bar\kappa = \kappa$.

η Let $\eta = \bigcup k''\lambda$. By definition of $X_{\nu+1}$, we have $k(\nu) \in X_{\nu+1}$, so $k(\nu) < \alpha_{\nu+1}$ for all $\nu < \lambda$. Thus $\eta \leqslant \bar\alpha$.

Since $\alpha_\nu < g \circ k(\nu) < \alpha_{\nu+1}$ for all $\nu < \lambda$, we have

(2) $$\bar\alpha = \bigcup_{\nu<\lambda} g \circ k(\nu).$$

Now by clause (iv) in the definition of $m(\nu + 1)$, $g_\lambda \upharpoonright k''\lambda = g \upharpoonright k''\lambda$. Thus by (2), we have

(3) $$\bar\alpha = \bigcup_{\nu<\lambda} g_\lambda \circ k(\nu).$$

Since k is monotone increasing, we have $k''\nu \subseteq k(\nu)$ for all $\nu < \lambda$. Thus $g_\lambda''(k''\nu) \subseteq g_\lambda''k(\nu)$ for all $\nu < \lambda$, i.e. $g_\lambda \circ k''\nu \subseteq g_\lambda''k(\nu)$ for all $\nu < \lambda$. So from (3) we have

(4) $$\bar\alpha = \bigcup_{\nu<\lambda} (\bar\alpha \cap g_\lambda''k(\nu)).$$

This is the same as

(5) $$\bar\alpha = \bigcup(\bar\alpha \cap g_\lambda''\eta).$$

So by (1) and (5) (noting that $\eta \leqslant \bar\alpha$) we have

(6) $$\bar\alpha = \bigcup(\bar\alpha \cap \bar g''\eta).$$

Now by definition of k we have $k(0) > \kappa$, so $\eta = \bigcup k''\lambda > \kappa$. So as $\bar\kappa = \kappa$ we have $\kappa < \eta \leqslant \bar\alpha$. So by choice of $\bar\kappa$ we have $\bigcup(\bar\alpha \cap \bar g''\eta) > \eta$. Thus by (6) we have $\bar\alpha > \eta$. But (6) also tells us that $\bar g$ maps a subset of η cofinally into $\bar\alpha$. Thus, in particular, $\bar\alpha \in S$.

6.11 Lemma. $\bar\beta = \beta(\bar\alpha)$.

Proof. By definition, $\bar{g}$ is $\Sigma_1(\langle J_{\bar{\varrho}}, \bar{A}\rangle)$. So $\bar{g} \cap (\bar{\alpha} \times \bar{\alpha})$ is $\Sigma_1(\langle J_{\bar{\varrho}}, \bar{A}\rangle)$. But $\bar{\varrho} = \varrho_{\bar{\beta}}^{n-1}$, $\bar{A} = A_{\bar{\beta}}^{n-1}$. Thus $\bar{g} \cap (\bar{\alpha} \times \bar{\alpha})$ is $\Sigma_n(J_{\bar{\beta}})$. By (6) above, $\bar{g} \cap (\bar{\alpha} \times \bar{\alpha})$ maps a subset of $\eta < \bar{\alpha}$ cofinally into $\bar{\alpha}$. Hence $\bar{\alpha}$ is Σ_n-singular over $J_{\bar{\beta}}$. Thus $\beta(\bar{\alpha}) \leqslant \bar{\beta}$.

Suppose that $\beta(\bar{\alpha}) < \bar{\beta}$. Then there is an $f \in J_{\bar{\beta}}$ and a $\delta < \bar{\alpha}$ such that f maps δ cofinally into $\bar{\alpha}$. Now, $\tilde{\pi} \upharpoonright \bar{\alpha} = \mathrm{id} \upharpoonright \bar{\alpha}$, so we have $\tilde{\pi}(\delta) = \delta$ and $f \subseteq \tilde{\pi}(f)$. But $\models_{J_{\bar{\beta}}}$ "$\mathrm{dom}(f) = \delta$", so applying $\tilde{\pi}\colon J_{\bar{\beta}} \prec_{n-1} J_\beta$ we have $\models_{J_\beta}$ "$\mathrm{dom}(\tilde{\pi}(f)) = \delta$". Thus we must have $\tilde{\pi}(f) = f$. But $\models_{J_{\bar{\beta}}}$ "$\bigcup f''\delta = \bar{\alpha}$", so applying $\tilde{\pi}$, $\models_{J_\beta}$ "$\bigcup f''\delta = \tilde{\pi}(\bar{\alpha})$". Since $\tilde{\pi}(\bar{\alpha}) \geqslant \alpha > \bar{\alpha}$, this is impossible. Hence $\beta(\bar{\alpha}) = \bar{\beta}$. □

6.12 Lemma. $n = n(\bar{\alpha})$.

Proof. By the properties of $\bar{g} \cap (\bar{\alpha} \times \bar{\alpha})$ mentioned above we have $n(\bar{\alpha}) \leqslant n$. So if $n = 1$ we are done. Assume that $n > 1$.

Let $\bar{f}$ be a $\Sigma_{n-1}(J_{\bar{\beta}})$ function from a bounded subset of $\bar{\alpha}$ into $\bar{\alpha}$. We shall show $\bar{f}$
that $\bar{f}''\bar{\alpha}$ is bounded in $\bar{\alpha}$, thereby proving that $n(\bar{\alpha}) = n$. Let $u = \mathrm{dom}(\bar{f})$. Let u
$\bar{\pi} = \tilde{\pi} \upharpoonright J_{\varrho_{\bar{\beta}}^{n-2}}$. By 5.6 we know that

$$\bar{\pi}\colon \langle J_{\varrho_{\bar{\beta}}^{n-2}}, A_{\bar{\beta}}^{n-2}\rangle \prec_1 \langle J_{\varrho_{\beta}^{n-2}}, A_{\beta}^{n-2}\rangle$$

and

$$\bar{\pi}(p_{\bar{\beta}}^{n-1}) = p_{\beta}^{n-1}.$$

Since $\bar{\varrho} = \varrho_{\bar{\beta}}^{n-1}$, we can find an $x \in J_{\bar{\varrho}}$ such that $\bar{f}$ is $\Sigma_1^{\langle J_{\varrho_{\bar{\beta}}^{n-2}}, A_{\bar{\beta}}^{n-2}\rangle}(\{x, p_{\bar{\beta}}^{n-1}\})$. Let x
f be defined over $\langle J_{\varrho_{\beta}^{n-2}}, A_{\beta}^{n-2}\rangle$ by means of the same Σ_1 definition in parameters f
$\bar{\pi}(x), p_{\beta}^{n-1}$.

Since $\bar{f} \subseteq \bar{\alpha} \times \bar{\alpha}$ and $\pi \upharpoonright \bar{\alpha} = \mathrm{id} \upharpoonright \bar{\alpha}$, we have $\bar{f} \subseteq f$. Again, u is a $\Sigma_{n-1}(J_{\bar{\beta}})$ subset of $\bar{\alpha} \leqslant \bar{\varrho} = \varrho_{\bar{\beta}}^{n-1}$, so by 4.6, $\langle J_{\bar{\varrho}}, u\rangle$ is amenable. But u is bounded in $\bar{\alpha}$. Hence $u \in J_{\bar{\varrho}}$. Thus $\pi(u)$ is defined. Since u is a bounded subset of $\bar{\alpha}$ and $\pi \upharpoonright \bar{\alpha} = \mathrm{id} \upharpoonright \bar{\alpha}$, we have $\pi(u) = u$. But the statements

"$\bar{f}$ is a function" and "$\mathrm{dom}(f) \subseteq u$"

are $\Pi_1^{\langle J_{\varrho_{\bar{\beta}}^{n-2}}, A_{\bar{\beta}}^{n-2}\rangle}(\{x, p_{\bar{\beta}}^{n-1}, u\})$. Hence as $\bar{\pi}$ is Σ_1-elementary, f is a function and $\mathrm{dom}(f) \subseteq u$. Thus $f = \bar{f}$.

This shows that $\bar{f}$ is $\Sigma_1^{\langle J_{\varrho_{\beta}^{n-2}}, A_{\beta}^{n-2}\rangle}(\{\pi(x), p_{\beta}^{n-1}\})$. But $\pi(x) \in X_\lambda$. So by 6.9, f is bounded in $\bar{\alpha}$, and we are done. □

6.13 Lemma. $\bar{\varrho} = \varrho(\bar{\alpha})$ *and* $\bar{A} = A(\bar{\alpha})$.

Proof. Directly from 6.11 and 6.12. □

6.14 Lemma. $\bar{p} = p(\bar{\alpha})$.

Proof. Directly from 6.13 and 6.10. □

6.15 Lemma. $\bar{g} \cap (\bar{\alpha} \times \bar{\alpha}) = g_\lambda \cap (\bar{\alpha} \times \bar{\alpha}) = g^{(\bar{\alpha})} \cap (\bar{\alpha} \times \bar{\alpha})$ *and* $\kappa^{(\bar{\alpha})} = \kappa^{(\alpha)} = \kappa$.

Proof. By our previous results. □

6.16 Lemma. $\bar{\alpha}$ falls under Case 5 in the definition of $C_{\bar{\alpha}}$.

Proof. Since $\bar{\alpha} > \omega_1$, $\bar{\alpha}$ does not fall under Case 1. Since $\bar{\alpha} \in Q$, $\bar{\alpha}$ does not fall under Case 2. Since $\bar{\alpha}$ is a limit point of Q (by definition of the function t) $\bar{\alpha}$ does not fall under Case 3. If $n > 1$, then by 6.12, $\bar{\alpha}$ does not fall under Case 4. And if $n = 1$, then $\bar{\beta} = \bar{\varrho}$, so as $\pi: J_{\bar{\varrho}} \prec_1 J_{m(\lambda)}$ and $\lim(\lambda)$, $\bar{\beta}$ is a limit ordinal, so by 6.11, $\bar{\alpha}$ still does not fall under Case 4. Hence $\bar{\alpha}$ must fall under Case 5. □

6.17 Corollary. $\bar{\alpha} \notin E$.

Proof. Since all members of E fall under Case 1 or Case 4. □

6.18 Lemma. $C_{\bar{\alpha}} = \bar{\alpha} \cap C_\alpha$.

Proof. Define $\bar{k}: \bar{\theta} \to \bar{\gamma}$, $\bar{m}: \bar{\theta} \to \bar{\varrho}$, $(\bar{X}_\nu \mid \nu < \bar{\theta})$, $(\bar{\alpha}_\nu \mid \nu < \bar{\theta})$ from $\bar{\alpha}$ just as k, m, $(X_\nu \mid \nu < \theta)$, $(\alpha_\nu \mid \nu < \theta)$ were defined from α. Since $\bar{\alpha}$ is a limit point of C_α, we clearly have $\bar{\theta} = \lambda$ here. And a straightforward induction proof shows that for $\nu < \lambda$, $\bar{k}(\nu) = k(\nu)$, $\pi(\bar{m}(\nu)) = m(\nu)$, $\pi''\bar{X}_\nu = X_\nu$, $\bar{\alpha}_\nu = \alpha_\nu$.

Now define $\bar{t}$ from $\bar{\alpha}$ as t was defined from α. For some $\bar{\lambda}$, we will have $\lambda = t(\bar{\lambda})$. By induction on $\nu < \bar{\lambda}$, we get $\bar{t}(\nu) = t(\nu)$. Hence

$$C_{\bar{\alpha}} = \{\bar{\alpha}_{\bar{t}(\nu)} \mid \nu < \bar{\lambda}\} = \{\alpha_{t(\nu)} \mid \nu < \bar{\lambda}\} = \bar{\alpha} \cap C_\alpha. \quad \square$$

The proof of 6.1 is finally complete.

Exercises

1. Strong Embeddings

This exercise is concerned with establishing a sort of "dual" to theorem 5.6. This result says that if there is an embedding

$$\sigma: \langle J_{\bar{\varrho}}, \bar{A} \rangle \prec_1 \langle J_{\varrho^n_\beta}, A^n_\beta \rangle,$$

then $\langle J_{\bar{\varrho}}, \bar{A} \rangle$ must have the form $\bar{\varrho} = \varrho^n_{\bar{\beta}}$, $\bar{A} = A^n_{\bar{\beta}}$, and the embedding σ can be extended to an embedding

$$\tilde{\sigma}: J_{\bar{\beta}} \prec_{n+1} J_\beta.$$

In the result proved below, the roles of $\langle J_{\bar{\varrho}}, \bar{A} \rangle$ and $\langle J_{\varrho^n_\beta}, A^n_\beta \rangle$ in the above are interchanged.

Let $\langle J_{\bar{\varrho}}, \bar{A} \rangle$, $\langle J_\varrho, A \rangle$ be amenable structures. We say that an embedding

$$\sigma: \langle J_{\bar{\varrho}}, \bar{A} \rangle \prec_1 \langle J_\varrho, A \rangle$$

is *strong* iff, whenever $\varphi(x, y)$ is a Σ_0 formula of $\mathscr{L}(A)$, if

$$\{(x, y) \in J_{\bar{\varrho}} \mid \models_{\langle J_{\bar{\rho}}, \bar{A} \rangle} \varphi(\mathring{x}, \mathring{y})\}$$

is well-founded, then

$$\{(x, y) \in J_\varrho \mid \models_{\langle J_\rho, A \rangle} \varphi(\mathring{x}, \mathring{y})\}$$

is well-founded. (Notice that in describing this property as an attribute of σ, we are really using the fact that in order to specify a mapping it is necessary to specify the domain and the range. The actual behaviour of σ plays no part in the definition of strongness.)

We shall prove that, for any $n > 0$, if $\langle J_\varrho, A\rangle$ is amenable and

$$\sigma\colon \langle J_{\varrho^n_{\bar\beta}}, A^n_{\bar\beta}\rangle \prec_1 \langle J_\varrho, A\rangle$$

is strong, then there is a unique ordinal β such that $\varrho = \varrho^n_\beta$, $A = A^n_\beta$, and a (strong) embedding

$$\tilde\sigma\colon J_{\bar\beta} \prec_{n+1} J_\beta$$

such that $\sigma \subseteq \tilde\sigma$.

It suffices to prove the following: Let $n, i > 0$, and suppose that

$$\sigma\colon \langle J_{\varrho^n_{\bar\beta}}, A^n_{\bar\beta}\rangle \prec_i \langle J_\varrho, A\rangle$$

is strong, where $\langle J_\varrho, A\rangle$ is amenable. Then there are η, B, $\tilde\sigma$, such that $\sigma \subseteq \tilde\sigma$ and

(i) $\varrho = \varrho^1_{\eta,B}$, $A = A^1_{\eta,B}$, $\tilde\sigma(p^{n-1}_{\bar\beta}) = p^1_{\eta,B}$;

(ii) $\tilde\sigma\colon \langle J_{\varrho^{n-1}_{\bar\beta}}, A^{n-1}_{\bar\beta}\rangle \prec_{i+1} \langle J_\eta, B\rangle$ is strong.

Set: $\quad \bar\varrho = \varrho^n_{\bar\beta}$, $\bar A = A^n_{\bar\beta}$, $\bar\eta = \varrho^{n-1}_{\bar\beta}$, $\bar B = A^{n-1}_{\bar\beta}$, $\bar p = p^{n-1}_{\bar\beta}$.

Note that: $\quad J_{\bar\eta} = h^*_{\bar\eta,\bar B}(J_{\bar\varrho} \times \{\bar p\})$.

Define: $\quad \bar h((i, x)) \simeq h_{\bar\eta,\bar B}(i, (x, p)) \quad (x \in J_{\bar\varrho})$.

Define relations $\bar D$, $\bar E$, $\bar I$, $\bar B'$ on $J_{\bar\varrho}$ by:

$$\bar D = \operatorname{dom}(\bar h);$$
$$\bar E = \{(x, y) \in \bar D^2 \mid \bar h(x) \in \bar h(y)\};$$
$$\bar I = \{(x, y) \in \bar D^2 \mid \bar h(x) = \bar h(y)\};$$
$$\bar B' = \{x \in \bar D \mid \bar h(x) \in \bar B\}.$$

Since $\bar D$, $\bar E$, $\bar I$, $\bar B'$ are $\Sigma_1^{\langle J_{\bar\eta}, \bar B\rangle}(\{\bar p\})$, they are $\Sigma_0^{\langle J_{\bar\varrho}, \bar A\rangle}$. Let D, E, I, B' have the same Σ_0 definitions over $\langle J_\varrho, A\rangle$. Since σ is strong, E is well-founded. Let

$$\bar M = \langle \bar D, \bar I, \bar E, \bar B'\rangle,$$
$$M = \langle D, I, E, B'\rangle.$$

Let $\bar T$ be the Σ_1 satisfaction relation for the structure $\bar M$. Then

$$\bar T(\varphi, (\vec x)) \leftrightarrow \models^{\Sigma_1}_{\langle J_{\bar\eta}, \bar B\rangle} \varphi(\bar h(\vec x)^\circ).$$

Since $\bar T$ is $\Sigma_1^{\langle J_{\bar\eta}, \bar B\rangle}(\{\bar p\})$, it is $\Sigma_0^{\langle J_{\bar\varrho}, \bar A\rangle}$. Let T have the same Σ_0 definition over $\langle J_\varrho, A\rangle$.

1A. Prove that T is the Σ_1 satisfaction relation for the structure M.

Since the satisfaction relations $\bar{T}$, T are Σ_0 in $\langle J_{\bar{\varrho}}, \bar{A}\rangle$, $\langle J_\varrho, A\rangle$, respectively, by the same definition, and σ is Σ_i-elementary, we have

$$(\sigma \restriction \bar{D})\colon \bar{M} \prec_{i+1} M.$$

Thus M satisfies the identity axioms (for I) and the Axiom of Extensionality. So we may define the factor models

$$\bar{M}^* = \bar{M}/\bar{I} = \langle \bar{D}^*, \bar{E}^*, \bar{B}^*\rangle$$

$$M^* = M/I = \langle D^*, E^*, B^*\rangle.$$

Let $\bar{k}\colon \bar{M} \to \bar{M}^*$ and $k\colon M \to M^*$ be the natural projections. Since $\bar{M}^*$, M^* are well-founded and extensional, let $\bar{l}$, l be their transitivisation isomorphisms, respectively. Clearly,

$$\bar{l}\colon \bar{M}^* \cong \langle J_{\bar{\eta}}, \bar{B}\rangle, \qquad \bar{h} = \bar{l} \circ \bar{k}.$$

Let

$$l\colon M^* \cong \langle J_\eta, B\rangle,$$

and set

$$h = l \circ k.$$

Define $\sigma^*\colon \bar{M}^* \prec_{i+1} M^*$ by $\sigma^* \circ \bar{k} = k \circ \sigma$, and define

$$\tilde{\sigma}\colon \langle J_{\bar{\eta}}, \bar{B}\rangle \prec_{i+1} \langle J_\eta, B\rangle$$

by $\tilde{\sigma} \circ \bar{h} = h \circ \sigma$. We have the following commutative diagram of the situation.

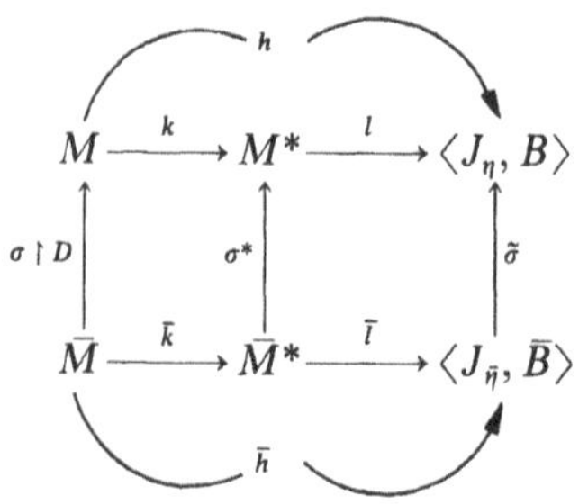

1B. Prove that $\tilde{\sigma} \restriction J_{\bar{\varrho}} = \sigma$.

Set $p = \tilde{\sigma}(\bar{p})$.

1C. Prove that

$$(i, x) \in D \to h((i, x)) = h_{\eta, B}(i, (x, p)).$$

1D. Prove that

$$A = \{(i, x) \mid x \in J_\varrho \wedge \models_{\langle J_\eta, B\rangle} \varphi_i(\mathring{x}, \mathring{p})\},$$

where $(\varphi_i \mid i < \omega)$ is as usual.

1E. Prove that $\varrho = \varrho^1_{\eta, B}$.

1F. Prove that $p = p^1_{\eta, B}$.

1G. Conclude that $A = A^1_{\eta, B}$.

1H. Prove that $\tilde{\sigma}$ is strong. (Hint. Pull back to $\bar{D}$ and D, and use the fact that σ is strong.)

That completes the proof.

The result just proved may be used to give a proof of the Covering Lemma (Chapter V) different from the one given in this book. This alternative proof may be found in *Devlin and Jensen* (1975).

2. *The Combinatorial Principle* $\square^\kappa(E)$

For each infinite cardinal κ, let

$$S_\kappa = \{\alpha \in S \mid \mathrm{cf}(\alpha) \leqslant \kappa\}.$$

Let $\square^\kappa(E)$ denote the following assertion. There is a sequence $(C_\alpha \mid \alpha \in S_\kappa)$ such that:

(i) C_α is a club subset of α;
(ii) if $\mathrm{cf}(\alpha) < \kappa$, then $\mathrm{otp}(C_\alpha) < \kappa$;
(iii) if $\bar{\alpha} < \alpha$ is a limit point of C_α, then $\bar{\alpha} \in S_\kappa$, $\bar{\alpha} \notin E$, and $C_{\bar{\alpha}} = \bar{\alpha} \cap C_\alpha$.

2A. Prove that $\square^\kappa(E)$ implies $\square_\kappa(F)$, where $F = E \cap (\kappa^+ - \kappa)$. (Hint: Let $(C_\alpha \mid \alpha \in S_\kappa)$ be as in $\square^\kappa(E)$. For $\kappa < \alpha < \kappa^+$, let $C'_\alpha = C_\alpha \cap (\kappa^+ - \kappa)$. For $\alpha \leqslant \kappa$, define C'_α in two cases. If κ is regular, let $C'_\alpha = \alpha$. If κ is singular, and if $\delta = \mathrm{cf}(\kappa)$, let C'_κ be a club subset of κ of type δ. If $\alpha < \kappa$ is a limit point of C'_κ, let $C'_\alpha = \alpha \cap C'_\kappa$. If $\alpha < \kappa$ is such that $\mu < \alpha \leqslant \nu$, where $\mu, \nu \in C'_\kappa$ are such that ν is the least element of C'_κ above μ, let $C'_\alpha = \alpha - \mu$. If $\alpha < \min(C'_\kappa)$, let $C'_\alpha = \alpha$. Then $(C'_\alpha \mid \alpha < \kappa^+$ & $\lim(\alpha))$ is a $\square_\kappa(F)$-sequence.)

2B. Prove that $\square(E)$ implies that $\square^\kappa(E)$ holds for any infinite cardinal κ. (Hint: Since the case $\kappa = \omega$ is trivial, assume $\kappa > \omega$. First define $(C'_\alpha \mid \alpha \in S_\kappa)$ to satisfy:

(i) C'_α is a club subset of α;
(ii) $\mathrm{otp}(C'_\alpha) \leqslant \kappa$;
(iii) if $\bar{\alpha} < \alpha$ is a limit point of C'_α, then $\bar{\alpha} \in S_\kappa$, $\bar{\alpha} \notin E$, and $C'_{\bar{\alpha}} = \bar{\alpha} \cap C'_\alpha$.

This is done as follows. Let $(C_\alpha \mid \alpha \in S)$ satisfy $\square(E)$ with the additional assumption $C_\alpha \subseteq \alpha - \kappa$ for $\alpha > \kappa$. (For a fixed κ this is trivially arranged.) For α singular, set $\xi_\alpha = \mathrm{otp}(C_\alpha)$, and let $f_\alpha: \xi_\alpha \to C_\alpha$ be the monotone enumeration of C_α. Define C'_α by recursion on α. For $\alpha \in S_\kappa$ such that $\xi_\alpha \leqslant \kappa$, let $C'_\alpha = C_\alpha$. Now suppose $\alpha \in S_\kappa$ and we wish to define C'_α. Thus $\xi_\alpha > \kappa$. Since $\mathrm{cf}(\xi_\alpha) = \mathrm{cf}(\alpha) \leqslant \kappa < \kappa < \xi_\alpha$, ξ_α is singular, so $\xi_\alpha \in S_\kappa$. By (ii) of $\square(E)$, $\xi_\alpha < \alpha$, so C'_{ξ_α} is defined. Set $C'_\alpha = f''_\alpha C'_{\xi_\alpha}$. Then $(C'_\alpha \mid \alpha \in S_\kappa)$ satisfies (i)–(iii) above. If κ is regular, $(C'_\alpha \mid \alpha \in S_\kappa)$ satisfies $\square^\kappa(E)$ already. Suppose κ is singular and let $\delta = \mathrm{cf}(\kappa)$. Let $(\delta_\nu \mid \nu < \delta)$ be a normal sequence cofinal in κ with $\delta_0 = 0$. Define $(\tilde{C}_\alpha \mid \alpha \in S_\kappa)$ as follows. Let $g_\alpha : \theta_\alpha \to C'_\alpha$ be the monotone enumeration of C'_α. If $\delta_\nu < \theta_\alpha \leqslant \delta_{\nu+1}$, set $\tilde{C}_\alpha = g''_\alpha(\theta_\alpha - (\delta_\nu + 1))$. If $\theta_\alpha = \sup\{\delta_\nu \mid \delta_\nu < \theta_\alpha\}$, set $\tilde{C}_\alpha = g''_\alpha\{\delta_\nu \mid \delta_\nu < \theta_\alpha\}$. Then $(\tilde{C}_\alpha \mid \alpha \in S_\kappa)$ is as required.)

2C. Prove that if $V = L$, then for any uncountable regular cardinal κ, there is a sequence $(X_\xi \mid \xi < \kappa^+)$ of classes such that for each closed set $X \subseteq \mathrm{On}$ of order-type κ:

(i) for all $\xi < \kappa^+$, $X \cap X_\xi$ is stationary in X;
(ii) if $\xi < \eta < \kappa^+$, then $X \cap X_\xi \cap X_\eta$ is not stationary in X.

(Hint: First use $\diamondsuit_\kappa$ to show that there are stationary sets $Y_\xi \subseteq \kappa$, $\xi < \kappa^+$, such that $Y_\xi \cap Y_\eta$ is not stationary whenever $\xi < \eta < \kappa^+$. Now let $(C_\alpha \mid \alpha \in S_\kappa)$ be as in $\square^\kappa(\emptyset)$. Let $(\varrho^\alpha_\xi \mid \xi < \eta_\alpha)$ be the monotone enumeration of C_α. Let

$$X_\delta = Y_\delta \cup \{\alpha \in \bigcup_{\nu < \kappa} S_\nu - \kappa \mid (\exists \xi \in Y_\delta)(\exists \beta \in S_\kappa)[\lim(\xi) \wedge \alpha = \varrho^\beta_\xi]\}.)$$

2D. Prove that if $V = L$, then for any uncountable regular cardinal κ there is a sequence $(X_\xi \mid \xi < \kappa)$ of pairwise disjoint classes such that for any closed set $X \subseteq \mathrm{On}$ of order-type κ, $X \cap X_\xi$ is stationary in X for every $\xi < \kappa$. (Hint: Use 2C.)

Deduce that, if $V = L$, then for each cardinal κ there is a set $A \subseteq \kappa$ such that neither A nor $\kappa - A$ contains a closed set of order-type ω_1. (See also the Notes on this chapter.)

3. *The Failure of $\square_\kappa$ and Large Cardinals*

Show that if κ^+ is not Mahlo in L, then $\square_\kappa$ holds. (Hint: Let $C \in L$ be a club subset of κ^+ consisting of singular cardinals in L. By 6.1, $\square$ holds in L, so there is a "$\square$-sequence" on C. Using the ideas from the proof of 6.2, modify this sequence to a $\square_\kappa$-sequence.)

Deduce that if $\square_\kappa$ fails, then κ^+ is Mahlo in L.

Notice that the above result provides an alternative solution to Exercise IV.5.

4. *The Principles $\square(E)$*

Prove Theorem VI.6.1′. (Use the argument of IX.2 as a starting point.)

Chapter VII
Trees and Large Cardinals in L

In this chapter we concentrate on the notion of a κ-tree in the case where κ is an inaccessible cardinal. In this case, assuming $V = L$, both the notion of a κ-Souslin tree and of a κ-Kurepa tree turn out to be closely related to large cardinal properties. Thus this chapter extends both Chapter IV, where we studied κ^+-trees, and (parts of) Chapter V, where we dealt with large cardinals.

1. Weakly Compact Cardinals and κ-Souslin Trees

The notion of a weakly compact cardinal has already been introduced in V.1, and we refer the reader back there for basic definitions. In particular, V.1.3 gives several equivalent definitions of weak compactness, and V.1.5 proves the result, relevant to us here, that if κ is a weakly compact cardinal, then $[\kappa$ is weakly compact$]^L$.

Assuming $V = L$, we shall prove that if κ is an inaccessible cardinal, then κ is weakly compact iff there is no κ-Souslin tree. This extends V.1.3 (viii), which says that, in ZFC, an inaccessible cardinal κ is weakly compact iff there is no κ-Aronszajn tree. We shall also show that under $V = L$, V.1.3 (ii) may be extended.

We shall require the following characterisation of weak compactness, which is really just a $V = L$ analogue of Π^1_1-indescribability (V.1.3 (iv)).

1.1 Lemma. *Assume $V = L$. Let κ be an inaccessible cardinal. Then κ is weakly compact iff, whenever $\varphi(\mathring{U}, \mathring{A}_1, \ldots, \mathring{A}_n)$ is a sentence of the language $\mathscr{L}(U, A_1, \ldots, A_n)$, if $A_1, \ldots, A_n \subseteq J_\kappa$ are such that*

$$(\forall U \subseteq J_\kappa)\,[\langle J_\kappa, \in, U, A_1, \ldots, A_n\rangle \models \varphi],$$

then for some $\alpha < \kappa$,

$$(\forall U \subseteq J_\alpha)\,[\langle J_\alpha, \in, U, A_1 \cap J_\alpha, \ldots, A_n \cap J_\alpha\rangle \models \varphi]. \quad \square$$

There are various ways of proving 1.1. One way is to make minor modifications to the proof that Π^1_1-indescribability characterises weak compactness in ZFC (V.1.3 (iv)). Another way is to prove that under the assumption $V = L$, the property in 1.1 is actually equivalent to the Π_1^1-indescribability condition, by noting that if λ is inaccessible, then $J_\lambda = V_\lambda$. (This requires a lemma that the $\alpha < \kappa$ of 1.1 can always be assumed to be an inaccessible cardinal. The proof of this fact involves adding a conjunct to the sentence φ which ensures this.) In any event, the proof of 1.1 is of no direct relevance to our work here, being essentially a part of large cardinal theory itself, rather than constructibility theory. So we do not give a full proof.

Now, by V.1.3(viii), if κ is a weakly compact cardinal, then there is no κ-Aronszajn tree, so certainly there can be no κ-Souslin tree. We shall prove that if $V = L$, then if κ is not weakly compact, there is a κ-Souslin tree. As usual when dealing with trees, we are assuming that κ is regular here. In fact, since we know from IV.2.4 that (if $V = L$) there is a κ-Souslin tree whenever κ is a successor cardinal, we need only consider the case where κ is inaccessible. Our construction of a κ-Souslin tree closely resembles that of IV.2.4. Indeed, since $\diamondsuit_\kappa(E)$ is valid for any stationary set $E \subseteq \kappa$ (assuming $V = L$), by examining the proof of IV.2.4 we see that it is sufficient, in order to show that there is a κ-Souslin tree for inaccessible, non-weakly compact κ, to prove the following combinatorial result:

1.2 Theorem. *Assume $V = L$. Let κ be an inaccessible cardinal which is not weakly compact. Then there is a stationary set $E \subseteq \kappa$ and a sequence $(C_\alpha \mid \alpha < \kappa \wedge \lim(\alpha))$ such that:*

(i) $\alpha \in E \to \mathrm{cf}(\alpha) = \omega$;

(ii) *C_α is a club subset of α;*

(iii) *if $\bar{\alpha} < \alpha$ is a limit point of C_α, then $\bar{\alpha} \notin E$ and $C_{\bar{\alpha}} = \bar{\alpha} \cap C_\alpha$.* □

By means of a slightly different argument, depending on VI.6.1′ rather than VI.6.1, it is possible to prove the following more general form of 1.2.

1.2′ Theorem. *Assume $V = L$. Let κ be an inaccessible cardinal which is not weakly compact. Let $A \subseteq \kappa$ be a stationary set of limit ordinals. Then there is a stationary set $E \subseteq A$ and a sequence $(C_\alpha \mid \alpha < \kappa \wedge \lim(\alpha))$ such that:*

(i) *C_α is a club subset of α;*

(ii) *if $\bar{\alpha} < \alpha$ is a limit point of C_α, then $\bar{\alpha} \notin E$ and $C_{\bar{\alpha}} = \bar{\alpha} \cap C_\alpha$.* □

(See Exercise 4.)

Before we turn to the proof of 1.2, we obtain some consequences of this result.

1.3 Theorem. *Assume $V = L$. Let κ be an inaccessible cardinal. Then the following are equivalent:*

(i) *κ is weakly compact;*

(ii) *if $E \subseteq \kappa$ is stationary in κ, then for some regular cardinal $\lambda < \kappa$, $E \cap \lambda$ is stationary in λ;*

(iii) *there is no κ-Souslin tree;*

(iv) *for all n, λ such that $1 < n < \omega$ and $1 < \lambda \leqslant \kappa$, the partition property*

$$\kappa \to [\kappa]^n_\lambda$$

(see below) is valid;

(v) *for some λ such that $1 < \lambda \leqslant \kappa$, the partition property*

$$\kappa \to [\kappa]^2_\lambda$$

(see below) is valid.

Proof. (i) → (ii). This is a simple application of Π^1_1-indescribability, and is left to the reader. $V = L$ is not required for this implication.

(ii) → (i). This follows from 1.2. If κ is not weakly compact, then the set $E \subseteq \kappa$ of 1.2 is stationary in κ, but if $\lambda < \kappa$ is regular, the set of limit points of C_λ is a club subset of λ which is disjoint from $E \cap \lambda$, so $E \cap \lambda$ is not stationary in λ.

(i) → (iii). This is a consequence of V.1.3 (viii) (there are no κ-Aronszajn trees). This part does not require $V = L$.

(iii) → (i). If κ is not weakly compact, then, using 1.2 we may repeat the argument of IV.2.4.

(i) → (iv). Condition (iv) involves a new partition relation. We write

$$\kappa \to [\mu]^n_\lambda$$

iff, whenever $f\colon [\kappa]^n \to \lambda$, there is a set $X \subseteq \kappa$, $|X| = \mu$, such that $f''[X]^n \neq \lambda$. Provided that $\lambda > 2$, this would seem to be much weaker than the condition

$$\kappa \to (\kappa)^n_\lambda,$$

which requires that the set X satisfy $|f''[X]^n| = 1$. An indeed, it is known that the two partition relations are not provably equivalent in ZFC. But as the theorem shows, in L these two relations are equivalent.

Since $\kappa \to (\kappa)^n_\lambda$ is a consequence of weak compactness (V.1.3 (ii)), the implication (i) → (iv) is provable in ZFC.

(iv) → (v). This is trivial, since (v) is a special case of (iv).

(v) → (i). It suffices to prove $\neg$ (iii) → $\neg$ (v). So let $\mathbf{T} = \langle \kappa, \leqslant_{\mathbf{T}} \rangle$ be a κ-Souslin tree. By discarding levels of $\mathbf{T}$ we may assume that for every $x \in \mathbf{T}$ the set $S(x)$ of all immediate successors of x in $\mathbf{T}$ has cardinality at least $|x|$. Let f_x be a map from $S(x)$ onto $x (= \{y \mid y < x\})$ for each $x \in \mathbf{T}$. Define $f\colon [\kappa]^2 \to \kappa$ as follows. If x, y are incomparable in $\mathbf{T}$, let $f(\{x, y\}) = 0$. Suppose $x, y \in \mathbf{T}$ are such that $x <_{\mathbf{T}} y$. Let $\bar{y}$ be the unique predecessor of y in $S(x)$. Let $f(\{x, y\}) = f_x(\bar{y})$. We show that f witnesses $\kappa \not\to [\kappa]^2_\kappa$.

Assume that $X \in [\kappa]^\kappa$ and $\alpha < \kappa$ are given. For each $x \in X - (\alpha + 1)$, let $y_x \in S(x)$ be such that $f_x(y_x) = \alpha$. Since $\mathbf{T}$ is κ-Souslin, there must be $x, x' \in X - (\alpha + 1)$ such that $y_x <_{\mathbf{T}} y_{x'}$. Then by definition, $f(\{x, x'\}) = \alpha$. The proof is complete. □

We turn now to the proof of 1.2. We assume $V = L$ from now on. We fix κ an inaccessible cardinal which is not weakly compact. By 1.1 there is a sentence φ of $\mathscr{L}(\mathring{B}, \mathring{D})$ and a set $B \subseteq \kappa$ such that

(a) $(\forall D \subseteq \kappa)\,[\langle J_\kappa, \in, B, D\rangle \models \varphi]$;

(b) $(\forall \alpha < \kappa)\,(\exists D \subseteq \alpha)\,[\langle J_\alpha, \in, B \cap \alpha, D\rangle \models \neg\, \varphi]$.

(We have made some simplifications here. In 1.1 we allowed any finite number of predicate letters in φ. But by using pairing functions we can always replace a finite number of predicates by a single predicate. Also, we have only considered predicates on ordinals in the above. But since there is a uniformly J_α-definable map from α onto J_α for all ordinals closed under the Gödel Pairing Function (see VI.3.19), and since we can always add a conjunct to φ to ensure that α is closed under the Gödel function, this also causes no loss of generality.)

Our proof of 1.2 depends heavily upon the proof of the global $\square$ principle in VI.6. We begin by recalling the definition of the class E of VI.6.

E is the class of all limit ordinals α such that for some ordinal $\beta > \alpha$:

(i) α is regular over J_β; and
(ii) there is a $p \in J_\beta$ such that whenever $p \in X \prec J_\beta$ and $X \cap \alpha$ is transitive, then $X = J_\beta$.

We define $\bar{E} \subseteq \kappa$ to be the set of all limit cardinals (note: *cardinals*) $\alpha < \kappa$ such that $\alpha \in E$ and for some $\beta > \alpha$ satisfying (i) and (ii) above, it is the case that:

(iii) $B \cap \alpha \in J_\beta$;
(iv) if $D \in \mathscr{P}(\alpha) \cap J_\beta$, then $\langle J_\alpha, \in, B \cap \alpha, D\rangle \models \varphi$.

Since $\bar{E} \subseteq E$, by VI.6.4, $\alpha \in \bar{E}$ implies $\mathrm{cf}(\alpha) = \omega$.

By VI.6.3 we know that $E \cap \kappa$ is stationary in κ. By modifying the proof of VI.6.3 slightly, we prove:

1.4 Lemma. *$\bar{E}$ is stationary in κ.*

Proof. Let $C \subseteq \kappa$ be club. We prove that $\bar{E} \cap C \neq \emptyset$. Since the set of all limit cardinals $\alpha < \kappa$ is club in κ, we may assume that all members of C are limit cardinals. Much as in VI.6.3, let N be the smallest $N \prec J_{\kappa^+}$ such that $(B, C) \in N$ and $N \cap \kappa$ is transitive. Let $\alpha = N \cap \kappa$. Let $\pi\colon J_\beta \cong N$. Then $\pi \restriction \alpha = \mathrm{id} \restriction \alpha$ and $\pi(\alpha) = \kappa$. Moreover, $\pi(B \cap \alpha) = B$ and $\pi(C \cap \alpha) = C$.

Exactly as in VI.6.3, we may prove that α, β are as in conditions (i) and (ii) above, with $p = (B \cap \alpha, C \cap \alpha)$. Moreover, we know that $B \cap \alpha \in J_\beta$, so (iii) holds. Finally, by choice of φ and absoluteness,

$$\models_{J_{\kappa^+}} (\forall D \subseteq \kappa)\,[\langle J_\kappa, \in, B, D\rangle \models \varphi].$$

Applying π^{-1},

$$\models_{J_\beta} (\forall D \subseteq \alpha)\,[\langle J_\alpha, \in, B \cap \alpha, D\rangle \models \varphi].$$

So by absoluteness,

$$(\forall D \in \mathscr{P}(\alpha) \cap J_\beta)[\langle J_\alpha, \in, B \cap \alpha, D\rangle \models \varphi].$$

But $\alpha \in C$ (as in VI.6.3), so α is a limit cardinal. Thus $\alpha \in \bar{E}$, and so $\alpha \in \bar{E} \cap C$, and we are done. □

We shall let S, $(C_\alpha \mid \alpha \in S)$ be as in VI.6. So, in particular, $(C_\alpha \mid \alpha \in S)$ is a $\square(E)$-sequence. We define a sequence $(\bar{C}_\alpha \mid \alpha < \kappa \wedge \lim(\alpha))$ to satisfy 1.2 for the stationary set $\bar{E} \subseteq \kappa$. That is, we shall define the sets $\bar{C}_\alpha$ so that $\bar{C}_\alpha$ is a club subset of α and whenever $\bar{\alpha} < \alpha$ is a limit point of $\bar{C}_\alpha$, then $\bar{\alpha} \notin \bar{E}$ and $\bar{C}_{\bar{\alpha}} = \bar{\alpha} \cap \bar{C}_\alpha$. There are several cases to consider. First a trivial case: set $\bar{C}_\omega = \omega$. From now on we shall assume $\alpha > \omega$. S, C_α

Case 1. α is not a limit cardinal.

In this case, let τ be the largest limit cardinal less than α, and set $\bar{C}_\alpha = \alpha - (\tau + 1)$. Since $\bar{E}$ consists only of limit cardinals, no limit point of $\bar{C}_\alpha$ can be in $\bar{E}$. Moreover, if $\bar{\alpha} < \alpha$ is a limit point of $\bar{C}_\alpha$, then $\tau < \bar{\alpha} < \alpha$, so $\bar{\alpha}$ falls under Case 1 as well, and $\bar{C}_{\bar{\alpha}} = \bar{\alpha} - (\tau + 1) = \bar{\alpha} \cap \bar{C}_\alpha$. There is nothing further to check in this case.

In order to describe the next case we require some preliminary notions.

Let U be the set of all limit cardinals $\alpha < \kappa$ such that for some $\beta > \alpha$: U

(i) α is regular over J_β;
(ii) $B \cap \alpha \in J_\beta$;
(iii) there is a $D \in \mathscr{P}(\alpha) \cap J_\beta$ such that $\langle J_\alpha, \in, B \cap \alpha, D\rangle \models \neg\varphi$.

We shall say that any β as above *testifies* that $\alpha \in U$.

1.5 Lemma. $U \cap \bar{E} = \emptyset$.

Proof. Let $\alpha \in \bar{E}$ and let $\beta > \alpha$ satisfy the definition for $\alpha \in \bar{E}$. Thus, in particular,

$$(\forall D \in \mathscr{P}(\alpha) \cap J_\beta)[\langle J_\alpha, \in, B \cap \alpha, D\rangle \models \varphi].$$

Now suppose that $\alpha \in U$, and let $\beta' > \alpha$ testify this fact. Thus, in particular,

$$(\exists D \in \mathscr{P}(\alpha) \cap J_{\beta'})[J_\alpha, \in, B \cap \alpha, D\rangle \models \neg\varphi].$$

Hence $\beta < \beta'$. But by VI.6.4, α is Σ_1-singular over $J_{\beta+1}$. Hence α is not regular over $J_{\beta'}$. Contradiction, since β' testifies $\alpha \in U$. Thus $\alpha \notin U$, and the lemma is proved. □

Now let W be the set of all $\alpha \in U$ such that if $\beta > \alpha$ is the least to testify $\alpha \in U$, then whenever $p \in J_\beta$ there is an $X \prec J_\beta$ such that $p \in X$ and $X \cap \alpha \in \alpha$. W

1.6 Lemma. *$U - W \subseteq E$. Moreover, if $\alpha \in U - W$ and $\beta > \alpha$ is the least to testify $\alpha \in U$, then β satisfies the definition for $\alpha \in E$.*

Proof. Let α, β be as above. Since $\alpha \notin W$ there is a $p \in J_\beta$ such that whenever $X \prec J_\beta$ is such that $p \in X$ and $X \cap \alpha$ is transitive, then $X \cap \alpha = \alpha$. Let p be in fact the $<_J$-least such element of J_β. Since $\alpha \in U$, let $D \in \mathscr{P}(\alpha) \cap J_\beta$ be the $<_J$-least subset of α such that $\langle J_\alpha, \in, B \cap \alpha, D\rangle \models \neg \varphi$. Let $q = (p, \alpha, B \cap \alpha, D)$. We prove the lemma by showing that if $q \in X \prec J_\beta$ and $X \cap \alpha$ is transitive, then $X = J_\beta$. It suffices to prove this for the smallest X for which $q \in X \prec J_\beta$ and $X \cap \alpha$ is transitive.

Let $\pi\colon X \cong J_{\bar\beta}$, $\bar\beta \leqslant \beta$. Since $p \in X$, we have $X \cap \alpha = \alpha$, so $\pi \restriction \alpha = \mathrm{id} \restriction \alpha$, $\pi(\alpha) = \alpha$, $\pi(B \cap \alpha) = B \cap \alpha$, $\pi(D) = D$. Moreover, since α is regular over J_β, α is regular over $J_{\bar\beta}$. Thus $\bar\beta$ testifies that $\alpha \in U$. So by the minimality of β, we have $\bar\beta = \beta$.

Suppose now that $\pi(p) \in Y \prec J_\beta$ and $Y \cap \alpha$ is transitive. Let $\bar Y = \pi^{-1}{}'' Y$. Then $\bar Y \cap \alpha = Y \cap \alpha$, so, as $\pi^{-1}\colon J_\beta \prec J_\beta$, we have $p \in \bar Y \prec J_\beta$ and $\bar Y \cap \alpha$ is transitive. Thus by choice of p, $\bar Y \cap \alpha = \alpha$. Thus $Y \cap \alpha = \alpha$. But Y was arbitrary here. Hence $\pi(p)$ has the same property as p. So as p was chosen $<_J$-minimally and $\pi(p) \leqslant_J p$ (because π is a collapsing map) we have $\pi(p) = p$. It follows at once that $\pi(q) = q$.

Now by choice of X, every element of X is definable from parameters in $\alpha \cup \{q\}$ in J_β. (Because the set of all elements of J_β which are so definable is an elementary submodel of J_β containing q which is transitive on α, and X is the smallest such.) But we have $X \prec J_\beta$, $\pi\colon X \cong J_\beta$, $\pi \restriction \alpha = \mathrm{id} \restriction \alpha$, $\pi(q) = q$. Hence $\pi = \mathrm{id} \restriction X$. Thus $X = J_\beta$, and we are done. □

Case 2. $\alpha \in W$.

Let $\beta > \alpha$ be the least to testify $\alpha \in U$, and let $D \in \mathscr{P}(\alpha) \cap J_\beta$ be $<_J$-least such that

$$\langle J_\alpha, \in, B \cap \alpha, D\rangle \models \neg \varphi.$$

Since $\alpha \in W$, we can define submodels $X_\nu \prec J_\beta$, $\nu < \theta$ (some θ), as follows:

$$X_0 = \text{the smallest } X \prec J_\beta \text{ such that } (\alpha, B \cap \alpha, D) \in X \text{ and } X \cap \alpha \text{ is transitive;}$$

$$X_{\nu+1} = \text{the smallest } X \prec J_\beta \text{ such that } (\alpha, B \cap \alpha, D, \alpha_\nu) \in X \text{ and } X \cap \alpha \text{ is transitive;}$$

$$X_\lambda = \bigcup_{\nu<\lambda} X_\nu, \quad \text{if } \lim(\lambda) \text{ and } \sup_{\nu<\lambda} \alpha_\nu < \alpha \quad \text{(otherwise undefined),}$$

where for each ν we set

$$\alpha_\nu = X_\nu \cap \alpha.$$

Since $\alpha \in W$, the definition proceeds until a limit ordinal θ is reached for which $\sup_{\nu<\theta} \alpha_\nu = \alpha$. Thus the set

$$\bar C_\alpha = \{\alpha_\nu \mid \nu < \theta\}$$

is a club subset of α.

1.7 Lemma. *Let $\alpha \in W$. Let $\bar{\alpha} < \alpha$ be a limit point of $\bar{C}_\alpha$. Then $\bar{\alpha} \in W$, $\bar{\alpha} \in \bar{E}$, and $\bar{C}_{\bar{\alpha}} = \bar{\alpha} \cap \bar{C}_\alpha$.*

Proof. Let $\bar{\alpha} = \alpha_\lambda$, $\lim(\lambda)$, and let $\pi: J_{\bar{\beta}} \cong X_\lambda$. Thus, $\pi \upharpoonright \bar{\alpha} = \mathrm{id} \upharpoonright \bar{\alpha}$, $\pi(\bar{\alpha}) = \alpha$, $\pi(B \cap \bar{\alpha}) = B \cap \alpha$, $\pi(D \cap \bar{\alpha}) = D \cap \alpha$. So, as $\pi: J_{\bar{\beta}} \prec J_\beta$, it is immediate that $\bar{\beta}$ testifies that $\bar{\alpha} \in U$, and moreover that $\bar{\beta}$ is the least such ordinal. In particular, by 1.5, we have $\bar{\alpha} \notin \bar{E}$.

Let $\bar{p} \in J_{\bar{\beta}}$. Then $p = \pi(\bar{p}) \in X_\lambda$, so $p \in X_\nu$ for some $\nu < \lambda$. Let $X = \pi^{-1}{}'' X_\nu$. Then $\bar{p} \in X \prec J_{\bar{\beta}}$ and $X \cap \bar{\alpha} = \alpha_\nu < \bar{\alpha}$. Thus $\bar{\alpha} \in W$.

Define $\bar{D}$, $(\bar{X}_\nu \mid \nu < \bar{\theta})$, $(\bar{\alpha}_\nu \mid \nu < \bar{\theta})$ from $\bar{\alpha}$, $\bar{\beta}$ just as D, $(X_\nu \mid \nu < \theta)$, $(\alpha_\nu \mid \nu < \theta)$ were defined from α, β. Thus, in particular, $\bar{C}_{\bar{\alpha}} = \{\bar{\alpha}_\nu \mid \nu < \bar{\theta}\}$. It is easily seen that $\bar{\theta} = \lambda$ and $X_\nu = \pi'' \bar{X}_\nu$ for all $\nu < \bar{\theta}$. Hence $\bar{\alpha}_\nu = \alpha_\nu$ for all $\nu < \bar{\theta}$, and we have $\bar{C}_{\bar{\alpha}} = \bar{\alpha} \cap \bar{C}_\alpha$. The lemma is proved. □

That completes the discussion in Case 2. Notice that this case includes all regular $\alpha > \omega$, since if α is regular, then $\beta = \alpha^+$ testifies $\alpha \in U$, and $\alpha \in W$ holds by regularity. From now on we assume that $\alpha > \omega$ does not fall under either of Cases 1 or 2. Hence α is a singular limit cardinal. We now make use of the sequences C_α, $\alpha \in S$, from VI.6.

Let C'_α be the set of all limit cardinals $\lambda < \alpha$ which are limit points of C_α. Then C'_α is closed in α, and if $\mathrm{cf}(\alpha) > \omega$, C'_α is also unbounded in α. C'_α

Case 3. C'_α is bounded in α.

Then we must have $\mathrm{cf}(\alpha) = \omega$. Let $\bar{C}_\alpha$ be any ω-sequence cofinal in α. Since $\bar{C}_\alpha$ has no limit points, there is nothing to check in this case.

Now, if $\alpha \in E$, then in the definition of C_α in VI.6, α falls under either Case 1 ($\alpha < \omega_1$) or else Case 4 ($n(\alpha) = 1$ and $\mathrm{succ}(\beta(\alpha))$), so C_α is an ω-sequence cofinal in α. Hence $C'_\alpha = \emptyset$ for all $\alpha \in E$. Thus Cases 1 through 3 above include all $\alpha \in E$. So by 1.6, Cases 1 through 3 include all $\alpha \in U - W$. So if we assume from now on that $\alpha > \omega$ does not fall under any of cases 1 through 3, then $\alpha \notin U$ and C'_α is unbounded in α. We shall take $\bar{C}_\alpha$ to be a certain club subset of C'_α.

In the definition of C_α in VI.6, in Case 1 ($\alpha < \omega_1$), Case 2 ($\alpha \notin Q$), and Case 3 ($\alpha \in Q$ and $\sup(Q \cap \alpha) < \alpha$), α is not a cardinal, and hence falls under our present Case 1 above. And in Case 4 of VI.6 ($n(\alpha) = 1$ and $\mathrm{succ}(\beta(\alpha))$) we have $C'_\alpha = \emptyset$. All of these possibilities are covered by our present Cases 1 through 3. Since we are assuming now that α does not fall under any of these three cases, it follows that in the definition of C_α in VI.6, α falls under Case 5. In particular, by VI.6.17, if $\bar{\alpha}$ is a limit point of C_α (*a fortiori*: of $\bar{C}_\alpha$, when it has been defined), then $\bar{\alpha} \notin E$, so $\bar{\alpha} \notin \bar{E}$, and hence we need only concern ourselves with the proof that $\bar{C}_{\bar{\alpha}} = \bar{\alpha} \cap \bar{C}_\alpha$.

Let $\beta = \beta(\alpha)$, $n = n(\alpha)$ be as in VI.6. Let $(\alpha_\nu \mid \nu < \theta)$ be the monotone enumeration of C'_α, and set $\beta_\nu = \beta(\alpha_\nu)$. If $\bar{\alpha} = \alpha_\nu$, then our β_ν is just the $\bar{\beta}$ of VI.6 and we have (VI.6.12) $n(\alpha_\nu) = n$; moreover there is a map $\tilde{\pi}: J_{\bar{\beta}} \prec_{n-1} J_\beta$ such that $\tilde{\pi} \upharpoonright \bar{\alpha} = \mathrm{id} \upharpoonright \bar{\alpha}$, and, in case $\bar{\alpha} < \bar{\beta}$, such that $\tilde{\pi}(\bar{\alpha}) \geqslant \alpha$. (See just prior to VI.6.10.) Let π_ν denote this embedding. Thus for each $\nu < \theta$ we have an embedding $\pi_\nu: J_{\beta_\nu} \prec_{n-1} J_\beta$ such that $\pi_\nu \upharpoonright \alpha_\nu = \mathrm{id} \upharpoonright \alpha_\nu$, and in case $\alpha_\nu < \beta_\nu$, $\pi_\nu(\alpha_\nu) \geqslant \alpha$. β, n, θ α_ν, β_ν π_ν

1.8 Lemma. *Let $\bar{\alpha} < \alpha$ be a limit point of C'_α. Then $\bar{\alpha}$ does not fall under either of Cases 1 and 3 above.*

Proof. Since $\bar{\alpha}$ is a limit cardinal, it cannot fall under Case 1. Since $C'_{\bar{\alpha}} = \bar{\alpha} \cap C'_{\alpha}$, $\bar{\alpha}$ is a limit point of $C'_{\bar{\alpha}}$, so $\bar{\alpha}$ cannot fall under Case 3. $\square$

Case 4. $\alpha = \beta$.

Set $\bar{C}_\alpha = C'_\alpha$ in this case. Suppose that $\bar{\alpha} < \alpha$ is a limit point of $\bar{C}_\alpha$. Note that by 1.8, $\bar{\alpha}$ cannot fall under either of Cases 1 or 3. For some limit ordinal $\lambda < \theta$, we have $\bar{\alpha} = \alpha_\lambda$. If $\alpha_\lambda < \beta_\lambda$, then we would have $\pi_\lambda(\alpha_\lambda) \geqslant \alpha$, which is impossible since $\alpha = \beta \notin J_\beta \supseteq \mathrm{ran}(\pi_\lambda)$. (Since α is a cardinal, $\omega\alpha = \alpha$, of course, so $\omega\beta = \beta$.) Thus $\alpha_\lambda = \beta_\lambda$, i.e. $\beta(\bar{\alpha}) = \bar{\alpha}$. It follows that $\bar{\alpha}$ does not fall under Case 2 above, because any $\bar{\beta} > \bar{\alpha}$ which would testify $\bar{\alpha} \in U$ would need to be less than $\beta(\bar{\alpha})$ in order for $\bar{\alpha}$ to be regular over $J_{\bar{\beta}}$. Hence as $\alpha_\lambda = \beta_\lambda$, $\bar{\alpha}$ falls under Case 4. Thus

$$\bar{C}_{\bar{\alpha}} = C'_{\bar{\alpha}} = \bar{\alpha} \cap C'_\alpha = \bar{\alpha} \cap \bar{C}_\alpha,$$

and we are done in this case.

We assume now that α does not fall under any of Cases 1 through 4. For
$\tau_{\nu\tau}$ $\nu \leqslant \tau < \theta$, set $\pi_{\nu\tau} = \pi_\tau^{-1} \circ \pi_\nu$. Thus $\pi_{\nu\tau}\colon J_{\beta_\nu} \prec_{n-1} J_{\beta_\tau}$ and $\pi_{\nu\tau} \restriction \alpha_\nu = \mathrm{id} \restriction \alpha_\nu$ for all $\nu \leqslant \tau < \theta$. Clearly, $\langle (J_{\beta_\nu})_{\nu<\theta}, (\pi_{\nu\tau})_{\nu\leqslant\tau<\theta} \rangle$ is a directed Σ_{n-1}-elementary system. (See V.5 for the relevant definitions.) What is its direct limit? We claim that it is $\langle J_\beta, (\pi_\nu)_{\nu<\theta} \rangle$. Clearly, what we must prove is that $J_\beta = \bigcup_{\nu<\theta} \mathrm{ran}(\pi_\nu)$. We do this below.

ϱ_ν Let $\varrho_\nu = \varrho(\alpha_\nu) = \varrho_{\beta_\nu}^{n-1}$ $(\nu < \theta)$. By VI.6.7 (iii),

$$\bigcup_{\nu<\theta} \pi_\nu'' J_{\varrho_\nu} = J_\varrho. \tag{1}$$

Suppose now that $\nu < \theta$, and let $\bar{\beta} = \beta_\nu$, $\pi = \pi_\nu$. By VI.5.6 we know that

$$(1)' \quad \begin{array}{l} (\pi \restriction J_{\varrho_{\bar{\beta}}^{n-2}})\colon \langle J_{\varrho_{\bar{\beta}}^{n-2}}, A_{\bar{\beta}}^{n-2} \rangle \prec_1 \langle J_{\varrho_\beta^{n-2}}, A_\beta^{n-2} \rangle, \\ \pi(p_{\bar{\beta}}^{n-1}) = p_\beta^{n-1}; \end{array}$$

. . . .

. . . .

$$(n-1)' \quad \begin{array}{l} (\pi \restriction J_{\varrho_{\bar{\beta}}^1})\colon \langle J_{\varrho_{\bar{\beta}}^1}, A_{\bar{\beta}}^1 \rangle \prec_{n-2} \langle J_{\varrho_\beta^1}, A_\beta^1 \rangle, \\ \pi(p_{\bar{\beta}}^2) = p_\beta^2; \end{array}$$

$$(n)' \quad \begin{array}{l} \pi\colon J_{\bar{\beta}} \prec_{n-1} J_\beta, \\ \pi(p_{\bar{\beta}}^1) = p_\beta^1. \end{array}$$

Now, by definition of p_β^{n-1}, every element of $J_{\varrho_\beta^{n-2}}$ is Σ_1-definable from elements of $J_\varrho \cup \{p_\beta^{n-1}\}$ in $\langle J_{\varrho_\beta^{n-2}}, A_\beta^{n-2} \rangle$. So by (1),

$$\bigcup_{\nu<\theta} \pi_\nu'' J_{\varrho_{\beta_\nu}^{n-2}} = J_{\varrho_\beta^{n-2}}. \tag{2}$$

Repeating the same argument, using (2) in place of (1) and (2)′ in place of (1)′ now yields

$$\bigcup_{\nu<\theta} \pi_\nu'' J_{\varrho_{\beta_\nu}^{n-3}} = J_{\varrho_\beta^{n-3}}. \tag{3}$$

Continuing in this fashion, we obtain, eventually,

$$(n-1)\quad \bigcup_{\nu<\theta} \pi_\nu{}''J_{\varrho^1_{\beta_\nu}} = J_{\varrho^1_\beta}.$$

$$(n)\quad \bigcup_{\nu<\theta} \pi_\nu{}''J_{\beta_\nu} = J_\beta.$$

This last equality is the one we require.

If $\lambda < \theta$ is a limit ordinal now and we set $\bar{\alpha} = \alpha_\lambda$, $\bar{\beta} = \beta_\lambda$, and if we define $\bar{\beta}_\nu, \bar{\alpha}_\nu, \bar{\pi}_\nu, \bar{\pi}_{\nu\tau}$, for $\nu \leqslant \tau < \bar{\theta}$, from $\bar{\alpha}, \bar{\beta}$ as $\beta_\nu, \alpha_\nu, \pi_\nu, \pi_{\nu\tau}$, for $\nu \leqslant \tau < \theta$, were defined from α, β, then (clearly) $\bar{\theta} = \lambda$ and for $\nu \leqslant \tau < \bar{\theta}$, $\bar{\beta}_\nu = \beta_\nu$, $\bar{\alpha}_n = \alpha_\nu$, $\bar{\pi}_\nu = \pi_{\nu\lambda}$, $\bar{\pi}_{\nu\tau} = \pi_{\nu\tau}$. We utilise these observations below. $\lambda, \bar{\alpha}, \bar{\beta}$ $\bar{\beta}_\nu, \bar{\alpha}_\nu, \bar{\pi}_\nu,$ $\bar{\pi}_{\nu\tau}, \bar{\theta}$

Case 5. $B \cap \alpha \in J_\beta$.

Since $J_\beta = \bigcup_{\nu<\theta} \pi_\nu{}''J_{\beta_\nu}$, we may pick ν_α to be the least $\nu < \theta$ such that α, $B \cap \alpha \in \pi_\nu{}''J_{\beta_\nu}$. Set

$$\bar{C}_\alpha = \{\alpha_\nu \mid \nu_\alpha \leqslant \nu < \theta\}.$$

Let $\bar{\alpha} < \alpha$ be a limit point of $\bar{C}_\alpha$. Thus $\bar{\alpha} = \alpha_\lambda$ for some limit ordinal λ, $\nu_\alpha < \lambda < \theta$. By 1.8, $\bar{\alpha}$ cannot fall under either of Cases 1 and 3. Moreover, $\bar{\alpha}$ cannot fall under Case 4, since $\alpha \in \pi_\lambda{}''J_{\beta_\nu}$, which implies that $\alpha_\lambda \in J_{\beta_\nu}$ and $\pi_\lambda(\alpha_\lambda) = \alpha$. (Recall that $\pi_\lambda \restriction \alpha_\lambda = \mathrm{id} \restriction \alpha_\lambda$.) We show that $\bar{\alpha}$ also cannot fall under Case 2. Indeed, not only do we have $\bar{\alpha} \notin W$, but the stronger condition $\bar{\alpha} \notin U$. For suppose that $\bar{\beta} > \bar{\alpha}$ were to testify that $\bar{\alpha} \in U$. Since $\bar{\alpha}$ must be regular over $J_{\bar{\beta}}$, we have $\bar{\beta} < \beta_\lambda$. Now, $B \cap \bar{\alpha} \in J_{\bar{\beta}}$, so as $\lambda > \nu_\alpha$, we must have $\pi_\lambda(B \cap \bar{\alpha}) = B \cap \alpha$. Thus $B \cap \alpha \in J_{\pi_\lambda(\bar{\beta})}$. Again, we can pick $\bar{D} \in \mathscr{P}(\bar{\alpha}) \cap J_{\bar{\beta}}$ so that $\langle J_{\bar{\alpha}}, \in, B \cap \bar{\alpha}, \bar{D}\rangle \models \neg\varphi$. Let $D = \pi_\lambda(\bar{D})$. Since $\pi_\lambda\colon J_{\beta_\lambda} \prec_0 J_\beta$, we have $D \in \mathscr{P}(\alpha) \cap J_{\pi_\lambda(\bar{\beta})}$ and $\langle J_\alpha, \in, B \cap \alpha, D\rangle \models \neg\varphi$. Thus as $\pi_\lambda(\bar{\beta}) < \beta$, $\pi_\lambda(\bar{\beta})$ testifies that $\alpha \in U$. But α falls under Case 5, so $\alpha \notin U$. Contradiction! Hence $\bar{\alpha}$ does not fall under any of Cases 1 through 4. But $\lambda > \nu_\alpha$, so $B \cap \bar{\alpha} \in J_{\beta_\lambda} = J_{\beta(\bar{\alpha})}$. Hence $\bar{\alpha}$ falls under Case 5. But it is clear from the remarks we made just prior to Case 5, together with the facts that $\pi_\lambda(\bar{\alpha}) = \alpha$ and $\pi_\lambda(B \cap \bar{\alpha}) = B \cap \alpha$ (which are valid because $\pi_\lambda \restriction \bar{\alpha} = \mathrm{id} \restriction \bar{\alpha}$ and α, $B \cap \alpha \in \pi_\lambda{}''J_{\beta_\lambda}$), that $\nu_{\bar{\alpha}} = \nu_\alpha$. Hence

$$\bar{C}_{\bar{\alpha}} = \{\alpha_\nu \mid \nu_{\bar{\alpha}} \leqslant \nu < \bar{\theta}\} = \{\alpha_\nu \mid \nu_\alpha \leqslant \nu < \lambda\} = \bar{\alpha} \cap \bar{C}_\alpha.$$

That completes the proof in this case.

Case 6. Otherwise.

In particular, in this case we have $B \cap \alpha \notin J_\beta$. Suppose that $\nu < \theta$ were such that $B \cap \alpha_\nu \in J_{\beta_\nu}$. Then there must be a $\tau > \nu$ such that $\pi_{\nu\tau}(B \cap \alpha_\nu) \neq B \cap \alpha_\tau$, since otherwise we would have

$$B \cap \alpha = \bigcup_{\nu<\tau<\theta} \pi_{\nu\tau}(B \cap \alpha_\nu) = \pi_\nu(B \cap \alpha_\nu) \in J_\beta.$$

So we can define a normal sequence $(\nu(\iota) \mid \iota < \bar{\theta})$, for some $\bar{\theta} \leqslant \theta$, as follows.

$$\nu(0) = 0;$$

$$\nu(\iota + 1) = \text{the least } \nu > \nu(\iota) \text{ such that}$$
$$B \cap \alpha_{\nu(\iota)} \in J_{\beta_{\nu(\iota)}} \rightarrow \pi_{\nu(\iota), \nu}(B \cap \alpha_{\nu(\iota)}) \neq B \cap \alpha_{\nu};$$

$$\nu(\lambda) = \sup_{\iota < \lambda} \nu(\iota), \quad \text{if this is less than } \theta \text{ (otherwise undefined), for } \lim(\lambda).$$

$\bar{\theta}$ The definition proceeds until an ordinal $\bar{\theta}$ is reached for which $\sup_{\iota < \bar{\theta}} \nu(\iota) = \theta$. (Clearly, $\lim(\bar{\theta})$.) Set

$$\bar{C}_\alpha = \{\alpha_{\nu(\iota)} \mid \iota < \bar{\theta}\}.$$

Let $\bar{\alpha} < \alpha$ be a limit point of $\bar{C}_\alpha$. Thus $\bar{\alpha} = \alpha_{\nu(\lambda)}$ for some limit ordinal $\lambda < \bar{\theta}$. As in Case 5, 1.8 implies that $\bar{\alpha}$ cannot fall under Cases 1 and 3, and since $\bar{\alpha} = \alpha_{\nu(\lambda)} < \beta_{\nu(\lambda)} = \beta(\bar{\alpha})$, $\bar{\alpha}$ cannot fall under Case 4. We show that $\bar{\alpha}$ cannot fall under Case 2. In fact, as in Case 5 we show that $\bar{\alpha} \notin U$. Suppose, on the contrary, that $\alpha \in U$. Thus, in particular, $B \cap \bar{\alpha} \in J_{\beta_{\nu(\lambda)}}$. (Clearly, the least $\bar{\beta} > \bar{\alpha}$ which testifies $\bar{\alpha} \in U$ has to be less than $\beta(\bar{\alpha}) = \beta_{\nu(\lambda)}$.) But (as we proved earlier for β)

$$J_{\beta_{\nu(\lambda)}} = \bigcup_{\iota < \lambda} \pi_{\nu(\iota), \nu(\lambda)}{}'' J_{\beta_{\nu(\iota)}},$$

so for some $\iota < \lambda$, $B \cap \bar{\alpha} \in \pi_{\nu(\iota), \nu(\lambda)}{}'' J_{\beta_{\nu(\iota)}}$. Thus $B \cap \bar{\alpha} = \pi_{\nu(\iota), \nu(\lambda)}(B \cap \alpha_{\nu(\iota)})$. But this implies that $\pi_{\nu(\iota), \nu(\iota+1)}(B \cap \alpha_{\nu(\iota)}) = B \cap \alpha_{\nu(\iota+1)}$, contrary to the choice of $\nu(\iota + 1)$. Hence $\bar{\alpha}$ does not fall under any of Cases 1 through 4. But the above argument shows that $\bar{\alpha}$ does not fall under Case 5 either. Thus $\bar{\alpha}$ falls under Case 6, and we have

$$\bar{C}_{\bar{\alpha}} = \{\alpha_{\nu(\iota)} \mid \iota < \lambda\} = \bar{\alpha} \cap \bar{C}_\alpha.$$

The proof of 1.2 is complete.

2. *Ineffable Cardinals and κ-Kurepa Trees*

Ineffability is a large cardinal property which strengthens weak compactness. By definition, an uncountable, regular cardinal κ is said to be weakly compact iff, whenever $f: [\kappa]^2 \rightarrow 2$, there is an *unbounded* set $X \subseteq \kappa$ such that $|f''[X]^2| = 1$. We say that an uncountable, regular cardinal κ is *ineffable* iff, whenever $f: [\kappa]^2 \rightarrow 2$, there is a *stationary* set $X \subseteq \kappa$ such that $|f''[X]^2| = 1$.

Clearly, all ineffable cardinals are weakly compact. The converse is not true, and indeed, as we shall show presently, ineffability is a much stronger notion than weak compactness. It should be said that the notion of ineffability is a rather specialised one, not covered in many of the standard texts dealing with large cardinals. (For instance, it is not covered in *Drake* (1974) or *Jech* (1978).) Consequently we give here a few of the basic results concerning ineffable cardinals.

2.1 Theorem. *Let $\kappa > \omega$ be regular. Then κ is ineffable iff, whenever $(A_\alpha \mid \alpha < \kappa)$ is such that $A_\alpha \subseteq \alpha$ for all $\alpha < \kappa$, there is a set $A \subseteq \kappa$ such that the set $\{\alpha \in \kappa \mid A \cap \alpha\}$ is stationary in κ.*

Proof. $(\rightarrow)$ Let $(A_\alpha \mid \alpha < \kappa)$ be given, $A_\alpha \subseteq \alpha$ for all $\alpha < \kappa$. For each $\alpha < \kappa$, let $f_\alpha: \alpha \to 2$ be the characteristic function of A_α. If we can find a function $f: \kappa \to 2$ such that $\{\alpha \in \kappa \mid f \upharpoonright \alpha = f_\alpha\}$ is stationary, then $A = f^{-1}{}''\{1\}$ will be as required.

Let $\dashv$ be the lexicographic ordering on the set $\{f_\alpha \mid \alpha < \kappa\}$. Define a function $h: [\kappa]^2 \to 2$ by

$$h\{\{\alpha, \beta\}\} = 0 \quad \text{iff } f_\alpha \dashv f_\beta \quad (\alpha < \beta < \kappa).$$

By assumption there is a stationary set $X \subseteq \kappa$ such that $|h''[X]^2| = 1$. Suppose, for definiteness, that $h''[X]^2 = \{0\}$. (The other case is similar.) Thus

$$\alpha, \beta \in X \quad \text{and} \quad \alpha < \beta \quad \text{implies } f_\alpha \dashv f_\beta.$$

For each $\nu < \kappa$, let α_ν be the least member of X such that $\alpha_\nu \geqslant \nu$ and

$$(\forall \beta \in X)\,(\beta \geqslant \alpha_\nu \to f_\beta \upharpoonright \nu = f_{\alpha_\nu} \upharpoonright \nu).$$

By choice of X, this definition is always possible. Let

$$C = \{\gamma \in \kappa \mid (\forall \nu)\,(\nu < \gamma \to \alpha_\nu < \gamma\}.$$

Clearly, C is a club subset of κ. Thus the set

$$Y = X \cap C \cap \{\nu \in \kappa \mid \lim(\nu)\}$$

is stationary in κ. Now, if $\lim(\nu)$, α_ν is the first member of X not less than $\sup_{\eta<\nu}\alpha_\eta$. So, if $\nu \in Y$, we will have $\alpha_\nu = \nu$. Hence

$$\alpha \in Y \quad \text{implies } (\forall \beta \in Y)\,(\beta \geqslant \alpha \to f_\beta \upharpoonright \alpha = f_\alpha).$$

Define $f: \kappa \to 2$ by

$$f = \bigcup_{\alpha \in Y} f_\alpha.$$

Since $Y \subseteq \{\alpha \in \kappa \mid f \upharpoonright \alpha = f_\alpha\}$, we are done.

$(\leftarrow)$ Let $f: [\kappa]^2 \to 2$ be given. For $\alpha < \kappa$, define $f_\alpha: \alpha \to 2$ by

$$f_\alpha(\nu) = f(\{\nu, \alpha\}) \quad (\nu < \alpha).$$

By assumption there is a function $\bar{f}: \kappa \to 2$ such that the set

$$X = \{\alpha \in \kappa \mid f_\alpha = \bar{f} \upharpoonright \alpha\}$$

is stationary in κ. (Consider the sets $A_\alpha \subseteq \alpha$ for which f_α is the characteristic function.) Now, $\bar{f}$ is regressive on $X - 2$, so by Fodor's Theorem (III.3.1) there is a stationary set $Y \subseteq X$ and an integer $i \in 2$ such that

$$\alpha \in Y \to \bar{f}(\alpha) = i.$$

For $\nu, \alpha \in Y$, $\nu < \alpha$, we have

$$f(\{\nu, \alpha\}) = f_\alpha(\nu) = (\bar{f} \upharpoonright \alpha)(\nu) = \bar{f}(\nu) = i.$$

Hence $|f''[Y]^2| = 1$. □

Strengthening the notion of Π_1^1-indescribability, which we have already noted as being equivalent to weak compactness (V.1.3), is that of Π_2^1-indescribability. An inaccessible cardinal κ is said to be *Π_2^1-indescribable* if, whenever $\varphi(\mathring{X}, \mathring{Y}, \mathring{U}_1, \ldots, \mathring{U}_n)$ is a sentence of $\mathscr{L}(X, Y, U_1, \ldots, U_n)$ and $U_1, \ldots, U_n \subseteq V_\kappa$ are such that

$$(\forall X \subseteq V_\kappa)(\exists Y \subseteq V_\kappa)[\langle V_\kappa, \in, X, Y, U_1, \ldots, U_n\rangle \models \varphi(\mathring{X}, \mathring{Y}, \mathring{U}_1, \ldots, \mathring{U}_n)],$$

then for some $\alpha < \kappa$,

$$(\forall X \subseteq V_\alpha)(\exists Y \subseteq V_\alpha)[\langle V_\alpha, \in, X, Y, U_1 \cap V_\alpha, \ldots, U_n \cap V_\alpha\rangle \models \varphi(\mathring{X}, \mathring{Y}, \mathring{U}_1, \ldots, \mathring{U}_n)].$$

Clearly, if κ is Π_2^1-indescribable, it must be Π_1^1-indescribable, i.e. weakly compact. The converse is not true. Indeed, we have:

2.2 Theorem. *If κ is Π_2^1-indescribable, then the set*

$$\{\lambda \in \kappa \mid \lambda \text{ is weakly compact}\}$$

is unbounded in κ.

Proof. (Sketch) There is a sentence $\varphi(\mathring{X}, \mathring{Y})$ of $\mathscr{L}(X, Y)$ such that an ordinal α is weakly compact iff

$$(\forall X \subseteq V_\alpha)(\exists Y \subseteq V_\alpha)[\langle V_\alpha, \in, X, Y\rangle \models \varphi(\mathring{X}, \mathring{Y})].$$

(Simply consider the defining property $\alpha \to (\alpha)_2^2$.) Given $\gamma < \kappa$ now, apply Π_2^1-indescribability for the structure $\langle V_\kappa, \in, X, Y, \{\gamma\}\rangle$ and the sentence $\varphi(\mathring{X}, \mathring{Y}) \wedge \exists x(x \in \mathring{U})$. □

2.3 Theorem. *If κ is ineffable, then κ is Π_2^1-indescribable.*

Proof. Let $\varphi(\mathring{X}, \mathring{Y}, \mathring{U}_1, \ldots, \mathring{U}_n)$ be a sentence of $\mathscr{L}(X, Y, U_1, \ldots, U_n)$, and let $U_1, \ldots, U_n \subseteq V$ be such that

$$(\forall X \subseteq V_\kappa)(\exists Y \subseteq V_\kappa)[\langle V_\kappa, \in, X, Y, U_1, \ldots, U_n\rangle \models \varphi].$$

Let

$$C = \{\lambda \in \kappa \mid |V_\lambda| = \lambda\}.$$

Clearly, C is a club subset of κ. We claim that for some $\lambda \in C$,

$$(\forall X \subseteq V_\lambda)(\exists Y \subseteq V_\lambda)[\langle V_\lambda, \in, X, Y, U_1 \cap V_\lambda, \ldots, U_n \cap V_\lambda\rangle \models \varphi],$$

thereby proving the theorem.

Suppose not. Then for each $\lambda \in C$ we can pick a set $X_\lambda \subseteq V_\lambda$ such that for all $Y \subseteq V_\lambda$,

$$\langle V_\lambda, \in, X, Y, U_1 \cap V_\lambda, \ldots, U_n \cap V_\lambda\rangle \models \neg\varphi.$$

Since $|V_\kappa| = \kappa$ and $|V_\lambda| = \lambda$ for all $\lambda \in C$, we may apply ineffability using 2.1 to conclude that there is a set $X \subseteq V_\kappa$ such that the set

$$A = \{\lambda \in C \mid X_\lambda = X \cap V_\lambda\}$$

is stationary in κ.

By assumption, we can find a set $Y \subseteq V_\kappa$ such that

$$\langle V_\kappa, \in, X, Y, U_1, \ldots, U_n\rangle \models \varphi.$$

Let

$$E = \{\lambda \in \kappa \mid \langle V_\lambda, \in, X \cap V_\lambda, Y \cap V_\lambda, U_1 \cap V_\lambda, \ldots, U_n \cap V_\lambda\rangle \prec \langle V_\kappa, \in, X, Y, U_1, \ldots, U_n\rangle\}.$$

Clearly, E is club in κ. Hence we can find a $\lambda \in E \cap A$. But then we have

$$\langle V_\lambda, \in, X_\lambda, Y \cap V_\lambda, U_1 \cap V_\lambda, \ldots, U_n \cap V_\lambda \rangle \models \varphi,$$

contrary to the choice of X_λ.

The theorem is proved. □

We shall show presently that if $V = L$, then ineffability is closely related to the Kurepa Hypothesis. Indeed, as we shall see, it plays the same role for Kurepa trees as does weak compactness for Souslin trees. But first it is of interest (though of no use to us here) to present the following result, which, it should be emphasised, is a theorem of ZFC.

2.4 Theorem. *If κ is an ineffable cardinal, then $\diamondsuit_\kappa$ holds.*

Proof. For each $\alpha < \kappa$, let (S_α, C_α) be, if possible, any pair of subsets of α such that C_α is club in α and $(\forall \gamma \in C_\alpha)(\gamma \cap S_\alpha \neq S_\gamma)$. In case no such pair exists, define

$S_\alpha = C_\alpha = \emptyset$. This defines $((S_\alpha, C_\alpha) \mid \alpha < \kappa)$ by recursion. We show that $(S_\alpha \mid \alpha < \kappa)$ is a $\diamondsuit_\kappa$-sequence.

Let $S \subseteq \kappa$. Suppose that the set

$$\{\alpha \in \kappa \mid S \cap \alpha = S_\alpha\}$$

were not stationary in κ. Then we could find a club set $C \subseteq \kappa$ such that

$$(\forall \alpha \in C)(S \cap \alpha \neq S_\alpha).$$

By the ineffability of κ and 2.1, together with some simple coding device, we can find sets $\bar{S}, \bar{C} \subseteq \kappa$ such that the set

$$A = \{\alpha \in \kappa \mid \bar{S} \cap \alpha = S_\alpha \wedge \bar{C} \cap \alpha = C_\alpha\}$$

is stationary in κ. Pick $\alpha, \beta \in A \cap C$, $\alpha < \beta$. Then

$$(*) \qquad S_\beta \cap \alpha = \bar{S} \cap \alpha = S_\alpha, \quad \text{and}$$

$$(**) \qquad C_\beta \cap \alpha = \bar{C} \cap \alpha = C_\alpha.$$

Since C_α is club in α is club in β, using $(**)$ we have

$$\alpha = \sup(C_\alpha) = \sup(\alpha \cap C_\beta) \in C_\beta.$$

Thus by choice of (S_β, C_β) we must have $\alpha \cap S_\beta \neq S_\alpha$. But this contradicts $(*)$. Thus the set $\{\alpha \in \kappa \mid S \cap \alpha = S_\alpha\}$ is stationary in κ, and the theorem is proved. $\square$

We turn now to the study of ineffable cardinals in L. As was the case with weakly compact cardinals (V.1.5), we can prove that ineffability relativises to L.

2.5 Lemma. *If κ is ineffable, then $[\kappa$ is ineffable$]^L$.*

Proof. We make use of 2.1. In L, let $(A_\alpha \mid \alpha < \kappa)$ be such that $A_\alpha \subseteq \alpha$ for all $\alpha < \kappa$. By absoluteness, this set is such a sequence in V, so by ineffability using 2.1, there is a set $A \subseteq \kappa$ such that

$$X = \{\alpha \in \kappa \mid A_\alpha = A \cap \alpha\}$$

is stationary in κ. Now, for each $\alpha \in X$, $A \cap \alpha = A_\alpha \in L$. Hence as X is cofinal in κ, $A \cap \gamma \in L$ for all $\gamma < \kappa$. But κ is weakly compact, so by V.1.4 this implies that $A \in L$. Hence $X \in L$ as well. But, by absoluteness, in L, X is stationary and $X = \{\alpha \in \kappa \mid A_\alpha = A \cap \alpha\}$. Thus by 2.1 applied inside L, we conclude that κ is ineffable in the sense of L. $\square$

We shall prove that if $V = L$, an inaccessible cardinal κ will be ineffable iff there is no κ-Kurepa tree. But what exactly do we mean by a "κ-Kurepa tree" for inaccessible κ? For if κ is inaccessible, the κ-tree consisting of all binary sequences of lengths less than κ, ordered by inclusion, has 2^κ many κ-branches, and we surely do not want such a trivial example to be a "Kurepa tree". The only reason this tree is a κ-tree at all is because the inaccessibility of κ keeps the cardinality of each level less than κ. A more interesting notion is supplied by the following considerations.

A κ-tree **T** is said to be *slim* if $|T_\alpha| \leqslant |\alpha|$ for all infinite α. By a *κ-Kurepa tree* we shall mean a slim κ-tree with at least κ^+ many κ-branches. In the case where κ is a successor cardinal, this is at variance with the definition of IV.1, but the distinction is clearly unimportant in this case, as it is the *cofinal* behaviour of trees that is of interest to us. Let us agree to adopt the new definition for all κ from now on. Likewise for the definition of a "κ-Kurepa family", given below.

The restriction that our trees be slim could also be applied to the notion of a κ-Souslin tree. In fact it is easily seen that the κ-Souslin trees constructed (in L) in 1.3 and in IV.2.4 are slim. Consequently there would have been no loss if we had required *all* of our κ-trees to be slim.

By a *κ-Kurepa family* we shall mean a family, $\mathscr{F}$, of subsets of κ such that $|\mathscr{F}| \geqslant \kappa^+$ but for all infinite $\alpha < \kappa$, $|\{x \cap \alpha \mid x \in \mathscr{F}\}| \leqslant |\alpha|$. The same argument as in III.2.1 shows that the existence of a (slim) κ-Kurepa tree is equivalent to the existence of a κ-Kurepa family.

The following result is a theorem of ZFC.

2.6 Theorem. *If κ is ineffable, then there is no κ-Kurepa tree.*

Proof. Let $\mathscr{F} \subseteq \mathscr{P}(\kappa)$ be such that $|\{x \cap \alpha \mid x \in \mathscr{F}\}| \leqslant |\alpha|$ for all infinite $\alpha < \kappa$. Assuming that κ is ineffable, we show that $|\mathscr{F}| \leqslant \kappa$, so that $\mathscr{F}$ cannot be a κ-Kurepa family.

For each $\alpha \geqslant \omega$, let $(f_\nu^\alpha \mid \nu < \alpha)$ enumerate $\{x \cap \alpha \mid x \in \mathscr{F}\}$. Set

$$R_\alpha = \{(\tau, \nu) \mid \tau \in f_\nu^\alpha\}.$$

Thus $R_\alpha \subseteq \alpha \times \alpha$. By ineffability (using 2.1 and a simple coding device) there is a set $R \subseteq \kappa \times \kappa$ such that the set

$$E = \{\alpha \in \kappa \mid R \cap (\alpha \times \alpha) = R_\alpha\}$$

is stationary in κ. For each $\nu < \kappa$, set $f_\nu = R''\{\nu\}$. We shall prove that $\mathscr{F} \subseteq \{f_\nu \mid \nu < \kappa\}$.

Let $f \in \mathscr{F}$, and suppose that $f \neq f_\nu$ for all $\nu < \kappa$. Since κ is regular we can find a club set $C \subseteq \kappa$ such that

$$\alpha \in C \rightarrow (\forall \nu < \alpha)(f \cap \alpha \neq f_\nu \cap \alpha).$$

Pick $\alpha \in C \cap E$. Then for $\nu < \alpha$,

$$f \cap \alpha \neq f_\nu \cap \alpha = \alpha \cap R''\{\nu\} = \alpha \cap R_\alpha''\{\nu\} = \{\tau \mid \tau \in f_\nu^\alpha\} = f_\nu^\alpha.$$

Hence

$$f \cap \alpha \notin \{f_\nu^\alpha \mid \nu < \alpha\} = \{x \cap \alpha \mid x \in \mathscr{F}\}.$$

Since $f \in \mathscr{F}$, this is absurd. This contradiction proves our result. □

Using $V = L$, we now prove the converse to the above theorem.

2.7 Theorem. *Assume $V = L$. Let κ be an uncountable regular cardinal which is not ineffable. Then there is a κ-Kurepa tree.*

Proof. The proof is very similar to that of IV.3.3 (the construction of a κ^+-Kurepa tree). As there, it is more convenient to construct a κ-Kurepa family.

By 2.1, let $(A_\alpha \mid \alpha < \kappa)$ be the $<_J$-least sequence such that $A_\alpha \subseteq \alpha$ for all α, and whenever $A \subseteq \kappa$, the set $\{\alpha \in \kappa \mid A \cap \alpha = A_\alpha\}$ is not stationary in κ. Notice that $(A_\alpha \mid \alpha < \kappa)$ is a definable element of J_{κ^+}.

For each $\alpha < \kappa$, let M_α be the smallest $M \prec J_\kappa$ such that $(\alpha + 1) \cup \{(A_\nu \mid \nu \leqslant \alpha)\} \subseteq M$, and let $\sigma_\alpha \colon M_\alpha \cong J_{f(\alpha)}$. Notice that for infinite α, $|f(\alpha)| = |\alpha|$. It is clear that the function $f \colon \kappa \to \kappa$ so defined is a definable element of J_{κ^+}.

Let

$$\mathscr{F} = \{x \subseteq \kappa \mid (\forall \alpha < \kappa)(x \cap \alpha \in J_{f(\alpha)})\}.$$

If we can show that $|\mathscr{F}| \geqslant \kappa^+$, then $\mathscr{F}$ will be a κ-Kurepa family, and we shall be done. We assume $|\mathscr{F}| \leqslant \kappa$ and derive a contradiction.

Let $X = (x_\nu \mid \nu < \kappa)$ be the $<_J$-least enumeration of $\mathscr{F}$. Notice that both $\mathscr{F}$ are X are definable elements of J_{κ^+}.

By recursion, define submodels $N_\nu \prec J_{\kappa^+}$, for $\nu < \kappa$, as follows.

$$N_0 = \text{the smallest } N \prec J_{\kappa^+} \text{ such that } N \cap \kappa \in \kappa;$$

$$N_{\nu+1} = \text{the smallest } N \prec J_{\kappa^+} \text{ such that } N_\nu \cup \{N_\nu\} \subseteq N \text{ and } N \cap \kappa \in \kappa;$$

$$N_\delta = \bigcup_{\nu < \delta} N_\nu, \quad \text{if } \lim(\delta).$$

Set

$$\alpha_\nu = N_\nu \cap \kappa.$$

Then $(\alpha_\nu \mid \nu < \kappa)$ is a normal sequence in κ. Set

$$x = \{\alpha_\nu \mid \nu < \kappa \wedge \alpha_\nu \notin x_\nu\}.$$

Then $x \subseteq \kappa$ and $x \neq x_\nu$ for all $\nu < \kappa$, so $x \notin \mathscr{F}$. We obtain our contradiction by showing that $x \cap \alpha \in J_{f(\alpha)}$ for all $\alpha < \kappa$. We argue much as in IV.3.3.

Let $\alpha < \kappa$ be given. Let η be the largest limit ordinal such that $\alpha_\eta \leqslant \alpha$. Since $x \cap \alpha$ differs from $x \cap \alpha_\eta$ by at most a finite set, in order to show that $x \cap \alpha \in J_{f(\alpha)}$ it suffices to show that $x \cap \alpha_\eta \in J_{f(\alpha_\eta)}$. (The function f is clearly non-decreasing.)

Since $x \cap \alpha_\eta = \{\alpha_\nu \mid \nu < \eta \wedge \alpha_\nu \notin x_\nu\}$, it is in fact enough to show that $(\alpha_\nu \mid \nu < \eta)$ and $(x_\nu \cap \alpha_\eta \mid \nu < \eta)$ are elements of $J_{f(\alpha_\eta)}$.

Let $\pi \colon N_\eta \cong J_\beta$. Then $\pi \restriction \alpha_\eta = \text{id} \restriction \alpha_\eta$, $\pi(\kappa) = \alpha_\eta$, and $\pi(x) = (x_\nu \cap \alpha_\eta \mid \nu < \alpha_\eta)$. In particular, $(x_\nu \cap \alpha_\eta \mid \nu < \eta) \in J_\beta$. And by an argument just as in IV.3.3, we see that $(\alpha_\nu \mid \nu < \eta)$ is ZF^--definable from J_β. It thus suffices to show that $\beta < f(\alpha_\eta)$.

Suppose, on the contrary, that $f(\alpha_\eta) \leqslant \beta$. Since $\alpha_\eta + 1 \subseteq M_{\alpha_\eta}$, we have $\sigma_{\alpha_\eta}((A_\nu \mid \nu \leqslant \alpha_\eta)) = (A_\nu \mid \nu \leqslant \alpha_\eta)$, so $(A_\nu \mid \nu \leqslant \alpha_\eta) \in J_{f(\alpha_\eta)} \subseteq J_\beta$. Let

$$E = \{\gamma \in \alpha_\eta \mid A_\gamma = \gamma \cap A_{\alpha_\eta}\}.$$

Then $E \in J_\beta$. Suppose that

$$\models_{J_\beta} \text{“}E \text{ is stationary in } \alpha_\eta\text{”}.$$

Setting $\tilde{E} = \pi^{-1}(E)$ and applying $\pi^{-1}: J_\beta \prec J_{\kappa^+}$, we get

$$\models_{J_{\kappa^+}} \text{“}\tilde{E} \text{ is stationary in } \kappa\text{”}.$$

Hence $\tilde{E}$ really is stationary in κ (by absoluteness). But $(A_\nu \mid \nu < \kappa) \in N_\eta$ (by definability), so $\pi^{-1}((A_\nu \mid \nu < \alpha_\eta)) = (A_\nu \mid \nu < \kappa)$. Hence, setting $\tilde{A} = \pi^{-1}(A_{\alpha_\eta})$, we have $A \subseteq \kappa$ and

$$\models_{J_{\kappa^+}} \text{“}\tilde{E} = \{\gamma \in \kappa \mid A_\gamma = \gamma \cap \tilde{A}\}\text{”}.$$

This is contrary to the choice of $(A_\gamma \mid \gamma < \kappa)$, because the above sentence is absolute. Hence,

$$\models_{J_\beta} \text{“}E \text{ is not stationary in } \alpha_\eta\text{”}.$$

Thus for some $C \in J_\beta$ we have

$$\models_{J_\beta} \text{“}C \text{ is a club subset of } \alpha_\eta \text{ and } (\forall \gamma \in C)(A_\gamma \neq \gamma \cap A_{\alpha_\eta})\text{”}.$$

Setting $\tilde{C} = \pi^{-1}(C)$ we get, applying $\pi^{-1}: J_\beta \prec J_{\kappa^+}$,

$$\models_{J_{\kappa^+}} \text{“}\tilde{C} \text{ is a club subset of } \kappa \text{ and } (\forall \gamma \in \tilde{C})(A_\gamma \neq \gamma \cap \tilde{A})\text{”}.$$

Since $\pi^{-1} \restriction \alpha_\eta = \mathrm{id} \restriction \alpha_\eta$, we have $\tilde{C} \cap \alpha_\eta = C$. Hence as C is unbounded in α_η and $\tilde{C}$ is closed in κ (by absoluteness), $\alpha_\eta \in \tilde{C}$. Thus $A_{\alpha_\eta} \neq \alpha_\eta \cap \tilde{A}$. But $\tilde{A} = \pi^{-1}(A_{\alpha_\eta})$, so in fact we do have $\tilde{A} \cap \alpha_\eta = A_{\alpha_\eta}$, because $\pi^{-1} \restriction \alpha_\eta = \mathrm{id} \restriction \alpha_\eta$. Contradiction! The proof is complete. □

3. Generalised Kurepa Families and the Principles $\diamondsuit^+_{\kappa,\lambda}$

The following natural generalisation of the notion of a κ-Kurepa family was put forward by C. C. Chang. Let κ, λ denote uncountable cardinals, with κ regular[8]

8 The principle $\mathrm{KH}(\kappa,\kappa)$ is of some interest in the case where κ is singular. This is considered in Exercise 3.

and $\lambda \leqslant \kappa$. We define

$$\mathscr{P}_\lambda(\kappa) = \{x \subseteq \kappa \mid \omega \leqslant |x| < \lambda\}.$$

The (κ, λ)-*Kurepa Hypothesis*, $\mathrm{KH}(\kappa, \lambda)$, is the assertion that there is a family $\mathscr{F} \subseteq \mathscr{P}(\kappa)$ such that $|\mathscr{F}| \geqslant \kappa^+$ and for every $x \subseteq \mathscr{P}_\lambda(\kappa)$,

$$|\{f \cap x \mid f \in \mathscr{F}\}| \leqslant |x|.$$

Clearly, $\mathrm{KH}(\kappa, \kappa)$ implies the existence of a κ-Kurepa family. Hence by 2.6, we have

3.1 Theorem. *If κ is ineffable, then* $\mathrm{KH}(\kappa, \kappa)$ fails. $\square$

We shall prove that if $V = L$, the converse to 3.1 holds, a result which strengthens 2.7. We shall also prove that $V = L$ implies that $\mathrm{KH}(\kappa, \lambda)$ holds for all uncountable regular κ and all uncountable $\lambda < \kappa$. We do this by introducing a two cardinal version of the combinatorial principle $\diamondsuit_\kappa^+$.

We assume throughout that κ, λ are as stated at the outset of this section.

$\diamondsuit_{\kappa,\lambda}^+$ asserts the existence of a function $(S_x \mid x \in \mathscr{P}_\lambda(\kappa))$ such that:

(i) $S_x \subseteq \mathscr{P}(\bigcup x)$;

(ii) $|S_x| \leqslant |x|$;

(iii) if $X \subseteq \kappa$, then there is an unbounded set $B \subseteq \kappa$ with the property that whenever $x \in \mathscr{P}_\lambda(\kappa)$ has no largest element and is such that $B \cap x$ is cofinal in x, then $X \cap \alpha, B \cap \alpha \in S_x$, where $\alpha = \bigcup x$.

3.2 Theorem. $\diamondsuit_{\kappa,\lambda}^+$ *implies* $\mathrm{KH}(\kappa, \lambda)$.

Proof. Recall that H_κ is a model of ZF^-. Fix some set of skolem functions for H_κ. Let $(S_x \mid x \in \mathscr{P}_\lambda(\kappa))$ satisfy $\diamondsuit_{\kappa,\lambda}^+$, and for each $x \in \mathscr{P}_\lambda(\kappa)$, let M_x be the smallest (with respect to the chosen skolem functions) $M \prec H_\kappa$ such that $x \cup \{x\} \subseteq M$ and $(\forall \alpha \leqslant \bigcup x)(S_{x \cap \alpha} \subseteq M)$. Notice that $|M_x| = |x|$. Set

$$\mathscr{F} = \{f \subseteq \kappa \mid (\forall x \in \mathscr{P}_\lambda(\kappa))(f \cap x \in M_x)\}.$$

In order to prove $\mathrm{KH}(\kappa, \lambda)$, it clearly suffices to show that $|\mathscr{F}| \geqslant \kappa^+$, since in that case $\mathscr{F}$ will satisfy $\mathrm{KH}(\kappa, \lambda)$. We shall assume that $|\mathscr{F}| \leqslant \kappa$ and derive a contradiction. Notice that $\kappa \in \mathscr{F}$, so $\mathscr{F} \neq \emptyset$.

Let $(f_\nu \mid \nu < \kappa)$ enumerate all unbounded members of $\mathscr{F}$. (This enumeration need not be one-one.) For each $\nu < \kappa$, let C_ν be the set of all limit points of f_ν. Let X be the diagonal intersection of the sequence $(C_\nu \mid \nu < \kappa)$, i.e.

$$X = \{\alpha \in \kappa \mid (\forall \nu < \alpha)(\alpha \in C_\nu)\}.$$

Each set C_ν is club in κ, so X is club in κ. For each $\alpha \in X$, α is a limit ordinal and for any $\nu < \alpha$, $f_\nu \cap \alpha$ is unbounded in α.

By $\diamondsuit^+_{\kappa,\lambda}$, let $B \subseteq \kappa$ be unbounded and such that whenever $x \in \mathscr{P}_\lambda(\kappa)$ is such that $\alpha = \bigcup x$ is a limit point of $B \cap x$, then $X \cap \alpha, B \cap \alpha \in S_x$. Let $(\alpha_\nu \mid \nu < \kappa)$ be the monotone enumeration of the set

$$\{\alpha \in X \mid \alpha \text{ is a limit point of } B\}.$$

For $\nu < \kappa$, set

$$\beta_\nu = \min(B - \alpha_\nu).$$

Notice that

$$\alpha_\nu \leqslant \beta_\nu < \alpha_{\nu+1}.$$

Set

$$f = \{\beta_\nu \mid \nu < \kappa\}.$$

Then f in an unbounded subset of κ. Since $f_\nu \cap \alpha_{\nu+1}$ is unbounded in $\alpha_{\nu+1}$, but $f \cap \alpha_{\nu+1} \subseteq \beta_\nu + 1 < \alpha_{\nu+1}$ for each $\nu < \kappa$, we have $f \neq f_\nu$ for all $\nu < \kappa$. We obtain our contradiction by showing that $f \in \mathscr{F}$.

Let $x \in \mathscr{P}_\lambda(\kappa)$. We prove that $f \cap x \in M_x$. Let β be the greatest limit point of $f \cap x$. Then

$$f \cap x = (f \cap x \cap \beta) \cup (f \cap x - \beta),$$

where $f \cap x - \beta$ is finite. Being a finite subset of x, $f \cap x - \beta$ must be an element of M_x, since $x \subseteq M_x$ and $M_x \models \mathrm{ZF}^-$. So in order to show that $f \cap x \in M_x$, it suffices to show that $f \cap x \cap \beta \in M_x$.

Now, β is a limit point of $f \cap x$. But $f \subseteq B$. Thus β is a limit point of $B \cap (x \cap \beta)$. Hence $X \cap \beta$, $B \cap \beta \in S_{x \cap \beta} \subseteq M_x$. But clearly, $f \cap \beta$ is ZF^--definable from $X \cap \beta$, $B \cap \beta$ in exactly the same way that f was defined from X and B. Hence $f \cap \beta \in M_x$. Thus $f \cap \beta \cap x \in M_x$, and we are done. □

3.3 Theorem. *Assume $V = L$. If $\lambda < \kappa$, then $\diamondsuit^+_{\kappa,\lambda}$ is valid.*

Proof. For each $x \in \mathscr{P}_\lambda(\kappa)$, let M_x be the smallest $M \prec J_\kappa$ such that $x \cup \{x\} \cup \{\lambda\} \subset M$, and set $S_x = \mathscr{P}(\bigcup x) \cap M_x$. We prove that $(S_x \mid x \in \mathscr{P}_\lambda(\kappa))$ satisfies $\diamondsuit^+_{\kappa,\lambda}$. M_x S_x

Suppose otherwise, and let $X \subseteq \kappa$ be the $<_J$-least set such that there is no unbounded set $B \subseteq \kappa$ as in $\diamondsuit^+_{\kappa,\lambda}$. Note that both $(S_x \mid x \in \mathscr{P}_\lambda(\kappa))$ and X are definable from λ in J_{κ^+}. X

By recursion on $\nu < \kappa$, define a chain of submodels

$$N_0 \prec N_1 \prec \ldots \prec N_\nu \prec \ldots \prec J_{\kappa^+}$$ N_ν

as follows.

$$N_0 = \text{the smallest } N \prec J_{\kappa^+} \text{ such that } \lambda \in N \cap \kappa \in \kappa;$$

$$N_{\nu+1} = \text{the smallest } N \prec J_{\kappa^+} \text{ such that } N_\nu \cup \{N_\nu\} \subseteq N \text{ and } N \cap \kappa \in \kappa;$$

$$N_\delta = \bigcup_{\nu<\delta} N_\nu, \quad \text{if } \lim(\delta).$$

It is easily seen that this causes no difficulties. In particular, $|N_\nu| < \kappa$ for all $\nu < \kappa$. Moreover

$$\nu < \tau < \kappa \to N_\nu \prec N_\tau \prec J_{\kappa^+}.$$

For each $\nu < \kappa$, set

α_ν $$\alpha_\nu = N_\nu \cap \kappa.$$

Clearly, $(\alpha_\nu \mid \nu < \kappa)$ is a normal sequence in κ.

For each $\nu < \kappa$, let

$\sigma_\nu, \beta(\nu)$ $$\sigma_\nu : N_\nu \cong J_{\beta(\nu)}.$$

Clearly,

$$\sigma_\nu \restriction \alpha_\nu = \mathrm{id} \restriction \alpha_\nu, \qquad \sigma_\nu(\kappa) = \alpha_\nu, \qquad \sigma_\nu(X) = X \cap \alpha_\nu.$$

Set

B $$B = \{\beta(\nu) \mid \nu < \kappa\}.$$

B is an unbounded subset of κ. We shall obtain the desired contradiction by showing that B satisfies the requirements of $\diamondsuit^+_{\kappa,\lambda}$ for X.

x Fix x an arbitrary element of $\mathscr{P}_\lambda(\kappa)$ such that $\alpha = \bigcup x$ is a limit point of $B \cap x$. We shall show that $X \cap \alpha, B \cap \alpha \in M_x$, thereby completing the proof.

For each $\nu < \kappa$, we have $N_\nu \in N_{\nu+1} \prec J_{\kappa^+}$, and hence $\sigma_\nu, \beta(\nu) \in N_{\nu+1}$. But $|N_\nu| < \kappa$. Thus $\beta(\nu) \in N_{\nu+1} \cap \kappa = \alpha_{\nu+1}$. Also, since $\sigma_{\nu+1}(\kappa) = \alpha_{\nu+1}$ we have $\alpha_{\nu+1} < \beta(\nu+1)$. Thus for all $\nu < \kappa$ we have

(1) $$\beta(\nu) < \alpha_{\nu+1} < \beta(\nu+1).$$

But α is a limit point of $B = \{\beta(\nu) \mid \nu < \kappa\}$. Thus we must have $\alpha = \alpha_\eta$ for some

η limit ordinal $\eta < \kappa$.

Now, as we remarked earlier, X is J_{κ^+}-definable from λ. But

$$\sigma_\eta^{-1} : J_{\beta(\eta)} \prec J_{\kappa^+}, \qquad \sigma_\eta^{-1}(\lambda) = \lambda, \qquad \sigma_\eta^{-1}(X \cap \alpha_\eta) = X.$$

Thus $X \cap \alpha_\eta$ is $J_{\beta(\eta)}$-definable from λ.

Similarly, $B \cap \alpha_\eta$ is ZF$^-$-definable from $J_{\beta(\eta)}$ and λ in exactly the same way that B was defined from J_{κ^+} and λ. (This uses the fact that $\sigma_\eta^{-1} \restriction \alpha_\eta = \mathrm{id} \restriction \alpha_\eta$.)

Since $\lambda \in M_x$ and $M_x \models \mathrm{ZF}^-$, it follows that in order to prove that $X \cap \alpha$, $B \cap \alpha \in M_x$ it is sufficient to show that $\beta(\eta) \in M_x$. This will take some time, and requires some considerable extra machinery before we can even motivate the argument.

To avoid confusion between ordinals and sequences of ordinals, from now on

$\vec{\beta}, \vec{\beta} \restriction \tau$ we shall use $\vec{\beta}$ to denote $(\beta(\nu) \mid \nu < \kappa)$, and for any $\tau < \kappa$ we shall write $\vec{\beta} \restriction \tau$ for $(\beta(\nu) \mid \nu < \tau)$.

For $\nu < \mu < \kappa$, set

$\sigma_{\nu\mu}$ $$\sigma_{\nu\mu} = \sigma_\mu \circ \sigma_\nu^{-1}.$$

Thus

$$\sigma_{\nu\mu}: J_{\beta(\nu)} \prec J_{\beta(\mu)}.$$

Note that $\langle (J_{\beta(\nu)})_{\nu<\kappa}, (\sigma_{\nu\mu})_{\nu<\mu<\kappa}\rangle$ is a directed elementary system. We write $\vec{\sigma}$ for $(\sigma_{\nu\mu} \mid \nu < \mu < \kappa)$, and for any $\tau < \kappa$ we write $\vec{\sigma} \upharpoonright \tau$ for $(\sigma_{\nu\mu} \mid \nu < \mu < \tau)$. $\vec{\sigma}$ $\vec{\sigma} \upharpoonright \tau$

The following result is central to our entire argument.

(2) If $\gamma \in \eta \cap M_x$, then $\alpha_\gamma, \beta(\gamma), \vec{\beta} \upharpoonright (\gamma + 1), \vec{\sigma} \upharpoonright (\gamma + 1) \in M_x$.

To prove (2), let $\gamma \in \eta \cap M_x$. Since $\alpha = \alpha_\eta$ is a limit point of $B \cap x$ we can find a $\tau < \eta$ such that $\tau > \gamma$ and $\beta(\tau) \in x \subseteq M_x$. γ

Define a sequence $(N'_\nu \mid \nu < \theta)$, for some θ, as follows.

$$\begin{aligned} N'_0 &= \text{the smallest } N \prec J_{\beta(\tau)} \text{ such that } \lambda \in N \cap \alpha_\tau \in \alpha_\tau; \\ N'_{\nu+1} &= \text{the smallest } N \prec J_{\beta(\tau)} \text{ such that } N'_\nu \cup \{N'_\nu\} \subseteq N \text{ and } N \cap \alpha_\tau \in \alpha_\tau; \\ N'_\delta &= \bigcup_{\nu<\delta} N'_\nu, \quad \text{if } \lim(\delta). \end{aligned}$$

The definition will break down at some stage θ when $\sup(\bigcup_{\nu<\theta} N'_\nu \cap \alpha_\tau) = \alpha_\tau$.

We have

$$(*) \qquad \sigma_\tau^{-1}: J_{\beta(\tau)} \prec J_{\kappa^+}, \quad \sigma_\tau^{-1}(\lambda) = \lambda, \quad \sigma_\tau^{-1}(\alpha_\tau) = \kappa, \quad \sigma_\tau^{-1} \upharpoonright \alpha_\tau = \mathrm{id} \upharpoonright \alpha_\tau.$$

So by induction on ν we see that

$$\nu < \tau \rightarrow N'_\nu \text{ is defined and } \sigma_\tau^{-1}{}'' N'_\nu = N_\nu.$$

It follows that $\theta = \tau$, of course, since $\sup_{\nu<\tau}(N'_\nu \cap \alpha_\tau) = \sup_{\nu<\tau}\alpha_\nu = \alpha_\tau$. For each $\nu < \tau$, let

$$\sigma'_\nu: N'_\nu \cong J_{\beta'(\nu)}.$$

Since $N'_\nu \cong N_\nu$ (by $\sigma_\tau^{-1} \upharpoonright N'_\nu$). We have $\beta'(\nu) = \beta(\nu)$ for all $\nu < \tau$. Thus $(\beta'(\nu) \mid \nu < \tau) = \vec{\beta} \upharpoonright \tau$. This shows that $\vec{\beta} \upharpoonright \tau$ is ZF$^-$-definable from $\beta(\tau), \lambda, \alpha_\tau$. Now, $\beta(\tau), \lambda \in M_x$. And by $(*)$ above,

$$\alpha_\tau = [\text{the largest cardinal}]^{J_{\beta(\tau)}},$$

so $\alpha_\tau \in M_x$ as well. Thus $\vec{\beta} \upharpoonright \tau \in M_x$. Since $\gamma \in M_x$ and $\gamma < \tau$ it follows that $\beta(\gamma) = (\vec{\beta} \upharpoonright \tau)(\gamma) \in M_x$ and $\vec{\beta} \upharpoonright (\gamma + 1) = (\vec{\beta} \upharpoonright \tau) \upharpoonright (\gamma + 1) \in M_x$. Also, $\alpha_\gamma =$ [the largest cardinal]$^{J_{\beta(\gamma)}} \in M_x$. It remains to prove that $\vec{\sigma} \upharpoonright (\gamma + 1) \in M_x$.

Now, in the definition of $(N_\nu \mid \nu < \kappa)$, if we replace J_{κ^+} by N_μ and κ by α_μ, we will obtain the sequence $(N_\nu \mid \nu < \mu)$, as is easily seen. So, as $\sigma_\mu: N_\mu \cong J_{\beta(\mu)}$ and $\sigma_\mu \upharpoonright \alpha_\mu = \mathrm{id} \upharpoonright \alpha_\mu$, the same definition with parameters $J_{\beta(\mu)}$ and α_μ will produce the sequence $(\sigma''_\mu N_\nu \mid \nu < \mu)$. But it is easily seen that $\sigma_{\nu\mu}^{-1}$ is the collapsing isomorphism for $\sigma''_\mu N_\nu$. Since $\alpha_\mu =$ [the largest cardinal]$^{J_{\beta(\mu)}}$ for all μ, this shows that $(\sigma_{\nu\mu} \mid \nu < \mu < \tau)$ is ZF$^-$-definable from $\vec{\beta} \upharpoonright \tau$. But $\vec{\beta} \upharpoonright \tau \in M_x$. Thus $\vec{\sigma} \upharpoonright \tau \in M_x$ and it follows at once that $\vec{\sigma} \upharpoonright (\gamma + 1) \in M_x$. So (2) is proved.

Two further results follow easily from the above.

(3) For $\tau < \eta$, $\tau \in M_x$ iff $\beta(\tau) \in M_x$.

By (2), if $\tau \in M_x$, then $\beta(\tau) \in M_x$. To prove the converse, assume that $\beta(\tau) \in M_x$. Thus $\alpha_\tau = [\text{the largest cardinal}]^{J_{\beta(\tau)}} \in M_x$. Now, from λ, $\beta(\tau)$, α_τ we may define the sequence $(N'_\nu \mid \nu < \theta)$ as in the proof of (2) above. As we observed then, we must have $\theta = \tau$. So this defines τ from λ, $\beta(\tau)$, α_τ in a ZF^- fashion. So as $\lambda, \beta(\tau), \alpha_\tau \in M_x$, we conclude that $\tau \in M_x$, and (3) is proved.

(4) $\sup(\eta \cap M_x) = \eta$.

Clearly, $\sup(\eta \cap M_x) \leqslant \eta$. To prove the opposite inequality, let $\nu < \eta$. Then $\alpha_\nu < \alpha_\eta$, so as $\alpha_\eta = \alpha = \sup(B \cap x \cap \alpha)$, we can find a $\tau < \eta$ such that $\alpha_\nu < \beta(\tau) \in x \subseteq M_x$. By (1), $\nu < \tau$. By (3), $\tau \in M_x$. So $\nu \leqslant \sup(\eta \cap M_x)$, and (4) follows at once.

Now let

π, δ $$\pi\colon M_x \cong J_\delta$$

and set

η^* $$\eta^* = \pi''(\eta \cap M_x).$$

By virtue of (2) we may define

$\vec{\beta}^*$ $$\vec{\beta}^* = \bigcup_{\gamma \in \eta \cap M_x} \pi(\vec{\beta} \upharpoonright (\gamma + 1)),$$

$\vec{\sigma}^*$ $$\vec{\sigma}^* = \bigcup_{\gamma \in \eta \cap M_x} \pi(\vec{\sigma} \upharpoonright (\gamma + 1)).$$

Since π is a collapsing isomorphism, the following are easily checked:

(5) η^* is an ordinal.

$\beta^*(\nu)$ (6) $\vec{\beta}^*$ is an η^*-sequence of ordinals, say $\vec{\beta}^* = (\beta^*(\nu) \mid \nu < \eta^*)$.

$\sigma^*_{\nu\mu}$ (7) $\vec{\sigma}^*$ is a system of maps of the form $\vec{\sigma}^* = (\sigma^*_{\nu\mu} \mid \nu < \mu < \eta^*)$.

(8) $\beta^*(\nu) = \pi(\beta(\pi^{-1}(\nu)))$ for all $\nu < \eta^*$.

(9) $\sigma^*_{\nu\mu} = \pi(\sigma_{\pi^{-1}(\nu), \pi^{-1}(\mu)})$ for all $\nu < \mu < \eta^*$.

We know that $\langle (J_{\beta(\nu)})_{\nu<\eta}, (\sigma_{\nu\mu})_{\nu<\mu<\eta} \rangle$ is a directed elementary system with direct limit $\langle J_{\beta(\eta)}, (\sigma_{\nu\eta})_{\nu<\eta} \rangle$. Using (8) and (9) it is easily checked that $\langle (J_{\beta^*(\nu)})_{\nu<\eta^*}, (\sigma^*_{\nu\mu})_{\nu<\mu<\eta^*} \rangle$ is a directed elementary system. Let $\langle \langle U, E \rangle, (\sigma^*_\nu)_{\nu<\eta^*} \rangle$ be a direct limit of this system. We may define an embedding

h $$h\colon \langle U, E \rangle \prec \langle J_{\beta(\eta)}, \in \rangle$$

by letting ν range over η^* in the following commutative diagram:

$$\begin{array}{ccc} J_{\beta(\pi^{-1}(\nu))} & \xrightarrow{\sigma_{\pi^{-1}(\nu),\eta}} & J_{\beta(\eta)} \\ \uparrow{\scriptstyle \pi^{-1}} & & \uparrow{\scriptstyle h} \\ J_{\beta^*(\nu)} & \xrightarrow{\sigma^*_\nu} & \langle U, E\rangle \end{array}$$

Thus $\langle U, E\rangle$ is well-founded, and we may take $\langle U, E\rangle$ to be of the form $\langle J_{\beta^*}, \in\rangle$ for some unique ordinal β^*. β^*

If $\nu < \eta^*$, then

$$(\sigma_{\pi^{-1}(\nu)})^{-1} \circ \pi^{-1}: J_{\beta^*(\nu)} \prec J_{\kappa^+},$$

so there is an $\alpha^*_\nu < \beta^*(\nu)$ such that

$$\alpha^*_\nu = [\text{the largest cardinal}]^{J_{\beta^*(\nu)}}.$$

Also,

$$\sigma_\eta^{-1} \circ h: J_{\beta^*} \prec J_{\kappa^+},$$

so there is an $\alpha^* < \beta^*$ such that

$$\alpha^* = [\text{the largest cardinal}]^{J_{\beta^*}}.$$ α^*

The following result is immediate:

(10) $\alpha^*_\nu = \pi(\alpha_{\pi^{-1}(\nu)})$ and $\sigma^*_\nu(\alpha^*_\nu) = \alpha^*$ for all $\nu < \eta^*$, and $h(\alpha^*) = \alpha_\eta$.

Moreover, as we show next:

(11) $\sigma^*_\nu \restriction \alpha^*_\nu = \mathrm{id} \restriction \alpha^*_\nu$ for all $\nu < \eta^*$.

Since $\langle J_{\beta^*}, (\sigma^*_\nu)_{\nu<\eta^*}\rangle$ is the transitive direct limit of $\langle (J_{\beta^*(\nu)})_{\nu<\eta^*}, (\sigma^*_{\nu\mu})_{\nu<\mu<\eta^*}\rangle$, it suffices to prove that $\sigma^*_{\nu\mu} \restriction \alpha^*_\nu = \mathrm{id} \restriction \alpha^*_\nu$ for all $\nu < \mu < \eta^*$. But this follows easily from (9) and the properties of the system $\vec{\sigma}$.

(12) $\alpha^* = \sup_{\nu<\eta^*} \alpha^*_\nu$.

Since $\sigma^*_\nu(\alpha^*_\nu) = \alpha^*$ for all $\nu < \eta^*$, we have $\sup_{\nu<\eta^*} \alpha^*_\nu \leqslant \alpha^*$. To prove the opposite inequality, suppose $\gamma < \alpha^*$. Pick $\nu < \eta^*$ so that $\gamma = \sigma^*_\nu(\bar{\gamma})$ for some $\bar{\gamma}$. Since $\sigma^*_\nu(\alpha^*_\nu) = \alpha^*$, we have $\bar{\gamma} < \alpha^*_\nu$. So by (11), $\gamma = \sigma^*_\nu(\bar{\gamma}) = \bar{\gamma}$. Thus $\gamma < \alpha^*_\nu$. This proves that $\alpha^* \leqslant \sup_{\nu<\eta^*} \alpha^*_\nu$, and completes the proof of (12).

We are now able to indicate the purpose of the above considerations. It is easily seen that $\vec{\beta}^*$ and $\vec{\sigma}^*$ are ZF^--definable from β^*, α^*, and λ in the same way that $\vec{\beta}$ and $\vec{\sigma}$ were defined from κ^+, κ, and λ. (See, in particular, the proof of (2) above and the definitions of $\vec{\beta}^*$, $\vec{\sigma}^*$, β^*, α^*.) Since $\alpha^* = [\text{the largest cardinal}]^{J_{\beta^*}}$, this means that $\vec{\beta}^*$ and $\vec{\sigma}^*$ are ZF$^-$-definable from β^*, λ.

Assume for the time being that $\beta^* \in J_\delta$. Since $J_\delta \models \text{Zf}^-$ (because $M_x \models \text{ZF}^-$), it follows that $\vec{\beta}^*, \vec{\sigma}^* \in J_\delta$. Thus $\pi^{-1}(\vec{\beta}^*)$ and $\pi^{-1}(\vec{\sigma}^*)$ are defined. Since $\pi\colon M_x \cong J_\delta$ is a collapsing isomorphism and $\eta^* = \pi''(\eta \cap M_x)$, it is a routine consequence of (6), (7), (8), (9) and the definition of $\vec{\beta}^*$ and $\vec{\sigma}^*$ that

$$\vec{\beta} \restriction \eta = \pi^{-1}(\vec{\beta}^*) \restriction \eta \quad \text{ and } \vec{\sigma} \restriction \eta = \pi^{-1}(\vec{\sigma}^*) \restriction \eta.$$

Suppose first that $\pi^{-1}(\vec{\beta}^*) = \vec{\beta} \restriction \eta$. Since J_{β^*} is the unique transitive limit of the system $\langle (J_{\beta^*(\nu)})_{\nu<\eta^*}, (\sigma^*_{\nu\mu})_{\nu<\mu<\eta^*} \rangle$, it follows that $J_{\pi^{-1}(\beta^*)}$ is the unique transitive limit of the system $\langle (J_{\beta(\nu)})_{\nu<\eta}, (\sigma_{\nu\mu})_{\nu<\mu<\eta} \rangle$. Thus $\pi^{-1}(\beta^*) = \beta(\eta)$. Hence $\beta(\eta) \in \operatorname{ran}(\pi^{-1}) = M_x$, and we are done.

Otherwise, $\pi^{-1}(\vec{\beta}^*)$ is a proper end-extension of $\vec{\beta} \restriction \eta$. Thus the directed elementary system determined by $\pi^{-1}(\vec{\beta}^*)$, $\pi^{-1}(\vec{\sigma}^*)$ is an end-extension of $\langle (J_{\beta(\nu)})_{\nu<\eta}, (\sigma_{\nu\mu})_{\nu<\mu<\eta} \rangle$. So $J_{\pi^{-1}(\beta^*)(\eta)}$ is the transitive direct limit of $\langle (J_{\beta(\nu)})_{\nu<\eta}, (\sigma_{\nu\mu})_{\nu<\mu<\eta} \rangle$, which means that $\pi^{-1}(\vec{\beta}^*)(\eta) = \beta(\eta)$. It follows that $\beta(\eta)$ is ZF$^-$-definable from $\pi^{-1}(\vec{\beta}^*)$ and α as the unique element γ of $\operatorname{ran}(\pi^{-1}(\vec{\beta}^*))$ such that $\alpha = [\text{the largest cardinal}]^{J_\gamma}$. (By (1), each $\beta(\nu)$ has a unique α_ν associated with it, so the same will be true for the members of $\pi^{-1}(\vec{\beta}^*)$. Since $\pi^{-1}(\vec{\beta}^*)(\eta) = \beta(\eta)$, the relevant "$\alpha_\nu$" here is $\alpha_\eta = \alpha$.) But $\alpha = \bigcup x$ and $x \in M_x$, so $\alpha \in M_x$. Also, $\pi^{-1}(\vec{\beta}^*) \in \operatorname{ran}(\pi^{-1}) = M_x$. Hence $\beta(\eta) \in M_x$, as required.

So we see that the proof boils down to showing that (as was assumed for the above discussion) $\beta^* \in J_\delta$. As a first step we prove:

(13) $\quad \mathscr{P}(\alpha_\eta) \cap M_x \nsubseteq J_{\beta(\eta)}$.

We know that $x \in \mathscr{P}(\alpha_\eta) \cap M_x$, so it suffices to show that $x \notin J_{\beta(\eta)}$. Well, we have $|x| < \lambda < \alpha_0 < \alpha_\eta$. Since λ is a cardinal, $|x|^{J_{\beta(\eta)}} < \lambda$. But $\sup(x) = \alpha_\eta$. Hence $\models_{J_{\beta(\eta)}}$ "α_η is singular". But this is a contradiction, since $\sigma_\eta^{-1}\colon J_{\beta(\eta)} \prec J_{\kappa^+}$ and $\sigma_\eta^{-1}(\alpha_\eta) = \kappa$. This proves (13). (Incidentally, this is the only point were we need the fact that $\lambda < \kappa$.)

We complete the proof by showing that if $\delta \leqslant \beta^*$, then, contrary to the above, $\mathscr{P}(\alpha_\eta) \cap M_x \subseteq J_{\beta(\eta)}$. First two results which do not require this assumption.

(14) $\quad$ If $z \in \mathscr{P}(\alpha_\eta) \cap M_x$, $\quad$ then $\pi(z) \in \mathscr{P}(\alpha^*) \cap J_\delta$.

Since $\pi(z) = \{\pi(\xi) \mid \xi \in z \cap M_x\}$, in order to prove (14) it suffices to show that if $\xi \in z \cap M_x$, then $\pi(\xi) \in \alpha^*$. Suppose $\xi \in z \cap M_x$. Then $\xi < \alpha_\eta$. Now, $\alpha_\eta = \sup_{\nu<\eta} \alpha_\nu$, so by (4) we can find a $\nu \in \eta \cap M_x$ such that $\xi < \alpha_\nu$. By (10), $\pi(\xi) < \pi(\alpha_\nu) = \alpha^*_{\pi(\nu)}$. But $\pi(\nu) < \eta^*$. So by (12), $\pi(\xi) < \alpha^*$. This proves (14).

(15) $\quad$ If $z \in J_{\beta^*}$ and z is a bounded subset of α^*, then $h(z) = \pi^{-1}(z)$.

Since z is a bounded subset of α^*, (12) tells us that we can pick $\nu < \eta^*$ sufficiently large so that $z \subseteq \alpha^*_\nu$. Since $z \in J_{\beta^*}$, we can assume that ν is chosen here so that $z = \sigma^*_\nu(\bar{z})$ for some $\bar{z} \subseteq \alpha^*_\nu$. By (11) and (10), $\sigma^*_\nu \restriction \alpha^*_\nu = \operatorname{id} \restriction \alpha^*_\nu$ and $\sigma^*_\nu(\alpha^*_\nu) = \alpha^*$, so

$z = z \cap \alpha^*_\nu = \sigma^*_\nu{}''\bar{z} = \bar{z}$. Thus $z \in J_{\beta^*(\nu)}$ and $\sigma^*_\nu(z) = z$. Since $z \in J_{\beta^*(\nu)}$, $\pi^{-1}(z)$ is defined. We have $\pi^{-1}(z) \subseteq \pi^{-1}(\alpha^*_\nu)$, so by (10), $\pi^{-1}(z) \subseteq \alpha_{\bar{\nu}}$, where $\bar{\nu} = \pi^{-1}(\nu)$. Now, $\sigma_{\bar{\nu}\eta} \upharpoonright \alpha_{\bar{\nu}} = \mathrm{id} \upharpoonright \alpha_{\bar{\nu}}$, so $\sigma_{\bar{\nu}\eta}{}''\pi^{-1}(z) = \pi^{-1}(z)$. But by choosing ν large enough below η^* we may assume that z is a bounded subset of α^*_ν, and hence that $\pi^{-1}(z)$ is a bounded subset of $\alpha_{\bar{\nu}}$. Thus $\sigma_{\bar{\nu}\eta}(\pi^{-1}(z)) = \sigma_{\bar{\nu}\eta}{}''\pi^{-1}(z) = \pi^{-1}(z)$. By definition of h now, we have

$$h(z) = \sigma_{\bar{\nu}\eta} \circ \pi^{-1} \circ \sigma^{*\,-1}_\nu(z) = \sigma_{\bar{\nu}\eta} \circ \pi^{-1}(z) = \pi^{-1}(z),$$

which proves (15).

To complete the proof of the theorem we now have:

(16) If $\delta \leqslant \beta^*$, then $\mathscr{P}(\alpha_\eta) \cap M_x \subseteq J_{\beta(\eta)}$.

Let $\bar{z} \in \mathscr{P}(\alpha_\eta) \cap M_x$. Let $z = \pi(\bar{z})$. By (14), $z \in \mathscr{P}(\alpha^*) \cap J_\delta$. Since $\delta \leqslant \beta^*$, $z \in J_{\beta^*}$. Thus $h(z) \in J_{\beta(\eta)}$. It suffices, therefore, to prove that $h(z) = \bar{z}$.

Now, $z \subseteq \alpha^*$, so, using (10), $h(z) \subseteq h(\alpha^*) = \alpha_\eta$. But by (4), $\alpha_\eta = \sup_{\nu \in \eta \cap M_x} \alpha_\nu$. Thus

$$h(z) = \bigcup_{\nu \in \eta \cap M_x} [h(z) \cap \alpha_\nu].$$

Likewise

$$\bar{z} = \bigcup_{\nu \in \eta \cap M_x} [\bar{z} \cap \alpha_\nu].$$

So we have

$$\begin{aligned}
h(z) &= \bigcup_{\nu \in \eta \cap M_x} [h(z) \cap \alpha_\nu] \\
&= \bigcup_{\nu < \eta^*} [h(z) \cap \alpha_{\pi^{-1}(\nu)}] && \text{(by definition of } \eta^*) \\
&= \bigcup_{\nu < \eta^*} [h(z) \cap \pi^{-1}(\alpha^*_\nu)] && \text{(by (10))} \\
&= \bigcup_{\nu < \eta^*} [h(z) \cap h(\alpha^*_\nu)] && \text{(by (15) applied to } \alpha^*_\nu) \\
&= \bigcup_{\nu < \eta^*} [h(z \cap \alpha^*_\nu)] && \text{(since } h \text{ is an isomorphism)} \\
&= \bigcup_{\nu < \eta^*} [\pi^{-1}(z \cap \alpha^*_\nu)] && \text{(by (15) applied to } z \cap \alpha^*_\nu) \\
&= \bigcup_{\nu < \eta^*} [\pi^{-1}(z) \cap \pi^{-1}(\alpha^*_\nu)] && \text{(since } \pi^{-1} \text{ is an isomorphism)} \\
&= \bigcup_{\nu < \eta^*} [\bar{z} \cap \alpha_{\pi^{-1}(\nu)}] && \text{(since } \pi(\bar{z}) = z \text{ and by (10), respectively)} \\
&= \bigcup_{\nu \in \eta \cap M_x} [\bar{z} \cap \alpha_\nu] && \text{(by definition of } \eta^*) \\
&= \bar{z}.
\end{aligned}$$

We are done. □

By 3.2 and 3.3, if $V = L$, then $\mathrm{KH}(\kappa, \lambda)$ is valid whenever $\lambda < \kappa$. By virtue of 3.1 and 3.2, our next result shows that if $V = L$, then $\mathrm{KH}(\kappa, \kappa)$ iff κ is not ineffable.

3.4 Theorem. *Assume $V = L$. Then $\diamondsuit^+_{\kappa,\kappa}$ holds iff κ is not ineffable.*

Proof. If κ is ineffable, then by 3.1, $\neg\, \mathrm{KH}(\kappa, \kappa)$, so by 3.2, $\neg\, \diamondsuit^+_{\kappa,\kappa}$.

Conversely, suppose κ is not ineffable. We prove $\diamondsuit^+_{\kappa,\kappa}$ by means of an argument very similar to that used in 3.3 above. Because of this similarity, we simply describe the changes that must be made to the proof in the present case. The idea is to modify the definition of the models M_x so that an analogue of (13) may be proved, since this is the only point in the proof of 3.3 where we made use of the fact that $\lambda < \kappa$. (At all other points where λ was mentioned, we may now simply omit all mention of λ, and everything proceeds as before.)

Let $(A_\alpha \mid \alpha < \kappa)$ be the $<_J$-least sequence such that $A_\alpha \subseteq \alpha$ for all $\alpha < \kappa$, but for any $A \subseteq \kappa$, the set $\{\alpha \in \kappa \mid A_\alpha = \alpha \cap A\}$ is not stationary in κ. (Such a sequence exists by 2.1.) For each $x \in \mathscr{P}_\kappa(\kappa)$, let M_x be the smallest $M \prec J_\kappa$ such that $x \cup \{x\} \cup \{A_{\cup x}\} \subseteq M$. Now define S_x, $x \in \mathscr{P}_\kappa(\kappa)$, as before, and proceed exactly as in 3.3 except for the verification of (13). At this point we argue as follows.

We wish to prove that $\mathscr{P}(\alpha_\eta) \cap M_x \nsubseteq J_{\beta(\eta)}$. We assume otherwise and derive a contradiction. By definition, we have $A_{\alpha_\eta} \in M_x$, so by our assumption, $A_{\alpha_\eta} \in J_{\beta(\eta)}$.

Now, $(A_\gamma \mid \gamma < \kappa)$ is J_{κ^+}-definable, so $(A_\gamma \mid \gamma < \kappa) \in N_0 \subseteq N_\eta$, so $(A_\gamma \mid \gamma < \eta) = \sigma_\eta((A_\gamma \mid \gamma < \kappa)) \in J_{\beta(\eta)}$. Thus

$$X = \{\gamma \in \alpha_\eta \mid A_\gamma = \gamma \cap A_{\alpha_\eta}\} \in J_{\beta(\eta)}.$$

Suppose that

$$\models_{J_{\beta(\eta)}} \text{"}X \text{ is stationary in } \alpha_\eta\text{"}.$$

Set $\bar{X} = \sigma_\eta^{-1}(X)$. Since $\sigma_\eta^{-1}: J_{\beta(\eta)} \prec J_{\kappa^+}$, we have

$$\models_{J_{\kappa^+}} \text{"}\bar{X} \text{ is stationary in } \kappa\text{"}.$$

Thus $\bar{X}$ really is stationary in κ. (Because $\mathscr{P}(\kappa) \subseteq J_{\kappa^+}$.)

Again, by absoluteness,

$$\models_{J_{\beta(\eta)}} \text{"}X = \{\gamma \in \alpha_\eta \mid A_\gamma = \gamma \cap A_{\alpha_\eta}\}\text{"}.$$

So if we set $A = \sigma_\eta^{-1}(A_{\alpha_\eta})$, we have, since $\sigma_\eta^{-1}: J_{\beta(\eta)} \prec J_{\kappa^+}$,

$$\models_{J_{\kappa^+}} \text{"}\bar{X} = \{\gamma \in \kappa \mid A_\gamma = \gamma \cap A\}\text{"}.$$

Thus it really is the case that

$$\bar{X} = \{\gamma \in \kappa \mid A_\gamma = \gamma \cap A\}.$$

But we assumed that no A exists for which such a set $\bar{X}$ is stationary. This contradiction proves that

$$\models_{J_{\beta(\eta)}} \text{“}X \text{ is not stationary in } \alpha_\eta\text{”}.$$

So there is a set $C \in J_{\beta(\eta)}$ such that

$$\models_{J_{\beta(\eta)}} \text{“}C \text{ is club in } \alpha_\eta \text{ and } (\forall \gamma \in C)(A_\gamma \neq \gamma \cap A_{\alpha_\eta})\text{”}.$$

Let $\bar{C} = \sigma_\eta^{-1}(C)$. By $\sigma_\eta^{-1}: J_{\beta(\eta)} \prec J_{\kappa^+}$, we get

$$\models_{J_{\kappa^+}} \text{“}\bar{C} \text{ is club in } \kappa \text{ and } (\forall \gamma \in \bar{C})(A_\gamma \neq \gamma \in A)\text{”}$$

By absoluteness, $\bar{C}$ is thus a club subset of κ such that $(\forall \gamma \in \bar{C})(A_\gamma \neq \gamma \cap A)$.

Now, $\bar{C} \cap \alpha_\eta = C$ (because $\sigma_\eta^{-1} \restriction \alpha_\eta = \text{id} \restriction \alpha_\eta$ and $\sigma_\eta^{-1}(\alpha_\eta) = \kappa$) and, by absoluteness from $J_{\beta(\eta)}$, C is club in α_η, so as $\bar{C}$ is closed in κ, we have $\alpha_\eta \in \bar{C}$. Thus $A_{\alpha_\eta} \neq \alpha_\eta \cap A$. But $\alpha_\eta \cap A = \sigma_\eta(A) = A_{\alpha_\eta}$ (by the two properties of σ_η^{-1} just mentioned), so we have a contradiction. The proof is complete. □

Exercises

1. Weakly Compact Cardinals and Set mappings

A *set mapping* is (for our purposes) a function $f: [\kappa]^n \to \kappa$ (for some $n \in \omega$) such that $f(\sigma) \notin \sigma$ for all $\sigma \in [\kappa]^n$. A set $X \subseteq \kappa$ is said to be *free* for such a set mapping if $f''[X]^n \cap X = \emptyset$. We write $(\kappa, n) \to \lambda$ if every set mapping $f: [\kappa]^n \to \kappa$ has a free set of cardinality λ.

1 A. Prove that if κ is weakly compact, then $(\kappa, n) \to \kappa$ for all $n \in \omega$.

1 B. Prove that if $V = L$, then κ is weakly compact iff κ is uncountable and regular and either $(\kappa, 2) \to \kappa$ or else $(\kappa, n) \to \kappa$ for all $n \in \omega$.

2. Weakly Compact Cardinals and Colourings of Graphs

A *graph* is a structure $\mathscr{G} = \langle G, E\rangle$, where G is a non-empty set, called the set of *vertices* of $\mathscr{G}$, and E is a set of pairs from G, called the set of *edges* of $\mathscr{G}$. If $\{x, y\} \in E$, we say that x and y are *joined* in $\mathscr{G}$. A *subgraph* of $\mathscr{G}$ is a substructure of $\mathscr{G}$ in the usual sense. If $H \subseteq G$, $\mathscr{G} \restriction H$ denotes the subgraph of $\mathscr{G}$ with domain H. We say $\mathscr{G} \restriction H$ is *small* if $|H| < |G|$.

Let $\mathscr{G} = \langle G, E\rangle$ be a graph, μ a cardinal. A mapping $h: G \to \mu$ is called a *μ-colouring* if $h(x) \neq h(y)$ whenever x and y are joined in $\mathscr{G}$. The least μ for which $\mathscr{G}$ has a μ-colouring is called the *chromatic number* of $\mathscr{G}$.

A basic question of graph theory is: how is the chromatic number of a graph $\mathscr{G}$ effected by the chromatic number of its small subgraphs? By $P(\kappa)$, let us mean the following assertion: if $\mathscr{G}$ is a graph of cardinality κ, all of whose small subgraphs have countable chromatic number, then $\mathscr{G}$ has countable chromatic number.

2A. Prove that if κ is weakly compact, then $P(\kappa)$ holds.

We shall prove that if $V = L$, then for uncountable regular κ, the converse to the above result is valid. Assume $V = L$ from now on. Let κ be an uncountable regular cardinal, not weakly compact. Let $E \subseteq \kappa$ be stationary and such that $E \cap \lambda$ is not stationary in λ for all limit ordinals $\lambda < \kappa$, with $\mathrm{cf}(\alpha) = \omega$ for all $\alpha \in E$. Assume that $\beta + \omega < \alpha$ whenever $\beta < \alpha \in E$. Let $(B_n^\alpha \mid n < \omega)$ be a partition of α, for each $\alpha \in E$, such that whenever $(B_n \mid n < \omega)$ is a partition of κ, the set

$$\{\alpha \in E \mid \mathrm{cf}(\alpha) = \omega \wedge (\forall n \in \omega)(B_n \cap \alpha = B_n^\alpha)\}$$

is stationary in κ. For $\alpha \in E$, let A_α be a cofinal ω-sequence in α, chosen so that

$$\forall n[B_n^\alpha \text{ unbounded in } \alpha \to A_\alpha \cap B_n^\alpha \neq \emptyset]\ .$$

Let $\mathscr{G}$ be the graph with domain κ, in which two points $\nu < \alpha$ are joined iff $\alpha \in E$ and $\nu \in A_\alpha$.

2B. Prove that $\mathscr{G}$ has chromatic number at least ω_1.

2C. Prove that, for any $\lambda < \kappa$, there is an enumeration $(x_\nu \mid \nu < \theta)$ of λ such that the set of all $\eta < \nu$ for which x_η is joined to x_ν is finite for all $\nu < \theta$, and use this to deduce that $\mathscr{G} \restriction \lambda$ has countable chromatic number.

3. KH(κ, κ) *for Singular* κ

3A. Assume GCH. Prove that if κ is singular and $\mathrm{cf}(\kappa) > \omega$, then whenever $\mathscr{F} \subseteq \mathscr{P}(\kappa)$ is such that the set $\{\lambda \in \kappa \mid |\{f \cap \lambda \mid f \in \mathscr{F}\}| \leqslant \lambda\}$ is stationary in κ, then $|\mathscr{F}| \leqslant \kappa$. (Hint: Work on a club subset of κ of order-type $\mathrm{cf}(\kappa)$ and use Fodor's Theorem.)

By the above, assuming GCH, KH(κ, κ) fails for all singular cardinals κ of uncountable cofinality. The following exercises show that if $V = L$, KH(κ, κ) is valid for all singular cardinals κ of cofinality ω. We fix κ a singular cardinal of cofinality ω from now on.

3B. Assume GCH. Show that if $\mathscr{F}$ is a set of ω-sequences cofinal in κ, then for any uncountable set $X \subseteq \kappa$.

$$|\{f \cap X \mid f \in \mathscr{F}\}| \leqslant |X|.$$

By virtue of the above, in order to prove KH(κ, κ) assuming $V = L$, it suffices to construct (from $V = L$) a family $\mathscr{F}$ of κ^+ many ω-sequences cofinal in κ such that $|\{f \cap X \mid f \in \mathscr{F}\}| \leqslant \omega$ for all countable sets $X \subseteq \kappa$. Assume $V = L$ from now on. For each $x \in \mathscr{P}_\kappa(\kappa)$, let M_x be the smallest M such that $x \cup \{x\} \subseteq M \prec J_{\kappa^+}$. Let

$$\mathscr{F} = \{f \subseteq \kappa \mid \mathrm{otp}(f) = \omega \ \&\ \sup(f) = \kappa \ \&\ (\forall x \in \mathscr{P}_{\omega_1}(\kappa))[f \cap x \in M_x]\}.$$

The aim is to prove that $|\mathscr{F}| = \kappa^+$, which at once establishes KH(κ, κ), of course.

Let $(\kappa_n \mid n < \omega)$ be the $<_J$-least ω-sequence of cardinals cofinal in κ such that $\kappa_0 > \omega$. For each $n < \omega$, let N_n be the smallest $N \prec J_{\kappa^{++}}$ such that $\kappa_n \subseteq N$, and let $N = \bigcup_{n<\omega} N_n$.

3C. Prove that $N_0 \prec N_1 \prec \ldots \prec N_n \prec \ldots \prec N \prec J_{\kappa^{++}}$.

3D. Prove that $\kappa \subseteq N$ and that $N \cap \kappa^+ \in \kappa^+$.

Let

$$j: N \cong J_\varrho,$$

and for each $n < \omega$, set

$$\bar{N}_n = j'' N_n.$$

3E. Prove that $\bar{N}_0 \prec \bar{N}_1 \prec \ldots \prec \bar{N}_n \prec \ldots \prec J_\varrho$.

Let

$$j_n: \bar{N}_n \cong J_{\varrho_n},$$

and set

$$R = \{\varrho_n \mid n < \omega\}.$$

3F. Show that $R \notin N$. (Hint: We know that $(\kappa_n \mid n < \omega) \in N$. Then if $R \in N$, we get $\langle (J_{\varrho_n})_{n<\omega}, (j_n \circ j_m^{-1})_{m<n<\omega} \rangle \in N$. Thus $J_\varrho \in N$, a contadiction.)

Now assume, by way of contadiction, that $|\mathscr{F}| \leqslant \kappa$.

3G. Show that (under the above assumption) $\mathscr{F} \in J_\varrho$ and $\mathscr{F} \subseteq J_\varrho$.

3H. Obtain a contradiction with 3F by proving that $R \in \mathscr{F}$. (This is the challenging part of the exercise, and you are on your own from now on. Good luck!)

4. More on Weak Compactness in L

Prove Theorem 1.2′. (See Exercise VI.4.)

Chapter VIII
Morasses and the Cardinal Transfer Theorem

By now it should be quite clear how it is that $V = L$ is of use when it is necessary to carry out intricate constructions by recursion. The uniform structure of L, and in particular the Condensation Lemma, enables us to take care of many "future" possibilities in a relatively small number of steps. For instance, in the construction of a Souslin tree, we need to take care of a collection of ω_2 potential uncountable antichains in ω_1 steps. As we shall see in this chapter, the uniformity of L enables us to do much more than this. In certain circumstances it is possible to construct structures of cardinality ω_2 in ω_1 steps. The idea is to use a sort of two-staged condensation principle, simultaneously approximating the final structure of size ω_2 by means of structures of size ω_1, and approximating each of these approximations by countable structures. In order to make this work, what is necessary is to investigate the way in which the two parts of such an approximation procedure must (and can) fit together. The essential combinatorial structure of L which is required is called a "morass". There is no need to stop there. We can go on to develop three-stage "morasses" which enable us to get up to ω_3 using only countable structures, and so on. In fact the subject of morasses is a vast area on its own, and would require an entire book of its own for a complete coverage. What we shall do in this chapter is look at the very simplest kind of morass, the one that gets us up to ω_2, in full detail, and then give little more than a glance at what comes after. In order both to motivate and illustrate the definition and use of a morass we take the problem which itself led to the development of morass theory, the Cardinal Transfer Problem of Model Theory.

1. Cardinal Transfer Theorems

Cardinals Transfer Theorems are generalised Löwenheim-Skolem Theorems. In its simplest form, the Löwenheim-Skolem Theorem says that if $\mathscr{A}$ is a model of a countable, first-order language, K, then there is a countable K-structure $\mathscr{B}$ such that $\mathscr{B} \equiv \mathscr{A}$. (More generally, given any infinite cardinal κ, there is a K-structure $\mathscr{B}$ of cardinality κ such that $\mathscr{B} \equiv \mathscr{A}$. Here and throughout it will be assumed that all structures considered are infinite, and that all first-order theories involved admit infinite models. This will exclude trivial special cases.) Now suppose that the language K contains a distinguished unary predicate symbol U. If κ, λ are cardinals (both infinite) we shall say that a K-structure $\mathscr{A}$ has *type* (κ, λ) iff $\mathscr{A}$ has the form

$$\mathscr{A} = \langle A, U^{\mathscr{A}}, \ldots \rangle$$

where $|A| = \kappa$ and $|U^{\mathscr{A}}| = \lambda$. The idea of a cardinal transfer theorem is to obtain a Löwenheim-Skolem Theorem which preserves the relationship between the cardinality of the domain and that of the distinguished subset. The simplest case is the so-called "Gap-1 Cardinal Transfer Property", which says that every K-structure of type (κ^+, κ) (for some infinite κ) is elementarily equivalent to a K-structure of type (ω_1, ω). As we shall see presently, this result is provable in ZFC. Assuming GCH, we may replace (ω_1, ω) by any type (λ^+, λ) where λ is regular. Assuming $V = L$ we may drop the requirement that λ be regular here. (These results are considered in the Exercises.) The "Gap-2 Property" says (in the simplest case) that every K-structure of type (κ^{++}, κ) is elementarily equivalent to one of type (ω_2, ω). More generally there is a "Gap-n Property" for every positive integer n. There are also more general types of Cardinal Transfer Property, which we shall not consider here. The reader may consult *Chang-Keisler* (1973) for further details (including applications) of Cardinals Transfer Theorems.

As mentioned above, the Gap-1 Property is (in its simplest form) provable in ZFC. The Gap-2 Property, and indeed the Gap-n Property for any $n \geqslant 2$, is provable in ZFC + $(V = L)$. Our aim here is to use the simple version of the Gap-2 Property to motivate and illustrate the notion of a morass. In order to do this it is convenient to begin with a brief account of the proof of the Gap-1 Theorem. (In particular we shall need all of the model theoretic notions developed for the Gap-1 Theorem in order to prove the Gap-2 Theorem.)

We fix, once and for all, a countable, first-order language, K, with a distinguished unary predicate symbol, U. We shall show that if $\mathscr{A}$ is a K-structure of type (κ^+, κ) for some infinite cardinal κ, there is a K-structure $\mathscr{B}$ of type (ω_1, ω) such that $\mathscr{B} \equiv \mathscr{A}$. We recall some basic notions of model theory. For further details the reader should consult, for example, *Chang-Keisler* (1973).

K' will denote an arbitrary, countable expansion of K. A particular example of an expansion of K is obtained by adjoining to K an individual constant $\mathring{x}$ for each x in a given set X. This expansion will be denoted by K_X. In this case, X may be uncountable: this is the only case where uncountable languages may be considered. If $\mathscr{A}$ is any K-structure, then $\langle \mathscr{A}, (a)_{a \in A} \rangle$ is a K_A-structure. (We adopt the usual convention that A is the domain of $\mathscr{A}$, B the domain of $\mathscr{B}$, etc.)

The first-order theory of a structure $\mathscr{A}$ is denoted by $Th(\mathscr{A})$. Thus if $\mathscr{A}$ is a K-structure,

$$Th(\mathscr{A}) = \{\varphi \mid \varphi \text{ is a sentence of } K \text{ such that } \mathscr{A} \models \varphi\}.$$

Let T be a K_X-theory. An *element-type* of T is a set $\Sigma(x)$ of K_X-formulas with free variable at most x, such that for some model $\mathscr{A}$ of T and some element a of $\mathscr{A}$, $\mathscr{A} \models \Sigma(a)$. (The notation is self-explanatory.) In this case we say that a *realises* $\Sigma(x)$ *in* $\mathscr{A}$.

Let κ be an infinite cardinal. A structure $\mathscr{A}$ is said to be κ-saturated iff for every set $B \subseteq A$ such that $|B| < \kappa$, every element-type of $Th(\langle \mathscr{A}, (b)_{b \in B} \rangle)$ is realised in the structure $\langle \mathscr{A}, (b)_{b \in B} \rangle$. A structure $\mathscr{A}$ is said to be *saturated* iff it is $|\mathscr{A}|$-saturated. The following theorem is standard.

1.1 Theorem. (i) *Let $\mathscr{A}, \mathscr{B}$ be saturated K'-structures of cardinality κ, $\mathscr{A} \equiv \mathscr{B}$. Then $\mathscr{A} \cong \mathscr{B}$.*

Moreover, if $A' \subseteq A$, $B' \subseteq B$, $|A'| = |B'| < \kappa$, and $h: A' \leftrightarrow B'$ are such that

$$\langle \mathscr{A}, (a)_{a \in A'} \rangle \equiv \langle \mathscr{B}, (ha)_{a \in A'} \rangle,$$

then there is an isomorphism $\tilde{h}: \mathscr{A} \cong \mathscr{B}$ such that $\tilde{h} \upharpoonright A' = h$.

(ii) *Assume* GCH. *Let κ be an uncountable regular cardinal. Then any K'-theory has a saturated model of cardinality κ.* □

A structure $\mathscr{A}$ is said to be *homogeneous* iff, whenever $B \subseteq A$, $|B| < |A|$, and $h: B \to A$ is such that

$$\langle \mathscr{A}, (b)_{b \in B} \rangle \equiv \langle \mathscr{A}, (hb)_{b \in B} \rangle,$$

there is an automorphism $\tilde{h}: \mathscr{A} \cong \mathscr{A}$ such that $\tilde{h} \upharpoonright B = h$.

It is immediate from 1.1 (i) that any saturated structure is homogeneous. By virtue of 1.1 (ii), this provides an existence proof for uncountable homogeneous structures of regular cardinality, assuming GCH. As far as countable homogeneous structures are concerned, the existence is provable in ZFC alone.

1.2 Theorem. *Let T be a K'-theory. Then T has a countable, homogeneous model.* □

We shall make use of countable, homogeneous structures in our proof of the Gap-1 Theorem. The following result will also be required.

1.3 Theorem. *If $\mathscr{A}_0 \prec \mathscr{A}_1 \prec \ldots \prec \mathscr{A}_n \prec \ldots$ $(n < \omega)$ is an elementary chain of countable, homogeneous structures, then $\mathscr{A} = \bigcup_{n < \omega} \mathscr{A}_n$ is a countable, homogeneous structure.* □

A structure $\mathscr{A}$ is said to be *special* if there is an elementary chain

$$\mathscr{A}_0 \prec \mathscr{A}_1 \prec \ldots \prec \mathscr{A}_n \prec \ldots \prec \mathscr{A}$$

such that:

(i) $\mathscr{A} = \bigcup_{n < \omega} \mathscr{A}_n$;

(ii) $|A_0| < |A_1| < \ldots < |A_n| < \ldots < |A|$;

(iii) $\mathscr{A}_{n+1}$ is $|A_n|$-saturated for every n.

The following result tells us all we need to know about special structures.

1.4 Theorem. (i) *Every K'-theory has a special model (of some cardinality).*

(ii) *If $\mathscr{A}, \mathscr{B}$ are special structures of the same cardinality such that $\mathscr{A} \equiv \mathscr{B}$, then $\mathscr{A} \cong \mathscr{B}$.* □

The key step in the proof of the Gap-1 Theorem is the following lemma.

1.5 Lemma. *Let $\mathscr{A} = \langle A, U, \dots \rangle$ be a K-structure of type (κ^+, κ) for some κ. Then there are countable, homogeneous K-structures $\mathscr{B}$, $\mathscr{C}$ such that:*

(i) $\mathscr{B} = \langle B, V, \dots \rangle$, $\mathscr{C} = \langle C, V, \dots \rangle$ (i.e. $U^{\mathscr{B}} = U^{\mathscr{C}}$);

(ii) $\mathscr{B} \equiv \mathscr{C} \equiv \mathscr{A}$;

(iii) $\mathscr{B} \prec \mathscr{C}$ *and* $\mathscr{B} \neq \mathscr{C}$;

(iv) $\mathscr{B} \cong \mathscr{C}$.

Proof. Let $\mathscr{A}_0 \prec \mathscr{A}$ be such that $U \subseteq A_0$ and $|A_0| = \kappa$. Pick $a \in A - A_0$, and let $\mathscr{A}_1 \prec \mathscr{A}$ be such that $A_0 \cup \{a\} \subseteq A_1$ and $|A_1| = \kappa$. Let $h\colon A_0 \leftrightarrow A_1$, and form the structure

$$\mathscr{A}^* = \langle \mathscr{A}_1, A_0, h \rangle .$$

By 1.4 (i), let $\mathscr{D}^*$ be a special structure such that $\mathscr{D}^* \equiv \mathscr{A}^*$, say

$$\mathscr{D}^* = \langle \mathscr{D}_1, D_0, k \rangle .$$

Let $\mathscr{D}_0$ be the restriction of $\mathscr{D}_1$ to domain D_0. Since $\mathscr{D}^* \equiv \mathscr{A}^*$, it is easily seen that $\mathscr{D}_0 \prec \mathscr{D}_1$. Since $\mathscr{D}^*$ is special, it is straightforward to check that both $\mathscr{D}_0$ and $\mathscr{D}_1$ are special. But $k\colon D_1 \leftrightarrow D_0$. Hence by 1.4 (ii), $\mathscr{D}_0 \cong \mathscr{D}_1$. Note also that $U^{\mathscr{D}_0} = U^{\mathscr{D}_1}$. Let $f\colon \mathscr{D}_1 \cong \mathscr{D}_0$ and consider the structure

$$\mathscr{D}^{**} = \langle \mathscr{D}_1, D_0, f \rangle .$$

By 1.2, let $\mathscr{C}^{**} \equiv \mathscr{D}^{**}$ be countable homogeneous, say

$$\mathscr{C}^{**} = \langle \mathscr{C}_1, C_0, g \rangle .$$

Let $\mathscr{C}_0$ be the restriction of $\mathscr{C}_1$ to domain C_0. It is routine to check that $\mathscr{C}_0, \mathscr{C}_1$ are both countable, homogeneous structures, that $\mathscr{C}_0 \equiv \mathscr{C}_1 \equiv \mathscr{A}$, $U^{\mathscr{C}_0} = U^{\mathscr{C}_1}$, $\mathscr{C}_0 \prec \mathscr{C}_1$, $\mathscr{C}_0 \neq \mathscr{C}_1$, and that $g\colon \mathscr{C}_1 \cong \mathscr{C}_0$. Thus $\mathscr{B} = \mathscr{C}_0$ and $\mathscr{C} = \mathscr{C}_1$ satisfy the lemma. □

The above lemma shows that it is quite possible to have structures $\mathscr{A}$, $\mathscr{B}$ such that $\mathscr{A} \prec \mathscr{B}$ and $\mathscr{A} \cong \mathscr{B}$. The following lemma also involves this situation.

1.6 Lemma. *Let $\mathscr{B}_0 \prec \mathscr{B}_1 \prec \dots \prec \mathscr{B}_n \prec \dots$ $(n < \omega)$ be an elementary chain of isomorphic, countable, homogeneous models. Then $\mathscr{B}_\omega = \bigcup_{n<\omega} \mathscr{B}_n$ is countable and homogeneous and $\mathscr{B}_\omega \cong \mathscr{B}_n$ for all $n < \omega$.*

Proof. By 1.3 we know that $\mathscr{B}_\omega$ is countable and homogeneous. We prove that $\mathscr{B}_\omega \cong \mathscr{B}_0$. The idea is to construct enumerations $(b_n^0 \mid n < \omega)$, $(b_n^\omega \mid n < \omega)$ of B_0, B_ω, respectively, so that

$$\langle \mathscr{B}_0, (b_n^0)_{n<\omega} \rangle \equiv \langle \mathscr{B}_\omega, (b_n^\omega)_{n<\omega} \rangle ,$$

which at once implies that $h\colon \mathscr{B}_0 \cong \mathscr{B}_\omega$, where we define $h(b_n^0) = b_n^\omega$ for all $n < \omega$.

Suppose that $b_0^0, \ldots, b_n^0, b_0^\omega, \ldots, b_n^\omega$ are defined and satisfy

(i) $$\langle \mathscr{B}_0, b_0^0, \ldots, b_n^0 \rangle \equiv \langle \mathscr{B}_\omega, b_0^\omega, \ldots, b_n^\omega \rangle .$$

(The definition of b_0^0, b_0^ω is a degenerate case of this definition, so we omit it.) Let $b_{n+1}^0 \in B_0$. We show that there is an element b_{n+1}^ω of B_ω such that

(ii) $$\langle \mathscr{B}_0, b_0^0, \ldots, b_n^0, b_{n+1}^0 \rangle \equiv \langle \mathscr{B}_\omega, b_0^\omega, \ldots, b_n^\omega, b_{n+1}^\omega \rangle .$$

Since $\mathscr{B}_0 \prec \mathscr{B}_\omega$, we have

$$\langle \mathscr{B}_0, b_0^0, \ldots, b_n^0 \rangle \equiv \langle \mathscr{B}_\omega, b_0^0, \ldots, b_n^0 \rangle ,$$

so by (i),

$$\langle \mathscr{B}_\omega, b_0^0, \ldots, b_n^0 \rangle \equiv \langle \mathscr{B}_\omega, b_0^\omega, \ldots, b_n^\omega \rangle .$$

Thus as $\mathscr{B}_\omega$ is homogeneous and $b_{n+1}^0 \in B_\omega$ there is a $b_{n+1}^\omega \in B_\omega$ such that

$$\langle \mathscr{B}_\omega, b_0^0, \ldots, b_n^0, b_{n+1}^0 \rangle \equiv \langle \mathscr{B}_\omega, b_0^\omega, \ldots, b_n^\omega, b_{n+1}^\omega \rangle .$$

Since $\mathscr{B}_0 \prec \mathscr{B}_\omega$, this at once yields (ii).

To complete the proof that $\mathscr{B}_\omega \cong \mathscr{B}_0$ we show that if we are given $b_0^0, \ldots, b_n^0$, $b_0^\omega, \ldots, b_n^\omega$ as in (i) and $b_{n+1}^\omega \in B_\omega$ is given, we can find an element b_{n+1}^0 of B_0 to satisfy (ii). (The required enumerations $(b_n^0 \mid n < \omega)$, $(b_n^\omega \mid n < \omega)$ can then be defined by recursion using a "back and forth" procedure to ensure that all elements of B_0, B_ω are included in these sequences.) Well, we can pick an integer $m < \omega$ such that $b_0^\omega, \ldots, b_n^\omega, b_{n+1}^\omega \in B_m$. Since $\mathscr{B}_m \prec \mathscr{B}_\omega$, we have

(iii) $$\langle \mathscr{B}_m, b_0^\omega, \ldots, b_n^\omega, b_{n+1}^\omega \rangle \equiv \langle \mathscr{B}_\omega, b_0^\omega, \ldots, b_n^\omega, b_{n+1}^\omega \rangle .$$

Since $\mathscr{B}_m \cong \mathscr{B}_0$ there are elements $b_0, \ldots, b_n, b_{n+1}$ of B_0 such that

(iv) $$\langle \mathscr{B}_m, b_0^\omega, \ldots, b_n^\omega, b_{n+1}^\omega \rangle \equiv \langle \mathscr{B}_0, b_0, \ldots, b_n, b_{n+1} \rangle .$$

By (i), (iii) and (iv),

$$\langle \mathscr{B}_0, b_0, \ldots, b_n \rangle \equiv \langle \mathscr{B}_0, b_0^0, \ldots, b_n^0 \rangle .$$

So as $b_{n+1} \in B_0$ and $\mathscr{B}_0$ is homogeneous, there is an element b_{n+1}^0 of B_0 such that

$$\langle \mathscr{B}_0, b_0, \ldots, b_n, b_{n+1} \rangle \equiv \langle \mathscr{B}_0, b_0^0, \ldots, b_n^0, b_{n+1}^0 \rangle .$$

Combining this with (iii) and (iv) we get (ii), as required. This completes the proof that $\mathscr{B}_\omega \cong \mathscr{B}_0$, and with it the proof of the lemma. □

We are now able to prove the Gap-1 Theorem.

1.7 Theorem. *Let $\mathscr{A} = \langle A, U, \ldots \rangle$ be a K-structure of type (κ^+, κ). Then there is a K-structure $\mathscr{B}$ of type (ω_1, ω) such that $\mathscr{B} \equiv \mathscr{A}$.*

Proof. By 1.5 there are countable, homogeneous structures $\mathscr{B}_0, \mathscr{B}_1$ such that $\mathscr{B}_0 \equiv \mathscr{B}_1 \equiv \mathscr{A}$, $\mathscr{B}_0 \prec \mathscr{B}_1$, $U^{\mathscr{B}_0} = U^{\mathscr{B}_1}$, $\mathscr{B}_0 \neq \mathscr{B}_1$, $\mathscr{B}_0 \cong \mathscr{B}_1$. The idea of the proof is to define, by recursion, a strictly increasing elementary chain

$$\mathscr{B}_0 \prec \mathscr{B}_1 \prec \ldots \prec \mathscr{B}_\nu \prec \ldots \qquad (\nu < \omega_1)$$

of countable, homogeneous structures such that for all $\nu < \omega_1$, $\mathscr{B}_\nu \cong \mathscr{B}_0$ and $U^{\mathscr{B}_\nu} = U^{\mathscr{B}_0}$, so that $\mathscr{B} = \bigcup_{\nu < \omega_1} \mathscr{B}_\nu$ is as required by the theorem.

$\mathscr{B}_0, \mathscr{B}_1$ are already defined. Suppose we have defined $\mathscr{B}_\nu$. Since $\mathscr{B}_\nu \cong \mathscr{B}_0$, we may let $\mathscr{B}_{\nu+1}$ be related to $\mathscr{B}_\nu$ as $\mathscr{B}_1$ is related to $\mathscr{B}_0$. This leaves us with the case where $\delta < \omega_1$ is a limit ordinal and $\mathscr{B}_\nu$, $\nu < \delta$, are all defined. In this case we let $\mathscr{B}_\delta = \bigcup_{\nu < \delta} \mathscr{B}_\nu$. By 1.6, $\mathscr{B}_\delta$ is as required. The proof is complete. □

The above proof depended upon the countability of the structures $\mathscr{B}_\nu$ in a significant way. Consequently, there seems to be no possibility of extending the chain $(\mathscr{B}_\nu \mid \nu < \omega_1)$ to an ω_2-chain and thereby produce a model of type (ω_2, ω). In fact it is easily seen that it is not possible to increase the size of a "gap" in a Cardinal Transfer Theorem. But when it comes to trying to prove the Gap-2 Cardinal Transfer Theorem we have some extra initial information: we start with a structure of type (κ^{++}, κ). How can we make use of this fact to obtain an elementarily equivalent structure of type (ω_2, ω)? Ideally we would like to utilise the methods developed in order to prove the Gap-1 Theorem. Thus, the idea is to construct the desired (ω_2, ω)-model as a limit of some system of countable *approximations* to it, in the sense that each of the structures $\mathscr{B}_\nu$, $\nu < \omega_1$, of 1.7 is an *approximation* to the sought-after (ω_1, ω)-model $\mathscr{B}$. But if we are to obtain a model of type (ω_2, ω) as a limit of countable models, there is no point in trying to use an elementary chain of structures. Rather we require some kind of elementary directed system of models. We commence by considering a "naive" approach to this problem.

We wish to construct a model $\mathscr{B}$ of type (ω_2, ω). We may regard this model as a union of a chain

$$\mathscr{B}_0 \prec \mathscr{B}_1 \prec \ldots \prec \mathscr{B}_\nu \prec \ldots \prec \mathscr{B} \qquad (\nu < \omega_2)$$

of structures of type (ω_1, ω), all having the same distinguished subset, U. Each of these (ω_1, ω)-structures $\mathscr{B}_\nu$ can itself be represented as the union of a chain

$$\mathscr{B}_{\nu 0} \prec \mathscr{B}_{\nu 1} \prec \ldots \prec \mathscr{B}_{\nu\tau} \prec \ldots \prec \mathscr{B}_\nu \qquad (\tau < \omega_1)$$

of countable structures, all with the same U. Thus the structure $\mathscr{B}$ is a sort of limit of the system of countable structures $\mathscr{B}_{\nu\tau}$, $\nu < \omega_2, \tau < \omega_1$. The question is, can we construct such a system *from below* in order to determine the limit structure $\mathscr{B}$? It turns out that if we assume $V = L$, this can be done, though it is by no means an easy matter, and relies heavily upon the Fine Structure Theory. The central point is the construction of a framework upon which a suitable directed elementary system can be built. This framework is known as a *morass*. In the next section we shall give a precise definition of a morass and show how such a structure can be built in L. In a sense, when we write down the axioms for a morass we are simply stating some properties of the Fine Structure Theory of a certain hierarchy of structures of the form $\langle J_\varrho, A\rangle$. It is thus not altogether surprising to discover that the structure so defined is somewhat "richer" than is required to prove the Gap-2 Theorem. In order to prove this (and many other applications of morasses) a simpler structure suffices. This "simplified morass" will be described in section 4. In section 3 we shall give a proof of the Gap-2 Theorem using the "standard" morass constructed in L. Section 4 contains an alternative proof of the Gap-2 Theorem using the simplified morass structure. The reason for this duplication is that the proof in section 3, using the standard morass, illustrates just how the Fine Structure Theory enables this theorem to be proved (which is, of course, the main aim of this book), whereas the (simpler) proof in section 4 serves as a prototype for other applications of morasses. Thus the reader who symply wants to learn how to use a morass may go straight on to section 4 from this point. (Though some acquaintance with section 2 is necessary if the reader wishes to find out just where the simplified morass comes from.)

2. Gap-1 Morasses

We can obtain a structure of cardinality κ^+ as a limit of a κ^+-chain of structures of cardinality κ. In order to determine a structure of cardinality κ^{++} as a limit of κ^+ many structures each of cardinality κ, a chain of structures will not work, and we must define instead some sort of directed system of structures. The underlying set-theoretic problem then is to establish some sort of framework upon which such a system can be built, corresponding to the well-ordered set κ^+ used as the domain of κ^+-chains. Such a framework (or indexing system) is called a *morass*. or more precisely a $(\kappa^+, 1)$-morass. For definiteness, we shall present our development for the case $\kappa = \omega$. The general case is entirely similar. So what we shall describe is a $(\omega_1, 1)$-morass. (We shall then say a few words about (κ, n)-morasses for $n > 1$.)

In order to formulate the notion of a morass, let us fix some sort of schematic representation of what we require. We want to determine, by means of countable structures, all the structures of cardinality ω_1 which lie in an increasing ω_2-chain determining a structure of cardinality ω_2. Let $\mathscr{A}$ denote the structure of cardinality ω_2 we are aiming at, and let $\mathscr{A}_\nu$, $\omega_1 < \nu < \omega_2$ be the increasing chain of length ω_2, where $|\mathscr{A}_\nu| = \omega_1$ for each ν. We can represent this as in Fig. 1.

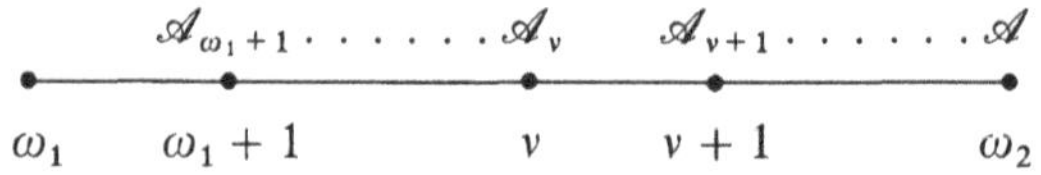

Fig. 1

For each ν, $\omega_1 < \nu < \omega_2$, we have a chain of models with limit $\mathscr{A}_\nu$. Each member of this chain will be a countable structure. We shall index each such by a countable ordinal τ, so $\mathscr{A}_\nu$ will be the limit of the structures $\mathscr{A}_{\nu\tau}$ for certain τ. We shall not use all countable ordinals τ here, just a certain collection associated with ν. It will turn out to be both natural and convenient to specify the ordinals τ associated with ν by defining a well-founded relation $\dashv$ on ω_2 so that $\{\tau \mid \tau \dashv \nu\}$ is the set of ordinals for which $\mathscr{A}_{\nu\tau}$ is defined, with $\{\tau \mid \tau \dashv \nu\}$ being totally ordered by $\dashv$ and $\bar{\tau} \dashv \tau \dashv \nu$ implying that, in some sense $\mathscr{A}_{\nu\tau}$ "extends" $\mathscr{A}_{\nu\bar{\tau}}$. However, just being able to determine the structures $\mathscr{A}_\nu$ piecemeal will not be enough. We need to determine the *sequence* $(\mathscr{A}_\nu \mid \omega_1 < \nu < \omega_2)$. In order to do this, we do not simply approximate the models $\mathscr{A}_\nu$, but rather the initial segments $(\mathscr{A}_{\bar{\nu}} \mid \omega_1 < \bar{\nu} \leqslant \nu)$ of our final chain.

Dropping our reference to the models $\mathscr{A}_\nu$, $\mathscr{A}_{\nu\tau}$ now, let us concentrate on the indexing system upon which we shall define the model system: this will be our "morass". We have seen that we need to be able to obtain each interval $[\omega_1, \nu]$, $\omega_1 < \nu < \omega_2$, as a limit of intervals $[\alpha, \tau]$, $\alpha < \tau < \omega_1$, in order that we shall never have to consider uncountable models $\mathscr{A}_{\nu\tau}$ during the course of our eventual construction. Just what the ordinals α, τ here are will clearly be unimportant: what counts is how these intervals fit together to form the indexing system. Hence we may assume that all of the small approximating intervals are disjoint: i.e. if $[\alpha, \tau]$, $[\alpha', \tau']$ are part of our morass, and if $\alpha < \alpha'$, then $\tau < \alpha'$. (This is not a misprint!) In point of fact, when we come to give the formal definition of a morass, we shall not use the entire interval $[\alpha, \tau]$ but rather a certain closed subset of it. This does not effect the combinatorial properties of the morass at all, but will make matters a little simpler when we come to construct a morass in L.

If $[\bar{\alpha}, \bar{\tau}]$, $[\alpha, \tau]$ are intervals in the morass with $\bar{\tau} \dashv \tau$, there will be an embedding $\pi_{\bar{\tau}\tau}$ of $[\bar{\alpha}, \bar{\tau}]$ into $[\alpha, \tau]$. And if $\omega_1 < \tau < \omega_2$, $[\omega_1, \tau]$ will be a direct limit of all the intervals $[\bar{\alpha}, \bar{\tau}]$ in the morass with $\bar{\tau} \dashv \tau$, under the π-embeddings. All of this is indicated in Fig. 2, where we adopt the usual convention that the relationship $\bar{\tau} \dashv \tau$ is indicated by a line drawn from τ downwards to $\bar{\tau}$. In connection with Fig. 2, notice that we draw each of the morass intervals $[\alpha, \tau]$ horizontally, to emphasise how they all fit together. In reality, by the disjointness of the intervals, the whole thing could be drawn as a single straight line, and indeed that is what it really is. But it is clearer to draw each morass interval horizontally as shown, so we shall continue to do so.

The problem is how do we set this up so that it works? In particular, the ω_1 many countable intervals must all fit together neatly so that the limits on the top level do indeed give us the chain of intervals $\langle [\omega_1, \nu] \mid \omega_1 < \nu < \omega_2 \rangle$. In order to arrange this, we shall have to make matters somewhat more complicated than we have indicated so far. One aspect of this is that some morass intervals will be initial parts of other morass intervals, so that the disjointness of intervals that we spoke of a few moments ago will not be true for all pairs of intervals (though it will be the case that the only two possibilities are disjointness or initial segments).

We shall say that an ordinal α is *adequate* iff it is either admissible or else is a limit of admissible ordinals. The adequate ordinals thus form a proper class of ordinals which is closed and unbounded in every uncountable cardinal. Moreover, each adequate ordinal has strong closure properties under definability.

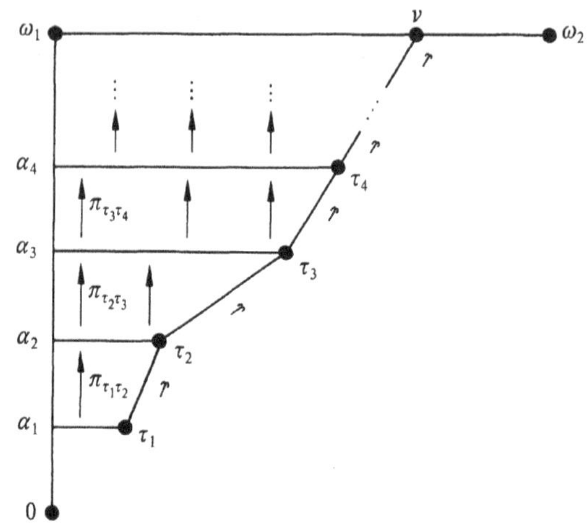

Fig. 2

Let $\mathscr{S}$ be a set of ordered pairs (α, ν) of adequate ordinals such that $\alpha < \nu < \omega_2$, $\alpha \leqslant \omega_1$, and whenever $(\alpha, \nu), (\alpha', \nu') \in \mathscr{S}$, then

$$\alpha < \alpha' \to \nu < \alpha'.$$

Define:

$$S^0 = \{\alpha \in \omega_1 + 1 \mid \exists \nu[(\alpha, \nu) \in \mathscr{S}]\};$$

S^1 $$S^1 = \{\nu \in \omega_2 \mid \exists \alpha[(\alpha, \nu) \in \mathscr{S}]\};$$

S $$S = S^0 \cup S^1;$$

S_α $$S_\alpha = \{\nu \in S^1 \mid (\alpha, \nu) \in \mathscr{S}\}, \qquad \text{for } \alpha \in S^0;$$

α_ν $$\alpha_\nu = \text{the unique ordial } \alpha \in S^0 \text{ such that } (\alpha, \nu) \in \mathscr{S}, \text{ for } \nu \in S^1.$$

Intuitively, S_{ω_1} is the ω_2-chain we are trying to determine, whilst each S_α, $\alpha < \omega_1$, is a countable approximation to S_{ω_1}. Fig. 3 illustrates the notation.

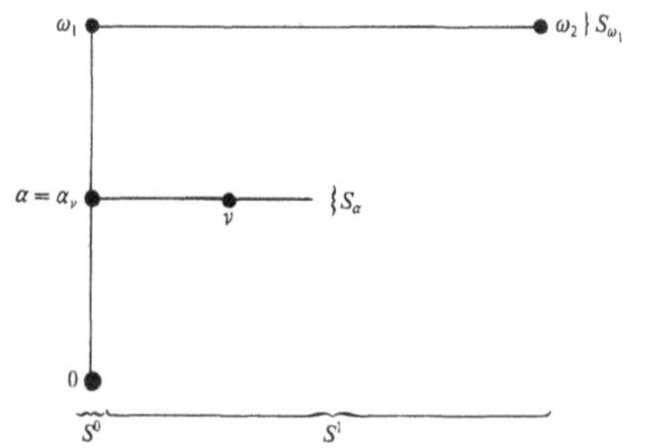

Fig. 3

Let $\dashv$ be a tree ordering on S^1 such that

$$\nu \dashv \tau \to \alpha_\nu < \alpha_\tau .$$

Let $(\pi_{\nu\tau} \mid \nu \dashv \tau)$ be a commutative system of maps

$$\pi_{\nu\tau} \colon (\nu + 1) \to (\tau + 1) .$$

Let

$$\mathcal{M} = \langle S, \mathcal{S}, \dashv, (\pi_{\nu\tau})_{\nu \dashv \tau} \rangle .$$

We say that $\mathcal{M}$ is an $(\omega_1, 1)$-*morass* (*morass*, from now on) iff the following axioms (M 0) through (M 7) are satisfied.

(M 0) (a) S_α is closed in $\sup(S_\alpha)$ for all $\alpha \in S^0$, and if $\alpha < \omega_1$, then $\sup(S_\alpha) \in S_\alpha$;

(b) $\omega_1 = \max(S^0) = \sup(S^0 \cap \omega_1)$ and $\omega_2 = \sup(S_{\omega_1})$.

(M 1) If $\nu \dashv \tau$, then

$$\pi_{\nu\tau} \restriction \alpha_\nu = \mathrm{id} \restriction \alpha_\nu, \qquad \pi_{\nu\tau}(\alpha_\nu) = \alpha_\tau, \qquad \pi_{\nu\tau}(\nu) = \tau$$

and $\pi_{\nu\tau}$ maps $S_{\alpha_\nu} \cap (\nu + 1)$ into $S_{\alpha_\tau} \cap (\tau + 1)$ in an order-preserving fashion so that:

(i) if γ is the first member of S_{α_ν}, then $\pi_{\nu\tau}(\gamma)$ is the first member of $S_{\alpha\tau}$;
(ii) if γ immediately succeeds β in $S_{\alpha_\nu} \cap (\nu + 1)$, then $\pi_{\nu\tau}(\gamma)$ immediately succeeds $\pi_{\nu\tau}(\beta)$ in S_{α_τ};
(iii) if γ is a limit point in $S_{\alpha_\nu} \cap (\nu + 1)$, then $\pi_{\nu\tau}(\gamma)$ is a limit point in S_{α_τ}.

Thus what (M 1) says is that the maps $\pi_{\nu\tau}$ embed each morass "interval" $S_{\alpha_\nu} \cap (\nu + 1)$ into the morass "interval" $S_{\alpha_\tau} \cap (\tau + 1)$ in a structure preserving fashion.

(M 2) If $\bar{\tau} \dashv \tau$ and $\bar{\nu} \in S_{\alpha_{\bar{\tau}}} \cap \bar{\tau}$, and we set $\nu = \pi_{\bar{\tau}\tau}(\bar{\nu})$, then

$$\bar{\nu} \dashv \nu \quad \text{and} \quad \pi_{\bar{\nu}\nu} \restriction \bar{\nu} = \pi_{\bar{\tau}\tau} \restriction \bar{\nu} .$$

What (M 2) says is that the morass embeddings $\pi_{\bar{\nu}\nu}$ fit together nicely as we move right along each row S_α. Figure 4 provides the picture.

(M 3) $\{\alpha_\nu \mid \nu \dashv \tau\}$ is closed in α_τ for every $\tau \in S^1$.

(M 3) tells us that as we move up along a branch $\{\nu \mid \nu \dashv \tau\}$ of the morass tree, all limit points exist on this branch, or more precisely, the limit points are on the morass "levels" they "ought" to be.

(M 4) If τ is not maximal in S_{α_τ}, then the set $\{\alpha_\nu \mid \nu \dashv \tau\}$ is unbounded in α_τ.

(M 4) tells us that any point which is not at the extreme right hand end of its morass level is a limit point in the morass tree $\dashv$. This has the rather surprising

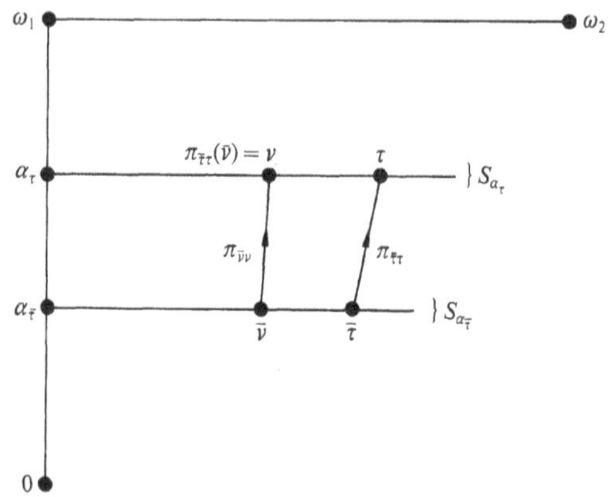

(M 2) asserts that $\bar{\nu} \dashv \nu$ and $\pi_{\bar{\nu}\nu} \upharpoonright \bar{\nu} = \pi_{\bar{\tau}\tau} \upharpoonright \bar{\nu}$.

Fig. 4

consequence that if α is a successor point in S^0, then S_α has only one member. Thus the approximations S_α, $\alpha \in S^0$, to S_{ω_1} do not "get better" monotonically as α increases: only at limit stages is there any chance of some progress in this sense.

(M 5) If $\{\alpha_\nu \mid \nu \dashv \tau\}$ is unbounded in α_τ, then

$$\tau = \bigcup_{\nu \dashv \tau} \pi_{\nu\tau}{}''\nu .$$

Used in conjunction, (M 3), (M 4), and (M 5) tell us that if τ is not the maximal point in its level, then the entire structure up to τ, in particular the morass interval $S_{\alpha_\tau} \cap \tau$, is the limit of the lower structure. For by these three axioms, together with (M 0), we see that if τ is not the maximal point of S_{α_τ}, then

$$S_{\alpha_\tau} \cap \tau = \bigcup_{\bar{\tau} \dashv \tau} \pi_{\bar{\tau}\tau}{}''(S_{\alpha_{\bar{\tau}}} \cap \bar{\tau}) .$$

By (M 0), this applies in particular to any point $\tau \in S_{\omega_1}$. Thus the entire top level of the morass is determined by the structure below.

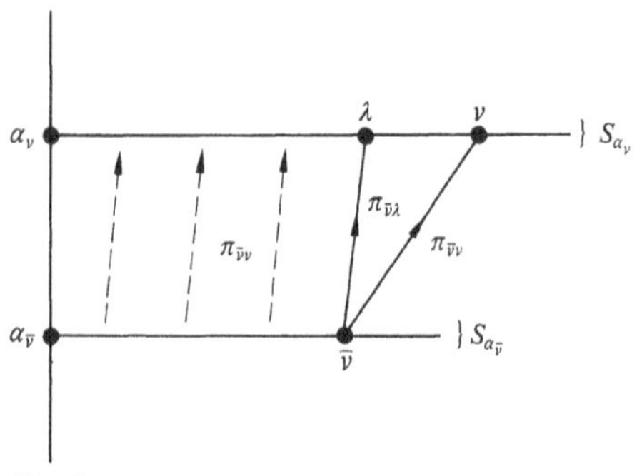

Fig. 5

(M 6) If $\bar{\nu}$ is a limit point of $S_{\alpha_{\bar{\nu}}}$ and $\bar{\nu} \dashv \nu$, and if we set $\lambda = \sup(\pi_{\bar{\nu}\nu}''\bar{\nu})$, then $\bar{\nu} \dashv \lambda$ and $\pi_{\bar{\nu}\lambda} \restriction \bar{\nu} = \pi_{\bar{\nu}\nu} \restriction \bar{\nu}$.

Loosely speaking, (M 6) says that although $\pi_{\bar{\nu}\nu}$ need not map $\bar{\nu}$ cofinally into ν, all of the morass maps $\pi_{\bar{\nu}\nu}$ are nevertheless cofinal maps in some sense. Figure 5 illustrates the situation.

(M 7) If $\bar{\nu}$ is a limit point of $S_{\alpha_{\bar{\nu}}}$, $\bar{\nu} \dashv \nu$, $\nu = \sup(\pi_{\bar{\nu}\nu}''\bar{\nu})$, and if

$$\alpha \in \bigcap_{\bar{\tau} \in S_{\alpha_{\bar{\nu}}} \cap \bar{\nu}} \{\alpha_\eta \mid \bar{\tau} \dashv \eta \dashv \pi_{\bar{\nu}\nu}(\bar{\tau})\},$$

then

$$(\exists \tau \in S_\alpha)(\bar{\nu} \dashv \tau \dashv \nu).$$

(Notice that by (M 2), if $\bar{\tau} \in S_{\alpha_{\bar{\nu}}} \cap \bar{\nu}$, then $\bar{\tau} \dashv \pi_{\bar{\nu}\nu}(\bar{\tau})$.)

(M 7) says, more or less, that a level "intermediate" between two levels cannot peter out at some limit point. See Fig. 6.

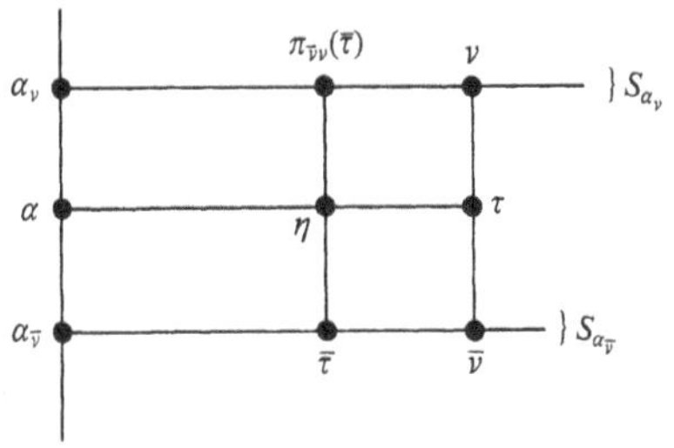

Fig. 6

That defines the notion of a morass. We should point out that it is known that such a structure cannot be constructed in ZFC. But if we assume $V = L$, as we do from now on, then we can construct a morass, though it will take some time to do so. We start with a simple model-theoretic notion.

If ν is a limit ordinal and $X \subseteq J_\nu$, we write

$$X \prec_Q J_\nu,$$

and say that X is a *Q-submodel* of J_ν, iff, for all Σ_0-formulas $\varphi(v_0, v_1)$ of $\mathscr{L}_X$,

$$\models_X (\forall \alpha)(\exists \beta > \alpha)\, \varphi(\beta, J_\beta) \quad \text{iff } \models_{J_\nu} (\forall \alpha)(\exists \beta > \alpha)\, \varphi(\beta, J_\beta).$$

Clearly, if $X \prec_Q J_\nu$ then $X \prec_1 J_\nu$, since by $\lim(\nu)$ we can bind any existential quantifier by some J_β for $\beta < \nu$. Conversely, if $X \prec_1 J_\nu$ is such that $X \cap \nu$ is cofinal in ν, then $X \prec_Q J_\nu$. Hence the notion of a Q-submodel lies between the notions of a Σ_1-submodel and a cofinal Σ_1-submodel. We shall use the notion of a Q-submodel, or rather the associated notion of a Q-embedding, when we define the morass tree relation $\dashv$.

As a first step to the construction of a morass, we define

$\mathscr{S}$
$$\mathscr{S} = \{(\alpha, \nu) \mid \alpha < \nu < \omega_2 \wedge \omega < \alpha \leqslant \omega_1 \wedge \nu \text{ is adequate} \wedge \models_{J_\nu} \text{“}\alpha \text{ is regular and is the largest cardinal”}\}.$$

Notice that if $(\alpha, \nu) \in \mathscr{S}$, then α is admissible, hence adequate.

Define S^0, S^1, S, S_α, α_ν now as before. Notice that $S_{\alpha_\nu} \cap \nu$ is uniformly $\Sigma_1^{J_\nu}(\{\alpha_\nu\})$ for $\nu \in S^1$. Notice also that S_α is always closed in $\sup(S_\alpha)$ and that $\sup(S_\alpha) \in S_\alpha$ for $\alpha < \omega_1$.

If $\nu \in S^1$ now, then clearly $\nu \neq \omega_1$, so we may define

$\beta(\nu)$ $\quad$ $\beta(\nu) =$ the least $\beta \geqslant \nu$ such that ν is singular over J_β;
$n(\nu)$ $\quad$ $n(\nu) =$ the least $n \geqslant 1$ such that ν is Σ_n-singular over $J_{\beta(\nu)}$;
$\varrho(\nu)$ $\quad$ $\varrho(\nu) = \varrho_{\beta(\nu)}^{n(\nu)-1}$;
$A(\nu)$ $\quad$ $A(\nu) = A_{\beta(\nu)}^{n(\nu)-1}$.

Notice that as ν is Σ_{n-1}-regular over $J_{\beta(\nu)}$, we have $\varrho(\nu) \geqslant \nu$, and that if $\varrho(\nu) > \nu$, then, since $\varrho(\nu) \leqslant \beta(\nu)$, $\models_{J_{\varrho(\nu)}}$ “ν is regular”. Notice also that if $\nu \in S_\alpha$, then $\models_{J_{\beta(\nu)}}$ “α is regular”. For if $\beta(\nu) = \nu$, this is true because $\nu \in S_\alpha$, whilst if $\beta(\nu) > \nu$ it follows from the two facts $\models_{J_{\beta(\nu)}}$ “ν is regular” and $\models_{J_\nu}$ “α is regular”. Also, if $\tau \in S_\alpha \cap \nu$, where $\nu \in S_\alpha$, then as α is the largest cardinal in J_ν, τ cannot be a cardinal in J_ν, so $\varrho(\tau) \leqslant \beta(\tau) < \nu \leqslant \varrho(\nu) \leqslant \beta(\nu)$.

We turn now to the definition of the tree relation $\dashv$ and the maps $\pi_{\nu\tau}$, $\nu \dashv \tau$. The idea is as follows. To each $\nu \in S^1$ we shall associate a certain parameter $p(\nu) \in J_{\varrho(\nu)}$, so that, in particular, every element of $J_{\varrho(\nu)}$ is Σ_1-definable from parameters in $\alpha_\nu \cup \{p(\nu)\}$ in the structure $\langle J_{\varrho(\nu)}, A(\nu)\rangle$. We shall then set $\nu \dashv \tau$ iff $\alpha_\nu < \alpha_\tau$ and there is an embedding

$$\sigma\colon \langle J_{\varrho(\nu)}, A(\nu)\rangle \prec_1 \langle J_{\varrho(\tau)}, A(\tau)\rangle$$

such that

$$(\sigma \restriction J_\nu)\colon J_\nu \prec_Q J_\tau, \qquad \sigma \restriction \alpha_\nu = \mathrm{id} \restriction \alpha_\nu, \qquad \sigma(p(\nu)) = p(\tau).$$

The definability property of $p(\nu)$ just mentioned will ensure that σ is unique here, and we shall set

$$\pi_{\nu\tau} = (\sigma \restriction \nu) \cup \{(\tau, \nu)\}.$$

Let us remark right away that the requirement that $(\sigma \restriction J_\nu)\colon J_\nu \prec_Q J_\tau$ in the above definition is a minor technical matter connected with morass axiom (M 1), and otherwise plays no role in our development. So the reader can for the most part ignore this point.

The definition of the parameter $p(\nu)$ depends upon the nature of ν. There are two cases to consider. We partition S^1 into two sets thus:

P $\quad$ $P = \{\nu \in S^1 \mid n(\nu) = 1 \wedge \mathrm{succ}(\beta(\nu))\}$;
R $\quad$ $R = S^1 - P$.

$\gamma(\nu)$ $\quad$ In case $\nu \in P$, let $\gamma(\nu)$ denote that ordinal γ such that $\beta(\nu) = \gamma + 1$.

Notice that if $\nu \in P$, then $\varrho(\nu) = \beta(\nu) = \gamma(\nu) + 1$, $A(\nu) = \emptyset$, and (since $\nu \leqslant \beta(\nu)$ and ν is adequate) $\nu \leqslant \gamma(\nu)$. Whilst if $\nu \in R$, then $\lim(\varrho(\nu))$.

Let $\nu \in S^1$ now, and set $\alpha = \alpha_\nu$, $\beta = \beta(\nu)$, $n = n(\nu)$, $\varrho = \varrho(\nu)$, $A = A(\nu)$, and if $\nu \in P$, $\gamma = \gamma(\nu)$. $\alpha, \beta, n, \varrho, A$ γ

2.1 Lemma. $\varrho_\beta^n \leqslant \alpha$.

Proof. Since α is the largest cardinal in J_ν, the definition of β, n ensures that there is a $\Sigma_n(J_\beta)$ map $f \subseteq \nu \times \alpha$ such that $f''\alpha$ is cofinal in ν. Since either $\beta = \nu$ or else $\models_{J_\beta}$ "ν is regular", $f \notin J_\beta$. But if $\varrho_\beta^n > \nu$, then by amenability, $f = f \cap (\nu \times \nu) \in J_{\varrho_\beta^n} \subseteq J_\beta$. Hence $\varrho_\beta^n \leqslant \nu$. It follows that there is a $\Sigma_n(J_\beta)$ map g such that $g''\nu = J_\beta$. Let $(A_\xi \mid \xi < \alpha)$ be a partition of α into α sets of cardinality α in J_ν, and let f_ξ be the $<_J$-least map from A_ξ onto $f(\xi)$ for each $\xi \in \mathrm{dom}(f)$. Since α is the largest cardinal in J_ν, we have $f_\xi \in J_\nu$ for all ξ. Let

$$k = \bigcup \{f_\xi \mid \xi \in \mathrm{dom}(f)\}.$$

Then k is a $\Sigma_n(J_\beta)$ function such that $k''\alpha = \nu$. Hence $g \circ k$ is a $\Sigma_n(J_\beta)$ function such that $g \circ k''\alpha = J_\beta$. Thus $\varrho_\beta^n \leqslant \alpha$. $\square$

Case 1. $\nu \in P$.

2.2 Lemma. *There is a $q \in J_\gamma$ such that every $x \in J_\gamma$ is J_γ-definable from parameters in $J_\alpha \cup \{q\}$.*

Proof. By 2.1 there is a $p \in J_\beta$ such that every $x \in J_\beta$ is Σ_1-definable in J_β from parameters in $J_\alpha \cup \{p\}$. Since $J_\beta = \mathrm{rud}(J_\gamma)$, there is a rudimentary function f and an element $q \in J_\gamma$ such that $p = f(J_\gamma, q)$. We show that q is as in the lemma.

Let $x \in J_\gamma$. For some Σ_0 formula φ of $\mathscr{L}$ and some $\vec{z} \in J_\alpha$, x is the unique x in J_β such that

$$(\exists y \in J_\beta)[\models_{J_\beta} \varphi(\mathring{y}, \mathring{\vec{z}}, \mathring{p}, \mathring{x})].$$

Pick $y \in J_\beta$ so that

$$\models_{J_\beta} \varphi(\mathring{y}, \mathring{\vec{z}}, \mathring{p}, \mathring{x}).$$

For some rudimentary function g and some $u \in J_\gamma$, we have $y = g(J_\gamma, u)$. So

$$(*) \qquad \models_{J_\beta} \varphi(g(J_\gamma, u)^\circ, \mathring{\vec{z}}, f(J_\gamma, q)^\circ, \mathring{x}).$$

Since g, f are rudimentary, hence simple, the formula $\varphi(g(y, u), \vec{z}, f(y, q), x)$ is Σ_0 in variables $y, u, \vec{z}, q, x$. So by VI.1.18 there is an $\mathscr{L}$-formula ψ such that $(*)$ is equivalent to

$$(**) \qquad \models_{J_\gamma} \psi(\mathring{x}, \mathring{u}, \mathring{\vec{z}}, \mathring{q}).$$

It follows at once that x is the unique element of J_γ such that $\models_{J_\gamma} \exists u \psi(\mathring{x}, \mathring{u}, \mathring{\vec{z}}, \mathring{q})$, and the lemma is proved. $\square$

$q(\nu)$ Let $q(\nu)$ be the $<_J$-least $q \in J_\gamma$ such that every $x \in J_\gamma$ is J_γ-definable from parameters in $\alpha \cup \{q\}$. (Since α is adequate, α is closed under Gödel's Pairing Function, so by VI.3.17, there is a $\Sigma_1^{J_\alpha}$ map from α onto J_α. But $\gamma \geqslant \nu > \alpha$, so this map is an element of J_γ. Hence $q(\nu)$ exists by virtue of 2.2.)

Set

$p(\nu)$
$$p(\nu) = (q(\nu), \gamma(\nu), \nu, \alpha_\nu).$$

That defines $p(\nu)$ in Case 1. We check that $p(\nu)$ has the property we mentioned earlier, that every element of $J_{\varrho(\nu)}$ is Σ_1 definable from parameters in $\alpha_\nu \cup \{p(\nu)\}$ in $\langle J_{\varrho(\nu)}, A(\nu)\rangle$. In this case, what this says is that every element of J_β is Σ_1 definable from parameters in $\alpha \cup \{p(\nu)\}$ in J_β.

Let $x \in J_\beta$. Then for some rudimentary function f and some $u \in J_\gamma$, $x = f(J_\gamma, u)$. Since J_β is rud closed, $f \upharpoonright J_\beta$ is $\Sigma_1^{J_\beta}$. So x is Σ_1 definable from γ and u in J_β. Since $\gamma = (p(\nu))_1$, we are done if we can show that u is Σ_1 definable from parameters in $\alpha \cup \{p(\nu)\}$ in J_β. Well, by choice of $q(\nu)$ there is an $\mathscr{L}$-formula φ and elements $\vec{z} \in \alpha$ such that

$$u = \text{the unique } u \in J_\gamma \text{ such that } \models_{J_\gamma} \varphi(\mathring{u}, \vec{\mathring{z}}, \mathring{q}(\nu)).$$

But this defines u in a $\Sigma_1^{J_\beta}$ fashion from $\gamma, \vec{z}, q(\nu)$. Hence as $\gamma = (p(\nu))_1$ and $q(\nu) = (p(\nu))_0$, we see that u is Σ_1 definable from $\vec{z}, p(\nu)$ in J_β, and we are done.

Case 2. $\nu \in R$.

$q(\nu)$ Let $q(\nu)$ be the $<_J$-least $q \in J_\varrho$ such that every $x \in J_\varrho$ is Σ_1-definable from parameters in $\alpha \cup \{q\}$ in $\langle J_\varrho, A\rangle$. Since $\varrho^1_{\varrho, A} = \varrho^n_\beta$, this is possible by virtue of 2.1. (An argument as in Case 1 allows us to write α in place of J_α in this definition.) Set

$p(\nu)$
$$p(\nu) = \begin{cases} (q(\nu), \nu, \alpha_\nu), & \text{if } \nu < \varrho, \\ (q(\nu), \alpha_\nu), & \text{if } \nu = \varrho. \end{cases}$$

That defines $p(\nu)$ in Case 2.

Note that as $q(\nu) = (p(\nu))_0$, it follows from the definition of $q(\nu)$ that every element of $J_{\varrho(\nu)}$ is Σ_1 definable from parameters in $\alpha_\nu \cup \{p(\nu)\}$ in $\langle J_{\varrho(\nu)}, A(\nu)\rangle$ in this case also.

Having now defined the parameters $p(\nu)$, $\nu \in S^1$, we establish a series of lemmas which will enable us to construct a morass in the manner outlined earlier.

2.3 Lemma. *The sequences* $\langle (J_{\varrho(\eta)}, A(\eta), J_\eta, p(\eta)) \mid \eta \in S_{\alpha_\nu} \cap \nu\rangle$ *is uniformly* $\Sigma_1^{J_\nu}(\{\alpha_\nu\})$ *for all* $\nu \in S^1$.

Proof. Since ν is adequate, this follows easily from the fact that if $\eta \in S_{\alpha_\nu} \cap \nu$, then $\beta(\eta) < \nu$, mentioned earlier. □

2.4 Lemma. *Let* $\nu, \tau \in S^1$. *Suppose that*

$$\sigma\colon \langle J_{\varrho(\nu)}, A(\nu)\rangle \prec_1 \langle J_{\varrho(\tau)}, A(\tau)\rangle$$

is such that $\sigma(p(\nu)) = p(\tau)$. *Then* σ is uniquely determined by $\sigma \restriction \alpha_\nu$. *Moreover*:

(i) $\nu \in P \leftrightarrow \tau \in P$;

(ii) $\sigma(\alpha_\nu) = \alpha_\tau$;

(iii) $\nu < \varrho(\nu) \leftrightarrow \tau < \varrho(\tau)$;

(iv) $\nu < \varrho(\nu) \rightarrow \sigma(\nu) = \tau$;

(v) *if* $\nu \in P$, *then* $\sigma(\gamma(\nu)) = \gamma(\tau)$;

(vi) $\sigma(q(\nu)) = q(\tau)$.

Proof. The uniqueness of σ follows from the definability property of $p(\nu)$. The remaining assertions of the lemma all follow from the definitions of $p(\nu)$ and $p(\tau)$. □

2.5 Lemma. *Let* $\nu \in S^1$, $\bar{\varrho} \leqslant \varrho(\nu)$, $\bar{A} \subseteq J_{\bar{\varrho}}$. *Let*

$$\sigma\colon \langle J_{\bar{\varrho}}, \bar{A}\rangle \prec_1 \langle J_{\varrho(\nu)}, A(\nu)\rangle$$ σ

be such that $p(\nu) \in \operatorname{ran}(\sigma)$. *Then there is a (necessarily unique)* $\bar{\nu} \in S^1$ *such that* $\bar{\varrho} = \varrho(\bar{\nu})$, $\bar{A} = A(\bar{\nu})$. *Moreover,* $\sigma(p(\bar{\nu})) = p(\nu)$.

Proof. To commence, notice that $\langle J_{\bar{\varrho}}, \bar{A}\rangle$ is amenable. For if $\nu \in P$, then $A(\nu) = \emptyset$, so $\bar{A} = \emptyset$ and amenability is trivial, and if $\nu \in R$, then $\lim(\varrho(\nu))$, so $\lim(\bar{\varrho})$, and for each $\eta < \bar{\varrho}$, we have

$$\models_{\langle J_{\varrho(\nu)}, A(\nu)\rangle} \exists x\,[x = A(\nu) \cap J_{\sigma(\eta)}],$$

so

$$\models_{\langle J_{\bar{\varrho}}, \bar{A}\rangle} \exists x\,[x = \bar{A} \cap J_\eta].$$

Set $\alpha = \alpha_\nu$, $\beta = \beta(\nu)$, $n = n(\nu)$, $\varrho = \varrho(\nu)$, $A = A(\nu)$, $p = p(\nu)$, $q = q(\nu)$, and, if $\nu \in P$, $\gamma = \gamma(\nu)$. $\alpha, \beta, n, \varrho, A, p, q, \gamma$

Case 1. $\nu \in P$.

Thus $\beta = \varrho = \gamma + 1$, $\bar{A} = A = \emptyset$, and ν is regular over J_γ. Since $p \in \operatorname{ran}(\sigma)$, we have $q, \gamma, \nu, \alpha \in \operatorname{ran}(\sigma)$. Let $\bar{q} = \sigma^{-1}(q)$, $\bar{\gamma} = \sigma^{-1}(\gamma)$, $\bar{\alpha} = \sigma^{-1}(\alpha)$, $\bar{\nu} = \sigma^{-1}(\nu)$. Let $\bar{\sigma} = \sigma \restriction J_{\bar{\gamma}}$. Clearly, $\bar{\sigma}\colon J_{\bar{\gamma}} \prec J_\gamma$ and $q \in \operatorname{ran}(\bar{\sigma})$. $q, \bar{\gamma}, \bar{\alpha}, \bar{\nu}$, $\bar{\sigma}$

Claim A. $\bar{\nu} \in S^1$ and $\alpha_{\bar{\nu}} = \bar{\alpha}$.

Since $\sigma\colon J_{\bar{\varrho}} \prec_1 J_\varrho$ and $\sigma(\bar{\nu}) = \nu$, we have $(\sigma \restriction J_{\bar{\nu}})\colon J_{\bar{\nu}} \prec J_\nu$. Hence $\bar{\nu}$ is adequate. Moreover, since $\sigma(\bar{\alpha}) = \alpha$,

$$\models_{J_{\bar{\nu}}} \text{“}\bar{\alpha} \text{ is regular and is the largest cardinal”}.$$

Thus $\bar{\nu} \in S_{\bar{\alpha}}$. Claim A is proved.

Claim B. $\bar{\sigma}''\,\bar{\nu}$ is cofinal in ν.

For each $m < \omega$, set

$$X_m = \{x \in J_\gamma \mid x \text{ is } \Sigma_{m+1}\text{-definable from parameters in } \alpha \cup \{q\} \text{ in } J_\gamma\}.$$

Then $X_m \prec_m J_\gamma$, and there is a J_γ-definable map from α onto X_m. Since α is the largest cardinal in J_ν and $\alpha \subseteq X_m \prec_1 J_\gamma$, $X_m \cap \nu$ is transitive, so set $\nu_m = X_m \cap \nu$. Since ν is regular over J_γ and there is a J_γ-definable map from α onto ν_m, we must have $\nu_m < \nu$. But by choice of q,

$$\bigcup_{m<\omega} X_m = J_\gamma .$$

Thus $\sup_{m<\omega} \nu_m = \nu$. But for each m, ν_m is J_γ-definable from q, so $\{\nu_m \mid m < \omega\} \subseteq \operatorname{ran}(\bar{\sigma})$. This proves Claim B.

For later use, we point out that the sequence $(\nu_m \mid m < \omega)$ is clearly $\Sigma_1^{J_{\gamma+1}}(\{p\})$.

Claim C. $\bar{\nu}$ is regular over $J_{\bar{\gamma}}$.

We know that ν is regular over J_γ. But $\bar{\sigma}$: $J_{\bar{\gamma}} \prec J_\gamma$ and

$$\bar{\nu} = \bar{\gamma} \rightarrow \nu = \gamma,$$
$$\bar{\nu} < \bar{\gamma} \rightarrow \bar{\sigma}(\bar{\nu}) = \nu,$$

so Claim C is immediate.

Claim D. $\bar{q}$ is the $<_J$-least element of $J_{\bar{\gamma}}$ such that every $x \in J_{\bar{\gamma}}$ is $J_{\bar{\gamma}}$-definable from parameters in $\bar{\alpha} \cup \{\bar{q}\}$.

Let $x \in J_{\bar{\gamma}}$. Then $\sigma(x) \in J_\gamma$, so for some $\vec{\delta} \in \alpha$, $\bar{\sigma}(x)$ is J_γ-definable from q, $\vec{\delta}$. Set $s = (\vec{\delta})$, and let φ be a formula of $\mathscr{L}$ such that:

(i) $\models_{J_\gamma} \forall z \exists y \forall y' [y' = y \leftrightarrow \varphi(y', z, \mathring{q})]$;
(ii) $\models_{J_\gamma} \forall z \forall y [\varphi(y, z, \mathring{q}) \rightarrow (\exists \vec{\xi})(z = (\vec{\xi}))]$;
(iii) $(\forall y \in J_\gamma)[y = \bar{\sigma}(x) \leftrightarrow \models_{J_\gamma} \varphi(\mathring{y}, \mathring{s}, \mathring{q})]$.

Let t be the $<_J$-least element of J_γ such that $\models_{J_\gamma} \varphi(\mathring{\bar{\sigma}(x)}, \mathring{t}, \mathring{q})$. Then t is J_γ-definable from $\bar{\sigma}(x)$, q. But $\bar{\sigma}(x), q \in \operatorname{ran}(\bar{\sigma}) \prec J_\gamma$. Hence $t \in \operatorname{ran}(\bar{\sigma})$. By choice of t, $t \leqslant_J s$, so $t \in J_\alpha$. Thus $t = (\vec{\xi})$ for some $\vec{\xi} \in \alpha$. By (i) above,

$$(\forall y \in J_\gamma)[y = \bar{\sigma}(x) \leftrightarrow \models_{J_\gamma} \varphi(\mathring{y}, \mathring{t}, \mathring{q})].$$

Applying $\bar{\sigma}^{-1}$ and setting $\bar{t} = \bar{\sigma}^{-1}(t) = (\vec{\zeta})$, we get,

$$(\forall y \in J_{\bar{\gamma}})[y = x \leftrightarrow \models_{J_{\bar{\gamma}}} \varphi(\mathring{y}, \mathring{\bar{t}}, \mathring{\bar{q}}).$$

But $\vec{\zeta} \in \bar{\alpha}$. Thus x is $J_{\bar{\gamma}}$-definable from parameters in $\alpha \cup \{\bar{q}\}$.

Now suppose that $\bar{q}' <_J \bar{q}$ also has the property that every element of $J_{\bar{\gamma}}$ is $J_{\bar{\gamma}}$-definable from parameters in $\bar{\alpha} \cup \{\bar{q}'\}$. Then in particular there are $\vec{\xi} \in \bar{\alpha}$ such that $\bar{q}$ is $J_{\bar{\gamma}}$-definable from $\vec{\xi}$, $\bar{q}'$. Applying $\bar{\sigma}$ and setting $q' = \bar{\sigma}(\bar{q}')$, $\vec{\xi} = \bar{\sigma}(\vec{\zeta})$, we see that $q' <_J q$ and that q is J_γ-definable from $\vec{\xi}$, q'. Hence every element of J_γ is J_γ-definable from parameters in $\bar{\alpha} \cup \{q'\}$, contrary to the choice of q. That completes the proof of Claim D.

Claim E. $\bar{\nu}$ is Σ_1-singular over $J_{\bar{\gamma}+1}$.

Using Claims A, C, D we may define $(\bar{v}_m | m < \omega)$ from $J_{\bar{\gamma}}, \bar{\alpha}, \bar{q}, \bar{v}$ exactly as we defined $(v_m | m < \omega)$ from J_γ, α, q, v in the proof of Claim B. Then $(\bar{v}_m | m < \omega)$ is a $\Sigma_1(J_{\bar{\gamma}+1})$ sequence which is cofinal in $\bar{v}$, proving Claim E.

Claim F. $\beta(\bar{v}) = \bar{\gamma} + 1$, $n(\bar{v}) = 1$, $\bar{v} \in P$, $\varrho(\bar{v}) = \bar{\gamma} + 1$, $\gamma(\bar{v}) = \bar{\gamma}$.

By Claims C and E.

Claim G. $q(\bar{v}) = \bar{q}$, $\sigma(p(\bar{v})) = p(v)$.

By Claims F and D, $q(\bar{v}) = \bar{q}$. Thus by Claims A and F, $p(\bar{v}) = (\bar{q}, \bar{\gamma}, \bar{v}, \bar{\alpha})$. Hence $\sigma(p(\bar{v})) = p(v)$. Claim G is proved.

That completes the proof of the lemma in Case 1.

Case 2. $v \in R$.

Set $\bar{q} = \sigma^{-1}(q)$, $\bar{\alpha} = \sigma^{-1}(\alpha)$. Set $\bar{v} = \sigma^{-1}(v)$ if $v < \varrho$ and set $\bar{v} = \bar{\varrho}$ if $v = \varrho$. $\bar{q}, \bar{\alpha}, \bar{v}$

By VI.5.6 there is a unique $\bar{\beta} \geq \bar{\varrho}$ such that $\bar{\varrho} = \varrho_{\bar{\beta}}^{n-1}$, $\bar{A} = A_{\bar{\beta}}^{n-1}$, and an embedding $\tilde{\sigma}: J_{\bar{\beta}} \prec_n J_\beta$, $\sigma \subseteq \tilde{\sigma}$. $\bar{\beta}$ $\tilde{\sigma}$

Claim H. $\bar{v} \in S^1$ and $\alpha_{\bar{v}} = \bar{\alpha}$.

If $v = \varrho$, then $\bar{v} = \bar{\varrho}$ and $\sigma: J_{\bar{v}} \prec_1 J_v$. And if $v < \varrho$, then $\sigma(\bar{v}) = v$, so $(\sigma \restriction J_{\bar{v}}): J_{\bar{v}} \prec J_v$. In either case, $\bar{v}$ is adequate. Since we always have $(\sigma \restriction J_{\bar{v}}): J_{\bar{v}} \prec_1 J_v$ and $\sigma(\bar{\alpha}) = \alpha$, $\bar{\alpha}$ is regular inside $J_{\bar{v}}$ and is the largest cardinal inside $J_{\bar{v}}$. Hence $\bar{v} \in S_{\bar{\alpha}}$. Claim H is proved.

Claim I. $\bar{v}$ is Σ_{n-1}-regular over $J_{\bar{\beta}}$.

Suppose not. Then, since $\bar{\alpha}$ is the largest cardinal in $J_{\bar{v}}$, we can find a $\Sigma_{n-1}(J_{\bar{\beta}})$ map f such that $f''\bar{\alpha}$ is cofinal in $\bar{v}$. There are now two cases to consider.

Suppose first that $v < \varrho$. Thus $\bar{v} < \bar{\varrho}$ and $\sigma(\bar{v}) = v$. If $f \in J_{\bar{\beta}}$, then by applying $\tilde{\sigma}: J_{\bar{\beta}} \prec_n J_\beta$, we see that $\tilde{\sigma}(f)$ maps a subset of α cofinally into v, contrary to v being regular inside J_β. Hence $f \notin J_{\bar{\beta}}$. But by using Gödel's pairing function we can code f as a $\Sigma_{n-1}(J_{\bar{\beta}})$ subset of $\bar{v}$. Thus $\mathscr{P}(\bar{v}) \cap \Sigma_{n-1}(J_{\bar{\beta}}) \nsubseteq J_{\bar{\beta}}$. Thus $\varrho_{\bar{\beta}}^{n-1} \leq \bar{v}$. But $\varrho_{\bar{\beta}}^{n-1} = \bar{\varrho} > \bar{v}$, so we have a contradiction. That proves the claim in the first case.

Now suppose that $v = \varrho$. Thus $\bar{v} = \bar{\varrho}$. In $J_{\bar{v}}$, let $(A_\xi | \xi < \bar{\alpha})$ be a partition of $\bar{\alpha}$ into $\bar{\alpha}$ many sets of cardinality $\bar{\alpha}$. For each $\xi \in \mathrm{dom}(f)$, let $k_\xi \in J_{\bar{v}}$ be the $<_J$-least map from A_ξ onto $f(\xi)$. (Since $\bar{\alpha}$ is the largest cardinal in $J_{\bar{v}}$, k_ξ is well-defined here.) Set

$$k = \bigcup \{k_\xi | \xi \in \mathrm{dom}(f)\}.$$

Clearly, k is a $\Sigma_{n-1}(J_{\bar{\beta}})$ function such that $k''\bar{\alpha} = \bar{v}$. But $\bar{v} = \bar{\varrho} = \varrho_{\bar{\beta}}^{n-1}$ and $\bar{\alpha} < \bar{v}$, so this contradicts the definition of the Σ_{n-1}-projectum. Claim I is proved.

Claim J. $\bar{q}$ is the $<_J$-least element of $J_{\bar{\varrho}}$ such that every element of $J_{\bar{\varrho}}$ is Σ_1-definable from parameters in $\bar{\alpha} \cup \{\bar{q}\}$ in $\langle J_{\bar{\varrho}}, \bar{A} \rangle$.

Let $x \in J_{\bar{\varrho}}$. Then $\sigma(x) \in J_\varrho$ so for some $\vec{\delta} \in \alpha$, $\sigma(x)$ is Σ_1-definable from $q, \vec{\delta}$ in $\langle J_\varrho, A \rangle$. Set $s = (\vec{\delta})$. Let φ be a Σ_0-formula of $\mathscr{L}$ such that:

(i) $\models_{\langle J_\rho, A \rangle} \forall z \, \exists y \, \forall y' [y' = y \leftrightarrow \exists u \varphi(u, y', z, \mathring{q})]$;

(ii) $\models_{\langle J_\rho, A \rangle} \forall z \, \forall y [\exists u \varphi(u, y, z, \mathring{q}) \rightarrow (\exists \vec{\xi})(z = (\vec{\xi}))]$;

(iii) $(\forall y \in J_\varrho)[y = \sigma(x) \leftrightarrow \models_{\langle J_\rho, A \rangle} \exists u \, \varphi(u, \mathring{y}, \mathring{s}, \mathring{q})]$.

Let $<^*$ be the lexicographic ordering on $L \times L$ induced by $<_J$. Clearly, $<^*$ is Σ_1 definable, and $<^* \cap (J_\varrho \times J_\varrho)$ is uniformly $\Sigma_1^{J_\rho}$ for all limit $\varrho > 0$. Let (t, u_0) be the $<^*$-least pair such that

$$\models_{\langle J_\rho, A\rangle} \varphi(\mathring{u}_0, \mathring{\sigma}(x) \mathring{t}, \mathring{q}).$$

Then (t, u_0) is Σ_1-definable from $\sigma(x), q$ in $\langle J_\varrho, A\rangle$. Hence $t, u_0 \in \operatorname{ran}(\sigma)$. Since $t \leqslant_J s$, $t \in J_\alpha$. Thus $t = (\vec{\xi})$ for some $\vec{\xi} \in \alpha$. By (i) above,

$$(\forall y \in J_\varrho)[y = \sigma(x) \leftrightarrow \models_{\langle J_\rho, A\rangle} \exists u \varphi(u, \mathring{y}, \mathring{t}, \mathring{q})].$$

Applying σ^{-1}, we obtain

$$(\forall y \in J_{\bar{\varrho}})[y = x \leftrightarrow \models_{\langle J_{\bar{\beta}}, A\rangle} \exists u \varphi(u, \mathring{y}, \mathring{\bar{t}}, \mathring{\bar{q}})],$$

where $\bar{t} = \sigma^{-1}(t) = (\vec{\zeta})$, $\vec{\zeta} \in \bar{\alpha}$. Hence x is Σ_1-definable from parameters in $\alpha \cup \{\bar{q}\}$ in $\langle J_{\bar{\varrho}}, \bar{A}\rangle$. The rest of the proof of Claim J is entirely similar to the argument used in proving the minimality of $\bar{q}$ in Claim D (for Case 1). So Claim J is established.

Claim K. $\bar{\nu}$ is Σ_n-singular over $J_{\bar{\beta}}$.

By Claim J,

$$J_{\bar{\varrho}} = h^*_{\bar{\varrho}, \bar{A}}(J_{\bar{\alpha}} \times \{q\}).$$

So there is a $\Sigma_1(\langle J_{\bar{\varrho}}, \bar{A}\rangle)$ map from a subset of $\bar{\alpha}$ onto (in particular) $\bar{\nu}$. Since $\bar{\varrho} = \varrho_{\bar{\beta}}^{n-1}$, $\bar{A} = A_{\bar{\beta}}^{n-1}$, this map is $\Sigma_n(J_{\bar{\beta}})$. Claim K is proved.

Claim L. $\bar{\beta} = \beta(\bar{\nu})$, $n = n(\bar{\nu})$, $\bar{\nu} \in R$, $\bar{\varrho} = \varrho(\bar{\nu})$, $\bar{A} = A(\bar{\nu})$.

Directly from Claims I and K. (For $\bar{\nu} \in R$, notice that as $\tilde{\sigma}$: $J_{\bar{\beta}} \prec_1 J_\beta$, $\lim(\bar{\beta})$ follows from $\lim(\beta)$.)

Claim M. $\bar{q} = q(\bar{\nu})$ and $\sigma(p(\bar{\nu})) = p(\nu)$.

By Claims L and J, $\bar{q} = q(\bar{\nu})$. If $\nu < \varrho$ now, then $\bar{\nu} < \bar{\varrho}$ and we have $p(\bar{\nu}) = (q(\bar{\nu}), \bar{\nu}, \alpha_{\bar{\nu}})$, so $\sigma(p(\bar{\nu})) = p(\nu)$ by Claim H and the equality $\sigma(\bar{\nu}) = \nu$. If $\nu = \varrho$, then $\bar{\nu} = \bar{\varrho}$ and $p(\bar{\nu}) = (q(\bar{\nu}), \alpha_{\bar{\nu}})$, so again $\sigma(p(\bar{\nu})) = p(\nu)$.

That completes the proof of the lemma. □

$\bar{\nu}, \nu, \bar{\alpha}$ **2.6 Lemma.** *Let $\bar{\nu} \in S_{\bar{\alpha}}$, $\nu \in S_\alpha$, $\bar{\alpha} < \alpha$, and suppose that $\bar{\nu}$ is a limit point of $S_{\bar{\alpha}}$. Let*

σ
$$\sigma: \langle J_{\varrho(\bar{\nu})}, A(\bar{\nu})\rangle \prec_1 \langle J_{\varrho(\nu)}, A(\nu)\rangle$$

be such that $\sigma(p(\bar{\nu})) = p(\nu)$.

ν' *Let $\nu' = \sup(\sigma''\bar{\nu})$. Then $\nu' \in S_\alpha$ and there is an embedding*

$$\sigma': \langle J_{\varrho(\bar{\nu})}, A(\bar{\nu})\rangle \prec_1 \langle J_{\varrho(\nu')}, A(\nu')\rangle$$

such that $\sigma \restriction J_{\bar{\nu}} \subseteq \sigma'$, $(\sigma' \restriction J_{\bar{\nu}})$: $J_{\bar{\nu}} \prec_Q J_{\nu'}$, and $\sigma'(p(\bar{\nu})) = p(\nu')$.

Proof. Case 1. $\nu \in P$.

Let $(\nu_m | m < \omega)$ be the sequence defined in Claim B of 2.5. This sequence is $\Sigma_1^{J_{\rho(\nu)}}(\{p(\nu)\})$ and is cofinal in ν. Hence $\{\nu_m | m < \omega\} \subseteq \mathrm{ran}(\sigma)$, and so $\nu' = \nu$. There is nothing to prove in this case.

Case 2. $\nu \in R$.

Set $\beta = \beta(\nu)$, $n = (\nu)$, $\varrho = \varrho(\nu)$, $A = A(\nu)$, $q = q(\nu)$, $p = p(\nu)$, $\bar\varrho = \varrho(\bar\nu)$, $\bar A = A(\bar\nu)$, $\bar q = q(\bar\nu)$, $\bar p = p(\bar\nu)$. $\beta, n, \varrho, A, q, p, \bar\varrho, \bar A, \bar q, \bar p$

Now, $S_{\bar\alpha} \cap \bar\nu$ is $\Sigma_1^{J_{\bar\nu}}(\{\bar\alpha\})$ and $S_\alpha \cap \nu$ is $\Sigma_1^{J_\nu}(\{\alpha\})$ by the same definition. And by 2.4, $\sigma(\bar\alpha) = \alpha$. Hence as $S_{\bar\alpha} \cap \bar\nu$ is cofinal in $\bar\nu$ and $\nu' = \sup(\sigma''\bar\nu)$, applying $(\sigma \restriction J_{\bar\nu})\colon J_{\bar\nu} \prec_1 J_\nu$ gives $\nu' = \sup(S_\alpha \cap \nu')$. But S_α is closed in $\sup(S_\alpha)$ and $\nu' \leqslant \nu \in S_\alpha$. Hence $\nu' \in S_\alpha$.

Set $\eta = \sup(\sigma''\bar\varrho)$, $\tilde A = A \cap J_\eta$. Since $\mathrm{ran}(\sigma) \subseteq J_\eta$, $p, \alpha \in J_\eta$. By Σ_0-absoluteness, $\eta, \tilde A$

$$\sigma\colon \langle J_{\bar\varrho} \bar A\rangle \prec_0 \langle J_\eta, \tilde A\rangle .$$

But σ is cofinal in η. Hence

$$\sigma\colon \langle J_{\bar\varrho}, \bar A\rangle \prec_1 \langle J_\eta, \tilde A\rangle .$$

Set

$$X = h^*_{\eta, \tilde A}(J_\alpha \times \{q\}).$$

Let

$$\pi\colon \langle J_\gamma, B\rangle \cong \langle X, \tilde A \cap x\rangle . \qquad \pi, \gamma, B$$

Thus

$$\pi\colon \langle J_\gamma, B\rangle \prec_1 \langle J_\eta, \tilde A\rangle .$$

Claim A. $\mathrm{ran}(\sigma) \subseteq X$.

Let $x \in \mathrm{ran}(\sigma)$. Then $x \in J_\varrho$, so x is Σ_1-definable from parameters in $\alpha \cup \{q\}$ in $\langle J_\varrho, A\rangle$. Let $\bar x = \sigma^{-1}(x)$. An argument as used in the proof of Claim J in 2.5 shows that $\bar x$ is Σ_1-definable from parameters in $\bar\alpha \cup \{\bar q\}$ in $\langle J_{\bar\varrho}, \bar A\rangle$. Hence for some $i \in \omega$, $\bar z \in J_{\bar\alpha}$,

$$\bar x = h_{\bar\varrho, \bar A}(i, (\bar z, \bar q)).$$

Applying $\sigma\colon \langle J_{\bar\varrho}, \bar A\rangle \prec_1 \langle J_\eta, \tilde A\rangle$, and setting $z = \sigma(\bar z)$,

$$x = h_{\eta, \tilde A}(i, (z, q)).$$

Hence $x \in X$, which proves Claim A.

Claim B. $X \cap \nu = \nu'$.

Let $\xi \in X \cap \nu$. Then for some $z \in J_\alpha$ and some $i \in \omega$,

$$\xi = h_{\eta, \tilde A}(i, (z, q)).$$

Since $\lim(\eta)$, there is a $\tau < \eta$ such that

$$\xi = h_{\tau, \tilde{A} \cap J_\tau}(i, (z, q)).$$

Since $\eta = \sup(\sigma''\bar{\varrho})$, we can pick τ here so that $\tau = \sigma(\bar{\tau})$ for some $\bar{\tau} < \bar{\varrho}$. Set

$$\theta = \sup[\nu \cap h^*_{\tau, \tilde{A} \cap J_\tau}(J_\alpha \times \{q\})],$$
$$\bar{\theta} = \sup[\bar{\nu} \cap h^*_{\bar{\tau}, \bar{A} \cap J_{\bar{\tau}}}(J_{\bar{\alpha}} \times \{\bar{q}\})].$$

Now, $\tilde{A} \cap J_\tau = A \cap J_\eta \cap J_\tau = A \cap J_\tau$, so $h_{\tau, \tilde{A} \cap J_\tau} \in J_\varrho$ by amenability. Since $\alpha < \nu$ and ν is regular inside J_ϱ, it follows that $\theta < \nu$. Similarly, $\bar{\theta} < \bar{\nu}$. But clearly, $\sigma(\bar{\theta}) = \theta$. Hence

$$\xi < \theta = \sigma(\bar{\theta}) < \sup(\sigma''\bar{\nu}) = \nu'.$$

Thus $X \cap \nu \subseteq \nu'$.

Now let $\xi \in \nu'$. For some $\bar{\delta} < \bar{\nu}$, $\xi \in \delta = \sigma(\bar{\delta})$. Since $\bar{\delta} < \bar{\nu}$, there is an $\bar{f} \in J_{\bar{\nu}}$ $\bar{f}\colon \bar{\alpha} \xrightarrow{\text{onto}} \bar{\delta}$. Since $(\sigma \restriction J_{\bar{\nu}})\colon J_{\bar{\nu}} \prec_0 J_{\nu'}$, $f = \sigma(\bar{f}) \in J_{\nu'}$ and $f\colon \alpha \xrightarrow{\text{onto}} \delta$. But by Claim A, $f \in X$. So as $\alpha \subseteq X$, $\delta = f''\alpha \subseteq X$. Hence $\xi \in X$. This shows that $\nu' \subseteq X \cap \nu$. Claim B is proved.

We have

$$\operatorname{ran}(\sigma) \prec_1 \langle J_\eta, \tilde{A}\rangle,$$
$$X \prec_1 \langle J_\eta, \tilde{A}\rangle,$$
$$\operatorname{ran}(\sigma) \subseteq X.$$

Thus

$$\operatorname{ran}(\sigma) \prec_1 \langle X, \tilde{A} \cap X\rangle.$$

So, if we set $\sigma' = \pi^{-1} \circ \sigma$, we have

$$\sigma'\colon \langle J_{\bar{\varrho}}, \bar{A}\rangle \prec_1 \langle J_\gamma, B\rangle.$$

By Claim B, $\pi^{-1} \restriction \nu' = \mathrm{id} \restriction \nu'$, so $\sigma' \restriction \bar{\nu} = \sigma \restriction \bar{\nu}$. Hence $\sigma \restriction J_{\bar{\nu}} \subseteq \sigma'$. Moreover, $(\sigma \restriction J_{\bar{\nu}})\colon J_{\bar{\nu}} \prec_0 J_{\nu'}$ cofinally, so $(\sigma' \restriction J_{\bar{\nu}})\colon J_{\bar{\nu}} \prec_Q J_{\nu'}$. So in order to complete the proof of the lemma it remains to show that $\gamma = \varrho(\nu')$, $B = A(\nu')$, and $\pi^{-1}(p) = p(\nu')$.

By Σ_0-absoluteness,

$$\pi\colon \langle J_\gamma, B\rangle \prec_0 \langle J_\varrho, A\rangle.$$

β' So by VI.5.6 there is a unique β' such that $\gamma = \varrho^{n-1}_{\beta'}$, $B = A^{n-1}_{\beta'}$, and a mapping
$\tilde{\pi}, q', p'$ $\tilde{\pi} \supseteq \pi$ such that $\tilde{\pi}\colon J_{\beta'} \prec_{n-1} J_\beta$. Set $q' = \pi^{-1}(q)$, $p' = \pi^{-1}(p)$. Notice that if $\nu < \varrho$, then $\bar{\nu} < \bar{\varrho}$ and $p = (q, \nu, \alpha)$, $\bar{p} = (\bar{q}, \bar{\nu}, \bar{\alpha})$, so $p' = (q', \pi^{-1}(\nu), \pi^{-1}(\alpha))$. But by Claim B, $\pi^{-1}(\nu) = \nu'$. And since $\pi^{-1} \restriction \nu' = \mathrm{id} \restriction \nu'$, $\pi^{-1}(\alpha) = \alpha$. Thus $p' = (q', \nu', \alpha)$, and $\nu' < \gamma$. Again, if $\nu = \varrho$, then $\bar{\nu} = \bar{\varrho}$ and $p = (q, \alpha)$, $\bar{p} = (\bar{q}, \bar{\alpha})$, so $p' = (q', \alpha)$, and (using Claim B) $\nu' = \gamma$. Hence in order to show that $\pi^{-1}(p) = p(\nu')$, it suffices to show that $\pi^{-1}(q) = q(\nu')$, i.e. that $q' = q(\nu')$.

Claim C. ν' is Σ_{n-1} regular over $J_{\beta'}$.

This is proved exactly as in Claim I in the proof of 2.5, so we do not give any details here.

Claim D. ν' is Σ_n singular over $J_{\beta'}$.

By Claim B,

$$\nu' = \nu' \cap h^*_{\eta, \tilde{A}}(J_\alpha \times \{q\}).$$

So as $\pi: \langle J_\gamma, B\rangle \prec_1 \langle J_\eta, \tilde{A}\rangle$ and $\nu' \subseteq X = \mathrm{ran}(\pi)$, we have

$$\nu' = \nu' \cap h^*_{\gamma, B}(J_\alpha \times \{q'\}).$$

So there is a $\Sigma_1(\langle J_\gamma, B\rangle)$ map from α onto ν'. But $\gamma = \varrho^{n-1}_{\beta'}$, $B = A^{n-1}_{\beta'}$, so this map is $\Sigma_n(J_{\beta'})$. Claim D is proved.

Claim E. $\beta' = \beta(\nu')$, $n = n(\nu')$, $\nu' \in R$, $\gamma = \varrho(\nu')$, $B = A(\nu')$.

By Claims C and D. (For $\nu' \in R$, notice that as $\tilde{\pi}: J_{\beta'} \prec_{n-1} J_\beta$, $\lim(\beta')$ follows from $\lim(\beta)$.)

Claim F. $q' = q(\nu')$.

By definition,

$$X = h^*_{\eta, \tilde{A}}(J_\alpha \times \{q\}).$$

So, applying π^{-1},

$$J_\gamma = h^*_{\gamma, B}(J_\alpha \times \{q'\}).$$

Hence every member of J_γ is Σ_1-definable from parametes in $\alpha \cup \{q'\}$ in $\langle J_\gamma, B\rangle$. An argument as in the proof of Claim D od 2.5 now completes the proof of Claim F.

The lemma is proved. □

We are now in a position to commence the construction of our morass. For $\nu, \tau \in S^1$, set $\nu \dashv \tau$ iff $\alpha_\nu < \alpha_\tau$ and there is an embedding

$$\sigma: \langle J_{\varrho(\nu)}, A(\nu)\rangle \prec_1 \langle J_{\varrho(\tau)}, A(\tau)\rangle$$

such that

(i) $\sigma \restriction \alpha_\nu = \mathrm{id} \restriction \alpha_\nu$;
(ii) $\sigma(p(\nu)) = p(\tau)$;
(iii) $(\sigma \restriction J_\nu): J_\nu \prec_Q J_\tau$.

Clearly, $\dashv$ is a partial ordering on S^1. And since $\nu \dashv \tau$ implies $\alpha_\nu < \alpha_\tau$, $\dashv$ is well-founded. We show that $\dashv$ is a tree. It suffices to show that if $\tau \in S_\alpha$ and $\bar{\alpha} < \alpha$, there is at most one $\nu \dashv \tau$ with $\alpha_\nu = \bar{\alpha}$. Let $\nu \dashv \tau$, $\alpha_\nu = \bar{\alpha}$. Then there is an embedding

$$\sigma: \langle J_{\varrho(\nu)}, A(\nu)\rangle \prec_1 \langle J_{\varrho(\tau)}, A(\tau)\rangle$$

as above. Now, $J_{\varrho(\nu)} = h^*_{\varrho(\nu), A(\nu)}(J_{\bar\alpha} \times \{p(\nu)\})$, so by applying σ and using properties (i) and (ii) above,

$$\operatorname{ran}(\sigma) = h^*_{\varrho(\tau), A(\tau)}(J_{\bar\alpha} \times \{p(\tau)\}).$$

Thus ran (ϱ) is entirely determined by τ and $\bar\alpha$. Since σ^{-1} is a collapsing map, it follows that $\varrho(\nu)$ is completely determined by τ and $\bar\alpha$. But if $\nu_1, \nu_2 \in S_{\bar\alpha}$ and $\nu_1 < \nu_2$, then $\varrho(\nu_1) < \nu_2 \leqslant \varrho(\nu_2)$. Hence $\nu \in S_{\bar\alpha}$ is unique here. Thus $\dashv$ is a tree on S^1. (It should be noted that the morass levels S_α are *not* the levels of this tree.)

By 2.4, if $\nu \dashv \tau$, the map σ testifying this fact is unique, so it may be denoted by
$\sigma_{\nu\tau}$ $\sigma_{\nu\tau}$. Clearly, the system of embeddings $(\sigma_{\nu\tau} | \nu \dashv \tau)$ is commutative. Set

$\pi_{\nu\tau}$ $$\pi_{\nu\tau} = (\sigma_{\nu\tau} \upharpoonright \nu) \cup \{(\tau, \nu)\}.$$

Then $(\pi_{\nu\tau} | \nu \dashv \tau)$ is a commutative system of maps

$$\pi_{\nu\tau}\colon (\nu + 1) \to (\tau + 1).$$

We show that the structure

$$\mathcal{M} = \langle S, \mathcal{S}, \dashv, (\pi_{\nu\tau})_{\nu \dashv \tau}\rangle$$

so defined is a morass.

(M 0) This is immediate. (To show that $\sup(S^0 \cap \omega_1) = \omega_1$ only requires a simple application of the Condensation Lemma. All other parts of (M 0) really *are* immediate.)

(M 1) If $\nu \dashv \tau$, then $S_{\alpha_\nu} \cap \nu$ is $\Sigma_1^{J_\nu}(\{\alpha_\nu\})$ and $S_{\alpha_\tau} \cap \tau$ is $\Sigma_1^{J_\tau}(\{\alpha_\tau\})$ by the same definition. But $(\sigma_{\nu\tau} \upharpoonright J_\nu)\colon J_\nu \prec_Q J_\tau$ and $\sigma_{\nu\tau}(\alpha_\nu) = \alpha_\tau$, so the assertions of (M 1) are immediate. (It is precisely in order to obtain (M 1) that we introduced the notion of a Q-embedding. And we only need this notion in order to prove (ii) and (iii) of (M 1) in the particular case $\gamma = \nu$.)

(M 2) Let $\bar\tau \in S_{\bar\alpha}$, $\tau \in S_\alpha$, $\bar\tau \dashv \tau$, $\bar\nu \in S_{\bar\alpha} \cap \bar\tau$, $\nu = \pi_{\bar\tau\tau}(\bar\nu)$. We must show that $\bar\nu \dashv \nu$ and $\pi_{\bar\nu\nu} \upharpoonright \bar\nu = \pi_{\bar\tau\tau} \upharpoonright \bar\nu$.

Let $\sigma = \sigma_{\bar\tau\tau} \upharpoonright J_{\bar\tau}$. Thus $\sigma\colon J_{\bar\tau} \prec_1 J_\tau$, $\sigma(\bar\alpha) = \alpha$, $\sigma(\bar\nu) = \nu$. By 2.3,

$$(\sigma \upharpoonright J_{\varrho(\bar\nu)})\colon \langle J_{\varrho(\bar\nu)}, A(\bar\nu), J_{\bar\nu}, \{p(\bar\nu)\}\rangle \prec \langle J_{\varrho(\nu)}, A(\nu), J_\nu, (p(\nu))\rangle.$$

Hence $\bar\nu \dashv \nu$ and $\pi_{\bar\nu\nu} \upharpoonright \bar\nu = (\sigma \upharpoonright J_{\bar\nu}) \upharpoonright \bar\nu = \sigma \upharpoonright \bar\nu = \sigma_{\bar\tau\tau} \upharpoonright \bar\nu = \pi_{\bar\tau\tau} \upharpoonright \bar\nu$.

(M 3) Let $\tau \in S_\alpha$. Let $\bar\alpha < \alpha$ be a limit ordinal such that the set $\{\alpha_\nu | \nu \dashv \tau \wedge \alpha_\nu < \bar\alpha\}$ is unbounded in $\bar\alpha$. We must show that there is a $\nu \dashv \tau$ such that $\bar\alpha = \alpha_\nu$.

For each $\eta \dashv \tau$, let $X_\eta = \operatorname{ran}(\sigma_{\eta\tau})$. Thus $(X_\eta | \eta \dashv \tau)$ is an increasing sequence of Σ_1 submodels of $\langle J_{\varrho(\tau)}, A(\tau)\rangle$. Set

$$X = \bigcup \{X_\eta | \eta \dashv \tau \wedge \alpha_\eta < \bar\alpha\}.$$

Then $X \prec_1 \langle J_{\varrho(\tau)}, A(\tau)\rangle$. Let

$$\sigma\colon \langle J_\varrho, A\rangle \cong \langle X, A(\tau)\cap X\rangle .$$

By 2.5 there is a unique $\nu \in S^1$ such that $\varrho = \varrho(\nu)$, $A = A(\nu)$, $\sigma(p(\nu)) = p(\tau)$. Since $\bar\alpha = \sup\{\alpha_\eta \mid \eta \dashv \tau \wedge \alpha_\eta < \bar\alpha\}$, it is easily seen that we must have $\nu \in S_{\bar\alpha}$. And since $X_\eta \cap J_\tau \prec_Q J_\tau$ for all $\eta \dashv \tau$, we have $X \cap J_\tau \prec_Q J_\tau$. It follows that $\nu \dashv \tau$ and $\sigma_{\nu\tau} = \sigma$.

(M 4) Let $\tau \in S_\alpha$, and suppose that τ is not maximal in S_α. We must show that the set $\{\alpha_\nu \mid \nu \dashv \tau\}$ is unbounded in α (i.e. that τ is a limit point in $\dashv$).

Pick $\lambda \in S_\alpha$, $\lambda > \tau$, λ admissible. Let $\theta < \alpha$ be given. Let X be the smallest Σ_1 elementary submodel of $\langle J_{\varrho(\tau)}, A(\tau)\rangle$ which contains $p(\tau)$ and θ and is such that $X \cap \alpha$ is transitive. Now, $\langle J_{\varrho(\tau)}, A(\tau)\rangle$ is an element of J_λ, and λ is admissible, so $X \in J_\lambda$. But α is regular inside J_λ. Hence $\bar\alpha = X \cap \alpha \in \alpha$. Let

$$\sigma\colon \langle J_\varrho, A\rangle \cong \langle X, A(\tau)\cap X\rangle .$$

Using 2.5 we see that there is a unique $\nu \in S_{\bar\alpha}$ such that $\varrho = \varrho(\nu)$, $A = A(\nu)$, $\nu \dashv \tau$, and $\sigma = \sigma_{\nu\tau}$. Since $\theta < \bar\alpha = \alpha_\nu$, we are done.

(M 5) Let $\tau \in S_\alpha$, and suppose that $\{\alpha_\nu \mid \nu \dashv \tau\}$ is unbounded in α. We must show that $\tau = \bigcup_{\nu \dashv \tau} [\pi_{\nu\tau}{}''\nu]$.

In fact we show that $J_\tau = \bigcup_{\nu \dashv \tau} [\sigma_{\nu\tau}{}''J_\nu]$. Since $\supseteq$ is trivial, we only have to worry about $\subseteq$. Let $x \in J_\tau$. Then for some $\vec\delta \in \alpha$, x is Σ_1-definable from $p(\tau)$, $\vec\delta$ in $\langle J_{\varrho(\tau)}, A(\tau)\rangle$. Pick $\nu \dashv \tau$ such that $\vec\delta \in \alpha_\nu$. Then, since $\vec\delta, p(\tau) \in \operatorname{ran}(\sigma_{\nu\tau}) \prec_1 \langle J_{\varrho(\tau)}, A(\tau)\rangle$, we have

$$x \in \operatorname{ran}(\sigma_{\nu\tau}) \cap J_\tau = \sigma_{\nu\tau}{}''J_\nu ,$$

as required.

(M 6) Let $\bar\nu$ be a limit point of $S_{\alpha_{\bar\nu}}$, $\bar\nu \dashv \nu$, $\nu' = \sup(\pi_{\bar\nu\nu}{}''\bar\nu)$. We must show that $\bar\nu \dashv \nu'$ and $\pi_{\bar\nu\nu'} \upharpoonright \bar\nu = \pi_{\bar\nu\nu} \upharpoonright \bar\nu$.

But this is immediate by 2.6.

(M 7) Let $\bar\nu$ be a limit point of $S_{\bar\alpha}$, $\nu \in S_\alpha$, $\nu \dashv \nu$, $\nu = \sup[\pi_{\bar\nu\nu}{}''\bar\nu]$. Let $\bar\alpha < \theta < \alpha$ be such that for each $\bar\tau \in S_{\bar\alpha} \cap \bar\nu$, S_θ contains an $\eta \in S^1$ such that $\bar\tau \dashv \eta \dashv \pi_{\bar\nu\nu}(\bar\tau)$. We must show that S_θ contains an η such that $\bar\nu \dashv \eta \dashv \nu$.

For each $\bar\tau \in S_{\bar\alpha} \cap \bar\nu$, set $\tau = \pi_{\bar\nu\nu}(\bar\tau)$ and let $\eta(\bar\tau)$ denote the unique $\eta \in S_\theta$ such that $\bar\tau \dashv \eta \dashv \tau$. Note that the function η is monotone increasing. Set $\eta = \sup\{\eta(\bar\tau) \mid \bar\tau \in S_{\bar\alpha} \cap \bar\nu\}$. Since S_θ is closed, $\eta \in S_\theta$. We show that $\bar\nu \dashv \eta \dashv \nu$. Since $\dashv$ is a tree and $\bar\nu \dashv \nu$, it suffices to show that $\eta \dashv \nu$. This will take some time. $\eta(\bar\tau)$ η

By the verification of (M 2), if $\bar\tau, \bar\tau' \in S_{\bar\alpha} \cap \bar\nu$, $\bar\tau < \bar\tau'$, we have:

$$\sigma_{\bar\tau,\eta(\bar\tau)} \upharpoonright J_{\bar\tau} = \sigma_{\bar\tau',\eta(\bar\tau')} \upharpoonright J_{\bar\tau} \quad \text{and} \quad \sigma_{\eta(\bar\tau),\tau} \upharpoonright J_{\eta(\bar\tau)} = \sigma_{\eta(\bar\tau'),\tau'} \upharpoonright J_{\eta(\bar\tau)} .$$

Hence we can define functions σ_0, σ_1 by

$$\sigma_0 = \bigcup\{\sigma_{\bar\tau,\eta(\bar\tau)} \upharpoonright J_{\bar\tau} \mid \bar\tau \in S_{\bar\alpha} \cap \bar\nu\}, \qquad \sigma_1 = \bigcup\{\sigma_{\eta(\bar\tau),\tau} \upharpoonright J_{\eta(\bar\tau)} \mid \bar\tau \in S_{\bar\alpha} \cap \bar\nu\} .$$ σ_0, σ_1

Clearly,

$$\sigma_0\colon J_{\bar\nu} \prec_0 J_\eta, \qquad \sigma_1\colon J_\eta \prec_0 J_\nu .$$

But σ_0, σ_1 are cofinal. Hence

$$\sigma_0\colon J_{\bar\nu} \prec_1 J_\eta, \qquad \sigma_1\colon J_\eta \prec_1 J_\nu .$$

2.7 Lemma. $\operatorname{ran}(\sigma_{\bar\nu\nu}) \cap J_\nu \subseteq \operatorname{ran}(\sigma_1)$.

Proof. If $\bar\tau \in S_{\bar\alpha} \cap \bar\nu$ and $x \in J_{\bar\tau}$, then $\sigma_1 \circ \sigma_0(x) = \sigma_{\eta(\bar\tau),\tau} \circ \sigma_{\bar\tau,\eta(\bar\tau)}(x) = \sigma_{\bar\tau\tau}(x)$, so $\sigma_1 \circ \sigma_0 \restriction J_{\bar\tau} = \sigma_{\bar\tau\tau} \restriction J_{\bar\tau} = \sigma_{\bar\nu\nu} \restriction J_{\bar\tau}$ (using the verification of (M 2)). But $S_{\bar\alpha} \cap \bar\nu$ is cofinal in $\bar\nu$. Hence $\sigma_1 \circ \sigma_0 = \sigma_{\bar\nu\nu} \restriction J_{\bar\nu}$. Thus we have

$$\operatorname{ran}(\sigma_{\bar\nu\nu}) \cap J_\nu = \sigma_{\bar\nu\nu}{}''J_{\bar\nu} = \sigma_1 \circ \sigma_0{}''J_{\bar\nu} \subseteq \operatorname{ran}(\sigma_1).$$

Since $\sigma_1 | \theta = \mathrm{id} \restriction \theta$ and (by confinality) $\sigma_1\colon J_\eta \prec_Q J_\nu$, we obtain $\eta \dashv \nu$ (and hence the verification of (M 7)) as an immediate consequence of the next lemma.

2.8 Lemma. *There is a* $\tilde\sigma \supseteq \sigma_1$ *such that*

$$\tilde\sigma\colon \langle J_{\varrho(\eta)}, A(\eta)\rangle \prec_1 \langle J_{\varrho(\nu)}, A(\nu)\rangle$$

and $\tilde\sigma(p(\eta)) = p(\nu)$.

$\sigma, \beta, n, \varrho, A, q, p, \gamma$ *Proof.* Set $\sigma = \sigma_1$, $\beta = \beta(\nu)$, $n = n(\nu)$, $\varrho = \varrho(\nu)$, $A = A(\nu)$, $q = q(\nu)$, $p = p(\nu)$, and if $\nu \in P$, $\gamma = \gamma(\nu)$.

Case 1. $\nu \in P$.

Thus $\varrho = \beta = \gamma + 1$ and $A = \emptyset$. Set

M $$M = \{x \in J_\gamma \mid x \text{ is } J_\gamma\text{-definable from parameters in } \operatorname{ran}(\sigma) \cup \{q\}\}.$$

Thus $M \prec J_\gamma$.

Claim A. $M \cap J_\nu = \operatorname{ran}(\sigma)$.

Let $x \in M \cap J_\nu$. Thus for some r and some Σ_r formula φ, and for some $y \in \operatorname{ran}(\sigma)$, x is the unique $x \in J_\gamma$ such that $\models_{J_\gamma} \varphi(\mathring{x}, \mathring{y}, \mathring{q})$. Define $X_m \prec_m J_\gamma$ just as in the proof of Claim B of 2.5, and, as there, set $\nu_m = X_m \cap \nu$. Let $\pi_m\colon X_m \cong J_{\gamma_m}$, and set $\pi_m(q) = q_m$. Since there is a J_γ-definable map from α onto γ_m, $\gamma_m < \nu$. Hence $(\gamma_m, q_m) \in J_\nu$. Now, (γ_m, q_m) is clearly Σ_1-definable from p in J_ϱ. So as $p \in \operatorname{ran}(\sigma_{\bar\nu\nu}) \prec_1 J_\varrho$, $(\gamma_m, q_m) \in \operatorname{ran}(\sigma_{\bar\nu\nu})$. So by 2.7, $(\gamma_m, q_m) \in \operatorname{ran}(\sigma)$. Pick $m \geqslant r$ so that $x, y \in J_{\nu_m}$. Since $\pi_m^{-1}\colon J_{\gamma_m} \prec_m J_\gamma$ and $\pi_m^{-1} \restriction J_{\nu_m} = \mathrm{id} \restriction J_{\nu_m}$, x is the unique $x \in J_{\gamma_m}$ such that $\models_{J_{\gamma_m}} \varphi(\mathring{x}, \mathring{y}, \mathring{q}_m)$. This provides us with a Σ_1 definition of x from γ_m, y, q_m in J_ν. But $\gamma_m, y, q_m \in \operatorname{ran}(\sigma) \prec_1 J_\nu$. Hence $x \in \operatorname{ran}(\sigma)$, which proves Claim A.

Let

$\tilde\sigma, \bar\gamma$ $$\tilde\sigma\colon J_{\bar\gamma} \cong M .$$

Thus

$$\tilde\sigma\colon J_{\bar\gamma} \prec J_\gamma .$$

$\bar q$ By Claim A, $\sigma \subseteq \tilde\sigma$. In particular, $\tilde\sigma(\theta) = \alpha$. Set $\bar q = \tilde\sigma^{-1}(q)$.

Claim B. (i) $\nu < \gamma \to \eta < \bar{\gamma} \wedge \tilde{\sigma}(\eta) = \nu$;
(ii) $\nu = \gamma \to \eta = \bar{\gamma}$.

Suppose $\nu < \gamma$. Then, since $\nu \in S_\alpha$ and $\gamma = \gamma(\nu)$, we have $\nu = [\alpha^+]^{J_\nu}$. But $\alpha = \sigma(0) \in \operatorname{ran}(\sigma) \subseteq M \prec J_\gamma$. Hence $\nu \in M$. But $\sigma''\eta$ is cofinal in ν. Hence $\eta = \tilde{\sigma}^{-1}(\nu)$. Thus $\eta < \bar{\gamma}$ and $\tilde{\sigma}(\eta) = \nu$.

Now suppose $\eta \neq \bar{\gamma}$. Thus $\eta < \bar{\gamma}$, and so $\tilde{\sigma}(\eta)$ is defined. Since $\sigma''\eta$ is cofinal in ν, $\tilde{\sigma}(\eta) \geqslant \nu$. Thus $\gamma > \nu$. Hence $\nu = \gamma \to \eta = \bar{\gamma}$. Claim B is proved.

Claim C. η is regular over $J_{\bar{\gamma}}$.

We know that ν is regular over J_γ. But $\tilde{\sigma}$: $J_{\bar{\gamma}} \prec J_\gamma$ and $\tilde{\sigma}''\eta$ is cofinal in ν, so this claim follows from Claim B.

Claim D. $\bar{q}$ is the $<_J$-least element of $J_{\bar{\gamma}}$ such that every element of $J_{\bar{\gamma}}$ is $J_{\bar{\gamma}}$-definable from parameters in $\theta \cup \{\bar{q}\}$.

Argue just as in Claim D of 2.5.

Claim E. η is Σ_1-singular over $J_{\bar{\gamma}+1}$.

By Claims C and D we may define $(\eta_m \mid m < \omega)$ from $J_{\bar{\gamma}}, \theta, \bar{q}, \eta$ exactly as $(\nu_m \mid m < \omega)$ was defined from J_γ, α, q, ν in Claim B of 2.5, thereby obtaining a $\Sigma_1(J_{\bar{\gamma}+1})$ ω-sequence cofinal in η, which proves Claim E.

Claim F. $\eta \in P$, $\gamma(\eta) = \bar{\gamma}$, $\varrho(\eta) = \beta(\eta) = \bar{\gamma} + 1$, $q(\eta) = \bar{q}$, $p(\eta) = (\bar{q}, \bar{\gamma}, \eta, \theta)$.

By Claims C, D, E.

Since $\tilde{\sigma}$: $J_{\bar{\gamma}} \prec J_\gamma$, by VI.1.19 there is a unique extension of $\tilde{\sigma}$ to an embedding $\tilde{\sigma}$: $J_{\varrho(\eta)} \prec_1 J_\varrho$. Using Claims B and F, $\tilde{\sigma}(p(\eta)) = p$. That completes the proof in this case.

Case 2. $\nu \in R$.

Let $h = h_{\varrho, A}$, and for $\tau < \varrho$, set $h_\tau = h_{\tau, A \cap J_\tau}$. Let $\delta = \delta(\nu) =$ the least $\delta < \varrho$ such that $q \in J_\delta$ and $\alpha \in h_\delta^*(J_\alpha \times \{p\})$, and such that $\nu \in h_\delta^*(J_\alpha \times \{p\})$ in case $\nu < \varrho$. Since $\lim(\varrho)$ and $J_\varrho = h^*(J_\alpha \times \{p\})$, such a δ can always be defined. $h, h_\tau, \delta, \delta(\nu)$

For $\delta \leqslant \tau < \varrho$, let

$$X_\tau = h_\tau^*(J_\alpha \times \{p\}).$$ X_τ

Then $X_\tau \prec_1 \langle J_\tau, A \cap J_\tau \rangle$. Moreover, by choice of p, $\bigcup_{\delta \leqslant \tau < \varrho} X_\tau = J_\varrho$. Since α is the largest cardinal inside J_ν, $X_\tau \cap \nu$ is transitive, so set $\nu_\tau = X_\tau \cap \nu$. Let ν_τ

$$\pi_\tau \colon \langle X_\tau, A \cap X_\tau \rangle \cong \langle J_{\gamma_\tau}, A_\tau \rangle.$$ $\pi_\tau, \gamma_\tau, A_\tau$

Set $\pi_\tau(p) = p_\tau$. Notice that $\pi_\tau \upharpoonright \nu_\tau = \mathrm{id} \upharpoonright \nu_\tau$, and that if $\nu < \varrho$, then $\pi_\tau(\nu) = \nu_\tau$. p_τ

Claim G. Let $\delta \leqslant \tau < \varrho$. Then $\langle J_{\gamma_\tau}, A_\tau \rangle \in J_\nu$.

Since $\langle J_\varrho, A \rangle$ is amenable, $h_\tau \in J_\varrho$. Hence $X_\tau \in J_\varrho$, and there is an $f \in J_\varrho$ such that f: $\alpha \leftrightarrow X_\tau$. Set

$$E = \{(\xi, \zeta) \subset \alpha^2 \mid f(\xi) \in f(\zeta)\},$$
$$B = \{\xi \in \alpha \mid f(\xi) \in A\}.$$

Thus $(\alpha, E, B) \in J_\varrho$. So if $v = \varrho$, then $(\alpha, E, B) \in J_\gamma$. If $v < \varrho$, then v is a cardinal inside J_ϱ, so $\mathscr{P}^{J_\rho}(\alpha) \subseteq J_v$, by applying II.5.5 within J_v, so again $(\alpha, E, B) \in J_v$. But v is adequate, so the transitive realisation of the well-founded, extensional structure $\langle \alpha, E, B \rangle$ is also in J_v. In other words, $\langle J_{\gamma_\tau}, \in, A_\tau \rangle \in J_v$, which proves Claim G.

By Claim G, $v_\tau < v$ for all τ. Hence $(v_\iota \mid \delta \leqslant \tau < \varrho)$ is a cofinal sequence in v.

Claim H. $((\gamma_\tau, A_\tau, v_\tau, p_\tau) \mid \delta \leqslant \tau < \varrho)$ is $\Sigma_1^{\langle J_\rho, A \rangle}(\{p\})$.

This is immediate from the definition.

Define

M $$M = \{x \in J_\varrho \mid x \text{ is } \Sigma_1\text{-definable from parameters in } (\operatorname{ran}(\sigma) \cap J_\alpha) \cup \{p\} \text{ in } \langle J_\varrho, A \rangle\}.$$

Of course, $\operatorname{ran}(\sigma) \cap J_\alpha = \sigma'' J_\theta = J_\theta$ here, but we have given the definition of M in the form required for the proof. We have

$$\langle M, A \cap M \rangle \prec_1 \langle J_\varrho, A \rangle.$$

Claim I. $M \cap J_v = \operatorname{ran}(\sigma)$.

Let $x \in M \cap J_v$. Then for some Σ_0-formula φ of $\mathscr{L}$ and some $y \in \operatorname{ran}(\sigma) \cap J_\alpha$, x is the unique $x \in J_\varrho$ such that $\models_{\langle J_\rho, A \rangle} \exists u \varphi(u, \mathring{x}, \mathring{y}, \mathring{p})$. Pick τ so that $x, y \in X_\tau$, $x \in J_{v_\tau}$, and for some $u \in J_\tau$, $\models_{\langle J_\tau, A \cap J_\tau \rangle} \varphi(\mathring{u}, \mathring{x}, \mathring{y}, \mathring{p})$. Then x is the unique $x \in J_\tau$ such that $\models_{\langle J_\tau, A \cap J_\tau \rangle} \exists u \varphi(u, \mathring{x}, \mathring{y}, \mathring{p})$. But $x, y, p \in X_\tau \prec_1 \langle J_\tau, A \cap J_\tau \rangle$, so applying π_τ and noting that $\pi_\tau \upharpoonright J_{v_\tau} = \operatorname{id} \upharpoonright J_{v_\tau}$, we see that x is the unique $x \in J_{\gamma_\tau}$ such that $\models_{\langle J_{\gamma_\tau}, A_\tau \rangle} \exists u \varphi(u, \mathring{x}, \mathring{y}, \mathring{p})$. This gives us a Σ_1 definition of x from $\gamma_\tau, A_\tau, y, p_\tau$ in J_v.

We may assume that τ was chosen above so that $\tau \in \operatorname{ran}(\sigma_{\bar{v}v})$. To see this, pick $\xi \in \sigma_{\bar{v}v}'' \bar{v}$ large enough so that whenever τ is such that $v_\tau \geqslant \xi$, then τ has the properties used above. Since $\sigma_{\bar{v}v}'' \bar{v}$ is cofinal in v, such a ξ can be found. The smallest τ with $v_\tau \geqslant \xi$ is now Σ_1-definable from ξ, p in $\langle J_\varrho, A \rangle$, by virtue of Claim H. So as $\xi, p \in \operatorname{ran}(\sigma_{\bar{v}v}) \prec_1 \langle J_\varrho, A \rangle$, we have $t \in \operatorname{ran}(\sigma_{\bar{v}v})$.

By Claim H, it now follows that $\gamma_\tau, A_\tau, p_\tau \in \operatorname{ran}(\sigma_{\bar{v}v})$. So by 2.7, $\gamma_\tau, A_\tau, p_\tau \in \operatorname{ran}(\sigma)$. Hence $\gamma_\tau, A_\tau, y, p_\tau \in \operatorname{ran}(\sigma)$. But $\operatorname{ran}(\sigma) \prec_1 J_v$. Thus $x \in \operatorname{ran}(\sigma)$.

Now let $x \in \operatorname{ran}(\sigma)$. Then $x \in J_v$. We show that $x \in M$. Since $x \in J_\varrho$, x is Σ_1-definable from parameters in $\alpha \cup \{p\}$ in $\langle J_\varrho, A \rangle$. So for some Σ_0 formula φ of $\mathscr{L}$ and some $y \in J_\alpha$, x is the unique $x \in J_\varrho$ such that

$$\models_{\langle J_\rho, A \rangle} \exists u \varphi(u, \mathring{x}, \mathring{y}, \mathring{p}).$$

Let y be the $<_J$-least such parameter. We show that $x \in M$ by proving that $y \in \operatorname{ran}(\sigma)$. Pick $\tau < \varrho$ such that $x, y \in X_\tau$, $x \in J_{v_\tau}$, so that for some $u \in J_\tau$,

$$\models_{\langle J_\tau, A \cap J_\tau \rangle} \varphi(\mathring{u}, \mathring{x}, \mathring{y}, \mathring{p}).$$

Then

$$\models_{\langle J_\tau, A \cap J_\tau \rangle} \exists u \varphi(u, \mathring{x}, \mathring{y}, \mathring{p}),$$

so applying π_τ, much as before, we see that x is the unique $x \in J_{\gamma_\tau}$ such that

$$\models_{\langle J_{\gamma_\tau}, A_\tau \rangle} \exists u \varphi(u, \mathring{x}, \mathring{y}, \mathring{p}_\tau).$$

Moreover, y is the $<_J$-least such parameter. This provides us with a Σ_1 definition of x from $\gamma_\tau, A_\tau, y, p_\tau$ in J_ν. As above, we can assume that τ has been chosen so that $\gamma_\tau, A_\tau, p_\tau \in \operatorname{ran}(\sigma)$. Hence as $x \in \operatorname{ran}(\sigma)$ and $\operatorname{ran}(\sigma) \prec_1 J_\nu$, the minimality of y gives $y \in \operatorname{ran}(\sigma)$, as required. Claim I is proved.

Let

$$\tilde{\sigma}: \langle J_{\bar{\varrho}}, \bar{A} \rangle \cong \langle M, A \cap M \rangle.$$ $\tilde{\sigma}, \bar{\varrho}, \bar{A}$

Thus

$$\tilde{\sigma}: \langle J_{\bar{\varrho}}, \bar{A} \rangle \prec_1 \langle J_\varrho, A \rangle.$$

By Claim I, $\sigma \subseteq \tilde{\sigma}$. In particular, $\tilde{\sigma}(\theta) = \alpha$. Set $\bar{p} = \tilde{\sigma}^{-1}(p)$. $\bar{p}$

By VI.5.6 there is a $\bar{\beta} \geqslant \bar{\varrho}$ such that $\bar{\varrho} = \varrho_{\bar{\beta}}^{n-1}$, $\bar{A} = A_{\bar{\beta}}^{n-1}$, and an extension $\tilde{\sigma}'$ of $\tilde{\sigma}$ such that $\tilde{\sigma}': J_{\bar{\beta}} \prec_n J_\beta$. $\bar{\beta}$ $\tilde{\sigma}'$

Claim J. η is Σ_{n-1} regular over $J_{\bar{\beta}}$.

This follows immediately from the fact that $\tilde{\sigma}': J_{\bar{\beta}} \prec_n J_\beta$ and $\sup[\tilde{\sigma}'{}''\eta] = \nu$.

Claim K. η is Σ_n singular over $J_{\bar{\beta}}$.

By definition of M, we have

$$J_{\bar{\varrho}} = h^*_{\bar{\varrho}, \bar{A}}(J_\theta \times \{\bar{p}\}).$$

Hence there is a $\Sigma_1(\langle J_{\bar{\varrho}}, \bar{A} \rangle)$ map from θ onto η. Since $\bar{\varrho} = \varrho_{\bar{\beta}}^{n-1}$, $\bar{A} = A_{\bar{\beta}}^{n-1}$, this map is $\Sigma_n(J_{\bar{\beta}})$.

Claim L. $\bar{\beta} = \beta(\eta)$, $n = n(\eta)$, $\eta \in R$, $\bar{\varrho} = \varrho(\eta)$, $\bar{A} = A(\eta)$.

By Claims J and K.

Hence

$$\tilde{\sigma}: \langle J_{\varrho(\eta)}, A(\eta) \rangle \prec \langle J_\varrho, A \rangle.$$

But $p \in \operatorname{ran}(\tilde{\sigma})$, so by 2.5, $\tilde{\sigma}(p(\eta)) = p$. The lemma is proved now. □

3. The Gap-2 Cardinal Transfer Theorem

In this section we prove the following theorem.

3.1 Theorem. *Assume there is a morass (i.e. an $(\omega_1, 1)$-morass). Let $\mathscr{A}$ be a K-structure of type (κ^{++}, κ) for some uncountable cardinal κ. Assume that $2^\kappa = \kappa^+$. Then there is a K-structure $\mathscr{B}$ of type (ω_2, ω) such that $\mathscr{B} \equiv \mathscr{A}$.* □

By virtue of the results of the last section, this implies that the Gap-2 Property is valid in L. In the exercises we indicate how the result may be extended to cover any type (λ^{++}, λ) in place of (ω_2, ω).

We fix $\mathcal{M} = \langle S, \mathscr{S}, \dashv, (\pi_{\nu\tau})_{\nu\dashv\tau}\rangle$ a morass from now on. We are given a K-structure $\mathscr{A}$ of type (κ^{++}, κ). We may assume that $\mathscr{A}$ has the form

$$\mathscr{A} = \langle \kappa^{++}, \kappa, <, \dots \rangle,$$

where $\kappa = U^{\mathscr{A}}$ and $<$ is the usual ordering of κ^{++}. If $\mathscr{B} \equiv \mathscr{A}$ and $e \in B$, we shall denote by $\mathrm{Pr}^{\mathscr{B}}(e)$ the set of all $<$-predecessors of e in the sense of $\mathscr{B}$, i.e.

$$\mathrm{Pr}^{\mathscr{B}}(e) = \{b \in B \mid \mathscr{B} \models b < e\}.$$

The key model-theoretic fact required for our proof is supplied by the following lemma.

3.2 Lemma. *Assume $2^{\kappa} = \kappa^{+}$. Then there are K-structures $\mathscr{B}, \mathscr{C}$ such that:*

(i) $\mathscr{B} \equiv \mathscr{C} \equiv \mathscr{A}$;

(ii) $\mathscr{B} \prec \mathscr{C}$ *and* $U^{\mathscr{B}} = U^{\mathscr{C}}$;

(iii) *there is an embedding* $\sigma\colon \mathscr{B} \prec \mathscr{C}$ *and an element* $e \in B$ *such that*:

(a) $U^{\mathscr{B}} \subseteq \mathrm{Pr}^{\mathscr{B}}(e)$;

(b) $\sigma \restriction \mathrm{Pr}^{\mathscr{B}}(e) = \mathrm{id} \restriction \mathrm{Pr}^{\mathscr{B}}(e)$;

(c) $B \subseteq \mathrm{Pr}^{\mathscr{C}}(\sigma(e))$.

Proof. For those familiar with the term, we remark that the proof is by means of a "Δ-system" argument.

For each $\alpha < \kappa^{++}$, let $\mathscr{A}_\alpha = \langle A_\alpha, \kappa, <, \dots\rangle$ be the smallest $\mathscr{A}_\alpha \prec \mathscr{A}$ such that $\kappa \cup \{\alpha\} \subseteq A_\alpha$. (Since $<$ well-orders $\mathscr{A}$, this definition makes sense.) We can clearly find a cofinal set $X \subseteq \kappa^{++}$ such that $A_\alpha \neq A_\beta$ whenever $\alpha, \beta \in X$, $\alpha \neq \beta$. Since $|A_\alpha| = \kappa$ for all $\alpha \in X$, we may assume that $\mathrm{otp}(A_\alpha) = \theta$ for all $\alpha \in X$, where θ is a fixed ordinal, $\kappa < \theta < \kappa^{+}$. Let $(a_\nu^\alpha \mid \nu < \theta)$ be the monotone enumeration of A_α for each $\alpha \in X$. Since $\alpha \in A_\alpha$, there is a least ordinal $\varrho < \theta$ such that $(a_\varrho^a \mid a \in X)$ is cofinal in κ^{++}. Since $(\kappa^{+})^{\kappa} = \kappa^{+}$, we may assume that $(a_\nu^\alpha \mid \nu < \varrho) = (a_\nu^\beta \mid \nu < \varrho)$ for all $\alpha, \beta \in X$. We may further assume that for all $\alpha, \beta \in X$, if $\alpha < \beta$ then $a_\varrho^\beta > a_\nu^\alpha$ for all $\nu < \theta$. Thus if we set

$$Y = \{a_\nu^\alpha \mid \nu < \varrho\} \qquad \text{(for any } \alpha \in X\text{)},$$
$$Z_\alpha = \{a_\nu^\alpha \mid \varrho \leqslant \nu < \theta\} \qquad \text{(each } \alpha \in X\text{)},$$

we have:

$$A_\alpha = Y \cup Z_\alpha \qquad \text{(all } \alpha \in X\text{)},$$
$$Y \cap Z_\alpha = \emptyset \qquad \text{(all } \alpha \in X\text{)},$$
$$Y < Z_\alpha < Z_\beta \qquad \text{(all } \alpha, \beta \in X,\ a < \beta\text{)}.$$

(i.e. $\{A_\alpha \mid \alpha \in X\}$ forms a Δ-*system*.)

Now there are at most $\kappa^\kappa = \kappa^+$ non-isomorphic K-structures of cardinality κ. So we can find $\alpha, \beta \in X$, $\alpha < \beta$, such that $\mathscr{A}_\alpha \cong \mathscr{A}_\beta$. It is clear that the only possible isomorphism $\sigma\colon \mathscr{A}_\alpha \cong \mathscr{A}_\beta$ is the unique order-isomorphism of A_α onto A_β (as sets of ordinals). Thus if we take

$$\mathscr{B} = \mathscr{A}_\alpha, \qquad \mathscr{C} = \mathscr{A}, \qquad e = a_\varrho^\alpha,$$

then $\mathscr{B}, \mathscr{C}, \sigma, e$ are clearly as required for the lemma. □

By means of an argument almost identical to that used in 1.5, we can use 3.2 in order to prove the following sharper result.

3.3 Lemma. *Assume $2^\kappa = \kappa^+$. Then there are countable homogeneous K-structures $\mathscr{B}_0, \mathscr{C}_0$ such that:*

(i) $\mathscr{B}_0 \equiv \mathscr{C}_0 \equiv \mathscr{A}$;
(ii) $\mathscr{B}_0 \prec \mathscr{C}_0$ *and* $U^{\mathscr{B}_0} = U^{\mathscr{C}_0}$;
(iii) *there is an embedding* $\sigma_0\colon \mathscr{B}_0 \prec \mathscr{C}_0$ *and an element* $e_0 \in B_0$ *such that:*
  (a) $U^{\mathscr{B}_0} \subseteq \mathrm{Pr}^{\mathscr{B}_0}(e_0)$;
  (b) $\sigma_0 \restriction \mathrm{Pr}^{\mathscr{B}_0}(e_0) = \mathrm{id} \restriction \mathrm{Pr}^{\mathscr{B}_0}(e_0)$;
  (c) $B_0 \subseteq \mathrm{Pr}^{\mathscr{C}_0}(\sigma_0(e_0))$;
(iv) $\langle \mathscr{B}_0, e_0\rangle \cong \langle \mathscr{C}_0, \sigma_0(e_0)\rangle$.

Proof. Since the proof is virtually the same as in 1.5 we give only a brief sketch. Commence with $\mathscr{B}, \mathscr{C}, \sigma, e$ as in 3.2. By replacing $\mathscr{C}$ by its skolem hull around $B \cup \sigma''B$ if necessary, we may assume that $|C| = |B|$. Let $h\colon C \leftrightarrow B$. Let

$$\langle \mathscr{C}', \mathscr{B}', \sigma', e', \sigma'(e'), h'\rangle \equiv \langle \mathscr{C}, \mathscr{B}, \sigma, e, \sigma(e), h\rangle$$

be special. Then, in particular, $\langle \mathscr{C}', \sigma'(e')\rangle$ and $\langle \mathscr{B}', e'\rangle$ are special structures of the same cardinality, so let

$$k'\colon \langle \mathscr{C}', \sigma'(e')\rangle \cong \langle \mathscr{B}', e'\rangle.$$

Let

$$\langle \mathscr{C}_0, \mathscr{B}_0, \sigma_0, e_0, \sigma_0(e_0), k_0\rangle \equiv \langle \mathscr{C}', \mathscr{B}', \sigma', e', \sigma'(e'), k'\rangle$$

be countable and homogeneous. Then $\mathscr{B}_0, \mathscr{C}_0, \sigma_0, e_0$ are as required by the lemma. □

We are now ready to commence our construction of an (ω_2, ω)-model $\mathscr{B} \equiv \mathscr{A}$. We shall obtain $\mathscr{B}$ as a limit of a certain directed, elementary system.

To each $\tau \in S^1$ we shall attach a K-structure $\mathscr{B}_\tau \equiv \mathscr{A}$ and an element $e_\tau \in B_\tau$. If $\nu, \tau \in S^1$ and $\nu < \tau$ (as ordinals) we shall have $\mathscr{B}_\nu \prec \mathscr{B}_\tau$. We shall set, for each $\tau \in S^1$,

$$\mathscr{B}_\tau^0 = \bigcup_{\nu < \tau} \mathscr{B}_\nu; \qquad \mathscr{B}_\tau^+ = \bigcup_{\nu \in S_{\alpha_\tau}} \mathscr{B}_\nu,$$

it being understood that ν, τ, etc. vary over S^1 in such situations. The directed system we construct will be called an $\mathscr{M}$-*complex*. We begin with an axiomatic description of the system. We fix $\mathscr{B}_0, \mathscr{C}_0, \sigma_0, e_0$ as in 3.3.

An $\mathcal{M}$*-complex* (for $\mathcal{A}$) is a structure

$$\mathcal{D} = \langle(\mathcal{B}_\tau)_{\tau\in S^1}, (e_\tau)_{\tau\in S^1}, (\sigma_{\bar\tau\tau})_{\bar\tau\dashv\tau}\rangle$$

such that:

(C 1) $\tau \in S^1 \to \mathcal{B}_\tau \equiv \mathcal{A}$ & $e_\tau \in B_\tau$;

(C 2) $\tau \in S^1 \cap \omega_1 \to \mathcal{B}_\tau$ is countable homogeneous and $\langle\mathcal{B}_\tau, e_\tau\rangle \cong \langle\mathcal{B}_0, e_0\rangle$;

(C 3) $\nu, \tau \in S^1$ & $\nu < \tau \to \mathcal{B}_\nu \prec \mathcal{B}_\tau$ & $U^{\mathcal{B}_\nu} = U^{\mathcal{B}_\tau}$;

(C 4) $\tau \in S^1 \to B_\tau^0 \subseteq \mathrm{Pr}^{\mathcal{B}_\tau}(e_\tau)$;

(C 5) the embeddings $\sigma_{\bar\tau\tau}\colon \mathcal{B}_{\bar\tau}^+ \prec \mathcal{B}_\tau$ for $\bar\tau \dashv \tau$, form a commutative system;

(C 6) $\bar\tau \dashv \tau \to \sigma_{\bar\tau\tau}(e_{\bar\tau}) = e_\tau$;

(C 7) $\bar\tau \dashv \tau$ & $\bar\nu \in S_{\alpha_\tau} \cap \bar\tau$ & $\nu = \pi_{\bar\tau\tau}(\bar\nu) \to \sigma_{\bar\tau\tau} \restriction B_{\bar\nu} = \sigma_{\bar\nu\nu} \restriction B_{\bar\nu}$;

(C 8) $\bar\tau \dashv \tau$ & $\nu \in S^1$ & $\alpha_\nu < \alpha_{\bar\tau} \to \sigma_{\bar\tau\tau} \restriction B_\nu^+ = \mathrm{id} \restriction B_\nu^+$;

(C 9) if τ is a limit point of $\dashv$, then $\mathcal{B}_\tau = \bigcup_{\bar\tau\dashv\tau} \sigma_{\bar\tau\tau}{}''\mathcal{B}_{\bar\tau}$.

Given an $\mathcal{M}$-complex as above, the Gap-2 Theorem follows at once. For if we set

$$\mathcal{B} = \bigcup_{\nu\in S_{\omega_1}} \mathcal{B}_\beta,$$

then $\mathcal{B} \equiv \mathcal{A}$ and by (C 3) and (C 4), $\mathcal{B}$ has type (ω_2, ω).

The construction of an $\mathcal{M}$-complex proceeds by recursion on $\tau \in S^1$. To commence, if τ_0 is the least ordinal in S^1 we take $\mathcal{B}_{\tau_0} = \mathcal{B}_0$, $e_{\tau_0} = e_0$. The induction step in the construction splits into three cases.

Case 1. τ is minimal in $\dashv$.

By morass axiom (M 4), since τ is not a limit point in $\dashv$, we must have $\alpha_\tau \neq \omega_1$. So $\mathcal{B}_\tau^0$ is a union of a countable elementary chain of countable homogeneous structures. Thus, using 1.6 in case this chain is of limit length, $\mathcal{B}_\tau^0$ is countable homogeneous and $\mathcal{B}_\tau^0 \cong \mathcal{B}_0$. Let $e_\tau^0 \in B_\tau^0$ correspond to $e_0 \in B_0$ under such an isomorphism. Then, by the properties of $\mathcal{B}_0, \mathcal{C}_0, \sigma_0, e_0$ we can find a structure $\langle\mathcal{B}_\tau, e_\tau\rangle$ such that the relationship between $\langle\mathcal{B}_\tau, e_\tau\rangle$ and $\langle\mathcal{B}_\tau^0, e_\tau^0\rangle$ is the same as that between $\langle\mathcal{C}_0, \sigma_0(e_0)\rangle$ and $\langle\mathcal{L}_0, e_0\rangle$. In particular, we have:

$$\mathcal{B}_\tau^0 \prec \mathcal{B}_\tau; \quad U^{\mathcal{B}_\tau^0} = U^{\mathcal{B}_\tau}; \quad \langle\mathcal{B}_\tau, e_\tau\rangle \cong \langle\mathcal{B}_\tau^0, e_\tau^0\rangle; \quad B_\tau^0 \subseteq \mathrm{Pr}^{\mathcal{B}_\tau}(e_\tau).$$

Thus $\mathcal{B}_\tau, e_\tau$ satisfy (C 1)–(C 4), whilst no new cases of (C 5)–(C 9) arise.

Case 2. τ is a limit point of $\dashv$.

Consider the directed elementary system

$$\langle\mathcal{B}_\nu^+)_{\nu\dashv\tau}, (\sigma_{\bar\nu\nu})_{\bar\nu\dashv\nu\dashv\tau}\rangle.$$

Let

$$\langle \mathscr{C}, (\sigma_\nu)_{\nu \dashv \tau} \rangle$$

be its direct limit. We may define an embedding

$$j \colon \mathscr{B}^0_\tau \prec \mathscr{C}$$

as follows. Let $x \in B^0_\tau$. For some $\nu < \tau$, $x \in B_\nu$. Suppose first that $\nu \in S_{\alpha_\tau}$. By (M 5) we have

$$S_{\alpha_\tau} \cap \tau = \bigcup_{\bar\tau \dashv \tau} \pi_{\bar\tau\tau}{}''(S_{\alpha_{\bar\tau}} \cap \bar\tau).$$

So we can find a $\bar\tau \dashv \tau$ and a $\bar\nu \in S_{\alpha_{\bar\tau}} \cap \bar\tau$ such that $\nu = \pi_{\bar\tau\tau}(\bar\nu)$. By (M 4), ν is a limit point of $\dashv$, so (C 9) tells us that

$$B_\nu = \bigcup_{\bar\nu \dashv \nu} \sigma_{\bar\nu\nu}{}''\mathrm{B}_{\bar\nu}.$$

So we can find a $\bar\tau \dashv \tau$ sufficiently high in $\dashv$ so that $x = \sigma_{\bar\nu\nu}(\bar x)$ for some $\bar x \in B_{\bar\nu} \subseteq B_{\bar\tau}$. Set $j(x) = \sigma_{\bar\tau}(\bar x)$ in this case. On the other hand, if $\alpha_\nu < \alpha_\tau$, then if we pick $\bar\tau \dashv \tau$ so that $\alpha_{\bar\tau} > \alpha_\nu$, we have $x \in B_{\bar\tau}$, and we can set $j(x) = \sigma_{\bar\tau}(x)$. Using (C 5), (C 7), and (C 8) it is routine to verify that j is well-defined and elementary from $\mathscr{B}^0_\tau$ into $\mathscr{C}$, and that for any $\bar\tau \dashv \tau$, $\bar\nu \in S_{\alpha_{\bar\tau}} \cap \bar\tau$, $\nu = \pi_{\bar\tau\tau}(\bar\nu)$, the following diagram commutes:

$$\begin{array}{ccc} \mathscr{B}^0_\tau & \xrightarrow{\;j\;} & \mathscr{C} \\ {\scriptstyle \mathrm{id}}\uparrow & & \uparrow{\scriptstyle \sigma_{\bar\tau} \restriction B_{\bar\nu}} \\ \mathscr{B}_\nu & \xrightarrow[\sigma_{\bar\nu\nu} \restriction B_{\bar\nu}]{} & \mathscr{B}_{\bar\nu} \end{array}$$

We may thus choose $\mathscr{C}$ specifically so that $j = \mathrm{id} \restriction B^0_\tau$. Let $\mathscr{B}_\tau = \mathscr{C}$ and set $\sigma_{\bar\tau\tau} = \sigma_{\bar\tau}$ for all $\bar\tau \dashv \tau$. By (C 6) below τ, there is a unique element $e_\tau \in B_\tau$ so that $\sigma_{\bar\tau\tau}(e_{\bar\tau}) = e_\tau$ for all $\bar\tau \dashv \tau$. We check that (C 1)–(C 9) hold for $\mathscr{B}_\tau$, e_τ under these definitions. The only one that is not immediate is (C 2).

Using an obvious notation we have:

$$\langle \mathscr{B}_\tau, e_\tau \rangle = \bigcup_{\bar\tau \dashv \tau} \langle \sigma_{\bar\tau\tau}{}''\mathscr{B}_{\bar\tau}, e_\tau \rangle.$$

Since $(\sigma_{\bar\tau\tau} \restriction B_{\bar\tau}) \colon \langle \mathscr{B}_{\bar\tau}, e_{\bar\tau} \rangle \cong \langle \sigma_{\bar\tau\tau}{}''\mathscr{B}_{\bar\tau}, e_\tau \rangle$, the structures $\langle \sigma_{\bar\tau\tau}{}''\mathscr{B}_{\bar\tau}, e_\tau \rangle$, for $\bar\tau \dashv \tau$, form an elementary chain of isomorphic, countable homogeneous structures. Thus by 1.6, $\langle \mathscr{B}_\tau, e_\tau \rangle$ is countable homogeneous and for any $\bar\tau \dashv \tau$ we have $\langle \mathscr{B}_\tau, e_\tau \rangle \cong \langle \sigma_{\bar\tau\tau}{}''\mathscr{B}_{\bar\tau}, e_\tau \rangle \cong \langle \mathscr{B}_{\bar\tau}\, e_{\bar\tau} \rangle \cong \langle \mathscr{B}_0, e_0 \rangle$. This proves (C 2).

Before we commence Case 3, we observe the following consequence of the axioms for an $\mathscr{M}$-complex:

(C 10) $\nu, \tau \in S^1 \cap \omega_1 \;\&\; \nu < \tau \rightarrow \langle \mathscr{B}_\tau, e_\nu \rangle \cong \langle \mathscr{B}_\tau, e_\tau \rangle.$

To see this, we first note that by (C 3),

$$\langle \mathscr{B}_\nu, e_\nu \rangle \prec \langle \mathscr{B}_\tau, e_\nu \rangle .$$

Also by (C 2),

$$\langle \mathscr{B}_\nu, e_\nu \rangle \cong \langle \mathscr{B}_\tau, e_\tau \rangle .$$

Thus

$$\langle \mathscr{B}_\tau, e_\nu \rangle \equiv \langle \mathscr{B}_\tau, e_\tau \rangle .$$

So as $\mathscr{B}_\tau$ is homogeneous,

$$\langle \mathscr{B}_\tau, e_\nu \rangle \cong \langle \mathscr{B}_\tau, e_\tau \rangle ,$$

which is (C 10).

We shall make use of (C 10) in dealing with Case 3.

Case 3. τ immediately succeeds $\bar{\tau}$ in $\dashv$.

Note that by (M 4), $\alpha_\tau \neq \omega_1$, so $\tau < \omega_1$. There are three subcases to consider.

Case 3.1. τ is minimal in S_{α_τ}.

Thus $\bar{\tau}$ is minimal in $S_{\alpha_{\bar{\tau}}}$ (by (M 1)). Using (C 10) and (possibly) Lemma 1.6, $\langle \mathscr{B}_\tau^0, e_{\bar{\tau}} \rangle \cong \langle \mathscr{B}_0, e_0 \rangle$. Thus there is a countable homogeneous structure $\mathscr{B}_\tau$ and an embedding $\sigma\colon \mathscr{B}_\tau^0 \prec \mathscr{B}_\tau$ such that:

$$\mathscr{B}_\tau^0 \prec \mathscr{B}_\tau, \quad U^{\mathscr{B}_\tau^0} = U^{\mathscr{B}_\tau}, \quad \sigma \restriction \mathrm{Pr}^{\mathscr{B}_\tau^0}(e_{\bar{\tau}}) = \mathrm{id} \restriction \mathrm{Pr}^{\mathscr{B}_\tau^0}(e_{\bar{\tau}}), \quad B_\tau^0 \subseteq \mathrm{Pr}^{\mathscr{B}_\tau}(\sigma(e_{\bar{\tau}})) .$$

Let $\sigma_{\bar{\tau}\tau} = \sigma$, $e_\tau = \sigma(e_{\bar{\tau}})$. It is routine to check (C 1)–(C 9) for τ.

Case 3.2. τ immediately succeeds η in S_{α_τ}.

Thus by (M 1), $\bar{\tau}$ immediately succeeds $\bar{\eta}$ in $S_{\alpha_{\bar{\tau}}}$, where $\pi_{\bar{\tau}\tau}(\bar{\eta}) = \eta$. Moreover, we have $\mathscr{B}_\tau^0 = \mathscr{B}_\eta$. Let $b = \sigma_{\bar{\eta}\eta}(e_{\bar{\tau}})$.

By (C 10) and (C 2) we have

$$\langle \mathscr{B}_{\bar{\eta}}^+, e_{\bar{\tau}} \rangle \cong \langle \mathscr{B}_{\bar{\tau}}, e_{\bar{\tau}} \rangle \cong \langle \mathscr{B}_0, e_0 \rangle .$$

Thus, applying $\sigma_{\bar{\eta}\eta}\colon \mathscr{B}_{\bar{\eta}}^+ \prec \mathscr{B}_\eta$,

$$\langle \sigma_{\bar{\eta}\eta}{}'' \mathscr{B}_{\bar{\eta}}^+, b \rangle \cong \langle \mathscr{B}_0, e_0 \rangle .$$

So as $\sigma_{\bar{\eta}\eta}{}'' \mathscr{B}_{\bar{\eta}}^+ \prec \mathscr{B}_\eta$,

$$\langle \mathscr{B}_\eta, b \rangle \equiv \langle \mathscr{B}_0, e_0 \rangle .$$

Thus by (C 2),

$$\langle \mathscr{B}_\eta, b \rangle \equiv \langle \mathscr{B}_\eta, e_\eta \rangle .$$

So as $\mathscr{B}_\eta$ is homogeneous,

$$\langle \mathscr{B}_\eta, b \rangle \cong \langle \mathscr{B}_\eta, e_\eta \rangle .$$

Thus by (C 2)$'$,

$$\langle \mathscr{B}_\eta, b\rangle \cong \langle \mathscr{B}_0, e_0\rangle .$$

It follows that we can find a countable homogeneous structure $\mathscr{B}_\tau$ and an embedding σ: $\mathscr{B}_\tau^0 \prec \mathscr{B}_\tau$ such that

$$\mathscr{B}_\tau^0 \prec \mathscr{B}_\tau, \quad U^{\mathscr{B}_\tau^0} = U^{\mathscr{B}_\tau}, \quad \sigma \upharpoonright \mathrm{Pr}^{\mathscr{B}_\tau^0}(b) = \mathrm{id} \upharpoonright \mathrm{Pr}^{\mathscr{B}_\tau^0}(b), \quad B_\tau^0 \subseteq \mathrm{Pr}^{\mathscr{B}_\tau}(\sigma(b)),$$
$$\langle \mathscr{B}_\tau^0, b\rangle \cong \langle \mathscr{B}_\tau, \sigma(b)\rangle .$$

Let $e_\tau = \sigma(b)$, $\sigma_{\bar{\tau}\tau} = \sigma \circ \sigma_{\bar{\eta}\eta}$.

It is immediate that (C 1)–(C 6) are preserved by this definition. Also, (C 9) does not apply in this case, and (C 8) follows easily from (C 7) (and the induction hypothesis). So we need to check (C 7) for τ. It clearly suffices to prove this for the case $v = \eta$ only, i.e. we show that

$$\sigma_{\bar{\tau}\tau} \upharpoonright B_{\bar{\eta}} = \sigma_{\bar{\eta}\eta} \upharpoonright B_{\bar{\eta}} .$$

Well, we have

$$B_{\bar{\eta}} \subseteq \mathrm{Pr}^{\mathscr{B}_{\bar{\tau}}}(e_{\bar{\tau}}) \subseteq \mathrm{Pr}^{\mathscr{B}_{\bar{\eta}}^+}(e_{\bar{\tau}}) .$$

So, applying $\sigma_{\bar{\eta}\eta}$,

$$\sigma_{\bar{\eta}\eta}{}'' B_{\bar{\eta}} \subseteq \mathrm{Pr}^{\mathscr{B}_\eta}(b) .$$

But

$$\sigma \upharpoonright \mathrm{Pr}^{\mathscr{B}_\eta}(b) = \mathrm{id} \upharpoonright \mathrm{Pr}^{\mathscr{B}_\eta}(b) .$$

Thus

$$\sigma \circ \sigma_{\bar{\eta}\eta}{}'' B_{\bar{\eta}} = \sigma_{\bar{\eta}\eta}{}'' B_{\bar{\eta}} ,$$

i.e.

$$\sigma_{\bar{\tau}\tau} \upharpoonright B_{\bar{\eta}} = \sigma_{\bar{\eta}\eta} \upharpoonright B_{\bar{\eta}} .$$

For future use, we note that for any K-formula $\varphi(\vec{y}, \vec{x})$:

(**) if $\vec{y} \in B_\eta^0$, $\vec{x} \in B_{\bar{\tau}}^+$, then $\mathscr{B}_\eta \models \varphi(\vec{y}, \sigma_{\bar{\eta}\eta}(\vec{x}))$ iff $\mathscr{B}_\tau \models \varphi(\vec{y}, \sigma_{\bar{\tau}\tau}(\vec{x}))$.

To see this, apply σ to the left-hand side and note that as $e_{\bar{\eta}} < e_{\bar{\tau}}$ in $\mathscr{B}_{\bar{\eta}}^+$, an application of $\sigma_{\bar{\eta}\eta}$ yields $e_\eta < b$ in $\mathscr{B}_\eta$, so $B_\eta^0 \subseteq \mathrm{Pr}^{\mathscr{B}_\tau^0}(b)$.

Case 3.3. τ is a limit point in S_{α_τ}.

Thus $\bar{\tau}$ is a limit point in $S_{\alpha_{\bar{\tau}}}$. There are two subcases to consider.

Case 3.3.1. $\lambda = \sup_{\bar{v} < \bar{\lambda}} \pi_{\bar{\tau}\tau}(\bar{v}) < \tau$.

In this case, $\langle \mathscr{B}_\tau^0, e_\lambda\rangle$ is the union of the elementary chain

$$(\langle \mathscr{B}_v, e_\lambda\rangle \mid v \in S_{\alpha_\tau} \;\&\; \lambda \leqslant v < \tau) .$$

By (C 10), this is a chain of isomorphic, countable homogeneous structures. So by 1.6,

$$\langle \mathscr{B}_\tau^0, e_\lambda \rangle \cong \langle \mathscr{B}_\lambda, e_\lambda \rangle \cong \langle \mathscr{B}_0, e_0 \rangle .$$

Thus we can find a countable homogeneous structure $\mathscr{B}_\tau$ and an embedding $\sigma\colon \mathscr{B}_\tau^0 \prec \mathscr{B}_\tau$ such that:

$$\mathscr{B}_\tau^0 \prec \mathscr{B}_\tau,\ U^{\mathscr{B}_\tau^0} = U^{\mathscr{B}_\tau},\ \sigma \upharpoonright \mathrm{Pr}^{\mathscr{B}_\tau^0}(e_\lambda) = \mathrm{id} \upharpoonright \mathrm{Pr}^{\mathscr{B}_\tau^0}(e_\lambda),\quad B_\tau^0 \subseteq \mathrm{Pr}^{\mathscr{B}_\tau}(\sigma(e_\lambda)),$$
$$\langle \mathscr{B}_\tau^0, e_\lambda \rangle \cong \langle B_\tau, \sigma(e_\lambda) \rangle .$$

Let $e_\tau = \sigma(e_\lambda)$, $\sigma_{\bar\tau\tau} = \sigma \circ \sigma_{\bar\tau\lambda}$. Much as in Case 3.2, we see that (C 1)–(C 9) continue to hold, and that (for later use):

if $\varphi(\vec y, \vec x)$ is any K-formula, then

(**) if $\vec y \in B_\lambda^0$, $\vec x \in B_{\bar\tau}^+$, then $\mathscr{B}_\lambda \models \varphi(\vec y, \sigma_{\bar\tau\lambda}(\vec x))$ iff $\mathscr{B}_\tau \models \varphi(\vec y, \sigma_{\bar\tau\tau}(\vec x))$.

The final case is by far the most complicated one, though as will be seen, we have already "done all of the work" for this case, in the sense that our construction is a "limit" construction.

Case 3.3.2. $\sup_{\bar\nu < \bar\tau} \pi_{\bar\tau\tau}(\bar\nu) = \tau$.

For each $\bar\nu \in S_{\alpha_{\bar\tau}} \cap \bar\tau$, let $\eta(\bar\nu)$ be the $\dashv$-least η such that $\bar\nu \dashv \eta \dashv \pi_{\bar\tau\tau}(\bar\nu)$. [Notice that as $\pi_{\bar\tau\tau}(\bar\nu)$ is not maximal in S_{α_τ}, there is no possibility that $\eta = \pi_{\bar\tau\tau}(\bar\nu)$ here.] Clearly, $(\alpha_{\eta(\bar\nu)} \mid \bar\nu \in S_{\alpha_{\bar\tau}} \cap \bar\tau)$ is non-decreasing. Also, by morass axiom (M 4), it is in fact strictly increasing. Set

$$\alpha = \sup \{\alpha_{\eta(\bar\nu)} \mid \bar\nu \in S_{\alpha_{\bar\tau}} \cap \bar\tau\} .$$

By (M 3), $\alpha \in S^1$, and in fact whenever $\bar\nu \in S_{\alpha_{\bar\tau}} \cap \bar\tau$, there is a $\nu' \in S_\alpha$ such that $\bar\nu \dashv \nu' \,\underline{\dashv}\, \pi_{\bar\tau\tau}(\bar\nu)$. So by (M 7) and the fact that τ immediately succeeds $\bar\tau$ in $\dashv$, we see that $\alpha = \alpha_\tau$. We shall define $\mathscr{B}_\tau$ as a "diagonal limit" of the structures $\mathscr{B}_{\eta(\bar\nu)}$, for $\bar\nu \in S_{\alpha_{\bar\tau}} \cap \bar\tau$.

For $\bar\nu, \bar\gamma \in S_{\alpha_{\bar\tau}} \cap \bar\tau$, $\bar\nu \leqslant \bar\gamma$, let $\eta(\bar\nu, \bar\gamma) = \pi_{\bar\gamma, \eta(\bar\gamma)}(\bar\nu)$. Thus $\eta(\bar\nu, \bar\gamma)$ is the unique $\eta \in S_{\alpha_{\eta(\bar\gamma)}}$ such that $\bar\nu \dashv \eta \dashv \pi_{\bar\tau\tau}(\bar\nu)$. (See Fig. 7.)

Notice that:

(i) $\bar\nu \dashv \eta(\bar\nu) \dashv \eta(\bar\nu, \bar\gamma) \dashv \pi_{\bar\tau\tau}(\bar\nu)$;

(ii) $\sigma_{\eta(\bar\nu), \eta(\bar\nu, \bar\gamma)}{}''B_{\eta(\bar\nu)}^0 \subseteq B_{\eta(\bar\nu, \bar\gamma)}^0$.

Also, since $\pi_{\bar\tau\tau}$ is cofinal in τ on $\bar\tau$ and the sequence $(\alpha_{\eta(\bar\nu)} \mid \bar\nu \in S_{\alpha_{\bar\tau}} \cap \bar\tau)$ is cofinal in α_τ, we have, setting $\nu = \pi_{\bar\tau\tau}(\bar\nu)$:

(iii) $\mathscr{B}_\tau^0 = \bigcup \{\sigma_{\eta(\bar\nu), \nu}{}''\mathscr{B}_{\eta(\nu)}^0 \mid \bar\nu \in S_{\alpha_{\bar\tau}} \cap \bar\tau\}$.

Claim. If $\bar\nu, \bar\gamma \in S_{\alpha_{\bar\tau}} \cap \bar\tau$, $\bar\nu \leqslant \bar\gamma$, $\vec y \in B_{\eta(\bar\nu)}^0$, $\vec x \in B_{\bar\tau}^+$, then for any K-formula φ:

$$\mathscr{B}_{\eta(\bar\nu)} \models \varphi(\vec y, \sigma_{\bar\nu, \eta(\bar\nu)}(\vec x)) \quad \text{iff } \mathscr{B}_{\eta(\bar\gamma)} \models \varphi(\sigma_{\eta(\bar\nu), \eta(\bar\nu, \bar\gamma)}(\vec y), \sigma_{\bar\gamma, \eta(\bar\gamma)}(\vec x)) .$$

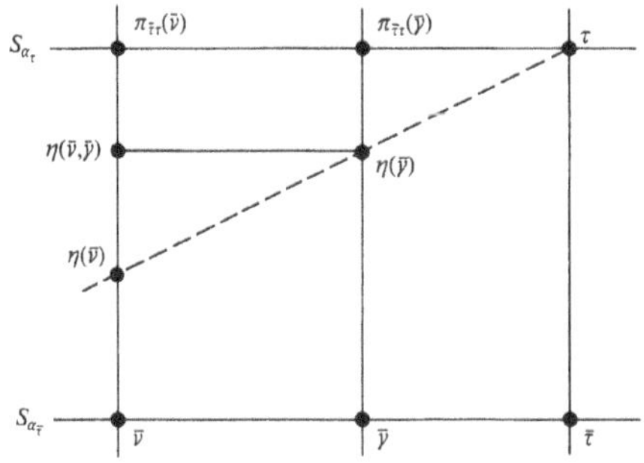

Fig. 7

We postpone the proof of this claim for a moment and complete the definition of $\mathscr{B}_\tau$.

Let X be an arbitrary, countable, infinite set disjoint from B_τ^0, and let

$$h\colon (B_{\bar\tau}^+ - B_{\bar\tau}^0) \leftrightarrow X .$$

Let $B_\tau = B_\tau^0 \cup X$, and define a function $\sigma\colon B_{\bar\tau}^+ \to B_\tau$ by

$$\sigma(x) = \begin{cases} h(x), & \text{if } x \notin B_{\bar\tau}^0 \\ \sigma_{\bar\nu\nu}(x), & \text{if } x \in B_{\bar\nu}, \text{ where } \bar\nu \in S_{\alpha_{\bar\tau}} \cap \bar\tau \text{ and } \nu = \pi_{\bar\tau\tau}(\bar\nu). \end{cases}$$

By (C 7) and the induction hypothesis, this clearly defines a function on $B_{\bar\tau}^+$.

We define a K-structure $\mathscr{B}_\tau$ on B_τ so that $\sigma\colon \mathscr{B}_{\bar\tau}^+ \prec \mathscr{B}_\tau$. We can do this in a unique way so that for all K-formulas φ and all $\bar\nu \in S_{\alpha_{\bar\tau}} \cap \bar\tau$, $\vec y \in B_{\eta(\bar\nu)}$, $\vec x \in B_{\bar\tau}^+$, we have

$$\text{(iv)} \quad \mathscr{B}_\tau \models \varphi(\sigma_{\eta(\bar\nu),\nu}(\vec y), \sigma(\vec x)) \quad \text{iff } \mathscr{B}_{\eta(\bar\nu)} \models \varphi(\vec y, \sigma_{\bar\nu,\eta(\bar\nu)}(\vec x)),$$

where $\nu = \pi_{\bar\tau\tau}(\bar\nu)$. By (iii) above, the $\vec y$'s take care of the whole of B_τ^0, and the $\vec x$'s take care of X, so all of B_τ is covered. By the claim there is no conflict between different choices of $\bar\nu$. Hence $\mathscr{B}_\tau$ is uniquely determined. And by means of an argument as in Case 2 there is an embedding $j\colon \mathscr{B}^0 \prec \mathscr{B}_\tau$. By choosing X suitably we can clearly ensure that $j = \mathrm{id} \restriction B_\tau^0$. Thus $\mathscr{B}_\tau^0 \prec \mathscr{B}_\tau$. We set $\sigma_{\bar\tau\tau} = \sigma$, $e_\tau = \sigma_{\bar\tau\tau}(e_{\bar\tau})$.

Now, by equivalence (iv) above, $\langle \mathscr{B}_\tau, e_\tau \rangle$ is a sort of limit of the isomorphic (to $\langle \mathscr{B}_0, e_0 \rangle$) countable homogeneous structures $\langle \mathscr{B}_{\eta(\bar\nu)}, \sigma_{\bar\nu,\eta(\bar\nu)}(e_{\bar\tau}) \rangle$, for $\bar\nu \in S_{\alpha_\tau} \cap \bar\tau$. By means of arguments which are in essence the same as those used to prove 1.3 and 1.6, it is easily seen that $\mathscr{B}_\tau$ is countable homogeneous and that $\langle \mathscr{B}_\tau, e_\tau \rangle \cong \langle \mathscr{B}_0, e_0 \rangle$. (In fact we did not give the proof of 1.3, since this is a "standard" result of model theory, but the details are easily worked out. The idea is to construct the desired homogeneity automorphism by means of a "back and forth" procedure as used in the proof of 1.6. This is similar to, but a little easier than the argument in 1.6 itself.) Thus (C 2) is preserved. The verification of (C 1) and (C 3)–(C 9) is routine.

There remains the verification of the claim. This is done by induction on $\bar{\gamma} \in S_{\alpha_\xi} \cap \bar{\tau}$. For $\bar{\gamma}$ the minimal member of S_{α_ξ}, the claim is trivially valid. Suppose next that $\bar{\gamma}$ immediately succeeds $\bar{v}$ in S_{α_ξ}. Then by induction it suffices to prove the claim for this one pair $\bar{v}, \bar{\gamma}$. We have:

$$\sigma_{\bar{v},\eta(\bar{v})}(\vec{x}),\ \vec{y} \in B_{\eta(\bar{v})},$$
$$\sigma_{\eta(\bar{v}),\eta(\bar{v},\bar{\gamma})}\colon \mathscr{B}^+_{\eta(\bar{v})} \prec \mathscr{B}_{\eta(\bar{v},\bar{\gamma})},$$
$$\sigma_{\eta(\bar{v}),\eta(\bar{v},\bar{\gamma})} \circ \sigma_{\bar{v},\eta(\bar{v})} = \sigma_{\bar{v},\eta(\bar{v},\bar{\gamma})}.$$

So for any φ, we clearly have

$$\mathscr{B}_{\eta(\bar{v})} \models \varphi(\vec{y}, \sigma_{\bar{v},\eta(\bar{v})}(\vec{x})) \quad \text{iff } \mathscr{B}_{\eta(\bar{v},\bar{\gamma})} \models \varphi(\sigma_{\eta(\bar{v}),\eta(\bar{v},\bar{\gamma})}(\vec{y}), \sigma_{\bar{v},\eta(\bar{v},\bar{\gamma})}(\vec{x})).$$

Now, $\eta(\bar{\gamma})$ immediately succeeds γ in $\dashv$ and $\eta(\bar{\gamma})$ immediately succeeds $\eta(\bar{v},\bar{\eta})$ in $S_{\alpha_{\eta(\gamma)}}$. Hence Case 3.2 applies to $\eta(\bar{\gamma})$. Note that as $\vec{y} \in B^0_{\eta(\bar{v})}$, (ii) above gives

$$\sigma_{\eta(\bar{v}),\eta(\bar{v},\bar{\gamma})}(\vec{y}) \in B^0_{\eta(\bar{v},\bar{\gamma})}.$$

So by (*) above,

$$\mathscr{B}_{\eta(\bar{v},\bar{\gamma})} \vdash \varphi(\sigma_{\eta(v),\eta(v,\gamma)}(\vec{y}), \sigma_{\bar{v},\eta(\bar{v},\bar{\gamma})}(\vec{x})) \quad \text{iff } \mathscr{B}_{\eta(\bar{\gamma})} \models \varphi(\sigma_{\eta(\bar{v}),\eta(\bar{v},\bar{\gamma})}(\vec{y}), \sigma_{\bar{\gamma},\eta(\bar{\gamma})}(\vec{x})).$$

The claim follows from the above two equivalences.

Finally, suppose that $\bar{\gamma}$ is a limit point in S_{α_ξ}. Let

$$\lambda = \sup_{\bar{v}<\bar{\gamma}} \pi_{\bar{\gamma},\eta(\bar{\gamma})}(\bar{v}).$$

Either $\lambda = \eta(\bar{\gamma})$ or else $\lambda < \eta(\bar{\gamma})$. Thus, either by identity or else by (**), respectively, we have, for all $\vec{y} \in B^0_\lambda$, $\vec{x} \in B^+_{\bar{\gamma}}$,

$$\text{(v)} \quad \mathscr{B}_{\eta(\bar{\gamma})} \models \varphi(\vec{y}, \sigma_{\bar{\gamma},\eta(\bar{\gamma})}(\vec{x})) \quad \text{iff } \mathscr{B}_\lambda \models \varphi(\vec{y}, \sigma_{\bar{\gamma},\lambda}(\vec{x})).$$

Let $\alpha = \sup_{\bar{v}<\bar{\gamma}} \alpha_{\eta(\bar{v})}$. Let $\bar{\eta}$ be the unique $\bar{\eta} \in S_\alpha$ such that $\bar{\gamma} \dashv \bar{\eta} \sqsupseteq \lambda$. Then $\bar{\eta}$ immediately succeeds $\bar{\gamma}$ in $\dashv$. For if $\bar{\gamma} \dashv \bar{\eta} \dashv \bar{\eta}$, then for $\bar{v} \in S_{\alpha_\xi} \cap \bar{\tau}$, we would have $\alpha_{\eta(\bar{v})} \leqslant \alpha_{\bar{\eta}}$, contrary to $\bar{\eta} \in S_\alpha$ and the choice of α. So, recalling the definition of λ we see that Case 3.3.2 applies to $\bar{\eta}$. So by induction, for $\vec{y} \in B^0_{\eta(\bar{v})}$, $\bar{v} \in S_{\alpha_\xi} \cap \bar{\gamma}$, $\vec{x} \in B^+_{\bar{\gamma}}$, we have, by (iv),

$$\mathscr{B}_{\eta(\bar{v})} \models \varphi(\vec{y}, \sigma_{\bar{v},\eta(\bar{v})}(\vec{x})) \quad \text{iff } \mathscr{B}_{\bar{\eta}} \models \varphi(\sigma_{\eta(\bar{v}),v}(\vec{y}), \sigma_{\bar{\gamma}\bar{\eta}}(\vec{x})),$$

where $v = \pi_{\bar{\gamma},\bar{\eta}}(\bar{v})$. Applying $\sigma_{\bar{\eta}\lambda}\colon \mathscr{B}_{\bar{\eta}} \prec \mathscr{B}_\lambda$ to the right hand side we obtain:

$$\mathscr{B}_{\eta(\bar{v})} \models \varphi(\vec{y}, \sigma_{\bar{v},\eta(\bar{v})}(\vec{x})) \quad \text{iff } \mathscr{B}_\lambda \models \varphi(\sigma_{\eta(\bar{v}),v}(\vec{y}), \sigma_{\bar{\gamma}\lambda}(\vec{x})),$$

where this time $v = \pi_{\bar{\gamma},\lambda}(\bar{v})$. But $\vec{y} \in B^0_{\eta(\bar{v})}$, so $\sigma_{\eta(\bar{v}),v}(\vec{y}) \in B^0_\lambda$ here. Thus by (v) we obtain

$$\mathscr{B}_{\eta(\bar{v})} \models \varphi(\vec{y}, \sigma_{\bar{v},\eta(\bar{v})}(\vec{x})) \quad \text{iff } \mathscr{B}_{\eta(\bar{\gamma})} \models \varphi(\sigma_{\eta(\bar{v}),\eta(\bar{v},\bar{\gamma})}(\vec{y}), \sigma_{\bar{\gamma},\eta(\bar{\gamma})}(\vec{x})),$$

as required.

The claim is proved, and with it the Gap-2 Theorem.

4. Simplified Morasses

The morass defined in section 2 and used in section 3 provides us with some insight into the structure of the constructible hierarchy. But as we have seen, it is not particularly easy to use, requiring the consideration of many separate cases (five in the proof of the Gap-2 Theorem), one of them (Case 3.3.2) quite complicated. If one's main interest is simply to use morasses to prove theorems like the Gap-2 Theorem, this complexity is a nuisance. In this section we show that it is an avoidable nuisance. We shall describe a "simplified morass" structure. The existence of a simplified morass is provably (in ZFC) equivalent to the existence of a full morass in the sense of section 2. (We shall give one half of the proof, the half of relevance to us here.) And as we shall see, it is considerably easier to prove the Gap-2 Theorem using a simplified morass.

As with the morass of section 2, the motivation is the approximation of a structure of cardinality ω_2 by means of a system of ω_1 many countable structures. (As before, we consider the case of an ω_1 morass for definiteness, but everything generalises quite easily to and arbitrary uncountable regular cardinal κ.)

A simplified morass (morass precisely, a simplified $(\omega_1, 1)$-morass) consists of a structure

$$\mathcal{M} = \langle (\theta_\alpha \mid \alpha \leqslant \omega_1), (\mathcal{F}_{\alpha\beta} \mid \alpha < \beta \leqslant \omega_1) \rangle$$

satisfying the following six conditions (which we examine below):

(P 0) (a) $\theta_0 = 1$, $\theta_{\omega_1} = \omega_2$, $(\forall \alpha < \omega_1)(0 < \theta_\alpha < \omega_1)$;

(b) $\mathcal{F}_{\alpha\beta}$ is a set of order-preserving functions $f\colon \theta_\alpha \to \theta_\beta$;

(P 1) $|\mathcal{F}_{\alpha\beta}| \leqslant \omega$ for all $\alpha < \beta < \omega_1$;

(P 2) if $\alpha < \beta < \gamma$, then $\mathcal{F}_{\alpha\gamma} = \{f \circ g \mid f \in \mathcal{F}_{\beta\gamma} \;\&\; g \in \mathcal{F}_{\alpha\beta}\}$;

(P 3) if $\alpha < \omega_1$, then $\mathcal{F}_{\alpha,\alpha+1} = \{\mathrm{id} \restriction \theta_\alpha, f_\alpha\}$, where f_α is such that for some $\delta < \theta_\alpha$, $f_\alpha \restriction \delta = \mathrm{id} \restriction \delta$ and $f_\alpha(\delta) \geqslant \theta_\alpha$;

(P 4) if $\alpha \leqslant \omega_1$ is a limit ordinal, if $\beta_1, \beta_2 < \alpha$, and if $f_1 \in \mathcal{F}_{\beta_1\alpha}$, $f_2 \in \mathcal{F}_{\beta_2\alpha}$, then there is a $\gamma < \alpha$, $\gamma > \beta_1, \beta_2$, and there are $f_1' \in \mathcal{F}_{\beta_1\gamma}$, $f_2' \in \mathcal{F}_{\beta_2\gamma}$, $g \in \mathcal{F}_{\gamma\alpha}$, such that $f_1 = g \circ f_1'$, $f_2 = g \circ f_2'$.

(P 5) for all $\alpha > 0$, $\theta_\alpha = \bigcup \{f'' \theta_\beta \mid \beta < \alpha \;\&\; f \in \mathcal{F}_{\beta\alpha}\}$.

The idea of the above definition is this. We approximate $\theta_{\omega_1} = \omega_2$ by means of the countable ordinals θ_α, $\alpha < \omega_1$. To do this we need to know how the intervals θ_α "fit inside" θ_{ω_1}. $\mathcal{F}_{\alpha\beta}$ consists of a set of order-preserving maps from θ_α into θ_β. Each map f in $\mathcal{F}_{\alpha\beta}$ gives one way in which θ_α "fits inside" θ_β as an approximation to it. (P 1) tells us that there are not too many ways in which this can happen for any given pair $\alpha, \beta < \omega_1$. (P 2) is self-explanatory. (P 3) (together with (P 0)(b)) says that at successor steps in the approximation procedure there are just two ways in which θ_α fits inside $\theta_{\alpha+1}$, both very simple (P 4) tells us that the "approximation tree" going up to θ_{ω_1} does not have branches which "split" at limit levels.

A particular consequence of (P 5) is that θ_{ω_1} is entirely determined by the countable parts of the simplified morass.

We shall use the simplified morass to prove the Gap-2 Theorem. We are given a K-structure $\mathscr{A} = \langle A, U, \ldots \rangle$ of type (κ^{++}, κ) and wish to construct a K-structure $\mathscr{B}$ of type (ω_2, ω) such that $\mathscr{B} \equiv \mathscr{A}$. We commence as in section 3. In particular, let $\mathscr{B}_0, \mathscr{C}_0, e_0, \sigma_0$ be as in lemma 3.3. We construct (instead of an $\mathscr{M}$-complex) sequences

$$(\mathscr{B}_\alpha \mid \alpha \leqslant \omega_1), \quad (h_\alpha \mid \alpha \leqslant \omega_1), \quad (f^* \mid f \in \mathscr{F}_{\beta\alpha}, \beta < \alpha \leqslant \omega_1)$$

so that:

(C 1) $\mathscr{B}_\alpha \equiv \mathscr{A}$;

(C 2) $h_\alpha \colon \theta_\alpha \to B_\alpha$ is order-preserving (where B_α is ordered by the linear ordering which is part of $\mathscr{B}_\alpha$);

(C 3) $f^* \colon \mathscr{B}_\beta \prec \mathscr{B}_\alpha$ and $U^{\mathscr{B}_\alpha} \subseteq \operatorname{ran}(f^*)$ for $\beta < \alpha \leqslant \omega_1$, $f \in \mathscr{F}_{\beta\alpha}$;

(C 4) if $\alpha < \omega_1$, then $\langle \mathscr{B}_\alpha, h_\alpha(\delta) \rangle \cong \langle \mathscr{B}_0, e_0 \rangle$ for all $\delta < \theta_\alpha$;

(C 5) $(f \circ g)^* = f^* \circ g^*$, whenever $f \in \mathscr{F}_{\gamma\alpha}$, $g \in \mathscr{F}_{\beta\gamma}$, $\beta < \gamma < \alpha$;

(C 6) $h_\alpha \circ f = f^* \circ h_\beta$ for each $f \in \mathscr{F}_{\beta\alpha}$;

(C 7) if $f \in \mathscr{F}_{\beta\alpha}$ and $\operatorname{ran}(f) \subseteq \delta < \theta_\alpha$, then $\operatorname{ran}(f^*) \subseteq \operatorname{Pr}^{\mathscr{B}_\alpha}(h_\alpha(\delta))$.

Provided we can carry out this construction we shall be done, since then $\mathscr{B}_{\omega_1}$ is of type (ω_2, ω) as required. ($|B_{\omega_1}| = \omega_2$ by (C 2) and $|U^{\mathscr{B}_{\omega_1}}| = \omega$ by (C 3).) We construct the above sequences by recursion on α.

$\mathscr{B}_0$ has been defined already. We set $h_0(0) = e_0$. Now suppose that we are at a successor step, $\alpha + 1$. By (P 3), $\mathscr{F}_{\alpha, \alpha+1} = \{\operatorname{id} \restriction \theta_\alpha, f\}$, where for some $\delta < \theta_\alpha$, $f_\alpha \restriction \delta = \operatorname{id} \restriction \delta$ and $f_\alpha(\delta) \geqslant \theta_\alpha$. By (C 4), $\langle \mathscr{B}_\alpha, h_\alpha(\delta) \rangle \cong \langle \mathscr{B}_0, e_0 \rangle$, so we can find $\mathscr{B}_{\alpha+1}, \sigma$ such that:

$$\mathscr{B}_\alpha \prec \mathscr{B}_{\alpha+1}, \quad U^{\mathscr{B}_\alpha} = U^{\mathscr{B}_{\alpha+1}}, \quad \sigma \colon \mathscr{B}_\alpha \prec \mathscr{B}_{\alpha+1}, \quad B_\alpha \subseteq \operatorname{Pr}^{\mathscr{B}_{\alpha+1}}(\sigma(h_\alpha(\delta))),$$
$$\sigma \restriction \operatorname{Pr}^{\mathscr{B}_\alpha}(h_\alpha(\delta)) = \operatorname{id} \restriction \operatorname{Pr}^{\mathscr{B}_\alpha}(h_\alpha(\delta)), \quad \langle \mathscr{B}_{\alpha+1}, \sigma(h_\alpha(\delta)) \rangle \cong \langle \mathscr{B}_\alpha, h_\alpha(\delta) \rangle.$$

Set $(\operatorname{id} \restriction \theta_\alpha)^* = \operatorname{id} \restriction B_\alpha$, $f_\alpha^* = \sigma$.

Suppose now that $h \in \mathscr{F}_{\beta, \alpha+1}$, $\beta < \alpha$. To define h^*, choose $f \in \mathscr{F}_{\alpha, \alpha+1}$, $g \in \mathscr{F}_{\beta\alpha}$, so that $h = f \circ g$ (by (P 2)) and let $h^* = f^* \circ g^*$. Now, g is clearly uniquely determined by h here, but if $\operatorname{ran}(h) \subseteq \delta$, then f is not. However, by (C 7) we have $\operatorname{ran}(g^*) \subseteq \operatorname{Pr}^{\mathscr{B}_\alpha}(h_\alpha(\delta))$, so by choice of σ, h^* does not depend upon the choice of f. Hence h^* is well-defined in all cases.

Define $h_{\alpha+1} \colon \theta_{\alpha+1} \to B_{\alpha+1}$ by

$$h_{\alpha+1}(\nu) = \begin{cases} h_\alpha(\nu), & \text{if } \nu < \theta_\alpha, \\ \sigma(h_\alpha(\bar{\nu})), & \text{if } \nu = f_\alpha(\bar{\nu}) \geqslant \theta_\alpha. \end{cases}$$

(Using (P 5), it is easy to see that $h_{\alpha+1}$ is well-defined on $\theta_{\alpha+1}$.)

We must check that (C 1)–(C 7) are preserved. (C 1) is clear. (C 2) holds because $B_\alpha \subseteq \operatorname{Pr}^{\mathscr{B}_{\alpha+1}}(\sigma(h_\alpha(\delta)))$. For (C 3), note that since $\langle \mathscr{B}_\alpha, h_\alpha(\delta) \rangle \cong \langle \mathscr{B}_0, e_0 \rangle$, we have

$U^{\mathscr{B}_\alpha} \subseteq \mathrm{Pr}^{\mathscr{B}_\alpha}(h_\alpha(\delta))$, and that by choice of σ, $\sigma \restriction \mathrm{Pr}^{\mathscr{B}_\alpha}(h_\alpha(\delta)) = \mathrm{id} \restriction \mathrm{Pr}^{\mathscr{B}_\alpha}(h_\alpha(\delta))$, so $U^{\mathscr{B}_{\alpha+1}} = U^{\mathscr{B}_\alpha} \subseteq \mathrm{ran}((\mathrm{id} \restriction \theta_\alpha)^*)$ and $U^{\mathscr{B}_{\alpha+1}} = U^{\mathscr{B}_\alpha} \subseteq \mathrm{ran}(f_\alpha^*)$. (C 4) is a simple consequence of the fact that $\mathscr{B}_{\alpha+1}$ is countable homogeneous (cf. the corresponding arguments in section 3, in particular the proof of (C 10) there). (C 5) holds by definition. (C 6) need only be verified for $\mathscr{F}_{\alpha,\alpha+1}$, i.e. it must be shown that if $f \in \mathscr{F}_{\alpha,\alpha+1}$, then $h_{\alpha+1} \circ f = f^* \circ h_\alpha$. But this is immediate. Finally, (C 7) also only requires verification for $\mathscr{F}_{\alpha,\alpha+1}$, which is a triviality.

There remains the limit case (i.e. lim (α)). Let

$$\mathscr{F} = \bigcup_{\beta<\alpha} \mathscr{F}_{\beta\alpha}$$

For each $f \in \mathscr{F}$, let $d(f)$ be that β such that $f \in \mathscr{F}_{\beta\alpha}$. For $f, f' \in \mathscr{F}$, set $f <^* f'$ iff $d(f) < d(f')$ and there is a $g \in \mathscr{F}_{d(f),d(f')}$ such that $f = f' \circ g$. Note that, if it exists, the g here is uniquely determined by f, f'. Hence for $f <^* f'$ we may define an embedding

$$\pi_{ff'}: \mathscr{B}_{d(f)} \prec \mathscr{B}_{d(f')}$$

by

$$\pi_{ff'} = g^*,$$

where $f = f' \circ g$. By (P 2) and (P 4), $<^*$ is a transitive, directed relation on $\mathscr{F}$. Clearly,

$$\langle (\mathscr{B}_{d(f)})_{f \in \mathscr{F}}, (\pi_{ff'})_{f <^* f'} \rangle$$

is a commutative, directed elementary system. Let

$$\langle \mathscr{B}_\alpha, (f^*)_{f \in \mathscr{F}} \rangle$$

be a direct limit. Using (C 6) we may define $h_\alpha: \theta_\alpha \to B_\alpha$ by requiring commutativity of the following diagram for all $\beta < \alpha$:

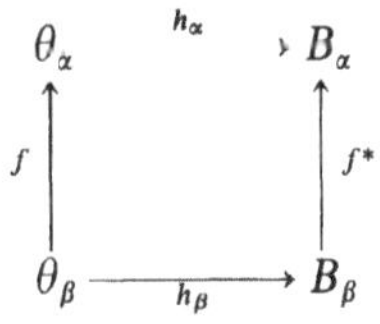

(By (P 5), this does define h_α on all of θ_α.)

We must verify (C 1)–(C 7). The only one that is not entirely trivial is (C 4). But if α is a countable limit ordinal, then $\langle \mathscr{F}, <^* \rangle$ has a cofinal subset of order-type ω, so (C 4) follows from lemma 1.6. That completes the proof of the Gap-2 Theorem using a simplified morass.

We turn now to the question of the existence of a simplified morass. It should be stressed that the definition of the simplified morass is designed to make applications easy. The simplified morass structure is not particularly closely related to the constructible hierarchy in the way that the "standard" morass is. In fact, in order to construct a simplified morass, what we shall in fact do is start with a standard morass and use it to construct the new morass, rather than the fine structure theory. This construction is not at all intuitive, and is motivated solely by the aim of obtaining the various properties of a simplified morass.

The general idea is to define the ordinals θ_α of the simplified morass as the order-types of certain well-ordered sets $(W_\alpha, <_\alpha)$ of finite tuples of elements of the standard morass, and to obtain the embeddings in $\mathscr{F}_{\alpha\beta}$ as compositions of some specific maps from W_α into W_β. In order to make this work we first of all have to add some extra points to the morass to enable us to "smooth out" the irregularities in the morass structure which manifested themselves in the large number of cases required to prove the Gap-2 Theorem using the standard morass.

Beyond this very rough outline, the rest is, unfortunately, highly technical, so you may expect a somewhat rough ride. Best of luck!

We fix a standard morass

$$\mathscr{M} = \langle S, \mathscr{S}, \dashv, (\pi_{\nu\tau})_{\nu \dashv \tau} \rangle$$

as in section 2. We shall write

$$\bar{\nu} \dashv_* \nu \quad \text{iff } \nu \text{ immediately succeeds } \bar{\nu} \text{ in } \dashv;$$

and (see Fig. 8)

$$\mu \,\text{-}\!\!\dashv \nu \quad \text{iff there are } \bar{\nu}, \bar{\mu} \text{ such that } \bar{\nu} \dashv_* \nu,\ \bar{\mu} \in S_{\alpha_{\bar{\nu}}} \cap \bar{\nu}, \text{ and } \bar{\mu} \dashv_* \mu \dashv \pi_{\bar{\nu}\nu}(\bar{\mu}).$$

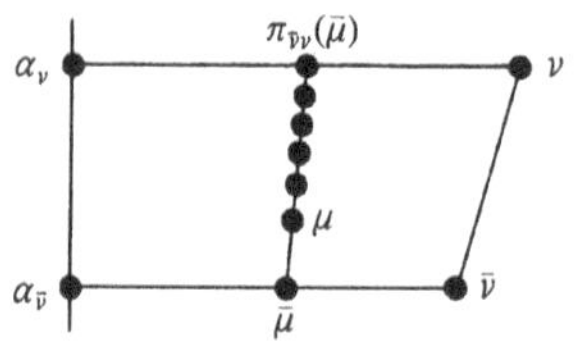

Fig. 8

We may (and shall) assume that if $\bar{\alpha} < \alpha$ and $\bar{\nu}$ is minimal in $S_{\bar{\alpha}}$ and ν is minimal in S_α, then $\bar{\nu} \dashv \nu$. (Simply extend $\dashv$ to achieve this. None of the morass axioms are effected by this.)

For each $\alpha \in S^0 \cap \omega_1$, we set $\nu_\alpha = \max(S_\alpha)$. Let

$$\begin{aligned} A &= \{\alpha \in S^0 \cap \omega_1 \mid \nu_\alpha \text{ is a successor in } \dashv\}; \\ A_0 &= \{\alpha \in A \mid \nu_\alpha \text{ is a successor in } S_\alpha\}; \\ A_1 &= \{\alpha \in A \mid \nu_\alpha \text{ is a limit in } S_\alpha \text{ but } \pi_{\bar{\nu}\nu_\alpha} \restriction \bar{\nu} \text{ is not cofinal in } \nu_\alpha, \\ &\qquad \text{where } \bar{\nu} \dashv_* \nu_\alpha\}. \end{aligned}$$

We now add some more points to the morass. For $\alpha \in A$, set

$$S_\alpha^+ = \{\nu_a + \tau \mid \tau \in S_\alpha \cap \nu_\alpha\}.$$

Extend the relation $\triangleleft$ to $\triangleleft'$ by setting, for $\nu_\alpha + \tau \in S_\alpha^+$,

$$\bar{\tau} \triangleleft' \nu_\alpha + \tau \quad \text{iff } \bar{\tau} \triangleleft \tau,$$

and then extending to obtain transitivity.

Let

$$\bar{S} = S^0 \cup \{\alpha + 1 \mid \alpha \in A_0 \cup A_1\}.$$

For $\alpha \in \bar{S}$, let

$$\bar{S}_\alpha = \begin{cases} S_\alpha, & \text{if } \alpha \in S^0 - A, \\ S_\alpha \cup S_\alpha^+, & \text{if } \alpha \in A - (A_0 \cup A_1), \\ S_\alpha - \{\nu_\alpha\}, & \text{if } \alpha \in A_0 \cup A_1, \\ S_{\bar{\alpha}} \cup S_{\bar{\alpha}}^+, & \text{if } \alpha = \bar{\alpha} + 1,\ \bar{\alpha} \in A_0 \cup A_1. \end{cases}$$

For $\nu \in \bigcup_{\alpha \in \bar{S}} \bar{S}_\alpha$, let

$$\alpha'_\nu = \text{the largest } \alpha \text{ such that } \nu \in \bar{S}_\alpha,$$
$$\alpha''_\nu = \text{the smallest } \alpha \text{ such that } \nu \in \bar{S}_\alpha.$$

(Notice that there are at most two α such that $\nu \in \bar{S}_\alpha$.)

Let $(\gamma_\nu \mid \nu \leqslant \omega_1)$ be the monotone enumeration of $\bar{S}$. For $\nu \leqslant \omega_1$, let W_ν be the set of all finite tuples $(\eta_0, \ldots, \eta_n)$ such that n is odd and:

(i) $\eta_0 \in \bar{S}_{\gamma_\nu}$;

(ii) $\eta_{2k+1} \triangleleft' \eta_{2k}$;

(iii) $\eta_{2k+2} \in S_{\alpha'_{\eta_{2k+1}}}$, $\eta_{2k+2} > \eta_{2k+1}$.

Let $<_\nu$ be the Kleene-Brouwer ordering on W_ν; that is, $(\vec{\eta}) <_\nu (\vec{\mu})$ iff $\vec{\mu}$ is an initial segment of $\vec{\eta}$ or else $\vec{\eta}$ precedes $\vec{\mu}$ lexicographically. It is easily seen that W_ν is well-ordered by $<_\nu$, so let $\theta_\nu = \operatorname{otp}(W_\nu, <_\nu)$. In the following we shall identify θ_ν with $(W_\nu, <_\nu)$.

Clearly, $0 < \theta_\nu < \omega_1$ for $\nu < \omega_1$, and $\theta_{\omega_1} = \omega_2$. As a prelude to defining the sets $\mathscr{F}_{\alpha\beta}$ of embeddings we define some special maps.

First of all, let $\bar{\nu} \triangleleft \nu$, $\alpha'_{\bar{\nu}} = \gamma_\alpha$, $\alpha''_\nu = \gamma_\beta$. We define $\tilde{\pi}_{\bar{\nu}\nu}\colon W_\alpha \to W_\beta$ by

$$\tilde{\pi}_{\bar{\nu}\nu}((\eta_0, \ldots, \eta_n)) = \begin{cases} (\pi_{\bar{\nu}\nu}(\eta_0), \eta_1, \ldots, \eta_n), & \text{if } \eta_0 \leqslant \bar{\nu}, \\ (\nu, \bar{\nu}, \eta_0, \ldots, \eta_n), & \text{if } \eta_0 > \nu. \end{cases}$$

Now suppose $\eta \dashv \nu$. Then there are $\bar{\nu}, \bar{\eta}$ such that $\bar{\nu} \triangleleft_* \nu$, $\bar{\eta} \in S_{\alpha_\nu} \cap \bar{\nu}$ and $\bar{\eta} \triangleleft_* \eta \triangleleft \pi_{\bar{\nu}\nu}(\bar{\eta})$. Let $\eta' = \pi_{\bar{\nu}\nu}(\bar{\eta})$. For some α, β, we have $\alpha'_\eta = \gamma_\alpha$, $\alpha''_\nu = \gamma_\beta$. Let $\varrho = \alpha'_{\bar{\nu}}$. Notice that by (M 4), $\eta = \max(S_{\gamma_\alpha})$, $\nu = \max(S_{\gamma_\beta})$, and (hence) $\gamma_\alpha, \gamma_\beta \in A$. We define $\sigma_{\eta\nu}\colon W_\alpha \to W_\beta$ by:

$$\sigma_{\eta\nu}((\eta_0, \ldots, \eta_n)) = \begin{cases} (\pi_{\eta\eta'}(\eta_0), \eta_1, \ldots, \eta_n), & \text{if } \eta_0 < \eta \text{ or } (\eta_0 = \eta \text{ and } \eta_1 \neq \bar{\eta}), \\ (\nu + \pi_{\eta\eta'}(\tau), \eta_1, \ldots, \eta_n), & \text{if } \eta_0 = \eta + \tau,\ \tau > 0, \\ \tilde{\pi}_{\bar{\nu}\nu}((\eta_2, \ldots, \eta_n)), & \text{if } (\eta_0, \eta_1) = (\eta, \bar{\eta}) \text{ and } \eta_2 \in \bar{S}_\varrho, \\ (\nu, \bar{\nu}), & \text{if } (\eta_0, \ldots, \eta_n) = (\eta, \bar{\eta}). \end{cases}$$

Again, let $\alpha \in A_0$. Then we can find $\bar{v}, \varrho$ so that $\bar{v} \dashv_* v_\alpha$ and v_α immediately succeeds ϱ in S_α. Let $v = v_\alpha$, $\pi_{\bar{v}v}(\bar{\varrho}) = \varrho$, $\alpha = \gamma_\delta$. Define $g_0^\delta\colon W_\delta \to W_{\delta+1}$ and $g_1^\delta\colon W_\delta \to W_{\delta+1}$ by $g_0^\delta = \mathrm{id} \restriction W_\delta$ and

$$g_1^\delta((\eta_0, \ldots, \eta_n)) = \begin{cases} (\eta_0, \ldots, \eta_n), & \text{if } \eta_0 < \varrho \text{ or } (\eta_0 = \varrho \text{ and } \eta_1 < \bar{\varrho}), \\ (v, \bar{v}, \eta_2, \ldots, \eta_n), & \text{if } \eta_0 = \varrho, \eta_1 = \bar{\varrho}, \text{ and either } \eta_2 \text{ does not} \\ & \qquad \text{exist or else } \eta_2 \neq \bar{v}, \\ (v, \eta_3, \ldots, \eta_n), & \text{if } \eta_0 = \varrho, \eta_1 = \bar{\varrho}, \eta_2 = \bar{v}, \\ (v + \varrho, \eta_1, \ldots, \eta_n), & \text{if } \eta_0 = \varrho \text{ and } \eta_1 > \bar{\varrho}. \end{cases}$$

Finally, let $\alpha \in A_1$. Then there are $\bar{v}, \lambda$ such that $\bar{v} \dashv_* v_\alpha$ and $\lambda = \sup \pi_{\bar{v}v}{}''\bar{v} < v_\alpha$. By (M 6), $\bar{v} \dashv \lambda$. Let $\alpha = \gamma_\delta$. Define $g_0^\delta\colon W_\delta \to W_{\delta+1}$ and $g_1^\delta\colon W_\delta \to W_{\delta+1}$ by $g_0^\delta = \mathrm{id} \restriction W_\delta$ and

$$g_1^\delta((\eta_0, \ldots, \eta_n)) = \begin{cases} (\eta_0, \ldots, \eta_n), & \text{if } \eta_0 < \lambda \text{ or } (\eta_0 = \lambda \text{ and } \eta_1 < \bar{v}), \\ (v, \eta_1, \ldots, \eta_n), & \text{if } \eta_0 = \lambda \text{ and } \eta_1 = \bar{v}, \\ (v + \eta_0, \eta_1, \ldots, \eta_n), & \text{if } \eta_0 > \lambda \text{ or } (\eta_0 = \lambda \text{ and } \eta_1 > \bar{v}). \end{cases}$$

The proofs of all parts of the following lemma are routine (and hence omitted).

4.1 Lemma.

(i) *$\pi_{\bar{v}v}, \sigma_{\eta v}, g_i^\delta$ are all order-preserving.*

(ii) *For some $\alpha < \delta$, $g_1^\delta \restriction \alpha = \mathrm{id} \restriction \alpha$ and $g_1^\delta(\alpha) \geqslant \theta_\delta$. (Recall that we identify θ_δ with $(W_\delta, <_\delta)$.)*

(iii) $\bar{v} \dashv v' \dashv v \to \pi_{\bar{v}v} = \tilde{\pi}_{v'v} \circ \tilde{\pi}_{\bar{v}v'}$.

(iv) $\eta \dashv \varrho \dashv v \to \sigma_{\eta v} = \sigma_{\varrho v} \circ \sigma_{\eta\varrho}$

(v) *If $\eta \dashv v$, where $\bar{v} \dashv_* v$ and $\bar{\eta} \dashv_* \eta$, then $\tilde{\pi}_{\bar{v}v} = \sigma_{\eta v} \circ \tilde{\pi}_{\bar{\eta}\eta}$.*

(vi) *Let $\bar{v} \dashv_* v = v_\alpha$, $\alpha \in A_0 \cup A_1$, $\alpha = \gamma_\delta$, and let $\lambda = \sup \pi_{\bar{v}v}{}''(S_{\alpha_v} \cap \bar{v})$. Let $\bar{\lambda} \dashv \lambda$ be such that $\alpha_{\bar{\lambda}} = \alpha_{\bar{v}}$. (Thus either $\bar{v} = \bar{\lambda}$ or else $\bar{v}$ immediately succeeds $\bar{\lambda}$ in $S_{\alpha_{\bar{v}}}$.) Let $\bar{\lambda} \dashv_* \lambda' \dashv \lambda$. Then $\tilde{\pi}_{\bar{v}v} = g_1^\delta \circ \tilde{\pi}_{\bar{\lambda}\lambda}$ and for all $\eta \dashv v$ we have either $\sigma_{\eta v} = g_1^\delta \circ \tilde{\pi}_{\lambda\lambda'} \circ \sigma_{\eta\lambda'}$ or else $\eta = \lambda'$ and $\sigma_{\eta v} = g_1^\delta \circ \tilde{\pi}_{\eta\lambda}$.*

(vii) *Let $\tau \in S_\alpha \cap v_\alpha$, $\alpha \in A_0 \cup A_1$, $\alpha = \gamma_\delta$. Then:*

$$\bar{\tau} \dashv \tau \to \tilde{\pi}_{\bar{\tau}\tau} = g_0^\delta \circ \tilde{\pi}_{\bar{\tau}\tau},$$
$$\eta \dashv v \to \sigma_{\eta v} = g_0^\delta \circ \sigma_{\eta v}. \quad \square$$

Let $\bar{\mathscr{F}}$ be the set of all the maps $\pi_{\bar{v}v}, \sigma_{\eta v}, g_i^\delta$, and let $\mathscr{F}$ be the closure of $\bar{\mathscr{F}}$ under finite compositions. For $\alpha < \beta \leqslant \omega_1$, set

$$\mathscr{F}_{\alpha\beta} = \{f \in \mathscr{F} \mid \mathrm{dom}(f) = W_\alpha \text{ and } \mathrm{ran}(f) \subseteq W_\beta\}.$$

The structure

$$\mathscr{M}_0 = \langle (\theta_\alpha \mid \alpha \leqslant \omega_1), (\mathscr{F}_{\alpha\beta} \mid \alpha < \beta \leqslant \omega_1) \rangle$$

is not yet a simplified morass, but as we show below it already has most of the properties we require.

4.2 Lemma. *$\mathcal{M}_0$ has property* (P 1), *i.e.* $|\mathcal{F}_{\alpha\beta}| \leqslant \omega$ *for* $\alpha < \beta < \omega_1$.

Proof. Clear. □

4.3 Lemma. *$\mathcal{M}_0$ has property* (P 2), *i.e. if* $\alpha < \beta < \gamma \leqslant \omega_1$, *then*

$$\mathcal{F}_{\alpha\gamma} = \{f \circ g \,|\, f \in \mathcal{F}_{\beta\gamma} \;\&\; g \in \mathcal{F}_{\alpha\beta}\}.$$

Proof. ($\supseteq$) By definition.

($\subseteq$). By induction on γ. There are many different cases. As an example we deal with the case $\bar{\nu} \dashv_* \nu$, $\bar{\nu}$ a limit point in $S_{\alpha_{\bar{\gamma}}}$, $\pi_{\bar{\nu}\nu}$ is cofinal. Then γ is a limit ordinal. We must use the induction hypothesis to show that (where $\mathrm{dom}(\tilde{\pi}_{\bar{\nu}\nu}) = W_\alpha$):

(*) for cofinally many $\beta < \gamma$, there are $f \in \mathcal{F}_{\beta\gamma}$, $g \in \mathcal{F}_{\alpha\beta}$ such that $\tilde{\pi}_{\bar{\nu}\nu} = f \circ g$.

By 4.1 (v) we have

$$\tilde{\pi}_{\bar{\nu}\nu} = \sigma_{\eta\nu} \circ \tilde{\pi}_{\bar{\eta}\eta} \quad \text{for } \eta \dashv \nu.$$

By morass properties (see the argument used in handling Case 3.3.2 in the proof of the Gap-2 Theorem in section 3) we have:

$$\alpha_\nu = \sup\{\alpha_\eta \,|\, \eta \dashv \nu\}.$$

Thus (*) follows.

For the other cases, use 4.1 (iii), (iv), (vi), (vii). □

4.4 Lemma. *Let* $\alpha < \omega_1$. *Then* $\mathcal{F}_{\alpha,\alpha+1}$ *is either a singleton or else consists of a pair* $\{\mathrm{id} \restriction \theta_\alpha, f_\alpha\}$ *such that for some* $\delta < \theta_\alpha$, $f_\alpha \restriction \delta = \mathrm{id} \restriction \delta$ *and* $f_\alpha(\delta) \geqslant \theta_\alpha$.

Proof. If $\gamma_\alpha \in A_0 \cup A_1$, then we are done by 4.1 (ii). If $\gamma_\alpha \notin A_0 \cup A_1$, then $\bar{\gamma} = \gamma_{\alpha+1}$ is a successor in S^0, so by morass properties $S_{\bar{\gamma}} = \{\nu\}$ for some ν. By our initial special assumption on the morass tree, we have $\bar{\nu} \dashv_* \nu$ for some $\bar{\nu}$. Thus $\mathcal{F}_{\alpha,\alpha+1} = \{\tilde{\pi}_{\bar{\nu}\nu}\}$, and again we are done. □

4.5 Lemma. *$\mathcal{M}_0$ has property* (P 4).

Proof. Let $\alpha \leqslant \omega_1$ be a limit ordinal. We define a certain subset $\mathcal{G}_\alpha \subseteq \bigcup_{\gamma<\alpha} \mathcal{F}_{\gamma\alpha}$ and leave it to the reader to check that it is always possible to find a $g \in \mathcal{G}_\alpha$ which verifies (P 4). Let $\gamma = \gamma_\alpha$, and if $\alpha < \omega_1$, let $\nu = \nu_\alpha$.

Case 1. $\gamma \in A_1$ or $\alpha = \omega_1$.
Set $\mathcal{G}_\alpha = \{\tilde{\pi}_{\bar{\tau}\tau} \,|\, \bar{\tau} \dashv \tau, \tau \in S_\gamma, \tau < \sup(S_\gamma)\}$.

Case 2. $\gamma \in A_0$.
Let ν immediately succeed τ in S_{γ_α}. Set $\mathcal{G}_\alpha = \{\tilde{\pi}_{\bar{\tau}\tau} \,|\, \bar{\tau} \dashv \tau\}$.

Case 3. $\gamma \in (S^0 \cap \omega_1) - A$.
Set $\mathcal{G}_\alpha = \{\tilde{\pi}_{\bar{\nu}\nu} \,|\, \bar{\nu} \dashv \nu\}$.

Case 4. $\gamma \in A - (A_0 \cup A_1)$.
Set $\mathcal{G}_\alpha = \{\sigma_{\eta\nu} \,|\, \eta \dashv \nu\}$. □

4.6 Lemma. $\omega_2 = \bigcup \{f''\theta_\alpha \mid \alpha < \omega_1 \;\&\; f \in \mathscr{F}_{\alpha\omega_1}\}$.

Proof. Obvious. □

Our task now is to modify $\mathscr{M}_0$ so that (P 4), (P 5) and (P 0) are satisfied, as well as (P 1), (P 2) and (P 3). The only part of (P 0) that we do not have so far is $\theta_0 = 1$. Lemma 4.4 tells us that we are part way to having (P 3) already. And lemma 4.6 gives us (P 5) for the case $\alpha = \omega_1$.

By 4.6,

$$\omega_2 = \bigcup_{\gamma < \omega_1} \left(\bigcup \{f''\theta_\gamma \mid f \in \mathscr{F}_{\gamma\omega_1}\}\right).$$

So we can find a $\gamma < \omega_1$ such that

$$|\bigcup \{f''\theta_\gamma \mid f \in \mathscr{F}_{\gamma\omega_1}\}| = \omega_2.$$

For $\gamma < \alpha \leqslant \omega_1$, let

$$S'_\alpha = \bigcup \{f''\theta_\gamma \mid f \in \mathscr{F}_{\gamma\alpha}\}.$$

Notice that by choice of γ,

$$|S'_{\omega_1}| = \omega_2.$$

4.7 Lemma. *If $\gamma < \beta < \alpha \leqslant \omega_1$, then*

$$S'_\alpha = \bigcup \{f''S'_\beta \mid f \in \mathscr{F}_{\beta\alpha}\}.$$

Proof.
$$\begin{aligned} S'_\alpha &= \bigcup \{h''\theta_\gamma \mid h \in \mathscr{F}_{\gamma\alpha}\} = \bigcup \{(f \circ g)''\theta_\gamma \mid f \in \mathscr{F}_{\beta\alpha}, g \in \mathscr{F}_{\gamma\beta}\} \\ &= \bigcup \{f''(\bigcup \{g''\theta_\gamma \mid g \in \mathscr{F}_{\gamma\beta}\}) \mid f \in \mathscr{F}_{\beta\alpha}\} = \bigcup \{f''S'_\beta \mid f \in \mathscr{F}_{\beta\alpha}\}. \quad \square \end{aligned}$$

Before we state our next lemma, we note that if $\alpha < \beta \leqslant \omega_1$ and $\tau_1, \tau_2 < \theta_\alpha$, $f_1, f_2 \in \mathscr{F}_{\alpha\beta}$, and if $f_1(\tau_1) = f_2(\tau_2)$, then $\tau_1 = \tau_2$ and $f_1 \upharpoonright \tau_1 = f_2 \upharpoonright \tau_2$. (This is easily proved by induction on β, using 4.4 for the initial step $\beta = \alpha + 1$, 4.3 for the successor step, and 4.5 for the limit step.)

4.8 Lemma. *Let $\gamma < \beta < \alpha \leqslant \omega_1$. The following are equivalent:*

(i) $(\exists f \in \mathscr{F}_{\beta\alpha})(f''S'_\beta = S'_\alpha)$;

(ii) $(\forall f, g \in \mathscr{F}_{\beta\alpha})(f \upharpoonright S'_\beta = g \upharpoonright S'_\beta)$;

(iii) $(\forall f \in \mathscr{F}_{\beta\alpha})(f''S'_\beta = S'_\alpha)$.

Proof. (i) → (ii). Choose $f \in \mathscr{F}_{\beta\alpha}$ such that $f''S'_\beta = S'_\alpha$. Suppose there is a $g \in \mathscr{F}_{\beta\alpha}$ such that $g(\tau) \neq f(\tau)$ for some $\tau \in S_\beta$. By (i), $g''S'_\beta \subseteq S'_\alpha = f''S'_\beta$, so we can choose $\tau' \in S'_\beta$ such that $f(\tau') = g(\tau)$. But clearly, $\tau' \neq \tau$, so this contradicts the observation made above.

(ii) → (iii). By 4.7.

(iii) → (i). Trivial. □

We shall call an ordinal $\alpha \leqslant \omega_1$ *redundant* if there are $\beta < \alpha$, $f \in \mathscr{F}_{\beta\alpha}$ such that $\beta > \gamma$ and $f'' S'_\beta = S'_\alpha$. Let

$$N = \{\alpha \leqslant \omega_1 \mid \alpha > \gamma \ \&\ \alpha \text{ is not redundant}\}.$$

Clearly, $\omega_1 \in N$.

4.9 Lemma. *$N \cap \omega_1$ is a club subset of ω_1.*

Proof. To prove closure, suppose that α is a limit point of $N \cap \omega_1$ and that α is redundant. Choose $\beta < \alpha$ and $f \in \mathscr{F}_{\beta\alpha}$ such that $\beta > \gamma$ and $f'' S'_\beta = S'_\alpha$. Since α is a limit point of $N \cap \omega_1$ we can find a $\delta \in N \cap \omega_1$ such that $\beta < \delta < \alpha$. Choose $g \in \mathscr{F}_{\delta\alpha}$ and $h \in \mathscr{F}_{\beta\delta}$ such that $f = g \circ h$. Since δ is not redundant, $h'' S'_\beta \subset S'_\delta$. But then $f'' S'_\beta = (g \circ h)'' S'_\beta \subset g'' S_\delta \subseteq S_\alpha$, contrary to the choice of β and f.

To prove unboundedness, let $\gamma < \beta < \omega_1$. We find the least element of $N \cap \omega_1$ greater than β. Let α be the least ordinal such that $\beta < \alpha \leqslant \omega_1$ and $(\forall f \in \mathscr{F}_{\beta\alpha})(f'' S'_\beta \neq S'_\alpha)$. Such an ordinal exists, since we clearly have $(\forall f \in \mathscr{F}_{\beta\omega_1}) \cdot (f'' S'_\beta \neq S'_{\omega_1})$. By 4.8 we can choose $f_1, f_2 \in \mathscr{F}_{\beta\alpha}$ such that $f_1 \restriction S'_\beta \neq f_2 \restriction S'_\beta$. If α is a limit ordinal, then by applying 4.5 we can get a counterexample to the minimality of α. Thus $\alpha = \delta + 1 < \omega_1$ for some $\delta \geqslant \beta$. If $\delta > \beta$, then by the minimality of α there is an $f \in \mathscr{F}_{\beta\delta}$ such that $f'' S'_\beta = S'_\delta$. If $g \in \mathscr{F}_{\delta\alpha}$ and $g'' S'_\delta = S'_\alpha$, then $g \circ f \in \mathscr{F}_{\beta\alpha}$ and $(g \circ f)'' S'_\beta = S_\alpha$, contradicting the choice of α. Thus $(\forall g \in \mathscr{F}_{\delta\alpha}) \cdot (g'' S'_\delta \neq S'_\alpha)$. Clearly, the same conclusion holds if $\delta = \beta$. It follows easily that α is not redundant, so $\alpha \in N \cap \omega_1$. Note that we have shown that if $\gamma < \beta < \omega_1$, then the least element of $N - \beta$ is a successor ordinal $\delta + 1$ for some $\delta \geqslant \beta$, and if $\delta > \beta$ then $(\exists f \in \mathscr{F}_{\beta\delta})(f'' S'_\beta = S'_\delta)$. □

Now let $(\eta_\nu \mid \nu \leqslant \omega_1)$ be the monotone enumeration of N, and for $\nu \leqslant \omega_1$, let $\theta'_\nu = \operatorname{otp}(S'_{\eta_\nu})$. Note that for $\nu < \omega_1$, $\eta_\nu < \omega_1$, so $0 < \theta'_\nu < \omega_1$. Also, $\eta_{\omega_1} = \omega_1$ and $|S'_{\omega_1}| = \omega_2$, so $\theta'_{\omega_1} = \omega_2$. We identify S'_{η_ν} with θ'_ν from now on. Subject to this identification, let $\mathscr{F}'_{\nu\tau}$ denote $\mathscr{F}_{\eta_\nu\eta_\tau}$.

4.10 Lemma. *Except for the fact that θ'_0 may not equal* 1, *the structure*

$$\mathscr{M}_1 = \langle(\theta'_\nu \mid \nu \leqslant \omega_1), (\mathscr{F}'_{\nu\tau} \mid \nu < \tau \leqslant \omega_1)\rangle$$

is a simplified morass.

Proof. Most of this is quite straightforward, and is left for the reader to check. To prove (P 4), use the fact that N is closed. For (P 3), note that for any $\nu < \omega_1$, $\eta_{\nu+1}$ is the least element of $N - \eta_\nu$, so by the proof of 4.9, $\eta_{\nu+1} = \delta + 1$ for some $\delta \geqslant \eta_\nu$, $(\forall f \in \mathscr{F}_{\delta\eta_{\nu+1}})(f'' S'_\delta \neq S'_{\eta_{\nu+1}})$, and if $\delta > \eta_\nu$ then $(\exists f \in \mathscr{F}_{\eta_\nu\delta})(f'' S'_{\eta_\nu} = S'_\delta)$, so by 4.8, $(\forall f, g \in \mathscr{F}_{\eta_\nu\delta})(f \restriction S'_{\eta_\nu} = g \restriction S'_{\eta_\nu})$. Clearly, $\mathscr{F}_{\delta\alpha_{\nu+1}} = \mathscr{F}_{\delta,\delta+1}$ is as in (P 3), and it is now not hard to show that $\mathscr{F}'_{\nu,\nu+1}$ is too. (P 5) follows easily from 4.7. □

Finally, if $\theta'_0 > 1$, we simply add an initial segment to the structure obtained above and reindex. All that is required is to build up to the existing θ'_0 by a simplified morass-like structure consisting of θ'_0 levels, starting with $\{0\}$. This is easily achieved. We are done.

5. *Gap-n Morasses*

In order to prove the Gap-$(n+1)$ Cardinal Transfer Theorem, we need a morass structure which enables us to construct a model of cardinality ω_{n+1} using only countable structures. The morass required to do this is a gap-n morass, or more precisely a (ω_1, n)-morass. Assuming $V = L$, such morasses can be developed, and thus, assuming $V = L$ the Gap-n Theorem is valid for all n. Unfortunately, for $n > 1$, the definition and construction of a gap-n morass is little short of horrendous, and would require for a reasonable treatment a volume comparable to the present one. However, although it is not possible to even give the definition of a gap-n morass here (for $n > 1$), it is possible to indicate why one might expect that such a structure exists, and what it should look like.

The simplest type of system for building models is an elementary chain, which we may regard as a one-dimensional system. Then come gap-1 morasses (together with their associated model complexes), which we may think of as two-dimensional systems. A gap-2 morass would then be a three-dimensional system, and in general a gap-n morass would be an $n+1$-dimensional system. The formal definition of a gap-n morass would then proceed in the "obvious fashion". Just as a gap-1 morass was defined on a set, $\mathscr{S}$, of ordered pairs (α, ν) of ordinals, with $\alpha \leqslant \omega_1, \nu < \omega_2$, so a gap-2 morass is defined on a set, $\mathscr{S}$, of ordered triples (α, τ, ν) of ordinals such that $\alpha \leqslant \omega_1, \tau \leqslant \omega_2, \nu < \omega_3$, and so on. Indeed, the construction of such a structure is, in principal the same as in the gap-1 case, using the fine structure theory. Unfortunately, matters rapidly become very complicated, and so we must end our rather brief account at this point.

Exercises

1. Morasses and the Kurepa Hypothesis

Prove that the existence of a κ^+-morass implies $KH(\kappa^+)$. (Hint. For each $\nu \in S_{\kappa^+}$, let $X_\nu = \{\bar{\nu} \in S^1 \cap \kappa^+ \mid \bar{\nu} \dashv \nu\}$, and show that the family $\mathscr{F} = \{X_\nu \mid \nu \in S_{\kappa^+}\}$ is a κ^+-Kurepa family.)

2. Morasses and the Combinatorial Principle $\square$

Prove that if there is an ω_1-morass, then $\square_{\omega_1}$ is valid. (Hint. For each limit point ν of S_{ω_1}, let $C_\nu = \{\sup(\pi_{\bar{\nu}\nu}{}''\bar{\nu}) \mid \bar{\nu} \dashv \nu\} \cap \nu$, and investigate the properties of the sets C_ν.) Does this generalise to arbitrary successor cardinals κ^+ in place of ω_1?

3. Cardinal Transfer Theorems

The first result to be proved is that, assuming GCH, if $\mathscr{A}$ is a K-structure of type (ω_1, ω), then for any uncountable regular cardinal κ there is a K-structure $\mathscr{B}$ of type (κ^+, κ) such that $\mathscr{B} \equiv \mathscr{A}$. The general idea is to proceed much as in 1.7, using saturated structures of size κ instead of countable homogeneous structures. 1.1 (i) guarantees that all of the structures in the chain are isomorphic. The difficulty lies in ensuring that limit stages preserve saturation. This requires the use of a clever trick. Fix $\mathscr{A}$ as above now.

A K-structure $\mathscr{B}$ is said to be *U-saturated* if it satisfies the definition of saturatedness for all element types $\Sigma(x)$ which contain the formula $U(x)$.

3 A. Show that if $\mathscr{B}$ is a saturated K-structure of cardinality κ such that $\mathscr{B} \equiv \mathscr{A}$, then there is a saturated K-structure $\mathscr{B}'$ of cardinality κ such that $\mathscr{B} \prec \mathscr{B}'$, $\mathscr{B} \neq \mathscr{B}'$, $U^{\mathscr{B}} = U^{\mathscr{B}'}$, $\mathscr{B} \cong \mathscr{B}'$.

3 B. Show that if $\mathscr{B}$ is a U-saturated K-structure of cardinality κ, there is a saturated K-structure $\mathscr{B}'$ of cardinality κ such that $\mathscr{B} \prec \mathscr{B}'$ and $U^{\mathscr{B}} = U^{\mathscr{B}'}$.

There is no less of generality in assuming that the given model $\mathscr{A}$ has a binary predicate E with the property that for each finite set $H \subseteq U^{\mathscr{A}}$, there is an element $a \in U^{\mathscr{A}}$ such that $H = \{x \mid x E^{\mathscr{A}} a\}$. Using this assumption, the following key step of the proof can be established.

3 C. Show that if $\lambda < \kappa^+$ and $(\mathscr{B}_\nu \mid \nu < \lambda)$ is an elementary chain of U-saturated structures elementarily equivalent to $\mathscr{A}$, each of cardinality κ and all having the same distinguished subset U, then $\bigcup_{\nu<\lambda} \mathscr{B}_\nu$ is U-saturated.

3 D. Show that there is a K-structure $\mathscr{B}$ of type (κ^+, κ) such that $\mathscr{B} \equiv \mathscr{A}$.

The second result to be proved is that, assuming $V = L$, if $\mathscr{A}$ is a K-structure of type (ω_1, ω), then for any singular cardinal κ there is a K-structure $\mathscr{B}$ of type (κ^+, κ) such that $\mathscr{B} \equiv \mathscr{A}$. (What we actually require is $\square_\kappa$ together with GCH.)

Fix κ a singular cardinal from now on, and let $\mu = \mathrm{cf}(\kappa)$. Let $G: \mu \to \kappa$ be an increasing sequence of regular cardinals such that $G(0) = 0$, $G(1) > \omega$, and $\sup(G''\mu) = \kappa$. By $\square_\kappa$, let $(S_\alpha \mid \alpha < \kappa^+ \wedge \lim(\alpha))$ be such that:

(i) S_α is a closed subset of α;

(ii) if $\mathrm{cf}(\alpha) > \omega$, then S_α is unbounded in α;

(iii) $|S_\alpha| < \kappa$;

(iv) if $\gamma \in S_\alpha$, then $S_\gamma = \gamma \cap S_\alpha$.

Modifying the previously defined notion of "special" a little, let us now agree to call a K-structure $\mathscr{B}$ of cardinality κ *special* iff there is an elementary chain $(\mathscr{B}_\alpha \mid \alpha < \mu)$ of saturated structures such that $\mathscr{B} = \bigcup_{\alpha<\mu} \mathscr{B}_\alpha$ and $|B_\alpha| = G(\alpha)$ for each $\alpha < \mu$. A mapping $r: B \to \mu$ is called a *ranking* of $\mathscr{B}$ iff there is such a chain with

$$r(x) = \text{the least } \alpha \text{ such that } x \in B_{\alpha+1}$$

for all $x \in B$. Similarly we define the notions of *U-special* and *U-ranking* by replacing "saturated" by "U-saturated".

3 E. Show that if $\mathscr{A}$ is any K-structure, there is a special structure $\mathscr{B}$ such that $\mathscr{B} \equiv \mathscr{A}$. (We are assuming GCH throughout.)

3 F. Show that if $\mathscr{A}$, $\mathscr{B}$ are special structures with rankings r, s, respectively, then $\mathscr{A} \cong \mathscr{B}$ and there is an isomorphism $f: \mathscr{A} \cong \mathscr{B}$ such that $s(f(x)) = r(x)$ for all $x \in A$.

3G. Let $\mathscr{A}$ be U-special with U-ranking r, and let $\mathscr{B}$ be special with ranking s, and suppose that $\mathscr{A} \equiv \mathscr{B}$. Show that there is an embedding $f\colon \mathscr{A} \prec \mathscr{B}$ such that $U^{\mathscr{B}} \subseteq \operatorname{ran}(f)$ and $s(f(x)) = r(x)$ for all $x \in A$.

3H. Let $\mathscr{A}$ be a K-structure of type (ω_1, ω). Let $\mathscr{B}$ be a U-special structure with U-ranking r, $\mathscr{B} \equiv \mathscr{A}$. Show that there is a special structure $\mathscr{B}'$ with ranking r' such that $\mathscr{B} \prec \mathscr{B}'$, $\mathscr{B} \neq \mathscr{B}'$, $U^{\mathscr{B}} = U^{\mathscr{B}'}$, and $r \subseteq r'$.

Given U-special structures $\mathscr{B}, \mathscr{B}'$ with U-ranking r, r', respectively, we write:

$$\begin{aligned}
(\mathscr{B}, r) \lhd (\mathscr{B}', r') \quad & \text{iff } \mathscr{B} \prec \mathscr{B}' \ \& \ r \subseteq r' \ \& \ U^{\mathscr{B}} = U^{\mathscr{B}'};\\
(\mathscr{B}, r) \propto_{\gamma} (\mathscr{B}', r') \quad & \text{iff } \mathscr{B} \prec \mathscr{B}' \ \& \ r \restriction U^{\mathscr{B}} = r' \restriction U^{\mathscr{B}} \ \& \ U^{\mathscr{B}} = U^{\mathscr{B}'}\\
& \& \, (r(x) < \gamma \to r'(x) < \gamma) \ \& \ (r(x) \geqslant \gamma \to r(x) = r'(x));\\
(\mathscr{B}, r) \propto (\mathscr{B}', r') \quad & \text{iff } (\exists \gamma)[(\mathscr{B}, r) \propto_{\gamma} (\mathscr{B}', r')].
\end{aligned}$$

Fix $\mathscr{A}$ a given K-structure of type (ω_1, ω) now. As before, assume that $\mathscr{A}$ has a binary predicate E which codes the finite subsets of $U^{\mathscr{A}}$ by elements of $U^{\mathscr{A}}$. To obtain a K-structure $\mathscr{B}$ of type (κ^+, κ) such that $\mathscr{B} \equiv \mathscr{A}$, the idea is to construct an elementary chain $(\mathscr{B}_\nu \mid \nu < \kappa^+)$ of U-special structures, all having the same U-set, and a sequence $(r_\nu \mid \nu < \kappa^+)$ such that r_ν is a U-ranking of $\mathscr{B}_\nu$, with $\mathscr{B}_0 \equiv \mathscr{A}$. The construction is carried out to preserve the following conditions:

(A) $\quad \alpha < \beta \to (\mathscr{B}_\alpha, r_\alpha) \propto (\mathscr{B}_\beta, r_\beta)$;

(B) $\quad \alpha \in S_\beta \to (\mathscr{B}_\alpha, r_\alpha) \lhd (\mathscr{B}_\beta, r_\beta)$;

(C) $\quad$ if $\alpha = S_\gamma(G(\beta))$, then $x \in A_\gamma - A_\alpha \to r_\gamma(x) \geqslant \beta$.

The only difficulty lies in the limit step in the definition. In case S_α is cofinal in α (for a limit stage α), set:

$$\mathscr{B}_\alpha = \bigcup_{\beta \in S_\alpha} \mathscr{B}_\beta, \qquad r_\alpha = \bigcup_{\beta \in S_\alpha} r_\beta.$$

In case S_α is not cofinal in α, in which case $\operatorname{cf}(\alpha) = \omega$, of course, pick a sequence $(\alpha_n \mid n < \omega)$ cofinal in α with $\alpha_0 = \sup(S_\alpha)$, let ψ be least such that $G(\psi) > \operatorname{otp}(S_\alpha)$, and pick a monotone sequence $(\varphi_n \mid n < \omega)$ of ordinals such that $\varphi_0 = 0$, $\varphi_1 > \psi$, and $\varphi_n < \mu$, with

$$(\mathscr{B}_{\alpha_i}, r_{\alpha_i}) \propto_{\varphi_{i+1}} (\mathscr{B}_{\alpha_{i+1}}, r_{\alpha_{i+1}})$$

for all $i < \omega$. Then set

$$\mathscr{B}_\alpha = \bigcup_{i<\omega} \mathscr{B}_{\alpha_i}.$$

For $x \in B_\alpha$, let $i(x)$ be the least i such that $x \in B_{\alpha_i}$, and set

$$r_\alpha(x) = \max(\varphi_{i(x)}, r_{\alpha_i}(x)).$$

3I. Check that the above definitions can indeed be carried out, and that they define a sequence $((\mathscr{B}_\alpha, r_\alpha) \mid \alpha < \kappa^+)$ as stated, to prove the Gap-1 Theorem for singular κ.

Further details of the above results can be found in *Chang-Keisler* (1973), *Devlin* (1973), and *Jensen* (1972).

Finally we consider the Gap-2 Theorem.

3 J. Assume $V = L$. Show that for any infinite cardinal κ, if $\mathscr{A}$ is a K-structure of type (ω_2, ω) there is a K-structure $\mathscr{B}$ of type (κ^{++}, κ) such that $\mathscr{B} \equiv \mathscr{A}$. (Hint. If κ is regular, the proof is a straightforward modification of the proof given in this chapter, using the ideas from exercises 3 A through 3 D above. If κ is singular, a mixture of the methods used in the chapter and those of exercises 3 E through 3 I is required, but in this case the proof is quite tricky. In particular, the $\square_\kappa$-sequence used must be obtained from the morass. (More precisely, from the fine structure construction of the morass.)

4. Morasses and the Combinatorial Principle $\diamondsuit_{\omega_2}$

4 A. Show that it is possible to define transitive structures M_ν, for each $\nu \in S^1$ in an ω_1-morass, such that:

(i) $\nu \in \mathrm{dom}(M_\nu)$, and $\mathrm{dom}(M_\nu) \cap \mathrm{On}$ is less than any element of S_{α_ν} and any element of S^0 above α_ν. (So, in particular, M_ν is countable for all $\nu \in S^1 \cap \omega_1$.)

(ii) For $\nu \dashv \tau$, there is an embedding $\sigma_{\nu\tau}$: $M_\nu \prec_1 M_\tau$ such that all of the following conditions are satisfied:

(iii) $(\sigma_{\nu\tau} \mid \nu \dashv \tau)$ is a commutative system.

(iv) $\sigma_{\nu\tau} \upharpoonright \nu = \pi_{\nu\tau} \upharpoonright \nu$.

(v) if $\bar{\tau} \dashv \tau$, $\bar{\nu} \in S_{\alpha_\tau} \cap \bar{\tau}$, $\nu = \pi_{\bar{\tau}\tau}(\bar{\nu})$, then $\sigma_{\bar{\nu}\nu} = \sigma_{\bar{\tau}\tau} \upharpoonright M_{\bar{\nu}}$.

(vi) if $\tau \in S^1$ is a limit point of $\dashv$, then $M_\tau = \bigcup_{\nu \dashv \tau} \sigma_{\nu\tau}'' M_\nu$.

(vii) $(\mathscr{P}(\nu) \cap M_\nu \mid \nu \in S_{\omega_1})$ is a $\diamondsuit_{\omega_2}$-sequence.

(viii) $(\mathscr{P}(\nu) \cap \Sigma_1(M_\nu) \mid \nu \in S_{\omega_1})$ is a $\diamondsuit^+_{\omega_2}$-sequence.

(Hint. Consider the structures $\langle J_{\varrho(\nu)}, A(\nu) \rangle$ used to construct the morass.)

4 B. Use 4 A to construct an ω_2-Souslin tree by means of a morass-like system of countable trees and embeddings between them.

5. A Coarse Morass

We investigate what kind of morass structure can be constructed using only elementary properties of L.

Let us call an ordinal ν *special* iff:

(i) either $L_\nu \models \mathrm{ZF}^-$ or else $\{\tau \in \nu \mid L_\tau \models \mathrm{ZF}^-\}$ is unbounded in ν;

(ii) $L_\nu =$ "there is exactly one uncountable cardinal".

For example, ω_2 is special. We define

$$S^1 = \{\nu \in \omega_2 \mid \nu \text{ is special}\};$$
$$\alpha_\nu = \omega_1^{L_\nu}, \quad \text{for } \nu \in S^1;$$
$$S^0 = \{\alpha_\nu \mid \nu \in S^1\};$$
$$S_\alpha = \{\nu \in S^1 \mid \alpha_\nu = \alpha\}, \quad \text{for } \alpha \in S^0.$$

5 A. Prove that $S_{\alpha_\nu} \cap \nu$ is uniformly definable in L_ν for $\nu \in S^1$.

5 B. Prove the following. Let $\nu \in S_\alpha$. Then there is an admissible ordinal $\beta = \beta(\nu) > \nu$ such that for some element p of L_β, every member of L_β is definable from parameters in $\alpha \cup \{p\}$ in L_β. Moreover, if τ succeeds ν in S_α, then there is such a $\beta < \tau$.

For each $\nu \in S^1$, let $\beta(\nu)$ be the least ordinal as above, and let $p(\nu)$ be the $<_L$-least such parameter.

5 C. Prove that ν is uniformly definable in $L_{\beta(\nu)}$.

5 D. Let $\nu \in S_\alpha$, and set $\beta = \beta(\nu)$, $p = p(\nu)$. Let $\bar{\alpha} < \alpha$, and suppose that X is the smallest $X \prec L_\beta$ such that $X \cap \alpha = \bar{\alpha}$ and $p \in X$. Let $\pi^{-1}: X \cong L_{\bar{\beta}}$, and set $\bar{\nu} = \pi^{-1}(\nu)$, $\bar{p} = \pi^{-1}(p)$. Show that $\bar{\alpha} \in A$, $\bar{\nu} \in S_{\bar{\alpha}}$, $\bar{\beta} = \beta(\bar{\nu})$, $\bar{p} = p(\bar{\nu})$.

For $\nu, \tau \in S^1$, define $\nu \dashv \tau$ iff $\alpha_\nu < \alpha_\tau$ and there is an embedding $\sigma: L_{\beta(\nu)} \prec L_{\beta(\tau)}$ such that $\sigma \restriction \alpha_\nu = \mathrm{id} \restriction \alpha_\nu$ and $\sigma(p(\nu)) = p(\tau)$.

5 E. Prove that if $\nu \dashv \tau$, the map σ above is unique.

Denote the unique map σ in the above by $\sigma_{\nu\tau}$. Note that by 5 C. above, $\sigma_{\nu\tau}(\nu) = \tau$. Hence $(\sigma_{\nu\tau} \restriction L_\nu): L_\nu \prec L_\tau$. Note also that $\sigma_{\nu\tau} \restriction \alpha_\nu = \mathrm{id} \restriction \alpha_\nu$, and $\sigma_{\nu\tau}(\alpha_\nu) = \alpha_\tau$. Let $\pi_{\nu\tau} = \sigma_{\nu\tau} \restriction (\nu + 1)$.

5 *F*. Prove that $\dashv$ is a tree ordering on S^1.

5 G. Verify that the system just constructed satisfies morass axioms (M 0) through (M 5), and investigate what happens when you try to prove (M 6) and (M 7).

The structure defined above is sometimes referred to as a *coarse morass*.

6. *Morasses and Large Cardinal Axioms*

Prove that if $V = L[A]$, where $A \subseteq \omega_1$, then there is a morass. Deduce that if ω_2 is not inaccessible in L, then there is a morass in the real world.

Chapter IX
Silver Machines

1. Silver Machines

Silver machines are a device for avoiding the use of the fine structure theory in proving results such as $\square_\kappa$. The idea is as follows. In proving, say $\square_\kappa$, as we did in Chapter IV, the main tool was the hierarchy of skolem functions $h_{\varrho_\alpha^n, A_\alpha^n}$. Of course, these functions, and the properties of them that we made use of, were obtained by our fine structure theory. But the fine structure theory itself was not used in the proof of $\square_\kappa$. Any hierarchy of functions with similar properties would suffice. As we shall see, it is possible to construct such a functioinal hierarchy without using the fine structure theory. The idea is as follows.

We shall say that an ordinal α is *-definable from a class X of ordinals iff there is an $\mathscr{L}$-formula $\varphi(v_0, \ldots, v_n)$ and elements $\beta_1, \ldots, \beta_n, \gamma$ of X such that α is the unique ordinal for which

$$\models_{L_\gamma} \varphi(\mathring{\alpha}, \mathring{\beta}_1, \ldots, \mathring{\beta}_n).$$

The idea behind the machine concept is this. Suppose we were to define a **-skolem function* for L as a function h such that $\mathrm{dom}(h) \subseteq \omega \times \mathrm{On}^{<\omega}$, $\mathrm{ran}(h) \subseteq \mathrm{On}$, and whenever α is *-definable from $X \subseteq \mathrm{On}$, then $\alpha \in h''(\omega \times X^{<\omega})$, where we use $X^{<\omega}$ to denote $\bigcup_{n<\omega} X^n$. In order to construct, say, a $\square$-sequence, we might then go on to define a hierarchy of (set) functions convering to h, possessing some kind of condensation property. And to a point, this is the idea behind the definition of a Silver machine. But there are some differences. For instance, we shall not work with a single skolem function h but rather an infinite family of functions h_i, $i < \omega$. Although h_i will, in some sense, correspond to the function $h(i, -)$ of the above sketch, the index i will not be the Gödel number of a formula as was the case with the skolem functions of the fine structure theory, and for different indices i the functions h_i may be quite different in structure. (Hence there is no point in trying to combine them into one function.)

One remark concerning the use of the word "machine". This stems from the motivation which led Silver to develop the concept in the first place. "Silver hierarchy" would be a more suitable term for the structure we shall develop here (which is not quite the same as the original), but we shall, of course, stick to the established usage.

A structure

$$N = \langle X, <, (h_i)_{i<\omega}\rangle$$

is said to be *eligible* iff:

(i) $X \subseteq \mathrm{On}$;

(ii) $<$ is the usual ordering on X;

(iii) for each i, h_i is a partial function from $X^{k(i)}$ into X, for some integer $k(i)$.

If N is as above and λ is an ordinal, we set

$$N_\lambda = \langle X \cap \lambda, <, (h_i \cap \lambda^{k(i)+1})_{i<\omega}\rangle.$$

We sometimes write N_∞ instead of N.

If N, λ are as above and $A \subseteq X \cap \lambda$, $N_\lambda[A]$ denotes the closure of A under the functions of N_λ.

Let $N^j = \langle X^j, \langle, (h_i^j)_{i<\omega}\rangle$ be eligible structures of the same similarity type, for $j = 1, 2$. We write $N^1 \lhd N^2$ iff $X^1 \subseteq X^2$ and for all $i < \omega$ and all $x_1, \ldots, x_{k(i)} \in X^1$,

$$h_i^1(x_1, \ldots, x_{k(i)}) \simeq h_i^2(x_1, \ldots, x_{k(i)}).$$

A *machine* is an eligible structure of the form

$$M = \langle \mathrm{On}, <, (h_i)_{i<\omega}\rangle,$$

which satisfies the following three conditions:

I. *Condensation Principle.* If $N \lhd M_\lambda$, there is an α such that $N \cong M_\alpha$.

II. *Finiteness Principle.* For each λ there is a finite set $H \subseteq \lambda$ such that for any set $A \subseteq \lambda + 1$,

$$M_{\lambda+1}[A] \subseteq M_\lambda[(A \cap \lambda) \cup H] \cup \{\lambda\}.$$

III. *Skolem Property.* If α is $*$-definable from the set $X \subseteq \mathrm{On}$, then $\alpha \in M[X]$; moreover there is an ordinal $\lambda < [\sup(X) \cup \alpha]^+$, uniformly Σ_1 definable from $X \cup \{\alpha\}$, such that $\alpha \in M_\lambda[X]$.

Some explanatory comments are perhaps in order here. In the light of our introductory remarks, the inclusion of the Condensation Principle and of the Skolem Property in this definition should come as no surprise. But why the Finiteness Principle? This says that the hierarchy $(M_\lambda \mid \lambda \in \mathrm{On})$ grows very slowly, with only finitely many new ordinals being calculated at each stage. Hence events of set theoretic interest will occur only at limit levels of the hierarchy. This fact will be of considerable use to us, much as we used the fact that the structures L_α are only easily handled when α is a limit ordinal (as we saw in Chapter II).

There are several ways to construct a machine, but in essence the idea is that the machine should code the truth definition associated with $*$-definability. The following devices are introduced in order to facilitate our proof of the Condensation Property for the machine.

Suppose $N = \langle X, <, (h_i)_{i<\omega}\rangle$ is an eligible structure, where X is a set. Let $*$ denote $\sup(X)$, and set $X^* = X \cup \{*\}$. Define partial functions h_i^*, $i < \omega$, on X^* as follows:

(a) if $s \in X^{k(i)}$ and $h_i(s)$ is defined, set $h_i^*(s) = h_i(s)$;
(b) if $s \in X^{k(i)}$ and $h_i(s)$ is undefined, set $h_i^*(s) = *$;
(c) if $s \in (X^*)^{k(i)}$ contains $*$, set $h_i^*(s) = *$.

Let $N^* = \langle X^*, <, (h_i^*)_{i<\omega}, \{*\}\rangle$. Though essentially the same as N, N^* has the advantage (for us) that all of its functions are total, which is the reason for its introduction.

Suppose that S is a first-order language. The infinitary language $S^\#$ is obtained from S by allowing the formation of countably infinite conjunctions and disjunctions of quantifier free formulas. A *universal* sentence of $S^\#$ is a sentence of the form

$$\forall v_0 \ldots \forall v_n \varphi(v_0, \ldots, v_n)$$

where φ is quantifier free. A *universal theory* in $S^\#$ is a consistent set of universal sentences of $S^\#$.

Suppose that S is the first-order language of some eligible structure, and T is a theory in $S^\#$. We say that T is *α-categorical* if the structure $\langle \alpha, <\rangle$ has exactly one expansion to an S-structure satisfying T. (The definition of satisfaction for $S^\#$ is quite straightforward.)

The following lemma indicates how the above concepts can assist us in proving that our machine has the Condensation Property.

1.1 Lemma. *Let $M = \langle \mathrm{On}, <, (h_i)_{i<\omega}\rangle$ be an eligible structure. Let S be the language of the structures M_λ^*. Suppose there is a universal theory T in $S^\#$ such that:*

(i) *T is $(\alpha + 1)$-categorical for all α;*
(ii) *$M_\alpha^* \models T$ for all α.*

Then M has the Condensation Property.

Proof. Let $N \lhd M_\alpha$. Since $M_\alpha^* \models T$ and T is universal, we clearly have $N^* \vdash T$. The domain of N is a set of ordinals, so there is a unique ordinal $\bar{\alpha}$ and a unique isomorphism $\pi: N \cong \bar{N}$, where $\bar{N} = \langle \bar{\alpha}, <, (\bar{h}_i)_{i<\omega}\rangle$ is eligible. But $\bar{N}^* \models T$, so as T is $(\bar{\alpha} + 1)$-categorical, $\bar{N}^* = M_{\bar{\alpha}}^*$. Hence $\bar{N} = M_{\bar{\alpha}}$, and we are done.

We are now ready to commence the construction of our machine. As a first step we define a certain well-ordering of $\mathrm{On}^{<\omega}$.

It is easily seen that the following rules do define a well-ordering of $\mathrm{On}^{<\omega}$. Let $s, t \in \mathrm{On}^{<\omega}$.

(i) If s is a proper subsequence of t, then any permutation of s precedes any permutation of t.
(ii) If s is a permutation of t, then s and t are ordered lexicographically.
(iii) If $s = (\alpha_1, \ldots, \alpha_n)$, $t = (\alpha_1, \ldots, \alpha_{n-1}, \beta_1, \ldots, \beta_m)$ and $\beta_1, \ldots, \beta_m < \alpha_n$, then any permutation of t precedes any permutation of s. (In case $n = 1$ here, $\alpha_1, \ldots, \alpha_{n-1}$ is interpreted as the empty string.)

We denote by $<^*$ the well-ordering of $\mathrm{On}^{<\omega}$ so defined. For later use, we note that for $s, t \in \mathrm{On}^{<\omega}$, $\max(s) < \max(t)$ implies $s <^* t$. (Proving this should help the reader to understand the definition of $<^*$ more fully.)

For $s \in \mathrm{On}^{<\omega}$, we denote by $\hat{s}$ the ordinal corresponding to s in the ordering $<^*$, that is

$$\hat{s} = \mathrm{otp}(\langle\{t \mid t <^* s\}, <^*\rangle).$$

Define functions $P_n\colon \mathrm{On}^n \to \mathrm{On}$ by setting $P_n(s) = \hat{s}$. These are the "*pairing functions*". Note that $P_n((\alpha_1, \ldots, \alpha_n)) \geqslant \max(\alpha_1, \ldots, \alpha_n)$.

Define partial functions Q_i from On to On by

$$Q_i(\alpha) = \begin{cases} \alpha_i, & \text{if } \alpha = P_n((\alpha_1, \ldots, \alpha_n)) \text{ and } i \leqslant n; \\ \text{otherwise undefined.} \end{cases}$$

These are the "*pairing inverses*".

Notice that

$$P = \langle \mathrm{On}, <, (P_n)_{n<\omega}, (Q_i)_{i<\omega} \rangle$$

is an eligible structure. Indeed, P satisfies two of the machine axioms, as we prove next.

1.2 Lemma. *P has the Finiteness Property.*

Proof. Let λ be given. For some n and some $\alpha_1, \ldots, \alpha_n \leqslant \lambda$,

$$\lambda = P_n((\alpha_1, \ldots, \alpha_n)).$$

Set

$$H = \{\alpha_1, \ldots, \alpha_n\} \cap \lambda.$$

Clearly, if $A \subseteq \lambda + 1$,

$$P_{\lambda+1}[A] \subseteq P_\lambda[(A \cap \lambda) \cup H] \cup \{\lambda\}. \quad \square$$

1.3 Lemma. *P has the Condensation Property.*

Proof. We use 1.1. Let S be the language of the structures P_λ^*. It is clear that, for fixed n, m, there is a first-order, quantifier free formula $\varphi_{n,m}(v_1, \ldots, v_n, v_{n+1}, \ldots, v_m)$ of S which says

$$\text{"}(v_1, \ldots, v_n) <^* (v_{n+1}, \ldots, v_m)\text{"}.$$

Let T_0 be the following universal $S^\#$ theory:

$$\begin{aligned} T_0 = {} & \{\forall x(x = * \vee x < *)\} \cup \{Q_n(*) = * \mid n \in \omega\} \\ & \cup \{\forall x_1 \ldots x_n(x_1 = * \vee \ldots \vee x_n = * \to P_n(\vec{x}) = *) \mid n \in \omega\} \\ & \cup \{\forall x_1 \ldots x_n y_1 \ldots y_m[\varphi_{n,m}(\vec{x}, \vec{y}) \to (P_n(\vec{x}) < P_m(\vec{y}) \vee P_m(\vec{y}) = *) \mid n, m \in \omega\} \\ & \cup \{\forall x \bigvee_{n<\omega} [x = P_n(Q_1(x), \ldots, Q_n(x))]\}. \end{aligned}$$

It is easily seen that T_0 is $(\alpha + 1)$-categorical for all α. And clearly, $P_\alpha^* \models T_0$ for all α. So by 1.1, P has the Condensation Property. □

In order to incorporate the Skolem Property into our machine, we introduce a first-order language, Γ, appropriate for *-definability.

The basic symbols of Γ are as follows:

variables: v_n $(n \in \omega)$;

connectives: $\wedge, \neg$;

predicates: $=, \in$;

constants: t_α^φ (for certain φ, α described below);

quantifiers: $\exists^\alpha$ $(\alpha \in \mathrm{On})$.

If φ is a formula of Γ, the *rank* of φ, $\varrho(\varphi)$, is the least α such that:

(i) if $\exists^\gamma$ occurs in φ, then $\gamma \leqslant \alpha$;
(ii) if t_γ^ψ occurs in φ, then $\gamma < \alpha$.

For each α and each Γ-formula φ of rank α, the language Γ has a constant t_α^φ, and this is the only occasion on which such a constant is defined.

The definitions of the language Γ and of the rank function ϱ thus proceed by means of a simultaneous recursion, which is easily seen to be well-defined.

The language Γ is interpreted in L as follows. The interpretation of t_α^φ is the set $\{x \in L_\alpha \mid \models_{L_\alpha} \varphi(\mathring{x})\}$, and the interpretation of $\exists^\alpha v_n$ is $(\exists x \in L_\alpha)$. Clearly, each member of L is denoted by a constant t_α^φ, and elements of $L_{\alpha+1}$ are just the interpretations of the constants t_α^φ as φ varies. For each α, L_α has a canonical name in Γ, namely $t_\alpha^{(v_0 = v_0)}$. This name is denoted by l_α. Similarly, α has a canonical name, $t_\alpha^{\mathrm{On}(v_0)}$, which we denote by o_α.

The formal definition of Γ in set theory is as follows.

$$v_n = \langle n + 6 \rangle;$$

$$t_\alpha^\varphi = \langle 0 \rangle ^\frown \varphi ^\frown \langle \omega + \alpha + 1 \rangle;$$

$$(x = y) = \langle 1 \rangle ^\frown x ^\frown y \qquad (x, y \text{ variables or constants});$$

$$(x \in y) = \langle 2 \rangle ^\frown x ^\frown y \qquad (x, y \text{ variables or constants});$$

$$(\varphi \wedge \psi) = \langle 3 \rangle ^\frown \varphi ^\frown \psi \qquad (\varphi, \psi \text{ formulas});$$

$$(\neg \varphi) = \langle 4 \rangle ^\frown \varphi \qquad (\varphi \text{ a formula});$$

$$(\exists^\alpha v_n \varphi) = \langle 5 \rangle ^\frown \langle \alpha \rangle ^\frown \langle n \rangle ^\frown \varphi \qquad (\varphi \text{ a formula}).$$

Thus each formula of Γ is a finite sequence of ordinals. Using the pairing functions P_n, we may now associate with each Γ-formula φ a unique single ordinal $\hat{\varphi}$. Similarly, each constant c of Γ is assigned an ordinal $\hat{c}$. If the ordinal α denotes a formula or a constant of Γ, we denote that formula/constant by $\ulcorner \alpha \urcorner$.

1.4 Lemma.

(i) *If* φ *is a subformula of* ψ, *then* $\hat{\varphi} < \hat{\psi}$.

(ii) *If* φ is $(\exists^{\alpha} v)\, \psi(v)$ *and* t *is* t^{θ}_{γ}, *where* $\gamma < \alpha$, *then* $\hat{\psi}(t) < \hat{\varphi}$.

(iii) *If* φ *is* $(t^{\theta_1}_{\alpha_1} = t^{\theta_2}_{\alpha_2})$ *or* $(t^{\theta_1}_{\alpha_1} \in t^{\theta_2}_{\alpha_2})$, *and if* $\varrho(\psi) \leqslant \max(\alpha_1, \alpha_2)$, *then* $\hat{\psi} < \hat{\varphi}$.

Proof. (i) If φ is a subformula of ψ, then as sequences of ordinals, φ is a subsequence of ψ, so $\varphi <^* \psi$. Hence $\hat{\varphi} < \hat{\psi}$.

(ii) This is a direct application of clause (iii) in the definition of $<^*$.

(iii) For definiteness, suppose $\alpha_1 \leqslant \alpha_2$. Thus $\max(\varphi) = \omega + \alpha_2 + 1$. Since $\varrho(\psi) \leqslant \alpha_2$, $\max(\psi) \leqslant \omega + \alpha_2$. Hence $\max(\psi) < \max(\varphi)$. So, as we remarked earlier, $\psi <^* \varphi$, giving $\hat{\psi} < \hat{\varphi}$. □

Our machine will need to be able to handle the elementary syntax of Γ. Accordingly, we make the following definitions.

For $\nu \leqslant \omega . \omega$, let k_ν be the constant unary function with value ν. Let I be the unary function $I(\alpha) = \omega + \alpha$. Let J be the unary function $J(\alpha) = \alpha + 1$.

Let

$$N = \langle \text{On}, <, (P_n)_{n<\omega}, (Q_i)_{i<\omega}, (k_\nu)_{\nu \leqslant \omega . \omega}, I, J \rangle.$$

The eligible structure N can clearly handle the bcsic syntax of the language Γ.

1.5 Lemma. *N has the Finiteness Property.*

Proof. Given λ, pick $n, \alpha_1, \ldots, \alpha_n$ so that $\lambda = P_n((\alpha_1, \ldots, \alpha_n))$ and let $H = \{\alpha_1, \ldots, \alpha_n\} \cap \lambda$. H is uniquely defined, and if $A \subseteq \lambda + 1$, then

$$N_{\lambda+1}[A] \subseteq N_\lambda[(A \cap \lambda) \cup H] \cup \{\lambda\}. \quad \square$$

1.6 Lemma. *N has the Condensation Property.*

Proof. If $\lambda \leqslant \omega . \omega$ and $X \lhd N_\lambda$, then $X = N_\lambda$ and there is nothing to prove. For the case $\lambda > \omega . \omega$, we use 1.1. Let S be the language of the structures N^{**}_λ. Let T_0 be as in the proof of 1.3. T_0 will take care of the P-part of N, so what we must do is extend T_0 to a universal $S^\#$ theory which uniquely characterises the functions k_ν, I, J. Let T_1 be the following universal $S^\#$ theory.

$$\begin{aligned} T_1 = T_0 &\cup \{\forall x \forall y [k_\nu(x) = k_\nu(y) \mid \nu \leqslant \omega . \omega) \\ &\cup \{\forall x \forall y [[(x = *) \wedge (\bigwedge_{\nu \leqslant \omega . \omega} k_\nu(x) = *)] \vee [(x < *) \\ &\wedge \bigwedge_{\tau < \nu \leqslant \omega . \omega} (k_\tau(x) < k_\nu(x)) \wedge [\bigvee_{\nu \leqslant \omega . \omega} (k_\nu(x) = y) \vee (k_{\omega . \omega}(x) < y)]]]]\} \\ &\cup \{\forall x \forall y [[x = * \wedge I(x) = * \wedge J(x) = *] \vee [(x < *) \\ &\wedge [\bigvee_{\nu \leqslant \omega . \omega} ((x = k_\nu(x)) \wedge (I(x) = k_{\omega+\nu}(x))) \\ &\vee (x \geqslant k_{\omega . \omega}(x) \wedge I(x) = x)] \wedge [(x < J(x)) \wedge (y \leqslant x \vee J(x) \leqslant y)]]]\}. \end{aligned}$$

Clearly, T_1 is $(\alpha+1)$-categorical for all $\alpha \geqslant \omega.\omega$. Moreover, $N_\alpha^* \models T_1$ for $\alpha \geqslant \omega.\omega$. Hence by 1.3, if $X \lhd N_\alpha$, where $\alpha \geqslant \omega.\omega$, then $X \cong N_{\bar\alpha}$ for some unique $\bar\alpha$. □

Now, L-truth for Γ-sentences is clearly definable. So we may define a function F from ordinals to $\{0, 1\}$ as follows:

$$F(\alpha) = \begin{cases} 1, & \text{if } \alpha = \hat\varphi \text{ where } \varphi \text{ is a true sentence of } \Gamma; \\ 0, & \text{if } \alpha = \hat\varphi \text{ where } \varphi \text{ is a false sentence of } \Gamma; \\ \text{otherwise undefined.} \end{cases}$$

We may now define functions G, H from ordinals to ordinals by:

$$G(\alpha) = \begin{cases} \widehat{t_\gamma^\psi}, & \text{if } \alpha = (\exists^\delta v\varphi(v))^\wedge \text{ and } \exists^\delta v\varphi(v) \text{ is true and } \widehat{t_\gamma^\psi} \text{ is least} \\ & \text{such that } \gamma < \delta \text{ and } \varphi(t_\gamma^\psi) \text{ is true;} \\ \text{otherwise undefined.} \end{cases}$$

$$H(\alpha) = \begin{cases} \beta, & \text{if } \alpha = \varphi(v)^\wedge \text{ and } \beta \text{ is least such that } \varphi(o_\beta) \text{ is true;} \\ \text{otherwise undefined.} \end{cases}$$

Set

$$M = \langle \mathrm{On}, <, (P_n)_{n<\omega}, (Q_i)_{i<\omega}, (k_\nu)_{\nu \leqslant \omega.\omega}, I, J, F, G, H \rangle.$$

Clearly, M is an eligible structure. We show that M is a machine.

1.7 Lemma. *M has the Finiteness Property.*

Proof. The proof of 1.5 is still valid. □

1.8 Lemma. *M has the Skolem Property.*

Proof. Let α be *-definable from $X = \{\beta_1, \ldots, \beta_n, \gamma\}$, α being the unique ordinal such that $\models_{L_\gamma} \varphi(\mathring{\alpha}, \mathring{\beta}_1, \ldots, \mathring{\beta}_n)$ where φ is some $\mathscr{L}$-formula. Obtain the formula $\psi(v_0)$ of Γ from $\varphi(v_0, \ldots, v_n)$ by replacing v_i by o_{β_i}, for $i = 1, \ldots, n$, and each quantifier $\exists v$ by $\exists^\gamma v$. Clearly, if t is a constant of Γ, $\psi(t)$ will be a true sentence of Γ (in L) iff the interpretation of t in L is α.

Let $\delta = \psi(v_0)^\wedge$. Notice that δ is computable from $\beta_1, \ldots, \beta_n, \gamma$ using the functions of M (in fact the functions of N). Let

$$\lambda = \sup\{\delta, \beta_1, \ldots, \beta_n, \gamma, \alpha\}.$$

Clearly, $\lambda < \sup(X \cup \{\alpha\})^+$, and λ is (uniformly) Σ_1 definable from $X \cup \{\alpha\}$. Then $\delta \in M_\lambda[X]$. But $H(\delta) = \alpha$. Hence $\alpha \in M_\lambda[X]$, as required. □

1.9 Lemma. *M has the Condensation Property.*

Proof. If $\lambda \leqslant \omega.\omega$ and $X \lhd M_\lambda$, then by virtue of the functions k_ν, we have $X = M_\lambda$, so there is nothing more to prove. For the case $\lambda \geqslant \omega.\omega$ we use 1.1. Let S be the language of the structures M_λ^*. Let T_1 be the universal $S^\#$ theory defined in the proof of 1.6. As we saw in 1.6, T_1 will take care of the N-part of M. What we must do now is extend T_1 to a universal theory T in $S^\#$ which characterises uniquely the remaining functions of M.

Notice first that the functions F, G have the following recursive definitions (by simultaneous recursion for F and G):

$$[F(*) = *] \wedge \forall x[F(x) = 0 \vee F(x) = 1 \vee F(x) = *]$$

$$\wedge \forall \alpha[F(\alpha) = 1 \leftrightarrow [\alpha = (t_\nu^\varphi = t_\nu^\psi)^\wedge \wedge F((\forall^\nu v)(\varphi(v) \leftrightarrow \psi(v))^\wedge) = 1]$$

$$\vee [\alpha = (t_\nu^\varphi = t_\tau^\psi)^\wedge \wedge \nu < \tau \wedge F((\forall^\tau v)(\psi(v) \leftrightarrow v \in l_\nu \wedge \varphi(v))^\wedge) = 1]$$

$$\vee [\alpha = (t_\tau^\psi = t_\nu^\varphi)^\wedge \wedge \nu < \tau \wedge F((\forall^\tau v)(\psi(v) \leftrightarrow v \in l_\nu \wedge \varphi(v))^\wedge) = 1]$$

$$\vee [\alpha = (t_\nu^\varphi \in t_\nu^\psi)^\wedge \wedge F((\exists^\nu v)(\psi(v) \wedge (\forall^\nu w)(w \in v \leftrightarrow \varphi(w)))^\wedge) = 1]$$

$$\vee [\alpha = (t_\nu^\varphi \in t_\tau^\psi)^\wedge \wedge \nu < \tau \wedge F((\exists^\tau v)(\psi(v)$$

$$\wedge (\forall^\tau w)(w \in v \leftrightarrow w \in l_\nu \wedge \varphi(w)))^\wedge) = 1]$$

$$\vee [\alpha = (t_\tau^\varphi \in t_\nu^\psi)^\wedge \wedge \nu < \tau \wedge F((\exists^\nu v)(\psi(v)$$

$$\wedge (\forall^\tau w)(w \in v \leftrightarrow \varphi(w)))^\wedge) = 1]$$

$$\vee [\alpha = (\varphi \wedge \psi)^\wedge \wedge F(\hat{\varphi}) = 1 \wedge F(\hat{\psi}) = 1]$$

$$\vee [\alpha = (\neg \varphi)^\wedge \wedge F(\hat{\varphi}) = 0]$$

$$\vee [\alpha = ((\exists^\nu v_n) \varphi)^\wedge \wedge G(\alpha) \neq *]]$$

$$\wedge \forall \alpha[F(\alpha) = 0 \leftrightarrow . \; . \; . \; . \; . \; . \; . \; . \; . \; .];$$

$$G(*) = * \wedge \forall \alpha \forall \beta[G(\alpha) = \beta \leftrightarrow \alpha = (\exists^\nu v \varphi(v))^\wedge \wedge \beta = (t_\tau^\psi)^\wedge \wedge \tau < \nu$$

$$\wedge F(\varphi(t_\tau^\psi)^\wedge) = 1 \wedge (\forall \gamma < \beta)(\gamma = t_\iota^\theta \wedge \iota < \nu \to F(\varphi(t_\iota^\theta)^\wedge) = 0].$$

Using 1.4, it is easily seen that the above definitions are sound. The function H has the following definition:

$$[H(*) = *] \wedge \forall \alpha \forall \beta[H(\alpha) = \beta \leftrightarrow \alpha = \varphi(v)^\wedge \wedge F(\varphi(o_\beta)^\wedge) = 1$$

$$\wedge (\forall \gamma < \beta)(F(\varphi(o_\gamma)^\wedge) = 0)].$$

Roughly speaking, T will consist of T_1 together with the above definitions of F, G, H. That T will be $(\alpha + 1)$-categorical for all $\alpha \geqslant \omega . \omega$ and that M_α^* will be a model of T for all $\alpha \geqslant \omega . \omega$ is clear. What we need to check, though, is that it is possible to write the above definitions as universal sentences of $S^\#$.

The appearance of the constants 0, 1 causes no problems, since the functions k_0, k_1 yield these values for all $x \neq *$. And the functions of N also enable us to handle the passage from formulas to ordinals and back again. For the passage from formulas to ordinals this is clear. For the reverse passage, considering the definition of F as an example, we may commence the $F(\alpha) = 1$ clause thus:

$$F(\alpha) = 1 \leftrightarrow \bigvee_{n < \omega} [\alpha = P_n(Q_1(\alpha), \ldots, Q_n(\alpha)) \wedge \ldots].$$

For each α which denotes a formula there will be a unique n such that $\alpha = P_n(Q_1(\alpha), \ldots, Q_n(\alpha))$, and $(Q_1(\alpha), \ldots, Q_n(\alpha))$ will be $\ulcorner\alpha\urcorner$, so the relevant disjunct in the above will deal with $\ulcorner\alpha\urcorner$. Allied to this is the classification of a formula into its logical type. But there are only a finite number of types, and so we may form a disjunction over these. We leave it to the reader to check the fine details now, and declare the lemma proved. □

That completes the proof that M is a machine.

2. The Combinatorial Principle □

We use the machine constructed above in order to prove the combinatorial principle □ from $V = L$. More precisely, we prove the following theorem:

2.1 Theorem. *Assume $V = L$. Let A be a class of limit ordinals. Then there is a class $E \subseteq A$ such that:*

(i) *if $\kappa > \omega$ is regular and $A \cap \kappa$ is stationary in κ, then $E \cap \kappa$ is stationary in κ;*

(ii) *$\square(E)$ holds.* □

We recall that $\square(E)$ says that there is a sequence $(C_\alpha \mid \alpha \in S)$, where S is the class S
of all singular limit ordinals, such that:

(i) C_α is a closed unbounded subset of α;

(ii) $\mathrm{otp}(C_\alpha) < \alpha$;

(iii) if $\bar{\alpha} < \alpha$ is a limit point of C_α, then $\bar{\alpha} \in S$, $\bar{\alpha} \notin E$, and $C_{\bar{\alpha}} = \bar{\alpha} \cap C_\alpha$.

Our proof of 2.1 using the machine M will be closely modelled upon the fine structure proof in VI.6. In one aspect the machine proof is better: it is uniform on α, avoiding the necessity of looking separately at different cases, which was a feature of the fine structure proof. With the machine, the analogue of the most difficult case in VI.6 works in all cases.

We assume $V = L$ from now on. M denotes the machine constructed in section 1. When we use the machine, a finite set of ordinals will often be referred to as a *parameter*. Since we may identify finite sets of ordinals with members of $\mathrm{On}^{<\omega}$ in a canonical manner, the well-ordering $<^*$ of $\mathrm{On}^{<\omega}$ gives us a well-ordering of all parameters.

Let α be a limit ordinal, $\beta \geqslant \alpha$. We say that α is *singular at* β if there is a parameter $p \subseteq \beta$ and a $\gamma < \alpha$ such that $M_\beta[\gamma \cup p] \cap \alpha$ is cofinal in α.

2.2 Lemma. *If $\alpha \in S$ there is a $\beta < \alpha^+$ such that α is singular at β.*

Proof. Let $\gamma = \mathrm{cf}(\alpha)$, and let f be the $<_L$-least map from γ cofinally into α. Let $\delta < \alpha^+$ be such that $f \in L_\delta$. Set $p = \{\alpha, \delta\}$.

For $\xi < \gamma$, $f(\xi)$ is the unique ordinal ζ such that $\models_{L_\delta}$ "ζ is the value at ξ of the $<_L$-least map from $\mathrm{cf}(\alpha)$ cofinally into α". So for each $\xi < \alpha$, $f(\xi)$ is *-definable from $\{\xi, \alpha, \delta\}$, and by the Skolem Property there is a $\beta(\xi) < [\max(f(\xi), \xi, \alpha, \delta)]^+$ such that $f(\xi) \in M_{\beta(\xi)}[\{\alpha, \delta, \xi\}]$. Let $\beta = \sup_{\xi<\gamma} \beta(\xi)$. Since $\beta(\xi) < \alpha^+$ for all $\xi < \gamma$ and $\gamma < \alpha$, we have $\beta < \alpha^+$. Also, for each $\xi < \gamma$, $f(\xi) \in M_\beta[\{\xi, \alpha, \delta\}]$. Hence $\operatorname{ran}(f) \subseteq M_\beta[\gamma \cup p]$. Thus $M_\beta[\gamma \cup p] \cap \alpha$ is cofinal in α, and we are done. □

Let α be a limit ordinal, $\beta \geqslant \alpha$. Let us say that α is *semi-singular at* β iff there is a parameter $p \subseteq \beta$ such that whenever $p \subseteq X \lhd M_\beta$ and $X \cap \alpha$ is transitive, then $X \cap \alpha = \alpha$.

2.3 Lemma.

(i) *If α is singular at β, then α is semi-singular at β.*

(ii) *If $\operatorname{cf}(\alpha) > \omega$ and α is semi-singular at β (with parameter p), then α is singular at β (with parameter p).*

Proof. (i) Let $p \subseteq \beta$ be a parameter and let $\gamma < \alpha$ be such that $M_\beta[\gamma \cup p] \cap \alpha$ is cofinal in α. Set $p' = p \cup \{\gamma\}$. We show that α is semi-singular at β with parameter p'. Let $p' \subseteq X \lhd M_\beta$ be such that $X \cap \alpha$ is transitive. Since $\gamma \in X$, we have $\gamma \subseteq X$. So as $X \lhd M_\beta$, we have $M_\beta[\gamma \cup p] \subseteq X$. Hence $X \cap \alpha$ is confinal in α. Thus as $X \cap \alpha$ is transitive, we must have $X \cap \alpha = \alpha$.

(ii) Let α be semi-singular at β with parameter p. By recursion, define substructures $X_n \lhd M_\beta$ and ordinals $\alpha_n \leqslant \alpha$ is follows.

$$X_0 = M_\beta[p]; \qquad \alpha_0 = \sup(X_0 \cap \alpha);$$

$$X_{n+1} = M_\beta[\alpha_n \cup p]; \qquad \alpha_{n+1} = \sup(X_{n+1} \cap \alpha).$$

Set

$$X_\omega = \bigcup_{n<\omega} X_n, \qquad \alpha_\omega = \sup_{n<\omega} \alpha_n.$$

Clearly, $X_\omega \lhd M_\beta$ and $X_\omega \cap \alpha = \alpha_\omega$. Since $p \subseteq X_\omega$ therefore, we must have $X_\omega \cap \alpha = \alpha$, i.e. $\alpha_\omega = \alpha$. Since $\operatorname{cf}(\alpha) > \omega$, it follows that $\alpha_n = \alpha$ for some $n < \omega$. Let n be the least such. If $n = 0$, then $M_\beta[0 \cup p] \cap \alpha$ is cofinal in α, and if $n > 0$, then $\alpha_{n+1} < \alpha_n = \alpha$ and $M_\beta[\alpha_{n-1} \cup p] \cap \alpha$ is cofinal in α, so in either case α is singular at β (with parameter p). □

Let $\gamma < \alpha \leqslant \beta$. Let $p \subseteq \beta$ be a parameter. We shall say that (γ, p) *jumps below* α *in* M_β iff $M_\beta[\gamma \cup p] \cap \alpha \neq \gamma$.

2.4 Lemma. *Let $\alpha \in S$, $\beta \geqslant \alpha$, $p \subseteq \beta$ a parameter. The following are equivalent:*

(i) *α is semi-singular at β with parameter p;*

(ii) *for all $\gamma < \alpha$, (γ, p) jumps below α in M_β.*

Proof. (i) → (*ii*). Let $\gamma < \alpha$ and set $X = M_\beta[\gamma \cup p]$. Suppose $X \cap \alpha = \gamma$. Then since $p \subseteq X \lhd M_\beta$ and γ is transitive, we have $\gamma = X \cap \alpha = \alpha$, which is absurd. Hence $X \cap \alpha \neq \gamma$, proving (ii).

(ii) → (i). Let $p \subseteq X \lhd M_\beta$ be such that $X \cap \alpha$ is transitive. Set $\gamma = X \cap \alpha$. Suppose $\gamma < \alpha$. Then $M_\beta[\gamma \cup p] \cap \alpha \neq \gamma$. But $\gamma \cup p \subseteq X \lhd M_\beta$, so $M_\beta[\gamma \cup p] \subseteq X$, and we have $X \cap \alpha \neq \gamma$, a contradiction. Hence $\gamma = \alpha$, proving (i). □

The class E of 2.1 consists of all ordinals $\alpha \in A$ such that for some $\gamma \geqslant \alpha$ and some parameter $q \subseteq \gamma$: E

(i) α is not semi-singular at γ;

(ii) if $\bar{\alpha} \in A \cap \alpha$, then either $(\bar{\alpha}, q)$ jumps below α in M_γ or else $\bar{\alpha}$ is semi-singular at γ with a parameter in $M_\gamma[\bar{\alpha} \cup q]$.

2.5 Lemma. *Let $\kappa > \omega$ be a regular cardinal, and assume that $A \cap \kappa$ is stationary in κ. Then $E \cap \kappa$ is stationary in κ.*

Proof. Let C be a club subset of κ. We show that $C \cap E \neq \emptyset$. Let $f: \kappa \to \kappa$ be defined by C

$$f(\alpha) = \text{the least element of } C \text{ greater than } \alpha. \qquad f$$

Then $f \in L_{\kappa^+}$ so for some ordinal $\theta < \kappa^+$, f is the θ-th element of L in the well-ordering $<_L$. Let $\varrho < \kappa^+$ be such that $\varrho > \theta$ and (say) L_ϱ is a model of ZF$^-$. By absoluteness, θ ϱ

$$f = [\text{the } \theta\text{-th element of } L \text{ in the ordering } <_L]^{L_\varrho}.$$

Let α be, if possible, the least ordinal in $A \cap \kappa$ such that α

(i) $M_{\kappa^+}[\alpha \cup \{\theta, \varrho\}] \cap \kappa = \alpha$;

(ii) if $p \subseteq M_{\kappa^+}[\alpha \cup \{\theta, \varrho\}]$ is a parameter, then α is not semi-singular at κ^+ with parameter p.

We show that α is well-defined here. Define a chain

$$X_0 \lhd X_1 \lhd \ldots \lhd X_\nu \lhd \ldots \lhd M_{\kappa^+} \ (\nu < \kappa)$$

by recursion, as follows. Let $X_0 \lhd M_{\kappa^+}$ be such that $\theta, \varrho \in X_0$ and $\alpha_0 = X_0 \cap \kappa \in \kappa$. If $X_\nu \lhd M_{\kappa^+}$ is defined and $\alpha_\nu = X_\nu \cap \kappa \in \kappa$, let $X_{\nu+1} \lhd M_{\kappa^+}$ be such that $\alpha_\nu \in \alpha_{\nu+1} = X_{\nu+1} \cap \kappa \in \kappa$. If $\lim(\nu)$ and X_η is defined for all $\eta < \nu$ and such that $\alpha_\eta = X_\eta \cap \kappa \in \kappa$ for all $\eta < \nu$, let $X_\nu = \bigcup_{\eta < \nu} X_\eta$, $\alpha_\nu = \sup_{\eta < \nu} \alpha_\eta$. Since $\kappa > \omega$ is regular, this definition causes no difficulty Since $\{\alpha_\nu \mid \nu < \kappa\}$ is club in κ, we can find a $\nu < \kappa$ such that $\lim(\nu)$ and $\alpha_\nu \in A \cap \kappa$. X_ν, α_ν

Since $\alpha_\eta \cup \{\theta, \varrho\} \subseteq X_\eta \lhd X_\nu \lhd M_{\kappa^+}$ for all $\eta < \nu$, we have

$$M_{\kappa^+}[\alpha_\nu \cup \{\theta, \varrho\}] \subseteq X_\nu.$$

But $X_\nu \cap \kappa = \alpha_\nu$. Thus

$$M_{\kappa^+}[\alpha_\nu \cup \{\theta, \varrho\}] \cap \kappa = \alpha_\nu.$$

Thus α_ν satisfies condition (i) above.

Now suppose that $p \subseteq M_{\kappa^+}[\alpha_\nu \cup \{\theta, \varrho\}]$ is a parameter. For some $\eta < \nu$, $p \subseteq M_{\kappa^+}[\alpha_\eta \cup \{\theta, \varrho\}]$. Thus $p \subseteq X_\eta \lhd M_{\kappa^+}$, and

$$X_\eta \cap \alpha_\nu = X_\eta \cap \kappa \cap \alpha_\nu = \alpha_\eta < \alpha_\nu.$$

Thus α_ν is not semi-singular at κ^+ with parameter p. This shows that α_ν satisfies condition (ii) above.

It follows that α is well-defined, and indeed that $\alpha \leqslant \alpha_\nu$.

Now let

π, γ $$\pi\colon M_\gamma \cong M_{\kappa^+}[\alpha \cup \{\theta, \varrho\}].$$

Then

$$\pi \upharpoonright \alpha = \mathrm{id} \upharpoonright \alpha \quad \text{and} \quad \pi(\alpha) = \kappa.$$

$\bar{\theta}, \bar{\varrho}, q$ Let $\bar{\theta} = \pi^{-1}(\theta)$, $\bar{\varrho} = \pi^{-1}(\varrho)$, and set $q = \{\bar{\theta}, \bar{\varrho}\}$. We show that γ, q satisfy the definition of E for α. Notice that $\gamma = M_\gamma[\alpha \cup q]$.

Suppose that α were semi-singular at γ. Then for some parameter $p \subseteq \gamma$, α will be semi-singular at γ with parameter p. Let $\delta < \alpha$. Then by 2.4, $M_\gamma[\delta \cup p] \cap \alpha \neq \delta$. Applying π, and using the fact that $\pi \upharpoonright \alpha = \mathrm{id} \upharpoonright \alpha$, we have $M_{\kappa^+}[\delta \cup \pi(p)] \cap \alpha \neq \delta$. Thus, whenever $\delta < \alpha$, $(\delta, \pi(p))$ jumps below α in M_{κ^+}. So by 2.4, α is semi-singular at κ^+ with parameter $\pi(p)$. But $\pi(p) \subseteq \pi''\gamma = M_{\kappa^+}[\alpha \cup \{\theta, \varrho\}]$, so this contradicts the choice of α. Hence α is not semi-singular at γ.

Now let $\bar{\alpha} \in A \cap \alpha$ be such that $(\bar{\alpha}, q)$ does not jump below α in M_γ. Thus

$$M_\gamma[\bar{\alpha} \cup q] \cap \alpha = \bar{\alpha}.$$

Applying π and using the fact that $\pi \upharpoonright \alpha = \mathrm{id} \upharpoonright \alpha$, we get

$$M_{\kappa^+}[\bar{\alpha} \cup \{\theta, \varrho\}] \cap \alpha = \bar{\alpha}.$$

Using property (i) of α we get

$$M_{\kappa^+}[\bar{\alpha} \cup \{\theta, \varrho\}] \cap \kappa = \bar{\alpha}.$$

So by the minimality of α there is a parameter $p \subseteq M_{\kappa^+}[\bar{\alpha} \cup \{\theta, \varrho\}]$ such that $\bar{\alpha}$ is semi-singular at κ^+ with parameter p. Let $\delta < \bar{\alpha}$. By 2.4,

$$M_{\kappa^+}[\delta \cup p] \cap \bar{\alpha} \neq \delta.$$

Applying π^{-1},

$$M_\gamma[\delta \cup \pi^{-1}(p)] \cap \bar{\alpha} \neq \delta.$$

So as $\delta < \bar{\alpha}$ was arbitrary, 2.4 tells us that $\bar{\alpha}$ is semi-singular at γ with parameter $\pi^{-1}(p)$. Since $\pi^{-1}(p) \subseteq M_\gamma[\bar{\alpha} \cup q]$, this completes the proof that $\alpha \in E$.

We obtain the contradiction which proves the lemma by showing that $\alpha \in C$. Let $\nu < \alpha$. Then $f(\alpha)$ is definable from ν, θ in L_ϱ. Hence $f(\nu)$ is *-definable from $\{\nu, \theta, \varrho\}$. So by the Skolem Property for M, $f(\nu) \in M_{\kappa^+}[\alpha \cup \{\theta, \varrho\}] \cap \kappa = \alpha$. Hence $f''\alpha \subseteq \alpha$. Thus by definition of f, α is a limit point of C. Hence $\alpha \in C$, and we are done. □

As a first step towards the construction of a □(*E*)-sequence, we construct a sequence $(C_\alpha \mid \alpha \in S)$ such that:

(i) C_α is a club subset of α;
(ii) if $\bar{\alpha}$ is a limit point of C_α, then $\bar{\alpha} \in S$ and $C_{\bar{\alpha}} = \bar{\alpha} \cap C_\alpha$.

Let $\alpha \in S$. By 2.2 and 2.3, we may define $\beta(\alpha)$ as the least ordinal β such that α is semi-singular at β. Let $p(\alpha)$ be the $<^*$-least parameter p such that α is semi-singular at $\beta(\alpha)$ with parameter p. $\beta(\alpha)$ $p(\alpha)$

2.6 Lemma. *$\beta(\alpha)$ is a limit ordinal.*

Proof. Let $\beta = \beta(\alpha), p = p(\alpha)$. Suppose that $\beta = \lambda + 1$. By the Finiteness Property for M there is a finite set $H \subseteq \lambda$ such that for any set $A \subseteq \beta$.

$$(*) \qquad M_\beta[A] \subseteq M_\lambda[(A \cap \lambda) \cup H] \cup \{\lambda\}.$$

Set $q = p \cup H$. We show that α is semi-singular at λ with parameter q, thereby contradicting the definition of $\beta(\alpha)$ (which is greater than λ), and hence proving the lemma.

Let $q \subseteq X \lhd M_\lambda$, $X \cap \alpha$ transitive. Set $Y = M_\beta[X]$. Since $\lambda < \beta$, we have

$$M_\lambda[X] \subseteq M_\beta[X] = Y.$$

But by (*),

$$Y = M_\beta[X] \subseteq M_\lambda[X] \cup \{\lambda\}.$$

Hence either $Y = M_\lambda[X]$ or else $Y = M_\lambda[X] \cup \{\lambda\}$. In either case we have

$$Y \cap \alpha = M_\lambda[X] \cap \alpha.$$

But $X \lhd M_\lambda$, so $M_\lambda[X] = X$, and we therefore have $Y \cap \alpha = X \cap \alpha$, so $Y \cap \alpha$ is transitive. But $p \subseteq Y \lhd M_\beta$, so this means that $Y \cap \alpha = \alpha$. Hence $X \cap \alpha = \alpha$, as required. □

2.7 Lemma. *Let $\alpha \in S$, and set $\beta = \beta(\alpha), p = p(\alpha)$. For every $\gamma < \alpha$ there is a $\delta < \beta$ such that (γ, p) jumps below α in M_δ.*

Proof. Let $\gamma < \alpha$. By 2.4, (γ, p) jumps below α in M_β; i.e. $M_\beta[\gamma \cup p] \cap \alpha \neq \gamma$. So for some ordinal $\xi, \gamma < \xi < \alpha$, we have $\xi \in M_\beta[\gamma \cup p]$. Let $\xi_1, \dots, \xi_n$ be a finite sequence of ordinals such that $\xi_n = \xi$ and for each i, either $\xi_i \in \gamma \cup p$ or else ξ_i is obtained from $\xi_1, \dots, \xi_{i-1}$ by an application of an M-function. By 2.6, β is a limit ordinal, so we can find a $\delta < \beta$ such that $\delta > \max(p), \xi_1, \dots, \xi_n$. Clearly, $\xi \in M_\delta[\xi_1, \dots, \xi_n]$, so $M_\delta[\gamma \cup p] \cap \alpha \neq \gamma$, as required. □

2.8 Lemma. *Let $\alpha \in S$, and set $\beta = \beta(\alpha)$, $p = p(\alpha)$. Then $\beta = M_\beta[\alpha \cup p]$.*

Proof. Let $X = M_\beta[\alpha \cup p]$. Since $X \lhd M_\beta$, the Condensation Property for M gives us a unique π and a unique λ such that $\pi: X \cong M_\lambda$. Clearly, $\pi \restriction \alpha = \mathrm{id} \restriction \alpha$. Since $X = X[\alpha \cup p]$, we have, applying π and setting $q = \pi'' p$, $\lambda = M_\lambda[\alpha \cup q]$. But

$\lambda \leqslant \beta$ and by an easy isomorphism argument, α is semi-singular at λ, so $\lambda = \beta$. Again, the same easy isomorphism argument shows that α is semi-singular at λ with parameter q, so as $q \leq^* p$ we have $q = p$. Thus $\beta = M_\beta[\alpha \cup p]$, as stated. □

2.9 Lemma. *Let $\alpha \in S$, $\mathrm{cf}(\alpha) > \omega$, and set $\beta = \beta(\alpha)$, $p = p(\alpha)$. Then for some $\theta < \alpha$, $M_\beta[\theta \cup p]$ is cofinal in β.*

Proof. By 2.3(ii) there is a $\theta < \alpha$ such that $M_\beta[\theta \cup p] \cap \alpha$ is cofinal in α. We show that $M_\beta[\theta \cup p]$ is cofinal in β. Suppose not, and pick $\delta < \beta$ such that $M_\beta[\theta \cup p] \subseteq \delta$.

Let $\gamma < \alpha$. By 2.4, $M_\beta[\gamma \cup p] \cap \alpha \neq \gamma$. If $\gamma \leqslant \theta$, we have $M_\beta[\gamma \cup p] \subseteq M_\delta[\gamma \cup p]$, so $M_\delta[\gamma \cup p] \cap \alpha \neq \gamma$. And if $\gamma > \theta$, then $M_\beta[\theta \cup p] \subseteq M_\delta[\theta \cup p] \subseteq M_\delta[\gamma \cup p]$, so as $M_\beta[\theta \cup p] \cap \alpha$ is cofinal in α, $M_\delta[\gamma \cup p] \cap \alpha \neq \gamma$. In either case, therefore, (γ, p) jumps below α in M_δ. Since $\gamma < \alpha$ was arbitrary, 2.4 tells us that α is semi-singular at δ, contrary to $\delta < \beta$. □

We are now able to define C_α, $\alpha \in S$ to satisfy conditions (i) and (ii) specified above.

β, p $\delta(\nu), \alpha_\nu$ Fix $\alpha \in S$, and set $\beta = \beta(\alpha)$, $p = p(\alpha)$. We define increasing, continuous sequences of ordinals, $(\delta(\nu) \mid \nu \leqslant \lambda)$, $(\alpha_\nu \mid \nu \leqslant \lambda)$, for some limit ordinal $\lambda \leqslant \alpha$, by recursion, as follows.

$$\delta(0) = \alpha_0 = 0;$$

$$\delta(\nu + 1) = \text{the least } \delta \leqslant \beta \text{ such that } \alpha_\nu \cup p \subseteq \delta \text{ and } (\alpha_\nu, p) \text{ jumps below } \alpha \text{ in } M_\delta;$$

$$\alpha_{\nu+1} = \text{the least } \gamma \leqslant \alpha \text{ such that } (\gamma, p) \text{ does not jump below } \alpha \text{ in } M_{\delta(\nu+1)};$$

$$\delta(\eta) = \sup_{\nu<\eta} \delta(\nu), \quad \text{if } \lim(\eta);$$

$$\alpha_\eta = \sup_{\nu<\eta} \alpha_\nu, \quad \text{if } \lim(\eta).$$

λ The definition breaks down when an ordinal λ is reached for which $\delta(\lambda) \geqslant \beta$ or $\alpha_\lambda \geqslant \alpha$.

Note that by continuity, for limit η, (α_η, p) does not jump below α in $M_{\delta(\eta)}$.

We show that $(\delta(\nu) \mid \nu \leqslant \lambda)$ is increasing. Suppose $\delta(\nu + 1) \leqslant \delta(\nu)$. Since (α_ν, p) jumps below α in $M_{\delta(\nu+1)}$, it follows that (α_ν, p) jumps below α in $M_{\delta(\nu)}$. This contradicts the properties of α_ν. Hence $\delta(\nu) < \delta(\nu + 1)$.

Next we show that for limit η, α_η is the least $\gamma \leqslant \alpha$ such that (γ, p) does not jump below α in $M_{\delta(\eta)}$, just as is the case at successor stages. We prove this by induction on η. Suppose $\gamma < \alpha_\eta$ were such that (γ, p) does not jump below α in $M_{\delta(\eta)}$. Pick $\nu < \eta$ such that $\alpha_\nu > \gamma$. Then as $\delta(\nu) < \delta(\eta)$, $M_{\delta(\nu)}[\gamma \cup p] \cap \alpha = \gamma$. By definition if ν is a successor ordinal, and by induction hypothesis if ν is a limit ordinal, this implies that $\alpha_\nu \leqslant \gamma$, contrary to the choice of ν. This proves the result.

We now show that $(\alpha_\nu \mid \nu < \lambda)$ is increasing. Well, we clearly cannot have $\alpha_{\nu+1} = \alpha_\nu$. But if $\alpha_{\nu+1} < \alpha_\nu$, then by the properties of α_ν, $(\alpha_{\nu+1}, p)$ must jump below α in $M_{\delta(\nu)}$, and hence also in $M_{\delta(\nu+1)}$, contrary to the known properties of $\alpha_{\nu+1}$. Hence $\alpha_\nu < \alpha_{\nu+1}$.

Now, if $\delta(\nu) < \beta$ and $\alpha_\nu < \alpha$, then by 2.7, $\delta(\nu+1) < \beta$, so by 2.4, $\alpha_{\nu+1} < \alpha$. Hence $\lim(\lambda)$. Suppose $\delta(\lambda) < \beta$. Then by 2.4, $\alpha_\lambda < \alpha$, which contradicts the choice of λ. Thus $\delta(\lambda) = \beta$. It follows that $\alpha_\lambda = \alpha$. For if $\alpha_\lambda < \alpha$, then by 2.4, (α_λ, p) jumps below α in $M_{\delta(\lambda)}$, contrary to the properties of $\alpha_\lambda, \delta(\lambda)$.

We set

$$C_\alpha = \{\alpha_\eta \mid \eta < \lambda\},$$ C_α

a club subset of α. We shall show that if $\bar\alpha$ is a limit point of C_α, then $\bar\alpha \in S$ and $C_{\bar\alpha} = \bar\alpha \cap C_\alpha$. But before that, we note for later use that as $(\delta(\nu) \mid \nu < \lambda)$ is strictly increasing and cofinal in β and $(\alpha_\nu \mid \nu < \lambda)$ is strictly increasing and cofinal in α, we have:

2.10 Lemma. $\mathrm{cf}(\beta(\alpha)) = \mathrm{cf}(\alpha)$. □

For $\eta < \lambda$, now, set

$$Y_\eta = M_{\delta(\eta)}[\alpha_\eta \cup p].$$ Y_η

Since $Y_\eta \lhd M_{\delta(\eta)}$, the Condensation Property gives an isomorphism

$$\pi_\eta \colon M_{\psi(\eta)} \cong Y_\eta.$$ $\pi_\eta, \psi(\eta)$

Let $\pi_\eta^{-1}(p) = p_\eta$. Notice that $\pi_\eta \restriction \alpha_\eta = \mathrm{id} \restriction \alpha_\eta$. p_η

2.11 Lemma. *Let $\eta < \lambda$, $\lim(\eta)$. Then $\alpha_\eta \in S$ and $\beta(\alpha_\eta) = \psi(\eta)$, $p(\alpha_\eta) = p_\eta$.*

Proof. We show first that α_η is semi-singular at $\psi(\eta)$ with parameter p_η. By 2.4 it suffices to show that for all $\gamma < \alpha_\eta$, (γ, p_η) jumps below α_η in $M_{\psi(\eta)}$.

Let $\gamma < \alpha_\eta$. By the properties of α_η,

$$M_{\delta(\eta)}[\alpha_\eta \cup p] \cap \alpha = \alpha_\eta,$$
$$M_{\delta(\eta)}[\gamma \cup p] \cap \alpha \neq \gamma.$$

Combining these two facts gives

$$M_{\delta(\eta)}[\gamma \cup p] \cap \alpha_\eta \neq \gamma.$$

But $\alpha_\eta \cup p \subseteq Y_\eta \lhd M_{\delta(\eta)}$. So we get

$$Y_\eta[\gamma \cup p] \cap \alpha_\eta \neq \gamma.$$

Applying π_η^{-1} gives

$$M_{\psi(\eta)}[\gamma \cup p_\eta] \cap \alpha_\eta \neq \gamma,$$

as required.

Since α_η is semi-singular at $\psi(\eta)$, we must have $\alpha_\eta \in S$, of course, so the first part of the lemma is proved.

Suppose that $\beta(\alpha_\eta) \neq \psi(\eta)$. Then by the above, $\beta(\alpha_\eta) < \psi(\eta)$. Since $\delta(\eta) = \sup_{\nu<\eta} \delta(\nu)$ and $Y_\eta = \bigcup_{\nu<\eta} Y_\nu$, we can pick $\nu < \eta$ such that $\delta(\nu) > \pi_\eta(\beta(\alpha_\eta))$ and $\pi_\eta(p(\alpha_\eta)) \subseteq Y_\nu$.

Now, (α_ν, p) does not jump below α in $M_{\delta(\nu)}$, so

$$M_{\delta(\nu)}[\alpha_\nu \cup p] \cap \alpha_\eta = \alpha_\nu.$$

But $\pi_\eta(p(\alpha_\eta)) \subseteq Y_\nu = M_{\delta(\nu)}[\alpha_\nu \cup p]$, so it follows that

$$M_{\delta(\nu)}[\alpha_\nu \cup \pi_\eta(p(\alpha_\eta))] \cap \alpha_\eta = \alpha_\nu.$$

Thus as $\delta(\nu) > \pi_\eta(\beta(\alpha_\eta))$,

$$M_{\pi_\eta(\beta(\alpha_\eta))}[\alpha_\nu \cup \pi_\eta(p(\alpha_\eta))] \cap \alpha_\eta = \alpha_\nu.$$

But clearly,

$$\alpha_\eta \cup \pi_\eta(p(\alpha_\eta)) \subseteq \pi_\eta{}''M_{\beta(\alpha_\eta)} \lhd M_{\pi_\eta(\beta(\alpha_\eta))}.$$

Hence

$$(\pi_\eta{}''M_{\beta(\alpha_\eta)})\,[\alpha_\nu \cup \pi_\eta(p(\alpha_\eta))] \cap \alpha_\eta = \alpha_\nu.$$

Applying π_η^{-1},

$$M_{\beta(\eta)}[\alpha_\nu \cup p(\alpha_\eta)] \cap \alpha_\eta = \alpha_\nu.$$

But $\alpha_\nu < \alpha_\eta$. So by 2.4, α_η is not semi-singular at $\beta(\alpha_\eta)$ with parameter $p(\alpha_\eta)$. This is absurd, of course. Hence $\beta(\alpha_\eta) = \psi(\eta)$. It follows at once that $p(\alpha_\eta) \leqslant^* p_\eta$.

Suppose that $p(\alpha_\eta) <^* p_\eta$. Then $\pi_\eta(p(\alpha_\eta)) <^* p$. So by definition of p, α is not semi-singular at β with parameter $\pi_\eta(p(\alpha_\eta))$. So by 2.4, there is a $\gamma < \alpha$ such that

$$M_\beta[\gamma \cup \pi_\eta(p(\alpha_\eta))] \cap \alpha = \gamma.$$

Suppose first that $\gamma < \alpha_\eta$. By the above, we get

$$M_{\delta(\eta)}[\gamma \cup \pi_\eta(p(\alpha_\eta))] \cap \alpha_\eta = \gamma.$$

So as $\alpha_\eta \cup \pi_\eta(p(\alpha_\eta)) \subseteq Y_\eta \lhd M_{\delta(\eta)}$,

$$Y_\eta[\gamma \cup \pi_\eta(p(\alpha_\eta))] \cap \alpha_\eta = \gamma.$$

Applying π_η^{-1},

$$M_{\psi(\eta)}[\gamma \cup p(\alpha_\eta)] \cap \alpha_\eta = \gamma.$$

But $\psi(\eta) = \beta(\alpha_\eta)$, so by 2.4 we have a contradiction.

Now suppose that $\gamma \geqslant \alpha_\eta$. By 2.8 we have

$$p_\eta \subseteq \psi(\eta) = \beta(\alpha_\eta) = M_{\beta(\alpha_\eta)}[\alpha_\eta \cup p(\alpha_\eta)].$$

Applying π_η,

$$p \subseteq M_{\delta(\eta)}[\alpha_\eta \cup \pi_\eta(p(\alpha_\eta))].$$

Hence

$$p \subseteq M_\beta[\alpha_\eta \cup \pi_\eta(p(\alpha_\eta))].$$

So as $\alpha_\eta \leqslant \gamma$,

$$p \subseteq M_\beta[\gamma \cup \pi_\eta(p(\alpha_\eta))].$$

Thus

$$M_\beta[\gamma \cup p] \subseteq M_\beta[\gamma \cup \pi_\eta(p(\alpha_\eta))].$$

So by choice of γ,

$$M_\beta[\gamma \cup p] \cap \alpha = \gamma.$$

This contradicts 2.4. Hence we must have $p(\alpha_\eta) = p_\eta$, and the proof is complete. □

2.12 Lemma. *Let $\eta < \lambda$, $\lim(\eta)$. Set $\bar{\alpha} = \alpha_\eta$, $\bar{\beta} = \beta(\alpha_\eta)$, $\bar{p} = p(\alpha_\eta)$, and define $\bar{\lambda}$, $(\bar{\delta}(\nu) \mid \nu < \bar{\lambda})$, $(\bar{\alpha}_\nu \mid \nu < \bar{\lambda})$, $(\bar{Y}_\nu \mid \nu < \bar{\lambda})$ from $\bar{\alpha}$ just as λ, $(\delta(\nu) \mid \nu < \lambda)$, $(\alpha_\nu \mid \nu < \lambda)$, $(Y_\nu \mid \nu < \lambda)$ were defined from α above. Let $\pi = \pi_\eta$. Then for all $\nu < \bar{\lambda}$:* $\eta, \bar{\alpha}, \bar{\beta}, \bar{p}, \bar{\lambda}$ $\bar{\delta}(\nu), \bar{\alpha}_\nu, \bar{Y}_\nu$ π

(i) $\bar{\alpha}_\nu = \alpha_\nu$;
(ii) $\pi(\bar{\delta}(\nu + 1)) = \delta(\nu + 1)$;
(iii) $\pi'' M_{\bar{\delta}(\nu)} = M_{\delta(\nu)} \cap Y_\eta$;
(iv) $\pi'' \bar{Y}_\nu = Y_\nu$.

Proof. We first of all prove (i)–(iii) by a simultaneous induction on ν.

By 2.11 we have:

$$\pi\colon M_{\bar{\beta}} \cong Y_\eta = M_{\delta(\eta)}[\bar{\alpha} \cup p], \quad \pi(\bar{p}) = p, \quad \pi \upharpoonright \bar{\alpha} = \mathrm{id} \upharpoonright \bar{\alpha}.$$

Since $\bar{\alpha}_0 = \alpha_0 = 0$ and $\bar{\delta}(0) = \delta(0) = 0$, the first step in the induction is trivial. Limit stages are immediate by continuity. So assume now that the result holds at $\nu < \bar{\lambda}$. Set $\bar{\delta} = \bar{\delta}(\nu + 1)$, $\delta = \pi(\bar{\delta})$. We prove that $\delta = \delta(\nu + 1)$, $\bar{\alpha}_{\nu+1} = \alpha_{\nu+1}$. Our proof of the first of these equalities will also yield $\pi'' M_{\bar{\delta}} = M_\delta \cap Y_\eta$.

Note that by definition of δ.

$$(\pi \upharpoonright \bar{\delta})\colon M_{\bar{\delta}} \prec M_\delta.$$

Applying π to $\bar{\alpha}_\nu \cup \bar{p} \subseteq \bar{\delta}$ gives $\alpha_\nu \cup p \subseteq \delta$. Also, we have

$$M_{\bar{\delta}}[\bar{\alpha}_\nu \cup \bar{p}] \cap \bar{\alpha} \neq \bar{\alpha}_\nu; \qquad M_{\bar{\delta}}[\bar{\alpha}_{\nu+1} \cup \bar{p}] \cap \bar{\alpha} = \bar{\alpha}_{\nu+1},$$

so as $\bar{\alpha}_{\nu+1} < \bar{\alpha} = \alpha_\eta$, we conclude that

$$M_{\bar{\delta}}[\bar{\alpha}_\nu \cup \bar{p}] \cap \alpha_\eta \neq \bar{\alpha}_\nu.$$

Applying $\pi \upharpoonright \bar{\delta}$ gives

$$M_{\delta}[\alpha_{\nu} \cup p] \cap \alpha_{\eta} \neq \alpha_{\nu}.$$

Thus $\delta(\nu + 1) \leqslant \delta$. We show that $\delta \leqslant \delta(\nu + 1)$ as well. We have

$$M_{\delta(\nu+1)}[\alpha_{\nu} \cup p] \cap \alpha \neq \alpha_{\nu}; \qquad M_{\delta(\nu+1)}[\alpha_{\nu+1} \cup p] \cap \alpha = \alpha_{\nu+1},$$

so combining these two results gives

$$M_{\delta(\nu+1)}[\alpha_{\nu} \cup p] \cap \alpha_{\nu+1} \neq \alpha_{\nu}.$$

Thus for some $\xi \in \alpha_{\nu+1}$, $\xi > \alpha_{\nu}$, we have $\xi \in M_{\delta(\nu+1)}[\alpha_{\nu} \cup p]$. Hence we can find a finite sequence $\xi_1, \ldots, \xi_n$ of ordinals in $\delta(\nu + 1)$ such that $\xi_n = \xi$ and for all $i = 1, \ldots, n$, either $\xi_i \in \alpha_{\nu} \cup p$ or else ξ_i is the value of some M function at some members of $\{\xi_1, \ldots, \xi_{i-1}\}$. Now,

$$\xi_1, \ldots, \xi_n \in M_{\delta(\nu+1)}[\alpha_{\nu} \cup p] \subseteq M_{\delta(\eta)}[\alpha_{\eta} \cup p] = Y_{\eta},$$

so we can define $\bar{\xi}_i = \pi^{-1}(\xi_i)$ for $i = 1, \ldots, n$. Since $\alpha_{\nu} < \xi_n = \xi < \alpha_{\nu+1} < \alpha_{\eta}$, we have $\alpha_{\nu} < \bar{\xi}_n = \xi < \alpha_{\nu+1} < \alpha_{\eta}$. And for each i, either $\bar{\xi}_i \in \alpha_{\nu} \cup p$ or else $\bar{\xi}_i$ is the value of some M-function at members of $\{\bar{\xi}_1, \ldots, \bar{\xi}_{i-1}\}$. So, if we set $\bar{\varrho} = \max(\bar{\xi}_1, \ldots, \bar{\xi}_n)$, we have $\bar{\xi} \in M_{\bar{\varrho}+1}[\alpha_{\nu} \cup \bar{p}]$. Hence

$$M_{\bar{\varrho}+1}[\alpha_{\nu} \cup \bar{p}] \cap \bar{\alpha} \neq \alpha_{\nu}.$$

Thus by choice of $\bar{\delta}$, $\bar{\delta} \leqslant \bar{\varrho} + 1$. Now set $\varrho = \pi(\bar{\varrho})$. Since $\bar{\varrho} = \max(\bar{\xi}_1, \ldots, \bar{\xi}_n)$, we have $\varrho = \max(\xi_1, \ldots, \xi_n) < \delta(\nu + 1)$. Also,

$$\varrho \in M_{\delta(\nu+1)}[\alpha_{\nu} \cup p] \subseteq M_{\delta(\eta)}[\alpha_{\eta} \cup p] = Y_{\eta}.$$

But the function $J(\gamma) = \gamma + 1$ is an M-function, and $\delta(\eta)$ is a limit ordinal, so it follows that

$$\varrho + 1 \in M_{\delta(\eta)}[\alpha_{\eta} \cup p] = Y_{\eta}.$$

Hence $\pi(\bar{\varrho} + 1) = \varrho + 1$. Since $\bar{\delta} \leqslant \bar{\varrho} + 1$, applying π gives $\delta \leqslant \varrho + 1 \leqslant \delta(\nu + 1)$, as required.

We now have $\pi(\bar{\delta}(\nu + 1)) = \delta(\nu + 1)$. It follows at once that

$$\pi''M_{\bar{\delta}(\nu+1)} = M_{\delta(\nu+1)} \cap Y_{\eta}.$$

We prove that $\bar{\alpha}_{\nu+1} = \alpha_{\nu+1}$.

By definition,

(i) $M_{\delta}[\alpha_{\nu+1} \cup p] \cap \bar{\alpha} = \alpha_{\nu+1}$,

(ii) $\gamma < \alpha_{\nu+1} \rightarrow M_{\delta}[\gamma \cup p] \cap \alpha_{\nu+1} \neq \gamma$.

Since $(\pi \upharpoonright \bar{\delta})\colon M_{\bar{\delta}} \prec M_{\delta}$ and $\alpha_{\nu+1} \cup p \subseteq M_{\delta} \cap Y_{\eta} = \pi'' M_{\bar{\delta}}$, (i) and (ii) give

(i)′ $M_{\bar{\delta}}[\alpha_{\nu+1} \cup \bar{p}] \cap \bar{\alpha} = \alpha_{\nu+1}$,
(ii)′ $\gamma < \alpha_{\nu+1} \rightarrow M_{\bar{\delta}}[\gamma \cup \bar{p}] \cap \alpha_{\nu+1} \neq \gamma$.

Hence $\bar{\alpha}_{\nu+1} = \alpha_{\nu+1}$.

That completes the proof of (i)–(iii). We are left with (iv).

Using (iii) we have

$$\pi''\bar{Y}_{\nu} = \pi''(M_{\bar{\delta}(\nu)}[\bar{\alpha}_{\nu} \cup \bar{p}]) = (M_{\delta(\nu)} \cap Y_{\eta})\,[\alpha_{\nu} \cup p].$$

But

$$M_{\delta(\nu)}[\alpha_{\nu} \cup p] = Y_{\nu} \subseteq Y_{\eta},$$

so we have

$$(M_{\delta(\nu)} \cap Y_{\eta})\,[\alpha_{\nu} \cup p] = M_{\delta(\nu)}[\alpha_{\nu} \cup p] = Y_{\nu}.$$

Thus $\pi''\bar{Y}_{\nu} = Y_{\nu}$, proving (iv). $\square$

2.13 Corollary. *Let $\alpha \in S$. If $\bar{\alpha}$ is a limit point of C_{α}, then $\bar{\alpha} \in S$ and $C_{\bar{\alpha}} = \bar{\alpha} \cap C_{\alpha}$.*

Proof. Using the above notation, $C_{\alpha} = \{\alpha_{\nu} \mid \nu < \lambda\}$ and for some limit ordinal η, $\bar{\alpha} = \alpha_{\eta}$. By 2.12, $C_{\bar{\alpha}} = \{\alpha_{\nu} \mid \nu < \bar{\lambda}\}$. But $(\alpha_{\nu} \mid \nu < \lambda)$ is strictly increasing and $\sup_{\nu < \bar{\lambda}} \alpha_{\nu} = \bar{\alpha} = \sup_{\nu < \eta} \alpha_{\nu}$, so $\bar{\lambda} = \eta$ and $C_{\bar{\alpha}} = \bar{\alpha} \cap C_{\alpha}$. $\square$

Our next step in obtaining $\square(E)$ is to thin down the sets C_{α} to sets C'_{α} such that:

(i) C'_{α} is a closed subset of α;
(ii) if $\mathrm{cf}(\alpha) > \omega$, then C'_{α} is unbounded in α;
(iii) if $\bar{\alpha}$ is a limit point of C'_{α}, then $\bar{\alpha} \in S$ and $C'_{\bar{\alpha}} = \bar{\alpha} \cap C'_{\alpha}$;
(iv) $\mathrm{otp}(C'_{\alpha}) < \alpha$.

It will then be a fairly easy matter to turn $(C'_{\alpha} \mid \alpha \in S)$ into a $\square(E)$-sequence.

Let $\alpha \in S$, and set $\beta = \beta(\alpha)$, $p = p(\alpha)$. Define λ, $(\delta(\nu) \mid \nu < \lambda)$, $(\alpha_{\nu} \mid \nu < \lambda)$, β, p
$(Y_{\eta} \mid \eta < \lambda)$, $(\pi_{\eta} \mid \eta < \lambda)$, $(\psi(\eta) \mid \eta < \lambda)$, $(p_{\eta} \mid \eta < \lambda)$ as before.

2.14 Lemma. *Let $\eta_1 < \eta_2 < \lambda$ be limit ordinals. Then $\sup Y_{\eta_1} < \sup Y_{\eta_2}$.*

Proof. Since $\alpha_{\eta_1} < \alpha_{\eta_2}$, 2.4 gives

$$M_{\beta(\alpha_{\eta_2})}[\alpha_{\eta_1} \cup p(\alpha_{\eta_2})] \cap \alpha_{\eta_2} \neq \alpha_{\eta_1}.$$

Applying $\pi_{\eta_2}\colon M_{\beta(\alpha_{\eta_2})} \cong Y_{\eta_2}$ and using 2.11,

$$Y_{\eta_2}[\alpha_{\eta_1} \cup p] \cap \alpha_{\eta_2} \neq \alpha_{\eta_1}.$$

Hence

$$M_{\sup Y_{\eta_2}}[\alpha_{\eta_1} \cup p] \cap \alpha_{\eta_2} \neq \alpha_{\eta_1}.$$

But

$$M_{\delta(\eta_1)}[\alpha_{\eta_1} \cup p] \cap \alpha = \alpha_{\eta_1}.$$

So as $\sup Y_{\eta_1} \leqslant \delta(\eta_1)$, we have

$$M_{\sup Y_{\eta_1}}[\alpha_{\eta_1} \cup p] \cap \alpha_{\eta_2} = \alpha_{\eta_1}.$$

Thus $\sup Y_{\eta_1} < \sup Y_{\eta_2}$. □

In defining C'_α there are two cases to consider. Let γ be the least ordinal such that $M_\beta[\gamma \cup p]$ is cofinal in β.

Case 1. γ is a limit ordinal.
Set

$$C'_\alpha = \{\alpha_\eta \mid \lim(\eta) \wedge (\exists \xi \leqslant \alpha_\eta)\,[\sup Y_\eta = \sup M_\beta[\xi \cup p]]\}.$$

2.15 Lemma. *C'_α is closed in α.*

Proof. By the continuity of the sequence $(Y_\eta \mid \eta < \lambda)$. □

2.16 Lemma. *Let $\mathrm{cf}(\alpha) > \omega$. Then C'_α is unbounded in α.*

Proof. Let $H = \{\sup Y_\eta \mid \lim(\eta)\}$, $K = \{\sup M_\beta[\delta \cup p] \mid \delta < \gamma\}$. Clearly, H and K are club in β. (By 2.10, $\mathrm{cf}(\beta) = \mathrm{cf}(\alpha) > \omega$.) Hence $H \cap K$ is club in β. So we can pick arbitrarily large limit ordinals $\eta < \lambda$ so that $\sup Y_\eta \in K$. For any such η, $\sup Y_\eta = \sup M_\beta[\delta \cup p]$ for some $\delta < \gamma$. But

$$\sup M_\beta[\alpha_\eta \cup p] \geqslant \sup M_{\delta(\eta)}[\alpha_\eta \cup p] \geqslant \sup Y_\eta.$$

Hence we can find such a $\delta \leqslant \alpha_\eta$. Then $\alpha_\eta \in C'_\alpha$. □

2.17 Lemma. $\mathrm{otp}(C'_\alpha) < \alpha$.

Proof. Define $\theta: C'_\alpha \to \mathrm{On}$ by letting $\theta(\alpha_\eta)$ be the least $\xi \leqslant \alpha_\eta$ such that $\sup Y_\eta = \sup M_\beta[\xi \cup p]$. By 2.14, θ is order-preserving. But by definition of γ, $\mathrm{ran}(\theta) \subseteq \gamma$. Hence $\mathrm{otp}(C'_\alpha) \leqslant \gamma$. But by 2.9, $\gamma < \alpha$. □

2.18 Lemma. *Let $\bar\alpha$ be a limit point of C'_α. Then $\bar\alpha \in S$, $\bar\alpha$ falls under Case 1, and $C'_{\bar\alpha} = \bar\alpha \cap C'_\alpha$.*

Proof. Since $C'_\alpha \subseteq C_\alpha$, we know at once that $\bar\alpha \in S$. Let $\bar\gamma$ be least such that $M_{\bar\beta}[\bar\gamma \cup \bar p]$ is cofinal in $\bar\beta$. We must show that $\lim(\bar\gamma)$ and that $C'_{\bar\alpha} = \bar\alpha \cap C'_\alpha$.

Let $\bar\alpha = \alpha_\varrho$. Then ϱ is a limit of limit ordinals $\eta < \varrho$ for which $\alpha_\eta \in C_\alpha$. For each such η there is a least $\xi_\eta \leqslant \alpha_\eta$ such that $\sup Y_\eta = \sup M_\beta[\xi_\eta \cup p]$. Since the sequence $(Y_\eta \mid \eta < \lambda)$ is continuous, taking the supremum over all such η gives

$$\sup Y_\varrho = \sup M_\beta[\xi \cup p],$$

where $\xi = \sup_\eta \xi_\eta$. We show that $\bar\gamma = \xi$. Since $\lim(\xi)$, this proves $\lim(\bar\gamma)$.

Let $\tau = \sup Y_\varrho$. Then since $\tau = \sup M_\beta[\xi \cup p]$, we have

$$M_\beta[\xi \cup p] = M_\tau[\xi \cup p].$$

But $\tau \leqslant \delta(\varrho) < \beta$. Hence

$$M_\beta[\xi \cup p] = M_{\delta(\varrho)}[\xi \cup p].$$

So

$$\tau = \sup M_{\delta(\varrho)}[\xi \cup p].$$

But

$$\xi \cup p \subseteq Y_\varrho \lhd M_{\delta(\varrho)}.$$

So,

$$\tau = \sup Y_\varrho[\xi \cup p].$$

Thus

$$\sup Y_\varrho = \sup Y_\varrho[\xi \cup p].$$

Applying π_ϱ^{-1},

$$\bar{\beta} = \sup M_{\bar{\beta}}[\xi \cup \bar{p}].$$

Hence $\bar{\gamma} \leqslant \xi$. Suppose that $\bar{\gamma} < \xi$. Then for some η, $\bar{\gamma} < \xi_\eta$. Now,

$$\sup M_{\bar{\beta}}[\bar{\gamma} \cup \bar{p}] = \bar{\beta}.$$

So

$$\sup M_{\bar{\beta}}[\xi_\eta \cup \bar{p}] = \bar{\beta}.$$

Applying π_ϱ,

$$\sup Y_\varrho[\xi_\eta \cup p] = \sup Y_\varrho.$$

But

$$Y_\varrho[\xi_\eta \cup p] = M_{\delta(\varrho)}[\xi_\eta \cup p].$$

Hence

$$\sup M_{\delta(\varrho)}[\xi_\eta \cup p] = \sup Y_\varrho.$$

Thus

$$\sup M_\beta[\xi_\eta \cup p] \geqslant \sup Y_\varrho.$$

But

$$\sup M_\beta[\xi_\eta \cup p] = \sup Y_\eta < \sup Y_\varrho,$$

so we have a contradiction. Hence $\bar{\gamma} = \xi$.

Since $\bar{\alpha}$ is a limit point of C_α, we know that $C_{\bar{\alpha}} = \bar{\alpha} \cap C_\alpha$. So

$$C_{\bar{\alpha}} = \{\alpha_\eta \mid \eta < \varrho\}.$$

Hence,

$$C'_{\bar{\alpha}} = \{\alpha_\eta \mid \lim(\eta) \wedge \eta < \varrho \wedge (\exists \xi \leqslant \alpha_\eta)\, [\sup \bar{Y}_\eta = \sup M_{\bar{\beta}}[\xi \cup \bar{p}]]\}.$$

Using 2.12(iv) we get at once,

$$C'_{\bar{\alpha}} = \pi_\varrho{}'' C'_{\bar{\alpha}} = \bar{\alpha} \cap C'_\alpha. \quad \square$$

That completes the construction and study of C'_α in Case 1.

Case 2. Otherwise.

We commence by defining a descending sequence of ordinals $\eta_1, \ldots, \eta_n$ for some n. First let η be least ($< \alpha$) such that $\sup M_\beta[\eta \cup p] = \beta$. Since we are not in Case 1, η is a successor ordinal. Set $\eta_1 = \eta - 1$. Thus

$$\varphi_1 = \sup M_\beta[\eta_1 \cup p] < \beta.$$

Now suppose that $\eta_1, \ldots, \eta_{i-1}$ are defined, where $i > 1$. Let η be least ($< \alpha$) such that $\sup M_\beta[\eta \cup p \cup \{\eta_1, \ldots, \eta_{i-1}\}] = \beta$. If $\lim(\eta)$, then $n = i - 1$ and the definition stops. Otherwise set $\eta_i = \eta - 1$. Then

$$\varphi_i = \sup M_\beta[\eta_i \cup p \cup \{\eta_1, \ldots, \eta_{i-1}\}] < \beta.$$

Since $\eta_1 > \eta_2 > \eta_3 > \ldots$, the definition stops after finitely many steps. We set

$$q = q(\alpha) = \{\eta_1, \ldots, \eta_n\}, \qquad \varphi = \varphi(\alpha) = \max(\varphi_1, \ldots, \varphi_n).$$

Set

$$C'_\alpha = \{\alpha_\eta \mid \lim(\eta) \wedge q \subseteq \alpha_2 \wedge \sup Y_\eta > \varphi \\ \wedge (\exists \xi \leqslant \alpha_\eta)\,[\sup Y_\eta = \sup M_\beta[\xi \cup p \cup q]]\}.$$

Since we shall have no further need to refer to the γ of Case 1, we now define γ to be the least ordinal such that $M_\beta[\gamma \cup p \cup q]$ is cofinal in β. By definition of q, we have $\lim(\gamma)$.

2.19 Lemma.

(i) *C'_α is closed in α;*

(ii) *if* $\mathrm{cf}(\alpha) > \omega$, *then C'_α is unbounded in α;*

(iii) $\mathrm{otp}(C'_\alpha) < \alpha$.

Proof. Just replace p by $p \cup q$ in the proof of 2.15, 2.16, and 2.17, (also, γ has a new meaning now of course.) □

2.20 Lemma. *Let $\bar{\alpha}$ be a limit point of C'_α, say $\bar{\alpha} = \alpha_\varrho$. Then:*

(i) *$\bar{\alpha}$ falls under Case 2;*

(ii) $\pi_\varrho(q(\bar{\alpha})) = q$ *(i.e. $q(\bar{\alpha}) = q$);*

(iii) $\sup[\pi_\varrho{}''\varphi(\bar{\alpha})] = \varphi(\alpha)$;

(iv) $C'_{\bar{\alpha}} = \bar{\alpha} \cap C'_\alpha$.

Proof. We prove (i)–(iii); (iv) then follows easily, much as in 2.18. In fact (iii) itself follows from (i) and (ii) as we now prove. It suffices to show that for each $i = 1, \ldots, n$,

$$\sup[\pi_\varrho{}''\sup M_{\bar{\beta}}[\bar{\eta}_i \cup \bar{p} \cup \{\bar{\eta}_1, \ldots, \bar{\eta}_{i-1}\}]] = \\ \sup M_\beta[\eta_i \cup p \cup \{\eta_1, \ldots, \eta_{i-1}\}],$$

i.e.

$$\sup \pi_\varrho{}''M_{\bar{\beta}}[\bar{\eta}_i \cup \bar{p} \cup \{\bar{\eta}_1, \ldots, \bar{\eta}_{i-1}\}] = \sup M_\beta[\eta_i \cup p \cup \{\eta_1, \ldots, \eta_{i-1}\}].$$

In fact we prove that

$$\pi_\varrho''M_{\bar\beta}[\bar\eta_i \cup \bar p \cup \{\bar\eta_1, \ldots, \bar\eta_{i-1}\}] = M_\beta[\eta_i \cup p \cup \{\eta_1, \ldots, \eta_{i-1}\}].$$

We have

$$\begin{aligned}\pi_\varrho''M_{\bar\beta}[\bar\eta_i \cup \bar p \cup \{\bar\eta_1, \ldots, \bar\eta_{i-1}\}] &= Y_\varrho[\eta_i \cup p \cup \{\eta_1, \ldots, \eta_{i-1}\}]\\ &= M_{\delta(\varrho)}[\eta_i \cup p \cup \{\eta_1, \ldots, \eta_{i-1}\}],\end{aligned}$$

by definition of Y_ϱ. But $\delta(\varrho) \geqslant \sup Y_\varrho > \varphi$, so by definition of φ,

$$\delta(\varrho) > \sup M_\beta[\eta_i \cup p \cup \{\eta_1, \ldots, \eta_{i-1}\}].$$

Hence

$$M_\beta[\eta_i \cup p \cup \{\eta_1, \ldots, \eta_{i-1}\}] = M_{\delta(\varrho)}[\eta_i \cup p \cup \{\eta_1, \ldots, \eta_{i-1}\}].$$

That proves (iii), assuming (i) and (ii). We must therefore prove (i) and (ii) to be done. In fact the proof of (i) is contained in the proof of (ii) so we simply concentrate on (ii). We prove by induction on i that for each $i = 1, \ldots, n$, $\bar\eta_i$ is defined and $\bar\eta_i = \eta_i$, and that if η is least such that $M_{\bar\beta}[\eta \cup p \cup \{\bar\eta_1, \ldots, \bar\eta_n\}]$ is cofinal in $\bar\beta$, then lim (η).

Suppose we have proved that for all $j = 1, \ldots, i-1$, $\bar\eta_j$ is defined and $\bar\eta_j = \eta_j$. Since $\sup Y_\varrho > \varphi_i$, we have

$$\begin{aligned}\sup Y_\varrho &> \sup M_\beta[\eta_i \cup p \cup \{\eta_1, \ldots, \eta_{i-1}\}]\\ &\geqslant \sup M_{\delta(\varrho)}[\eta_i \cup p \cup \{\eta_1, \ldots, \eta_{i-1}\}]\\ &= \sup Y_\varrho[\eta_i \cup p \cup \{\eta_1, \ldots, \eta_{i-1}\}].\end{aligned}$$

Applying π_ϱ^{-1}, we get

$$\bar\beta > M_{\bar\beta}[\eta_i \cup \bar p \cup \{\bar\eta_1, \ldots, \bar\eta_{i-1}\}].$$

If we can show that $\bar\beta = \sup M_{\bar\beta}[(\eta_i + 1) \cup \bar p \cup \{\bar\eta_1, \ldots, \bar\eta_{i-1}\}]$, then by definition we shall have $\bar\eta_i = \eta_i$.

Since $\bar\alpha = \alpha_\varrho \in C'_\alpha$, there is a $\xi \leqslant \alpha$ such that

$$\sup Y_\varrho = \sup M_\beta[\xi \cup p \cup q].$$

Hence as $\delta(\varrho) \geqslant \sup Y_\varrho$,

$$M_\beta[\xi \cup p \cup q] = M_{\delta(\varrho)}[\xi \cup p \cup q].$$

But $\xi \cup p \cup q \subseteq Y_\varrho \lhd M_{\delta(\varrho)}$, so

$$M_{\delta(\varrho)}[\xi \cup p \cup q] = Y_\varrho[\xi \cup p \cup q].$$

Thus

$$\sup Y_\varrho = \sup Y_\varrho[\xi \cup p \cup q].$$

Applying π_ϱ^{-1},

$$\bar{\beta} = \sup M_{\bar{\beta}}[\xi \cup \bar{p} \cup q].$$

Now,

$$\sup M_\beta[\xi \cup p \cup q] = \sup Y_\varrho \leqslant \delta(\varrho) < \beta,$$

and

$$\sup M_\beta[(\eta_i + 1) \cup p \cup q] = \beta,$$

so we must have $\eta_i + 1 > \xi$. Thus

$$\bar{\beta} = \sup M_{\bar{\beta}}[(\eta_i + 1) \cup \bar{p} \cup q].$$

But $\{\eta_i, \eta_{i+1}, \dots, \eta_n\} \subseteq \eta_i + 1$. Hence

$$\begin{aligned}\bar{\beta} &= \sup M_{\bar{\beta}}[(\eta_i + 1) \cup \bar{p} \cup \{\eta_1, \dots, \eta_{i-1}\}] \\ &= \sup M_{\bar{\beta}}[(\eta_i + 1) \cup \bar{p} \cup \{\bar{\eta}_1, \dots, \bar{\eta}_{i-1}\}],\end{aligned}$$

as required.

For each limit ordinal $\eta < \varrho$ now, let $\xi_\eta < \alpha_\eta$ be least such that

$$\sup Y_\eta = \sup M_\beta[\xi_\eta \cup p \cup q].$$

By 2.14, if $\xi = \sup_{\eta<\varrho} \xi_\eta$, we have $\lim(\xi)$,

$$\sup Y_\varrho = \sup(\bigcup_{\eta<\varrho} Y_\eta) = \sup M_\beta[\xi \cup p \cup q],$$

and for each $\eta < \varrho$, $\sup M_\beta[\xi_\eta \cup p \cup q] < \sup Y_\varrho$. Since $\sup Y_\varrho \leqslant \delta(\varrho)$, it follows that ξ is the least ordinal such that $\sup M_{\delta(\varrho)}[\xi \cup p \cup q] = \sup Y_\varrho$, and hence that ξ is the least ordinal such that $\sup Y_\varrho[\xi \cup p \cup q] = \sup Y_\varrho$. Applying π_ϱ^{-1}, we see that ξ is the least ordinal such that $\sup M_{\bar{\beta}}[\xi \cup \bar{p} \cup q] = \bar{\beta}$. Since $\lim(\xi)$, this means that the definition of $\bar{q}$ stopped at stage $n + 1$, so $\bar{q} = q$. The proof is complete. □

To complete the proof of 2.1 now, we use the sequence $(C'_\alpha \mid \alpha \in S)$ to build a $\square(E)$-sequence. The following lemma sums up what we know about the sets C'_α.

2.21 Lemma.

(i) *C'_α is a closed subset of α;*

(ii) *if* $\mathrm{cf}(\alpha) > \omega$, *then C'_α is unbounded in α;*

(iii) $\mathrm{otp}(C'_\alpha) < \alpha$;

(iv) *if $\bar{\alpha}$ is a limit point of C'_α, then $\bar{\alpha} \in S$ and $C'_{\bar{\alpha}}$.* □

The following lemma will enable us to avoid the class E on limit points of the final $\square(E)$-sequence.

2.22 Lemma. *If $\alpha \in E$, then $C'_\alpha \cap A$ is bounded in α.*

Proof. Let $\beta = \beta(\alpha)$, $p = p(\alpha)$, and adopt the notation used in the definition of C_α and C'_α. Since $\alpha \in E$ there is a $\gamma \geqslant \alpha$ and a parameter $q \subseteq \gamma$ such that:

(a) α is not semi-singular at γ;

(b) if $\bar{\alpha} \in A \cap \alpha$, then either $(\bar{\alpha}, q)$ jumps below α in M_γ or else $\bar{\alpha}$ is semi-singular at γ with a parameter in $M_\gamma[\bar{\alpha} \cup q]$.

By (a), $\gamma < \beta$. Since $\sup_{\eta<\lambda} \delta(\eta) = \beta$ and $\sup_{\eta<\lambda} \alpha_\eta = \alpha$ and $\beta = M_\beta[\alpha \cup p]$, we can find an ordinal $\eta_0 < \lambda$ such that

$$q \cup \{\gamma\} \subseteq M_{\delta(\eta_0)}[\alpha_{\eta_0} \cup p].$$

Suppose that $C'_\alpha \cap A$ were unbounded in α. Then we could find a limit ordinal $\eta < \lambda$ such that $\eta \geqslant \eta_0$ and $\alpha_\eta \in A$. By the definition of α_η,

$$M_{\delta(\eta)}[\alpha_\eta \cup p] \cap \alpha = \alpha_\eta.$$

So as $q \subseteq M_{\delta(\eta)}[\alpha_\eta \cup p]$,

$$M_{\delta(\eta)}[\alpha_\eta \cup q] \cap \alpha = \alpha_\eta.$$

So as $\gamma < \delta(\eta)$,

$$M_\gamma[\alpha_\eta \cup q] \cap \alpha = \alpha_\eta.$$

Thus by (b) above (with $\bar{\alpha} = \alpha_\eta$), α_η must be semi-singular at γ with some parameter in $M_\gamma[\alpha_\eta \cup q]$. Consider the isomorphism

$$\pi_\eta\colon M_{\psi(\eta)} \cong M_{\delta(\eta)}[\alpha_\eta \cup p].$$

Let $\bar{\gamma} = \pi_\eta^{-1}(\gamma)$, $\bar{q} = \pi_\eta^{-1}(q)$. Using 2.4, we see easily that (since $\pi_\eta^{-1} \restriction \alpha_\eta = \mathrm{id} \restriction \alpha_\eta$) α_η is semi-singular at $\bar{\gamma}$ with a parameter in $M_{\bar{\gamma}}[\alpha_\eta \cup \bar{q}]$. But by 2.11, $\bar{\gamma} < \xi(\eta) = \beta(\alpha_\eta)$, so this is impossible. Hence C'_α must be bounded in α. □

2.23 Corollary. *If $\alpha \in S$ and $\bar{\alpha} < \alpha$ is a limit point of $C'_\alpha \cap A$, then $\bar{\alpha} \notin E$.*

Proof. Let $\bar{\alpha} < \alpha$ be a limit point of $C'_\alpha \cap A$. By 2.21 (iv), $C'_{\bar{\alpha}} = \bar{\alpha} \cap C'_\alpha$. But $\bar{\alpha}$ is a limit point of $C'_\alpha \cap A$ and hence of $C'_{\bar{\alpha}} \cap A$, so $\sup C'_{\bar{\alpha}} \cap A = \bar{\alpha}$. So by 2.22 $\bar{\alpha} \notin E$. □

Now define sets C''_α by;

$$C''_\alpha = \begin{cases} C'_\alpha - \sup(C'_\alpha \cap A), & \text{if } \sup(C'_\alpha \cap A) < \alpha, \\ \text{the closure of } (C'_\alpha \cap A), & \text{if } \sup(C'_\alpha \cap A) = \alpha. \end{cases}$$

Clearly, the sets C''_α have the following properties:

(i) C''_α is a closed subset of α;
(ii) if $\mathrm{cf}(\alpha) > \omega$, then C''_α is unbounded in α;
(iii) $\mathrm{otp}(C''_\alpha) < \alpha$;
(iv) if $\bar{\alpha} \in C''_\alpha$, then $\bar{\alpha} \in S$, $\bar{\alpha} \notin E$, and $C''_{\bar{\alpha}} = \bar{\alpha} \cap C''_\alpha$.

Define sets D_α, $\alpha \in S$, by recursion, thus:

$$D_\alpha = \begin{cases} \bigcup \{D_\gamma \mid \gamma \in C''_\alpha\}, & \text{if } \sup(C''_\alpha) = \alpha; \\ \bigcup \{D_\gamma \mid \gamma \in C''_\alpha\} \cup \{\alpha_n \mid n < \omega\}, & \text{if } \sup(C''_\alpha) < \alpha, \end{cases}$$

where $(\alpha_n \mid n < \omega)$ is any sequence cofinal in α with $\alpha_0 = \sup(C''_\alpha)$.

As in IV.5.1, it is easily seen that $(D_\alpha \mid \alpha \in S)$ is a $\square(E)$-sequence. The proof of 2.1 is complete.

Exercises

1. Using the argument from Chapter V.5 as a model, obtain a machine proof of the Covering Lemma.

2. Obtain machine proofs of the results in Chapter VII concerning trees and large cardinals in L.

Remarks and Historical Notes

Chapter I

Sections 1–6

This is all basic set theory. See the "Historical Notes" section of Jech (1978) for historical and bibliographical details.

Section 7

Theorems 7.1 and 7.2 are due (independently) to Mostowski and Shepherdson and appeared in Mostowski (1949).

Section 8

Theorems 8.1 and 8.2 are due to Montague and Lévy. See Montague (1961) and Lévy (1960). The Lévy Hierarchy was introduced in Lévy (1965). The concept of absoluteness is due to Gödel, who in principle proved 8.3.

Section 9

There are various (essentially equivalent) ways of defining a formal "language of set theory" within set theory. The precise definition chosen is our own, as is all of the development in this section. The Basic Set Theory (BS) has been formulated independently by various authors, among them Gandy, Jensen, and ourselves. The first published account of this seems to be in Gandy (1974).

Section 10

Again the material of this section is our own version of well-known, essentially "folklore" material.

Section 11

Admissible sets were first studied by S. Kripke and R. Platek. See Kripke (1964) and Platek (1966). Most of the results in this section now have the status of "folklore". For further details on admissible sets consult Barwise (1975).

Chapter II

Section 1

The notion of constructibility is due to K. Gödel. In Gödel (1939), the sets L_α were defined and used to define L. In his famous 1940 monograph (the first ever "book on L"), Gödel defined L using a collection of definability functions. All of the material in this section is due to Gödel.

Section 2

The material in this section was known implicitly to Gödel. The first explicit proof that "L is Σ_1" appeared in Karp (1967). The proof given here is our own version.

Section 3

The validity of AC in L was announced in Gödel (1938), and a proof was given in Gödel (1940). The present proof is our own version of Gödel's original argument.

Section 4

Theorem 4.3 is due to Shepherdson (1951, 1952, 1953).

Section 5

Again these results were all known to Gödel in one form or another, with the proof given here our own version.

Section 6

The material in this section is due mainly to R.B. Jensen. Various sets of notes on the results were in circulation for several years in the early to mid 1960s, but the material was never published until the appearance of Jensen (1972) and Devlin (1973).

Section 7

The remarks for Section 6 apply.

Exercises

1. The primitive recursive set functions were first studied in depth by Jensen and Karp (1971).
2. Relative constructibility was first investigated by Hajnal (1956) and Lévy (1957).
3. This was proved by various people, ourselves included.
4. Hajnal (1956).
5. Jensen (unpublished).

Chapter III

Section 1

The Souslin Problem was proposed in M. Souslin (1920). The first systematic investigation of trees was carried out by D. Kurepa (1935). This paper includes Theorems 1.1 and 1.4. Both of these were later rediscovered by other authors, among them F.B. Jones (Theorem 1.1) and E. Miller (Theorem 1.4). Theorem 1.5 is due to Jensen, who announced the result in Jensen (1968). The first construction of a model of set theory in which there are no Souslin trees is due to Solovay and Tennenbaum (1971). See Jech (1978) for further details.

Section 2

The Kurepa Hypothesis was first formulated in Kurepa (1935). The proof that KH is equivalent to the existence of a Kurepa tree is also due to Kurepa himself. The construction of a Kurepa tree from $V = L$ is due to Solovay (unpublished), shortly after Jensen's construction of a Souslin tree in L. The first construction of a model of set theory in which there are no Kurepa trees is due to Silver (1971). Silver started with a model in which there is an inaccessible cardinal. Solovay observed that since his construction of a Kurepa tree in L works (with minor changes) for any universe $L[A]$ where $A \subseteq \omega_1$ (in $L[A]$), Silver's assumption of the consistency of an inaccessible cardinal is necessary. For further details on Kurepa trees, consult Jech (1978).

Section 3

Fodor's Theorem (Theorem 3.1) appeared in Fodor (1956). Lemma 3.4 is due to K. Kunen. Everything else in this section is due to Jensen.

Exercises

1 D. This is due to Baumgartner (unpublished).
2. Solovay (unpublished).
3 A–3 D. These are due to Devlin (1979 a).
4. Jensen (unpublished).

Chapter IV

All Sections

Practically all of this is due to Jensen. The results were proved and circulated privately around 1967/68, but did not appear in print until the appearance of Jensen (1972) and Devlin (1973). See the notes for Chapter VI for more details along this line. Theorem 2.11 was first proved by J. Gregory (1976) using a direct argument. The proof given here, using 2.5 through 2.10, is a distillation of subsequent observations of Jensen, Gregory, Laver, and ourselves.

Exercises

1 A. Aronszajn.
1 B. Specker.
1 C. Jensen.
2. Jensen.
3. This observation was used by Rowbottom and Lévy (independently) to obtain a model of set theory in which there is a Kurepa tree.
4. Jensen.
5. Jensen.
6. The consistency result mentioned here is due to Solovay (unpublished).
7, 8. These results were formulated by Shelah, and sharpen previous results of Gregory, Jensen, Laver, and ourselves.

Chapter V

Section 1

Drake (1974) and Jech (1978) supply all of the relevant history, etc.

Section 2

In 1964, F. Rowbottom proved, in his PhD thesis (Wisconsin) that if there is a measurable cardinal ($\kappa(\omega_1)$ suffices), then $\mathscr{P}^L(\omega)$ is countable. This finally appeared in Rowbottom (1971). Rowbottom used the method of indiscernibles to prove this result. In the same year, H. Gaifman used the method of iterated ultrapowers to obtain the conclusion of Theorem 2.11, and with it various consequences, including Rowbottom's result, from the existence of a measurable cardinal. See Gaifman (1964). Solovay extracted $0^{\#}$ from the proof, and proved various result about $0^{\#}$. See Solovay (1967). In his PhD thesis (Berkeley, 1966), J. Silver subsequently obtained all of Gaifman's and Solovay's result using the method of indiscernibles, starting from the existence of $\kappa(\omega_1)$. Silver's work was finally published in Silver (1971).

The methods used throughout this section are essentially those developed by Silver (with one or two ideas from Rowbottom). See Jech (1978) for further details concerning $0^{\#}$.

Section 3

This is due to Solovay, who also proved that $0^{\#}$ is Δ^1_3-definable over the integers.

Section 4

The main result of this section are due to K. Kunen. The proofs given are due to Silver.

Section 5

The Covering Lemma is due to Jensen, and appeared in Devlin and Jensen (1975). The proof given here is a somewhat simpler version due to M. Magidor.

Exercises

1A. Silver.
1B. Various people.
1C. Harrington and Shelah.
2. Solovay.
3. Various people.
4B, 4C. Bukovski.

Chapter VI

All Sections

Everything in this chapter is due to Jensen. (Exception: Theorem 6.1′ is due to Beller and Litman (1980), following Jensen's proofs.) Jensen first worked out a "fine structure theory" for the L_α-hierarchy in the early to mid 1960s. His results were circulated in the form of a number of increasingly bulky sets of handwritten notes. Then, in 1970 he defined the J_α-hierarchy and reworked the fine structure theory for this hierarchy. Working from his notes, we then wrote three sets of (handwritten) notes which for a while became the "standard text" in the area. These notes were subsequently turned into the book Devlin (1973). In the meantime, Jensen (1972) came out, covering some of the same material. Until the appearance of this volume, these were essentially the only sources for this material.

Exercises

1. Jensen. See Devlin and Jensen (1975).
2. Prikry and Solovay (1975).
2D. Starting with a model with a Mahlo cardinal, Shelah has constructed a model of set theory in which every stationary subset of ω_2 consisting of ω-cofinal ordinals contains a closed set of type ω_1.
4. Beller and Litman. (Unpublished in this form.)

Chapter VII

All Sections

Theorem 1.2 is due to Jensen (1972). Theorem 1.2′ is due to Beller and Litman (1980). The rest of the chapter is due to Jensen.

Exercises

1. Devlin.
2. Shelah.
3A. Erdos, Hajnal and Milner.
3B. Prikry, Jensen (independently). More recently, S. Todorcevic has proved that $KH(\kappa,\kappa)$ for $\mathrm{cf}(\kappa) = \omega$ follows from $\Box_\kappa$.
4. Beller and Litman.

Chapter VIII

Section 1

Theorem 1.7 and its proof are due to R.L. Vaught. (See Morley & Vaught (1962).)

Section 2

Everything in this section is due to Jensen.

Section 3

Theorem 3.1 is due to Jensen. In his original proof, Jensen made use of some techniques developed by Keisler involving omitting types arguments. The proof given, using a Δ-system approach was later worked out by Donder. The construction of the $\mathcal{M}$-complex is effectively the same as in the original Jensen proof.

Section 4

Simplified morasses were developed by Velleman (see Velleman (1984, 1984a)), following some work on morasses by Shelah and Stanley (1983). The construction of a simplified morass from a standard morass is due to Donder (1983).

Another simplified morass-like structure is due to Silver, and is known as *W*. For details of this see Kanamori (1982).

The application of both the standard morass and the simplified morass given here uses the morass only as an "indexing system". Many of the other uses of morasses employ diagonalisation techniques using $\diamondsuit$-like properties that can be incorporated into morasses. Exercise 4 provides one example. Others can be found in the references given above.

Section 5

At present only reference to gap-*n* morasses is Stanley (1975).

Chapter IX

All Sections

The notion of a machine was worked out by Silver to provide an alternative to the fine structure theory for proving $\square_\kappa$ and similar results. The construction of a machine given here, though essentially that of Silver, owes much to an unpublished account of the subject written by A. Litman. The proof of $\square$ using a machine, whilst again Silver's proof at heart, follows closely the version given in Beller and Litman (1980).

Bibliography

Barwise, J.
1975 Admissible Sets and Structures. Springer-Verlag

Beller, A., Litman, A.
1980 A Strengthening of Jensen's □ Principles. J. Symbolic Logic *45*, 251–264

Chang, C. C., Keisler, H. J.
1973 Model Theory. North Holland

Devlin, K. J.
1973 Aspects of Constructibility. Springer-Verlag: Lecture Notes in Mathematics *354*
1979 Fundamentals of Contemporary Set Theory. Springer-Verlag
1979a Variations on ◇. J. Symbolic Logic *44*, 51–58

Devlin, K. J., Jensen, R. B.
1975 Marginalia to a Theorem of Silver. In "Logic Conference Kiel 1974". Springer-Verlag: Lecture Notes in Mathematics *499*, 115–142

Devlin, K. J., Johnbråten, H.
1974 The Souslin Problem. Springer-Verlag: Lecture Notes in Mathematics *405*

Donder, H. D.
1984 Another Look at Gap-1 Morasses. Proc. Cornell Meeting 1982

Drake, F. R.
1974 Set Theory: An Introduction to Large Cardinals. North Holland

Fodor, G.
1956 Eine Bemerkung zur Theorie der Regressiven Funktionen. Acta Sci. Math. (Szeged) *17*, 139–142

Gaifman, H.
1964 Measurable Cardinals and Constructible Sets. Notices Amer. Math. Soc. *11*, 771

Gandy, R.
1974 Set-Theoretic Functions for Elementary Syntax. In "Axiomatic Set Theory". Proc. Symp. in Pure Math. *13*, Part II (ed. A. Mathias), Amer. Math. Soc.

Gödel, K.
1938 The Consistency of the Axiom of Choice and of the Generalised Continuum Hypothesis. Proc. Natl. Acad. USA *24*, 556–557
1939 Consistency Proof for the Generalised Continuum Hypothesis. Proc. Natl. Acad. USA *25*, 220–224
1940 The Consistency of the Axiom of Choice and of the Generalised Continuum Hypothesis. Ann. Math. Studies *3*

Gregory, J.
1976 Higher Souslin Trees and the Generalised Continuum Hypothesis. J. Symb. Logic *14*, 663–671

Hajnal, A.
1956 On a Consistency Theorem Connected with the Generalised Continuum Problem. Zeit. Math. Logik *2*, 131–136

Jech, T.
1978 Set Theory. Academic Press

Jensen, R. B.
1968 Souslin's Hypothesis is Incompatible with $V = L$. Notices Amer. Math. Soc. *15*, 935
1972 The Fine Structure of the Constructible Hierarchy. Annals of Math. Logic *4*, 229–308

Jensen, R. B., Karp, C.
1971 Primitive Recursive Set Functions. In Axiomatic Set Theory: Proc. Symp. Pure Math. *13*, (ed. D. Scott), Amer. Math. Soc. 143–167

Kanamori, A.
1982 On Silver's and Related Principles. In "Logic Colloquium '80" (ed. van Dalen et al.), North Holland

Karp, C.
1967 A Proof of the Relative Consistency of the Continuum Hypothesis. In "Sets, Models and Recursion Theory" (ed. J. Crossley), North Holland, 1–32

Kripke, S.
1964 Transfinite Recursion on Admissible Ordinals, I, II. J. Symbolic Logic *29*, 161–162

Kurepa, D.
1935 Ensembles Ordonnés et Ramifiés. Publ. Math. Univ. Belgrade *4*, 1–138

Lévy, A.
1957 Indépendence Conditionnelle de $V = L$ et d'Axiomes qui se Rattachent au Systeme de M. Gödel. C.R. Acad. Sci. Paris *245*, 1582–1583
1960 Axiom Schemata of Strong Infinity in Axiomatic Set Theory. Pacific J. Math. *10*, 223–238
1965 A Hierarchy of Formulas in Set Theory. Memoirs Amer. Math. Soc. *57*
1979 Basic Set Theory. Springer-Verlag

Mitchell, W.
1972 Aronszajn Trees and the Independence of the Transfer Property. Annals Math. Logic *5*, 21–46

Montague, R. M.
1961 Fraenkel's Addition to the Axioms of Zermelo. In "Essays on the Foundations of Mathematics" (ed. Y. Bar-Hillel et al.). The Magnes Press (Jerusalem), 91–114

Morley, M., Vaught, R.
1962 Homogeneous Universal Models. Math. Scand. *11*, 37–57

Mostowski, A.
1949 An Undecidable Arithmetical Statement. Fund. Math. *36*, 143–164

Platek, R.
1966 Foundations of Recursion Theory. PhD thesis, Stanford University

Prikry, K. L., Solovay, R. M.
1975 On Partitions into Stationary Sets. J. Symbolic Logic *40*, 75–80

Rowbottom, F.
1971 Some Strong Axioms of Infinity Incompatible with the Axiom of Constructibility. Ann. of Math. Logic *3*, 1–44

Shelah, S., Stanley, L. J.
1983 *S*-Forcing, I: A "Black Box" Theorem for Morasses

Shepherdson, J.
1951/52 Inner Models of Set Theory. J. Symbolic Logic *16*, 161–190; *17*, 225–237;
1953 *18*, 145–167

Silver, J. H.
1971 Some Applications of Model Theory in Set Theory. Ann. of Math. Logic *3*, 45–110
1971 a The Independence of Kurepa's Conjecture and Two-Cardinal Conjectures in Model Theory. In "Axiomatic Set Theory". proc. Symp. in Pure Math. *13* (ed. D. Scott), Amer. Math. Soc. 383–390

Solovay, R. M.
1967 A Nonstructible Δ_3^1 Set of Integers. Trans. Amer. Math. Soc. *127*, 50–75

Solovay, R. M., Tennenbaum, S.
1971 Iterated Cohen Extensions and Souslin's Problem. Ann. of Math. *94*, 201–245

Souslin, M.
1920 Problème 3. Fund. Math. *1*, 223

Stanley, L. J.
1975 L-Like Models of Set Theory: Forcing, Combinatorial Principles, and Morasses. PhD Thesis (Berkeley)

Todorcevic, S.
1983 Trees and Linearly Ordered Sets. In "Topology Handbook"

Velleman, D.
1984 Simplified Morasses. J. Symbolic Logic
1984 a Simplified Morasses with Linear Limits. J. Symbolic Logic

Glossary of Notation

$\kappa \cdot \lambda$ 15
κ^{λ} 15
CH 16
GCH 16
$\mathrm{cf}(\alpha)$ 18
$\kappa^{<\lambda}$ 19
Φ^{M} 24
Σ_n, Π_n 27
$\Sigma_n^{\mathrm{ZF}}, \Pi_n^{\mathrm{ZF}}, \Delta_n^{\mathrm{ZF}}$ 27
$\Sigma_n^{T}, \Pi_n^{T}, \Delta_n^{T}$ 27
$\mathscr{L}, \mathscr{L}_X, \mathscr{L}_V$ 31
φ, ψ, θ 32
$\langle x \rangle$ 32
$\langle x_1, \ldots, x_n \rangle$ 32
$s \frown t$ 32
$\| s \|$ 32
$\mathrm{Vbl}(x)$ 32
$\mathring{x}$ 32
$\mathrm{Const}(x)$ 32
$\mathrm{PFml}(x)$ 33
$\mathrm{Finseq}(x)$ 33
$F_{\in}, F_{=}, F_{\wedge}, F_{\neg}, F_{\exists}$ 34
$\mathrm{Build}(\varphi, \psi)$ 35
BS 36
$\mathrm{Seq}(u, a, n)$ 36
$\mathrm{Fml}(x)$ 37
$\mathrm{Const}(x, u)$ 38
$\mathrm{PFml}(x, u)$ 38
$\mathrm{Fml}(x, u)$ 38
$\mathrm{Fr}(\varphi, x)$ 38
$\mathrm{Sub}(\varphi', \varphi, v, t)$ 39
$\mathrm{Sat}(u, \varphi)$ 40
$\models_u \varphi$ 41
$\mathrm{Fml}^{\Sigma_0}(\varphi)$ 42
$\mathscr{L}_u(\mathring{A}_1, \ldots, \mathring{A}_k)$ 43
$\mathring{A}(x_1, \ldots, x_k)$ 43
$\Sigma_n^{\mathbf{M}}(N), \Sigma_n^{\mathbf{M}}, \Sigma_n(\mathbf{M})$ 44
$\Pi_n^{\mathbf{M}}(N), \Pi_n^{\mathbf{M}}, \Pi_n(\mathbf{M})$ 44
$\Delta_n^{\mathbf{M}}(N), \Delta_n^{\mathbf{M}}, \Delta_n(\mathbf{M})$ 44
$R \wedge S, R \vee S, \neg R$ 46
$\exists x R, (\exists x \in y) R$, etc 46
$R((x)_0), R((x)_1)$, etc 47
$\mathbf{N} \prec \mathbf{M}, \mathbf{N} \prec_n \mathbf{M}$ 48, 78
$\pi\colon \mathbf{N} \prec \mathbf{M}, \pi\colon \mathbf{N} \prec_n \mathbf{M}$ 48, 78
$X \prec \mathbf{M}, X \prec_n \mathbf{M}$ 48, 78

KP 48
H_κ 49
$\mathrm{Def}(X)$ 57
L_α 58
$\mathrm{Seq}(y, x)$ 65
$\mathrm{Pow}(y, x)$ 67
$\dashv$ 72
$<^*$ 73, 91
$<_L$ 73
$\mathrm{WO}(y, x), \mathrm{wo}(y, x)$ 74
$\mathrm{pr}(x)$ 75
$\mathrm{Enum}(\alpha, x)$ 76
$f(x) \simeq g(x)$ 86
$\models_{\mathbf{M}}^{\Sigma_n}$ 88
$r_\alpha(i, x)$ 90
$h_\alpha(i, x)$ 90
$G(\alpha, \beta)$ 90
$L_\alpha[A], L[A]$ 102
$<_{L[A]}$ 104
SH 109
$\hat{x}$ 109
$ht(x), ht_{\mathbf{T}}(x)$ 109
T_α 109
$T \restriction \alpha, \mathbf{T} \restriction \alpha$ 109
KH 118
$\mathscr{F} \restriction \alpha$ 118
$\diamondsuit$ 124, 135
$\diamondsuit'$ 127
$\diamondsuit^*$ 128
$\diamondsuit^+$ 129, 135
$\diamondsuit_\kappa, \diamondsuit_\kappa(E)$ 138
$\square_\kappa, \square_\kappa(E)$ 141
J_α 153, 154
$\mathrm{rud}(X)$ 154
$h_{\alpha, A}, H_{\alpha, A}$ 156
ϱ_α^n 156
A_α^n 156
p_α^n 156
$\square_\kappa'$ 158
$\mathscr{A} \models \varphi$ 169
$t^{\mathscr{A}}(x_0, \ldots, x_m)$ 169
$[X]^\alpha, [X]^{<\alpha}, [X]^{\leqslant\alpha}$ 170
$\kappa \to (\lambda)_\mu^\alpha$ 170
$\kappa \to (\lambda)_\mu^{<\omega}$ 173
$\kappa(\lambda)$ 173
$0^{\#}$ 176, 182
BSL 176

$\mathscr{A} \restriction X$ 176
$M(\Sigma, \alpha)$ 181
H_κ 181
h_α^φ 198
$H_\alpha^n(A), H_\alpha^\omega(A)$ 198
$S_\delta^n(\eta)$ 201
$I_\delta^n(\eta)$ 201
$S_\delta^\omega(\eta)$ 202
$I_\delta^\omega(\eta)$ 202
$S_\delta^0(\eta)$ 202
$I_\delta^0(\eta)$ 202
$a^{\#}$ 220
SCH 220
$F_i(x, y)$ 236, 241
$t_{\varphi(x_0, \ldots, x_n)}(u)$ 236
$Sat^A(u, \alpha, \varphi)$ 241
S 249
Wo 249
J_α 251
S_α 252
$<_\alpha^{\mathrm{S}}$ 257
$<_\alpha$ 257
$<_J$ 257
$r_{\alpha, A}(i, x)$ 259
$h_{\alpha, A}(i, x)$ 259
ϱ_α^n 266, 273
$\Gamma(\alpha, n, \varrho)$ 267
A_α^n 275
p_α^n 275
$\varrho_{\alpha, A}^n$ 277
$A_{\alpha, A}^n$ 278
$p_{\alpha, A}^n$ 278
$\square(E)$ 284
$\square$ 285
$\square^\kappa(E)$ 301
$\kappa \to [\mu]_\lambda^n$ 305
$\mathscr{P}_\lambda(\kappa)$ 320
$\mathrm{KH}(\kappa, \lambda)$ 320
$\diamondsuit_{\kappa, \lambda}^+$ 320
(κ, λ) 333
K_X 333
$\mathrm{Th}(\mathscr{A})$ 333
$\mathscr{S}$ 340
S, S^0, S^1 340
S_α 340
α_ν 340
$\dashv$ 341

Index

For EU product safety concerns, contact us at Calle de José Abascal, 56–1°, 28003 Madrid, Spain or eugpsr@cambridge.org.

www.ingramcontent.com/pod-product-compliance
Lightning Source LLC
LaVergne TN
LVHW102009210926
848483LV00010B/239
* 9 7 8 1 1 0 7 1 6 8 3 5 0 *